The Medical Device Industry

SCIENCE, TECHNOLOGY, AND REGULATION IN A COMPETITIVE ENVIRONMENT

edited by

Norman F. Estrin

Health Industry Manufacturers Association
Washington, D.C.

MARCEL DEKKER, INC. New York and Basel

ISBN 0-8247-8268-2

This book is printed on acid-free paper.

MARCEL DEKKER, INC.
270 Madison Avenue, New York, New York 10016

Current printing (last digit):
10 9 8 7 6 5 4 3 2 1

PRINTED IN THE UNITED STATES OF AMERICA

To my parents, my family
(Mitzi, Laura, Melissa, Yale, and Lea),
for their love

Foreword

The challenge of competing in today's fast-paced scientific, technical, regulatory, and economic environment is truly formidable. Medical device manufacturers and others involved in ensuring the availability of quality health-care products must become experts in understanding the regulatory process. They must deliver products that are safe and effective and have instructions designed to help prevent product misuse. They must assure safety in the work place and minimize adverse effects on the environment. Today's manufacturers must address these issues not only on a domestic basis, but internationally as well.

The issues are multifaceted and complex and raise a number of questions: How will changing coverage and payment policies affect our marketing strategies? What will be the impact of regulatory harmonization in the European community? How will the dramatic changes in Eastern Europe and the Soviet Union affect our marketing strategies? Are we accessing the latest in technological advances by utilizing technology transfer mechanisms already in use by foregin competitors? Are we ready to make the most of changes in the domestic and international standards processes? Is our educational system adequate to keep us competitive?

This book, which brings together contributors from a wide variety of organizations and disciplines, attempts to answer these and other important questions and provides valuable insights on both the domestic and global aspects of medical device manufacture, regulation, and marketing.

Alan H. Magazine, *President*
Health Industry Manufacturers Association
Washington, D.C.

Preface

We live in a world of rapid technological innovation. As the new decade unfolds, manufacturers of health-care products will be surrounded by a constellation of challenges and opportunities. Complex and fluctuating regulatory approaches by the Food and Drug Administration, evolving requirements for occupational safety and health, growing concern over pollution of the environment, changing health-coverage policies, and the advent of the global market will profoundly affect the competitiveness of the U.S. medical device industry.

This volume is designed to give manufacturers of medical devices and diagnostic products, government officials, attorneys, academicians, consultants, and students practical knowledge about biomedical technology, regulation, and their relationship to quality health care.

Organized for simplicity, the text examines four areas of major interest and influence. Parts two through five cover the *regulatory environment*, describing device regulation by the Food and Drug Administration. Parts six and seven cover *occupational and environmental safety and health*, introducing issues related to regulation by the Environmental Protection Agency and the Occupational, Safety and Health Administration. The reader will understand the legal and marketing implications of initiatives promulgated by the Health Care Financing Administration through reading the sections on *coverage, payment, and marketing: staying competitive*, parts eight through ten. *International issues* are discussed in parts eleven through fourteen and provide the tools needed for operating successfully in a global market.

For each of the four areas covered, we begin with a historical overview and legal framework to provide the reader with the perspective necessary to comprehend and evaluate subsequent information. The chapters that follow delineate the role of organizations that have impact on the regulatory environment and offer pragmatic advice from experienced professionals on industry practices. This "how to" approach to responding to regulatory initiatives will be especially helpful to companies developing their own programs and procedures.

Current and Emerging Issues concludes each of the four areas. Here, the reader gets a sense of the dynamic forces shaping the medical device industry. Issues will change. Some of those cited in this section will turn out to be minor perturbations, while others may have major regulatory and economic repercussions. A review of these issues will help the reader to identify and evaluate the significance of future issues.

I am delighted to see authors from diverse disciplines and backgrounds unite to educate and advise all concerned with quality health care. Your comments and suggestions for making future editions of *The Medical Device Industry: Science, Technology, and Regulation in a Competitive Environment* more valuable to you are welcomed.

I am indebted to the authors, not only for their contributions, but for advice and suggestions that helped shape this volume. Howard Holstein, Ed Basile, Paul Campbell, and Gordon Schatz were especially helpful in providing names of experts and subject areas. Special appreciation is extended to Gretchen New, Sharon Sealey, and Teresa Frakes for their overall coordination and indexing.

Norman F. Estrin, Ph.D.

Contents

part three
RESPONDING AS ORGANIZATIONS TO THE REGULATORY ENVIRONMENT

part four
RESPONDING AS COMPANIES TO THE REGULATORY ENVIRONMENT

Contributors

Robert S. Adler Associate Professor of Legal Studies, Graduate School of Business Administration, University of North Carolina, Chapel Hill, North Carolina

June C. Abbey Associate Dean of Research and Evaluation, Vanderbilt University School of Nursing, Nashville, Tennessee

Edwin D. Allen, Esq. Vice President and General Counsel, Health Industry Manufacturers Association, Washington, D.C.

F. Alan Andersen, Ph.D. Acting Director, Office of Science and Technology, Center for Devices and Radiological Health, Food and Drug Administration, Rockville, Maryland

William D. Appler, J.D. Partner, McDermott, Will & Emory, Washington, D.C.

Joseph S. Arcarese Director, Office of Training and Assistance, Center for Devices and Radiological Health, Food and Drug Administration, Rockville, Maryland

Edward M. Basile, J.D. Partner, King & Spalding, Washington, D.C.

John V. Bergen, Ph.D. Executive Director, National Committee for Clinical Laboratory Standards, Villanova, Pennsylvania

Thomas A. Boardman, J.D. Office of General Counsel, 3M Company, St. Paul, Minnesotta

Russell Bodoff Vice President, External Relations, American National Standards Institute, Washington, D.C.

Jerome S. Bortman Head, Office of Research and Technology Applications, Naval Air Development Center, Warminster, Pennsylvania

Robert C. Brandys, Ph.D., C.I.H., C.S.P., P.E. President, Occupational & Environmental Health Consulting Services, Inc. Hinsdale, Illinois

John M. Bredehoft, Esq. Cleary, Gottleib, Steen & Hamilton, Washington, D.C.

Elizabeth A. Bridgman Assistant Executive Director, Association for the Advancement of Medical Instrumentation, Arlington, Virginia

Stephen L. Brown, Ph.D. Project Manager, Environ Corporation, Arlington, Virginia

Noel L. Buterbaugh President, Whittaker Bioproducts, Inc., Walkersville, Maryland

Kathleen A. Buto Acting Director, Bureau of Policy Development, Health Care Financing Administration, Baltimore, Maryland

Nancy E. Cahill McDermott, Will & Emory, Washington, D.C.

Paul M. Campbell Executive Consultant, The Lash Group, Washington, D.C.

Earle L. Canty Vice President, Quality Assurance, Ventritex, Inc., Sunnyvale, California

Arnold C. Celnicker, J.D. Assistant Professor, College of Business, and Adjunct Professor for Antitrust, College of Law, The Ohio State University, Columbus, Ohio

F. Michael Cole, Esq. Laxalt, Washington, Perito & Dubic, Washington, D.C.

Amiram Daniel, Ph.D. Director, Diagnostic Manufacturing and Biotechnology Programs, Health Industry Manufacturers Association, Washington, D.C.

Thomas J. DiPasquale, J.D. Corporate Counsel, Office of General Counsel, 3M Company, St. Paul, Minnesota

Robert A. Dormer, Esq. Partner, Hyman, Phelps & McNamara, Washington, D.C.

William H. Duffell, MBA, BS Chem, J.D. Director of Corporate Regulatory Compliance, Corporate Regulatory Affairs, Syntex (U.S.A.) Inc., Palo Alto, California

H. Neal Dunning, Ph.D. Director, Division of Small Manufacturers Assistance, Center for Devices and Radiological Health, Food and Drug Administration, Washington, D.C.

Norman F. Estrin, Ph.D. Vice President, Science and Technology, Health Industry Manufacturers Association, Washington, D.C.

Marcel P. Fages Director, Association for Manufacturing Excellence, Wyoming, Ohio

Robert C. Flink Director, International Regulatory Affairs and Standards, Corporate Regulatory Affairs, Medtronic, Inc., Minneapolis, Minnesota

Susan Bartlett Foote, J.D. Associate Professor, School of Business Administration, University of California, Berkeley, California

Glen Paul Freiberg Director, Regulatory Affairs and Quality Assurance, Boehringer Mannheim Diagnostics, Division of Boehringer Mannheim Corporation, Indianapolis, Indiana

Richard C. Fries, P.E. CRE Product Safety Manager, Anesthesia Product Assurance, Ohmeda Anesthesia Systems, Madison, Wisconsin

Michael C. Fuchs Medical Equipment Industry Specialist, Office of Microelectronics and Instrumentation, U.S. Department of Commerce, Washington, D.C.

Pamela M. Garside The Newhouse Group, New York, New York

Kathryn L. Gleason, Esq. Morgan, Lewis & Bockius, Counselors at Law, Washington, D.C.

Donald F. Grabarz DFG & Associates, Inc., Salt Lake City, Utah

Walter E. Gundaker Acting Deputy Director, Center for Devices and Radiological Health, Food and Drug Administration, Rockville, Maryland

Arthur Hull Hayes, Jr., M.D. President and CEO, EM Pharmaceuticals, Inc., Hawthorne, New York

Marilyn Hernandez Director, Public Relations, External Relations, American National Standards Institute, Washington, D.C.

S. Craig Holden Ober, Kaler, Grimes & Shriver, Baltimore, Maryland

Howard M. Holstein Partner, Hogan & Hartson, Washington, D.C.

W. Fred Hooten Director, Office of Compliance and Surveillance, Center for Devices and Radiological Health, Food and Drug Administration, Rockville, Maryland

Peter Barton Hutt, Esq. Partner, Covington & Burling, Washington, D.C.

Elizabeth D. Jacobson, Ph.D. Director, Office of Science and Technology, Center for Devices and Radiological Health, Food and Drug Administration, Rockville, Maryland

James F. Jorkasky Director, Environmental and Occupational Safety and Health Programs, Health Industry Manufacturers Association, Washington, D.C.

John H. Keene, Dr. Ph. President, Biohaztec Associates, Inc., Midlothian, Virginia

Eleanor D. Kinney Associate Professor of Law and Director, The Center for Law and Health, Indiana University School of Law, Indianapolis, Indiana

Edward R. Kimmelman, J.D. Manager, Quality Assurance Diagnostic and Biotechnology Systems, Medical Products Department, E. I. du Pont de Nemours & Company, Inc., Wilmington, Deleware

Arthur C. Kohler Division of Small Manufacturers Assistance, Center for Devices and Radiological Health, Food and Drug Administration, Washington, D.C.

Stephen C. Kolakowsky Director, Regulatory Affairs, Carter Products, Division of Carter-Wallace, Inc., Cranbury, New Jersey

Eugene H. LaBrec, Ph.D. President, E. H. LaBrec & Associates, Rockville, Maryland

C. Stephen Lawrence, Esq. Hogan & Hartson, Washington, D.C.

J. Thomas Lowe, DDS Deputy Director, Office of Health Affairs, Food and Drug Administration, Rockville, Maryland

Ted R. Mannen, Esq. Principal, Dickstein, Shapiro & Morin, Washington, D.C.

Gaile L. McMann, Esq. McDermott, Will & Emery, Washington, D.C.

Michael J. Miller, J.D. Executive Director, Association for the Advancement of Medical Instrumentation, Arlington, Virginia

Robert F. Morrissey, Ph.D. Director, Sterilization Science Group, Ethicon, Inc., Johnson & Johnson, New Brunswick, New Jersey

Rodney R. Munsey, Esq. Partner, Hogan & Hartson, Washington, D.C.

Joel J. Nobel, M.D. President, Emergency Care Research Institute, Plymouth Meeting, Pennsylvania

Leif E. Olsen Director, Regulatory Affairs and Quality Assurance, Whittaker Bioproducts, Inc., Walkerville, Maryland

John E. Olsson Manager, Safety and Regulatory Engineering, Medical Systems Group, General Electric Company, Milwaukee, Wisconsin

Kenneth Pearson Vice President, Technical Committee Operations Division, American Society for Testing and Materials, Philadelphia, Pennsylvania

G. Briggs Phillips, Ph.D. President, Petra, Inc., St. Michaels, Maryland

Robert G. Pinco, Esq. Partner, Food and Drug Department, Baker & Hostetler, Washington, D.C.

Ann K. Pollock, Esq. Baker & Hostetler, Washington, D.C.

Alexis J. Prease, Esq. King & Spalding, Washington, D.C.

James A. Rice Senior Vice President, Strategic Development, Health One Corporation, Minneapolis, Minnesota

Barbara A. Rich Executive Vice President, National Association for Biomedical Research, Washington, D.C.

Wayne I. Roe President, Health Technology Associates, Washington, D.C.

Jean Russotto, Esq. Managing Partner, Oppenheimer, Wolff & Donnelly, Brussels, Belgium

Gordon B. Schatz, Esq. Reed, Smith, Shaw & McClay, Washington, D.C.

Gary S. Schneider* Manager, Policy Evaluations, Pharmaceutical Manufacturers Association, Washington, D.C.

Sara D. Schotland, Esq. Partner, Cleary, Gottleib, Steen & Hamilton, Washington, D.C.

---

*Deceased

Robert T. Schwartz Executive Director, Industrial Designers Society of America, Great Falls, Virginia

Marvin D. Shepherd Program Manager, Office of Environmental Health and Safety, University of California Medical Center, San Francisco, California

Robert L. Sheridan Deputy Director, Office of Device Evaluation, Center for Devices and Radiological Health, Food and Drug Administration, Silver Spring, Maryland

Max Sherman Director, Regulatory Affairs, Zimmer, Inc., Bristol Myers Company, Warsaw, Indiana

Dee Simons Director, Computer and Biomedical Technology Programs, Health Industry Manufacturers Association, Washington, D.C.

John F. Stigi Deputy Director, Division of Small Manufacturers Assistance, Center for Devices and Radiological Health, Food and Drug Administration, Rockville, Maryland

LaWayne R. Stromberg, M.D. Vice President, Medical Affairs, Baxter Healthcare Corporation, Deerfield, Illinois

Sanford V. Teplitzky, Esq. Ober, Kaler, Grimes & Shriver, Baltimore, Maryland

Stephen D. Terman, Esq. Health Industry Manufacturers Association, Washington, D.C.

Frankie L. Trull President and Executive Director, National Association for Biomedical Research, Washington, D.C.

Joseph G. Valentino, J.D. Executive Associate, United States Pharmacopeial Convention, Inc., Rockville, Maryland

Marilyn M. Wilder Whitted & Buoscio, Merrillville, Indiana

George T. Willingmyre, P.E. Vice President, American National Standards Institute, Washington, D.C.

Robert E. Wren Director, Office of Coverage Policy, Bureau of Policy Development, Health Care Financing Administration, Baltimore, Maryland

part one

Introduction

1

Focus on Patient Care

SUSAN BARTLETT FOOTE *University of California, Berkeley, California*

I. INTRODUCTION

The patient is the ultimate consumer of all medical devices, from the simplest cotton swab to the most sophisticated monitoring device. However, with the exception of some over-the-counter products, the medical device manufacturer rarely has a direct relationship with the patient in the marketplace. Unlike many other consumer products, a host of intermediaries influence the demand for medical devices. These intermediaries include policymakers, providers, and payers of health care services.

The public sector is by far the most powerful intermediary. Government plays a variety of roles in the delivery of health care, many of which directly or indirectly impact upon the medical device producer and the patient-consumer. For example, the Food and Drug Administration (FDA) has the authority to review medical devices before they can be placed on the market [1]. An unapproved product will never reach the consumer. Even once FDA permits the marketing of medical devices, government and others in the private sector may make the key decisions. For example, health care providers such as hospitals, physicians, and other health professionals often choose what equipment and procedures will be used on patients. And the prevalence of third party payers, whether it is through a government program like Medicare or a private insurance plan, further removes many patients from basic purchasing decisions. Thus, although these intermediaries all serve the health care system, the patient is dependent on these decision makers in the choice of medical services and equipment he or she ultimately receives.

There are two quite clear reasons why these various intervening institutions have emerged to stand between the producer and the patient. First, the use of medical devices is deeply embedded in the delivery of health care services. The wide range of health policies, designed primarily to structure the delivery of services, inevitably affects the device market and patients as well. Thus, it is difficult to discuss medical devices without reference to medical technology generally, which has been defined to include drugs, devices, procedures, and the systems in which health services are delivered [2]. Additionally, medical devices can be distinguished from other consumer products like lawn mowers and vacuum cleaners. Medical devices serve an important social purpose because they are often life-saving or life-enhancing. Thus, pressing questions about their availability, safety, and cost often give rise to public intervention not warranted for other consumer goods.

The complex environment in which these products are consumed can lead producers to overlook the concerns of the patient. However, members of the medical device industry must not lose sight of the interests of the patient, nor forget the health care context in which products are consumed. These factors offer both a challenge to producers of medical devices and an opportunity to serve the public interest.

The purpose of this chapter is to focus on the relationship of the device industry to patient care. First, it will provide an overview of what patients want from the health care system. Second, it will discuss the rise of intermediaries and their impact on the relationship of industry to the public. Finally, as health care decisions have moved to the policy arena, the changing role of the patient in politics will be evaluated.

II. WHAT DO PATIENTS WANT?

It does not take much imagination to develop a common set of patient preferences for personal health care. As individuals, we all want access to "affordable" treatments. Americans, however, also want the care they receive to be absolutely safe and indisputably efficacious. In addition, we do not want run-of-the-mill care based upon last year's ideas; we want innovative, state-of-the-art technology. From the patients' perspective, medical devices are necessary tools for the attainment of their broader health goals. To the extent that devices ensure that safe, innovative, and accessible care is available, they are desirable.

It is important to recognize, despite these common social goals, that patients will not all agree about how best to achieve them or what priorities should be set along the way. Some may also hold unrealistic notions about safety, failing to appreciate that adverse reactions may not be anyone's fault. When resources are limited, decisions about priorities are likely to divide patient groups. For example, Oregon legislators recently decided to terminate payment for transplants under Medicaid, and instead 1500 pregnant women will receive regular prenatal exams with the same money. The funds would have paid for about 30 transplants [3].

Health care policy presents many hard questions. The challenge for all is to craft health policy that maximizes the achievement of the common goals of innovation, safety, and access.

III. THE RISE OF INTERMEDIARIES

A. From the Marketplace to Public Policy

For most of the first half of this century, patients sought health care in the marketplace. The transactions between the providers of services and products and the individual patient were direct economic exchanges. There was, however, little to buy. Until the 1930s, the marketplace had little to offer for the treatment of illness. Hospitals were generally shunned as places where destitute people without family went to die [4]. Doctors held out little hope for treatment; consumers often turned to folk remedies or miracle cures. Indeed, in regard to the device industry, the public perception in the period before the Second World War was that many products in this industry were merely quack devices promising what they could not deliver.

Progress in the medical sciences, including important breakthroughs such as aseptic surgery, sulfa drugs, and vaccines, as well as sophisticated diagnostic tools and laboratory equipment, increased the desirability of medical care. These advances also added to the cost. While we continued to rely on the private sector to produce technological innovations, including medical devices, there was increased political pressure for government to become involved in expanding the access of excluded groups to the health care system. And, as products became more complex, additional pressure grew to regulate the safety of such devices.

What has occurred has been a shift from the primacy of the individual to decide what medical products and services to purchase to increased decision making at a higher level. Many health-related choices are made by government agencies, health professionals, and insurance companies. The advent of these intermediaries has had many beneficial effects, such as providing funds for care and ensuring levels of safety for the increasingly complex medical technology. Another consequence, however, has been to complicate the environment for producers, who must sort their way through the maze of relevant players, and for consumers, who must try to maximize their individual welfare within a set of public and private constraints.

B. Promoting Access to Care

After the Second World War, it became clear that many Americans were excluded from the health care marketplace. Some gained better access to the system through private insurance programs. The number of individuals with some form of private insurance coverage grew from 6% in 1948, to 25% in 1966, to 32% in 1982 [5]. Private insurance benefitted the wealthy and the employed.

After the war, the federal government confined its involvement in health care to support for hospital construction. The Hill-Burton Act, passed in 1946, has financed nearly 15% of annual hospital construction through a program of grants and loans [6]. Hill-Burton funds helped expand the infrastructure of hospitals and distribute beds to underserved areas of the country. The program has indirectly supported access for those excluded by requiring participating hospitals to promise to provide some care to the indigent.

The public commitment to access greatly increased in 1966 when Medicare was put in place for the elderly and the disabled. The Medicare program involved the federal government in payment for health care services. Medicare funds now account for about 40% of all hospital purchases of technology [7]. The Medicaid program was enacted in 1965 and is supported by a combination of federal and state funds. Medicaid has extended health care benefits to more than 21 million eligible, low-income people [8].

The availability of public funds, and the growth of private insurance programs, greatly improved patient access to health care. National health expenditures reflected this trend, soaring from 40.46 billion dollars in 1965 (5.9% of GNP) to 322.3 billion dollars (10.5% of GNP) by 1982. Per capita expenditures increased more than fivefold, from $207 in 1965 to $1337 in 1982. Public funds accounted for 41% of the expenditures by 1982 [9]. However, there is by no means universal coverage in the United States. Recent estimates are that 38.8 million people are uninsured. Given the rising costs of high quality care, this sizable number are effectively denied access to medical services.

The infusion of funds into the marketplace has had an enormous impact on innovation in the device industry. The story of renal dialysis is illustrative. The diffusion of dialysis technology was hampered by the lack of a market, and also reduced incentives to innovate. As Rettig has noted: "There were always equipment vendors ready to sell artificial kidneys to prospective buyers. . . . The scarce resource was never machines but money. . . . Unit costs for kidney machines were high, the buyers were few, and the financial means for paying for treatment costs for the potential patient pool were uncertain in all but a few cases" [10]. Due to political and social pressure from patients and physicians, Congress extended Medicare coverage for transplantation and dialysis from the elderly to more than 90% of the United States population in 1972 [11]. Innovative activity in the field of renal dialysis greatly increased as a consequence of rapid market expansion. Dialysis patients now had access to the latest technology, but at a high price. Costs of coverage averaged $1.8 billion for treatment of 80,000 individuals in 1984 alone [12].

C. Promoting Safety

As sophisticated medical technology came to characterize modern medical care, the public became more aware of the risks as well as the benefits of medical devices. In 1970, a federal study revealed 10,000 injuries and 731 deaths associated with medical devices in the preceding decade [13]. The public became increasingly concerned with this problem when the harmful effects of the Dalkon Shield and defects in certain types of cardiac pacemakers came to light. After years of debate, Congress passed the Medical Device Amendments of 1976 that conferred on the Food and Drug Administration the authority to regulate all medical devices.

Other chapters in this volume discuss the law in much greater detail. There is considerable controversy about how well the law has been implemented and what its effect has been on device safety and device innovation. Recent studies have indicated that while the law does impose additional costs on some products and some producers, data do not support a conclusion that the stream of innovative products has been significantly affected.

In addition to regulation, product safety is one of the several goals of product liability. Ideally, tort law complements safety regulation by deterring the production of harmful products, along with its primary purpose to compensate injured individuals [14]. The impact of product liability judgments and the increase in insurance rates as a consequence have been highly controversial. While data are difficult to acquire, it is clear that new theories of liability have expanded the number of potential lawsuits and that there has been a trend toward larger compensatory and punitive damage awards. Medical devices have been the focus of several mass tort actions, including thousands of cases brought against the producers of tampons for toxic shock injuries and against A. H. Robins, the manufacturer of the Dalkon Shield. There is no question that product liability has increased the costs of doing business in some sectors of the medical device industry.

D. Promoting Innovation

The private sector accounted for virtually all the research and development funds for innovation until 1965, although the level of support was relatively low [15]. In the last two decades, firms have greatly increased R&D spending. The Office of Technology Assessment reports that R&D expenditures on medical devices as a percent of sales were almost twice the industry average in 1980 [16]. The federal government has provided some support for research as well. Congress created the Artificial Heart Program (AHP) in 1963, and federal funds were approved the following year. Housed in the National Institutes of Health (NIH), the AHP has negotiated targeted contracts with private firms to develop parts and materials for the artificial heart. This approach, which is modeled on the space program, supports government involvement in development of particular technologies. The Artificial Kidney-Chronic Uremia (AK-CU) Program, also in NIH, has supported engineering projects to develop medical devices [17].

The data on medical product sales corroborate the conclusion that federal spending expanded industry sales. The industry experienced a burst of growth from the 1960s to 1982, with the dollar value of product shipments increasing from $1.4 billion in 1965, to $6 billion in 1976, to $15 billion in 1982, an annual percentage increase of over 14% per year. In 1982, for example, hospitals purchased $6 billion of the $16.8 billion in sales of medical products, much of that due to the availability of federal funds. Innovation became the hallmark of the industry, and patients and providers became accustomed to innovative new medical technology.

IV. THE REALITY OF COST CONTAINMENT

As long as the cost of health care was subordinate to other goals, the medical device industry thrived and innovative products continued to arrive on the market. Public policy seemed to promote the values shared by patients—increased access to safe and innovative technology. However, pressure to contain or reduce medical expenditures in the 1980s has raised a potential threat to both the industry and the patients.

Between 1967 and 1982, Medicare expenditures for in-patient services rose at an average annual rate of over 19%. Concerned with ever-rising outlays, Congress initiated efforts to revise the Medicare payment system. The Social Security Amendments of 1983 established the prospective payment system (PPS), which classifies each patient into one of over 400 diagnostic-related groups (DRGs), and reimburses hospitals at prices set in advance for each DRG rather than the actual costs of treatment. The intent of this new system was to encourage the efficient delivery of services in the hospital sector, leading ultimately to a reduction in overall Medicare expenditures. Because payments under Medicare represent a substantial portion of all expenditures for medical technology, changes in this system affect the behavior of hospitals and private insurers as well. States have also felt the budget crunch, and have introduced cost-reducing strategies for Medicaid programs as well [18].

Of course, to the extent that these cost containment efforts reduce or eliminate waste and inefficiency, patients are not harmed, and may even benefit from the reform. In addition, cost containment may provide incentives for innovative, cost-reducing technologies that will improve care and lower cost. Price competition may be a constructive value to introduce into a market that has been largely insensitive to it because of the prevalence of third-party payers one step removed from the providers. However, it appears that cost containment may require policy makers to compromise the goals of access to safe, innovative care in some cases, and will inevitably affect the patients' interests in safety, innovation, and access.

A few illustrations dramatize the problem. The efforts of the legislators in Oregon, referred to above, to pay for prenatal care for 1500 women instead of transplants for 30 Medicaid recipients illustrate the painful rationing process that cost constraints require.

Cost containment may also limit the access of Medicare patients to expensive innovations. HCFA recently announced that it will cover some heart transplants for a limited number of recipients. It is clear that the system as presently funded will not be able to afford heart replacement procedures for all those who might need them. A recent NIH advisory panel, in reviewing the heart program there, asked whether society should spend its money on artificial hearts instead of other therapies or modes of prevention, particularly if the cost of artificial hearts restricts access to other goods and services [19].

Others ask if we can afford the regulatory and legal procedures that are intended to ensure that safe products get to market. Recent trends in tort law, particularly product liability, potentially conflict with cost containment directly, and indirectly affect the values of access and innovation. The result of the expansion of tort law has been rising insurance premiums for all producers. For example, Puritan-Bennett, a leading manufacturer of hospital equipment, including anesthesia devices, faced a 750% increase in premiums in 1986, with less coverage and higher deductibles [20]. In a cost-contained environment, it becomes harder for equipment manufacturers to pass along the increased costs of doing business.

Increases in potential liability may also affect innovation and access. The example of IUDs illustrates the difficulty of balancing safety over access. Following the rash of lawsuits against the makers of the Dalkon Shield and other intrauterine devices (IUDs), many producers left the marketplace. With sales of its two IUDs amounting to only $11 million, Searle recently pulled its devices off the market, although the FDA had approved them and the company had prevailed in eight of the ten product liability suits [21]. By the mid-1980s, only one producer remained in the market, and that company did not widely advertise or aggressively market its product. Women who preferred the IUD for a variety of reasons over other birth control alternatives were denied access to their choice. Stories surfaced of individuals going to Canada to obtain the device. In this case, our social mechanisms to insure safety and compensation foreclosed the opportunity of others for access to the products they desired despite the known risks.

Cost containment has highlighted the painful issues that surround the distribution of high-quality care. Cost containment may force compromises in access or safety, and may adversely impact the flow of innovations. It may also trigger proposals for reform of the present policy environment [22]. While some measures to make the health care system more cost efficient are both inevitable and necessary, the interests of patients in access, safety, and innovation are at risk. What role will the patient play in resolving those hard decisions requiring balancing of interests that affect the individual?

V. HEALTH CARE POLITICS

A. The Interest Groups

The level of expenditures attests to the fact that health care is big business. And because of the powerful role of government as regulator and provider, health care is now highly political. Private interest groups have traditionally dominated in the political arena, and groups that are well organized and well funded generally prevail. Hospitals, insurers, and medical associations have all been particularly active and effective in politics.

As Morone and Dunham have pointed out, government now plays an unique role in health politics. No longer is government simply a pliant state that distributes funds to powerful interest groups. As the interests of the private organizations become linked, it becomes more difficult for any one group to dominate the political environment. The stakes are so large that any policy changes can have enormous consequences for any one of the participants. The result is a dense political environment in which public administrators have a growing influence on health policy, particularly those bureaucrats who make payment and coverage decisions for Medicare [23].

Who serves the interests of patients? Each of the private interest groups would argue that they serve as surrogates for the patients and work on their behalf. Doctors and hospitals, the argument goes, are in business to provide services for patients, and insurers exist to facilitate payment for patient care. However, the interests of these groups are not necessarily coincident with patient needs. More funds for hospitals may be good for those institutions, but there may not be a need for more beds. Caps on malpractice awards in lawsuits may serve the doctors' interests rather than those of the patients. More technology doesn't automatically mean that it is necessary or desirable.

Nor are the interests of government agencies necessarily always in the public's best interests, at least as members of the public might define them. Indeed, cost containment seems to drive many of the decisions of the Health Care Financing Administration (HFCA), the agency that administers Medicare. Cost control may be achieved at the expense of access of the elderly to innovative procedures.

B. The Patient in Politics

What role do patients play in the politics of health? How might they influence the decisions that can enhance or limit their individual and collective goals?

Patient groups and consumer organizations are traditionally fragmented, underfunded in comparison to professional and trade organizations, and extremely diffuse. We are all patients at some time in our lives. However, no one is enthusiastic about being a patient; it is a classification we are all happy to abandon. Thus, despite the universality of the condition, people have not identified collectively as patients, at least in the way we identify with others by gender, race, age, or region. One generally enters the ranks of patient under stressful conditions at best, traumatic at worst [24].

When the issue is access to care, those most concerned have been generally people excluded from the system. These groups are traditionally low-income or disadvantaged people with severely limited political power and influence. Rarely are the interests of the poor represented with the same energy as concerns of physicians or hospitals. It is true that there are some general, organized interest groups, such as the American Association of Retired Persons and the Women's Health Collective, that advocate forcefully for their constituencies on health-related matters. However, the resources of these groups pale in comparison to the American Medical Association or the Pharmaceutical Manufacturers Association, to name only two active organizations.

Much patient activism has tended to cluster around specific illnesses that have brought individuals or members of their families into the political arena. There is a host of smaller groups that are disease or treatment specific, such as parents of children with adverse DPT reactions or parents of offspring with spina bifida, a neural tube defect. These groups are more likely to be effective on specific, narrowly defined issues.

More directly, patient activism is disease specific, not always device specific. When a device is the center of attention, it is usually for negative reasons. The examples include the toxic shock cases brought by injured women against the manufacturers of tampons and cases brought against the producers of IUDs. Recent revelations of misrepresentations by G. D. Searle on the safety of its Copper-7 IUD may rekindle public distrust of device producers. While patients may clamor for more access, or more research for a certain disease like cancer or AIDS, they are not likely to find themselves promoting a particular device or product.

This is not to say that patient groups have had no impact on health care legislation. Liberal reformers fought long and hard for Medicare. Public interest organizations, such as Nader's Health Research Group, have been a persistent presence in Washington politics. Consumer groups have also made felicitous alliances. The National Childhood Vaccine Act of 1986 was supported by pediatricians and well-organized parents of children who had suffered adverse vaccine reactions. However, these alliances inevitably shift depending on the issues.

In regard to the medical device industry, patient groups should recognize that the manufacturers often have similar interests. On occasion, some consumer groups have strongly opposed FDA approval of particular devices. However, when faced with deep cost containment measures favored by government bureaucrats, the consumer interests in wider access to care, including high technology, coincide with the device manufacturing interests in preventing limitations on their market.

Many health policy decisions are made at the administrative, not the legislative, level. There are some efforts to include consumers in administrative decision making. For example, the FDA has 35 public advisory committees that include consumer members. Their role is to present a consumer perspective on issues that come under the review of various committees and to provide a link to community-based organizations and the FDA.

In other areas of health policy, there is little or no input from patients. A recent lawsuit by a Medicare recipient against the Secretary of Health and Human Services, *Jameson v. Bowen*, challenged HCFA's process of determining standards for whether a treatment is covered by Medicare. In *Jameson*, the plantiff's doctor recommended and performed percutaneous transluminal angioplasty surgery for treatment of his coronary problems. Medicare refused to pay for his arterial balloon surgery on the grounds that it was listed in the Medicare Carriers Manual as experimental. Among other complaints, the plaintiff argued that neither the process nor the resulting rules are published in the Federal Register or are otherwise generally available to the public. Medicare beneficiaries receive no advance notice that these particular services will not be covered [25]. Although the case ultimately settled without a final decision on the merits, the settlement agreement included a commitment on the part of the government to open up the procedures to some public scrutiny.

Efforts to institutionalize public participation in the creation and implementation of health policy have been limited at best. Often patients are excluded because they are not considered sufficiently "expert" in areas of technical or scientific matters. Even when included, it is difficult for one individual or even one organization to represent the diffuse and often inchoate "public interest."

Patient interests will never have the organizational coherence of trade groups or professional associations. Many groups have emerged, however, and when they forge strategic alliances with other private interests, some victories have been achieved. In this complex policy environment, however, the interests of the recipients of medical care probably always will be underrepresented and will depend heavily on the commitment of public officials to represent them.

VI. CONCLUSION

Device manufacturers and patients stand at opposite ends of the health care delivery system. In order to reach the consumer, the device producer must master a complex political and regulatory environment that includes the Food and Drug Administration, the Health Care Financing Administration, a multitude of state health agencies, and the complex and rapidly changing hospital marketplace, as well as health professionals and the private insurance industry. Many of these groups seek to facilitate the diffusion of new devices; indeed, the third-party payers, both public and private, have supported the lucrative medical marketplace. Often, however, health policies may act as barriers to entry into the marketplace. These barriers can be intentional, such as FDA regulation, or indirect, such as HCFA's refusal to pay for some innovative procedures that involve the use of medical devices. As cost containment pressures increase, it is inevitable that the interests of some of these groups will collide.

Although producers and consumers are separated by a sea of intermediaries, there are several key issues that both groups must recognize and respect. First, producers and patients should explore the common interests that they share. Patients want innovative products; they depend on life-saving technologies. When government policies to contain costs, for example, threaten to constrict access to desirable technologies, both industry and individuals can join together to defend their common interests.

For their part, patients must realize that the goals of universal access to safe, effective, innovative technologies may not always be attainable. They must realize that absolute safety may be technically unfeasible, that some innovations are too costly to justify, and that difficult choices will inevitably have to be made that favor some patients over others.

For its part, the industry must remember that it is only a piece of a complex and critically important service. There are times when diffusion of technologies might lead to inefficient resource allocation. When other values are at stake, the goals of a particular firm or group of firms should give way to these other social interests. This may mean that the market for a particular set of products or for cost-raising technologies is reduced temporarily or permanently. Maintaining a broader social perspective is essential for producers who are truly dedicated to serving the national interest in health.

REFERENCES

1. 21 U.S.C. secs. 301-322.
2. U.S. Congress, Office of Technology Assessment, *Strategies for Medical Technology Assessment*, Washington, D.C., 1982.
3. Oregon Legislators Face Up to the Hard Job of 'Playing God,' *Washington Post National Weekly Edition*, February 15-21, 1988, p. 33.
4. C. Rosenberg, *The Care of Strangers: The Rise of America's Hospital System*, Basic Books, New York, 1987.
5. S. B. Foote, From Crutches to CT Scans: Business Government Relations and Medical Product Innovation, *Research in Corporate Social Performance and Policy*, vol. 8, Greenwich, Connecticut, JAI Press, 1986, pp. 3-28.
6. J. Lave and L. Lave, *The Hospital Construction Act: An Evaluation of the HIll-Burton Program, 1948-1973*, American Enterprise Institute, Washington, D.C., 1974.
7. S. B. Foote, Coexistence, Conflict and Cooperation: Public Policies Toward Medical Devices, *Journal of Health Politics, Policy and Law, 11*: 501-523 (1986).
8. M. E. Lewin, Financing Care for the Poor and Underinsured: An Overview, *The Health Policy Agenda: Some Critical Questions*, American Enterprise Institute, Washington, D.C., 1985.
9. U.S. Bureau fo the Census, 1975. Cited in S. B. Foote, From Crutches to CT Scans: Business-Government Relations and Medical Product Inno-

vation, *Research in Corporate Social Performance and Policy*, vol. 8, JAI Press, 1986, pp. 3-28

10. R. Rettig, Lessons Learned from the End-Stage Renal Disease Experience, in *Technology and the Quality of Health Care* (R. H. Egdahl, ed.), Germantown, MD, 1978.
11. P.L. 92-603.
12. U.S. Congress, Office of Technology Assessment, *Federal Policies and the Medical Devices Industry*, Washington, D.C., October 1984.
13. S. B. Foote, Loops and Loopholes: Hazardous Device Regulation Under the 1976 Medical Device Amendments to the Food, Drug and Cosmetic Act, *Ecology Law Quaterly, 7*:101-135 (1978).
14. H. A. Latin, Problem-Solving Behavior and Theories of Tort Liability, *California Law Review, 73*:677-746 (1985).
15. Peterson, R. D. and C. R. MacPhee, *Economic Organization in Medical Equipment and Supply*, Lexington Books, Lexington, MA, 1973.
16. U.S. Congress, Office of Technology Assessment, *Federal Policies and the Medical Devices Industry*, Washington, D.C., 1984.
17. S. B. Foote, From Crutches to CT Scan: Business-Government Relations and Medical Product Innovation, *Research in Corporate Social Policy and Performance*, vol. 8, JAI Press, Greenwich, Connecticut, 1986, pp. 3-28.
18. S. B. Foote, Coexistence, Conflict and Cooperation: Public Policies Toward Medical Devices, *Journal of Health Politics, Policy, and Law, 11*: 501-523 (1986).
19. National Heart, Lung, and Blood Institute, Working Group on Mechanical Circulatory Support, *Artificial Heart and Assist Devices: Direction, Needs, Costs, Society and Ethical Issues*, Washington, D.C., May 1985.
20. M. Brody, When Products Turn into Liabilities, *Fortune*, March 2, 1986, p. 20.
21. The costs of defending only four lawsuits amounted to $1.5 million dollars. Recently, a federal judge unsealed documents that allegedly show that Searle may have misled regulators about potential hazards associated with its controversial Copper-7 IUD. Monsanto's Searle May have Misled U.S. on Potential Hazards of IUM, Data Show, *The Wall Street Journal*, March 14, 1988, p. 2.
22. See, for example, S. B. Foote, Product Liability and Medical Device Regulation: Proposal for Reform, in *New Medical Devices: Factors Influencing Invention, Development and Use*, National Academy Press, Washington, D.C., 1988.
23. J. A. Morone and A. B. Dunham, Slouching Towards National Health Insurance: The New Health Care Politics, *Yale Journal on Regulation, 2*: 263-292 (1985).
24. M. Lipsky and M. Lounds, Citizen Participation and Health Care: Problems of Government Induced Participation, *Journal of Health Politics, Policy and Law, 1*:85-111 (1971).
25. *Jameson v. Bowen*, U.S. District Court, No. CV-F-83-547 REC.

part two

Historic Overview/Legal Framework of the Regulatory Environment

2

A History of Government Regulation of Adulteration and Misbranding of Medical Devices

PETER BARTON HUTT *Covington & Burling, Washington, D.C.*

I. INTRODUCTION

Although food, drug, and cosmetic products have been articles of commerce and subject to government regulation for centuries [1], medical devices have only recently been recognized as a separate and distinct category of health products and subjected to their own unique form of government regulation. This chapter traces the history of government concern about the need for regulation of medical devices, initial regulation under the Federal Food, Drug, and Cosmetic Act of 1938 [2], more comprehensive regulation under the Medical Device Amendments of 1976 [3], subsequent consideration of refinements in the existing law, and other related legislation that authorizes FDA to regulate medical devices.

II. THE INITIAL DEVELOPMENT OF MEDICAL DEVICES

Simple medical devices—wooden splints to hold broken bones in places, makeshift crutches, and homemade stretchers to carry the ill, to name just a few—have undoubtedly existed since before recorded history. Evidence of their existence is found in numerous archeological findings. Dental devices, for example, were used by the ancient Egyptians and Etruscans [4].

Perhaps the beginning of the association of exaggerated medical claims with mechanical and electrical devices can be traced to the arrival of Franz Anton Mesmer in Paris, France, in February 1778 [5-8]. Mesmer contended that "animal magnetism," the primary "agent of nature," was the source of all health. Those who were sick could be cured by recharing them with animal magnetism through the use of magnets and later large tubs in which were placed iron rods connecting patients to specially magnetized jars of water.

A Royal Commission convened by the medical establishment, consisting of such prominent scientists as Lavoisier and Benjamin Franklin, conducted tests and determined in 1784 that Mesmer's treatments were ineffective.

In the United States, the first renowned fraudulent medical device was marketed by Dr. Elisha Perkins in the late 1700s [9]. Perkins developed two rods of brass and iron, about three inches long. Called "Perkins' Patent Tractors," he sold them throughout the country for eliminating disease from the body. Even George Washington purchased a set for his family. After Perkins died in 1799, his son carried on the business. Within 10 years, however, it had been exposed as a fraud.

Throughout the 1800s, public and legislative attention in the United States was focused on the need for legislation to control the adulteration and misbranding of food and drugs, not medical devices [1]. Initially, these laws were enacted at the state and local levels. It was not until the early 1900s that Congress enacted broad nationwide legislation authorizing federal regulation of food and drugs, in the form of the Biologics Act of 1902 [10] and the Food and Drugs Act of 1906 [11]. When Congress itself enacted food and drug legislation for the District of Columbia in 1888 [12] and 1898 [13], for example, it covered only food, drugs, and cosmetics.

In January 1879, Dr. E. R. Squibb, in an address to the Medical Society of the State of New York, proposed the enactment of a national statute to regulate food and drugs [14]. Ten days later, Representative Wright introduced the first comprehensive legislation in Congress [15]. It was to take 27 years before such a law ultimately would be enacted by Congress as the Food and Drugs Act of 1906.

In those intervening 27 years, the Division (later Bureau) of Chemistry in the U.S. Department of Agriculture (USDA), under the leadership of Dr. Harvey W. Wiley, conducted investigation after investigation, and issued report after report, relating to the adulteration and misbranding of food and drugs. Research has uncovered no USDA report during that era on medical devices. The legislative history of the 1906 Act is, in fact, devoid of any mention of medical devices.

Other sources, however, reveal that the well-publicized problem of exaggerated health claims for patent medicines was beginning to extend to medical devices as well. One of the great crusaders of that era, Samuel Hopkins Adams, wrote a series of articles in *Collier's Weekly* during 1905 and 1906 on nostrums and quackery, which were later reprinted by the American Medical Association [16]. An article on "the specialist humbug" from the September 1, 1906, issue of *Collier's Weekly* describes an "electro-vibration" apparatus for the cure of deafness, "magneto-conservative garments" for the cure of a large number of ailments, and other mechanical "cures." Nonetheless, the 1906 Act did not cover medical devices.

Following enactment of the 1906 Act, regulation of fraudulent medical devices therefore fell to the U.S. Post Office, under the Postal Fraud Statutes: 18 U.S.C. 1341, which provides criminal penalties for mail fraud, and 39 U.S.C. 3005, which provides that any mail containing false or fraudulent representations may be declared unmailable by the Postmaster General [17,18]. The Post Office has enforced these statutes since they were first enacted in 1872 [19], with the assistance of FDA and its predecessor agencies, and continues to do so to this day. In 1929, for example, FDA collaborated with the Post Office Department on some 73 mail order fraud cases, including "a cap device alleged to grow hair, and a supposedly electric belt and insoles for the treatment of rheumatism and kidney ailments" [20].

Throughout this period, fraudulent medical devices flourished. A 1916 report by the American Medical Association solely devoted to "cures" for deafness cited several worthless devices [21]. One of the popular fraudulent devices of this era was the Abrams "dynamizer" machine [22]. By placing a blood sample in the machine, Abrams contended that he could diagnose the precise disease from which the individual was suffering and even the precise spot within the body where the disease had its focus. By the time that Abrams died in 1924, his machine was exposed as a fraud.

By 1917, it was clear to FDA that the law should be expanded to include authority over medical devices. In its Annual Report to Congress that year, FDA stated that the 1906 Act "has its serious limitations . . . which render it difficult to control . . . fraudulent mechanical devices used for therapeutic purposes [23]. The 1926 FDA Annual Report described exaggerated therapeutic claims made for products containing radium, including one that consisted of a glass bulb to be hung over the bed to cause the dispersion of "all thoughts and worry about work and troubles and brings contentment, satisfaction, and bodily comfort that soon results in peaceful, restful sleep" [24]. Thus, FDA monitored these products and assisted the Post Office in their regulatory actions, but could take no action of its own.

III. THE FEDERAL FOOD, DRUG, AND COSMETIC ACT OF 1938

A. The Legislative History

In June 1933, as part of the New Deal program, legislation was introduced to modernize and expand the 1906 Act [25]. Five years later, that legislation was enacted as the Federal Food, Drug, and Cosmetic Act of 1938 [2]. As introduced, the legislation covered medical devices as well as food, drugs, and cosmetics. This was accomplished by defining the term "drug" to include medical devices. The 1933 FDA Annual Report explained the need for this expansion of the law:

> Mechanical devices, represented as helpful in the cure of disease, may be harmful. Many of them serve a useful and definite purpose. The weak and ailing furnish a fertile field, however, for mechanical devices represented as potent in the treatment of many conditions for which there is no effective mechanical cure. The need for legal control of devices of this type is self-evident. Products and devices intended to effect changes in the physical structure of the body not necessarily associated with disease are extremely prevalent and, in some instances, capable of extreme harm. They are at this time almost wholly beyond the control of any Federal statute [26].

In reintroducing the Bill in 1934, Senator Royal Copeland, the chief sponsor of the legislation, provided this further explanation:

> The present law defines drugs as substances or mixtures of substances intended to be used for the cure, mitigation, or prevention of disease. This narrow definition permits escape from legal control of all therapeutic or curative devices like electric belts, for example. It also permits the escape of preparations which are intended to alter the structure or some function of the body, as, for example, preparations intended to reduce excessive weight. There are many worthless and some dangerous devices and preparations falling within these classifications. S. 2800 contains ample authority to control them [27].

During Senate debate on the legislation in April 1935, however, Senator Clark of Missouri contended that it was improper, as a matter of common English language, to classify medical devices as drugs:

> Mr. Clark. Mr. President, I should like to ask the Senator from New York how he can reconcile the language of this section and the language of the amendment with the common, ordinary acceptation of the English language. In other words, here he says it is proper to describe as a drug 'all substances, preparations, and devices intended for use in the diagnosis, cure, mitigation, treatment, or prevention of disease in man or other animals.' In other words, if a man has invented a shoulder brace, a purely mechanical device, which he claims will straighten a man's shoulders and expand his chest and make for his health, according to the definition contained in this paragraph it has to be described as a drug and treated in law as a drug.
>
> I should like to ask the Senator from New York to justify any such misuse of common, ordinary English terms.
>
> Mr. Copeland. The Senator from New York would have no objection to the proposal about the particular devices mentioned by the Senator. But there are on the market a great many devices which are offered for use, and citizens are exploited, believing that they can be cured of all sorts of ailments by the use of them. For example, there is such a thing as a radium belt carrying a disc alleged to contain radium; it is claimed that if the Senator from Missouri should wear that belt he would never have appendicitis or gall-bladder disease or perhaps any other ailment.
>
> Mr. Clark. The language the Senator from New York has employed in the bill is broad enough to cover any device of which the Food and Drug Bureau of the Agricultural Department chooses to take jurisdiction. The point I am making is that if the devices ought to be outlawed, they ought to be outlawed, and I have no objection to that; but to maintain that a purely mechanical device is a drug and to be treated as a drug in law and in logic and in lexicography is a palpable absurdity, in my opinion [28].

Senator Clark went on to say that calling a medical device a drug was like "calling a sheep's tail a leg" [28]. Although Senator Copeland continued to defend this choice of language [29], because of continued criticism of the definitional terminology [30] a separate definition of the term "device" was added to the legislation [31] and later agreed to without further discussion [32].

While the legislation that became the 1938 Act was pending in Congress, three influential books were published that documented some of the more flagrant examples of unproven medical devices: Kallet and Schlink's *100,000,000 Guinea Pigs* [33], Lamb's *American Chamber of Horrors* [34], and Cramp's *Nostrums and Quackery and Pseudo-Medicine* [35]. These added to the general support for a new medical device law.

Once the definition of "device" was separated from the definition of "drug," the device provisions of the statute paralleled the drug provisions except in one final respect. In November 1937, the Massengill Company marketed a new product, Elixir Sulfanilamide, using diethylene glycol as the solvent. During the next several weeks, more than 100 people died of diethylene glycol poisoning. After a 34-page report to Congress by the Secretary of Agriculture [36], Congress added a premarket notification requirement for new drugs to the pending legislation. There was no comparable provision added, however, for new medical devices. Thus, as the statute was enacted, the provisions relating to the adulteration and misbranding of drugs and devices were basically the same except for the premarket notification provisions that applied to new drugs.

B. The 1938 Act

The 1938 Act provided FDA with statutory authority to take formal or informal regulatory action against the adulteration or misbranding of medical devices. A device was deemed to be adulterated if it contained any filthy material, was prepared under unsanitary conditions, or differed from the quality represented in its own labeling. A device was deemed to be misbranded if its labeling was false or misleading in any particular, or it failed to bear a label containing the name and address of the manufacturer, the net quantity of contents, and adequate directions for use and warnings against misuse, or it was dangerous to health when used as recommended in its labeling. These provisions were, in short, very short and simple. They remain the law today.

C. Enforcement of the 1938 Act

FDA enforced the device provisions of the 1938 Act with vigor and determination. Paradoxically, just after FDA was given adequate statutory authority to police the safety and labeling of devices, a flood of fraudulent devices began to appear on the market. For the next 25 years, FDA's enforcement resources were strained to the limit to handle the regulatory action needed to deal with such widely distributed quack devices as the Drown Radio-therapeutic Instrument, the Zerret Applicator, the Vrilium Tube, the Reich Orgone Accumulator, the Hubbard Electropsychometer (or E-Meter), the Ghadiali Spectrochrome, the Relax-A-Cisor, and the Diapulse devices, to name only a few [7]. All of these devices were, in fact, the subject of successful regulatory action by FDA. They consumed such a large amount of Agency resources, however, that consideration was soon given to enactment of additional legislation to strengthen the FDA authority.

D. Consideration of Additional FDA Authority

Health, Education, and Welfare (HEW) Secretary Hobby appointed a Citizens Advisory on the FDA, which issued a 63-page report in June 1955 [37]. The Committee recommended "more extensive work on therapeutic devices making false claims of diagnostic and curative properties," while recognizing that additional resources were needed to implement this recommendation. A second Citizens Committee in 1962 [38] failed to focus specifically on medical devices, but the later Kinslow Report in 1969 [39] concluded that "FDA has completely inadequate resources and statutory authority to regulate this growing and highly sophisticated industry" and recommended interim measures until "a more comprehensive control system" could be enacted.

Beginning in 1959, Senator Estes Kefauver launched an extensive investigation into FDA regulation of the drug industry. Following the thalidomide tragedy in 1962, Congress enacted the Drug Amendments of 1962 [40] to strengthen the new drug regulatory system. The 1962 Amendments converted the new drug provisions of the law from premarket notification to premarket approval and required FDA to grant explicit approval of the safety and effectiveness of every new drug prior to marketing. While the 1962 Amendments were being considered in Congress, a companion bill was also being considered to require premarket approval of new medical device products under the same type of system as was applied to new drugs [41,42]. As part of the agreement that resulted in enactment of the 1962 Amendments, however, all provisions relating to medical devices were deleted from the final legislation [43]. It was the understanding, at the time, that Congress would return to the matter of device legislation within a matter of months. In fact, this process consumed another 15 years.

Although much of FDA's regulatory activities with respect to regulation of medical devices following enactment of the 1938 Act involved enforcement action against fraudulent devices, the revolution in biomedical technology following World War II resulted in the introduction of a wide variety of important new life-saving medical devices that also demanded attention from the Agency. Thus, FDA devoted more and more effort to assuring the safety and effectiveness of these devices as well as protecting the public against fraudulent devices.

During 1962 to 1969, a wide variety of bills were introduced in Congress to provide FDA with increased authority over medical devices, ranging from premarket approval to appointment of a national commission to study the matter. President Kennedy urged enactment of new medical device authority in his consumer message in March 1962 [44]. President Johnson made the same recommendation in his consumer messages in February 1964 [45], March 1966 [46], and February 1967 [47]. When President Nixon assumed office and did not immediately issue a consumer message, the Democratic Study Group Task Force on Consumer Affairs again pushed for a new medical device legislation [48].

The February 1967 consumer message of President Johnson contained the most detailed description of proposed legislation:

> 1. *Insuring the Safety and Effectiveness of Medical Devices*
>
> Under present law, dangerous and worthless devices may be marketed until the Government—sometimes by chance, sometimes by complaint—

discovers them and gathers the necessary evidence to establish that they are hazardous or ineffective. This is a laborious process. It requires many months. It is costly.

In the meantime, the elderly and the seriously ill suffer most. Improper treatment with worthless devices can be the cruelest hoax of all.

We want to foster continued research and development of life-saving devices. But we must be sure they have been adequately tested before they are put on the market. We cannot be sure today.

Congressional testimony has revealed that

—Defective nails and screws for bone repair have required repeated operations to correct the damage.
—Some artificial eyes have resulted in serious infection.
—Useless heating and vibrating devices have caused the ill to squander their money and delay the pursuit of effective treatment.
—X-ray machines, which could have been properly safeguarded at little cost, emitted excessive doses of radiation.

I recommend the Medical Device Safety Act of 1967.

Under this Act, the Food and Drug Administration would be required to pre-clear certain therapeutic materials—such as artificial organ transplants—used mainly on or in the body. In addition, the FDA will establish standards to assure the safety and performance of certain classes of widely used devices—bone pins, catheters, x-ray equipment, and diathermy machines.

In every case, the rights of the parties will be protected by fair hearings.

This new law will not apply to simple and ordinary patient care items which have withstood the test of time and are generally recognized as safe and reliable. It will not apply to an item specially ordered or designed by a surgeon or physician. Nor will it inhibit the research and development essential to the advancement of the medical arts. It will, however, protect physician and patient alike from devices which are dangerous and unreliable.

Nonethless, there were no further congressional hearings and thus no serious consideration of proposed new device legislation during the 1960s.

E. The Court Cases and the Cooper Committee

While the legislation was pending, FDA moved ahead with its routine enforcement activities. As a matter of conscious policy, moreover, the Agency sought to classify as drugs, rather than as medical devices, important new products that arguably could be placed under either category, in order to achieve maximum regulatory control. Because the definitions of "drug" and "device" in the 1938 Act were virtually identical, and thus overlapped in many respects, FDA concluded that there was a strong possibility that the courts would agree with their policy of strengthening regulatory control through this administrative action.

In February 1968, the first of two important court decisions was handed down, endorsing the FDA approach. In *AMP Inc. v. Gardner* [49], the U.S. Court of Appeals for the Second Circuit held that a product consisting of a disposable applicator, a nylon ligature loop, and a nylon locking disc, used to ligate severed blood vessels during surgery, was a drug. The court determined that the product was essentially a suture, and because sutures are listed in official compendia as drugs the product was properly regarded as a drug. The court gave an expansive interpretation to the statute to be consistent with the congressional purpose "to keep inadequately tested medical and related products which might cause widespread danger to human life out of interstate commerce."

A year later, in April 1969, the Supreme Court held in *United States v. An Article of Drug . . . Bacto-Unidisk* [50] that an antibiotic sensitivity disc used as a laboratory screening test to help determine the proper antibiotic drug to administer to patients is a drug rather than a device. The Supreme Court decision was based primarily on public policy grounds. The Court first noted "the well-accepted principle that remedial legislation such as the Food, Drug, and Cosmetic Act is to be given a liberal construction consistent with the Act's overriding purpose to protect the public health" and then concluded that:

> the legislative history, read in light of the statute's remedial purpose, directs us to read the classification "drug" broadly, and to confine the device exception as nearly as is possible to the types of items Congress suggested in the debates, such as electric belts, quack diagnostic scales, and therapeutic lamps, as well as bathroom weight scales, shoulder braces, air conditioning units, and crutches.

This definitive statement from the Supreme Court gave FDA a clear directive to begin to reclassify as drugs a large number of borderline medical products that had previously been regulated as devices.

Thus, the need for legislative resolution of the matter became clear to all interested groups. Indeed, a national conference on the need for new medical device legislation was convened in Washington in September 1969 and the results were summarized in the *Journal of the American Medical Association* in December 1969 [51].

In his October 1969 consumer message, President Nixon joined his predecessors in endorsing medical device legislation, but announced that a thorough study would be undertaken before specific legislation was proposed:

> Another important medical safety problem concerns medical devices—equipment ranging from contact lenses and hearing aids to artificial valves which are implanted in the body. Certain minimum standards should be established for such devices; the government should be given additional authority to require premarket clearance in certain cases. The scope and nature of any legislation in this area must be carefully considered, and the Department of Health, Education, and Welfare is undertaking a thorough study of medical device regulation. I will receive the results of that study early in 1970 [52].

The Secretary of HEW promptly established a Study Group on Medical Devices, chaired by Dr. Theodore Cooper, then Director of the National Heart and Lung Institute. The Cooper Committee consisted of ten government officials—two from FDA, five from various parts of the National Institutes of Health (NIH), and three from other parts of HEW. The report, issued in September 1970 [53,54], concluded that medical devices present entirely different issues from new drugs. It rejected the approach that had been included in most of the pending legislation of simply applying the new drug provisions of existing law to new medical devices. Instead, the report recommended a different regulatory approach, designed specifically to deal with the breadth and diversity of medical devices.

The Booper Committee Report recommended two immediate steps: an inventory of all medical devices already on the market, and an initial classification of those marketed devices to determine those that should be subject to premarket approval by FDA, those for which performance standards would be an adequate form of regulation, and those for which neither premarket approval nor standards are required. Where premarket approval is required, the report recommended the establishment of standing advisory review panels as an integral part of the approval system. To implement this broad new concept, the report also recommended a number of specific details, including requirements for records and reports, registration and inspection of establishment, good manufacturing practices, and related provisions.

IV. THE MEDICAL DEVICE AMENDMENTS OF 1976

A. The Legislative History

Immediately upon receipt of the Cooper Committee Report, both administrative and legislative activity began to take shape. These two independent activities became intertwined during the next 6 years, culminating in enactment and then implementation of the Medical Device Amendments of 1976 [3].

After receipt of the Cooper Committee Report, the Department of HEW drafted legislation to embody its recommendations and circulated it within the government for comment. It took a full year—largely to accommodate changes demanded by the Department of Commerce and the Department of Defense—before the Administration's bill was introduced in December 1971 [55].

To expedite the legislative process, FDA itself undertook administrative action designed to implement the Cooper Committee Report and to prepare for the pending legislation. In 1971, FDA Commissioner Charles C. Edwards transferred the Office of Medical Devices from the Bureau of Drugs to the Office of the Associate Commissioner for Medical Affairs [56]. Following receipt of the Cooper Committee Report, FDA created a Medical Device Advisory Committee, undertook an inventory of existing medical devices on the market, and began the process of classifying medical devices [57]. FDA had hoped that the pending legislation would be enacted before the final classification process actually began, but Congress did not move quickly enough. Thus, in 1973 the Agency began to establish 14 classification panels of experts and in May 1975 it published a general notice advising device manufacturers about the medical device classification procedures [58]. Classification reports were issued by these panels before the legislation was enacted.

Beginning in 1970, FDA established its first device performance standard for impact-resistant lenses in eyeglasses and sunglasses [59]. In 1972, FDA chose to regulate in vitro diagnostic products as devices rather than drugs, and to establish a comprehensive system of labeling and performance standards rather than to require new drug applications, in reliance on the pending legislation [60]. Finally, in 1974, FDA Commissioner Alexander M. Schmidt took the Office of Medical Devices out of the Office of the Associate Commissioner for Medical Affairs and created a new Bureau of Medical Devices and Diagnostic Products [61].

These administrative initiatives grated on members of Congress and their staff, who argued that FDA should wait for specific statutory authority [62]. Nonetheless, they served a very important function of keeping pressure on Congress to enact the new legislation.

In the Senate, hearings were held on medical device legislation in September 1973 [63], S. 2368 was favorably reported by the Senate Committee on Labor and Public Welfare in January 1974 [64], and the bill was passed by the Senate in February 1974 [65]. Because the legislation did not pass the House that year, it was again considered by the Senate in the next session of Congress. After hearings on the Dalkon Shield in January 1975 [66], the Senate again favorably reported device legislation in the form of S. 510 in March 1975 [67], which was passed by the Senate in April 1975 [68].

In the House, hearings were held on IUD devices during May and June 1973 [69], and specifically on the pending device legislation in October 1973 [70], but no further action was taken on the legislation that year. By late 1974, the Senate had twice passed the legislation and, as was described above, FDA had itself begun to take independent administrative action to force the hand of Congress.

In a marathon series of drafting sessions, four individuals (representing FDA, the device industry, the House Subcommittee, and the House Office of Legislative Counsel) revised the entire legislation during December 1974 and January 1975, to clarify the intended congressional policy. The revised legislation was introduced as H.R. 5545, on which hearings were held in July 1975 [71]. After markup in November 1975, this legislation was reported out of Committee in February 1976 [72], and was passed by the House in March 1976 [73]. The House and Senate conferees met in March 1976 and issued a Conference Report in May 1976 [74]. Both houses concurred, and the bill was signed into law by the President on May 28, 1976 [3].

B. The Provisions of the 1976 Amendments

The final provisions of the 1976 Amendments [3] are remarkably faithful to the Cooper Committee Report. The 1976 Amendments did require an inventory and classification of all medical devices into those that require general controls (Class I), performance standards (Class II), and premarket approval (Class III). For all medical devices, the general provisions of the 1938 Act were substantially strengthened. FDA was required to be notified of every medical device prior to marketing. Devices that were substantially equivalent to pre-1976 devices could be marketed immediately, subject to any existing or future requirements for that type of device. Pre-1976 Class III devices could be required to submit proof of safety and effectiveness to FDA, but no statutory timetable or deadlines were required for this process. FDA was authorized to require the development of performance standards for Class II devices, but this was not made mandatory and was left to the sole discretion of FDA according to Agency priorities and resources. Registration of device establishments was required. Good manufacturing practice regulations were authorized. FDA was given authority to deal with the notification, repair, replacement, and refund of defective devices, and was authorized to ban any device that presents a substantial deception or substantial unreasonable risk of injury or illness. Thus, the final law greatly strengthened FDA authority to regulate medical devices but retained the fundamental concept of the Cooper Committee Report that regulation should be carefully tailored to the type of device involved.

In general, FDA has implemented the 1976 Amendments efficiently and effectively, avoiding both overregulation and underregulation [75,76]. The Agency has, in short, adhered to the Cooper Committee Report and the Congressional mandate to regulate medical devices in proportion to the degree of risk presented by the device. There have been no reported medical catastrophes caused by devices that can reasonably be attributed to a failure to implement the 1976 Amendments properly. FDA's approach to medical device regulation has raised the standards of the entire device industry, and thus greatly benefited the public health, without imposing unnecessary cost. The kinds of quack medical devices that were still marketed in the 1960s have been eliminated. A 1984 Office of Technology Assessment (OTA) report concluded that, "while regulatory costs have been incurred, regulation has generally not had a significant negative impact on the industry" [77]. Thus, FDA has succeeded in exerting sufficient regulatory control to protect the public health while at the same time avoiding overregulation that would discourage medical device innovation and thus harm the public health.

There has, of course, been criticism of FDA's regulation of medical devices under the 1976 Amendments. Some in industry have contended that FDA has stifled innovation, some consumer activists have contended that FDA has failed to protect the public health, and some in Congress have contended that the legislation has not been implemented in the way that they would prefer. Much of this criticism, including an adverse congressional report [78], has come from individuals who were not involved with the enactment of the legislation, were not aware of the reasons why the final legislation reflected the policy that it embodies, or simply disagree with the final version. All of the suggestions for useful administrative changes and legislative clarification of the 1976 Amendments [79,80] could be accommodated by administrative action taken within the broad discretion and flexibility given to FDA under the statute. A 1983 General Accounting Office (GAO) report [81] identified a few areas where FDA could strengthen its administration of the law and suggested some areas where the law itself might be amended, and a 1988 GAO report [82] suggested better documentation of "substantial equivalence" decisions under Section 510(k). Like the 1984 OTA report [77], however, they did not reflect significant criticism of FDA's overall approach.

FDA has responded systematically to these criticisms. In accordance with FDA Action Plan I [83], 10 task forces were established in 1984, from which have emerged a number of important administrative improvements [84, 85] and additional new goals [86]. Not all of the criticisms can or should be accommodated by FDA, however, because of resource constraints, priority decisions, and valid policy judgments.

Since 1976, the possibility of revising the Amendments in various different ways has been continuously discussed. A brief House oversight hearing on medical device regulation was held in May 1987 [87]. In December 1987, relatively simple clarifying amendments were proposed by FDA in S. 1928 [88]. In 1988, H.R. 4640, the Medical Device Improvements Act of 1988, was reported [89] and passed by the House [90]. The legislation was not the subject of hearings in the House, or of hearings or committee consideration in the Senate. Because it was controversial, the bill was not enacted and will likely be further considered in future sessions of Congress.

The 1976 Amendments specifically provided that any product that was regulated as a new drug before 1976 but as a device after 1976 was a "transitional" device that would automatically be classified in Class III. Although FDA was authorized to reclassify transitional devices under the same procedures and criteria that apply to all other devices, the Agency has not done so, largely because of the statutory prohibition against the use of proprietary industry data to support any reclassification. Small contact lens manufacturers have pushed for legislation to open up the soft contact lens market by reclassifying these transitional devices into Class II. S. 1808, which was favorably reported out of committee in October 1988 [91], would have required FDA to review the classification of all transitional devices and specifically to reclassify soft contact lens in Class II unless FDA affirmatively retained them in Class III. The bill was not enacted and will undoubtedly be considered further by Congress in the future.

V. ADDITIONAL FDA STATUTORY AUTHORITY RELATING TO MEDICAL DEVICES

A. The Radiation Control for Health and Safety Act of 1968

In 1968, Congress enacted the Radiation Control for Health and Safety Act, to control harmful radiation emissions from electronic products [92]. Responsibility for administration of this statute was delegated to FDA in 1971 [93]. It was initially administered by the FDA Bureau of Radiological Health, which was combined with the Bureau of Medical Devices in 1982 to form the current Center for Devices and Radiological Health [94]. Under this statute, FDA has issued performance standards for X-ray systems, computed tomography (CT) equipment, laser products, sunlamps, ultrasonic therapy products, and other related products [95]. Since 1968, there has been no serious consideration of amending the statute.

B. Drug Price Competition and Patent Term Restoration Act of 1984

In 1984, Congress enacted legislation providing for stronger protection of drug patents, combined with easier entry of generic drugs into the market after the patent and any period of market exclusivity have expired [96]. As part of that legislation, Congress awarded up to 5 years of patent term restoration for medical devices that are subject to a regulatory review period that would otherwise reduce the effective patent life for the product [97]. Medical devices subject to premarket approval by FDA have, in fact, now received patent term restoration under this statute [98].

C. The Cardiac Pacemaker Registry Act of 1984

As part of the Deficit Reduction Act of 1984, FDA was required to establish a national registry of all cardiac pacemaker devices and leads implanted or removed, for which Medicare makes payment [99]. The purposes of the registry are both to assist the Health Care Financing Administration (HCFA) in determining proper Medicare payments and to assist FDA in determining the compliance of the devices with regulatory requirements. Regulations governing the registry have been promulgated [100,101].

D. The Orphan Drug Act

Congress initially enacted the Orphan Drug Act in 1983 [102] to provide additional incentives for industry to make the investment necessary to develop drugs for rare diseases. In 1984, the Act was amended specifically to provide that any drug with a target patient population of fewer than 200,000 people is automatically regarded as an orphan drug [103]. As part of the Orphan Drug Amendments of 1988, medical devices have been added to the provision of the 1983 Act that allows financial assistance to defray the cost of developing products for rare diseases or conditions [104]. The 1988 Amendments also require the Department of Health and Human Services (HHS) to conduct a study to determine whether the application of the other statutory incentives that apply to orphan drugs should also be applied to orphan medical devices in order to encourage the development of such devices. It is likely that, at some time in the future, all of the provisions that relate to orphan drugs will also be applied to orphan devices.

VI. CONCLUSION

The two major milestones in FDA regulation of medical devices occurred when devices were first brought under federal regulatory control in 1938 and when they were first subjected to intense premarket review in 1976. On both of these occasions, Congress sought to balance the need to protect the public against adulteration and misbranding of medical devices against the need to foster the development of innovative new life-saving medical devices. FDA, in turn, has sought to achieve the same balance. Where problems have occurred, FDA has stepped in to handle them. Where major technological advances have become feasible, such as the development of monoclonal antibodies for diagnostic use, on the other hand, FDA has found ways to permit their marketing with a minimum of regulatory requirements. In this way, Congress and FDA have forged a unique regulatory system that is truly designed to enhance the public health.

REFERENCES

1. P. B. Hutt and P. B. Hutt II, *A History of Government Regulation of Adulteration and Misbranding of Food*, 39 FDC L.J. 2 (1984).
2. 52 Stat. 1040 (1938, 21 U.S.C. 301 et seq.
3. 90 Stat. 539 (1976).
4. L. Kanner, *History of Dentistry: Folklore of the Teeth*, p. 205 (1936).
5. R. Darnton, *Mesmerism* (1968).
6. R. C. Fuller, *Mesmerism and the American Cure of Souls* (1982).
7. W. F. Janssen, *The Gadgeteers*, in S. Barrett and G. Knight, *The Health Robbers*, chapter 16 (1976).
8. G. Bolch, tr., *Mesmerism: A Translation of the Original Scientific and Medical Writings of F. A. Mesmer* (1980).
9. M. Fishbein, *The Medical Follies*, chapter 11 (1925).
10. 32 Stat. 728 (1902).
11. 34 Stat. 768 (1906).
12. 25 Stat. 549 (1888).
13. 30 Stat. 246 (1989).
14. E. R. Squibb, *Proposed Legislation on the Adulteration of Food and Medicine* (January 1879).
15. H.R. 5916, 45th Cong., 3d Sess. (1879).
16. S. H. Adams, *The Great American Fraud* (various editions, 1905-1912).
17. F. M. Hart, *The Postal Fraud Statutes: Their Use and Abuse*, 11 FDC L.J. 245 (1956).
18. D. G. Crumbaugh, *Survey of the Law of Mail Fraud*, 1975 U. Ill. L. Forum 237 (1975).
19. J. H. Young, *The Medical Messiahs*, chapters IV and XIII (1967).
20. FDA 1929 Annual Report at 17.
21. American Medical Association, *Deafness Cures* (1916).
22. M. Fishbein, *The Medical Follies*, chapter VI (1925).
23. FDA 1917 Annual Report at 16.
24. FDA 1926 Annual Report at 26.
25. S. 1944, 73d Cong., 1st Sess. (1933).
26. FDA 1933 Annual Report at 13.
27. 78 Cong. Rec. 8960 (May 16, 1934).
28. 79 Cong. Rec. 4841 (April 2, 1935).
29. 79 Cong. Rec. 4842 (April 2, 1935).
30. 79 Cong. Rec. 4842-4844 (April 2, 1935).
31. S. Rep. No. 646, 74th Cong., 1st Sess. (1935).
32. 79 Cong. Rec. 8351-8355 (May 28, 1935).
33. A. Kallet and F. J. Schlink, *100,000,000 Guinea Pigs: Dangers in Everyday Foods, Drugs, and Cosmetics* (1933).
34. R. Lamb, *American Chamber of Horrors: The Truth About Food and Drugs* (1936).
35. A. J. Cramp, *Nostrums and Quackery and Pseudo-Medicine* (1936).
36. "Elixir Sulfanilamide," S. Doc. No. 124, 75th Cong., 2d Sess. (1937).
37. "Report of the Citizens Advisory Committee on the Food and Drug Administration," H.R. Doc. No. 227, 84th Cong., 1st Sess. (1955).
38. *Report of the Citizens Advisory Committee on the Food and Drug Administration* (1962).

39. FDA, *Report from the Study Group on Food and Drug Administration Consumer Protection Objectives and Programs*, 36, 38 (July 1969).
40. 76 Stat. 780 (1962).
41. H.R. 11582, 87th Cong., 2d Sess. (1962).
42. "Drug Industry Act of 1962," Hearings before the Committee on Interstate and Foreign Commerce, House of Representatives, 87th Cong., 2d Sess. (1962).
43. 108 Cong. Rec. 21049, 21058-21062 (September 27, 1962).
44. "Strengthening of Programs for Protection of Consumer Interests," 108 Cong. Rec. 4167, 4169 (March 15, 1962).
45. "Consumer Interests," 110 Cong. Rev. 1958, 1959 (February 5, 1964).
46. "Consumer Interests," 2 Weekly Compilation of Presidential Documents 422, 426-427 (March 28, 1966).
47. "Protecting the American Consumer," 3 Weekly Compilation of presidential Documents 261, 263, 267-268 (February 20, 1967).
48. 115 Cong. Rev. 23149 (August 11, 1969).
49. 389 F.2d 825 (2d Cir. 1968).
50. 394 U.S. 784 (1969).
51. *Excerpts and Summary of a National Conference on Medical Devices*, 210 J.A.M.A. 1745 (December 1, 1969).
52. "Consumer Protection," 5 Weekly Compilation of Presidential Documents 1516, 1523 (November 3, 1969).
53. Study Group on Medical Devices, *Medical Devices: A Legislative Plan* (September 1970).
54. T. Cooper, *Device Legislation*, 26 FDC L.J. 165 (1971).
55. H.R. 12316, 92d Cong., 1st Sess. (1971).
56. D. M. Link and L. R. Pilot, *FDA's Medical Device Program*, 6 FDA Papers 24 (May 1972).
57. D. M. Link, *Cooper Committee Report and Its Effect on Current FDA Medical Device Activities*, 27 FDC L.J. 624 (1972).
58. 40 Fed. Reg. 21848 (May 19, 1975).
59. 21 C.F.R. 801.410.
60. 21 C.F.R. Part 809.
61. 39 Fed. Reg. 5812 (February 15, 1974).
62. P. G. Rogers, *Medical Device Law—Intent and Implementation*, 36 FDC L.J. 4 (1981).
63. "Medical Device Amendments, 1973," Hearings before the Subcommittee on Health of the Committee on Labor and Public Welfare, United States Senate, 93d Cong., 1st Sess. (1973).
64. S. Rep. No. 93-670, 93d Cong., 2d Sess. (1974).
65. 120 Cong. Rec. 1798 (February 1, 1974).
66. "Food and Drug Administration Practice and Procedure, 1975," Joint Hearings before the Subcommittee on Health of the Committee on Labor and Public Welfare and the Subcommittee on Administrative Practice and Procedure of the Committee on the Judiciary, United States Senate, 94th Cong., 1st Sess. (1975).
67. S. Rep. No. 94-33, 94th Cong., 1st Sess. (1975).
68. 121 Cong. Rec. 10701 (April 17, 1975).
69. "Regulation of Medical Devices (Intrauterine Contraceptive Devices)," Hearings before a Subcommittee of the Committee on Government Opera-

tions, House of Representatives, 93rd Cong., 1st Sess. (1973).
70. "Medical Devices," Hearings before the Subcommittee on Public Health and Environment of the Committee on Interstate and Foreign Commerce, House of Representatives, 93d Cong., 1st Sess. (1973).
71. "Medical Device Amendments of 1975," Hearings before the Subcommittee on Health and the Environment of the Committee on Interstate and Foreign Commerce, House of Representatives, 94th Cong., 1st Sess. (1975).
72. H. Rep. No. 94-853, 94th Cong., 2d Sess. (1976).
73. 122 Cong. Rec. 5875 (March 9, 1976).
74. H.R. Rep. No. 1090, 94th Cong., 2d Sess. (1976).
75. P. B. Hutt, *Medical Device Regulation: Reasonable, Workable*, 3 Legal Times of Washington, No. 48, at 12 (May 5, 1980).
76. P. B. Hutt, *Legal Aspects of Introducing New Biomaterials*, in J. W. Boretos, ed., *Contemporary Biomaterials: Material and Host Response, Clinical Applications, New Technology and Legal Aspects*, p. 645 (1984).
77. Office of Technology Assessment, *Federal Policies and the Medical Devices Industry* 126 (OTA-H-230, 1984).
78. "Medical Device Regulation: The FDA's Neglected Child," Committee Print 98-F, 98th Cong., 1st Sess. (1983).
79. Health Industry Manufacturers Association, *Report and Recommendations of the HIMA Device and Diagnostic Product Approval Task Force* (October 1985).
80. D. A. Kessler et al., *Federal Regulation of Medical Devices*, 317 N. Engl. J. Med. 357 (August 6, 1987).
81. General Accounting Office, *Federal Regulation of Medical Devices—Problems Still to be Overcome* (HRD-83-53, 1983).
82. General Acocunting Office, *Medical Devices: FDA's 510(k) Operations Could be Improved* (PEMD-88-14, 1988).
83. Food and Drug Administration, *A Plan for Action* 17-22 (July 1985).
84. J. S. Benson et al., *The FDA's Regulation of Medical Devices: A Decade of Change*, 43 FDC L.J. 495 (1988).
85. Food and Drug Administration, *Executive Summary of the Criticisms Task Forces' Reports* (1985).
86. Food and Drug Administration, *A Plan for Action: Phase II* 8-9 (May 1987).
87. "Medical Devices and Drug Issues," Hearings before the Subcommittee on Health and the Environment of the Committee on Energy and Commerce, House of Representatives, 100th Cong., 1st Sess. 331 (1987).
88. S. 1928, 100th Cong., 1st Sess. (1987).
89. H.R. Rep. No. 100-782, 100th Cong., 2d Sess. (1988).
90. 134 Cong. Rec. H5848 (July 26, 1988) (daily ed).
91. S. Rep. No. 100-588, 100th Cong., 2d Sess. (1988).
92. 82 Stat. 1173 (1968), 42 U.S.C. 263b et seq.
93. 36 Fed. Reg. 12803 (July 7, 1971).
94. 47 Fed. Reg. 44614 (October 8, 1982).
95. 21 C.F.R. subchapt. J.
96. 98 Stat. 1585 (1984).
97. 35 U.S.C. 156.
98. 51 Fed. Reg. 34143 (September 25, 1986).
99. 98 Stat. 494, 1068 (1984), 42 U.S.C. 1395y(h).

100. 52 Fed. Reg. 27756 (July 23, 1987).
101. 21 C.F.R. part 805.
102. 96 Stat. 2049 (1983).
103. 98 Stat. 2815, 2817 (1984).
104. 102 Stat. 90 (1988).

3

Medical Device Regulation: The Big Picture

WILLIAM D. APPLER and GAILE L. McMANN *McDermott, Will & Emery, Washington, D.C.*

Companies entering the medical device arena are confronted with a broad spectrum of statutes and regulations. Some of these legal requirements are applicable because a firm is producing a medical device. Other laws are triggered by the mere fact that a company is doing business in the United States. With respect to medical devices per se, the Federal Food, Drug and Cosmetic Act is the primary statute with which device firms must contend. Accordingly, we begin this introductory chapter with an overview of this law. The remainder of the chapter addresses some of the less obvious (and perhaps easily overlooked) statutes and regulatory agencies that can affect companies that market medical devices.

I. INTRODUCTION

Medical devices are an extraordinarily heterogeneous category of products. The term "medical device" includes such technologically simple articles as ice bags and tongue depressors. On the other end of the continuum, very sophisticated articles such as pacemakers and surgical lasers are also medical devices. Perhaps it is this diversity of products coupled with the sheer number of different devices that makes the development of an effective and efficient regulatory scheme a unique challenge for the Congress and the Food and Drug Administration.

Historically, medical devices have been neglected from a legislative and regulatory perspective. In the early 1900s, Congressional attention focused on foods and drugs. The Pure Food and Drugs Act of 1906 [1] was passed to prohibit the distribution of adulterated or misbranded food and drugs in interstate commerce. This legislation, however, did not include any provisions to enable the Food and Drug Administration (FDA) to regulate medical devices. Thus, legitimate and bogus medical devices were freely marketed without any effective check on the safety of these articles or the accuracy of their claims.

II. MEDICAL DEVICE REGULATION PRIOR TO 1976

It was not until 1938, when the Pure Food and Drugs Act of 1906 underwent extensive revision, that the Congress expressly empowered the federal government to regulate medical devices. The Federal Food, Drug and Cosmetic Act of 1938 [2] expanded the FDA's regulatory control over foods and drugs and extended the Agency's authority to include medical devices and cosmetics.

The most significant rationale for authorizing FDA to regulate medical devices was the mounting level of consumer fraud. In the years preceding the Act of 1938, medical devices were marketed that touted false therapeutic claims. Many of these devices were patently harmful; others, by virtue of their bogus therapeutic claims, delayed consumers from seeking proper medical attention. Thus, a growing concern evolved—the public welfare was in jeopardy unless a mechanism was established to regulate the safety and reliability of medical devices.

The FDA's regulatory authority over medical devices granted by the Federal Food, Drug and Cosmetic Act of 1938 remained unchanged until the mid-1970s. During the same time period, however, the regulatory environment for foods and drugs changed dramatically. For example, in 1951, the Durham Humphrey Amendment [3] established the criteria for determining which drugs are suitable for lay use and which drugs must be restricted to prescription use. The 1958 Food Additive Amendments [4] to the Federal Food, Drug and Cosmetic Act established a new scheme for regulating foods and substances added to food products. And of course, in the wake of the thalidomide disaster, the Federal Food, Drug and Cosmetic Act was significantly amended in 1962 [5] to require that new drugs satisfy stringent approval criteria, which include safety and efficacy determinations. Thus, from 1938 to 1976, food and drugs received the lion's share of legislative and regulatory attention.

In the late 1960s and early 1970s there was some interest expressed by administrative officials and members of Congress in improving the regulatory framework for medical devices. Although some of this interest culminated in device regulation bills, formal legislation was not enacted until May 1976. Accordingly, until 1976, medical devices remained subject to the relatively ineffective regulatory provisions established in the 1938 Act.

One of the primary impetuses for the Medical Device Amendments was FDA's inability to regulate devices effectively under the authority granted by the Federal Food, Drug and Cosmetic Act of 1938. In the 1938 Act, the term "device" was defined as "any instrument, apparatus, or contrivance, including any of its components, parts, or accessories, intended (1) for use in the diagnosis, cure, mitigation, treatment, or prevention of disease in man or other animals; or (2) to affect the structure or any function of the body of man or other animals." Moreover, under the 1938 Act, FDA's authority was limited to medical devices introduced, or offered for introduction, into interstate commerce that were deemed to be misbranded or adulterated.

The authority to prohibit adulterated and misbranded articles from entering interstate commerce applies to all products regulated by FDA. Generally, an article is adulterated if it is composed of any filthy, putrid, or decomposed substance, or if it was manufactured or packaged in an unsanitary environment. Misbranded articles are those products whose labeling is false or misleading in any particular.

The Agency's jurisdiction is limited to products that travel in interstate commerce (i.e., cross state borders). Thus, from a legal standpoint, FDA does not have authority over products that are manufactured, marketed, and used within the confines of a state's borders. This general limitation on FDA authority, however, does not apply to medical devices. There is a presumption that all medical devices or their components have, at sometime, crossed state lines and therefore are subject to FDA's federal regulatory authority.

The 1938 Act placed FDA in a "reactive" posture. FDA had no regulatory power to prevent a device from entering the market; rather, the Agency was required to initiate action against the product after it was determined to be misbranded or adulterated. FDA's power to remove a fraudulent or unsafe device from the market consisted of seeking a court order to seize the device, requesting an injunction, or pursuing a criminal prosecution. The shortcomings of this limited regulatory authority were demonstrated by the enormous expenditure of government time and resources to stop the marketing of a few bogus devices.

The Diapulse device was an illustrative example of the time-consuming and inefficient process in which FDA became entangled when trying to remove a fraudulent device from the market. Diapulse was a heat-generating device marketed to health care practitioners in the mid-1960s. This "wonder" device bore numerous therapeutic claims, none of which could be substantiated by the manufacturer with valid scientific data. FDA first seized the product in 1965. However, due to protracted court proceedings and appeals, many years elapsed before the Agency obtained a final injunction against the device manufacturer [6].

At about the same time that the extreme inefficiency of FDA's enforcement actions against medical devices was fully recognized, the Agency received a regulatory boost from the courts. In the late 1960s, two court cases upheld FDA's power to regulate as drugs products that were generally considered to be devices. In *AMP, Inc. v. Gardner* [7], the court determined that an apparatus containing sutures used to ligate severed blood vessels was a drug, and therefore subject to the premarket approval procedures of the 1962 Amendments to the Federal Food, Drug and Cosmetic Act. The court found that the product was essentially a suture, which could fall within the statutory definition of a drug or a device. The court then reasoned that since the product presented some of the same health concerns that are created by drugs, a liberal interpretation of the Federal Food, Drug and Cosmetic Act should be applied in order to protect the public.

One year later, the U.S. Supreme Court upheld a determination by FDA that a cardboard disc, impregnated with various antibiotics and used to assess a patient's sensitivity to these antibiotics, was a drug [8]. After reviewing the legislative history of the 1938 Act, the Court determined that the definition of a "drug" should be construed broadly. In contrast, the Court stated that the definition of a "device" should be narrowly interpreted to include patently quack contraptions and items of a purely mechanical nature (e.g., crutches). Undoubtedly, the Court's opinion demonstrated a desire to protect the public's health by expanding FDA's premarket approval authority for drugs to other articles that could potentially cause harm if not adequately regulated.

The cumbersome and inefficient regulatory authority provided to FDA under the 1938 Act, coupled with several restrictive interpretations of the term "device" by the courts, helped set the stage for new legislation specifically addressing medical devices. But, perhaps the most important factor that spurred the Congress into legislative action, was the growing sophistication and complexity of medical devices.

Early medical devices were generally rather simple articles that could be used safely and appropriately without detailed or extensive instructions to health care practitioners. By the 1960s, however, medical technology was advancing at a rapid pace. Sophisticated and complex devices such as cardiac pacemakers, artificial heart valves, intrauterine contraceptives, and intraocular lenses were being developed and used in patients. This escalation in device technology mandated the establishment of a new administrative mechanism to ensure that medical devices were properly tested and manufactured before they entered the market. Thus, a shift in regulatory emphasis began to appear—away from action targeted primarily at fraudulent or unsafe bogus devices, and toward regulatory activity focused on the safety and effectiveness of legitimate medical devices.

III. THE MEDICAL DEVICE AMENDMENTS OF 1976

The insulation from extensive government regulation that devices enjoyed ended with the enactment of the Medical Device Amendments to the Federal Food, Drug and Cosmetic Act in 1976 [9]. Indeed, the Amendments established an intricate statutory framework to enable FDA to regulate nearly every aspect of medical devices—from testing through marketing. The complexity of the Amendments and the extensiveness of FDA's authority over medical devices elicit different responses depending on an individual's personal perspective. This, of course, reflects the common paradox that the businessman's nightmare is the lawyer's dream.

The foundation for the Medical Device Amendments of 1976 was the determination by Congress that a regulatory mechanism was needed to provide the public with a reasonable assurance of the safety and effectiveness of medical devices. From this basic premise sprung the complex regulatory framework that currently governs devices marketed in the United States.

A. Definition of a "Device"

The most logical starting point for any overview of the Medical Device Amendments is the definition of a "device." The current definition parallels the previous definition in the 1938 Act. A device is defined as:

an instrument, apparatus, implement, machine, contrivance, implant, in vitro reagent, or other similar or related article, including any component, part, or accessory, which is:

(1) recognized in the official National Formulary, or the United States Pharmacopeia,
(2) intended for use in the diagnosis of disease or other conditions, or in the cure, mitigation, treatment, or prevention of disease, in man or other animals, or
(3) intended to affect the structure or any function of the body of man or other animals, and which does not achieve any of its principal intended purposes through chemical action within or on the body of man or other animals and which is not dependent upon being metabolized for the achievement of any of its principal intended purposes [10].

This definition evidences an express attempt by Congress to distinguish medical devices from drugs. Of particular significance is the language that restricts the definition of a device to those articles that are not metabolized or otherwise chemically active in or on the body. Undoubtedly, this definitional distinction was in response to the rather creative judicial opinions that held that products that were apparently devices could nevertheless be regulated as drugs.

While this revised and expanded device definition is an improvement over the definition in the 1938 Act, the distinction between a drug and a device is not always clear. Indeed, advances in medical technology since the mid-1970s, when the Amendments were enacted, have led to the development of "hybrid products"—articles with both drug and device characteristics. For example, the incorporation of medicaments into bandages and wound dressings, the development of contraceptives that provide both a mechanical and a chemical barrier to fertilization, and the evolution of unique drug delivery systems have once again blurred the distinction between drugs and devices.

B. Device Classifications

Congress did not intend to create a statutory framework that would subject all medical devices to the same regulatory requirements. The initial basis for separating the vast number of medical devices is their marketing status on the date when the Medical Device Amendments were enacted. This differentiation of products marketed before versus after the date legislation was enacted appears in several different contexts in the Federal Food, Drug and Cosmetic Act. For example, the 1962 Amendments to the Act, which require that drugs are effective as well as safe, made special provisions for those drugs marketed before 1962. All medical devices marketed before May 28, 1976, are distinguished from devices developed after this date. Pre-1976 devices, and those devices similar to them, are categorized into one of three classes based on the amount of regulation that is necessary to provide a reasonable assurance of safety and effectiveness. Class I represents the least restrictive level of regulation; Class III devices are subject to the most stringent level of regulatory control. Devices developed after 1976 that are not similar to pre-1976 devices are placed in Class III.

The categorization of pre-1976 devices into Class I, II, or III is a multiphased process [11]. Initially, FDA establishes separate panels for generic types of devices (e.g., ophthalmic, cardiovascular, surgical). The advisory panels are composed of individuals—independent of the Agency—with diverse backgrounds such as medicine, engineering, and other health-related fields. A consumer spokesperson and a representative from the device manufacturing industry are also appointed to the panels, but these individuals do not vote on a panel's classification recommendations. Based on the amount of regulatory control that the panel determines is required to ensure the safety and effectiveness of a particular type of device, a recommendation for placing the device in Class I, II, or III is made. FDA is not bound by this recommendation and may elect to include the device in a different class. Before a device is formally classified, however, a notice of the proposed classification must be published in the Federal Register and an opportunity for public comment provided.

C. Hierarchy of Device Regulation

1. *Class I*

As described previously, the extent of regulation differs between the three classes of medical devices. Class I, General Controls [12], consists of requirements for registering device manufacturing facilities; providing FDA with regularly updated lists of marketed devices; complying with good manufacturing practices; and maintaining records and filing reports of device-related injuries and malfunctions.

Additionally, FDA has the authority to limit the use or distribution of Class I devices; to detain Class I devices that allegedly violate the Medical Device Amendments until a court order to initiate legal proceedings is obtained; and to promulgate a formal regulation to ban deceptive devices or devices that present an unreasonable risk of injury to consumers. General controls also enable FDA to require that notice be given to purchasers or users of devices that create unreasonable risks, and to require that the device manufacturer initiates corrective actions in certain circumstances of device failure. Of course, FDA has the authority to take enforcement action against Class I devices that violate the broad adulteration and misbranding provisions in the Federal Food, Drug and Cosmetic Act. Class I devices cannot be marketed until a premarket notification is submitted to FDA, unless FDA has exempted a device or class of devices from this requirement. Recent classification decisions by the Agency have exempted over 100 devices from the premarket notification requirement.

2. *Class II*

Class II devices are subject to the same general controls as articles in Class I with the additional requirement that Class II devices must comply with any applicable performance standards. Under the Medical Device Amendments of 1976 [13], FDA is authorized to establish performance standards for a particular device or a group of devices (e.g., cardiac pacemakers) by promulgating formal regulations. Performance standards are intended to supplement the general controls and provide a reasonable assurance that the device will operate in a safe and effective manner. Accordingly, performance standards can address a variety of factors such as the construction or composition of a device, the testing of a device, and the labeling of a device to ensure proper installation, use, and maintenance.

To date, FDA has failed to exercise this authority to establish performance standards for any of the numerous devices placed in Class II. Lack of adequate personnel and lack of resources are commonly cited as reasons for the Agency's inactivity in this area. FDA's legislative proposals for change in the statute have included sections removing performance standards (Class II) from the statutory scheme. Thus, at the present time, there is no real difference between Class I and II devices from a regulatory standpoint. Like products in Class I, Class II devices cannot be lawfully marketed until a premarket notification is submitted to FDA.

3. *Class III*

Finally, if the combination of general controls and a performance standard are insufficient to provide a reasonable assurance of safety and effectiveness, and the device is used in sustaining life or preventing impairment of health, or presents a potentially unreasonable risk of illness/injury, the device is placed in Class III. Class III contains those devices described in the previous sentence, devices developed after 1976 that are not sufficiently similar to pre-1976 devices, and devices that were regulated as new drugs before 1976 [14]. Class III devices developed after 1976 are subject to FDA's premarket approval requirements before they can be legally marketed in the United States, unless the Agency grants a petition to place them in a lower class. However, a grace period is provided from the FDA approval requirement for Class III devices already on the market prior to classification. These devices, and their substantial equivalents, may remain on the market without FDA approval even though placed in Class III until (a) 30 months following the final classification or (b) 90 days after FDA promulgates a regulation specifically requiring that a (Pre)market Approval Application be filed (whichever event occurs later). Of course, Class III devices are also subject to the general controls applicable to Class I and II devices.

The amendments contain provisions for reclassifying devices that may be invoked by a device manufacturer or by the Agency [15]. Moreover, the amendments authorize FDA to develop criteria for exempting certain Class I devices from the premarket notification requirement. Such an exemption permits manufacturers to market affected devices without first notifying FDA (i.e., submitting a premarket notification). The Agency is required to promulgate formal regulations for Class I device exemptions.

D. The Premarket Notification Process

As previously indicated, devices in Classes I (unless exempted by regulation) and II are subject to premarket notification requirements before they may be legally distributed in the United States. The premarket notification requirement also applies to pre-1976 devices in Class III until FDA formally requires the submission of premarket approval applications for such devices. The premarket notification [16], also called a 510(k) after the pertinent section in the Federal Food, Drug and Cosmetic Act, is a unique administrative process.

From a legislative perspective, the requirement in section 510(k) that a manufacturer notifies the Agency of its intent to introduce a medical device into interstate commerce was originally intended as a surveillance mechanism for FDA. The Agency, however, through its authority to promulgate regulations to implement the Medical Device Amendments of 1976, has transformed the premarket notification [510(k)] provision into a mechanism for clearing medical devices before commercial distribution commences [17]. Inasmuch as subsequent chapters discuss the 510(k) process in great detail, we provide only a brief overview of premarket notifications as an introduction.

The cornerstone of the premarket notification process is the ability of a manufacturer to market a medical device based on a claim that such device is "substantially equivalent" to a device in commercial distribution before May 28, 1976 (the date the Medical Device Amendments were enacted). To determine whether device B is substantially equivalent to device A, many factors are considered, including the claims or indications, the technology, the intended patient or consumer populations, and safety and risk comparisons. Significantly, the premarket notification does *not* result in a determination that the device is safe or effective. Instead, FDA merely concludes that the device is substantially equivalent to a device in commercial distribution before 1976 and therefore does not require the filing of a premarket approval application. FDA's review of a premarket notification does not entail a detailed evaluation of the 510(k) device or of the pre-1976 device to which substantial equivalence is alleged. Indeed, FDA could later determine that either one or both of these devices are adulterated or misbranded and subject to regulatory enforcement action.

Submitting a 510(k) to FDA is generally a rather straightforward endeavor for device manufacturers. From a manufacturer's standpoint, the premarket notification process is preferable to the more stringent premarket approval requirements for Class III devices and for those devices that are found not substantially equivalent to pre-1976 devices. Critics of the 510(k) process, which include members of the U.S. Congress, argue that FDA's premarket notification process and the concept of "substantial equivalence" have become much too liberal. Consequently, unsafe or ineffective devices are slipping onto the market. For example, a phenomenon known as piggybacking has occurred where a post-1976 device C is marketed on the basis of substantial equivalence to a post-1976 device B, which is marketed on the basis of substantial equivalence to a pre-1976 device A (i.e., if C=B, and B=A, then C=A). Critics of the premarket notification process allege that the piggybacking phenomenon enables the chain of "substantial equivalence" to become too attenuated.

In response to this criticism, and to advancements in medical device technology, the 510(k) process is currently undergoing change. With increasing frequency, FDA is requiring that 510(k) submissions contain clinical data to establish that a device is, in fact, similar to a particular pre-1976 device [18]. The Agency is also requesting clinical data to demonstrate the safety and effectiveness of devices utilizing new technologies. Accordingly, many commentators predict that the 510(k) premarket notification may be gradually transformed into a mini-premarket approval process.

E. The Premarket Approval Process

The alternative to the premarket notification is the premarket approval application (PMAA). The medical device PMAA [19] is the cousin of the new drug application (NDA) and results in a determination by FDA that the device is safe and effective for its labeled indications. A PMAA is required for Class III devices, which include (1) pre-1976 devices that were subsequently classified into Class III, and devices that are substantially equivalent to these devices; (2) pre-1976 devices that were regulated as new drugs; and (3) post-1976 devices that are not substantially equivalent to devices commercially available before the Medical Device Amendments were enacted.

Subsequent chapters describe the PMAA process at length. Accordingly, only a brief overview of this topic and citations to the relevant statutory and regulatory provisions are provided. Generally, the PMAA is a lengthy document consisting of a substantial amount of animal and human data to establish the device's safety and effectiveness. Information on the device's components, the manufacturing process, and the labeling is also included in a PMAA. FDA has promulgated detailed regulations governing the premarket approval process [20]. Much of the data incorporated into a PMAA is obtained from investigational studies of the device. The Federal Food, Drug and Cosmetic Act [21], as amended by the Medical Device Amendments of 1976, authorizes the use of devices in humans for investigational purposes. To implement these provisions in the Act, FDA has promulgated comprehensive regulations for Investigational Device Exemptions (IDE) that enable manufacturers to conduct clinical investigations of their devices [22].

The IDE requirements are analogous to the investigational new drug (IND) requirements for drugs and apply to most clinical studies of medical devices. For example, clinical investigations of medical devices are subject to monitoring by an institutional review board (IRB), informed consent must be obtained from study participants, and clinical study sponsors and investigators must comply with prescribed reporting requirements. Chapter 19 provides a thorough discussion of FDA's IDE regulations.

F. Restricted Devices

Although the premarket notification and premarket approval requirements first come to mind when one considers FDA regulation of medical devices, the Medical Device Amendments of 1976 and FDA regulations contain several other important provisions of which device manufacturers should be aware. The Federal Food, Drug and Cosmetic Act [23] authorizes FDA to promulgate regulations to restrict the sale, distribution, or use of a device in order to provide a reasonable assurance of the safety and effectiveness of such device. The most common "restriction" placed on devices is the requirement that they be used pursuant to prescriptions from licensed practitioners. In other words, FDA can determine that devices are not suitable for lay use, and designate such devices as prescription products.

Medical devices that are restricted to prescription use are also subject to a greater level of FDA regulation. Advertisements for restricted devices fall within FDA's jurisdiction [24], while advertisements for nonrestricted devices are subject to the less stringent standards of the Federal Trade Commission (FTC). This division of regulatory authority for advertising of restricted versus nonrestricted devices is consistent with provisions for drugs whereby FDA regulates advertising of prescription drugs and the FTC regulates advertising of nonprescription (over-the-counter) drugs. In addition, FDA inspectors have greater access to the records, files, and facilities of manufacturers that market restricted devices as opposed to manufacturers of nonrestricted devices [25]. This difference in inspection authority parallels the statutory provisions that distinguish facilities that produce prescription drugs from those that do not.

G. Banned Devices

The Medical Device Amendments of 1976 also provide FDA with the authority to ban certain devices [26]. The standard for banning a device is a determination by FDA that such device is substantially deceptive or presents an unreasonable and substantial risk of illness or injury. FDA bans a device from commercial distribution by proposing a regulation to that effect. But before a device is banned, the manufacturer and other interested parties must be provided an opportunity for a hearing. Recognizing the checkered past of the device industry, this authority to ban products was undoubtedly designed to enable FDA to act quickly and forcefully against hazardous or bogus devices that jeopardize the welfare of patients and consumers. To date, however, FDA has invoked this power against only one device—artificial hair implants.

H. Additional Device Regulations

To round out this discussion of FDA's pervasive regulatory power over medical devices and the manufacturers of medical devices, a brief overview of the Agency's authority to require the registration of facilities, establish good manufacturing practices, and impose medical device reporting requirements on manufacturers is presented. Each of these important issues is discussed in great detail in later chapters.

1. Facility Registration and Device Listing

The Federal Food, Drug and Cosmetic Act expressly mandates that manufacturers of medical devices register their establishments with FDA and provide the Agency with updated lists of devices in commercial distribution [27]. In conjunction with this provision, FDA has also promulgated specific regulations for registering facilities and listing marketed devices [28]. The forms for registering facilities and listing devices are not complex, but rather necessitate orderly record keeping by the manufacturer.

2. Good Manufacturing Practices

The Medical Device Amendments of 1976 expressly incorporated into the Federal Food, Drug and Cosmetic Act language authorizing FDA to establish good manufacturing practices (GMPs) for the device industry [29]. FDA has exercised this power by promulgating regulations for nearly every aspect of manufacturing—device components, production processes and controls, packaging, storage, labeling, and installation [30]. These GMPs are intended to ensure that devices will be safe and effective, as well as comply with statutory requirements.

The statute provides for a unique mechanism whereby FDA must consult with advisory panels before establishing GMP regulations. Advisory panels are composed of officers of federal, state, or local governments; representatives from the device industry and the health professions; and public interest spokespersons. Thus, Congress has legislatively mandated a check on FDA's GMP rule-making authority.

Perhaps due to this requirement for input from an advisory panel, and the fact that formal rule-making is time-consuming and personnel-intensive, the Agency has, with an increasing level of frequency, issued "guidelines" on GMP issues.

Unlike formal GMP regulations, guidelines do not have the force of law [31]. Instead, GMP guidelines are merely the Agency's opinions on particular aspects of device manufacturing. Because they are not legally binding, device manufacturers are not required to adhere to FDA guidelines, although their manufacturing operations must comply with formal GMP regulations. The manufacturer that elects to comply with an FDA guideline, however, is assured that the Agency will not initiate regulatory action against the company on the particular issue addressed by the guideline.

This distinction between formal GMP regulations and FDA guidelines recently arose in a lawsuit [32]. In that case, the Agency attempted to enforce a sterility level adopted in an FDA guideline as a GMP requirement. FDA lost this suit on the basis that this sterility level was merely a guideline—not a formal GMP regulation promulgated pursuant to the required review by an advisory panel and an opportunity for comment by other interested parties. In short, the Agency cannot compel device manufacturers to comply with informal FDA guidelines.

However, there is no question that the Center for Devices and Radiological Health (CDRH) is substantially increasing its use of guidelines, which (absent the unusual event of a lawsuit) have had a strong tendency over time to harden into GMP standards [33]. Certainly two key documents for GMP compliance are the Center's May 1987 Guidelines on Preproduction Quality Assurance and Process Validation [34]. These documents are probably more important than the GMP regulations [35] in providing guidance on medical device product manufacturing today.

3. *Device Reporting*

The last major aspect of FDA regulation of which manufacturers should be aware concerns reporting injuries and malfunctions attributed to medical devices. Both the Federal Food, Drug and Cosmetic Act [36] and FDA regulations [37] require the reporting of certain incidents related to medical devices. FDA has further set forth its views in a number of articles, question and answer (Q & A) papers, and guidelines [38]. Several subsequent chapters explain the medical device reporting provisions and provide information to help manufacturers comply with these requirements.

IV. FDA ENFORCEMENT

FDA has a host of enforcement options that it can invoke against a medical device or a manufacturer that violates an applicable statutory or regulatory requirement [39]. In a "worst case" scenario, FDA can ban a device as discussed previously. Additional severe measures include prosecutions, injunctions, and product seizures. During the course of a facility inspection, FDA has the authority to invoke administrative detention orders against devices that are believed to be adulterated or misbranded [40]. (To date, FDA has only issued detention orders against a handful of devices, none recently.) The Agency is also empowered to require that the device manufacturer repair, replace, or refund the price of a violative device [41]; FDA has not yet exercised this "3R" authority.

At the other end of the enforcement spectrum are less severe regulatory tools such as a Notice of Adverse Findings letter (NAF) or a Regulatory Letter. These documents indicate that FDA believes that the manufacturer is in violation of the Act or the regulations, and request that the recipient responds with a plan to correct the violation. Device manufacturers are strongly advised to take an NAF or a regulatory letter seriously and respond promptly. Fortunately, FDA is usually amenable to resolving problems informally rather than progressing to a sterner enforcement measure.

V. MISCELLANEOUS AREAS OF DEVICE REGULATION BY FDA

A. Imports and Exports

Firms that import or export medical devices should be familiar with the pertinent provisions in the Federal Food, Drug and Cosmetic Act that address these activities [42]. FDA is authorized to regulate imports and exports of medical devices. With respect to imported devices, the Agency works in conjunction with U.S. Customs, a subunit of the Department of the Treasury. Companies involved in the importation of medical devices could, at some time, come in contact with U.S. Customs and its unique administrative bureaucracy.

B. Radiation Emission

Firms marketing medical devices that emit radiation must comply with the Radiation Control For Health and Safety Act [43]. The law is intended to control the release of electronic product radiation which includes ionizing and nonionizing electromagnetic or particulate radiation, as well as sonic, infrasonic, and ultrasonic waves. FDA is the administrative agency that enforces this statute.

C. Packaging and Labeling

Manufacturers should also be aware of the provisions in the Fair Packaging and Labeling Act [44] that are applicable to medical devices. Basically, the Fair Packaging and Labeling Act establishes standards for the content, form, and placement of product labeling. Enforced by FDA, this statute is designed to promote accurate packaging and labeling information for devices available to consumers.

VI. OTHER SOURCES OF DEVICE REGULATION

A. Federal Trade Commission

Although the bulk of regulation stems from the Federal Food, Drug and Cosmetic Act and FDA rules, medical devices do fall within the purview of other federal and state authorities. For example, as mentioned previously, the Federal Trade Commission (FTC) has jurisdiction over the advertising of nonrestricted medical devices. It is important to understand the distinction between advertising and labeling. FDA regulates all device labeling, which includes materials affixed to a device, and product literature that accompanies the device such as an instruction manual or a package insert; FTC regulates nonrestricted device advertising.

FTC's authority to regulate device advertising is found in the Federal Trade Commission Act [45]. In a nutshell, the FTC is a watchdog to prevent the distribution of false advertisements. To accomplish this mission, the FTC has developed the "reasonable basis requirement," which mandates that advertising claims for medical devices are substantiated *before* they are disseminated. For example, express claims that utilize phrases such as "studies show," or "doctors recommend" must be supported by at least the advertised level of claim substantiation. The FTC has also developed standards for "comparative claims," which compare the characteristics of two or more competing products. Although medical device advertising is not usually a target of FTC enforcement activity, companies should be cognizant of this Agency's requirements to avoid potential problems in the commercialization of their products.

B. Federal Communications Commission

Perhaps one of the least obvious government agencies that can affect the marketing of a medical device is the Federal Communications Commission (FCC). Pursuant to the Communications Act of 1934, the FCC has authority over certain types of medical equipment. Specifically, FCC regulates devices that emit electromagnetic energy on frequencies within the radiofrequency spectrum. The Commission's involvement is necessary to avoid interference with authorized radio communication services. FCC regulations [46] establish operating, reporting, importing, and other requirements for affected devices. Examples of devices subject to FCC's jurisdiction include diathermy and ultrasonic equipment.

C. State Authorities

The Federal Food, Drug and Cosmetic Act [47] expressly prohibits any state from creating or enforcing device requirements that are different from or in addition to federal requirements. This federal preemption, however, is not absolute—states may petition FDA for an exemption to federal preemption based on local conditions or special circumstances. And, as discussed below, California and other states are attempting to establish state-specific requirements for FDA-regulated products. Accordingly, device manufacturers should be aware of both state and federal requirements applicable to their devices.

An interesting situation concerning federal preemption versus state law has recently arisen in California. In 1986, a voter referendum titled "The Safe Drinking Water and Toxic Enforcement Act" (Proposition 65) was passed. Proposition 65 established special warning requirements for products (including medical devices) that contain certain carcinogens or reproductive toxicants. Proposition 65 also established prohibitions against the discharge of these hazardous substances into the drinking water supply.

One of the many problems created by Proposition 65 is the requirement that products bear additional warning statements (or otherwise notify consumers of product hazards), despite the fact that these products are lawfully marketed pursuant to federal safety and labeling standards. For example, California authorities have set a permissible level for ethylene oxide residues that would force many manufacturers to warn consumers of the chemical's potential for reproductive toxicity. At the time of this writing, FDA-regulated products (i.e., devices as well as foods and drugs) have been granted a temporary exemption from the warning requirements of Proposition 65. However, much to the dismay of the affected industries, this "regulatory tug-of-war" between state and federal authorities has not been definitively resolved.

D. Occupational Safety and Health Administration

Medical device companies must also be aware of the general legal responsibilities that exist between an employer and his employees. The legal duty to provide a safe and hazard-free workplace is one example of an employer's obligation that has received a heightened level of attention in recent years. In the early 1970s Congress passed the Occupational Safety and Health Act [48] to protect the health and safety of workers. Pursuant to this law, the federal Occupational Safety and Health Administration (OSHA) has created numerous workplace standards, such as employee injury reports and sanitation requirements. Moreover, OSHA is authorized to inspect medical device facilities and to impose fines for violations of the law.

In addition to the federal requirements for safe workplaces, states have also enacted legislation to protect the welfare of workers. For example, many states have passed comprehensive "right-to-know" laws that mandate disclosure of health and safety hazards to employees. "Right-to-know" programs routinely require cautionary labeling on containers, warning signs in work areas, extensive employee training, and centralized filing systems for information on the hazardous substances that are used in, or manufactured in, an employer's facility.

E. Consumer Product Safety Commission

Certain medical devices could be subject to the Federal Hazardous Substances Act (FHSA) and its implementing regulations [49], which are enforced by the Consumer Product Safety Commission (CPSC). Briefly, the law establishes labeling and warning requirements for products that are toxic, flammable, corrosive, sensitizers, or irritants. Although the Act expressly excludes other FDA-regulated products (foods, drugs, and cosmetics), devices are not exempt.

In practice, however, CPSC's jurisdiction under the FHSA over medical devices is very limited. CPSC's authority is generally restricted to "consumer" products that could cause substantial personal injury or illness as a result of customary use in the household. To date, very few medical devices have fallen under CPSC's regulations. One particular brand of children's sunglasses, however, did involve dual regulation by FDA and CPSC. Because of their physiological effect on the eyes, FDA asserted that the glasses were devices; CPSC exerted its regulatory authority based on a determination that the glasses were made of a flammable type of plastic.

Companies producing devices that possess any of the specified hazardous characteristics covered by the Act and CPSC's regulations should be familiar with the applicable labeling requirements. This is advisable since FDA may require a company to include cautionary statements (identical or similar to CPSC requirements) so that the device is not misbranded. Moreover, warning statements on devices may be prudent from a product liability perspective.

F. Environmental Protection Agency

Companies should be cognizant of laws that affect what substances generated from the production of medical devices can be released into the air, water, and soil. On the federal level, the Environmental Protection Agency (EPA) enforces several statutes intended to preserve, and in some cases restore, the quality of our environment. The Clean Air Act [50], the Clean Water Act [51], and the Solid Waste Disposal Act [52] are among the important laws that could potentially affect the day-to-day operations of device companies. Furthermore, most states have enacted legislation tailored to their own specific environmental concerns. Accordingly, device firms should ensure that their production or processing procedures do not run afoul of federal and state requirements.

G. Statutory Exemptions for Devices

To avoid creating the impression that device companies are subject to an endless litany of statutory and regulatory requirements, we conclude this section by addressing several laws from which medical devices are exempt. For instance, medical devices are expressly excluded from one of the EPA's most important sources of regulatory authority—the Toxic Substances Control Act (TSCA) [53].

In recent years, this statute has been construed very liberally in order to enlarge EPA's regulatory net. For example, although TSCA was initially enacted to cover "chemical substances," certain genetically altered microorganisms now fall under the statute. Thus, absent the express exemption for medical devices, manufacturers of diagnostics that utilize genetically engineered bacteria could be required to comply with TSCA.

Medical devices, as defined in the Federal Food, Drug and Cosmetic Act, are also exempt from the Consumer Product Safety Act [54]. Enforced by the Consumer Product Safety Commission, the Act is intended to protect the public from unsafe products and to ensure that consumers are adequately warned of potentially dangerous articles. The rationale for the exemption is that FDA regulations presumably address any safety concerns that would otherwise cause a medical device to fall within this statute.

VII. CONCLUSION

Companies are confronted with a host of federal and state laws established to regulate business in general, and medical devices in particular. For better or worse, the days of relatively unrestrained marketing of medical devices are gone. The Medical Device Amendments of 1976 established a statutory framework governing nearly every aspect of devices and the industry. Using the Amendments as a springboard, FDA markedly expanded its regulatory authority over medical devices. Indeed, the manufacturing and marketing of medical devices has become a complex and regulation-intensive endeavor for companies. (This fact is clearly evidenced by the existence of over 60 chapters in a book on the medical device industry!)

This chapter is intended to provide the reader with a broad overview of medical device regulation and various government authorities that can be involved in this process. It is often easier, at least in theory, to grasp the intricacies of a subject if one does not lose sight of the "big picture." On this note, we conclude this introduction and defer to our distinguished co-authors to explain the nuts and bolts of medical device regulation.

REFERENCES

1. 34 Stat. 768 (1906).
2. 52 Stat. 1040 (1938).
3. 65 Stat. 648 (1951); 21 U.S. Code (U.S.C.) § 353.
4. 72 Stat. 1784 (1958); 21 U.S.C. § 348.
5. 76 Stat. 780 (1962).
6. House Comm. on Interstate and Foreign Commerce, Medical Device Amendments of 1976, H.R. Rep. No. 853, 94th Cong., 2d Sess. 7 (1976).
7. 389 F.2d 825 (2d Cir. 1968), cert. denied, 393 U.S. 825 (1968).
8. *United States v. Bacto-Unidisk*, 394 U.S. 784 (1969).
9. Pub. L. No. 94-295, 90 Stat. 539 (1976).
10. 21 U.S.C. § 321(h).
11. 21 U.S.C. § 360c.
12. "General Controls" refer to the regulatory controls authorized in sections 501, 502, 510, 516, 518, 519, and 520 of the Federal Food, Drug and Cosmetic Act.
13. 21 U.S.C. § 360d.
14. 21 U.S.C. § 360j(l) addresses "transition devices," which are articles regulated as drugs before the enactment of the Medical Device Amendments of 1976.
15. 21 U.S.C. § 360c(e)
16. 21 U.S.C. § 360(k).
17. 21 Code of Federal Regulations (C.F.R.) § 807.81-.97.
18. For a discussion of 510(k) piggybacking and FDA's requirements for clinical data in premarket notifications, consult R. Cooper, *Clinical Data Under Section 510(k)*, 42 Food Drug Cosm. L.J. 192-202 (1987).
19. 21 U.S.C. § 360e.
20. 21 C.F.R. § 814.20-.84.

21. 21 U.S.C. § 360j(g).
22. 21 C.F.R. § 812.1-.150.
23. 21 U.S.C. § 360j(e).
24. 21 U.S.C. § 352(r).
25. 21 U.S.C. § 374(a).
26. 21 U.S.C. § 360f(a); 21 C.F.R. § 895.1-.101 address banned devices.
27. 21 U.S.C. § 360.
28. 21 C.F.R. § 807.3-.65.
29. 21 U.S.C. § 360j(f).
30. 21 C.F.R. § 820.1-.198.
31. 21 C.F.R. § 10.90.
32. *U.S. v. Bioclinical Systems, Inc.*, 666 F. Supp. 82 (D. Md. 1987).
33. W. Appler, *Guidelines Aren't Regulations (And Other Lessons From The Federal Courts)*, 10 Medical Device & Diagnostic Industry 10-11 (Feb. 1988); L. Oster, *Washington Wrap-Up: More About BioClinical*, 10 Medical Device & Diagnostic Industry 14-18 (Apr. 1988).
34. Preproduction Quality Assurance Planning: Recommendations for Medical Device Manufacturers; Guideline on General Principles of Process Validation (May 1987).
35. 21 C.F.R. Part 820.
36. 21 U.S.C. § 360i.
37. 21 C.F.R. § 803.1-.36.
38. W. Hooten and R. Bimonte, *GMP Complaint Files: How They Relate to Reports Required Under MDR*, (Center for Devices and Radiological Health—Division of Compliance Programs); Center for Devices and Radiological Health—Office of Compliance, *Medical Device Reporting Questions and Answers*, Compliance Guidance Series (Feb. 1988); Food and Drug Administration, *Enforcement of the Medical Device Reporting (MDR) Regulation*, Compliance Program Guidance Manual—Program 7382.011 (Sept. 1988).
39. J. Gibbs, *Medical Devices And Regulatory Letters: An Analysis of FDA Enforcement Actions*, 9 Medical Device & Diagnostic Industry 42-47 (Aug. 1987).
40. 21 U.S.C. § 334(g); 21 C.F.R. § 800.55.
41. 21 U.S.C. § 360h.
42. 21 U.S.C. § 381.
43. 42 U.S.C. § 263b et seq.
44. 15 U.S.C. § 1451 et seq.
45. 15 U.S.C. § 52-55.
46. 47 C.F.R. Part 18.
47. 21 U.S.C. § 360k.
48. 29 U.S.C. § 651 et seq.
49. 15 U.S.C. § 1261 et seq.; 16 C.F.R. Part 1500.
50. 42 U.S.C. § 7401 et seq.
51. 33 U.S.C. § 1251 et seq.
52. 42 U.S.C. § 6901 et seq.
53. 15 U.S.C. § 2601 et seq.
54. 15 U.S.C. § 2051 et seq.

4

FDA's Regulatory Program for Medical Devices and Diagnostics

WALTER E. GUNDAKER *Center for Devices and Radiological Health, Food and Drug Administration, Rockville, Maryland*

I. INTRODUCTION

A new manufacturer of medical devices faces a mountain of responsibility under the requirements of the Federal Food, Drug, and Cosmetics Act. If the product is an electronic product that emits radiation, there are even further requirements under the Radiation Control for Health and Safety Act. Assistance in overcoming that mountain (sometimes of fear) is available, particularly from the Center's Division of Small Manufacturers Assistance (see Chap. 8). This chapter is also intended for that new manufacturer to provide a further glimpse of insight into how the FDA operates, the types of regulatory guidance and where it is published, the importance of compliance to the medical device good manufacturing practice regulation, and the importance of standards. It is written for a person unfamiliar with the organization and jargon of the FDA and, as such, hopefully will not be too superficial for someone deeply involved in Agency interactions.

II. FDA FIELD ACTIVITIES

A. Role of the District Offices

Perhaps the most important thing for a new manufacturer of medical devices to recognize is that once a product is cleared for marketing, either through the 510(k) or premarket approval (PMA) process, the primary contact he will have with the Agency is the local FDA District Office. It is this component that schedules and performs routine inspections, investigates complaints by patients or hospitals, performs or arranges for laboratory testing of devices, and decides on the appropriate Agency response to the findings of its investigations. The District Office has broad regulatory authority, including the power to detain imports and commercially distributed domestic devices, and to initiate legal actions such as seizures and injunctions.

Most importantly, the District Office feels responsible for the manufacturers within its geographical boundaries, and this sense of responsibility must be recognized when a manufacturer, through perhaps no fault of its own, either violates one of the FDA's regulations or takes some action regarding the devices it manufactures that could affect the public health and safety. As will be discussed later in this chapter, communications with the District Office should be established before the need arises so that there is no perception, when an issue is discovered at a later time, that there was an intent to keep the information from the Agency. One situation that would clearly fall into this category is when devices are recalled because of a safety or efficacy reason.

The District Offices are generally composed of an investigations branch, which is truly the eyes and ears of the Agency; a compliance branch, which acts on the violations discovered by the investigations branch; and a laboratory, which performs sample analysis and provides technical support. For medical devices, one regional laboratory—the Winchester Engineering and Analytical Center, located in the Northeast Region—has particular significance since it specializes in medical device research and testing. The locations of all of the District Offices, arranged according to the respective Regional Office, are given in Table 1. The District Offices also have resident posts that are strategically located within their areas to better serve the commercial and public needs.

TABLE 1 FDA District Offices by Region

Region	District Office
Northeast	Brooklyn
	Buffalo
	Boston
Mid-Atlantic	Philadelphia
	Baltimore
	Newark
	Cincinnati
Southeast	Atlanta
	Orlando
	Nashville
	New Orleans
	San Juan
Mid-West	Chicago
	Detroit
	Minneapolis
Southwest	Dallas
	Denver
	Kansas City
Pacific	San Francisco
	Los Angeles
	Seattle

B. Good Manufacturing Practice Inspections

With that brief description of the District Offices, the next thing to be aware of is the regulation that will directly affect how a manufacturer produces a medical device.

The Food and Drug Administration has enforced good manufacturing practice (GMP) regulations on the key industries it regulates for more than 20 years. With the passage of the Food, Drug, and Cosmetic Act in 1938, FDA had, for the first time, the authority to establish and impose reasonable sanitation standards on the production of foods, drugs, and cosmetics. The first drug GMPs were published in 1963 and have undergone several revisions since then. In 1969, FDA promulgated the good manufacturing practice regulations for the manufacture of foods, and they, too, have been revised since then. GMPs for blood and blood components were published in 1975. The medical device GMPs became effective in 1978 after a lengthy public comment period. In the intervening 10 years, they have become the primary basis on which the Agency evaluates a manufacturer's ability to produce a safe and effective product consistently.

The purpose of the medical device GMPs is to provide a framework of manufacturing controls. The Agency believes that manufacturers who adhere to these controls increase the probability that their devices will conform to their established specifications. This means that if the medical device has been properly designed, following the GMPs will ensure that there is consistent manufacturing of that product with the appropriate quality control checks prior to its distribution and use.

One term that the FDA uses regarding these regulations is that they are "umbrella" controls. The term "umbrella" signifies that the GMPs apply to all medical device manufacturers, including manufacturers of *in vitro* diagnostic products. Since the industry is so diverse and the manufacturing processes vary significantly, FDA established a very general standard for manufacturing practices that took into account the complexity and diversity of the many products. The rule was designed so it would apply to all manufacturers and would avoid prescribing to each of them specifically the precise details of what must be done.

The scope of the regulation also encompasses the word "current," as in "current good manufacturing practices." Practices that are feasible and valuable are meant to be the same as current and good. Therefore, practices that are believed by experts to be feasible and valuable are, by definition, GMP requirements even if the practices are not followed by a majority of the manufacturers. Therefore, the definition of current good manufacturing practices is not the lowest common denominator. Also, as times change, the definition of current will vary depending on the state of the art, available test equipment, and other quality control practices. Thus, what is feasible and valuable today may change over time.

The Agency expends a considerable amount of effort to enforce this important regulation. At the present time, the District Offices are performing approximately 1800 inspections per year to determine if manufacturers are meeting this regulation. The total operational resources that it takes to conduct these inspections amounts to about 75 person-years.

The Agency has also carefully evaluated the results of these inspections to determine which sections of the regulation are most frequently found to be in violation. The results of this analysis and other information on how to comply with the specifics of the GMP regulation may be found in Chapter 22 by W. Fred Hooten.

C. Benefits of GMP Compliance

Compliance with the GMP regulation is, of course, mandatory. However, there are benefits of compliance, in addition to the obvious, that many manufacturers may not appreciate. One of the benefits relates to the government-wide quality assurance program in which the Department of Defense (DOD) and the Department of Veterans Affairs (VA) consult with the FDA prior to making major purchases. In essence, DOD and VA ask the FDA whether particular manufacturers are acceptable government vendors. To answer this question, FDA considers such factors as a firm's record of compliance with the GMP, registration, listing, premarket notification, and other regulations. If a firm has a poor record of compliance, FDA will advise DOD and VA that the firm may not be eligible to sell products to the government because of its violations of FDA regulations. It is not often that a firm is ruled ineligible, but GMP compliance can very definitely affect a manufacturer's ability to sell to the federal government.

Export certificates represent a second benefit. Many foreign countries require certification of a manufacturer's compliance with United States regulatory requirements. A firm needing such certification can write to the FDA and request such a certificate. We issue many export certificates from the Center's Office of Compliance—but not before we have checked with the appropriate District Office to verify that the firm is in compliance with GMPs.

Compliance with the GMP regulation also expedites premarket approval of products. Final approval of any premarket approval application is contingent upon a satisfactory GMP inspection. An inspection does not necessarily have to be performed while the approval is pending; we will sometimes allow final approval of a product if a recent GMP inspection showed the manufacturer to be in compliance. One way or another, however, compliance with the GMPs is an element of the premarket approval process.

A fourth benefit of compliance with the GMP regulation relates to selling in countries outside the United States. A "Memorandum of Understanding" (MOU) is an agreement between FDA and the regulatory agency in another country that may include the exchange of inspectional information. An MOU is in place between the United States and the United Kingdom in which the results of inspections performed in the United Kingdom are made available to FDA so that we can decide whether or not we need to conduct our own inspection of those manufacturers. The United Kingdom makes similar use of data for U.S. medical device manufacturers inspected by the FDA. Compliance with the GMPs is again the major factor in either country's deciding if another inspection is needed and whether the products should be allowed to be imported.

III. DEVELOPMENT OF REGULATORY POLICY

This section will summarize how FDA regulatory policy is established and where to look for the policy, in order to enable better understanding of what the Agency is thinking and how it reacts to different situations.

The Regional Offices and, in turn, the District Offices are supervised by the Associate Commissioner for Regulatory Affairs. This major Agency component, also known as the Office of Regulatory Affairs (ORA), has approximately a third of the FDA's total personnel, and as such is responsible for responding to the day-to-day field operations, as well as the crisis situations such as over-the-counter drug tampering, food contamination, and recall co-ordination. The Associate Commissioner for Regulatory Affairs and staff are also the Agency's key organization for establishing the compliance and regulatory policy. The practices common to all products regulated by the Agency, such as the inspectional philosophy and practices, the procedures for processing legal actions, and the overall interpretations of the Act and regulations, are established by the ORA.

The technical expertise for specific product areas is located with the major centers of FDA, which include the Centers for Drug Evaluation and Research, Biologics Evaluation and Research, Food Safety and Applied Nutrition, Veterinary Medicine, and Devices and Radiological Health. Each of these centers has an office that focuses on the compliance aspects of their particular product area, and it is here that specific guidance is developed and disseminated throughout the Agency.

Cooperation and joint resolution of problems between the policy and procedure developers in the ORA and the Centers works extremely well, and communication channels are well established. The Agency's Compliance Policy Council, chaired by the Associate Commissioner for Regulatory Affairs, with its membership of the compliance chiefs from each Center and ORA, meets routinely to discuss the major activities and the issues facing each product area and the associated enforcement issues. This attitude of working together on issues carries into each of the staff offices and is put into practice when a policy document is conceptualized, drafted, and finalized.

For manufacturers of medical devices, there is, unfortunately, no single place where you will find all of the answers on regulatory or compliance policy. The following list, however, covers the major sources of information:

1. *Federal Register Preambles*: The labeling, good manufacturing practice, and other regulations specifically directed to medical device manufacturers (contained in Subchapter H, Part 800 of Title 21 of the Code of Federal Regulations) are the most binding requirements and are easily found. However, to understand the Agency's interpretation and reasons for writing these regulations, one should read the preamble to the regulation published with the final publication in the Federal Register. The preamble contains valuable insight into the development of the regulation.
2. *Code of Federal Regulations*: The specific regulations governing medical devices may be found in Parts 800-895 of Title 21 of the CFR. The requirements including labeling, medical device reporting, *in vitro* diagnostic labeling, premarket approval, registration and listing, investigational device exemptions, and device classifications. A manufacturer should also be aware of 21 CFR Part 7 regarding policy, procedures, and industry responsibilities for recalls.
3. *Federal Register Notices of Availability*: In recent years the Agency has made extensive use of this mechanism to advise manufacturers of important policy decisions. One example is the recently updated list of critical devices that are subject to the critical device requirements of the good manufacturing practice regulations [1]. Another is the Agency's guidelines on how it believes sponsors of investigational devices should solicit investigators [2].

4. *FDA Guidelines*: 21 CFR 10.90 describes the differences between regulations, guidelines, recommendations, and agreements. It states that guidelines are "procedures or practices of general applicability that are not legal requirements but are acceptable to FDA for a subject matter . . . ." Medical device manufacturers should therefore be knowledgeable of the guidelines on process validation [3], validation of the limulus amebocyte lysate test [4], and similar publications. *In vitro* diagnostic manufacturers should review carefully the draft guideline for compliance with the GMP regulation that is currently being developed [5].
5. *FDA Recommendations*: Recommendations are a means FDA uses to disseminate information on matters on which it has the authority to act under the law but for which it currently has no specific regulations. For medical device manufacturers there is a specific set of recommendations on preproduction quality assurance planning [6].
6. *Compliance Policy Guides*: Compliance policy guides are intended primarily as an internal statement of FDA policy regarding a specific compliance situation so that all of our District Offices are working under the same ground rules. They are available to anyone requesting a copy, however, and many guides are now being disseminated to the public through the Federal Register Notice of Availability. The 28 guides related to medical devices include the following:

7124.01	Contamination of Devices Labeled as Sterile
7124.10	Oxygen Equipment—Emergency and OTC Use
7124.14	Registration of Assemblers of Diagnostic X-Ray Systems as Device Manufacturers
7124.15	Inspection of Manufacturers of Device Components
7124.16	Re-Use of Medical Disposable Devices
7124.19	Commercial Distribution with Regard to Premarket Notification [Section 510(k)]
7124.21	Condoms; Defects—Criteria for Direct Reference Seizure
7124.28	Reconditioners/Rebuilders of Medical Devices

7. *Compliance Programs*: The Agency uses compliance programs to direct the District Offices in specific program areas. These programs include instructions on how to conduct inspections and the policy on what specific actions are to be taken based on the results of the inspections.

IV. REGULATORY ACTIONS AND RECALLS

In spite of the best efforts by the industry trade associations, the FDA's dissemination of information and guidance through our manufacturers assistance program, and the understanding of our field investigators and managers, there have been and probably always will be some level of regulatory or legal actions that must be taken by the FDA to protect the public health. The numbers of such regulatory actions provide a snapshot view of where the medical device industry and the FDA is at regarding compliance with the applicable laws and regulations. Because of the many factors that could influence the numbers, however, an in-depth analysis has not been performed. The data are presented in Table 2 by fiscal year, to state the numbers of actions recommended by the District Offices and reviewed by the Center for Devices and Radiological Health.

TABLE 2 Medical Device Regulatory Action Recommendations

Actions	Fiscal year 1984	1985	1986	1987	1988
Regulatory letters	79	88	80	75	68
Seizures	24	36	22	25	25
Injunctions	15	14	14	11	6
Civil penalties	14	3	4	1	4
Citations	3	5	1	3	3
Prosecutions	2	5	3	1	4
Totals	137	153	124	116	110

Another indicator of the status of the medical device industry is the number of recalls that the Agency has classified in recent years. It is to the credit of device manufacturers that all of these recalls have been voluntary in nature. Only under very unusual circumstances has the Agency had to resort to the threat of using Section 518 of the Federal Food, Drug, and Cosmetic Act, which provides a procedure for mandatory notification and recall. An FDA District Office normally becomes aware of a recall action by a manufacturer through an inspection, upon being advised by the manufacturer, or through information supplied by the user of the product. The District Office forwards the pertinent recall information to the Center, where it is re-viewed and categorized. The number of recalls, by fiscal year, are given in Table 3.

V. STANDARDS ENFORCEMENT

A. Medical Devices

At the present time, no performance standards have been promulgated under the 1976 medical device amendments to the Federal Food, Drug, and Cosmetic Act. Thus, there is no enforcement program, per se, for medical devices

TABLE 3 Medical Device Recalls by Fiscal Year

Fiscal Year	Number of recalls
1984	367
1985	584
1986	762
1987	601
1988	664

under Section 514 of the Act. A performance standard is presently being developed for apnea monitors, however, and as described in the next section, the Agency's experience with the enforcement of radiation safety standards may provide a hint at how the apnea monitor and other medical device standards would be enforced in the future.

Voluntary standards are considered to play an important role in promoting safe and effective devices and, as such, are supported by FDA. A new program to assess conformance to these standards has been developed, which, while it is related to standards enforcement, is by no means intended to imply any enforcement authority. Standards organizations such as the National Committee for Clinical Laboratory Standards, the American National Standards Institute, the Association for the Advancement of Medical Instrumentation, and others have developed and published voluntary standards that apply to the devices FDA regulates. Such voluntary standards play an important role in solving or preventing many of the public health problems associated with the use of such equipment, and thus it is important for the Agency to be aware of the extent to which manufacturers are following these accepted standards.

In this new program, the Agency is first asking manufacturers if they are following the voluntary standard and what types of claims they may be making in their current labeling, promotional materials, or instructions for use. Consideration is also being given to obtaining samples of the products for laboratory testing, but this will depend on the cost of both the devices and the testing. The five devices initially selected for this program are Foley catheters, autotransfusion devices, inflatable tracheal tube cuffs, defibrillators, and sphygmomanometers. A second phase of the program will be to determine if the voluntary standard is sufficient and adequate to address the public health concerns that may have prompted the writing of the standards or may now be evident. This will be accomplished by looking at other sources of postmarket surveillance data and determining if the standard is assisting in solving such problems. Our conclusions on the adequacy of a voluntary standard will be taken back to the standards organization, where changes hopefully could be initiated if needed.

B. Radiation Safety Standards

Radiation safety standards have been promulgated under the Radiation Control for Health and Safety Act for the following medical devices (reference is to the regulation within 21 CFR):

Medical and Dental X-Ray Equipment, 1020.30-1020.33
Medical Lasers, 1040.10
Ultrasonic Therapy, 1050.10
Sunlamp Products, 1040.20

Thus, manufacturers of these products must also comply with the standard and the other regulations promulgated under this Act.

As a general rule, the enforcement of these standards is conducted and coordinated by the Division of Standards Enforcement within the Center's Office of Compliance. In this way, manufacturers of these products, in addition to dealing with the FDA District Offices for GMP inspections, will also have direct communications with the Center regarding compliance with the radiation safety performance standards.

The standards are enforced through laboratory testing, field or in-use testing, review of reports submitted by manufacturers, and inspections of the manufacturing facilities. Depending on the particular product, all or only several of these four methods of enforcement are utilized. For example, since medical X-ray equipment is very large and complex, the primary emphasis for determining compliance is through in-use testing of this equipment in physicians' offices and in hospitals. This medical X-ray testing is performed primarily through contracts that FDA has with 33 state agencies. The data are then returned to the Center for analysis and compilation to determine if a specific manufacturer's product is in compliance with the standard.

Manufacturers of products subject to a radiation safety standard must also submit initial, model change, and annual reports to the Center. These reports must describe how a manufacturer complies with the standard through an appropriate quality control and testing program. Currently, the Center is receiving approximately 5500 reports per year.

VI. ADVICE TO A MANUFACTURER

Several recommendations can be made regarding how to deal with the FDA and its regulatory process. None of these bits of advice are dramatic or new, but in the course of observing a firm's interaction with the Agency, it is amazing how many times the failure to think of these steps can result in significant difficulties.

The first recommendation is to *know your district office*. This may not be an easy thing to accomplish since, understandably, there is a great reluctance to walk into a regulatory agency such as the FDA or OSHA and say "I'm here to get to know you." As opportunities arise, however, they should not be overlooked. Situations such as responding to a notice of an investigator's observations at the conclusion of an inspection (receipt of an FDA Form 483) or a notice of adverse findings letter are excellent opportunities to hand deliver a reply instead of simply mailing it. The verbal discussion of the reply may make the content much more meaningful, and will allow both sides to learn more about the intent and seriousness with which the subject is being approached.

Another recommendation preached by many in the past but still not practiced by all is to *prepare for inspections*. When the FDA investigator walks into your manufacturing facility or corporate offices, there should be a procedure established that everyone is familiar with as to who is called, who escorts the investigator through the facility, who is available to make copies of records requested, etc. A corrollary to this suggestion is to be prepared to deal with adverse inspectional findings or other communications from the Agency that indicate that the FDA has found violations, a serious health hazard, or other information that requires high-level company knowledge and decision making.

The idea to *take seriously 483's and letters* may also sound very basic, but once again, it's surprising how many regulatory actions are processed with no apparent indication that a firm seriously considered the violations noted by the Agency. As an example, if a firm receives a regulatory letter, in most cases, it will have been reviewed in draft by the Center for Devices and Radiological Health, and if appropriate action is not taken, further action such as seizure or injunction may very well be the next step by FDA. Such letters should be taken very seriously and the company position discussed in detail with the District Office compliance officials.

At the risk of offending the legal profession, the next recommendation, that *lawyers are nice, but . . .*, should be viewed with having the right person represent your company at the right time. For example, if the Agency is questioning the methods used to validate a sterilization process, the person explaining your answer to the investigator or district managers should be your quality control manager or a process validation expert, not the firm's legal department. Obviously, the situation may call for lawyers to be included in those discussions, particularly if legal issues are involved, but as a general rule, a regulatory affairs professional with a thorough knowledge of a firm's manufacturing and quality control process will be viewed with greater respect. They will have the current view of the firm rather than a strictly legal view as voiced by many attorneys.

On a broader scale, a firm should also *keep up with current events and procedures of the FDA*. This will minimize the changes or surprise interpretations that could have an effect upon a firm's operations and will allow for advance planning for new FDA requirements. The Agency publishes much of its new program information in bulletins and other broad distribution documents, but much can be learned from obtaining copies of FDA Compliance Policy Guides and Compliance Programs.

Closely coupled with keeping up with current events is the necessity to let the FDA know of your firm's opinions on issues, whether they are in the development stage at the Agency or are policies or programs established and in operation. There is always a certain amount of risk in saying *tell us what you think*, but the Agency does recognize that the firms it regulates are the true experts in device manufacturing and distribution, and your views are important. The Agency also recognizes that the regulation of manufacturers is not the only bottom line—solving public health problems is equally or more important, and there are generally many ways to solve those problems.

REFERENCES

1. Food and Drug Administration, Center for Devices and Radiological Health, *Advisory List of Critical Devices—1988, Federal Register*, March 17, 1988.
2. Food and Drug Administration, Center for Devices and Radiological Health, *Guidelines for Preparing Notices of Availability for Investigational Devices, Federal Register*, April 4, 1986.
3. Food and Drug Administration, Center for Drugs and Biologics and Center for Devices and Radiological Health, *Guideline on General Principles of Process Validation*, FDA, Rockville, Maryland, May 1987.
4. Food and Drug Administration, Centers for Drug Evaluation and Research, Biologic Evaluation and Research, Devices and Radiological Health, and Veterinary Medicine, *Guideline on Validation of the Limulus Amebocyte Lysate Test as an End-Product Endotoxin Test for Human and Animal Parenteral Drugs, Biological Products, and Medical Devices*, FDA, Rockville, Maryland, December 1987.
5. Food and Drug Administration, Center for Devices and Radiological Health, *Draft GMP Guidelines for the Manufacture of In Vitro Diagnostic Products; Availability, Federal Register*, April 7, 1988, p. 11561.

6. Food and Drug Administration, Center for Devices and Radiological Health, Division of Compliance Programs, *Preproduction Quality Assurance Planning: Recommendations for Medical Device Manufacturers*, FDA, Rockville, Maryland, May 1987.

5

Overview of Current FDA Requirements for Medical Devices

EDWARD M. BASILE *King & Spalding, Washington, D.C.*

I. LEGAL FRAMEWORK

The Federal Food, Drug, and Cosmetic Act (the Act), as amended [1], a statute enacted by the U.S. Congress, provides the U.S. Food and Drug Administration (FDA) with the authority to regulate, among other things, medical devices. Congress first gave FDA authority to regulate medical devices in the 1938 amendment to the Pure Food and Drugs Act of 1906. That amendment gave FDA "after-the-fact" authority to take action against medical devices that were unsafe or contained false or misleading labeling claims. The 1938 amendment did not, however, contain any premarket review or approval authority for medical devices. Thus, FDA could only take action against an unsafe or ineffective product after that product was first marketed.

After a period of time, though, Congress recognized that this "after-the-fact" authority was inadequate to deal with the complex safety and effectiveness issues presented by the increasingly sophisticated medical technology coming to market in the 1960s and the 1970s. Consequently, on May 28, 1976, Congress enacted the Medical Device Amendments to the Federal Food, Drug, and Cosmetic Act [2].

To implement this law, FDA has promulgated regulations that are published in Title 21 of the Code of Federal Regulations (C.F.R.) [3]. Title 21 is composed of nine volumes, with the majority of the regulations concerning medical devices contained in Parts 800-1299. References to these regulations are given throughout this chapter.

If FDA deletes, adds, or amends a regulation, it is published in the *Federal Register*. The *Federal Register* also serves as the official public notice system. When a change to a regulation is proposed, FDA publishes a notice, called a proposed rule, or publishes a notice of a citizen petition to change a regulation. After allowing time for comments from the public, FDA will then publish a final version of the regulation, with a preamble summarizing the comments submitted and specifying the effective date. The final regulation is then codified in Title 21 of the Code of Federal Regulations.

II. THE DEVICE CLASSIFICATION SYSTEM

A central feature of the Medical Device Amendments of 1976 (the Amendments) was the creation of a three-tier approach to regulating medical devices. The Amendments instruct FDA to classify medical devices, by regulation, into one of three regulatory classes—Class I, Class II, or Class III—each of which would be subject to a different level of regulatory control. The goal of the classification system is to impose on each device class sufficient regulation to provide a reasonable assurance of the safety and effectiveness of the devices in that class.

Class I devices are those devices that do "not present a potential unreasonable risk of illness or injury [4] and where general controls "are sufficient to provide reasonable assurance of the safety and effectiveness of the device [5]. Examples of Class I medical devices include a hot water bottle, a hand-powered breast pump, a tongue depressor, and a medical examination table. Class I medical devices are regulated by FDA's general authority over adulterated and misbranded devices and general record-keeping and inspection authorities, that is, FDA's "general controls." These general controls are described in detail below.

Class II devices involve a somewhat greater health risk than devices in Class I, but not as great as devices in Class III. Class II devices are those that cannot be classified in Class I because insufficient information exists to determine that general controls are sufficient to assure the devices' safety and effectiveness and for which there is sufficient information to establish a performance standard for the device [6]. These devices are thus subject to both general controls and are supposed to be subject to performance standards.

A performance standard is a regulation prescribing performance and other criteria a medical device must meet in order to be marketed. Although Congress intended that a performance standard be established for every Class II medical device, to date, FDA has yet to issue a performance standard for any Class II devices. This lack of performance standards is discussed in more detail below. Examples of Class II medical devices include porcelain dental crowns, vascular clips used to occlude small blood vessels, eye movement monitors, and blood glucose monitors.

Class III devices are those for which premarket approval is required. This class consists of devices for which insufficient information exists to determine that "general controls" would provide reasonable assurance of safety and effectiveness, and for which (unlike Class II devices) there is *not* sufficient information to establish a performance standard [7]. The statute also sets Class III devices apart from the second category of Class I devices on the alternative grounds, either that they are intended for supporting or sustaining life or for a use of substantial importance in preventing impairment of health, or that they present a "potential unreasonable risk of illness or injury" [8]. Examples of Class III devices are pacemakers, heart valves, and fetal electroencephalograph monitors.

Class III also includes those devices, such as sutures and contact lenses, that were regulated as drugs prior to the enactment of the Medical Device Amendments of 1976. These "transitional devices" were automatically placed in Class III by the 1976 Amendments [9]. It was perceived that these devices presented potential "unreasonable risks" that should be reviewed by FDA prior to their marketing to assure the safety and effectiveness of the devices.

All medical devices on the market when the medical device amendments were enacted have been given a proposed or final classification by FDA [10]. The process by which these devices were classified began with the assignment of each device to an advisory panel of experts from outside of the government. They are composed of members from the research and medical communities. An additional member representing the industry and one representing consumer interests also sit on the panel. There are approximately 16 panels organized according to broad medical categories (e.g., ophthalmologics, orthopedics, cardiovascular products, etc.). Each panel discussed the products assigned to it and considered any information provided by FDA or device manufacturers regarding the appropriate classification of such products. The review process covers the conditions of use for the device, other persons for whose use the device is intended, the benefit to health versus the risk of injury from use or misuse of the device, and the reliability of the device.

The panelists started out with a few basic presumptions. All devices were Class I unless their safety and effectiveness could not be assured without greater regulation. Implanted devices were assumed to belong in Class III unless a less regulated class could assure safety. Devices regulated as drugs before the enactment of the Amendments ("transitional devices") were automatically assigned to Class III. Then, the panelists considered groups of similar devices together. After completing its review, the panel submitted recommendations to FDA. FDA either accepted the recommendation or stated the reasons why the classification recommendation was changed.

FDA considered these recommendations, and then published panel-by-panel proposals in the *Federal Register* to classify all medical devices. Public comments were sent to FDA either agreeing or disagreeing with each proposal. After reviewing these comments, FDA published final classification decisions.

As of the date of this publication, there are approximately 460 devices with a proposed or final Class I classification; 1087 with a proposed or final Class II classification; and 135 with a proposed or final Class III classification. The panels thus followed basic human nature and tended to place most devices in the middle, that is, in Class II.

III. GENERAL CONTROLS

General controls are one of the methods FDA uses to regulate Class I, Class II, and Class III products. These controls fall into two general categories: (1) controls such as general record-keeping and inspection provisions and the adulteration and misbranding provisions that apply to all FDA regulated products and (2) controls enacted as part of the 1976 Amendments that apply only to medical devices. A discussion of these provisions follows.

A. Controls Applicable to All FDA-Regulated Products

Adulteration

The adulteration provisions are a very powerful safety enforcement tool for FDA. Under Section 501 of the Act, a device is deemed adulterated if it is composed of any "filthy, putrid, or decomposed substance," or "if it contains any poisonous or deleterious substance which may render the [device] injurious to health" [11]. In addition to this basic safety provision, the Act requires that medical devices be produced, packaged, and held under safe, sanitary conditions [11]. If they are not manufactured according to current good manufacturing practices (GMPs), the product is deemed adulterated even if there is no actual safety hazard from a particular device [12].

All devices are subject to the safety and quality portion of the adulteration section. Additional requirements for safety conformance are imposed on certain devices. If, for example, a Class III device does not conform with the specification in the premarket approval application it is considered adulterated [13].

A device that is deemed adulterated is in violation of the Act, and FDA may take regulatory enforcement action against the individual device or a group of devices and the company or individuals marketing this device. Enforcement action is discussed more fully below.

Misbranding

Section 502 of the Act defines the circumstances under which a device will be considered "misbranded" and thus subject to regulatory action by FDA [14]. The types of misbranding specified in Section 502 relate to information required to be included in or omitted from the label or labeling of a medical device [14]. The primary purpose of the misbranding provisions is to require medical device manufacturers to provide accurate information about their products to the users of the product. Thus, a product is "deemed to be misbranded [if] its labeling is false or misleading" or if the label does not contain the descriptive information required by Section 502 and implementing regulations [15].

Section 502 provides that the label for a medical device must contain "(1) the name and place of business of the manufacturer, packer, or distributor and (2) an accurate statement of the quantity of the contents in terms of weight, measure, or numerical count" [16]. A label is the written or printed matter placed on a container for the purposes of the Act [17]. The information required to appear on the label must be on the outside container or package or be visible through the outside wrapper.

A label, however, is not the only form of medical device "labeling" regulated by FDA. Medical device "labeling" includes the outer wrapper or packaging in any promotion or informational flyer, any written advertising, instructional booklets, or any printed or graphic material accompanying the device [18]. FDA has authority to act (e.g., deem a device to be misbranded) if any of the foregoing types of labeling make misleading claims, even though the device itself has a proper label.

The misbranding prohibition for false or misleading information covers a wide range of statements. A statement may, for example, be literally true, but may nevertheless be misleading. Thus, a statement that a manufacturer has received reports that a medical device cures a certain disease may be literally true, but would nevertheless be considered misleading and constitute misbranding if in fact that product had not been shown by valid scientific evidence to cure such a disease [19]. The omission of a material fact may also be considered misleading. Thus, the failure to provide an adequate warning about a risk presented by a device may result in that product being misbranded.

Inspections and Enforcement

FDA enforces the misbranding and adulteration provisions of the Act and the other device specific enforcement remedies discussed below by conducting inspections of facilities where medical devices are manufactured, processed, packed, or held for introduction into commercial distribution. Section 704 of the Act gives FDA authority "to inspect, at reasonable times and within reasonable limits and in a reasonable manner" such facilities [20]. A "reasonable time" may be 3:00 a.m. if a facility conducts manufacturing at that time. And although FDA generally completes inspections in a few days, one inspection can take several months if the agency finds significant problems.

FDA inspections are unannounced and may be conducted for any reason. Generally, though, inspections are conducted to determine compliance with current good manufacturing practice (CGMP) requirements. Other reasons for FDA to conduct an inspection include:

1. As part of the process of reviewing a premarket approval application for a Class III device
2. As part of the government-wide quality assurance program FDA conducts inspections for other government agencies before those agencies enter into contracts to purchase medical devices
3. As a follow-up to a DEN (Device Experience Network), USP (U.S. Pharmacopoeia), or a medical device report on a product problem
4. In conjunction with a product recall

In the routine GMP inspection, FDA determines compliance by conducting fairly detailed inspections that examine manufacturing operations and records. A key part of any GMP inspection is a thorough review of complaint files. By looking at these files, the FDA inspector can quickly identify potential problem areas and see what a company has done to remedy or respond to those problems.

GMP inspections must be conducted at least every 2 years for Class II and Class III products [21], but FDA will conduct inspections more frequently where significant problems are found. There is no time requirement on the frequency of inspections for Class I products. FDA conducts approximately 1500 medical device GMP inspections per year.

GMP inspections generally last 2 days. Inspectors will examine such things as the methods of manufacture, assembly, storage, sterilization, and, if applicable, packaging. Inspections also include detailed record review, especially complaints. Investigators are within their legal rights to request copies of records required to be kept by FDA's GMP regulations or finished product samples.

At the beginning of an inspection, the investigator will present a notice of inspection or Form 482. At the conclusion of the inspection, the investigator will leave a list of observations or Form 483. When the investigator returns to his office, he will prepare a detailed write-up of the inspection called an establishment inspection report (EIR).

If significant violations of the Act are found during an inspection, the FDA has a variety of enforcement tools it may select to assure compliance with the Act. FDA's choice of enforcement remedy will vary depending on the severity of the violation. One of the less severe actions FDA can take is the sending of a notice of adverse findings. This is a letter informing a firm that FDA believes violations of the Act exist and asking how the firm intends to correct those violations. Another more serious letter FDA can send is a regulatory letter. This is a letter from the FDA identifying violations of the Act and promising legal enforcement action by the FDA if corrections are not made.

If warnings are unsuccessful in bringing about corrections of apparent violations of the Act or if immediate action is needed to eliminate an imminent public health risk, FDA is authorized to take court action. Without going to court, FDA can also temporarily detain devices suspected of violating the Act. And, after obtaining court approval, FDA can actually "seize" (i.e., take control of) potentially violative devices wherever they may be found [22]. FDA can also bring a civil injunction action to prohibit a company from making or shipping a violative product [23]. A violation of the Act may also constitute a criminal misdemeanor or felony, in which case the FDA can prosecute an individual or corporation [24]. In all court-related actions, FDA operates through and in cooperation with local U.S. attorneys' offices and the Department of Justice.

B. Device-Specific General Controls

The following general controls were all enacted as part of the 1976 Amendments and apply only to medical devices.

Registration and Listing

Under Section 510 of the Act, every person engaged in the "manufacture, preparation, propagation, compounding or processing of . . . a device shall register . . . his name, place of business and such establishment [25]. This includes manufacturers of devices and components, repackers, relabelers, as well as initial distributors of imported devices. Those not required to register include manufacturers of raw materials, licensed practitioners, manufacturers of devices for use solely in research or teaching, warehousers, manufacturers of veterinary devices, and those who only dispense devices (pharmacies) [26].

Upon registration, FDA issues a device registration number. A change in the ownership or corporate structure of the firm, the location, or person designated as the official correspondent must be communicated to the FDA device registration and listing branch within 30 days. Registration must be done when first beginning to manufacture medical devices and must be updated yearly. Forms for registration (FD2892) are available form FDA.

Section 510 of the Act also requires all manufacturers to list the medical devices they market [27]. Listing must be done when first beginning to manufacture a product and must be updated every 6 months. Listing includes not only informing FDA of products manufactured, but also providing the agency with copies of labeling and advertising. Forms for listing (FD2892) are available from FDA.

Foreign firms that market products in the United States are permitted—but not required—to register, and are required to list. Foreign devices that are not listed are not permitted to enter the country.

Registration and listing provides FDA with information about the identity of manufacturers and the products they make. This information is very important to FDA. It enables the agency to schedule inspections of facilities and also to follow up on problems. When FDA learns about a safety defect in a particular type of device, it can use the listing information to notify all manufacturers of those devices about that defect.

Good Manufacturing Practices

FDA is authorized, under Section 520(f) of the Act, to promulgate regulations detailing compliance with current good manufacturing practices [28]. GMPs include the methods used in, and the facilities and controls used for, the manufacture, packing, storage, and installation of a device. The GMP regulations were established as manufacturing safeguards to ensure the production of a safe and effective device and include all of the essential elements of a quality assurance program [29]. Because manufacturers cannot test every device, the GMPs were established as a minimum standard of manufacturing to ensure that each device produced would be safe. If a product is not manufactured according to GMPs, even if it is later shown not to be a health risk, it is in violation of the Act and subject to FDA enforcement action.

The general objectives of the GMPs, not specific manufacturing methods, are found in Part 820 of the Code of Federal Regulations. The GMPs apply to the manufacture of every medical device. Rather than developing separate standards for each class of device, the GMPs designate two device categories. "Critical" devices are those intended to be surgically implanted and used to support or sustain human life, and whose failure could reasonably be expected to result in significant injury (Class III devices for the most part) [30]; all other devices are considered "noncritical." All device manufacturers must meet general GMP requirements. Critical device manufacturers have additional GMP requirements.

The GMP regulations cover most aspects of manufacturing: from the organization and qualifications of the quality assurance personnel to the physical building design and equipment. The GMPs also include requirements for raw material inspection and handling, product and processing control, packaging and labeling controls, and warehouse distribution procedures. Moreover, GMPs specify record-keeping requirements—these records are subject to review by FDA during inspection.

Notification and Repair, Replace, or Refund Authority

Under Section 518 of the Act, FDA may order manufacturers to notify device users of a risk presented by a device, require repair or replacement of a device, or require refund of the purchase price of a device where the device presents an unreasonable risk of substantial harm to human health [33].

FDA can invoke this "notification" requirement only if (1) there is an unreasonable risk of substantial harm to human health about which FDA has found it necessary to inform the users and (2) there is no other more practicable means of protecting the public health [32]. In those rare situations where providing the user of the device with a warning is insufficient, the Act allows FDA to require that a manufacturer notify medical professionals.

Where notification is not "sufficient to eliminate the unreasonable risk," FDA may exercise its repair, replacement, or refund authority [33]. Before doing so, however, FDA is required to hold an informal hearing. FDA may order the preparation of a plan for repair, replacement, or refund under Section 518(b) only if each of the following four tests is satisfied:

1. A device intended for human use presents an unreasonable risk of substantial harm to the public health.
2. There are reasonable grounds to believe that the device was not properly designed and manufactured with reference to the state of the art as it existed at the time of its design and manufacture.

3. There are reasonable grounds to believe that the unreasonable risk was not caused by failure of a person other than the manufacturer, importer, distributor, or retailer of the device to exercise due care in the installation, maintenance, repair, or use of the device.
4. The notification authorized by Section 518(a) would not by itself be sufficient to eliminate the unreasonable risk, and repair, replacement, or refund is necessary to eliminate such risk [33].

To date, FDA has never found it necessary to use its repair, replace, or refund authority under Section 518(b).

Medical Device Reporting

The 1976 Amendments also provide FDA with basic authority to require a manufacturer to document, follow up, and, on occasion, report to FDA problems reported to the manufacturer concerning its device. The medical device reporting (MDR) regulation, a mandatory adverse experience reporting system for medical devices, became effective on December 14, 1984. The authority for the MDR reporting system stems from Section 519 of the Act, which authorizes FDA to issue regulations to require device manufacturers, importers, and distributors to "maintain such records, make such reports, and provide such information as the [FDA] may by regulation reasonably require to assure that such device is not adulterated or misbranded and to otherwise assure its safety and effectiveness [34].

According to the MDR regulations implemented pursuant to this authority, an MDR report is required to be submitted to FDA whenever a manufacturer becomes aware of information that reasonably suggests that one of its devices (1) may have caused or contributed to a death or seroius injury, or (2) has malfunctioned, and, if the malfunction recurrs, is likely to cause or contribute to a death or serious injury [35].

When a death or serious injury has occurred, a telephone report ot FDA is required as soon as possible, but in no event later than 5 calendar days after receipt of the information [36]. The telephone report is required to contain a detailed description of the incident, including the identity of the device and the manufacturer. This telephone report must be followed by a written report within 15 working days of initial receipt of the information [36]. The written report is required to contain all information submitted in the telephone report and any additional relevant information received subsequent to the telephone report.

Reportable malfunctions are required to be reported to FDA in writing as soon as the necessary information for making the report is obtained, but in no event later than 15 working days after the initial receipt of the information [37]. A manufacturer is required to report to FDA each time it becomes aware of a reportable event, even if an event of the same or similar nature has been reported previously, or if it is believed to be the result of user error, faulty service, or maintenance.

The failure to make a required MDR report subjects the company responsible for the violation to FDA enforcement action. A detailed description of the MDR reporting requirements is contained in Chapter

Restricted Devices

Section 520(e) of the Act authorizes FDA to restrict the sale, distribution, or use of a device only upon the written or oral authorization of a practitioner licensed by law to administer or use such device, or upon such other conditions as FDA may prescribe in such regulation [38].

If, because of the potential for a device to cause a harmful effect or because of the collateral measures necessary to its use, the FDA determines that there cannot otherwise be reasonable assurance of its safety and effectiveness, FDA is authorized to limit the use of such a device to persons with specific training or experience or to persons in certain facilities. Such restrictions, however, cannot be limited only to persons who have board certification by the American Board of Medical Specialties.

Banned Devices

Section 516 of the Act authorizes FDA to ban a product if there is an unreasonable and substantial risk of injury to the public health [39]. Unreasonable risk may be caused by the device or the procedure in which it is used. A device may also be banned if it is substantially deceptive to the public [40]. FDA is not required to show that the manufacturer intentionally sought to mislead the public or harm consumers, just that the deception or unreasonable risk exists. If the health risk is low and the deception can be reduced or eliminated by changes in the labeling, FDA can allow corrective labeling.

Proceedings to ban a device may be instituted by FDA or in response to a citizen's petition. The procedure begins with the issuance of a proposed regulation to ban a device or a group of devices. After the public is given an opportunity to comment on the proposal, FDA issues the final regulation, which takes effect on publication in the *Federal Register*.

IV. BRINGING A MEDICAL DEVICE TO MARKET

A. Overview of the Premarket Approval and Review Process

Two statutory provisions govern the introduction of new medical devices into the marketplace: Sections 515 and 510(k) [40]. Section 515 establishes a premarket approval process for Class III devices. This process entails submitting a premarket approval application (PMA) containing data and information demonstrating the safety and effectiveness of the device. Section 510(k) establishes a premarket notification process. Under this process, a manufacturer is required to file with FDA, 90 days before a new device is to be marketed, a premarket notification demonstrating that the device in question is substantially equivalent to a device that was on the market before enactment of the 1976 Amendments—and, therefore, marketable without formal FDA approval.

In enacting Section 510(k), Congress, in effect, divided medical devices into two broad categories: (1) preamendment—those introduced into commercial distribution prior to May 28, 1976, the enactment date of the 1976 Amendments—and (2) postamendment-those introduced into commercial distribution after May 28, 1976.

New postamendment devices are automatically placed in Class III by the terms of the Act [41]. However, a postamendment device can be brought to market under the 510(k) process, if FDA determines that the device is "substantially equivalent" to a preamendment device. If the postamendment device is identical to a preamendment device, it is substantially equivalent.

The 510(k) is then accepted by FDA and the postamendment device is placed in the same class as the preamendment device to which it is substantially equivalent. The 510(k) or premarket notification process thus serves as a classification system for postamendment devices. For example, a postamendment elastic bandage or bed board that is identical to a preamendment Class I bandage or bed board would be found substantially equivalent to the preamendment predicate and classified in Class I.

Of course, not all postamendment devices are identical to a preamendment device. Most have minor changes in materials, design, or energy source. If the postamendment device is not identical to a preamendment device, FDA may nevertheless permit marketing under the 510(k) process if the agency concludes that the postamendment device is "substantially equivalent" to the preamendment device. Thus, a grade #1 stainless steel surgical blade can be found substantially equivalent to a Class I grade #4 stainless steel surgical blade. Similarly, even though a postamendment device has changes in design or materials, it may be substantially equivalent to a preamendment device. When deciding an issue of "substantial equivalence," FDA generally will look at whether the device has the same intended use as the preamendment device and whether the changes in material, design, or energy source raise any new questions of safety or effectiveness.

A Class III postamendment device may also be found to be substantially equivalent to a Class III preamendment device and marketed under a premarket notification. There are, however, some differences that apply to Class III devices. Remember, Class III devices are those for which a premarket approval application is or will be required. Thus, FDA will eventually require filing of premarket approval applications for all Class III preamendment devices. Until FDA calls for PMAs for such Class III devices, the law allows FDA to accept a 510(k) for a postamendment Class III device that is substantially equivalent to a Class III preamendment device. When FDA requires the filing of PMAs for a type of Class III device, for example, heart valves, manufacturers of both preamendment and postamendment versions of that Class III device—all heart valve manufacturers—will be required to submit a PMA application to FDA.

All postamendment devices that are not substantially equivalent to a preamendment device are classified in Class III. These devices thus cannot be marketed without a premarket approval application. It is important to bear in mind that FDA does not consider the risk presented by a device in determining whether the device is substantially equivalent to a preamendment device. Thus, it is theoretically possible that FDA could find that a postamendment device is not substantially equivalent and therefore requires a PMA application even though that device does not present a significant risk.

B. Reclassification

The 1976 Amendments contain five separate provisions for the reclassification of medical devices from Class III to Class II or I or Class II to I [42]. These provisions apply to preamendment Class III devices [43], "new" postamendment Class III devices [44], transitional devices [45], devices for which a performance standard is being promulgated [46], and preamendment Class III devices for which FDA is requiring the filing of PMA applications [47].

The procedure by which FDA will reclassify a device varies somewhat, depending on the statutory provision under which reclassification is sought.

Generally, though, the process will be initiated by the filing of a petition, although FDA can initiate a reclassification on its own initiative. FDA will generally refer the petition to an advisory panel. A public meeting will be held at which the advisory panel will hear presentations supporting or opposing the petition, discuss the merits of the petition, and then vote to recommend for or against reclassification.

If FDA accepts the positive recommendation of a panel, FDA will follow one of two procedures depending on the reclassification provision being used. FDA will either announce the reclassification in a final decision without publishing a notice in the *Federal Register* or publish a notice in the *Federal Register* proposing the reclassification. If a *Federal Register* proposal is issued, after public notice and comment, a final regulation will be issued establishing the new classification of the device.

C. Premarket Notification

The premarket notification or 510(k) provision, which is discussed above, is perhaps the most used, and one of the most critical provisions of the 1976 Amendments. Section 510(k) is used to notify the FDA in two situations: (1) when first marketing a device that is "substantially equivalent" to a preamendment device and (2) when changing a device in a way that could significantly affect safety or effectiveness. Examples of significant changes would be changes in design, material, composition, energy source, manufacturing process, or intended use.

A premarket notification must be filed at least 90 days before going to market with a new or substantially changed device. A premarket notification is generally a relatively short document that provides the name of the device, a description of the device, and a description of the device to which it is substantially equivalent, and includes proposed labeling that describes the device, its intended use, and its directions for use. FDA may, however, ask for clinical data to support a premarket notification where the description of the device is inadequate to show substantial equivalence. FDA generally reviews 510(k) applications within 90 days and tells the company whether the agency considers the product to be substantially equivalent. If so, the product is treated the same as the product to which it is equivalent. If not, FDA will require a company to file a PMA for the product.

A premarket notification is not an approval. A letter from FDA clearing a premarket notification is simply a determination by FDA that the proposed device is substantially equivalent to a preamendment device and thus will be placed in the same class as the preamendment device. FDA strictly prohibits any manufacturer from representing that a 510(k) is an "FDA approval" [48].

D. Premarket Approval

Anyone intending to introduce a "new" postamendment Class III medical device into commercial distribution is required to obtain FDA approval of a premarket approval (PMA) application before marketing a product. A premarket approval application is a voluminous document that includes clinical data and much greater detail than a premarket notification.

The PMA process begins with the submission of the PMA application. Although the Act only allows FDA 180 days to review a PMA, review time averages between 11 and 13 months. Approximately 45 days following the submission of the PMA, FDA makes a threshold decision on whether to "file" the PMA. This determination involves a threshold finding that the PMA contains on its face all of the information necessary to provide for review of the PMA. Once the PMA is filed, substantive review of the PMA by FDA begins. Following FDA substantive review, the PMA is referred to an advisory committee, which makes a recommendation on whether to approve or deny the PMA. FDA then either accepts or rejects the advisory committee recommendation and issues an approvable or not approvable letter. This is a letter informing the applicant that the application is approvable with certain minor revisions, such as final approved labeling and a GMP inspection, or it can be a letter stating that the PMA is not approvable without a major amendment, such as an additional clinical study. The final step in the process is the final FDA decision to approve or deny approval of the PMA. The content requirements for a PMA application and the procedure for filing PMAs and FDA's processing of these applications are described in detail in another chapter.

PMAs will also be required for all preamendment Class II devices and for all postamendment Class III devices that have been found to be substantially equivalent to a preamendment Class III under the 510(k) process. The procedure by which FDA will require PMA applications for pre- and postamendment Class III devices begins with the publication of a proposed regulation setting forth information about the device, such as the degree of risk presented by the device and the condition or conditions the device is intended to treat or diagnose [49]. This notice provides an opportunity for comment and an opportunity to request a reclassification of the device to Class II or I. If reclassification is requested, FDA is required to refer the reclassification request to an advisory committee for a recommendation on the request. The agency is then required to rule on the reclassification request prior to proceeding further with the call for PMAs. If there is no reclassification request, after reviewing the comments on the proposal to call for PMAs, FDA publishes a final rule setting forth the specific requirements for a PMA [49].

Manufacturers of pre- and postamendment Class III devices of that type are then required to file PMAs within 90 days from the date of publication of the final regulation. Failure to file the PMA within the required time results in the product being adulterated and subject to regulatory action [50]. If a PMA is filed within the 90-day period, the manufacturer can continue to market the product pending the FDA's review of the PMA. If FDA approves the PMA, the product can continue to be marketed. FDA's denial of such a PMA, however, means that the product can no longer be marketed.

E. Investigational Device Exemptions

The Act prohibits the introduction into interstate commerce of any unapproved medical device, even if that introduction is for an investigational use [51]. The Act provides, though, for an investigational device exemption (IDE) to this prohibition for clinical investigations involving one or more human subjects to determine the safety or effectiveness of a device. Clinical data collected under an IDE are usually used to support a PMA application for a Class III device, but may also be used to support a premarket notification for Class I or Class II device or a preamendment Class III device [52].

The general rule is that an IDE is required any time a study is conducted on an unapproved device that involves humans. There are, however, a few exceptions. An IDE is not required for a device undergoing consumer preference testing, testing of a modification, or testing of a combination of two or more devices in commercial distribution, if the testing is not for the purpose of determining safety or effectiveness and does not put subjects at risk [53]. What constitutes consumer preference testing or other activities that are exempt from the IDE requirements is not well defined. Companies who believe a proposed test of a device falls under this exemption would be well advised to seek the advice of legal counsel.

For those studies that do fall under the IDE requirements, there are two basic types of IDE studies—those involving significant risk devices, and nonsignificant risk devices. FDA approval of an IDE is required before testing of a significant risk device can be started. A significant risk device is one that:

1. Is intended as an implant and presents a potential for a serious health risk;
2. Is represented for use in supporting or sustaining life and presents a potential for a serious risk to health;
3. Is for a use of substantial importance in diagnosing, curing, medicating, or treating disease and presents a potential for a serious risk to health; or
4. Otherwise presents a potential for a serious risk to health [54].

Prior to beginning a clinical study for a significant risk device, a company is required to file an IDE application with FDA. FDA then has 30 days in which to respond to the IDE application, and actual review times are within this 30 days. The contents of and the procedures for filing an IDE application are described in detail in Chapter 19.

A clinical study of a nonsignificant risk device does not require submission or FDA approval of an IDE application if:

1. The device is labeled as an investigational device;
2. Institutional Review Board (IRB) approval is obtained;
3. Each patient is given informed consent; and
4. The sponsor monitors the study and the investigator maintains all records required by the IDE regulations [55].

Although the FDA has guidelines on the determination of whether a device is a significant or nonsignificant device, the criteria are somewhat vague. Even where a company obtains an opinion from a hospital's Institutional Review Board that a proposed study of a device involves a "nonsignificant" risk, FDA may nevertheless disagree with that decision. In such a situation FDA could take regulatory action or simply put the clinical investigation on hold requiring the filing of an IDE application. It is therefore useful for companies to seek the advice of legal counsel where there are questions about whether the use of the device to be studied presents a "nonsignificant risk."

The Act and its implementing regulations prohibit the promotion of an investigational device, although companies are permitted to recover costs of the device during the course of a clinical investigation [56]. The prohibition on commercialization or promotion of an investigational device applies to all unapproved devices, regardless of their classification or whether they are the subject of a clinical investigation. FDA prohibits any representation that an investigational device is safe or effective for the purposes for which it is being investigated.

An unapproved device may, however, be advertised or displayed at an educational or promotional symposium. The following precautions should be taken, though, in such cases:

1. Make clear that the device has not yet been approved by FDA by stating that it is an investigational device or that a 510(k) or PMA is pending or is about to be filed for the device.
2. Do not take purchase orders for the device or represent that you are prepared to take purchase orders or quote prices for the device.
3. Do not state or imply that the device is safe or effective.
4. Do not compare your device to other devices or suggest that your device is better than other devices.
5. A factual description of the device and how it operates is permitted. This description will necessarily include a description of the features of the device that distinguish it from other devices.
6. It is important that the description of the device and all other aspects of the advertisement present a fair balance of the device's strengths and weaknesses [57].

Development of advertising copy for an unapproved device is another area where the assistance of legal counsel can often be of help.

V. THE PERFORMANCE STANDARD SETTING PROCESS

In addition to general controls, Class II products are subject to the additional control of performance standards. Performance standards provide a detailed description of the construction, components, ingredients, and properties of the device and its compatibilities with power systems used in connection with such systems, and also include provisions for testing the device or measuring its performance characteristics.

The procedure by which a standard is promulgated includes a "Section 514(b) notice" that is published in the *Federal Register* and offers an opportunity to seek reclassification of the device to Class I [58]. If there are not requests to reclassify, or if the agency denies those requests, FDA then publishes a "Section 514(c) notice," which invites organizations to submit existing standards or develop new standards. Following consideration of any submission, FDA selects someone to write a standard or decides to write it itself.

After development of a proposed regulatory standard, FDA publishes the proposal in the *Federal Register* for public comment. Comments are reviewed and the agency can then issue a final standard. The process is expected to take several years—per product—to complete. Although performance standards are required for the 1087 devices in Class II, FDA has thus far only initiated the performance standards setting procedures for a handful of devices and has failed to complete the process for any device. Part of the difficulty in establishing performance standards can be attributed to the elaborate procedural requirements that must be followed to establish a performance standard. The problem has also been aggravated by the fact that the classification panels and FDA placed too many devices in Class II. Many of these devices belong in Class I. Finally, FDA has a manpower shortage that has required the agency to set priorities. Consequently, the review of PMAs, IDEs, and reclassification petitions has taken precedence over the establishment of performance standards.

One might ask whether this deficiency in implementing the law has resulted in any significant public health problem. To date, it has not. The FDA has used the 510(k) process and general controls such as GMPs, the MDR reporting system, and other enforcement tools to assure that medical devices are safe and effective. The real solution to the problem presented by requiring the establishment of performance standards would be to amend the law to make standards for Class II products discretionary. With such a change, FDA could focus on those devices that truly need performance standards and would not have the impossible burden of establishing such a large number of standards.

VI. OTHER MEDICAL DEVICE STATUTORY PROVISIONS

A. Custom Devices

The premarket notification and premarket approval requirements of Section 510 and Section 515, respectively, do not apply to a device that (1) is manufactured to the order of a individual physician or dentist; (2) is not generally available in finished form for purchase or for dispensing upon prescription; and (3) is not offered through labeling or advertising by the manufacturer, importer, or distributor thereof for commercial distribution [59]. To fall under this custom device exemption, the device must also (1) be intended for use by an individual patient named in the order of the physician or dentist; or (2) be intended to meet the special needs of such physician or dentist in the course of their professional practice; and (3) not be generally available to or generally used by other physicians or dentists.

The most common question asked about the custom device exemption is how many of a particular type of custom device may be manufactured before that device becomes "generally available or generally used." While the number is probably greater than 1 and less than 50, there is no specific guidance on this question. Companies are, however, advised to be cautious about the quantity of a custom device manufactured—otherwise they may find that FDA will take the position that the device is generally available and therefore not subject to the custom device exemption.

B. Trade Secrets

In enacting the Medical Device Amendments in 1976, Congress placed a substantial burden on manufacturers of Class III devices to develop scientific evidence required to prove the safety and effectiveness of their products. In exchange, the statute grants what amounts to a private license to market the product: FDA's approval of a PMA for a Class III device permits the marketing only of a particular product produced by a particular manufacturer.

Congress also recognized the economic value of the safety and effectiveness data developed to obtain such a private license and enacted the trade secret and confidentiality provisions aimed specifically at protecting the "competitive advantage" inherent in ownership of such data [60]. These provisions protect safety and effectiveness data from public disclosure and prohibit reliacne by one manufacturer on data submitted by another to gain approval of the same or similar device.

Section 520(c) of the Act provides the basic protection against public disclosure [61]. It prohibits disclosure of information submitted under the premarket approval provision and other provisions of the medical device amendments if those data qualify for the disclosure exemption in the Freedom of Information Act (FOIA) applicable to "trade secrets and commercial or financial information obtained from a person and privileged or confidential" [62]. Legislative history makes it clear that Congress considered safety and effectiveness data to fall within this exemption and thus to be protected from disclosure by Section 520(c) [47-50].

This protection has been codified further in FDA's regulations, which provide expressly that the safety and effectiveness data in a PMA file, as well as the file's very existence, shall not be disclosed to the public [63]. Under Part 814.9, the confidential "PMA file" includes all data and information submitted with the PMA. The confidentiality of such data has also been confirmed by the courts [64]. In this case, the court recognized that safety data on a medical device qualifies for the FOIA's "(b)(4)" exemption as confidential commercial information if there is a likelihood of substantial competitive injury resulting from disclosure [65].

The Act does provide for limited disclosure of a summary of the safety and effectiveness information contained in a PMA. Section 520(h) provides that, upon approval of a PMA, a summary of information respecting the safety and effectiveness of the device—but not the safety and effectiveness data itself—will be prepared and made public [66]. Even the summary, however, is expressly excluded by the statute for consideration by FDA in support of another party's application, and its release remains subject to Section 520(c). The purpose of Section 520(h) is to permit public scrutiny of FDA's decisions without sacrificing the confidentiality—and thus economic value—of data submitted in support of Class III medical devices [47 at 51].

The trade secret protection established in Section 520(c) [67]—as qualified by Section (b)(4) of the FOIA [68]—is broad. It appiles not only to information submitted as part of a PMA application, but also to "any information reported to or otherwise obtained by [FDA]" as part of a classification proceeding, a standard setting proceeding, a proceeding to ban a device, a proceeding to require notification of users of a device about a risk under Section 518(a), a proceeding to require repair, replacement or refund of a device under Section 518(b), a medical device report submitted under Section 519, an inspection for good manufacturing practices, or an investigational device exemption application.

C. Export

Although an unapproved medical device may not be entered into commercial distribution in the United States or subject to clinical investigation without an IDE exemption in the United States, such an unapproved device may be exported under certain limited conditions. Section 801(d) of the Act provides that a medical device may be exported if it:

1. Accords to the specifications of a foreign purchaser;
2. Is not in conflict with the laws of the country to which it is intended for export;
3. Is labeled on the outside of the shipping package that it is intended for export;
4. Is not offered for sale in domestic commerce; and
5. The Secretary has determined that the exportation of the device is not contrary to public health and safety and has the approval of the country to which it is intended for export [69].

FDA has established an administrative procedure for obtaining export permits from the agency. This is usually a fairly brief filing, which contains the basic information described above. FDA does not perform a detailed safety and effectiveness evaluation, and generally will approve a permit if the product is of a type that FDA is familiar with and has already approved a 510(k) or PMA for, or a product that has an approved IDE and is undergoing clinical investigation.

D. Exemptions

FDA may exempt a device from certain regulatory requirements under Section 513 of the Act [70]. A manufacturer may be exempt from the registration, listing, and premarket notification portions of Section 510 of the Act [71]. FDA has determined that for certain devices it is not necessary for the protection of public health to conduct inspections as often and receive premarket notification [510(k)] [72].

FDA may also exempt device manufacturers from certain GMP and recordkeeping requirements of Sections 519 and 520(f) of the Act [73]. An example is the Class I devices not labeled as sterile. Many of the GMP requirements are directed specifically at the manufacture of a sterile product and are unduly burdensome to a nonsterile Class I manufacturer.

VII. CONCLUSION

This has been a very brief overview of the Act as it applies to medical devices. More detail about all of these subjects can be found in the remaining chapters of this book.

ACKNOWLEDGMENTS

I would like to express my appreciation to Fred Degnan, Sherill LaPrade, and Maureen English for their assistance in preparing this chapter.

REFERENCES

1. Pub. L. No. 75-717, 52 Stat. 1040 (1938), as amended 21 U.S.C. §§ 301-392.
2. Medical Device Amendments of 1976, Pub. L. No. 94-295 (1976), 90 Stat. 539 (codified at 15 U.S.C. 55 at 21 U.S.C. *passim*).
3. 21 C.F.R. *passim*.
4. 21 U.S.C. 360c(a)(1)(A)(ii)(II).
5. 21 U.S.C. 360c(a)(1)(A)(ii).
6. 21 U.S.C. 360c(a)(1)(B).
7. 21 U.S.C. 360c(a)(1)(C).
8. 21 U.S.C. 360c(a)(1)(C)(ii)(II); 21 C.F.R. 860.3(c)(3).
9. 21 U.S.C. 360j(l).
10. See 21 U.S.C. 360c.
11. 21 U.S.C. 351(a).

12. 21 U.S.C. 351(h).
13. 21 U.S.C. 351(f).
14. 21 U.S.C. 352.
15. 21 U.S.C. 352(a).
16. 21 U.S.C. 352(b).
17. 21 U.S.C. 321(k).
18. 21 U.S.C. 321(m).
19. See *U.S. v. John J. Fulton Co.*, 33 F. 2d 506 (9th Cir. 1929).
20. 21 U.S.C. 374(a)(1).
21. 21 U.S.C. 360(h).
22. 21 U.S.C. 304(a)(2)(D).
23. 21 U.S.C. 332.
24. 21 U.S.C. 333.
25. 21 U.S.C. 360(c).
26. 21 U.S.C. 360(g).
27. 21 U.S.C. 360(j).
28. 21 U.S.C. 360j(f).
29. 21 C.F.R. 820.
30. 21 C.F.R. 820.3(f).
31. 21 U.S.C. 360(h).
32. 21 U.S.C. 360h(a)(2).
33. 21 U.S.C. 360h(b).
34. 21 C.F.R. 803.
35. 21 C.F.R. 803.1(a).
36. 21 C.F.R. 803.24(b)(1).
37. 21 C.F.R. 803.24(b)(2).
38. 21 U.S.C. 360j(e)(1).
39. 21 U.S.C. 360f.
40. 21 U.S.C. 360e, 360(k).
41. 21 U.S.C. 360c(f)(1).
42. 21 U.S.C. 360c(e), 360c(f), 360j(e), 360d(b), 360e(b).
43. 21 U.S.C. 360c(e).
44. 21 U.S.C. 360c(f).
45. 21 U.S.C. 360j(e).
46. 21 U.S.C. 360d(b).
47. 21 U.S.C. 360e(b).
48. See 21 C.F.R. 807.97.
49. 21 U.S.C. 360e(b).
50. 21 U.S.C. 351(f).
51. 21 U.S.C. 352(f) and (o).
52. 21 U.S.C. 360j(g).
53. 21 C.F.R. 812.2(c)(4).
54. 21 C.F.R. 812.3(m).
55. 21 C.F.R. 812.2(b)(1).
56. See C.F.R. 812.7.
57. See FDA, "Guideline for Preparing Notices of Availability of Investigational Medical Devices" (November 1985).
58. 21 U.S.C. 360d(b).
59. 21 U.S.C. 360, 360e.
60. See generally H.R. Rep. No. 94-853, 94th Cong., 2nd Sess. 48-5 (1976).
61. 21 U.S.C. 360j(c).
62. 21 U.S.C. 552(b)(4).

63. 21 C.F.R. 814.9.
64. See *Public Citizen Health Research Group v. Food and Drug Administration*, 704 F. 2d 1280, 1290-91 (D.C. Cir. 1983).
65. 5 US..C. 552(b)(4).
66. 21 U.S.C. 360j(h).
67. 21 U.S.C. 360j(c).
68. 5 U.S.C. 552(b)(4).
69. 21 U.S.C. 381(d).
70. 21 U.S.C. 360c.
71. 21 U.S.C. 360.
72. 21 U.S.C. 360(k).
73. 21 U.S.C. 360i, 360j(k).

part three

Responding as Organizations to the Regulatory Environment

6

CDRH's Medical Device Laboratories

ELIZABETH D. JACOBSON and F. ALAN ANDERSEN *Center for Devices and Radiological Health, Food and Drug Administration, Rockville, Maryland*

I. INTRODUCTION

Among the Food and Drug Administration's laboratories, the least well known may very well be those of the Center for Devices and Radiological Health (CDRH). This chapter describes the laboratory program and provides a sense of the relationship between it and the medical device industry.

The origin of the current laboratory facility can be traced to both the radiological health and the medical devices antecedents to CDRH. Beginning in the early 1970s, the medical device program maintained a small laboratory with capability to evaluate *in vitro* diagnostic device performance and mechanical and electrical engineering aspects of medical device hardware. The radiological health laboratories developed engineering, physics, and biology expertise over a period of over 25 years and became the largest component of the radiation protection program.

When the medical device and radiological health organizations were merged, an increased overall emphasis was placed on medical devices. The laboratories mirrored that pattern. Many of the radiological health programs were trimmed and the combined laboratory began to redirect its efforts toward medical devices. Today, the laboratory maintains its commitment to CDRH radiation protection goals, while putting most of this effort toward medical devices.

It is the medical device focus that begs questions from the industry: "What exactly do you do?" "Why do you do it?" "How do you do it?" "Why should we care?" These questions and more will be addressed below.

II. WHY DOES CDRH HAVE MEDICAL DEVICE LABORATORIES?

The Food and Drug Administration's Center for Devices and Radiological Health (CDRH) expends significant resources and effort to maintain a research-based science group as part of its medical device and radiological health organization. CDRH is making these expenditures at a time when resource constraints face all public health institutions. Why are scarce resources being used to support a research base?

The practical answer is that the research base provides benefits that are not available from any other source.

CDRH medical device laboratories provide this crucial support and input for Center programs because they are maintained by scientists and engineers whose personal expertise, publications, and analyses are predicated on their laboratory or field research. The laboratories have impact on such "premarket" programs as the review of premarket submissions, reclassification actions, development of manufacturer and reviewer guidelines and guidance documents, and laboratory analyses of selected devices prior to approval. Laboratory "postmarket" activities include failure analyses of problem products and review of corrective action plans, standards support, educational activities, and postmarket surveillance. The laboratories also provide the center with independent means for addressing fundamental public health issues, both in forecasting upcoming problems and in developing technological "fixes," that is, technological improvements that solve key problems. The laboratory program includes a mix of both forward-looking developmental projects and work of immediate applicability to current problems. This work includes the analysis of phenomena that cut across broad device areas, such as understanding the properties of the various types of materials used in devices, and problems of equipment automation.

III. WHAT KINDS OF MEDICAL DEVICE LABORATORY FACILITIES DOES CDRH HAVE, AND WHERE ARE THEY LOCATED?

The medical device laboratories are contained in the Office of Science and Technology (OST), one of nine offices within CDRH. OST is a multidisciplinary group, organized internally along disciplinary lines into five operating divisions—Biometric Sciences, Life Sciences, Physical Sciences, Mechanics and Materials Science, and Electronics and Computer Science. Division laboratories occupy a total of about 80,000 square feet, including 5000 square feet for animal care, in several buildings in Rockville, MD. The laboratories are linked by interactive computer facilities and centralized administrative support, and include physical science and engineering facilities (digital and analog electronics labs, materials lab, mechanics lab, electro-optics lab, computer science lab, medical imaging lab, ionizing and nonionizing radiation calibration facilities, shop facilities), life science laboratories (toxicology, molecular biology, and radiation bioeffects), and statistics and epidemiologic research facilities.

In-house efforts are greatly enhanced through collaborations with field laboratories and the Winchester Engineering and Analytical Center in FDA's Office of Regulatory Affairs, as well as through numerous collaborations with outside scientific groups (see Table 1).

TABLE 1 Examples of Collaborative Projects with Other Laboratories

	Institution	Project
Government	Clinical Center, National Institutes of Health (NIH)	Clinical evaluation/optimized mammography
	Centers for Disease Control and FDA Office of Regional Affairs	Disinfectant effect on dialyzer membranes
	National Center for Health Statistics	Population-based data on implants
	National Institute of Standards and Technology	Standard reference material for polyurethane
	Environmental Protection Agency	*In vitro* screening methods for toxicology
	Walter Reed Army Medical Center	Laser beam transmission properties
	Office of Naval Research	Microwave absorption by biomaterials
	National Cancer Institute, NIH	Metal toxicity and carcinogenicity
Academia	Johns Hopkins University	Nondestructive evaluation of devices
	Cleveland Clinic	Microelectronic circuitry for biosensors
	University of Kansas	Image analysis/tissue characterization
	Duke University	Reduction of ultrasound speckle noise/signal processing
	University of Pennsylvania	Ultraviolet-A radiation carcinogenesis
	Case Western Reserve University	Statistical evaluation of clinical laboratory measurements
	University of North Carolina	Inactivation of slow viruses
Industry	Eastman Kodak	Quantitative evaluation of medical imaging systems
	Richards Medical Co., University of Maryland, and Johns Hopkins University	Porous-coated implants
	International Acoustics, Inc.	Evaluation of heart valve acoustic signals

IV. WHAT IS THE ROLE OF THE CDRH MEDICAL DEVICE LABORATORIES?

In the fewest words possible, OST's role is to develop and maintain scientific and technical expertise in order to address critical public health needs. This expertise is applied to current Center programs as described above, and presented here in more detail.

A. Support of Center Programs

OST performs a variety of tasks that cross-cut the CDRH organization. For example, in the review of premarket submissions, OST performs two types of paper reviews of device applications for the Office of Device Evaluation (ODE): statistical review of the clinical data supplied in every PMA, as well as selected 510(k) and IDE applications, and technical reviews of selected applications, especially for first-of-a-kind devices, or for devices in which OST has special expertise.

Second, OST staff spend a significant amount of time in "front-end" activities that attempt to influence the types, quantity, and quality of the applications that are submitted to CDRH. Examples of this include important roles in various reclassification efforts, and in the drafting of guidance for manufacturers to aid them in providing quality submissions to CDRH. The latest wrinkle here is an OST-initiated Bulletin-Board Service (301-443-7496) where people outside the Center can gain access to selected guidance documents and provide the Center with immediate feedback.

Finally, OST does hands-on laboratory analyses of selected devices, before approval, at the request of ODE.

In enforcement, OST supports the Office of Compliance (OC) in a variety of ways. OST provides a great deal of the technical expertise used in the enforcement of the various performance standards for radiation-emitting devices promulgated under CDRH's "other" law, P.L. 90-602, the Radiation Control for Health and Safety Act (RCHSA). OST scientists develop and calibrate instruments, train FDA field personnel, and develop compliance test procedures and guidelines. This is of relevance to the medical device industry for two reasons: (1) some medical devices must meet certain of the RCHA performance standards (e.g., surgical lasers, sunlamp products, diagnostic X-ray equipment, ultrasound therapy equipment), and (2) OST will certainly play a similar role in the development and envorcement of any "514 standards" promulgated under P.L. 94-295, the Medical Device Amendments to the Food, Drug, and Cosmetic Act. OST staff currently serve as the technical resource in the Center's effort to develop a mandatory standard for apnea monitors.

OST also evaluates problem recalls and does laboratory analyses of selected devices as the request of OC and the field.

OST contributes to the Center's postmarketing surveillance activities through the conduct of epidemiologic studies. A current effort involves an interagency agreement with the National Center for Health Statistics to study the number, frequency of replacement, associated morbidity, associated complications, and mortality risks of selected device implants. In addition to generating scientific information for use in postmarketing product management, OST staff works with OC staff to improve the quality and ease of use of information collected in our mandatory reporting system. For example, OST provided project leadership for the development of automated trend analyses of Medical Device Reporting data reports.

The educational programs of the Center are the responsibility of the Office of Training and Assistance (OTA), and OST contributes to these primarily by acting as a source of expert consultants on various projects. OST provided and analyzed technical information for OTA in the recently completed survey of defibrillator use and maintenance practices in five states, and has developed a laboratory demonstration project in battery maintenance as a result of one of the survey recommendations. The laboratories also provide direct training to field personnel in a variety of enforcement activities.

Finally, OST staff actively support the development of performance standards in the voluntary standards arena, with representation on committees of such groups as the American Standards for Testing and Materials, American National Standards Institute, American Association for Medical Instumentation, National Committee for Clinical Laboratory Standards, and the International Standards Organization.

B. Product Evaluation: Research and Technical Innovation

It is product evaluation that enables the laboratories to make their unique contribution to the Center, and that creates the pool of scientific expertise with which OST executes the support functions described above. Product evaluation is a broad term, and refers to all those activities relating to the scientific evaluation of medical devices, or medical device technologies, including problem definition, statistical, engineering, physics, and biological risk analysis, and option development. (See the chapter on risk assessment for further development of this topic.) Product evaluation includes the laboratories' most basic evaluation of products and technologies, including getting on top of new technologies as they apply to medical devices.

A few examples serve to illustrate product evaluation. OST has an active research program in surgical lasers and fiber optic tips and probes. Current studies are directed at the characterization of the optical properties of sapphire surgical tips and quartz sphere tips, and at understanding laser spatial beam profiles, focusing properties, and transmission properties in various media.

Another research area of interest involves understanding the biophysics of electrical stimulation: to what extent can the potential for damage to tissue from electrical stimulation be assessed? Current work is directed toward the assessment of auditory nerve damage from cochlear stimulators, assessment of neural injury using neural models, and assessment of efforts of electrical stimulation on cellular signal transmission.

OST has also expanded laboratory capabilities in the area of biotechnology. In one project, scientists are constructing a DNA probe for a virus of medical significance to developing countries.

The laboratory materials program has used polyurethane as a model material because of its wide use in medical devices. Research projects include the development of an atlas of biomedical polyurethane infrared spectra, explant pathology studies of polyurethane trileaflet heart valve and blood pumping diaphragms, electric field effects and other stresses (e.g., thermal and solvent processing effects) on polyurethane degradation, and effects of mechanical stress on conductance and dielectric constants of polyurethane.

Finally, in OST's medical imaging research programs, sophisticated statistical image analysis techniques are being used to extract information from ultrasound images that the clinician is unable to comprehend upon visual inspection and that may relate to disease state in addition to physical structure.

New measurement technologies and instrumentation for CDRH's rather specialized needs are also the responsibility of OST. Not surprisingly, laboratory involvement in the technical aspects of medical device problems sometimes leads to technical innovations, or technical "fixes" to device problems that CDRH encounters in its regulatory role. For example, previous work on a hyperthermia applicator led to the development of an electromagnetic rewarming coil. The coil has been used to rewarm animals gradually after surgery done at low temperature, with the advantage of even, all-over heating. Potential applications exist for the rewarming of blood and other organs.

In a different arena, optimization of X-ray imaging systems has been a long-term objective of the Center. Comparison between computational and laboratory results on an optimized mammography design confirmed a one-third exposure reduction with equivalent, or even improved, image quality. The next step in this proejct, the fabrication of a system meeting the optimized design and a clinical evaluation of its performance, was begun this year through the establishment of an interagency agreement with the Clinical Center at the National Institutes of Health. Clinical trials are expected to begin in FY89.

In general, research results are presented at scientific meetings and prepared for publication in the open literature. In fiscal year 1988, 115 presentations were made at scientific meetings and 62 manuscripts were published. Technical innovation, on the other hand, finds its outlet in patent applications. Currently, OST staff hold 24 patents, with several additional patents pending.

V. HOW ARE RESEARCH PRIORITIES SET?

Given the breadth of disciplines involved in understanding the science behind all medical devices, it should be obvious immediately that OST laboratories cannot provide the Center with comprehensive expertise across the board. Resources are limited, and the Center must decide on research priorities. First, and most pragmatically, areas for research are identified that have application to many device problems. In this way, we try to maximize our chances for developing and maintaining expertise that will be of greatest relevance to the public health issues that the Center faces. For example, research programs have been developed in toxicology (what the device does to the body) and in materials, particularly materials degradation (what the body does to the device). The goal for these research programs is *not* to solve all of the Center's toxicology or materials problems in the laboratory. Rather, the goal is to maintain a cadre of "hands-on" experts who will better understand and be able to cope with real world device problems that involve, in this example, toxicology issues or materials questions.

In setting reasonable project priorities within these broad research areas, OST must also consider other criteria. For example, most of our research projects must fall into one of four categories: (1) things others cannot do (for lack of expertise, equipment, etc.); (2) things others will not do (for lack of an adequate incentive); (3) things others should not do (e.g., maintaining the only source of information on product failures); and (4) things other have not done (i.e., to fill crucial information gaps). Relevance to the mission of CDRH also is of paramount importance in selecting a research project. However, relevance is measured on a scale that ranges from immediate programmatic impact to importance for problems that will be seen in the coming decades.

Within this framework, the mechanics of the steps from proposal of research projects through completion of research projects are straightforward, and occur within the context of the CDRH planning cycle, which follows the federal fiscal year. Project proposals are made in the spring, are reviewed within OST, and are then presented to the rest of the Center in midsummer, along with other OST proposed activities. The target for a finished operating plan is October 1 of each year. Since many OST research projects span more than one year, there is a mid-year program review within the office, at which time progress and problems are discussed, and necessary redirections of effort are determined.

VI. WHY SHOULD INDUSTRY BE INTERESTED IN CDRH's MEDICAL DEVICE LABORATORIES?

The Center's medical device laboratories have a large impact on the medical device industry because of the particular kind of support given to the rest of the Center in a variety of cross-cutting ways (see above). Because of its research base, OST provides the Center with an additional strategy that can be used in support of solutions to public health problems. Of course, OST often is able to provide hard data or expert analysis to support the traditional statutory authority the Center exercises in its enforcement and premarket review activities. We also support the less traditional, but highly effective, educational strategies CDRH uses for problem-solving. The additional strategy, and OST's unique contribution to CDRH, is the analytical and laboratory-based ("hands-on") expertise the Center can use to identify potential medical device problems before people are affected and to develop solutions to problems where none were thought possible.

This analytical and laboratory-based expertise can, and often does, have an impact on manufacturers during negotiations with CDRH. It is harder for manufacturers to negotiate from a position of strength when Center staff share the manufacturer's technical understanding of a device, or of a potential adverse effect. When our staff knows how to fix the problem, a manufacturer's claim that it can't be fixed will not be received well by the Center.

On the positive side, however, laboratory-based technical expertise can often be utilized to the advantage of both the Center and the medical device industry. For example, the expertise of OST in radiofrequency and microwave radiation measurement and bioeffects, as well as medical imaging, put OST in an excellent position to shepherd the recent successful reclassification of magnetic resonance (MR) devices through the Center and Agency. We think industry benefits if we do know how to fix a problem that industry has not been able to resolve. This expertise also frequently translates into cooperation with industry to establish consensus standards that can resolve problems and obviate regulation.

The final reason why industry should be interested in CDRH's medical device laboratories relates to the Federal Technology Transfer Act of 1986 (P.L. 99-502). This act amended the Stevenson-Wydler Technology Innovation Act of 1980 to authorize a federal agency to permit directors of the agency's laboratories to enter into cooperative research and development agreements on behalf of the agency. The implementation of that authority in a federal regulatory agency such as FDA has been the topic of a great deal of discussion over the past 2 years. At least a potential conflict of interest exists when one portion of the agency is working jointly with industry and the resulting data or product is to be evaluated by another portion of the agency. FDA has concluded that sufficient safeguards can be placed such that its laboratories can make use of the new authority.

In July of 1988, the Commissioner of FDA delegated authority to implement the new technology transfer provisions to the Director of CDRH, the other FDA Centers, and the FDA Office of Regulatory Affairs. As a practical matter, the exercise of this new authority is, in CDRH, not intended to impact greatly on the collaborations that were described in Table 1; that is, what the medical device laboratories are now able to do in collaboration with industry, etc., won't be made more difficult. It is likely, therefore, that the extra effort to establish a cooperative agreement under the new authority will be made only when there is an opportunity to supplement federal funds for research, under terms and conditions that meet the needs of all parties involved. The possibilities are endless, or they are nonexistent.

As a result of the discussion in this chapter, the areas in which the medical device laboratories have expertise should be better known. Industry, therefore, is in a position to consider its own needs in relation to our capabilities and to suggest cooperative work. In CDRH, each example of potential collaboration will be examined on its own merits, with particular emphasis on the expected public health impact.

VII. CONCLUSION

The value of maintaining a laboratory facility in support of CDRH's regulatory programs and its overall public health mission has been proven for over 25 years. With an improved understanding of the role of the laboratory, the specific programs underway in it, and the facility itself, the medical device industry should be better prepared to deal with CDRH in general. Because the laboratory already has a history of collaboration with outside groups, and new technology transfer authority is now in place, the opportunity certainly exists for enhanced cooperation with the medical device industry to develop new data and to undertake technical innovation.

7

The FDA's Role in Assisting Small Manufacturers

H. NEAL DUNNING *Center for Devices and Radiological Health, Food and Drug Administration, Rockville, Maryland*

I. INTRODUCTION

Compliance with regulatory requirements is apt to cause greater difficulties for small companies than for large ones. Moreover, regulations involving public safety and health cannot be modified to favor small manufacturers—as can some federal actions, which provide for tax incentives, loan guarantees, or liberal grants. These issues are especially critical to the medical device industry, an area in which approximately 95% of all manufacturers have fewer than 500 employees (and, in fact, in which about 65% have fewer than 50 employees).

Recognizing the special vulnerability of small device firms to regulatory action, Congress incorporated a brief but significant paragraph—section 10—in the Medical Device Amendments of 1976. The following passage, excerpted from the report of the House Committee on Interstate and Foreign Commerce on the medical device amendments, explains the intentions of Congress in formulating section 10: "During hearings and development of the reported bill, the Committee became cognizant of the potentially detrimental economic impact [that] implementation of [the medical device amendments] might have on small device manufacturers. For this reason, several provisions of the bill are designed to avoid unnecessary regulation of medical devices . . . . In addition, the Committee has taken the unusual action of requiring that the [HEW] Secretary establish an identifiable office to provide technical and other nonfinancial assistance to small manufacturers of medical devices to assist them in complying with the requirements of the Act" [1]. The result of this action was the establishment of the Division (formerly Office) of Small Manufacturers Assistance (DSMA) within the Center for Devices and Radiological Health.

II. STRATEGIC CONSIDERATIONS REGARDING ASSISTANCE EFFORTS

A. Regulatory Requirements

The first consideration in developing an assistance program for "small manufacturers" is to define "small"—"manufacturers" already are defined by registration regulations developed to affect the Medical Device Amendments [2]. The Regulatory Flexibility Act requires that "small" entities must be defined as those having fewer than 500 employees, unless there are persuasive arguments against this defined level. In the medical device industry, a company having 500 employees is in the top 5% as far as size goes. Even so, it does not seem worthwhile to try to establish a different definition with the Small Business Administration.

DSMA uses the Small Business Administration's definition of "small" and takes a pragmatic stance that those manufacturers who seek our help or advice are our legitimate clients.

It also is desirable for DSMA to interact with large medical device manufacturers. The medical device industry is unique in many regards. One of the unique aspects is that the larger manufacturers often help the smaller manufacturing sector, either directly or indirectly through their trade and professional associations. Thus, DSMA commonly receives as much or more guidance than it gives to the larger companies. For example, the good manufacturing practices (GMP) manual is replete with "hands on" examples of how to handle the myriad problems in implementing a company's GMP documentation. Many of these examples were provided by medium- or large-sized companies.

Both "responsive" and "outreach" efforts are required for an effective assistance program for small manufacturers. The requirements for developing an effective "outreach" program in this area are remarkably similar to the requirements for a successful marketing effort by any major company. We need to know which companies, of what kind, at what location, and of what size and capabilities there are that (1) need our services; (2) will accept our services; and (3) can use the services that we provide. Then we must determine which of these services we can provide efficiently.

As DSMA became better acquainted with the industry, it became obvious that differing outreach efforts are appropriate for different audiences. For example, it often is not reasonable to spend time trying to contact the very smallest of companies except by mass mailings of general brochures. Conferences or workshops may require more expertise than the smallest companies have or more time than they can spend in attendance. These companies are best served by providing the specific information and explanations that they request. However, there obviously are thousands of companies that find workshops profitable—as has been observed in DSMA's highly successful series of nationwide workshops. There also undoubtedly are companies having characteristics that require still other treatments. In order to provide the needs of special companies, or specialized sectors of the industry, DSMA needs special data.

Unlike most federal efforts to help small businesses, DSMA has the advantage of the registration and listing information that is required of medical device manufacturers—by means of which established clientele and some of their characteristics can be determined. Also, DSMA has adopted modern methods of management to ensure maximum effectiveness from limited resources.

As an example, Dun & Bradstreet data and marketing services are used to help determine the characteristics of "small" manufacturers.

B. Industry Characteristics

At the time DSMA (then OSMA) was established, there was little knowledge, regulatory, economic, or structural, about the medical device industry.

DSMA wished to know more about the organization of the companies with which it dealt for several reasons. A major use of company organization information would be to sharpen the aim of DSMA programs. For example, in some companies, the main GMP authority is vested in the quality assurance manager. In others, it is vested in the vice president of operations, or regulatory affairs. Further, the small divisions of a large company that receive little or no legal or regulatory help from the central office are in effect "small manufacturers." By the same token, a division of equal size that receives all the support of a major company from its head offices should not be treated like a "small manufacturer." These data were not available from FDA sources. Likewise, there was no reasonably available commercial source of such data.

DSMA contracted with Louis Harris and Associates to conduct a national survey of medical device manufacturing establishments [3]. The purpose of the survey was to establish trends relating to sales, capital investment, employment, and innovation; to determine the impact of FDA regulations on industry; and to study the reactions of manufacturers to the medical device regulations. This survey, conducted in the fall of 1981, provided the first systematic assessment of the regulated manufacturers' experiences and attitudes toward the medical device regulations. It also evaluated early efforts of FDA to assist small manufacturers, and provided guidance for additional programs. In-depth interviews of senior company officials responsible for regulatory matters were conducted in a statistically valid number of establishments.

1. Survey Response

The survey revealed a young, vigorous, and successful medical device industry, whose growth and expansion seem to be largely unaffected by the introduction of the Medical Device Amendments of 1976. This appeared true despite the dire prophecies that the Medical Device Amendments would cripple the industry. The establishments that have entered the medical device industry included a disproportionate share of research and development-intensive establishments.

In the past few years, several studies of the nature of the medical device industry have been made by DSMA and by FDA's Office of Planning and Evaluation [4-7]. These studies showed that the industry's basic nature remains unchanged from that observed by Harris [3]. The industry is still young, growing, and aggressive, with an unusually large number of very small manufacturers and an unusually high input of advanced technology. An unusual interest in educational materials on regulatory matters also was shown.

Each year, the International Trade Administration publishes a summary of economical and trade factors involving "medical and dental" instruments and supplies [8]. Areas of the medical device industry are reported under the following Standard Industrial Classification (SIC) codes: X-Ray and Electromedical Equipment, SIC 3693; Surgical and Medical Instruments, SIC 3841; Surgical Appliances and Supplies, SIC 3842; Dental Equipment and Supplies,

SIC 3843; and Ophthalmic Goods, SIC 3851. The rapidly increasing importance of *in vitro* diagnostic devices is emphasized by the fact that in 1987, they were given a four-digit descriptor, SIC 2835, together with *in vivo* diagnostic products.

2. Response to Regulation

A proper strategy for assisting small manufacturers of medical devices (SMMD) must also include recognizing the response of SMMD to regulations, and contrasting this with the response of large firms with effective regulatory staffs to the same regulations. Relative responses are illustrated qualitatively in Figure 1.

This difference in response has been discussed and confirmed with both industry and government regulatory experts. Large firms commonly anticipate regulations, develop responses to them, and may even help write them (if only by extensive comments). Therefore, they are ready to respond, and do so rapidly after the regulation is finalized. Then these larger companies adapt rapidly and settle into rather routine responses. After this time, action by the FDA is minimal, dealing mostly with exceptions. Experience shows, however, that smaller firms seldom anticipate regulations and have minor influence on their formation. Because of their limited anticipation of regulations, and their limited staff resources, these small firms adapt more slowly to regulatory requirements and may need considerable help for years if the regulation is a complex one, like the GMP regulation, for example. In addition, interpretations and compliance programs of the Center for Devices and Radiological Health (CDRH) change from time to time, thus perpetuating the need for *all* companies to keep up to date.

The needs of SMMD for education and assistance also depend on the rate of formation and demise of companies. For example, even established areas

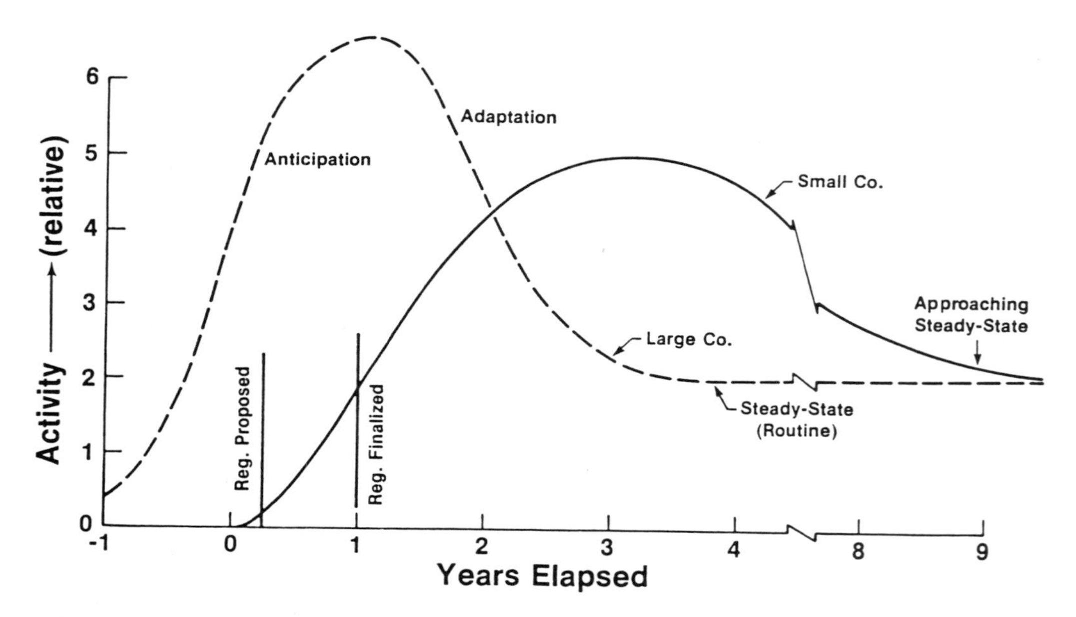

FIGURE 1 Comparative responses of large and small companies to a regulation.

such as registration, listing, and premarket notification are always new to companies with new product lines. Informational needs will therefore continue in all areas of regulation. The degree of magnitude of these needs depends on the rate of formation and demise of companies in the medical device area. Geiser [4] estimated that during the 1980-1985 period about 900 companies entered the industry and 300 left each year. Both of these numbers have increased according to FDA registration data. Table 1 shows that the number of registered establishments has increased at a rate of more than 10% compounded annually. This extensive tabulation also shows that the West and East Coasts continue to increase their lead over most of the rest of the United States as choice locations for new establishments.

As part of a more formal effort, DSMA received OMB clearance for a questionnaire to be administered at various locations. A major question was, "Please rank the subjects below for their importance in filling your needs for information or education." Answers to this question show that perceived needs for information and education on the various topics have evolved somewhat but in general remain remarkably steady. Results collated from workshops through the 1980s are summarized in Table 2.

It is apparent that the GMP regulation, and quality assurance in general, has been the favored subject, with the premarket notification process—510(k) —a close second. Both these subjects have a wide lead over all other subjects and, indeed, are favorite subjects of DSMA speakers. Quite recently CDRH and industry associations have begun to emphasize the need for preproduction quality assurance. Interestingly, this subject has been stressed in the GMP manual since the first edition in 1981 [9].

The most recent use of this questionnaire was at the 1987 Boston workshop series. There were a sufficient number of *in vitro* diagnostic device manufacturers to test if *their* priorities differed from general device manufacturers. The results are illustrated in Figure 2 and compared to the results from general device manufacturers.

The only significant differences in perceived educational needs between manufacturers of general devices and *in vitro* devices appear to be in "classification," where the interest of *in vitro* manufacturers is relatively quite low. This may not have much import since it has become apparent that although most *in vitro* products are in Class II, performance standards are not apt to be prepared for them.

III. DSMA ACTIVITIES

A. Responsive

1. Phone Calls and Correspondence

The "responsive" activities of the Division include response to questions on any subject relating to the regulation of medical devices, and providing the information that is requested by those who learn about the office and call, write, or visit. Logically, this was the first part of the assistance program that was initiated, and remains a vital and rapidly expanding activity.

Assistance to specific companies is generally more demanding on resources than group assistance efforts. However, an effective assistance program should supply as much personal or individual assistance as staff and resources permit without curtailing group efforts. Most of this assistance is provided by telephone and infrequently, by letter. Obtaining an "800" telephone num-

TABLE 1 Registered Device Manufacturers (by State)

June 1984		June 1985		June 1986		July 1987		August 1988	
California	1586	California	1758	California	1868	California	2088	California	2528
New York	1040	New York	1182	New York	1268	New York	1469	New York	1787
New Jersey	596	Illinois	649	Illinois	679	Florida	836	Florida	963
Illinois	570	New Jersey	645	New Jersey	658	Illinois	802	Illinois	927
Florida	465	Florida	552	Florida	654	New Jersey	755	New Jersey	857
Massachusetts	419	Massachusetts	460	Texas	472	Texas	592	Texas	718
Pennsylvania	368	Texas	432	Massachusetts	469	Massachusetts	536	Massachusetts	602
Texas	345	Pennsylvania	412	Pennsylvania	436	Pennsylvania	494	Pennsylvania	579
Ohio	263	Ohio	304	Ohio	338	Ohio	370	Ohio	440
Connecticut	217	Connecticut	251	Connecticut	262	Minnesota	306	Minnesota	356
Michigan	202	Michigan	227	Michigan	240	Connecticut	295	Michigan	353
Minnesota	183	Minnesota	196	Minnesota	229	Michigan	287	Connecticut	315
Missouri	175	Missouri	192	Missouri	211	Missouri	257	Washington	303
Colorado	163	Washington	191	Washington	199	Washington	245	Missouri	285
Wisconsin	161	Wisconsin	182	Wisconsin	194	Colorado	229	Wisconsin	253
Washington	154	Colorado	181	Colorado	192	Wisconsin	209	Colorado	248
Indiana	127	Maryland	145	Georgia	167	Georgia	202	Georgia	232
Maryland	126	Georgia	142	Indiana	154	Maryland	184	Maryland	212
Georgia	121	Indiana	139	Maryland	152	Indiana	175	Oregon	196
Oregon	111	Oregon	133	Virginia	149	Oregon	170	Indiana	193

North Carolina	109	Virginia	129	Oregon	145	North Carolina	160	Puerto Rico	189
Tennessee	107	Puerto Rico	125	North Carolina	135	Virginia	151	North Carolina	188
Virginia	105	Tennessee	123	Puerto Rico	129	Puerto Rico	149	Virginia	174
Puerto Rico	96	North Carolina	120	Tennessee	127	Tennessee	143	Tennessee	169
Arizona	82	Arizona	94	Hawaii	99	Hawaii	121	Hawaii	138
Kansas	70	Hawaii	87	Arizona	95	Arizona	115	Arizona	135
Hawaii	65	Utah	73	Kansas	83	Utah	100	Utah	113
Utah	64	Kansas	72	Utah	78	Kansas	96	Kansas	99
South Carolina	48	South Carolina	53	South Carolina	59	Alabama	72	Alabama	87
New Hampshire	43	Rhode Island	53	Alabama	59	South Carolina	65	Oklahoma	76
Rhode Island	43	New Hampshire	50	New Hamshire	55	Kentucky	63	South Carolina	73
Oklahoma	41	Oklahoma	48	Rhode Island	55	Rhode Island	62	Kentucky	70
Alabama	40	Alabama	47	Oklahoma	50	Oklahoma	62	New Hampshire	67
Kentucky	36	Kentucky	46	Kentucky	49	New Hampshire	60	Rhode Island	67
Nebraska	35	Iowa	42	Louisiana	45	Louisiana	54	Louisiana	64
Iowa	34	Louisiana	37	Iowa	41	Iowa	48	Iowa	60
Louisiana	30	Nebraska	36	Nebraska	38	Nevada	46	Nevada	57
Mississippi	26	Nevada	33	Nevada	37	Nebraska	42	Nebrasak	48
Arkansas	24	Mississippi	29	Arkansas	31	Arkansas	36	Arkansas	47
Nevada	21	Arkansas	29	Mississippi	30	Maine	36	Maine	40
Maine	20	Maine	27	Maine	28	Mississippi	35	Mississippi	36
New Mexico	19	Delaware	22	Delaware	25	New Mexiso	27	New Mexico	30

TABLE 1 (continued)

June 1984		June 1985		June 1986		July 1987		August 1988	
Vermont	17	New Mexico	21	New Mexico	20	Delaware	25	Washington, D.C.	30
Delaware	17	Vermont	17	Vermont	17	Vermont	22	Delaware	27
West Virginia	10	Montana	12	Idaho	15	Idaho	21	Idaho	27
Idaho	7	West Virginia	10	Washington, D.C.	14	Washington, D.C.	20	Vermont	24
Montana	7	Idaho	8	Montana	13	Montana	19	Montana	22
Washington, D.C.	6	Washington, D.C.	6	West Virginia	12	West Virginia	15	West Virginia	16
Wyoming	3	Wyoming	3	South Dakota	4	South Dakota	5	South Dakato	6
South Dakota	1	South Dakota	2	North Dakota	4	North Dakota	5	North Dakota	6
North Dakota	1	North Dakota	2	Wyoming	3	Wyoming	3	Wyoming	3
Alaska	1	Alaska	1	Alaska	1	Alaska	1	Alaska	1
Total	8620	Total	9830	Total	10,587	Total	12,376	Total	14,536

TABLE 2 Industry Interest in Information

	Relative ratings[a] by year				
Topic	1981	1982	1985	1986	1988
GMP	6.2	7.1	6.8	6.4	6.2
510(k)	4.9	6.8	7.1	5.8	6.0
Standards	4.9	4.1	4.8	4.6	4.3
PMA	4.3	4.7	5.0	4.8	5.6
Classification	4.0	4.3	3.5	4.8	3.6
Recall	3.1	3.7	3.5	3.8	3.7
IDE	3.1	3.6	4.4	3.3	4.0

[a]Higher rating indicates greater perceived need.

ber (800-638-2041) in 1981 has proved to be a major aid because it saves large amounts of time and money for both government and industry. These calls are increasing regularly, reaching about 16,000 in 1987; 25,000 in 1988; and a projected 40,000 in 1989.

Effective though a phone call may be, face-to-face communications often provide the most rapid route to understanding and to solution of problems. Therefore, as a result of specific invitations, DSMA conducts on-site visits with specific companies and arranges or provides office conferences with visitors to Washington.

2. *Office Conferences*

Representatives of many small manufacturers, both domestic and foreign, stop at the DSMA office for specific consultation. Many of these visitors have been referred by members of the review staff of the Office of Device Evaluation, or by trade associations. Although DSMA prefers to have office conferences prescheduled, most manufacturers have limited time to spend in Washington, D.C., so those who simply "drop by" are generally accommodated. As stated in an early editorial, "We are from the Government and We Are Here to Help you." Despite this now time-worn motto, the true basis of any assistance effort *is* a sincere desire to help.

In addition to these impromptu conferences DSMA arranges and participates, upon request, in meetings among small manufacturers and pertinent Center scientific and compliance personnel. Approximately 150 such meetings are conducted each year. These services coupled with the workshops may obviate or substantially reduce the need for small manufacturers to engage technical and legal consultants to comply with the simpler FDA requirements. However, DSMA staff *are* federal employees and therefore *cannot* act as advocates for a particular company section of industry. They can, however, insure that industry meets with the proper experts and *is* heard.

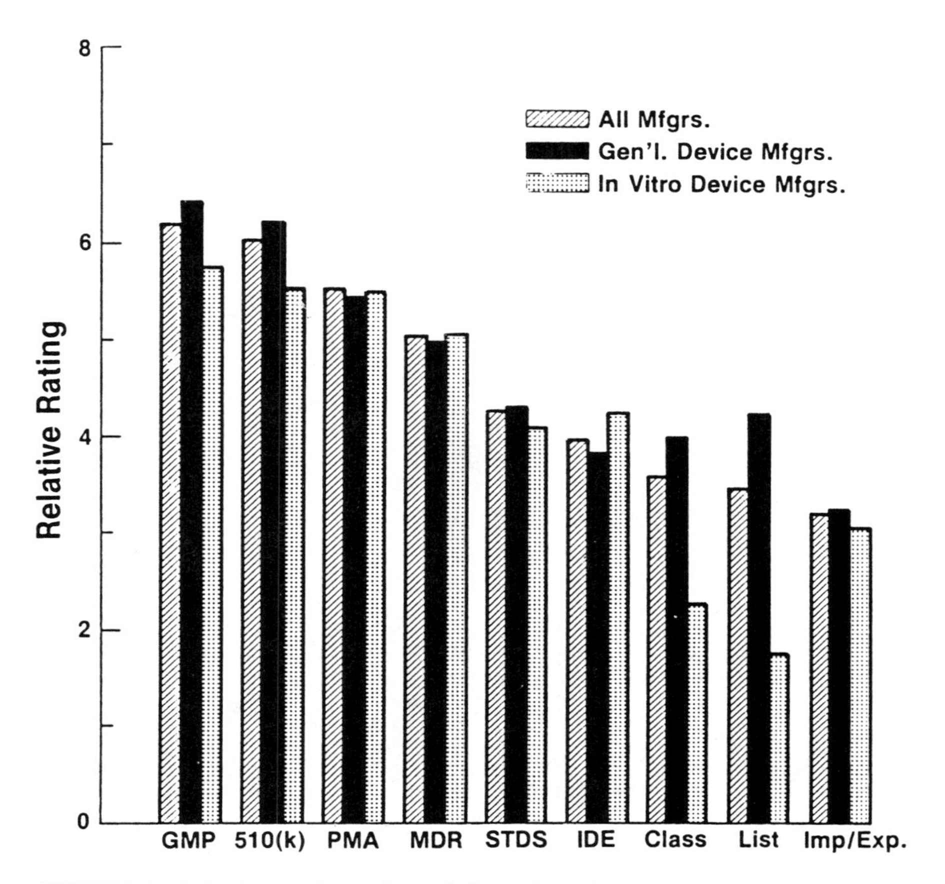

FIGURE 2 Priority rating of need for education.

3. *Basic Information Pamphlets*

DSMA tracks requests for information and, as certain areas begin to be the subject of increasing questions, attempts to obviate such requests by preparation of pamphlets or articles on the major questions. One of the best known of these was a relatively early product, "Everything You Always Wanted to Know About the Medical Device Amendments . . ." [10].

One of the most recent such attempts was published in the SMA NEWS section of *Medical Devices and Diagnostic Industry* [11]. Such publications, together with frequently used *Federal Register* documents and other information, are listed on an order form, "Printed Materials: Small Manufacturers Assistance." This form is distributed in all information and registration packs and at DSMA workshops.

B. Outreach Efforts

Various of the more important "outreach" activities developed to assist small manufacturers are discussed here to give a balanced view of these activities.

1. Workshops

In 1979, the Division of Small Manufacturers Assistance (DSMA) launched one of its most successful outreach programs, a new series of workshops at regional locations to help device firms meet FDA requirements.

DSMA's day-long workshops have received more notice and more favorable comment than most other DSMA programs. The workshops are focused on two general areas—good manufacturing practices, and regulatory requirements to market a device. Each of these general areas has spawned specialized workshops. These include the sterilized devices workshops in the GMP area, and import/export, *in vitro* diagnostics, and clinical concepts in the regulatory requirements area. As with other efforts, DSMA tries to balance supply with demand. Therefore, there have been more GMP workshops than any other kind, and in fact the one hundredth GMP workshop was held in Orange, CA, on March 16, 1988. The series was initiated in 1979, shortly after the inception of the small manufacturers assistance efforts. About 3000 people attended one or more DSMA workshops in 1987. The numbers and types of workshops for the past three years are summarized in Table 3. Attendance at these workshops has risen rapidly, with the result that preregistrations for almost all of them have been closed early because of room-size limitations. Despite their success, the number of workshops is limited by DSMA resources and budgetary considerations.

To meet increased demand, DSMA has cooperated with various professional societies to host workshop-like presentations at their functions. For example, such agreements have been made with the American Society for Quality Control, the Regulatory Affairs Professional Society, and the Food Drug Law Institute. Also, workshop sessions with DSMA staff have been featured at device expositions of Canon Communications, Inc., and other multiplier groups that have similar aims.

Other components of FDA participate fully in these workshops. The GMP workshop, for example, is conducted cooperatively with the Division of Compliance Programs (DCP). At regulatory sessions, speakers from the Office of Device Evaluation are prominent. In addition, all workshops involve the FDA District staff and the Director for the FDA District in which they are held. Regional Small Business Representatives of FDA also participate in workshops in their geographical areas.

TABLE 3 DSMA Workshops Conducted

	FY 1986	FY 1987	FY 1988
Regulatory	9	9	9
GMP	7	8	9
Sterilization	3	3	2
Import/export	3	5	2
Clinical concept	2	1	2
In vitro diagnostic	1	2	2
Total	25	28	26
Cities	9	8	9

This close involvement of various parts of FDA in DSMA workshops has several positive results. First, it lets the manufacturers become acquainted with many FDA personnel. Second, it lets FDA staff of all types become better acquainted with industry and its needs. Hopefully, better acquaintance will foster mutual respect for the sincerity and respectability of industry and government regulatory experts. These DSMA workshops also let industry voice its concerns so that DSMA can obtain information and understanding to carry out its assigned function of acting as an ombudsman for industry.

2. Ombudsman Activities

According to original Congressional guidance [12]:

> The second major function of the office was to serve as a ombudsman for the small manufacturer to deal with grievances which he may have concerning requirements and implementation of the law. This ombudsman role should not be construed as advocacy. The office should not be in conflict with the compliance efforts of the Bureau [now Center]. However, the office should ensure that the small manufacturer has been availed of the proce-dural safeguards to which he is entitled and is treated fairly and equitably.

This important function is much more difficult to quantify than most other assistance efforts. However, an effort of this kind has three basic elements. These are:

1. A real understanding of the needs of small manufacturers and an empathy for the problems besetting small, technical businesses.
2. A receptive FDA and CDRH management that is willing to listen to, and take account of, spokesmen for the industry.
3. An effective ombudsman to determine small manufacturers' needs, grievances, and problems (particularly those with FDA), and to interpret them into terms that allow the regulatory machinery to act on them. An effective effort of this type requires good rapport with both the industry and with FDA management.

DSMA's ombudsman role has been pursued assiduously, though quietly, especially with upper CDRH management, and has had some notable successes. For example, the GMP regulation calls for regular audits of the quality assurance program by "appropriately trained individuals not having direct responsibilities for the matters being audited." In 1979 and 1980 this requirement was being interpreted literally by small manufacturers, many of whom have *no* technically qualified person *not* involved in quality assurance, manufacturing, or supervision, and no one to perform such an audit. Therefore, several had gone so far as to hire independent auditors at considerable expense.

Fortunately, it was possible to clarify the intent and to have an interpretation prepared and printed, taking into account the interdependence of the main officers of a small company. This interpretation was published in a small readable pamphlet [13], which was distributed to manufacturers by DSMA. The interpretation was:

> Although an independent audit is preferred, FDA places more emphasis on the training and qualifications of auditors than on their independence

from the subject area of the audit. If a very small operation—particularly one in which everybody is directly involved in daily production activities—concludes that an independent audit would be unduly burdensome or impractical, the firm may perform its own audit, provided that it does so according to a written audit plan, such as a list of the areas to be audited and the aspects to be checked, and implements the plan according to a schedule.

Other ombudsman efforts have not always been as successful, but through the years, the Center management has developed a good understanding of the problems of small, regulated manufacturers.

The Center for Devices and Radiological Health undertook a major introspective effort in 1984 by establishing 10 criticism task forces to study various Center activity areas in depth, and to recommend remedies—legislative, regulatory, or policy—for problems encountered. An eleventh task force, the Small Business Criticism Task Force (SBCTF), was then established. The SBCTF had a representative from each of the 10 other task forces and was chaired by DSMA. The assignment of the SBCTF was to study recommendations of the other task forces, to ensure that the proposed remedies had optimal effects for small manufacturers, and to initiate their own recommendations. Top Center management took these recommendations very seriously. Many improvements have been made in the regulatory area—such as exempting 104 generic types of Class I devices from the premarket notification requirement, in addition to those exempted during the classification process. Other changes have been made in the policy area, such as expediting review of 510(k) applications for devices that are obviously substantially equivalent. Many of these are described in the Office of Device Evaluation Guidance Memoranda, the "Blue Book" [14]. Changes suggested that would require legislative action take more time, but several bills that contain changes related to suggestions of these criticism task forces have been submitted to the House or Senate.

3. *On-Site Visits*

The on-site visit program is demanding of DSMA time and resources and therefore is limited to special cases. Occasionally, however, DSMA staff are able to fill an invitation for a personal visit to a company. When conducted, such on-site visits are made to determine needs, views, and types of assistance most desired by small manufacturers and to provide assistance with specific problems that the firm wishes to discuss. During 1980, 80 on-site visits were conducted. Despite an increased DSMA staff, this number has decreased to about 40 per year in recent years to provide more staff resources for workshops. Such visits pay large dividends by keeping DSMA staff up-to-date, and by producing a better understanding of industry problems. This increases the ability and interest of DSMA staff in acting as ombudsman for industry.

4. *Coordination with Small Business Representatives*

DSMA also works closely with the small business representatives (SBRs) assigned to each FDA Region. These representatives advise and assist small businesses in complying with FDA rules and regulations. Unlike DSMA, whose efforts are restricted to medical and radiological devices, the SBRs try to respond to requests in all areas that FDA regulates—foods, drugs, devices, veterinary medicines, and cosmetics. Because of this broad range, the SBRs are limited in the technical aspects of their responses. Although various SBRs have varying interests, some of them spend nearly half their time on device-related matters. They assist in DSMA workshops (sometimes arranging them) and conduct several medical device on-site visits each year. As regional representatives, they also serve as excllent "listening posts" for problems that are bothering manufacturers. DSMA supplies SBRs with all the device brochures and materials used and helps coordinate the SBR efforts related to devices.

5. Mass and Sector Mailings

Because DSMA acts as an ombudsman for small manufacturers, it is vital that the staff be cognizant of Center activities and documents. Reprints of proposed and final rules, policy statements, guidelines, and other publications are described in several publications prepared by DSMA staff and are provided to manufacturers upon request. Materials of broad general interest (such as the medical device reporting requirements) are mass mailed to all registered device firms. Specific materials are mailed to interested sectors, for example, classification regulations are mailed to those firms that have listed devices subject to the particular *Federal Register* notice. Basic information packets are mailed to approximately 1000 newly registered firms each year and about 3000 complete introductory packages—including instructions on how to register a medical device establishment and how to prepare a Premarket Notification-510(k)—are mailed on request.

Another area of major concern for small manufacturers is their ability to keep up with FDA publications and regulations that affect their operations. The larger companies subscribe to the *Federal Register*, in which all formal regulatory statements are published. This is expensive—a few hundred dollars per year—and because it is published daily it is a real burden for anyone less than a full-time regulatory expert to track. Therefore, DSMA has assumed the load of tracking and stocking all *Federal Register* publications that pertain to medical devices. These are listed in the publication "Medical Device Federal Register Documents," which is updated semiannually and is available free from DSMA [15].

IV. QUALITY ASSURANCE

A federally funded assistance effort has heavy responsibility to both the industry and the Agency. Therefore, establishment of a system to ensure high quality responses is vital. Early congressional guidance [12] stressed the necessity for keeping assistance efforts out of and away from the compliance function of the agency. This has proved to be a highly valuable suggestion, and has been followed. Manufacturers must *know* that those who assist them will not "turn them in" for enforcement actions. This is an excellent example of the absolute need for government employees to avoid conflicts of interest.

Staff members involved in manufacturers assistance have a rewarding but difficult job. The nature of their work is advisory—yet they often must respond extemporaneously to questions that are expected to take weeks or months for formal response from regulatory staff or company lawyers. For example, manufacturers often need to know if their product is a medical device and thus subject to the Medical Device Amendments. If it is a device, they must know its classification number and classification name. This information can be obtained, formally, by preparing a Section 513(g) petition (a citizen's petition) and sending it to the Office of Compliance, CDRH. Responses are supposed to be available in 60 days, but this is not a legislated deadline so the press of business sometimes causes the period to extend for months. Alternatively, manufacturers may call DSMA and obtain the same information in a matter of minutes. Needless to say, this prompt response has a greater chance of error than does the carefully researched, written response complete with review by supervisors. Nevertheless, very few errors have occurred in this basic area.

As another example, DSMA members often spend one-third to one-half their time—on the phone, in meetings with industry, or at workshop sessions—on Premarket Notification discussions. The very valuable flexibility of this regulation leads to a great need for advice. Large companies have well-established policies and regulatory staff—often including legal input—to help make these decisions. Smaller companies often look to DSMA for assistance in this area regarding modified devices, even though DSMA staff stress that it is the company's decision whether or not to make such a submission. Advice of this kind could easily be wrong because of an improper communication of, for example, intended uses. Again, errors have been few because of a conservative approach by both DSMA and industry. Fortunately, the results of such advice also are subject to review by industry and by FDA. Further, when the purposes of this regulation and the views of the Office of Device Evaluation are well understood [14], the need for (or lack of) a 510(k) submission is usually apparent. Indeed the very few major incidents attributed to incorrect decisions in submitting 510(k) applications have involved major companies with large regulatory and legal staffs.

DSMA staff members try to give conservative advice, outlining the aims of the regulation involved and quoting precedent conclusions. They also are careful to emphasize that their words are *only advice* and do not represent a formal policy of the Agency. It is important to stress that assistance staff must not *be* in policy-making positions if they are to avoid conflicts of interest.

These examples illustrate the dilemma of an assistance staff—they are asked to give the kind of guidance that regulatory agencies often thrash about for months in resolving. But the assistance staff are expected to respond in minutes. This makes the assistance effort a trying one, but one with considerable personal reward. Veteran DSMA staff believe that the appeal of their job comes from a balance of the adrenaline-increasing promptness of response required coupled with the sincere appreciation of manufacturers who receive those responses.

The necessity to make prompt, accurate responses requires an established system of quality control. DSMA management has established both formal and informal systems. In the formal system, supervisors review the notes of staff members, discuss the questions and answers, and make suggestions for the future. Frequent staff meetings also allow dissemination and discussion of more general or generic problems and possible solutions. In addition, informational releases on new publications or decisions are circulated to all staff members, as are precedence letters showing medical device decisions. The most formal quality control would be exerted in the old-fashioned way—official disapproval, or approval, of a staff member's actions. Such "official" measures are of course far too late to correct errors: they can only motivate staff members to avoid them in the future.

Our informal quality assurance system probably is more effective and certainly more pleasant than the formal one. Staff members must feel free to discuss questions they have handled, or are handling, with their peers, subordinates, or supervisors. They must feel free to interchange questions and answers readily, to admit a need for more knowledge, or to argue their position if different from others. This is especially true because of the immense scope—from canes to defibrillating pacemakers—of the medical device industry. No single person can hope to know immediate answers to the 40,000 plus phone calls DSMA receives per year. However, he or she *can* be responsive, get the facts, give what advice they can confidently, and get back with detailed answers. Thus, working in a major assistance effort is a rewarding, but humbling, job.

The peculiarities of assistance efforts make this an area especially amenable to the latest quality effort—preproduction quality assurance. Here, the "production" occurs when a staff member provides answers. An effective preproduction QA system will go far to *assure* quality answers. Since the *quality* of an answer is very subjective, assistance efforts must be based on personnel who are broadly knowledgeable in the subject, confident enough in themselves to admit their limits, and perceptive enough to determine what the "client" is asking or should be asking to solve the problem.

Part of the "broadness" problem can be minimized by hiring or training staff members in various specialities. Specialists can be effectively hired by discipline—engineering, chemistry, health physics—and quite readily trained for specialties in regulations—510(k), PMAs, and IDEs. Some areas, such as the GMP regulation, really require incumbents with specialized background such as electronic engineering, *and* specialized experience and training in quality assurance.

The use of specialists requires close team work among the assistance staff so that each member *knows* who can handle the question and feels free to hand the phone to that specialist. In addition, a dedicated "call reception" staff can do much to forward calls to the correct staff member. Development of a staff that has this degree of internal ease among members is an exceedingly difficult goal. However, it is eminently worthwhile since it is the basis of a responsive and effective assistance staff.

Therefore, a solid responsive effort requires a special and dedicated staff—with members who are knowledgeable on a *wide* range of subjects, but know their own limits and are willing to admit them. In addition to a competent, dedicated staff, an effective assistance effort requires a responsive industry and supportive Agency management. A responsive industry is produced by years of *fair* treatment by the Agency that regulates the industry. Development of mutual respect and trust among industry and Agency management requires a deep understanding of each other's viewpoints.

Probably nowhere in government does the time-honored maxim, "We must be able to disagree, without being disagreeable," have closer application. Obviously, industry leaders like the least possible interference in their business of delivering safe and effective medical devices. Just as obviously, there are times that an effective regulatory agency must impose rules and regulations. Good communication, increasingly deep understanding of the other's viewpoint, and broad acquaintance between industry and the Agency are the key to reasonable regulation and to more innovative, advanced, safer, and more effective medical devices.

REFERENCES

1. Medical Device Amendments of 1976, Public Law 94-295, 94th Congress, U.S. Government Printing Office, Washington, D.C., 1976.
2. Establishment Registration and Device Listing for Manufacturers of Devices, 21 CFR Part 807, U.S. Government Printing Office, Washington, D.C., 1987.
3. Louis Harris and Associates, A Survey of Medical Device Manufacturers, FDA 82-4026, NTIS Accession No. PB-82-251331, Springfield, VA, 1982.
4. N. C. Gieser, The Evolving Medical Device Industry: 1976 to 1984, *Med. Device Diag. Ind.*, *8* (11): 50-54 (1986).
5. H. N. Dunning, Size and Employee Distributions in Medical Device Establishments, *Med. Device Diag. Ind.*, *9* (7): 26-30 (1987).
6. F. H. Dworkin and M. A. Stevan, IVD Industry Sends Messages to FDA, *Med. Device Diag. Ind.*, *8* (11): 38, 40-44 (1986)
7. H. N. Dunning and J. F. Stigi, Industry-FDA Interactions at Workshops for Medical Device Manufacturers (in press).
8. M. Fuchs, *Medical and Dental Instruments and Supplies*, Chapter 36, U.S. Industrial Outlook, International Trade Administration, Washington, D.C. (1988).
9. A. Lowery, *Device Good Manufacturing Practices Manual*, 4th ed., FDA 87-4179, Dept. of Health and Human Services, Washington, D.C. (1987).
10. H. Dunning, *Everything You Always Wanted to Know About the Medical Device Amendments. . .*, 2nd ed., FDA 84-4173, Dept. of Health and Human Services, Washington, D.C., 52 pp. (1984).
11. H. N. Dunning, Questions Most Often Asked of DSMA, *Med. Device Diag. Ind.*, *9* (11):22, 24-25 (1987).
12. A. Maguire (Cong. 7th Dist. N.J.), Letter to FDA Commissioner Alexander Schmidt, June 21, 1976.
13. Division of Compliance Programs (FDA), Device GMP: Applications and Interpretations for the Field Investigator, 48 pp., U.S. Government Printing Office, Washington, D.C. (1980).
14. K. Mohan, Office of Device Evaluation, Guidance Memoranda (The Blue Book), in press.
15. A. C. Kohler, *Medical Device Federal Register Documents*, 54 pp., FDA 86-4208, Dept. of Health and Human Services, Washington, D.C. (1986).

8

Precautionary Checklist for Addressing the Issue of Medical Device User Error and Product Misuse

KATHRYN L. GLEASON *Morgan, Lewis & Bockius, Washington, D.C.*

I. INTRODUCTION

Since enactment of the Medical Device Amendments in 1976, the U.S. healthcare system has witnessed exponential advancements in medical device technology. These technological advancements, while providing manifold contributions to public health, have also created new user concerns for the medical device industry to resolve. Medical device user errors and product misuse have become increasingly serious public health issues [1]. The Food and Drug Administration (FDA) recently reported that "a substantial percentage of device-related patient injuries are [now] caused by or involve errors in using . . . medical devices [2]." Given these significant new user concerns, manufacturers are well advised to consider carefully the regulatory and product liability implications of device user problems, and to implement precautionary measures designed specifically to address user errors and product misuse.

This chapter summarizes the regulatory and product liability issues that medical device manufacturers should consider when they receive information concerning improper device use. It also sets forth a precautionary program that manufacturers should undertake to reduce the risk of user problems and related regulatory, product liability, and patient safety concerns.

II. FDA REGULATORY OBLIGATIONS

When confronted with evidence of actual or potential medical device misuse or user error, manufacturers first should consider the FDA-related regulatory obligations that might be applicable to them. These obligations might include duties (1) to modify product design, (2) to change labeling, (3) to report the incidences to the FDA, and/or (4) to revise premarket submissions, as necessary, to reflect any product changes made in response to the user problem. Although practical applications of these regulatory considerations are incorporated into the precautionary checklist set forth later in this chapter, manufacturers should also be familiar with the specific regulatory requirements that might be triggered by reports of improper medical device use.

In considering product design changes, manufacturers will find no specific FDA requirement obligating them to modify product design in response to reports of user error and misuse. Manufacturers should be aware, however, that if the FDA receives evidence demonstrating that a device design is unsafe because it creates user problems, Agency officials could bring an action claiming that the device is "adulterated" in violation of the Federal Food, Drug, and Cosmetic Act (the Act) [3]. In addition, the FDA has authority to withdraw a premarket approval (PMA) if a product design is considered unsafe [4] or, for devices marketed pursuant to Section 510(k) of the Act, to request that a premarket approval application (PMAA) be filed [5]. In practice, however, the FDA ordinarily attempts to work informally with a company or an industry, to recommend any design changes that might be considered necessary to reduce the risk of user problems [6].

As with product design changes, there is no specific regulatory requirement for device manufacturers to revise labeling in response to reports of user error [7]. If failure to change labeling perpetuates the risk of user error, however, the product may be considered "misbranded" in violation of the Act [8]. A product is misbranded if it fails to bear adequate directions, which includes directions that will enable a person to use the device safely, without error, and for the purpose intended [9]. Given these misbranding concerns, manufacturers generally cooperate with the FDA's recommendations to improve labeling when it appears that labeling has caused or contributed to the risk of user errors [10].

In contrast to labeling obligations triggered by user errors, manufacturers are required by regulation to change labeling in response to reports of product misuse (i.e., when the product is being used for an unintended purpose or in an unintended manner) [11]. FDA regulation 21 C.F.R. § 801.4 provides that, if a manufacturer knows, or has knowledge of, facts that would give him notice that a device is being used for purposes other than those for which he offers it, "he is required to provide adequate labeling . . . which accords with such other uses [11]. Thus, for example, when dialyzer manufacturers became aware that their products labeled "single use only" were being reused, the FDA requested the "single use only" label to be revised to include adequate reprocessing instructions pursuant to 21 C.F.R. § 801.4 [12].

Although a full discussion of a manufacturer's Medical Device Reporting (MDR) obligations is provided elsewhere in this textbook, manufacturers should be aware that reports of "serious injuries" and "malfunctions" must be submitted to the FDA even in instances involving user error and product misuse [13]. Under the MDR regulations, a device manufacturer must report to FDA any information it receives or becomes aware of that "reasonably suggests that one of its devices (1) may have caused or contributed to a death or serious injury or (2) has malfunctioned," and, if the malfunction were to recur, would be "likely to cause or contribute to a death or serious injury [14]." The regulations make it clear that if the serious injuries or malfunctions are otherwise reportable, they should be filed with the Agency, "even if the manufacturer . . . believes that the event that requires a report is due to user error, the failure to service or maintain the device properly, the use of the device beyond its labeled useful life, or any other [improper use]

[13]." The circumstances that suggest user error, misuse, or abuse, however, should be included in the report to the Agency.

Finally, when manufacturers determine that design and/or labeling changes are necessary in response to user problems, they must consider their obligations to notify the FDA of these changes. For Class III devices, labeling and design changes generally require Supplemental PMAA review by the Agency prior to implementation [15]. If, however, the manufacturer adds warnings, deletes confusing references, or otherwise revises labeling to enhance the safe use of a device, these changes may be put into effect at the time the Supplement is submitted [16]. For those devices that were subject to Section 510(k) premarket notifications, a new notification would be required if the device has been "significantly changed or modified in design, components, method of manufacture, or intended use [17]." Manufacturers, however, may make their own determination as to whether design or labeling changes implemented in response to user problems are "significant" such that a new Section 510(k) notification would be necessary [18].

III. PRODUCT LIABILITY CONSIDERATIONS

In addition to presenting FDA regulatory concerns, incidences of user error and product misuse could trigger significant product liability exposure for medical device manufacturers. Consequently, when manufacturers receive reports of actual or potential product misuse or user error, they should consider the duties of care that courts have imposed on other companies in product liability challenges involving similar incidences. Specifically, manufacturers should consider their duty (1) to provide warnings in response to any foreseeable risk of user error, and to convey those warnings and related instructions clearly so that they do not confuse the user; (2) to provide adequate warnings of any foreseeable risk of product misuse; and (3) to design their product safely so that the design does not cause or contribute to user error or product misuse. These duties of care are summarized below.

In considering the product liability implications of user errors, companies should be aware that courts will determine liability for any resultant injury by examining both the foreseeability of the error [19,20] and the adequacy of the product warnings provided [20]. In instances where manufacturers reasonably could have anticipated the risk of user error, the courts have imposed a duty to provide adequate warnings that alert users to this potential problem [20,21]. Thus, for example, in one case, a surgical needle manufacturer was held liable for its failure to provide adequate warnings, when it was foreseeable that surgeons mistakenly would grasp the surgical needle in a weak area, causing it to break [22]. If user error is foreseeable, warnings and instructions must be conveyed unambiguously so as not to create further risks of user error. When labeling is misleading or otherwise confusing, causing readers to believe mistakenly that the product could be used for an unintended purpose or in an unintended manner, the manufacturer will be held liable for the resultant injury [23].

In some instances, user problems reasonably cannot be anticipated at the premarket stage of development and are discovered only after a product has been marketed. This is frequently the case with incidences of product misuse, when instructions have been stated clearly but subsequently are altered or ignored by the users. Although some courts view misuse as a defense to a product liability action [24], most do not when there is evidence that the misuse was foreseeable and properly could have been avoided by the issuance of new product warnings [25]. For example, in one case, a cystoscope manufacturer was held liable for a surgeon's eye injury because the court determined that the manufacturer had been aware that surgeons were using an incorrect eyepiece for direct viewing, but failed to warn against such use [26]. In another case, a manufacturer of an artificial breathing device was held liable for injury resulting from a nurse's improper connection of the device, when accidents caused by misconnection of similar devices had been reported in the literature [27].

The question of whether a new risk of user problems is reasonably foreseeable and requires corrective labeling is one of fact, to be considered on a case-by-case basis [22,26]. The determination will turn on the type of device involved, the scope of harm that might result from the product's improper use, and the type and scope of reports of improper use that have been received by the company [27]. Companies might become aware of improper product use from investigators or patients in premarket clinical trials, scientific literature, the FDA, the medical profession, and postmarket reports of injuries or device malfunctions received from the field. Whether at the premarket or postmarket stage, if a risk of improper product use has been identified from these sources, manufacturers have a duty to prepare adequate warnings addressing those risks [20,21], convey the warnings in a timely manner [28], and ensure that the warnings are "brought home" to product users [29].

Even when warnings are themselves adequate, if a product design may have caused or contributed to user error or misuse, courts in limited instances may hold a manufacturer liable for defective design [30]. Generally, in order for a manufacturer to be held liable for a defective design, the design must present an unreasonable danger, notwithstanding the warnings provided [31]. Courts employ different tests in determining whether a product's design is "unreasonably dangerous." The two tests primarily used are the "risk-benefit analysis" [31 at 655; 32] and the "consumer expectations test" [33]. The "risk-benefit analysis" balances product utility against product danger and determines whether the risk could have been reduced without affecting significantly the product's efficacy and cost [34]. For example, in one case, an FDA-regulated product was found to have a defective design, based on evidence that a safer alternative design existed at the time of the product's distribution [35]. Under the "consumer expectations test," a product is defectively designed if it fails to perform as safely as an ordinary consumer would expect when it is used in an intended or reasonably foreseeable manner [36]. For example, a manufacturer of a cardiac umbrella filter was held liable for defective design under the "consumer expectations test," because a patient would not expect the filter to present a risk of lung damage, impaired respiration or death, in light of safer existing technology [36 at 1255].

Failure, thus, to design medical devices safely so that they do not cause or contribute to improper product use, and failure to provide adequate and timely warnings of any foreseeable product misuse or user error, could present significant product liability exposure for medical device companies. Moreover, with the ever-increasing complexity and sophistication of medical device technology, it is reasonable to expect continued growth of product liability challenges in this area. In order to reduce these product liability risks, manufacturers are advised to refer to and implement the precautionary considerations set forth below.

IV. PRECAUTIONARY CHECKLIST FOR ADDRESSING THE ISSUE OF MEDICAL DEVICE USER ERROR AND PRODUCT MISUSE

Set forth below is a checklist of precautionary considerations to assist medical device companies in their efforts to reduce the risk of product misuse and user error. This checklist addresses (1) product design and other premarket activities, (2) labeling efforts, (3) training and user education activities, (4) monitoring of user problems and related document retention, and (5) issues that should be considered when a user problem has been identified and the company has begun to interact with FDA to resolve the problem.

A. Product Design and Other Premarket Activities

Companies should consider conducting premarket test(s) to assess users' ability to handle the device safely and effectively in actual-use settings [37]. These tests are particularly important for devices that present significant risks of user problems and/or are novel products. Test findings should be addressed as part of the company's premarket design efforts, and also should be incorporated into instruction manuals and other labeling materials as necessary.

If a company determines that premarket tests to evaluate user problems are unnecessary or inappropriate, the company should nonetheless request its technical experts to document how the device has been designed to minimize the risk of user error and misuse. This document should be provided to and approved by the company's counsel and "labeling review committee" (i.e., the group responsible for reviewing adequacy of product warnings and instructional materials).

The design of the product should be tailored to the potential user(s). For example, if devices are intended for lay users (e.g., home-use *in vitro* diagnostics), they should incorporate procedures that are as simple as possible, in order to reduce the risk of user confusion.

Finally, companies should generally be attentive to informal design standards that develop in the industry. For example, if a number of comparable devices in the industry have certain safety features to reduce the risk of user error and misuse, and one company does not, that company should consider incorporating the safety features into its device as well [38].

B. Labeling Considerations to Reduce the Risk of User Error and Misuse

A company's labeling review committee should include technical experts knowledgeable about design and actual use of the product. For example, it would be useful to include one of the scientists principally responsible for researching and developing the product, and an appropriate health-care professional who has applied the device in actual-use settings.

In order to reduce the risk of user error, labeling should be drafted to respond to the needs of all potential patient population groups. For example, if there is a risk of visually impaired patients using the device (e.g., elderly patients using glucose monitors), larger print to avoid user confusion is recommended. As another example, if the device is intended for lay users, it should be remembered that many adults have very low-grade reading capacity.

Consequently, labeling instructions should be as short and simple as possible. In drafting labeling generally, attempts at stylistic flair should be avoided; an approach that is simple, precise, and consistently worded will provide the greatest protection against liability concerns.

In order to reduce the risk of user error, labeling should address all reasonably foreseeable use conditions. This is an obvious precaution, but not always considered fully. For example, if a device might be used in dimly lit settings, larger print should be considered to enhance legibility of the instructions.

If a company has no previous experience in drafting labeling materials, there are literature references available to provide guidance on how to convey labeling claims and instructions effectively [39]. The FDA's Center for Devices and Radiological Health (CDRH), Office of Training and Assistance, should be able to provide further reference assistance on this issue, if needed.

Cautionary statements that are most important to prevent user errors and misuse should be considered significant warnings and, as such, should be conspicuously displayed (e.g., by color, bold type, or graphic highlighting). In providing color contrast to enhance a cautionary statement, pastel colors should not be used. (Eight percent of males have diminished ability to distinguish between pastel shades of red, green, and beige.) Suggested color contrasts include red-white, blue-white, yellow-black, green-white, and white-black [40].

Medical Device Reports (MDRs) provide a useful resource for determining the most likely causes of user errors. In advance of preparing labeling, companies may consider it useful to file a Freedom of Information Act request for all MDRs relating to malfunctions or serious injuries that involved user errors for similar devices. Findings from these reports should be incorpoated into labeling materials.

In drafting instruction manuals, companies should consider the following:

1. Many companies, prior to preparing instruction manuals, request FDA to provide them with instruction manuals for comparable devices, particularly if the device will be reviewed through the Section 510(k) process. Instruction manuals for comparable devices, however, might not be that well written. Consequently, particularly for Section 510(k) devices, poorly written instruction manuals tend to breed additional poorly written instruction manuals. Accordingly, companies should request guidance from appropriate officials within the CDRH's Office of Device Evaluation and/or Office of Training and Assistance, to help them identify those manuals that might be most appropriate for use as models. (For example, the CDRH Office of Training and Assistance has informally suggested that instructional materials relating to central venous catheters are generally well written and provide good guidance sources.)
2. Companies should realize that many of the Agency representatives that review labeling in premarket submissions have never observed actual use of the device. This is particularly true for technologically innovative products. Reviewers of Section 510(k) devices are also not charged with the responsibility of confirming the safety and effectiveness of those products as labeled; they need only determine the "substantial equivalence" of a device to predicate device product(s) [41]. Consequently, companies properly should not assume that Agency reviewers will identify all

instruction manual statements that present the risk of user error and misuse.

Companies should also bear in mind that FDA acceptance or approval of labeling/instructional materials will not provide an absolute shield of protection in product liability actions [42]. Even when labeling has met all governmental regulations, courts have held that such compliance is not a defense if a reasonable and prudent manufacturer would have taken added precautions [42 at 1442].

For these reasons, in addition to requesting a company's labeling review committee to review all labeling materials, it may also be useful for companies to request that an independent evaluation be undertaken of those materials. Such an evaluation could be conducted by consultants representing each of the professions or other groups intended to use the device.

3. Instructional materials should incorporate standardized terms readily identifiable by the professions or groups that will be using the device. Such terms should never be varied by synonyms or alternative phrases.
4. It is often useful to devote a separate portion of the instruction manual to common user errors, for quick reference use.
5. In order to clarify labeling, companies should use instructive illustrations to the extent possible. These instructive illustrations, however, should be precise; pictorial inaccuracies may actually add to user confusion.
6. Companies should include a statement on the manual cover page, noting that the manual should not be separated from the device. The CDRH's Office of Training and Assistance has identified unavailability of instructional materials as one of the significant problems leading to user error and misuse.
7. Many companies are concerned that device users do not take the time to read fine print. It might be useful, therefore, to include in the instruction manual a summary page of the manual's most important instructions.

   For product liability purposes, however, the summary should emphasize that the entire manual must be reviewed carefully, so that readers do not misinterpret the summary as a substitute reference source for the manual.

It may also be useful to prepare a laminated pre-use checklist that could either be affixed, or posted in close proximity, to the device. (Again, for product liability purposes, the checklist should emphasize that the entire manual must be reviewed.) According to the FDA, this checklist procedure has been used effectively with infusion pumps and anesthesia equipment [43, 44].

Finally, if a change has been made in labeling or instructional material to address a new risk of user error and misuse, this change should be "brought home" to the user [29]. In order to be considered adequate for product liability purposes, new warnings may need to be conveyed by a separate letter to the profession or group using the device, by label highlighting, and/or by use of sales representatives.

C. Training/User Education to Reduce the Risk of User Error and Misuse

In considering the proper role that training should play in reducing user error, some experts have concluded that training is the most effective method for conveying instructions and reducing the risk of product error and misuse [45]. Nonetheless, many companies express concerns about the inherent limitations of training programs to reduce the risk of user problems. For example, some companies have complained that doctors are occasionally unwilling to participate in training programs, although nurses and other support staff will typically attend. Training problems also arise as a result of medical staff changes. Other companies find it difficult to convey detailed and complex training instructions in a way that will be retained.

Problems endemic to training appear to decline, however, when companies or their relevant trade associations interact closely with relevant medical professional groups to deal with user concerns. A good example of industry/professional collaboration was seen with recent user education activities directed at anesthesia equipment [46]. As a result of increased industry-professional interaction, user errors appear to be occurring less frequently in that industry [44 at 9-10].

In dealing with the problems of training, some companies also consider it useful to raise the importance of training sessions directly with the hospital's administration (e.g., its risk management group). So as not to disrupt the company's relationships with doctors, however, any communications with the hospital's administration should not be presented as complaints about attendance at training sessions or as other grievances.

In order to reduce the risk of user error, companies should instruct their sales representatives or other training instructors:

1. To stress the importance of users reading carefully all instructional materials prior to use of the device.
2. Not to overpromote a product's ease of use or make any statements that would deemphasize product warnings or important product precautions.
3. Not to highlight, underline, or otherwise mark up instructional materials or promotional literature, because of the risk that it might negate warnings or diminish the perceived importance of other instructional information.
4. If a user of the device advises the sales representative that the device is being used for purposes other than that noted in the labeling and instructional materials, to note important precautions and limitations of the product.
5. Not to keep records of information conveyed to the device user. Although some companies request their sales force to maintain such records in an effort to demonstrate that important instructions for use and warnings have been discussed, because such records rarely are complete, they may later be more harmful than helpful to the company in a product liability action.
6. Not to provide incorrect or incomplete information, if the sales representative does not know the answer to a question. In such circumstances, sales representatives should simply state that they do not know the answer, but will attempt to obtain one from appropriate company sources.

D. Monitoring Activities

Appropriate personnel should be assigned to review applicable industry journals, scientific literature references, and other FDA policy statements issued by FDA or its expert panels, to identify any reports that might relate directly or indirectly to the safe use of the company's device(s). Such reports should be forwarded promptly to the company's labeling review committee [47].

Appropriate personnel should also be assigned to monitor receipt of all adverse reaction reports, in order to determine whether user errors and product misuse necessitate corrective actions. In order to facilitate this process, companies typically either prepare internal complaint status reports at regular intervals, or conduct more detailed trend analyses. In undertaking trend analyses, companies should be aware that the FDA is in the process of expanding its MDR data base so that it will track more effectively certain user problems. Companies should make sure, therefore, that their internal trend analysis efforts track user errors as well, so that they will be able to anticipate and prepare to respond to the FDA's concerns. All complaint status reports and trend analyses should be forwarded promptly to company counsel for review.

In determining whether evidence of user error and misuse should prompt corrective actions, companies should balance the number of reports received against the severity of adverse reactions that have resulted from the user problems. When user problems have resulted in serious injuries or death, fewer reports should prompt corrective activities. If the risk of use0 problems is stated on device labeling, companies should also conside0 whether such problems are occurring with greater frequency or severity than those risks referenced in the labeling; if so, labeling should be revised [48].

Companies should be attentive to the drafting of MDRs. Frequently, companies will cite "user error" as a possible attributing factor to a report of malfunction or serious injury. Companies should understand, however, that a record of "user errors" can also have product liability implications. If the reports suggest a trend involving the same type of error, corrective labeling or design measures may be necessary for product liability purposes.

Companies should also understand that all internal documentation relating to user error, misuse, or confusion may be discoverable in a product liability action. Such records may be probative in a court's review of the company's knowledge of events, as well as in its consideration of the adequacy and timeliness of warnings [49]. Given the significant product liability implications of internal documentation, therefore, all complaint and related records (e.g., MDR files, adverse experience reports, and good manufacturing practices complaint files) should be audited regularly by company's counsel.

If a company has received reports suggesting a trend of user problems, all documents addressing and responding to these problems should be drafted only at the request of company counsel and should be designated "Attorney-Client Work Product/Privileged Communication." As further protection of this privilege, there could be controlled and limited dissemination of documents through counsel. For example, company counsel may want to request that the recipients of sensitive documents not disseminate or copy the documents without prior authorization, and return all copies to counsel following their review and use.

Finally, if a decision has been made to change labeling materials or product design to address reports of user error and product misuse, companies should consider whether detail men need to notify their customers of these changes. If sales representatives are used as an integral part of selling a product and inducing its use, a court may hold that a new warning should be transmitted by sales representatives as well as in labeling [29].

E. Considerations Relating to Interacting with FDA, Once a User Problem Has Been Identified

The majority of FDA's generic user education/training activities have been received enthusiastically by participating manufacturers. From the manufacturers' perspective, this type of collaboration promotes good will between FDA and industry, as well as with the medical profession. More importantly, this collaboration may reasonably be expected to reduce ultimate product liability defense and related insurance costs for industry, if the programs actually accomplish their objective (i.e., to reduce the incidence of user error and product misuse) [44].

A few of FDA's efforts to review user problems, however, may present their own product liability implications. For example, FDA is in the process of reviewing design features on glucose monitors and, while the report findings are not expected to be presented as recommendations for specific design changes, they are likely to identify certain features on monitor equipment that might affect safe use of the machines [50]. Although not the intent of the Agency, failure of a manufacturer later to address FDA's design findings might support a claim of defective product design. As another example, FDA also recently recommended a pre-use checklist for anesthesia equipment, which recommendation was endorsed by the medical profession [51]. Failure now of a manufacturer to incorporate this checklist into its instructional materials is likely to be probative in a court's review of adequacy of warnings. Accordingly, if FDA officials suggest that they are considering certain design or labeling activities, it is incumbent upon companies to remind the Agency of these product liability implications and to emphasize that such implications properly should not be ignored in setting policy.

Additional business considerations that might also need to be raised with the FDA, depending upon the Agency's particular user concern and proposed initiative, include whether the FDA's efforts might have an adverse competitive effect (i.e., favor some companies over others), and whether the proposed activities might be so costly or present such anxiety for users of the device that, if implemented, availability of the device would be adversely affected.

Companies should seriously and objectively consider available evidence of user problems. If evidence clearly points to a problem, and others in the industry are working with the FDA to address the issue, it is advisable to cooperate with the Agency on these issues as well. If, however, the FDA undertakes an effort that appears to be an unduly harsh response to user problems, or appears not to be supported by available evidence, companies should consider the following:

1. If the FDA is imposing labeling or design changes as recommendations, rather than requirements, and a company determines that such suggestions are inapplicable or inappropriate, the company should document its determination along with appropriate justifications. For product liability purposes, the FDA's recommendations should not simply be ignored. The documentation might note (a) how existing labeling, design features, and other activities already undertaken by the company adequately address the Agency's user concerns; (b) how the company's position is not inconsistent with manufacturers of comparable devices; and (c) how available data relied on by the Agency is either flawed or insufficient (e.g., if MDR data are the sole source for the FDA's concern, such data often are insufficient to assess user problems adequately; it is for this reason

that the FDA will frequently undertake additional surveys or other similar data-compilation efforts as a first step to refining user concerns). If any of these three points cannot be made, companies are well advised to reconsider their position for product liability purposes.

2. To the extent possible, companies should rely on their relevant trade associations to raise objections with the FDA's corrective recommendations. Trade associations provide a necessary buffer between the individual companies and FDA representatives. If there is dissension within the trade association, however, and some companies agree to implement the FDA's recommendations, it would be advisable for other companies to consider undertaking the corrective activities as well, for purposes of product liability protection [38].
3. If the FDA's proposed corrective activities appear to be unduly harsh, companies should determine whether the Agency's recommendations might be able to be recast in a more positive and instructive fashion (e.g., recast from specific labeling warnings to a commitment by companies to expand user education).
4. If appropriate and useful, companies or their trade associations should consider requesting assistance from relevant medical professional groups to support their position. Medical professional support, for example, assisted the ultrasound industry in its efforts to recast warnings for nonfetal and multipurpose transducers [52].
5. If appropriate, companies should review whether devices that present comparable risks incorporate the corrective measures recommended by the FDA. The ultrasound industry, for example, raised the disparate treatment between the FDA's proposed warnings for ultrasound equipment and the omission of such warnings on other similar diagnostic x-ray equipment [52 at 3]. This finding was relied on to help persuade the FDA to consider less stringent measures to address ultrasound user concerns [52 at 3].
6. Finally, companies should consider timing in responding to the FDA's recommendations for device labeling and/or design changes. Delay tactics, from a product liability perspective, may be probative in a court's review of the timeliness or adequacy of response [53], and might support a claim of punitive damages as well [54].

V. CONCLUSIONS

As medical technology continues to evolve, it is a foregone conclusion that the medical device industry will continue to create new user challenges. If manufacturers adhere to the practical guide set forth in this chapter, they will be able to bring these new challenges to market in a way that will reduce the risk of user problems and related regulatory, product liability, and public health concerns.

REFERENCES

1. FDA records provide strong evidence of the toll of medical device user problems on human life. See, for example, FDA, Medical Device Report, No. M128709 (Sept. 26, 1986) (an anesthesia machine mistakenly operated

by a nurse caused a perforated esophagus and death for a young female); FDA Safety Alert: Pediatric Cribs (July 2, 1984) (improperly latched pediatric crib caused strangulation of an infant); Letter to apnea monitor user from John C. Villforth, Director, Center for Devices and Radiological Health, FDA (June 21, 1985) (apnea monitor mistakenly replugged electrocuted a baby while it slept).

2. See Joseph S. Arcarese, User Considerations for In Vitro Diagnostic Products, at 3 (Oct. 18, 1988) (remarks at Health Industry Manufacturers Association/FDA meeting in Rockville, MD).
3. Section 501(c) of the Federal Food, Drug, and Cosmetic Act ("Act"), 21 U.S.C. § 351(c) (1982) (a device is "adulterated . . . [if its] quality falls below that which it purports or is represented to possess").
4. Section 515(e)(1)(A),(B) of the Act, 21 U.S.C. § 360e(e)(1)(A),(B) (1982); 21 C.F.R. § 814.46 (1988).
5. Sections 513(a)(1)(C), 515(a) of the Act, 21 U.S.C. § 360c(a)(1)(C), 360e(a) (1982); see also 21 C.F.R. § 814.1 (1988).
6. See, for example, letter from Walter E. Gundaker, Director, Office of Compliance, Center for Devices and Radiological Health, FDA to manufacturers of breathing frequency monitors and heart rate monitors for home use (June 19, 1985) (requesting review of lead plug designs to reduce the risk of electrocution from product misuse).
7. See generally 21 C.F.R. Part 801 (1988).
8. See Section 502 of the Act, 21 U.S.C. § 352 (1982).
9. Section 502(f) of the Act, 21 U.S.C. § 352(f) (1982).
10. In 1984, for example, pediatric crib manufacturers cooperated with the FDA in making labeling changes to reduce the risk of improper latching procedures. See Post-Market Product Management, Office of Compliance, Center for Devices and Radiological Health, FDA, Internal Summary of Pediatric Crib Device Problem (July 2, 1986). In 1986, the contact lens industry cooperated with the Agency in making voluntary labeling changes with respect to extended-wear lenses, following evidence suggesting user inattention to lens maintenance instructions. See Letter to Mr. Cal Bowman, Regulatory Chairman, Contact Lens Institute, from Richard E. Lippman, O.D., Director, Division of Ophthalmic Devices, Center for Devices and Radiological Health, FDA (Nov. 19, 1986).
11. 21 C.F.R. § 801.4 (1988).
12. See "Dialyzer Processing is Maintenance Tool," Medical Devices, Diagnostics & Instrumentation ("The Gray Sheet"), Jan. 5, 1987, at 8.
13. 21 C.F.R. § 803.24(d)(1) (1988).
14. 21 C.F.R. § 803.1 (1988).
15. 21 C.F.R. § 814.39(a) (1988).
16. 21 C.F.R. § 814.39(d) (1988).
17. 21 C.F.R. § 807.81(a)(3) (1988).
18. Although there currently is no policy or other guidance clarifying manufacturers' obligations with respect to Section 510(k) modifications, the FDA may in the near future issue guidance specifying when design and labeling changes would trigger new Section 510(k) filings.
19. See, for example, *Thomas v. American Cystoscope Makers, Inc.*, 414 F. Supp. 255 (E.D. Pa. 1976).
20. See, for example, *Bristol-Myers Co. v. Gonzales*, 548 S.W. 2d 416 (Tex. Civ. App. 1976), *rev'd on other grounds*, 561 S.W. 2d 801 (Tex. 1978).

21. An adequate warning for FDA-regulated products has been defined as labeling that (1) adequately indicates the scope of the danger, (2) reasonably communicates the extent or seriousness of the harm that could result from misuse of the product, (3) includes physical aspects (color, print size) sufficient to alert a reasonably prudent person to the danger, (4) indicates the consequences that might result from failure to follow the warning, and (5) conveys the warning adequately (i.e., makes its meaning clear to the doctor and/or end-user). See *Richards v. Upjohn Co.*, 95 N.M. 675, 625 P.2d 1192 (Ct. App.), *cert. denied*, 94 N.M. 675, 615 P.2d 992 (1980). See also *Perfetti v. McGhan Medical*, 99 N.M. 645, 662 P.2d 646 (Ct. App.), *cert. denied*, 99 N.M. 644, 662 P.2d 645 (1983) (court upheld jury's finding that warning for mammary prosthesis inadequate in that it did not disclose nature and extent of danger).
22. *Ethicon, Inc. v. Parten*, 520 S.W.2d 527 (Tex. Civ. App. 1975).
23. *Bristol-Myers Co. v. Gonzales*, 548 S.W.2d 416 (Tex. Civ. App. 1976), *rev'd on other grounds*, 561 S.W.2d 801 (Tex. 1978) (court held that drug's labeling in the *Physicians Desk Reference* was misleading, causing readers mistakenly to believe that the drug could be used as a continuous irrigant).
24. See, for example, *Corbridge v. Clark Equipment Co.*, 730 P.2d 1005 (Idaho 1986) (product misuse recognized as an affirmative defense in product liability suit against forklift loader manufacturer because plaintiff used loader for towing in deep snow when it was not designed for such purpose); *Clark v. Boeing Co.*, 395 So. 2d 1226 (Fla. Dist. Ct. App. 1981) (in product liability action brought by flight attendant against aircraft manufacturer for defective engine design, court held that improper exposure of emissions was knowing misuse, which precluded claim).
25. See, for example, *Thomas v. American Cystoscope Makers, Inc.*, 414 F. Supp. 255 (E.D. Pa. 1976) (even though surgeon used improper eyepiece, court held this error foreseeable and correctable); *McCormack v. Hankscraft Company*, 154 N.W.2d 488 (Minn. 1967) (court held that manufacturer should have foreseen that children might tip over vaporizer and that this risk could have been avoided by an alternative design).
26. See *Thomas v. American Cystoscope Makers, Inc.*, 414 F. Supp. 255 (E.D. Pa. 1976).
27. See *Airco, Inc. v. Simmons First National Bank*, 276 Ark. 486, 638 S.W.2d 660 (1982) (in case holding manufacturer of artificial breathing machine liable for harm resulting from product's foreseeable misuse, court noted the irreversible brain damage to the patient and prior reports of similar user problems with such machines).
28. See *Miller v. Upjohn Co.*, 465 So. 2d 42, *cert. denied*, 467 So. 2d 533 (La. App. 1985) (drug manufacturer held liable for not warning that its drug stained teeth and pitted tooth enamel until four years after it became aware of these adverse reactions, even though manufacturer contended that FDA did not approve the warnings until just before their issuance).
29. See *Sterling Drug v. Yarrow*, 408 F.2d 978 (8th Cir. 1969) (because sales representatives were used as an integral part of selling an FDA-regulated product and inducing its use, court held that warning of new risk should be transmitted by sales representatives as well as by labeling).

30. Most prescription drug and device cases alleging defective design are brought under a theory of strict liability. See *Ortho Pharmaceutical Corp. v. Heath*, 722 P.2d 410 (Colo. 1986) (court upheld lower court verdict that oral contraceptive manufacturer strictly liable for defective design); *Coursen v. A.H. Robins Company, Inc.*, 764 F.2d 1329 (9th Cir. 1985) (IUD manufacturer held strictly liable for defective design); *Brochu v. Ortho Pharmaceutical Corp.*, 642 F.2d 652 (1st Cir. 1981) (court held oral contraceptive manufacturer could be strictly liable for defective design); *Thomas v. American Cystoscope Makers, Inc.*, 414 F. Supp. 255 (E.D. Pa. 1976) (manufacturer held strictly liable for improperly insulated resectoscope eyepiece, which burned surgeon's eye); *LeRay v. St. Paul Fire and Marine Insurance Co.*, 444 So. 2d 1252 (La. Ct. App. 1983) (cardiac umbrella filter manufacturer held strictly liable for plaintiff's brain damange due to filter's defective design).

    Even in defective design cases brought under a negligence theory, strict liability principles may be applied. See *Toner v. Lederle Laboratories*, 732 P.2d 297 (Idaho 1987) (court held that a vaccine manufacturer could be held negligent for improper design of vaccine); *McCormack v. Hankscraft Co.*, 154 N.W. 2d 488 (Minn. 1967) (court held evidence sufficient to find vaporizer manufacturer liable for child's burn injuries under theories of negligence and express warranty due to vaporizer's defective design).

    Some courts, however, will not apply strict liability to prescription drugs and devices. See, for example, *Brown v. Superior Court*, 751 P.2d 470 (Cal. 1988) (court held that strict liability standards should not be used to measure a drug manufacturer's liability for a defectively designed drug "because of the public interest in the development, availability, and reasonable price of drugs").
31. See, for example, *Brochu v. Ortho Pharmaceutical Corp.*, 642 F.2d 652 (1st Cir. 1981) ("When an unreasonable danger could have been eliminated without excessive cost or loss of product efficiency, liability may attach even though the danger was obvious or there was adequate warning").
32. See *Toner v. Lederle Laboratories*, 732 P.2d 297 (Idaho 1987).
33. See *LeRay v. St. Paul Fire and Marine Insurance Co.*, 444 So. 2d 1252, 1254-55 (La. Ct. App. 1983). In California, manufacturers must be attentive to both the "risk-benefit" and "consumer expectations" tests, since either standard of care may be used by the courts in that state. See, for example, *Ortho Pharmaceutical Corp. v. Heath*, 722 P.2d 410 (Colo. 1986), citing *Barker v. Lull Engineering Co.*, 24 Cal. 3d 413, 426-427, 143 Cal. Rptr. 225, 234, 573 P.2d 443, 452 (1978).
34. See, for example, *Brochu v. Ortho Pharmaceutical Corp.*, 642 F.2d 652, 655 (1st Cir. 1981) (evidence demonstrated that lower dosage of drug was safer and as effective as higher dose which caused injury).
35. *Ortho Pharmaceutical Corp. v. Heath*, 722 P.2d 410 (Colo. 1986) (stating that risks of higher dosage of drug outweighed any benefits).
36. See *LeRay v. St. Paul Fire and Marine Insurance Co.*, 444 So. 2d 1252 (La. Ct. App. 1983).
37. See Office of Compliance, Center for Devices and Radiological Health (CDRH), Food and Drug Administration (FDA), Draft *Preproduction Quality Assurance Planning: Recommendations for Medical Device Manufacturers* at 5 (May 1987) (general principles for conducting failure analyses).

38. See *LeRay v. St. Paul Fire and Marine Insurance Co.*, 444 So. 2d 1252 (La. Ct. App. 1983) (company held liable because cardiac umbrella filter could have been designed more safely).
39. See, for example, R. Rice and W. Paisley, *Public Communications Campaigns* Chs. 6, 12, 13, 14 (Sage Publ. 1981); Posavac, *Evaluations of Patient Education Programs: A Meta-Analysis*, 3 Eval. Health Prof. 47-62 (Mar. 1980).
40. Proposed ASTM Standard to Enhance Identification of Drug Names on Labels, submitted by Leslie Rendell-Baker, M.D. (Jan. 25, 1989).
41. See Section 510(k) of Act, 21 U.S.C. §360(k) (1982).
42. See, for example, *O'Gilvie v. International Playtex, Inc.*, 821 F.2d 1438 (10th Cir. 1987), *cert. denied*, 108 S.Ct. 2014 (1988) (holding that defendant's compliance with FDA warning requirements for tampons did not satisfy its duty to provide adequate warning).
43. Interview with James L. Morrison, Deputy Director, Office of Training and Assistance, Center for Devices and Radiological Health, FDA (Oct. 28, 1988).
44. See Joseph S. Arcarese, Controlling Risks from the Use of Medical Devices: The CDRH Perspective, at 4, 8-10 (Sept. 27-28, 1988) (remarks at the 12th Annual Meeting of the Regulatory Affairs Professionals Society, Alexandria, VA) (noting the effectiveness of pre-use checklists and that the cost of insurance for anesthesia equipment manufacturers has gone down as a result of recent industry-professional efforts to address user problems).
45. See Dr. Deborah Rugg, User Education Concerns (Sept. 20, 1988) (remarks at the Health and Education Council, Second National Symposium on AIDS Prevention, Baltimore, MD).
46. See Joseph S. Arcarese, Controlling Risks from the Use of Medical Devices: The CDRH Perspective, at 4-7 (Sept. 27-28, 1988) (remarks at the 12th Annual Meeting of the Regulatory Professionals Society, Alexandria, VA) (noting that FDA recently has been working with the Anesthesia Patient Safety Foundation (APSF) and the American Society of Anesthesiologists (ASA) to promote the development of anesthesia training simulators; participate in the development of voluntary performance and safety standards for anesthesia and respiratory devices; evaluate the effectiveness of recommended preanesthesia checklists; and develop a paper and poster on disconnect problems).
47. See generally *Lindsay v. Ortho Pharmaceutical Corp.*, 637 F.2d 87, 91 (2d Cir. 1980) (court held that manufacturer has duty to "keep abreast of the current state of knowledge of its products as gained through research, adverse reaction reports, scientific literature, and other available methods").
48. See generally 21 C.F.R. §803.24(c)(7), (d)(3)(iii)(D) (1988).
49. See *Miller v. Upjohn Co.*, 465 So. 2d 42 (La. Ct. App.), *cert. denied*, 467 So. 2d 533 (La. 1985) (manufacturer liable for not warning that drug stained teeth and pitted tooth enamel until FDA approved the warning, although it knew of the adverse reactions well before that time); *Airco, Inc. v. Simmons First National Bank*, 276 Ark. 486, 638 S.W.2d 660 (1982) (punitive damages appropriately considered because company continued to market breathing machines without labeling and design changes, despite prior reports that the devices could kill people because of foreseeable danger of human error).

50. See Post-Market Product Management, Office of Compliance, Center for Devices and Radiological Health, FDA, Internal Summary of Blood Glucose Monitor Problem (Sept. 21, 1988).
51. See 52 *Fed. Reg.* 5583 (Feb. 25, 1987) (notice of availability of anesthesia apparatus checkout recommendations).
52. See "FDA to Stay Ultrasound Transducer Labeling Requirement," Medical Devices, Diagnostics & Instrumentation ("The Gray Sheet"), Oct. 26, 1987, at 2-3.
53. See generally *Schenebeck v. Sterling Drug, Inc.*, 423 F.2d 919 (8th Cir. 1970) (drug manufacturer held liable for not providing warnings of adverse reactions until four years after it was aware of such reactions).
54. See *Airco, Inc. v. Simmons First National Bank*, 276 Ark. 486, 638 S.W.2d 660 (1982) (punitive damages appropriately considered for manufacturer's failure to warn against misconnection of breathing device, because similar accidents had been reported previously in the literature).

9

FDA's Role in Medical Device User Education

JOSEPH S. ARCARESE *Center for Devices and Radiological Health, Food and Drug Administration, Rockville, Maryland*

I. INTRODUCTION

A. The Issue of Medical Device Use

The progress of medicine in the past 30 or 40 years has been characterized in part by the rapid development and widespread use of increasingly sophisticated medical devices. This explosion of new diagnostic and therapeutic technology has made possible remarkable advances in the early recognition and successful treatment of many diseases and disabilities. The use of this technology in modern medicine, together with greater public consciousness about health and changes in diet and lifestyle to prevent disease, enables Americans to enjoy longer and more fulfilling lives than ever before.

But with the benefits of technology have also come some additional risks. As medical devices have increased in complexity and capability, and as the demands on the knowledge and technical skill of their human operators have increased, so too the opportunity has increased for unintended and unforeseen adverse consequences. Devices occasionally malfunction, and their users occasionally make mistakes in judgment or in action. As a consequence, the deaths and injuries of some patients occur—a tragic byproduct of the very effort of medical science to extend the benefits of modern health technology to more and more people.

A substantial percentage of device-related patient deaths and injuries is caused by or involves misuse of medical devices. This percentage varies dramatically between device categories; in the case of anesthesiology, 80% or more may be due to human error [1]. Typically, a smaller percentage of patient adverse outcomes are due to outright malfunction or catastrophic failure of devices themselves. In addition to the dolorous human toll of lives lost or harmed through avoidable accidents, user error leading to patient deaths and injuries also carries with it expensive financial liabilities. These use-mediated adverse incidents frequently result in malpractice suits for the health professionals involved, and also often result in product liability suits for the device manufacturer. These financial liabilities in turn affect the cost of medical care in this country. For example, the impact of professional liability has been estimated to be approximately 15% of the total U.S. expenditures on physicians' services [2].

B. FDA's Educational Roles

As a consequence of the Medical Device Amendments to the Federal Food, Drug and Cosmetic Act, the Food and Drug Administration (FDA) imposes many requirements regarding the manufacture, marketing and distribution of medical devices, in order to assure that devices are safe and effective. Most medical device manufacturers are aware of the FDA's regulatory role in their industry.

It is less widely known that FDA has educational programs addressing the needs of health professionals, lay consumers and patients, and manufacturers. It publishes educational documents, develops audiovisual materials, and conducts conferences and workshops, in order to provide these audiences with important medical device information. Many manufacturers have directly benefited from the work of FDA's Division of Small Manufacturers Assistance, which conducts workshops and publishes educational information to assist businesses in meeting FDA requirements and maintains a toll-free phone number to answer manufacturers' questions [3].

However, one of FDA's most important educational strategies consists not in merely distributing information to the public or to particular medical device users, but in its work with health professional organizations and device manufacturers to cooperatively solve problems involving the unsafe use of medical devices. FDA realizes that achieving device safety and effectiveness depends on educating and motivating medical device users in their proper operation. This chapter will focus on FDA's joint role with device manufacturers and the health professional organizations in a major educational effort to improve the performance of the device user.

II. THE SAFETY AND EFFECTIVENESS OF DEVICES

A. Device Design, Manufacture, and Use

The safety and effectiveness of medical devices depend not only on their design and manufacture, but also on the way in which they are used. It is axiomatic that the quality of a device's design and construction bears directly on both how safely it can be operated and how effective it will be in achieving its intended purpose. But the device itself does nothing without some sort of human intervention (even if only to start the machine or to implant it); how well the human users perform frequently determines whether or not the device's intended purpose will be safely and effectively achieved.

Even the most well-intentioned manufacturer's efforts to produce a safe and effective medical device can be frustrated when the device is improperly used or mishandled. This can happen when the device user either doesn't know how to use the device properly, or knows how to use it properly but is tired, aggravated, inattentive, or has some motivation not to follow the directions for its proper use. In an industry magazine editorial, Chairman and Chief Executive Officer of Bausch and Lomb Daniel Gill observed: "Everybody talks about quality these days, but in medical devices, quality can mean the difference between life and death . . . . But no matter how carefully we design and manufacture our products, they're going to be used by people. And people sometimes make mistakes. We've got to make sure that our high-quality products are used in a high-quality way" [4].

B. The Chain of Events in an Accident

The Medical Device Reporting System (MDR) is an FDA data base consisting of reports from device manufacturers of patient deaths and serious injuries allegedly involving their medical devices, and malfunctions of their devices that could cause death or serious injury. The data base, which is expanding at a rate of about 18,000 reports per year, contains numerous reports of incidents that are apparently related to health professional or lay user error rather than to some functional failure of a medical device. One typical example of health professional user error involves an MDR report of an injury to a patient undergoing hemodialysis:

> Patient was infused with air (air embolism) while on dialysis. Saline administration line had been cut and air entered system. Air detector alarm system was not armed at the time. The system is designed to detect air in the blood return line and clamp the line to prevent the air reaching the patient. This alarm system had been bypassed by the user. This is a potentially fatal complication.

In this example, an equipment safety feature, the air detector alarm system, had not been turned on, preventing it from functioning in the way in which it had been designed. The failure to turn the alarm on did not directly cause the injury, of course, but it set the stage allowing the accident to happen. Because warning systems sometimes alarm inappropriately, resetting them is an intrusive irritation for dialysis health care providers. Turning them off is seen as an apparently harmless way of avoiding the inconvenience, an unfortunate misunderstanding reinforced by the fact that discernible problems rarely result. The injury (or in other similar instances, patient death) happened when another event unexpectedly occurred (air accidentally introduced into the system through a severed saline line) and the inoperative air detector could not respond.

Medical devices are often misused, but often no harmful outcome ensues. Occasionally, however, another circumstance occurs in tandem, like an additional link added to a chain of events, which causes the misuse to result in tragedy. When this happens, the manufacturer of the device is often implicated in a product liability suit, even if the device user is totally at fault.

Since the user's behavior has such an intrinsic impact on the safety and effectiveness of medical devices, attention paid to the knowledge and motivation of users of medical devices by manufacturers, health professional organizations, and government is fundamentally important to patient welfare. In addition to the incalculable benefit of averting human death and suffering, efforts to decrease user error will also decrease the number of product liability and malpractice suits, and thus decrease the cost of health care delivery in this country.

III. ANALYZING THE CAUSES OF USER ERROR

A. Causes of User Error

There are essentially two fundamental categories of causes for user error. Either the user does not know how to use a device properly, and consequently makes errors in using it, or the user knows how, but does not correctly apply the knowledge for some reason.

B. User Lack of Awareness of Proper Device Operation

Devices are often packaged with instructional labeling, and in addition, many device manufacturers provide additional kinds of training opportunities for their clients, such as hospital in-service training, or training programs for lay users of the device at home. With this information available, it might seem puzzling that users could operate the device without possessing the requisite knowledge to do so correctly. Nevertheless, there are several reasons to account for this:

1. The information is sometimes unavailable to the end user even though the device is packaged with all the appropriate labeling materials. In a busy hospital, for example, equipment and operating manuals are quickly separated. To be useful, the information must be physically available to the user. In addition, some health professionals are hard to reach with training. For example, although nurses frequently go to in-service training programs, residents and staff physicians often do not. It also happens that new devices (but similar to previous models) or device upgrades are introduced without new training, without recognition that previous training may not be fully applicable to the new model or the upgraded device.
2. Even when information is available to a user, the user may not pay much attention to it. How many parents read the instructions on Christmas Eve *before* trying to assemble the doll houses or the bicycles? And how many computer hackers attempt to use a new piece of software before consulting the documentation manual? Some people believe they know enough about operating a device that they have little incentive to read the manufacturer's instructions carefully, or they enjoy the challenge of discovering its operation without aid. These same fundamental human motivations and behaviors also operate in the world of health care delivery. As a result, some medical personnel will not take the time and effort to carefully read and follow the manufacturer's instructions.
3. Even when the information is available, and even when the user *wants* to learn it, the information may simply be beyond the capability of the user to understand it. Instructional materials typically suffer from a host of problems. They are often vague, incomplete, confusing, inarticulate or erroneous, or they presuppose reading comprehension levels, information, or skills beyond the scope of many intended users. And what is true for written materials is also often true for live presentations. Unfortunately, *many manufacturers have not mastered the art of making instructional material and presentations understandable, memorable, and appropriate for the target audience,* and this defect has the serious consequence that the user may have no reasonable recourse but to use the device without adequate knowledge.

C. Other Factors Affecting Performance

User lack of familiarity about the proper use of medical devices, as prevalent as it may be, is only one category of reason why people use them incorrectly. There are many other reasons why users perform poorly, even when they *are* adequately informed:

1. The general environment in which the medical device operates may not be conducive to good use. The lighting may be too dim or too glaring for the user to see the screens, controls, and labels adequately, the ambient noise level may mask audible alarms, or critical features of the device may be hidden, for example by bedclothes or surgical drapes.
2. The user may be forced to respond to other incentives, such as unusually heavy workload, so that there is no time to perform critical pre-use check-out or important maintenance, or the facility management may be unsupportive of seemingly burdensome precautionary procedures which take additional time.
3. The typical user might occasionally be unusually fatigued, such as residents and medical students on duty for extended periods without adequate rest.
4. The user might be inattentive to proper device procedures, perhaps resulting from several years of accident-free work. In the misguided belief that he or she already knows everything there is to know about the device or can handle any unexpected complication, the user might be inclined to take critical and occasionally disastrous short-cuts in procedure.
5. The user may *intellectually* know how to properly operate a device, but may not be practiced enough to do it reliably; or the user may have some impairment of physical dexterity, making it difficult to perform some complex physical maneuvers.
6. The controls and labeling on the device itself might be presented in a confusing or illogical fashion, or might be designed without adequate attention to the real circumstances in which the device is intended to operate (poor "human factors engineering").

In the real world of health care delivery, many of these factors can interrelate. For example, in the midst of an emergency, when time is a critical factor and attention is diverted to other priorities, a tired user might become confused by a poorly designed and dimly illuminated operating panel and inadvertently select a wrong control, especially if he or she had used that particular model device only once or twice before. Several unanticipated links in a chain of events can sometimes be joined inadvertently and very rapidly, leading inexorably to disaster.

IV. THE PARTNERSHIP OF HEALTH PROFESSIONAL ORGANIZATIONS, MANUFACTURERS, AND GOVERNMENT

A. Systems Problem and Systems Solution

Regardless whether user error is caused by user ignorance, careless behavior, inattention, fatigue, environmental factors, user-unfriendly device design, poorly designed instructional materials, or any one of many other possible factors, user error is basically an educational systems problem. As such, it is often generic, affecting a whole category of devices regardless of manufacturer. Consequently it requires a systems solution that assures that health professionals and lay medical device users know how to use their medical devices appropriately, motivates them to put the knowledge into practice, and controls for environmental constraints and circumstances, such as heavy workload, fatigue, distractions, and conflicting priorities.

Dealing effectively with all those aspects of human behavior is exceedingly difficult for either manufacturers, health professional organizations, or government acting alone. But problems that one organization cannot solve alone might be successfully overcome when the three groups cooperate. Manufacturers, professional organizations, and government bear a mutual responsibility to work together to decrease user error and improve patient safety, and they each have unique resources to apply to these challenges.

Manufacturers have technical expertise about their equipment, and through their sales personnel, they sometimes have valuable observations about the user-related circumstances affecting their devices. For example, they may be aware of certain staff practices that are less than optimal, although they may be powerless to do anything about it directly. Many companies also have financial resources that they are willing to apply to solve problems, especially if this expense can be seen ultimately to offset liability losses. However, due to concern for restraint-of-trade difficulties and the competition between them, manufacturers often find it difficult to cooperatively address generic user problems themselves.

The health professional organizations have clinical expertise and credibility, and they have much insight into the generic advantages and disadvantages of devices under particular circumstances. They communicate regularly with their membership through newsletters, journals, and meetings, some of which provide continuing education credits. When professional organizations publish guidelines or standards of practice, these often have a great deal of influence on the practice patterns within the profession. However, these organizations generally do not intrude on the practice of specific individual members unless their behavior is flagrant. Some organizations sponsor data-gathering studies for particular issues, but they often rely on the clinical advice of their members as the grist with which to answer device-use questions.

FDA's Center for Devices and Radiological Health (CDRH) has responsibilities relating to the full gamut of medical devices; because of its broad scope, however, its staff resources generally available for any one specific problem are quite limited. It does have several data-gathering mechanisms and data bases from which it can draw useful information about generic problems, and it often performs in-house laboratory and bibliographic research. Within staff and budgetary constraints, it also sponsors specific data-gathering studies by which it can obtain statistically valid epidemiological information. FDA has an extensive capability for publication of information, using its own sources as well as the public and professional media. Also, it has the capability to sponsor meetings of health professonals and manufacturers to discuss health-related issues. Such a convocation under government aegis minimizes restraint-of-trade difficulties for manufacturers.

Readily acknowledging the unique capabilities and limitations of each other, manufacturers, health professional organizations, and CDRH have been cooperatively addressing a number of medical device problems involving user error. CDRH involvement in these joint projects is coordinated by its Office of Training and Assistance. The following sections describe several current examples of these projects.

B. Hemodialysis

Hemodialysis is a critical care technology upon which more than 80,000 Americans depend for their lives. It is also very complex, providing many opportunities for human error, such as the air embolism incident mentioned earlier. There are many different kinds of user problems in dialysis, involving all segments of the hemodialysis system, including water treatment, preparation and handling of dialysate concentrate, the dialysate delivery system, extracorporeal blood components, and dialyzers. Problems in hemodialysis each year result in patient deaths and injuries, malpractice suits against the health professionals, and product liability suits against the manufacturers.

Since 1986, the major U.S. dialysis manufacturers, under the aegis of the Health Industry Manufacturers Association, have met periodically with representatives of the American Nephrology Nurses Association and the Renal Physicians Association, and staff of CDRH. This committee was formed to develop an educational program for dialysis nurses and technicians, operating under the recognition that many dialysis misadventures were caused by human error, not by any inherent faults with the equipment.

The committee's first project was an educational videotape program, "Human Factors in Hemodialysis," a free copy of which was sent to every dialysis center in the country. It was developed as one important means of raising the awareness of dialysis health professionals about the importance of some fundamental dialysis quality assurance procedures. Although it provided information that some dialysis personnel might not know, its primary purpose was to motivate dialysis nurses and technicians to utilize precautionary procedures whose significance they might not fully appreciate. CDRH provided technical information about hemodialysis problems and produced a tape with its in-house production studio; the manufacturers paid for all the out-of-pocket costs of the production, including the duplication and mailing; and the health professions provided the clinical expertise necessary to make the production and obtain professional endorsement of its messages.

The committee's second and third videotape educational projects address "Water Treatment in Hemodialysis" and "Infection Control in Hemodialysis." With the encouragement of this committee, CDRH contracted for the development of a comprehensive water treatment manual to be distributed at no cost to dialysis facilities along with the videotapes.

C. Anesthesiology

Anesthesiology is another medical device area in which user errors account for a large number of avoidable patient deaths and injuries. Published estimates of avoidable deaths from anesthesia-related incidents range from 2000 to 10,000 per year [5,6], not to mention the toll of patients with serious brain injuries. This tragic litany has resulted in anesthesiology paying very high malpractice premiums. Product liability suits plague the manufacturers, so much so that one of the three major U.S. firms ceased manufacturing gas machines. The profession has faced this intolerable situation head on, and has solicited the help of manufacturers and government to reduce the number of avoidable anesthesia deaths and injuries. For many years, the three groups have cooperated in addressing various use problems, jointly supporting various educational seminars, and producing a variety of educational documents. They participate in the Anesthesia Patient Safety Foundation, a nonprofit organization composed of anesthesiologists, manufacturers, government representatives, biomedical engineers, hospital risk managers and liability insurers, all dedicated to finding ways of improving patient safety in anesthesia. Through the Foundation, they support patient safety research and education programs.

Working with the anesthesia professions and the anesthesia manufacturers, CDRH prepared and published a "Preanesthesia Checklist," which was distributed to every anesthesiologist, nurse anesthetist, and hospital risk manager in the country. The proper use of a pre-use checklist, common in the aviation world, is being promoted within the anesthesia professional community and by manufacturers. Many anesthesia mishaps can be avoided if pre-use check-out of equipment becomes routinely and conscientiously practiced [7].

One of the most highly acclaimed projects by this unique coalition has been a series of videotapes covering various critical topics of anesthesia patient safety. Produced by CDRH under the clinical oversight and financial assistance of the American Society of Anesthesiologists (ASA), and with the technical assistance of the anesthesia machine manufacturers, the series covers topics that relate to reducing human error in anesthesiology, such as record-keeping, mishaps, postanesthesia care, adverse outcomes, monitoring, and maintenance. Under a drug manufacturer's subsidy, ASA distributes the videotapes without charge to essentially all the hospital anesthesiology departments in the United States. Since they are endorsed and promoted by ASA, the content of these tapes constitutes a de facto standard of care.

The anesthesia machine manufacturers have made considerable strides in equipment redesign and automation to lessen the likelihood of human error. In addition, several malpractice insurance carriers have instituted a variety of premium incentives to encourage the use of patient physiological monitoring devices (such as pulse oximeters and carbon dioxide monitors) during anesthesiology, in an effort to stimulate additional reductions in adverse patient reactions, and thus reductions in claims.

No doubt as a result of the focused attention paid by the anesthesia professions to the issue of patient safety, along with the interest and aid of the manufacturers, government, and others, the frequency of adverse incident claims has lessened [8]. As a result of this lessening of risk exposure, malpractice insurance carriers have lowered the rates they charge anesthesiologists in comparison to other high-risk specialties. For example, during the period 1986-1989, several malpractice insurance carriers lowered the anesthesiologist rate relativity factor from 5 to 3.5, thus significantly reducing liability coverage expense [9].

D. Extended-Wear Contact Lenses

The increased occurrence and association of corneal ulcers with the use of extended-wear contact lenses also has been the subject of cooperative problem solving by manufacturers, professional societies, and government. CDRH became concerned with a steady accumulation of MDR reports about corneal ulcers, and called a meeting in December 1985 of extended-wear lens manufacturers and representatives of the relevant eye care professional associations to discuss the problem. All the groups agreed that, at least in part, the problem involved misuse of lenses, including wearing lenses too long between removals for cleaning and disinfection, as well as other improper lens care procedures. All groups agreed to participate in efforts to identify risk factors for contact lens-related corneal ulcers, correct the problem of user error, and improve extended-wear lens materials. Since that meeting, a number of significant steps have been taken. First, CDRH produced educational materials for consumers and health professionals, shared its analysis of the adverse incident data bases, and participated in planning for the corneal ulcer epidemiology study. CDRH is also working to develop recommendations about those features of labeling and user education that improve user compliance with instructions for appropriate eye and lens care procedures.

Second, working with CDRH, several major lens manufacturers evaluated their labeling and made important changes to increase user comprehension and compliance. They also developed new educational materials, such as videotapes for eye care professionals. In addition, manufacturers are conducting research on new and modified lens materials designed to increase the passage of oxygen to the cornea, which is vital to corneal health. Increased corneal oxygen flow may reduce the risk of corneal infection leading to corneal ulceration. Finally, the Contact Lens Institute, an industry organization that represents the major research-oriented lens manufacturers, has sponsored a large epidemiological study of extended-wear lens users and corneal ulcer patients, in order to assist CDRH in determining and evaluating the incidence of corneal ulcers and the factors that influence their occurrence in extended-wear contact lens patients.

Third, the eye care professional organizations implemented many educational programs for their membership about the corneal ulcer problem, including educational meetings, articles in professional journals, and special mailings to their members. These efforts greatly improved the awareness of the health care community about the potential problems associated with extended-wear lenses and increased their sensitivity to the importance of patient compliance with specific measures necessary to reduce risks associated with extended lens wear.

All of these steps help to elevate the awareness in the professional community of the importance of strong patient education. In response to the increased patient education and monitoring of patient compliance by eye care professionals, patients are becoming more deliberate in their lens care procedures. It is anticipated that these behavioral changes will help to reduce the incidence of corneal ulceration and other serious eye infections, benefiting patients, health care professionals, and manufacturers.

E. Other Cooperative Projects

In similar fashion, CDRH also has stimulated cooperative projects with health care professionals and manufacturers involved with home-use ventilators, self-monitoring blood glucose systems, defibrillators, and central venous catheters. All of these problem areas involve user errors, which contribute to deaths and injuries, and which might very well be solved with a concerted and cooperative effort.

V. CONCLUSION

This multidisciplinary approach, involving small-group meetings of manufacturers, health professionals, and the government, to analyze user problems and implement cooperative behavioral solution strategies, is built on the premise that the solution of user problems requires a coordinated and cooperative effort of the major parties involved. This approach recognizes that each of these groups has a different perspective on any one device problem, and that the solution strategy must recognize each perspective. Its strength lies in the fact that a solution strategy, once devised, is simultaneously owned and committed to by each group. Consequently, each has a stake in its success, and each group is willing to work to promote it.

It is hoped that the examples presented here will serve as models of a successful and mature approach to the solution of problems compromising patient safety involving the use of medical devices. As the representatives of industry, medicine, and government work together for medical device problem solving, fewer patients will die or be injured in avoidable medical device tragedies, and liability insurance costs will be further reduced. In addition, these groups will be able to take justifiable pride in their contribution to achieve an improvement in medical device safety and the betterment of this nation's health care.

Those interested in more information about problem-solving efforts already underway or planned in the near future are encouraged to write to the Office of Training and Assistance (HFZ-200), Center for Devices and Radiological Health, 5600 Fishers Lane, Rockville, MD 20857.

REFERENCES

1. C. B. Inlander, S. L. Lovell, and E. Weiner, *Medicine on Trial.* Prentice-Hall, New York, p. 51 (1988).
2. R. A. Reynolds, J. A. Rizzo, and M. L. Gonzalez, The Cost of Medical Professional Liability, *J. Am. Med. Assoc.*, *257*:2776-2781 (1987).
3. Manufacturers, distributors and importers needing information about FDA's medical devices requirements are encouraged to call 1-800-638-2041 or write to Division of Small Manufacturers Assistance, Office of Training and Assistance, CDRH (HFZ-220), 5600 Fishers Lane, Rockville, MD 20857.
4. D. Gill, *MD&DI Magazine*, March, p. 8 (1988).
5. J. B. Cooper, Toward Prevention of Anesthetic Mishaps, in *Analysis of Anesthetic Mishaps*, E. Pierce and J. Cooper, eds., Vol. 22, pp. 167-183, International Anesthesiology Clinics (1984).
6. J. B. Cooper, R. S. Newbower, and R. J. Kitz, An Analysis of Major Errors and Equipment Failures in Anesthesia Management, *Anesthesiology*, *60*:34-42 (1984).
7. P. Carstensen, FDA Issues Pre-Use Checkout, *Anesthesia Patient Safety Foundation Newsletter*, *1*:13-15 (1986).
8. M. D. Wood, Anesthesia Claims Decrease, *Anesthesia Patient Safety Foundation Newsletter*, *1*:21-23 (1986).
9. E. C. Pierce, Editorial: Good News and Bad News on Safety, *Anesthesia Patient Safety Foundation Newsletter*, *2*:22 (1987).

10

Industry's Role in Standards Development

GEORGE T. WILLINGMYRE *American National Standards Institute, Washington, D.C.*

I. INTRODUCTION

A. Standards Taxonomy

Standards are important. They can impact our industry in many ways. Most of the standards activity relevant to the medical device, diagnostic product, and health care information systems industry falls into one or more of the four following molds: regulatory, national voluntary consensus, foreign national, and international.

Regulatory standards are those that generally have some basis in law—for example, the standards that FDA might issue to assure the safety and efficacy of a medical device.

National voluntary standards are the work products of groups such as the National Committee for Clinical Laboratory Standards (NCCLS), the Association for the Advancement of Medical Instrumentation (AAMI), the American Society for Testing and Materials (ASTM), the Institute for Electrical and Electronic Engineers (IEEE), Underwriters Laboratories (UL), and the National Fire Protection Association (NFPA).

Foreign national standards are like our own national regulatory and voluntary standards except that they are for other countries.

International standards are the attempts by countries to try to reduce the differences in national standards through organizations such as the International Standards Organization (ISO) and the International Electrotechnical Commission (IEC).

Each of these is significant in its own way:

Regulatory standards are straightforward—it is against the law to sell a noncompliant product.

Voluntary standards are a little more complex. Voluntary standards represent what a particular group can reach a consensus upon, be it safety or merely a desirable performance feature. Many of the voluntary standards in the United States today were prepared in the hope of substituting a voluntary standard for a mandatory regulation. They are also important to manufacturers, since the documents often wind up in local ordinances or electric codes or purchase specifications. They may also be useful in marketing. Products may be labeled and advertised as meeting the requirements of a particular standard.

Foreign national standards are just like the regulatory and voluntary standards in the United States, except they are applicable in the foreign country. These are relevant to anyone who wants to do business in that country. Examples include the standards published in the United Kingdom by the British Standards Institution (BSI), or Department of Health and Social Security (DHSS), or in France by AFNOR or the French Government, or in Germany by the DIN.

International standards are significant to manufacturers when they begin to appear as "foreign national" standards. That an international standard exists or is in development may not have a practical impact until the standard starts to be referenced as a foreign national regulatory or voluntary requirement.

Standards may also take the form of definitions or guidelines, design specifications, or performance requirements. There are standards developed with a clear objective and those—more than might be expected—with no clear purpose.

B. Standards Principles

There are three fundamental principles that industry participants need to keep in mind during standards development.

The first is, *Understand why the standard is being written.* Make sure that a standard has a clear purpose. There should be a written statement of need for every standard with which a company is involved. A clear statement of the need for a standard will not only simplify the standards writing process, but will protect a company's interests.

The second point is, *Understand why the company is involved with the standards.* An industry representative on a standards committee has two responsibilities: to contribute his industry expertise and viewpoint and to represent his company's interests. The representative cannot possibly represent effectively the views of his company if he does not understand why the company is involved. Is he participating because of a personal interest in the subject or because the company has a bottom-line concern?

The third point is, *Understand the potential consequences of the standard.* An industry representative should appreciate that the standard may turn up in places not anticipated, places like purchasing specifications and product liability suits. He must understand also that the FDA will be watching the standard and may use it as a source document for a regulatory standard. If a company supports work on a standard, it should make sure that it could live with it not only as a voluntary standard, but also as a regulatory standard.

II. THE VOLUNTARY STANDARDS ENVIRONMENT

People who work on medical device standards have been conditioned to think only in terms of voluntary and regulatory standards. While that may be a useful distinction in law, in practice, the distinctions blur. In fact, most standards fit into a great gray area. From a practical point of view it does not matter very much if a standard is labeled mandatory or voluntary. There are examples of regulatory agency standards being promoted as guidelines and voluntary standards being used as mandatory requirements.

A. What Is a Standard?

In the present environment it is best to think of standards in the broadest sense. Dictionaries list many definitions for the word "standard." Almost always, however, the first definition is "anything taken as a basis of comparison," or "anything taken by general consent as the basis of comparison." This dictionary definition covers the most common manifestations of standards as they are seen by manufacturers: guidelines, accepted practices, approved models, reference methods and materials, and specifications.

Manufacturers of medical devices, diagnostics, and health care information systems are sensitive to comparisons of their products with "standards." These comparisons occur in many different forms with varying levels of influence on the business success of a product.

The journals are full of "letters to the editor" and peer-reviewed articles making product comparisons or comparisons to requirements contained in published standards documents. The effect of these comparisons varies, based on the weight of the evidence and the prestige of their authors.

Standards have been written to cover almost anyting that could be imagined. Conveniently, these standards fit into a fairly small number of categories.

Standard definitions attempt to create a common language. An example of this is the document "Nomenclature and Definitions for Use in the National Reference System in Clinical Chemistry." This was prepared by a group that sought, as one of their first tasks, to develop a common way of describing their area of concern.

The *recommended practice* is a type of standard that describes an accepted procedure for doing something. An example of this is the AAMI recommended practice on sterilization and sterility assurance.

A *testing method* is simply an agreed upon way of measuring something.

A *standard classification* describes categories of objects or concepts. An example of this is the white-cell differential test in which white cells are categorized in a standardized fashion based on cellular morphology.

A *specification*, which is perhaps the most common of all standards, sets limits on the characteristics of a product or material. Specifications can be of two types: design and performance.

Design specifications may specify the ingredients of a product—the amount of plasticizers, for example, the dimensions of the product, or other physical or chemical attributes. Design specifications are important because they promote uniformity and compatibility. Design specifications, however, have a major drawback. By dictating the design of the product they limit a manufacturer's flexibility and may impede innovation.

A *performance specification* sets limits on the performance of the product. For example, if the issue under consideration is the strength of a particular component, a design standard might specify the material it should be made of. A performance standard would require only that it withstand a specified degree of force. The advantage of the performance standard in this case is that when new materials are developed they can be substituted for the earlier material. This could not be done if the standard specified the type of materials. Performance standards can be more intellectually demanding than design standards. The standards developers have to know what performance objective they are writing the standard for and why they have chosen a particular performance limit.

Frequently a standard will include elements from several of these categories. It is not unusual to find standards that incorporate both design and performance specifications.

B. Who Writes Standards?

In the private sector, the American National Standards Institute (ANSI) has 11,000 standards. Thousands of individuals are working on these standards. Many kinds of organizations provide forums for this work.

1. Single Interest Groups

Single interest groups, whether made up of just manufacturers or just users, frequently contain extraordinary expertise. Standards developed by them can be extremely useful. Occasionally these standards may suffer from a narrow point of view and may cause problems.

Several years ago a professional association decided to develop a standard for hypodermic needles. The members of the committee, all of one profession, decided for the profession that only two or three sizes of hypodermic needles were needed and developed specifications along these lines. The first time the industry became aware of this standard was when it was published. It turned out that the practitioners actually purchased seven to nine other sizes in substantial quantity. Therefore, the limitation of two or three sizes was not appropriate. Considerable effort was expended to convince the association that their standard was unnecessarily restrictive. The argument offered by industry, and finally accepted by the association, was that the standard was unacceptable because people were in fact buying and using the sizes that had been slated for extinction. This example demonstrates how easy it is for standards writers to lose touch.

2. Consensus Groups

In contrast to single interest groups, consensus standards groups draw on professionals, users, regulators, manufacturers, and all identifiably concerned organizations. Consensus groups include ASTM and IEEE, NCCLS, and AAMI. Standards developed by groups like these are influential because of the credibility that arises from this interaction. Most of the voluntary standards for medical devices are prepared by consensus groups.

3. Independent Groups

Independent organizations such as Underwriter's Laboratories (UL) offer a gamut of third-party product review and certification. They develop standards, test products, offer certification services and listing marks, and conduct compliance inspections. The major focus is safety. A number of states and localities in the United States require that institutions installing certain types of equipment must have the equipment checked or certified by a nationally recognized testing laboratory, such as UL. Institutions promptly incorporate their local ordinance requirements into their purchasing practices.

4. *International Organizations*

Thirty or more years ago, international work on standards was regarded mainly as the concern of industry. Published international standards mainly dealt with nuts and bolts, roller bearings, electrical components, and test methods for steel and other materials. These standards were designed to encourage interchangeability of parts and to facilitate and reduce the cost of manufacturing.

Currently, the scope of international standards activity has expanded to professional and governmental groups. Standards in some countries have been used as nontariff barriers. This has brought the realization, on a worldwide basis, that standards should be international.

Recently 103 nations, including the United States, signed the General Agreements on Tariffs and Trade. The GATT agreements are designed to assure that technical regulations and standards do not create unnecessary obstacles and technical barriers to trade.

There are many international standards organizations, both governmental and nongovernmental, that are active in the clinical/medical field. They include the International Standards Organization (ISO), the International Electrotechnical Commission (IEC), the International Organization for Legal Metrology (IOML), and the International Federation of Clinical Chemists (IFCC).

III. THE REGULATORY STANDARDS ENVIRONMENT

The procedures for FDA promulgation of medical device regulatory standards were established by Congress in the Medical Devices Amendments of 1976. Industry must be prepared to participate in this process. Theoretically, the Center for Devices and Radiological Health could start the regulatory standards process tomorrow for any Class II devices. On the other hand, FDA has classified far too many products in the standards category and will as a practical matter never be able to write standards for them all. To date the FDA has initiated the process for just five of the more than 1000 products classified in the Class II standards category.

A. The Regulatory Process

The list below is a simplified overview of the Section 514 regulatory standards process. The first three steps are specific to the regulatory standards process. The fourth and fifth steps involve traditional FDA rule-making procedure:

1. Opportunity to request reclassification
2. Solicitation of standards
3. Acceptance or development of standard

4. Proposal of regulatory standard
5. Final rule

The first opportunity for industry involvement is during step 1, request for reclassification to Class I. Anyone can suggest at this point that the product does not really need a standard at all. Companies will have only 15 days to respond with a request for reclassification. Fifteen days is a very short time to develop a reclassification petition.

The next step—the invitation to submit or develop a standard—is designed to give voluntary standards organizations an opportunity to submit standards they have developed or to offer to develop a standard expressly for the FDA. Actually, anyone—including medical device companies—can participate in this step, but most of the responses will probably come from voluntary standards organizations.

Following a 60-day comment period, either the FDA will accept one of the offered standards, or it will decide to develop one itself. While there is no formal opportunity for comment at this stage, the regulation clearly indicates that standards developers, including any federal agency, "shall provide interested persons an opportunity to comment in the development of the standard."

The next two stages—the proposed and final rules—are the same as for other FDA regulations. That is, a proposed regulation is followed after a comment period by a final rule. The FDA must allow a year for the final rule to become effective, unless it can demonstrate that an earlier date is necessary to protect the public health.

At the time of writing this chapter, Congress is considering legislation that would remove the first three steps in the process noted above. FDA has maintained that the existing system is too cumbersome and is the reason why FDA has not issued any standards yet for Class II products.

B. FDA Product Classification

In response to the requirements of the 1976 Medical Device amendments to the Federal Food Drug and Cosmetic Act, the FDA has evaluated the need for performance standards for virtually all medical devices. While the FDA's work was unique in that it methodically considered one device after another, the actual decisions of the classification panels were based on the same information which underlies voluntary standards. The process resulted in over 1000 products in the Class II category for standards.

FDA's analysis of thermometers is instructive. The classification panel determined that a performance standard should be developed for thermometers "to assure the safety and effectiveness of the device." The FDA established that the risks associated with thermometers include misdiagnosis because of a faulty reading and mercury poisoning from ingestion of mercury. The data used to determine these risks came from "panel members' personal knowledge of and clinical experience with the device." Certainly, misdiagnosis and mercury poisoning are serious problems that should not be treated lightly. What is striking, however, is that no data are offered to support the panel recommendation. One would think the thermometer is such a common device that such serious risks would have been documented and that there would be estimates on the magnitude of the problem. Yet no such documentation is provided; the FDA based its judgment solely on the panel members' experiences.

The panel members had substantial experience in health care and one cannot dismiss their recommendations lightly. Given the findings of this panel of experts, one might agree, as the FDA did, that general controls would be inadequate to assure the safety and efficacy of thermometers and that mandatory standards would be required to keep unsafe thermometers from the market. On the other hand, one might conclude that the risks to public health from thermometers are not that serious and that there must be more risky products than thermometers that should receive higher priority for regulatory standards attention. HIMA believes that there are far too many products in the Class II category than deserve to be there.

IV. THE PROS AND CONS OF INVOLVEMENT

ANSI's 1987 Summary Annual Report of Medical Device Standards Board Activities identifies over 700 voluntary medical device standards completed or under development. These standards cover everything from needles, syringes, and thermometers to diagnostic test kits, electrical safety, and laboratory computers. Some of these standards are clearly defined and cover only a specific device. Others, however, are so broad—on sterilization, for example—that they cover whole classes of medical devices.

Why are these standards written? Standards do not just spring up out of nowhere. They come about because someone perceives a problem that can be solved by the adoption of a standard.

A. Why Get Involved?

The simple answer to the question of why companies get invovled with standards is that they consider it in their best interest to do so. They find that the benefits outweigh the costs.

The costs of standards are readily apparent. The travel and salary costs of personnel assigned to standards writing are easy to measure. To know how much a company is spending on standards, a manager need only add up the travel vouchers and factor in the direct salary expense. In addition, there is no uncertainty about expenses incurred to redesign a product to comply with an unanticipated standard. Once such expenses have been incurred, they have a direct and immediate impact on the company.

Unfortunately, the benefits can be very difficult to quantity. How much is it worth to prevent the incorproation of a totally unreasonable requirement in a voluntary standard? How much is it worth to have a good standard?

Another difficulty in evaluating the benefits is the uncertainty factor. There is no guarantee that any voluntary standard will actually get through the approval process and become a final approved document.

So, in deciding on whether to participate in standards activities a company must compare definite and measurable costs to uncertain and intangible benefits. Sometimes the case will be so overwhelming that a decision will be straightforward. Frequently, however, the decision will not be that simple.

B. The Advantages of Participation

1. Forestall Mandatory Standard

For many companies, averting a regulatory standard is one of the greatest incentives for participating in a voluntary standards project. If a reasonable voluntary standard can be developed, then it is less likely that the FDA will feel it is necessary to produce a regulatory standard. This is really a defensive approach—writing standards to prevent government intervention. This is a legitimate reason to participate and, in fact, is one that the FDA supports.

However, this can be a two-edged sword. The FDA can use a voluntary standard as the basis for a mandatory standard. That is, it can take a voluntary standard and add what ever requirements it thinks necessary and propose them as a mandatory standard.

2. Reduce Uncertainty

Standards can be very helpful in reducing uncertainty. By setting performance criteria or specifications that a product must meet, standards remove some of the options and make the design process simpler. An architect does not decide on the width of the doors on a case by case basis. There are certain standard widths, and he just uses whichever are appropriate. However, there is only a fine line between simplifying the design process and creating unnecessary obstacles to innovation by institutionalizing archaic concepts.

3. Develop Good Standard

Sometimes companies are faced with the prospect of a standards activity that will proceed regardless of whether they participate. Since industry many times has the best information on standards issues, its contribution is invaluable in developing a good standard. In situations like this, a company can be forced to participate simply to protect itself from unnecessary requirements.

4. International Considerations

International standards have assumed increasing importance in recent years. These standards can affect U.S. companies in two ways. If products are sold outside the United States, international and foreign national standards can directly limit a company's ability to sell if the products do not conform. Second, because of the General Agreement on Tariffs and Trade, the FDA will be obligated to use existing international standards when it wants to issue a domestic standard, unless it has a good reason not to.

5. Product Liability

Adherence to a reasonable standard may mitigate against unreasonable or punitive damages in a product liability suit. Participation in standards development activity can ensure that standards are reasonable. If standards can be used as a "sword" against a manufacturer, they might as well be reasonable swords. User standards can help a manufacturer by establishing the qualifications of, and care to be exercised by, the user.

6. Purchase Specifications

Requirements included in standards (even draft standards) often find their way into purchase specifications and can influence customers' purchasing decisions. It is frustrating but true that orders may be lost simply because a company's product did not meet one of the proposed requirements in a draft standard.

7. Responsive Industry Group

A hidden benefit from standards work can be the development of a responsive industry group. There have been several occasions when companies working on standards issues have been confronted by other regulatory issues that were unexpected when the standards work began. The companies were able to build on the relationships established in the standards process in coping with these problems.

C. The Consequences of Not Participating

Many of these reasons can just as easily be used to dissuade a company from participating. For every company that feels it important to help write standards that might be used as purchasing specifications, there are others that feel they would only be helping develop standards that might be used against them.

In many cases the question is not how will participating in a standard help the company, but rather, how will not participating in a standard hurt a company. Standards work can be viewed, at least in some fashion, like insurance. It may be costly to have, but it may be more costly not to have it.

Each company must make its own judgements about the relevance of these factors. These judgments will be made for standards in general, as the company develops a philosophy on standards participation, and also in every standard which comes their way. They must select those of most importance.

V. COMPANY PREPARATION AND OBLIGATIONS

The level of involvement in standards activities depends on each company. The minimum involvement should be to name someone in the company to coordinate standards activities. It would be wise to consider memberships in the voluntary standards groups that focus on the company's products. Such memberships will ensure that the company standards coordinator is informed of relevant standards activities.

A. Establishing a Policy, Approach, and Strategy

Each company, regardless of size, should establish a company policy with regard to standards activities. This policy should result from a conscious decision. The options are limitless. The simplest option is to keep informed but not participate. The minimum active option is to review each draft standard and return it with hand-written marginal comments. Other options vary from this to full participation in domestic and international standards activities, including chairing standards writing groups.

As each new standards project begins, a company should decide on a specific approach. For example, a company might elect to take a generally passive role in some projects, but when a standard is initiated involving the company's special expertise it might elect to actively participate.

The minimum active strategy is to review and comment on draft standards directly affecting a company's products. Comments should be prepared in the format prescribed by the standard writing group. In general, comments should reference the specific portion of the standard affected and indicate what should be added, deleted, or revised. Any recommendations for revision should be accompanied by suggested wording. And each comment should be followed by its rationale.

Very small companies may not be able to participate to this extent. For these companies, although it places a burden on the committee's comment coordinator, participation can be as minimal as reviewing the draft standard and marking it up with handwritten comments. The key is to make the company's views known in a credible way.

B. Selecting a Representative and Standards Development Tactics

If a company elects to actively participate in a standards writing group, it must select its representative with care. The representative must have a good knowledge of both the product and the company. The representative should be able to act in a professional manner with peers, customers, regulators, and other professionals. Finally, the representative must have the commitment of his company to support his involvement.

The better the representative knows the consensus process and the procedures of the standard group, the better chance there is for effective participation. Most standards groups have written procedure, which they will supply. They are not all the same.

Standards participants should understand the specific problem to be addressed by the standard. The scope of the standard and the rationale should be based on the problem to be solved. Problem definition studies, FDA Enforcement Reports, FDA Classification Panel Reports, and internal records like company complaint files are useful sources of information about problems with medical devices.

If a standards project has potential to affect significantly a company's business, the representative should be directly involved in writing the draft of the standard. Early input is very effective.

When the representative receives a draft standard, he or she should note the date by which comments are requested. The company response should be submitted before the deadline. When others in the company are also commenting, there should be follow-up to assure that their comments are available on time and are properly integrated into one response.

Proposed and final drafts of a standard will be circulated for ballot. Consider carefully any "no" vote. If the standard will affect a company's business unreasonably, a "no" vote is appropriate. If there are significant reservations, a "no" vote is reasonable. A "no" vote is a powerful tool and can delay or halt the standards process. All "no" votes should be explained. A "yes" vote "with comment" can be effective in handling reservations of less significance.

C. Effective Participation in Meetings

The first thing a company's representative to a standards committee meeting should do is to establish his own credibility and credentials and those of his company. A representative can do this informally by introducing himself to all the members of the committee and explaining why the company is interested in the activity.

People have many different perspectives and goals, and the more sensitive the representative is to them, the more successful he will be in not only resolving the problems but moving the discussion along efficiently. For example, it is not unusual in this era of product liability suits that a medical professional will be sensitive to statements that indicate the "cause of the problem" is medical technique or that the "problem can be corrected by proper technique" rather than by design specifications for the product. Government representatives and industry representatives have their own interests.

A fixed and rigid position that ignores reasonable arguments reduces credibility and makes it difficult to write a good standard. A certain amount of diplomacy is needed, as well as respect for others' viewpoints. While vigorous disagreement is sometimes appropriate, it should never be handled in a hostile manner.

An effective representative identifies the commitee's real leadership. While this may sound easy, often the chairman is not the real leader. This person or group can be recognized by the fact that the other members defer to them. Convincing the leadership of a position may sway the entire committee.

It is appropriate to form alliances with individuals having common viewpoints. Often this is not an alliance of only manufacturers or users but a mixture of these gorups. Alliances are sometimes based on personal credibility, company reputation, and people's familiarity with the company's product.

The representative should never promise to do something if he is not willing to fulfill the commitment. He must remember that if he has an assigned responsibility, the entire committee is relying on him.

Finally one should understand the ramifications of the standard being developed. It may have a significant effect, not only domestically but also internationally. It may affect future product development. The representative should know the difference between performance and design requirements, product versus user standards, and he must always remember that the words "may," "should," "shall," and "must" mean different things and that the phrase "when appropriate" is not only very useful but very often "appropriate."

The list of things to do to prepare for the standards process and the obligations involved is long. Fortunately, however, many of them can be included under the terms "common sense" and "courtesy."

VI. CONCLUSIONS

Harking back to the introduction of this chapter, it is clear that standards activities are important to the medical device, diagnostic product and health care information systems industry. Three key principles can guide industry's involvement in such projects to assure that the work is productive:

1. Understand why the standard is being developed.
2. Understand why the company is involved with the standard.
3. Understand the potential consequences of the standard.

This overview of the standards environment is intended to help industry participate most effectively in the standards development process. The concepts and principles embodied in this chapter are derived in large measure from a Standards Educational Program prepared by the Health Industry Manufacturers Association Standards Section (now Science and Technology) in 1984. The author acknowledges the key role of Erv Taylor, Mort Levin, Edward Kimmelman, John James, Richard Flaherty, Michael Miller, Jim Dugan, and Henry Wishinsky in contributing to the 1984 effort.

11

Role of NCCLS and ECCLS in the Voluntary Consensus Process

JOHN V. BERGEN *National Committee for Clinical Laboratory Standards, Villanova, Pennsylvania*

EDWARD R. KIMMELMAN *E. I. du Pont de Nemours & Company, Inc., Wilmington, Delaware*

I. INTRODUCTION—WHAT ARE NCCLS AND ECCLS?

The National Committee for Clinical Laboratory Standards (NCCLS) and its European counterpart, the ECCLS, are private-sector, nonprofit, educational organizations dedicated to the development and promotion of voluntary consensus standards and guidelines for clinical laboratory testing [1].

NCCLS is the only U.S. organization dedicated solely to clinical laboratory testing that is accredited for these purposes by the American National Standards Institute (ANSI). ECCLS employ roughly the same operating concept and procedures as the NCCLS and serves as an umbrella organization, coordinating the standards development work and encouraging adoption of its standards and guidelines through national "CCLS" groups and its members in the various countries in Europe. A third major "CCLS" is under development in Japan.

The standards, guidelines, and other related materials address technical matters in the scientific disciplines that make up clinical laboratory testing (clinical chemistry, hematology, immunology, microbiology, toxicology, and drug monitoring). They include descriptions of reference methods and materials, bench methods, performance standards, evaluation protocols for test methods and instruments, and procedures for specimen collection and quality control. In addition, guidelines have been published on various aspects of laboratory administration and safety, including labeling and inventory control, the shipment of potentially infectious specimens, laboratory design, and protection of laboratory workers from blood- and tissue-borne infection.

This chapter focuses primarily on NCCLS. Further information as to membership and operations of ECCLS and JCCLS can be obtained by contacting their headquarters.

A. Background and History of NCCLS

NCCLS was founded 20 years ago when clinical laboratory practice in the United States was in a state of profound change. Biomedical measurement systems were beginning to move from "strictly manual" methods to "automated instrument" methods, resulting in major redesign and relocation of clinical laboratory testing venues, and new types and levels of communication were required between the personnel doing laboratory testing and the clinicians using the test results. New relationships were emerging among the government agencies responsible for public health, the manufacturers of health-care products, and health-care practitioners, including laboratory professionals. An increasingly mobile population gave rise to the need for consistency of national laboratory performance to produce accurate biochemical measurements.

The changes, for the most part, were occurring in the absence of standards. Standards, where they existed, were developed by manufacturers or by laboratory professional societies, separated often by boundaries of discipline and potentially conflicting vested interests. An independent, national, interdisciplinary body that could assure broad scientific scrutiny and could confer the imprimatur of the entire clinical laboratory community did not exist.

In the fall of 1967, 31 clinicians and laboratory scientists representing 15 organizations met to discuss ways of "improving what we are doing for patients" by providing a formal structure for laboratory standardization, a national structure that would coordinate existing projects. Under the leadership of Dr. Russell Eilers, College of American Pathologists, the founders of NCCLS developed a statement of purposes that guides the organization to this day:

To promote the development, evaluation, approval, and implementation of national and international standards for the clinical laboratory field in order to maintain national laboratory performance at levels consistent with the needs of patient care.

To provide a mechanism for approval of clinical laboratory standards based upon a consensus of all parties interested in the scope, provision, and use of such standards.

To cooperate with other national and international organizations in developing and approving clinical laboratory standards, promoting their use, and evaluating the effectiveness of standards followed in practice.

To maintain a forum for communication and mutual education among professional, governmental, and industrial organizations concerned with the quality of clinical laboratory performance and the relationship of standards thereto [2].

In the 20 years since its founding, NCCLS has published more than 200 useful standards, guidelines, and other publications, and currently has over 130 active projects. NCCLS has grown from an original group of 31 interested people to an association with over 1000 member organizations and almost 1200 volunteers participating in its scientific projects and helping to manage the organization. Essentially all manufacturers, professional associations, and governmental agencies having a close interest in the quality of clinical laboratory testing in the United States are members of NCCLS or participate in development of NCCLS standards.

In 1977, NCCLS was accredited by the American National Standards Institute (ANSI) as a national standards organization for clinical laboratory sciences.

As a result of a Conference on a National Understanding for Development of Reference Materials and Methods for Clinical Chemistry, hosted by the U.S. Centers for Disease Control (CDC) in Atlanta in 1977, NCCLS became the home of the National Reference System for the Clinical Laboratory (NRSCL).

NCCLS served as the model for the ECCLS when it was founded in 1979. In 1985, NCCLS was designated a World Health Organization Collaborating Center for Clinical Laboratory Standards. In this capacity, NCCLS is disseminating information on its voluntary standards and guidelines to health-care organizations in the Pan American region.

B. Membership

NCCLS is an organization of organizations. There are no personal or individual memberships. Membership is divided into two basic categories: active and corresponding. There are subdivisions within the two basic categories (refer to Table 1).

C. Future Direction

As it has done in the past, NCCLS will anticipate and adjust to the environment affecting clinical laboratory testing. In 1988, NCCLS published a strategic plan based on an analysis of eight of these major environmental factors (health-care finances, laboratory organization, legal climate, new technologies, testing quality requirements, regulatory requirements, products and services desired from a consensus standards organization, and the societal implications of laboratory testing) [3].

The plan outlines 11 overall strategic objectives that recognize a significant percentage of clinical laboratory testing is moving from the central hospital or reference laboratory to decentralized, often less sophisticated testing sites. As a result, the plan calls for NCCLS to focus future attention on tailoring its current standards and guidelines to facilitate their use in these new settings, and to use these standards and guidelines as the basis of education and training programs designed to be presented at the local level.

With the heightened concern over the medical and societal implications associated with the transmission of infectious diseases, NCCLS will focus increased attention on guidelines designed to assure a safe laboratory work environment and the protection of the privacy of individuals being tested.

The strategic plan recognizes that NCCLS can assist laboratories by providing tools that facilitate efficient cost management.

The plan calls for NCCLS to look for alternative sources of income in addition to the traditional sources. Table 2 contains the complete list of strategic objectives.

II. HOW DOES NCCLS OPERATE?

A. Organization

The NCCLS is organized to facilitate member control of administrative policy and member management of the standards and guidleines development programs.

TABLE 1 Categories of NCCLS Membership

Active membership in NCCLS is open to any public or private academic, scientific, professional, governmental, or industrial organization, corporation, or agency based in the United States. Included in this category are professional associations, state and federal governmental agencies, corporations, and trade associations. Active members participate through their delegates in the balloting on NCCLS standards and the governing of the NCCLS organization.
In recent years, NCCLS has added a subcategory of *associate active member* for educational or health-care institutions that wish to vote on NCCLS scientific matters.
Corresponding membership is available to individual U.S. institutions such as hospitals, clinics, or medical organizations, state or local professional associations, and governmental agencies. International professional associations, governmental agencies, companies, and institutions may also become corresponding members. Corresponding members receive NCCLS standards on which they are encouraged to comment; the NCCLS UPDATE newsletter; Just for Corresponding Members letters; and special reports on clinical laboratory testing issues.
Sustaining membership is available to organizations that wish to provide financial support to NCCLS over and above their appropriate dues level. These organizations customarily provide a sustaining contribution of $5,000. The sustaining funds are accepted by NCCLS to assist the organization in carrying on its ongoing standards program or special projects. NCCLS encourages all organizations to consider whether sustaining membership is consistent with their long-range goals. Sustaining members are entitled to several additional copies of the NCCLS Library of Standards.

The NCCLS Board of Directors and its Executive Committee are comprised of a balanced representation from the three major constituencies of the organization: practicing professionals, government, and industry. The Board, with the assistance of various standing and advisory committees and special task forces, sets organization policy, approves major expenditures of funds, and reviews standards and guidelines prior to publication to ensure that the consensus process is being properly implemented and that the publication content meets NCCLS objectives.

The NCCLS area committees manage the standards and guidelines development programs: identifying new projects, setting priorities, reviewing developed documents to assure the validity of their content, and administering the costs associated with development of the projects under their control. Subcommittees and working groups reporting to the area committees actually write the standards and guidelines and revise them based on the broad review obtained during the consensus process.

The NCCLS National Office manages the day-to-day operations of the organization, facilitating the standards development process, the consensus process, and the production, printing, sale, and distribution of NCCLS publications. The National Office manages the daily finances, and the various communication media used to keep the membership and other interested parties abreast of NCCLS activities.

TABLE 2 NCCLS Strategic Objectives

- Adapt the consensus process to respond quickly to immediate needs or rapid changes in technology, while continuing to assure careful consideration of input from all interested parties. Balancing the need for timely response and careful review of input is a major concern of NCCLS.
- Modify NCCLS standards, guidelines, and other "products" for application in all testing sites. These standards, guidelines, and other "products" are intended to advise, assist, and serve as tools for professionals in the delivery of quality health care.
- Determine future new constituencies (e.g., the patient, the home tester) and their potential involvement in NCCLS.
- Broadly support education in the significance and use of clinical laboratory tests.
- Provide standards and guidelines that aid in cost accounting of clinical laboratory tests and serve as tools for measuring test utility.
- Provide guidelines to ensure the privacy of the individual being tested.
- Provide guidelines to ensure a safe work environment for laboratory personnel.
- Develop and implement alternative methods of presentation (in addition to printed documents) for NCCLS products, including computer programs and audiovisual formats.
- Identify, examine, and obtain funds from new sources for NCCLS projects and activities, recognizing that organizations and corporations not traditionally involved in NCCLS but concerned with health care will be interested in participation.
- Develop an organized, systematic approach to continuously review environmental factors that impact on NCCLS direction, operation, and strategy.
- Assure that accomplishment of these strategic objectives and the growth embodied in them will not alter the financial stability of NCCLS or jeopardize its ongoing operations.

Table 3 outlines the organization of the NCCLS.

B. Types of NCCLS Publications

NCCLS publications fall into three basic categories, depending on their intended use, the need for NCCLS "consensus," and the likelihood that NCCLS consensus can be achieved on the issues addressed in the publication:

1. *Standard*: A publication developed through the NCCLS consensus process that clearly identifies specific, essential requirements for materials, methods, or practices for use in an unmodified form. A standard may also contain clearly defined discretionary elements.

TABLE 3 NCCLS Organization

BOARD OF DIRECTORS

EXECUTIVE COMMITTEE. Composed of the NCCLS officers and the NCCLS Executive Director. The Executive Committee acts on behalf of the Board with the Board's specific authority.

AREA COMMITTEES. Composed of a numerically balanced membership of representatives of the three major constituencies of the NCCLS having expertise in a particular scientific area. There are area committees dealing with the following scientific and technical areas:

- Alternate Site Testing
- Clinical Chemistry
- Evaluation Protocols
- General Laboratory Practices
- Hematology
- Immunology and Ligand Assay
- Instrument Systems
- Microbiology
- Toxicology and Drug Monitoring

A council manages the National Reference System for the Clinical Laboratory.

STANDING COMMITTEES. Composed of experts in the consensus process, and other nontechnical areas valuable to the Board in managing the NCCLS. There are standing committees in the following areas:

- Constitution and Bylaws
- Finance
- International Relations
- Nominations
- Presidents
- Program
- Standards Management

TASK FORCES. Composed of experts in a specific technical or administrative area. The task force remains in existence only until the task is completed. Examples of task forces are:

- Task Force on the Protection of Laboratory Workers from Infectious Disease
- Task Force on Decentralized Laboratories
- Task Force on Document Use

EXECUTIVE DIRECTOR. The highest level employee of NCCLS, the executive director manages the operations of the National Office. The National Office staff includes personnel responsible for:

- Administrative liaison with area committees and subcommittees
- Member services
- Communications
- Financial systems
- Office management
- Secretarial support

2. *Guideline*: A publication developed through the NCCLS consensus process describing criteria for a general operating practice, procedure, or material. A guideline may be used as written or modified to fit specific needs.
3. *Committee report*: A publication that is not subjected to consensus review, but that the NCCLS Board of Directors agrees to publish for informational use.

C. Consensus

NCCLS "consensus" is the result of faithful adherence to a documented and established "notice and response" process. The objectives of the process are to assure the broadest possible review of an NCCLS publication by interested participants; complete, considered response to all substantive comments, either by modifying the publication or by explaining why a comment did not result in revision of the publication; and determination by vote that a strong majority of participants "approve" the publication.

Approval of a publication under the NCCLS consensus process connotes neither unanimous agreement with everything that is in the publication, nor unanimous acceptance of the publication by all the participants voting. It is more a statement that the publication meets its stated objectives closely enough that most of the people affected by it are "willing to live with it." Unanimous agreement and acceptance are in almost every case unlikely or impossible to achieve.

The complete consensus process involves publication at three levels, the "proposed," "tentative," and "approved" levels.

1. *Proposed level.* At the proposed level, the publication is intended to elicit as much comment as possible from the laboratory community. It is labeled for review and comment only. Comments are especially important at this level, because they alert the NCCLS to unforeseen difficulties in the application of the contents under laboratory conditions. Readers are strongly urged to send their comments in any form to NCCLS. The official comment period on a "proposed level" publication is 60 days from the date of publication, although comments are accepted up to the time the responsible committee meets to address the comments received. All substantive comments must be considered when the document is revised for publication at the tentative level.
2. *Tentative level.* At the tentative level, the publication is revised to address comments that the revision committee agrees are pertinent. Because all comments on the proposed level publication might not have been accepted, the tentative publication contains a summary of comments not leading to change and the committee's responses to them. Anyone who submitted a comment, therefore, should be able to find a change in the tentative level publication, or a negative response accompanied by an explanation from the committee. The tentative publication is intended for field testing in the laboratory and comment based on this testing. The official comment period is 1 year.
3. *Approved level.* After the 1-year comment period, a review committee considers the comments received on the tentative level publication and any new information that may affect the acceptability of the publication. Revisions are made where appropriate, and comments not leading to change

are again responded to. The publication is submitted to the NCCLS active membership for a vote to "approve or disapprove." At least two-thirds of the votes received must be "affirmative" for the document to be moved to the "approved" level. An "approved" publication is usually reevaluated after 3 years, or sooner if changes occur that affect the usefulness of the information contained in it.

There are variations on the full consensus process to ensure sufficient comment on publications that are targeted at a very limited segment of the clinical laboratory community, and to ensure efficient handling for publications that do not require the field testing obtained during the tentative stage of the full consensus process. These variations are described in the NCCLS administrative procedures [4].

D. Communication

The fact that NCCLS membership includes organizations from government, the practicing professions, and industry would be of little value unless it was coupled with a strong communication program. The consensus process outlined earlier is, of course, the heart of the communication effort. There are others also.

NCCLS UPDATE is a periodic newsletter designed to inform members and other interested people about NCCLS projects and the individuals participating in them. UPDATE also includes articles on NCCLS programs and those of other organizations that may have a significant effect on the clinical laboratory community. UPDATE has a circulation of over 2500.

Periodically, the members receive "Just for Delegates" and "Just for Corresponding Members" letters. These letters are designed to inform the active member and corresponding member organization of items and issues for which the NCCLS leadership wants feedback. In order to facilitate the feedback, the letters are accompanied by "Just from Delegates" and "Just from Corresponding Members" response forms. The letters are usually sent just after Board of Directors meetings. This ensures that the members are getting news while it is still fresh, and Board members are getting timely feedback.

During the year the NCCLS sponsors a number of regional breakfast or luncheon forums for delegates and other interested people. The forums, usually held in conjunction with major professional association national meetings, are led by members of the Executive Committee and the National Office staff. The attendees have the opportunity to hear the latest news, to respond to questions posed to them, and to provide project and program recommendations to NCCLS.

The NCCLS Annual Meeting in the early spring is another part of the organization's communication program. In addition to a major forum on a key issue affecting the clinical laboratory, the meeting features subject-specific breakout sessions, and communication forums targeted for delegates, corresponding members, and committee volunteers.

From time to time, NCCLS, itself or in conjunction with other interested organizations, sponsors meetings to discuss issues important to the clinical laboratory and to the health of the general public. Examples are meetings related to:

Standardization of enzyme measurement

Standardization of cholesterol measurement as a key element of the NIH national program to reduce the level of cholesterol in the general public
Development of a clear understanding about laboratory measurement accuracy and precision, and laboratory quality control

IV. WHAT ARE THE BENEFITS OF PARTICIPATION IN NCCLS?

A. General

NCCLS provides to the clinical laboratory field in general a communications and consensus-generating/review mechanism that is scrupulously administered and inherently free from "interest group" pressure.

That fact was clearly recognized in the fall of 1977, at a meeting of governmental, practicing professional, and industry representatives—the Conference on a National Understanding for the Development of Reference Methods and Materials for Clinical Chemistry, hosted by the Centers for Disease Control (CDC) in Atlanta, GA. The meeting had as its purpose to discuss the organization of a system for developing such reference methods and materials. The single major point of agreement was that NCCLS should serve as a home for the National Reference System for Clinical Chemistry [5].

The "National Reference System" was organized within the NCCLS structure and has, in fact, grown to include the development and "credentialing" of reference materials and methods for disciplines in addition to clinical chemistry. Its current name is the National Reference System for the Clinical Laboratory (NRSCL).

NCCLS provides the opportunity to be in the forefront of clinical laboratory standardization, to recommend and receive early information on new standards and guidelines to be developed, and to directly influence their content through participation (as a member, advisor, or observer) in the subcommittee writing the standard.

In June 1987, the NCCLS Board approved the start of a project to develop guidelines for the protection of laboratory workers from blood- and tissue-borne infection. Under the inspired and effective leadership of Dr. Stanley Bauer of Bronx Lebanon Hospital, a committee of over 30 experts in infectious diseases and infection control, representing all areas of the NCCLS membership, and a subcommittee drawn from this group, wrote M29-P, "Protection of Laboratory Workers from Infectious Disease Transmitted by Blood and Tissue." The guideline was published at the "proposed" level in less than 6 months after the authorization of the project—a record for a publication of this size and scope. The organizations involved in this effort unanimously praised Dr. Bauer and NCCLS for the quality and efficiency of this effort. The drafting technique used by Dr. Bauer and the committee was uniquely effective and will be used in the future on projects of similar breadth.

B. Government

NCCLS provides a forum for representatives of government to meet on "neutral ground" with industry and the practicing professions. For regulatory agencies, the NCCLS consensus process offers a reasonable alternative to the expensive process of developing mandatory standards. For all governmental agencies, NCCLS provides a mechanism for disseminating and encouraging the use of important educational and program information. Many Food and Drug Administration (FDA), Health Care Financing Adminsitration (HCFA), National Institutes of Health (NIH), National Institute of Standards and Technology (NIST, formerly the National Bureau of Standards), and CDC personnel participate directly on NCCLS scientific and administrative committees.

Development of the NCCLS M29-P (Protection of Laboratory Workers) guideline has assisted laboratories in implementing CDC's recommendations on universal precautions for handling specimens. The NRSCL Council's credentialing program has raised awareness of the work of NIST in developing definitive and reference methods, and providing Standard Reference Materials.

Several other NCCLS projects provide good examples of how governmental objectives are being achieved without imposing expensive and unnecessary restrictions on the groups being regulated.

M22-T, "Quality Assurance for Commercially Prepared Culture Media," developed with the input of participants from the federal government, along with experts from industry and professional laboratories, has allowed HCFA to reduce significantly the level of quality control testing of commercially prepared culture media. This reduced the cost of using these media without risk to the public.

Using standards, methods, and procedures described in M2-A3, M7-A, and M11-A, the NCCLS Subcommittee on Antimicrobial Susceptibility Testing administers a review process that results in an annual updating of quality control tables to include data on the latest approved drugs. The tables provide information on zones of inhibition and interpretive breakpoints. The subcommittee review takes the place of a process that previously was performed by the FDA. FDA now relies on the NCCLS information.

One of the key issues related to home-use diagnostics is labeling. Organizations representing both government and the general public have expressed concern about the understandability of labeling for these products. NCCLS is currently developing guidelines in this area, with the participation of FDA personnel and representation from manufacturers of home-use products.

Voluntary standards developed by NCCLS have helped maximize the beneficial effect of government efforts. Both manufacturers and practicing professionals realize that voluntary standards and guidelines can be used as the source documents for regulations. Both also understand that the voluntary consensus process gives them an opportunity for early and considered input; it would be very difficult for a regulatory agency to promulgate a regulation that differed in any significant way from a voluntary guideline on the same subject developed with such broad input.

C. Industry

As mentioned earlier, NCCLS standards and guidelines in some cases have eliminated the need for government regulation. In addition, these standards have resulted in the down-classification of certain diagnostic products from Class III, requiring premarket approval, to Class II, the standards classification. Recently, NCCLS developed a standard related to differential leukocyte counting systems. Based on this standard and the data NCCLS generated in developing it, the Health Industry Manufacturers Association (HIMA) convinced the FDA to make just that kind of down-classification for this product.

Participation in NCCLS provides industry representatives with a number of other benefits. It puts them in direct contact with experts in all aspects of the manufacture, use, and regulation of diagnostic products. In addition to the professional enrichment that results from such contacts, insights are gained on more effective premarket testing of new products and more efficient processing of regulatory reviews.

The NCCLS Area Committee on Evaluation Protocols has developed a series of documents that establish recommended protocols for manufacturers to test diagnostic products for the purpose of developing performance claims, and for purchasers of such products to confirm that product claims are being met in their labs. Uniformity in user testing helps manufacturers anticipate the nature and extent of such testing and helps the manufacturer advise its customers about valid and useful protocols.

D. Practicing Professionals

The practicing professional is the primary beneficiary of NCCLS standards development efforts. NCCLS "bench level" standards and guidelines are designed specifically to be effective tools for training laboratory personnel, evaluating new and existing methodologies in the laboratory, and assuring consistent, high-level laboratory performance.

NCCLS gives to practicing professionals an influential voice in developing the standards and guidelines for their practices. Through subcommittee participation and as reviewers and evaluators, practicing professionals provide the input that leads to practical, easily implementable consensus standards and guidelines.

Over the years, NCCLS has developed a series of 10 guidelines for collection and handling of blood specimens that have been used successfully for training laboratory personnel. Among the areas covered are collection by venipuncture, skin puncture, and arterial puncture; handling and transport of diagnostic specimens and etiologic agents; and collection, transport, and preparation of blood specimens for coagulation testing.

The series of evaluation protocols mentioned earlier is designed to assist laboratories of all levels of sophistication in evaluating the performance of products in their own settings. These protocols range form the confirmation of precision, accuracy, interference, and linearity claims with high levels of statistical confidence to "quick and dirty" preliminary evaluation of product performance. Additional projects are under consideration for evaluation of other performance variables.

There are NCCLS guidelines designed to be used as tools in the administrative management of the clinical laboratory. One provides guidance in the development of the laboratory's procedure manual. Another assists in the implementation of inventory control systems in the lab.

Recently, NCCLS published comprehensive physician's office laboratory guidelines, which provide a framework of recommended laboratory practices that have been specifically tailored to the office laboratory. The guidelines contain information on laboratory design, choosing a referral laboratory, labeling, handling and storage of specimens, procedure manuals, laboratory safety, quality assurance, and other subjects of interest to the physician who is interested in setting up an office laboratory or operating an existing one.

REFERENCES

1. The National Office of the NCCLS and the headquarters offices of the ECCLS and JCCLS are:

   National Committee for Clinical Laboratory Standards (NCCLS)
   771 E. Lancaster Avenue
   Villanova, PA 19085
   (215) 525-2435

   European Committee for Clinical Laboratory Standards (ECCLS)
   Central Office
   University Hospital
   S-221 85 Lund
   Sweden

   Japanese Committee for Clinical Laboratory Standards (JCCLS)
   2-2-14 Daita Setagaya-ku
   Tokyo, Japan 155
2. NCCLS Bylaws, Article II, Section 1.
3. NCCLS Strategic Plan, 1988-1993, "Preparing for Our Future."
4. NCCLS Adminsitrative Procedures, Sections 7.0 and 8.0.
5. "A National Understanding for the Development of Reference Materials and Methods for Clinical Chemsitry," ed. J. H. Boutwell, AACC, 1978. Library of Congress Catalog Card Number 78-60712, ISBN 0-91527.

12

Role of AAMI in the Voluntary Standards Process

MICHAEL J. MILLER and ELIZABETH A. BRIDGMAN *Association for the Advancement of Medical Instrumentation, Arlington, Virginia*

I. INTRODUCTION

Voluntary standards accomplish a number of important objectives in the medical device field, and the Association for the Advancement of Medical Instrumentation (AAMI) has contributed uniquely and materially to the accomplishment of these objectives.

Standards provide a common language for important communications among device developers, government, and users. This common language enhances device development, evaluation, improvements, safety, efficacy, and appropriate use.

Standards are an important medium of education and information in that they provide the user and manufacturer with fundamental elements of device safety, performance, and testing. They are an important medium for reconciling the medical technology expectations of the developer, user, and regulator.

II. WHO IS AAMI?

AAMI is a membership organization of 5000 technology developers, users, and managers who are united by a desire to share their considerable knowledge to improve their development, management, and use of medical technology. AAMI's members—engineers, technicians, physicians, scientists from many disciplines, managers, academicians, and researchers—represent a variety of institutions, including manufacturers, health care facilities, universities, government agencies, and research facilities.

AAMI serves its membership and the field by providing standards on equipment safety, performance, development, and use; other publications; educational programs; and a number of other services that foster effective equipment development, management, and use. AAMI initiated and now sponsors certification programs for health care engineers and technicians.

Although AAMI is well known for many of its services (such as its annual meeting and exhibit program) and publications (such as its journal, *Medical Instrumentation*, and magazine, *Biomedical Technology Today*), it has made a unique contribution to medical technology through its standards, which provide baseline safety and performance requirements for many types of widely used technologies. A listing of AAMI's standards is provided as an appendix to this chapter.

AAMI's efforts have many milestones and this chapter describes several of the most important.

III. 1969 BETHESDA CONFERENCE

Legal and regulatory action in the 1960s appeared to be leading to a regulatory structure that was inappropriate for medical devices. Court decisions confused the distinction between devices and drugs with respect to the authority that the Food and Drug Administration (FDA) had for regulating medical devices.

There was legitimate and significant concern that devices were to be regulated like drugs, which could have adversely affected device development. As is now widely recognized—in great part due to a 1969 AAMI conference—devices evolve from different types of developers and technologies and demand a very different development process and a user with a different background and experience than drugs. For these and other reasons, devices demand a different type of regulatory structure. We note in passing that due to the efforts of organizations such as AAMI, most devices are *not* now regulated like drugs (i.e., are not subjected to onerous premarket approval procedures).

The objective of the 1969 conference held in Bethesda, MD [sponsored and funded, in part, by the National Heart Institute of the National Institutes of Health (NIH), Department of Health, Education, and Welfare] was to define the important differences between device and drug development and use, to emphasize how these differences require a different regulatory scheme, and to outline a regulatory structure for devices or the means by which a structure could be developed. Over 100 prominent physicians, engineers, government officials and industry leaders participated in the conference.

The conference report [1], which was published in nearly 40 scientific journals, became a major basis for the Cooper Committee report and, as a result, the Medical Device Amendments of 1976. The Cooper Committee was established by the Department of Health, Education, and Welfare (now Health and Human Services, HHS) to evaluate the need for and to make recommendations on an appropriate structure for device regulation. It was chaired by Theodore Cooper, M.D., then Director of the National Heart Institute, NIH.

The 1970 Cooper Committee report recommended classification of devices into three regulatory categories: those subject to general controls, those subject to standards, and those subject to premarket approval of device safety and efficacy. Congress adopted this regulatory approach when it approved the Medical Device Amendments of 1976.

The 1969 AAMI conference report also provided information and recommendations on standards that became the foundation of much of the AAMI standards work of today. A major assumption and recommendation of the report was that while enforcement was the legitimate function of government, standards setting has been and should be the proper function of the professions and industry.

The landmark conference achieved all of its objectives and was an invaluable forum and resource for the consideration and promulgation of concepts fundamental to the Medical Device Amendments of 1976 and the development and use of voluntary and mandatory standards.

IV. 1972 AAMI STANDARDS CONFERENCE

In 1972, AAMI held a national conference on medical device standards in anticipation of FDA's reliance on voluntary and mandatory standards. The conference received financial support from the FDA and attracted the invited participation of representatives from some 100 medical, standards, engineering, trade, consumer, and governmental bodies.

One of the most important results of the conference was a recommendation to establish a national coordinating body for medical device standards. Specifically, the conference participants recommended "establishment of a national capability to accelerate and coordinate the development, promulgation, updating and utilization of medical device standards [and] . . . that FDA organize and fund a meeting, within 3 to 6 months, of organizations known to be engaged in, or to have the capability for, development, promulgation, or enforcement of medical devices standards in the U.S., for the purpose of establishing structure, membership, and management of such a National Capability . . . ." Further, they recommended "That the National Capability should serve as the executive or administrative agent for coordinating U.S. participation in international standards activities . . . ."

A meeting of interested and affected parties was held within a 6-month period, and this meeting of interested parties resulted in the creation of a medical device technical advisory board, later renamed the Medical Device Standards Board, or MDSB. This body has operated under the auspices of the American National Standards Institute (ANSI). The MDSB coordinates, in an advisory capacity, national standards activities and publishes an annual directory of national and international standards activities.

The 1972 conference and its report [2] were influential in compiling and promulgating basic concepts of the AAMI standards program. The conference report was published, in whole or part, in some 20 journals. During the conference, representatives outlined their perceptions of the benefits and risks of standards and what could be done to maximize benefits and reduce risk through appropriate policies and procedures. Some of these policies and procedures are described later in this chapter.

V. ELECTRICAL SAFETY CONCERNS

One of the areas in which AAMI made a major contribution was electrical safety education; its national standard on safe current limits is a baseline document for equipment design and purchase [3].

During the 1970s a number of scientific journals and popular magazines raised questions about electrical safety in hospitals. A controversy arose as to how to protect patients, who were vulnerable in a number of contexts, from electrical currents. This is a political issue as well as a scientific issue.

Many articles and studies dealt with protecting patients from fibrillation when they were vulnerable due to various electrical connections leading directly to the heart.

Serious questions have been raised as to whether the safety concerns of the 1970s were overstated and whether precautionary efforts were overdone. Many very important objectives were, however, accomplished, due in part to these and other legitimate safety concerns and activities, as the nation's political and health care leaders began to comprehend the benefits and challenges of the beginning of a large-scale introduction of medical technology into medical care.

First, the safety concerns of the 1970s enhanced the role of clinical engineers and other hospital equipment managers, giving them the recognition and authority to effect important changes in hospital equipment management, and thereby helping to enhance the role of this new health care professional and medical technology.

Second, sensitivity to equipment safety was heightened, which undoubtedly improved the safety of equipment in general.

Third, the electrical safety concerns provided the foundation for cost control and equipment management actions that placed major equipment issues in perspective (such as the cost/benefit tradeoffs for extremely expensive isolated power centers in hospitals).

Fourth, real or imagined, the dangers associated with perceived inadequate electrical safety precautions never materialized.

The electrical safety concern in many respects fostered clinical engineering in hospitals, which generated many benefits to medical technology development, management and use. Perhaps the most important contribution of the "safety" phase in the 1970s was the heightened concern in hospitals, which undoubtedly reduced adverse device experiences (i.e., device safety has been recognized by the Cooper Committee and by many other expert groups as depending as much on user knowledge and practice as it does on device design and development).

To date, there is little evidence that there are major problems with many medical devices. Some have even argued that the Medical Device Amendments of 1976 law is more stringent than it needs to be, considering the relatively limited documentation and justification provided at congressional hearings on adverse experiences.

VI. FDA'S IMPETUS TO STANDARDS

The FDA has provided an important impetus to medical device standards. After the passage of the Medical Device Amendments of 1976, the FDA published its "laundry" list of needed standards, and a number of organizations, but primarily AAMI, began work on many of the high-priority standards.

Another impetus to voluntary standards was a series of FDA contracts for standards development during the 1970s when the agency had a significant amount of funding for this type of activity.

Several of AAMI's most comprehensive standards are based on the contract work performed for the FDA, including AAMI standards for cardiac monitors, defibrillators, and electrosurgical devices. Much of the standards work that AAMI initiated and conducted in the 1970s and 1980s was based on needs expressed by the FDA, industry, and the professions for voluntary standards that could reduce the possibility of regulation.

VII. DEVICE LAW FOSTERED SAFEGUARDS

The 1969 and 1972 AAMI conferences and the medical device law fostered many of the policies on which AAMI standards are based. The perception existed in the 1970s that voluntary standards would likely become mandatory standards and should be written with this possibility in mind. In addition, many of the safeguards fostered by the law seemed very appropriate for voluntary standards, which are often de facto marketplace regulations.

These influences resulted in the recognition that the need for voluntary standards (and the possibility of resulting mandatory standards) must be clearly established before new standards development is undertaken. Similarly, each standard should have a rationale statement for important provisions to assure reasonable interpretation. The safeguards built into the AAMI standards process are as valuable in a voluntary context as in a mandatory context.

These policies and procedures protect the industry, professions, and patient by ensuring careful consideration of all viewpoints and interests and by ensuring that standards resources are utilized for constructive activities. This flurry of activity also resulted in an awareness of what adverse affects unneeded standards could produce and how standards could be misused. One of the best examples is the planned use of a draft AAMI pacemaker standard by a foreign government, although the standard was outdated and in some respects incomplete.

An important safeguard now built into AAMI standards is a clear explanation that the user has an important responsibility to make sure that the standard is still relevant and appropriate to the user's needs. AAMI carefully emphasizes to the potential user of a standard that although the standard represents a national consensus of experts in a general context, each provision of a specific standard must be carefully evaluated by the user to assure relevance to the user's specific needs. (This caveat is also important in the context of converting voluntary standards to regulatory standards, which is discussed later in this chapter. The regulator's needs are often quite different from the needs of the users and industry, and the development and "enforcement" process is also quite different.)

AAMI has been instrumental in promoting the concept of rationale in domestic and international standards. It is AAMI's policy to state why a standard is needed, why specific provisions are included, and why they are written the way they are.

Other national and international organizations have not adopted the concept of rationale because they fear that too much detail could increase their liability. AAMI's viewpoint is that once an organization decides to provide a service via standards, it should assume some risk of liability and should provide this service as effectively as it can. The nominal risk of increased liability to AAMI for its policy of clear rationale seems a very small price to pay for the increased level of service provided to its members.

This rationale has materially enhanced the value and use of standards, since the user of standards not only has safety, performance, and testing criteria, but also has the rationale or justification for the specific aspects of each requirement.

Voluntary standards have been characterized by many manufacturers as "de facto marketplace" regulations, and, as such, standards should not be undertaken unless there is a clearly stated need. Standards should be based on sound scientific data and rationale to assure that they will be useful and not restrictive of technology. This need statement and rationale should be an integral part of the standard. AAMI continues to urge other national and international standards developers to adopt the policy of incorporating rationale into their standards.

AAMI has insisted that all of its standards provide test methods to ensure that there are objective criteria for demonstrating adherence to performance requirements. Industry has supported the concept of test methods, since it provides manufacturers with predictability about how their devices will be tested by government and health care facilities. Predictability is an important factor in maintaining a stable marketplace and effective relationships between the industry and the users of medical devices. AAMI uses the concept of "referee" test methods in its standards, which permits manufacturers to employ other tests, provided that equivalence with the referee tests can be established in terms of comparability of test results.

Another important AAMI policy is to avoid design standards unless absolutely necessary (they are restrictive to device development), and to establish instead performance requirements, an approach that provides maximum flexibility to the device designer.

VIII. VOLUNTARY, NOT REGULATORY, STANDARDS

AAMI has also adopted the policy that generally voluntary standards do not make good regulatory standards, and AAMI discourages the use of its standards as regulations.

The reason for the policy is very sound. Voluntary standards are written with totally different objectives, and with no legal enforcement contemplated. AAMI's standards are educational and informational. Mandatory standards are to be used as enforcement tools, often with criminal sanctions. In addition, the process for writing a voluntary standard is quite different from the formal, legal process for writing a mandatory standard.

The major reason that the FDA's proposed policy of "endorsing" voluntary standards [4] was never adopted was the legitimate concern that voluntary standards conceived and intended to be used in a voluntary context should not generally be used by government agencies as regulations. There was proper concern that important safeguards provided by the law would be bypassed. Another important reason why AAMI discourages adoption of mandatory standards is that it believes standards are generally poor tools for regulation.

Nonetheless, AAMI standards have been adopted as regulations, a case in point being the Health Care Financing Administration's (HCFA) adoption of the AAMI *Recommended Practice for Reuse of Hemodialyzers* [5,6]. The circumstances of this adoption are unique: Congress required by law that HCFA either adopt a mandatory standard or cease dialyzer reuse cost reimbursement under Medicare. Congress imposed a 1-year deadline for the promulgation of a federal standard.

On a number of occasions AAMI has suggested that there was little or no need for a mandatory standard where existing voluntary standards and industry and user adherence were accomplishing all that the FDA wished to accomplish with a mandatory standard. AAMI formally communicated this position when the FDA proposed mandatory standards for such technologies as cardiac monitors and vascular prostheses where AAMI had already developed comprehensive standards that were being adhered to by the industry [7].

IX. CURRENT ACTIVITIES AND ISSUES

FDA and Congress are now unlikely to insist on any significant number of mandatory standards, although the medical device law contemplated many mandatory standards. There seems to be an increasing recognition by Congress and government agencies that the problems associated with medical devices are limited and that mandatory standards are generally an inappropriate form of regulation because of the amount of resources required for development, revision, and enforcement.

Additionally, AAMI has developed standards for many of the devices that need standards or is working on standards that FDA considers important. In other words, AAMI's standards work eliminated the need, in many instances, for regulation. AAMI has expended considerable effort convincing Congress and the FDA that standards are not good regulatory tools and that there are many other more appropriate ways to regulate medical devices.

Both FDA and Congress seem to be moving toward a national policy that recognizes that voluntary medical device standards are an acceptable and cost-effective approach (and a much better approach than mandatory standards) to the actual or potential device problems that can be resolved by standards. Congress is now considering an amendment to the law—proposed by the HHS—that would preclude FDA from developing and using mandatory standards if there are adequate voluntary standards being adhered to by the industry.

Although FDA has not used mandatory standards as a primary means of regulation, the agency does use voluntary standards in a number of regulatory contexts, including approving foreign devices for import, enforcing good manufacturing practices, administering the 510(k) process (the process by which the FDA concludes that a device is similar to another device and therefore needs less regulation), and evaluating premarket approval applications.

As noted above, voluntary standards are de facto regulations in the marketplace; AAMI standards are included in purchase specifications by most hospitals and other health care facilities. Thus, AAMI standards are a primary frame of reference in a regulatory or voluntary context wherever fundamental aspects of medical technology are specified.

X. CURRENT IMPETUS TO STANDARDS

The major incentives to voluntary standards activities now include:

1. The increasing importance of international standards in world trade. The United States has more reasons now than ever before to develop carefully thought out positions on domestic voluntary standards to assure an effective posture in international standards setting.
2. The need for basic safety and performance recommendations as a means of enhancing the manufacturer's marketplace and assuring the user of the quality of the product. Responsible manufacturers support standards because customers demand them and a stable marketplace must be based on them. The manufacturer and user need to establish a context

in which evaluations and communications about technology can occur, and the user needs to feel that there is agreement on the basic elements of safety, performance, and testing of devices before purchase and use.
3. The need to assure patients and users of safe environments for medical technology use.
4. Government agencies' willingness to defer to voluntary standards rather than utilize resources for mandatory standards.

XI. CHALLENGES

AAMI sees a number of challenges arising from voluntary standards activities. First, coordination of voluntary standards resources could be improved. Standards organizations sometimes initiate efforts where there is little need for the standards—or inadequate justification for the diversion of resources from other efforts. Before industry and the professions' resources are diverted from other health care activities, the proponent of a new standards effort should define (1) the need for the project, (2) the available and appropriate organizational capability for the effort, and (3) the level and type of resources required to complete the effort. The ANSI MDSB or a similarly representative coordinating body should review each such proposal before significant resources are committed.

Second, it is likely that government will use more standards for safety and efficacy regulation or quasi-regulation or as the basis for evaluating eligibility of providers for cost reimbursement. This will require careful coordination and communication between government and standards organizations to ensure that voluntary standards are used appropriately. HCFA, for example, is likely to continue using voluntary medical device standards to help establish eligibility for medical device cost reimbursement. Government agencies perceive standards as much less expensive than government development of standards, and much more effective. Government can and should rely on voluntary standards developers, since they have much more experience and latitude to develop needed standards.

Third, there could be increasing pressure for government or quasi-government management of voluntary standards efforts. This pressure could come from a number of sources, including:

1. Parties who feel that they have been aggrieved by standards.
2. Industries that feel that the increasing importance of international standards requires government intervention and coordination.
3. People who believe that better coordination and management of national resources can be accomplished by government or quasi-government management. The United States is one of the few developed countries that does not have central government or quasi-government management of standards efforts.
4. Organizations that oppose standards in general who perceive that their cost is not offset by benefit. This is an increasing factor in a cost-containment environment.

There has been strong pressure in the past for government coordination of all national standards efforts, and there will likely be pressure again.

XII. AAMI'S UNIQUE CONTRIBUTION

In the context of all of these issues, AAMI can, because of its commitment to better patient care technology, continue to provide unique services to the professions and industry. This service will be unique because:

1. AAMI is membership or constituent driven. AAMI is not in the business of *only* writing standards; AAMI is in the business of serving its members and medical technology. This materially affects AAMI's desire to avoid writing standards that are not needed and could in fact be harmful. AAMI has, for example, communicated to other organizations (on behalf of affected membership groups) its concern with efforts to write voluntary standards in areas where there is too little data available on the consequences of the standards, where the scope of activity has been inadequately defined, or where the need for standards has not been clearly established. (Fortunately, it is AAMI's experience that where there is not a legitimate need for a standard, a standard is very difficult to write. This does not mean, however, that valuable resources that could be used in other contexts are not wasted.)
2. AAMI has acknowledged that there are often better ways to solve problems than developing standards, and attempts to be very flexible about responses that are relevant to the perceived problems. For example, AAMI has developed educational documents in lieu of standards where the need for resolution was more rapidly and effectively managed by education and information rather than by stricter safety and performance requirements—which often cannot be written because there is inadequate data.

AAMI is not moving away from traditional standards setting, but is acknowledging that the challenges facing its membership require diverse services and responses. New areas of activity that initially require different responses than standards are medical software; clinical engineering management; and engineering or equipment management productivity, quality control, and risk management. Additional guidance is also being sought in the areas of lasers and anesthesiology, which will require more educationally oriented rather than standards related documents.

XIII. SUMMARY AND CONCLUSIONS

AAMI has played an important role in the development of policies and procedures for both voluntary and regulatory standards, nationally and internationally. It has also developed a body of standards that has had an important effect on the level of device safety and performance, education, and device regulation.

Although traditional voluntary standards will still play an important role in all of these areas, it is likely that new types of documents will evolve as the need for more rapid educational responses emerges.

Government, manufacturers, and users will rely more on standards in the future, as the benefits of voluntary standards over mandatory standards become more obvious.

Effective voluntary management of national standards resources, resulting in effective standards and effective representation of all interests, can enhance the safety and efficacy of medical devices. It can also help avoid government management of voluntary standards. As the participants of the 1969 AAMI conference emphasized, management of voluntary standards is the prerogative and responsibility of the professions and the industry.

REFERENCES

1. A National Conference on Medical Devices, Medical Instrumentation. November-December 1969, pp. 180-213, Part 1 of 2, Association for the Advancement of Medical Instrumentation, Arlington, VA.
2. "Proceedings of AAMI/FDA National Conference on Medical Device Standards," Medical Instrumentation, Volume 6, July-August 1972, pp. 288-315, Association for the Advancement of Medical Instrumentation, Arlington, VA.
3. "American National Standard, Safe Current Limits for Electromedical Apparatus" (ANSI/AAMI ESI-1985), Association for the Advancement of Medical Instrumentation, Arlington, VA.
4. FDA's Draft Standards Policy, *AAMI News*, January-February 1982, *17*(1): 5-6. Association for the Advancement of Medical Instrumentation, Arlington, VA.
5. *Recommended Practice for Reuse of Hemodialyzers* (AAMI ROH-1986), Association for the Advancement of Medical Instrumentation, Arlington, VA, 1986.
6. *Fed. Reg.*, *52*:36926 (October 2, 1987).
7. *Fed. Reg.*, *50*:48156 (November 21, 1985); *Fed. Reg.*, *51*:564 (January 6, 1986).

APPENDIX. STANDARDS, RECOMMENDED PRACTICES, AND TECHNICAL INFORMATION REPORTS PUBLISHED BY THE ASSOCIATION FOR THE ADVANCEMENT OF MEDICAL INSTRUMENTATION

Cardiovascular Monitoring and Surgery

Automatic External Defibrillators Design, Testing and Reporting Performance Results (AAMI Technical Information Report)	AAMI TIR No. 2-1987
Blood Pressure Transducers, General (American National Standard)	ANSI/AAMI BP22-1986
Blood Pressure Transducers, Interchangeability and Performance of Resistive Bridge Type (American National Standard)	ANSI/AAMI BP23-1986
Cardiac Defibrillator Devices (American National Standard)	ANSI/AAMI DF2-1989
Cardiac Monitors, Heart Rate Meters and Alarms (American National Standard)	ANSI/AAMI EC13-1983
Cardiac Valve Prostheses (American National Standard)	ANSI/AAMI CVP3-1981

Diagnostic Electrocardiographic Devices (American National Standard)	ANSI/AAMI EC11-1982
ECG Connectors (AAMI Standard)	AAMI ECGC-5/83
ECG Electrodes, Pregelled, Disposable (American National Standard)	ANSI/AAMI EC12-1983
Sphygmomanometers, Electronic or Automated (American National Standard)	ANSI/AAMI SP10-1987
Sphygmomanometers, Non-Automated (American National Standard)	ANSI/AAMI SP9-1985
Testing and Reporting Performance Results of Ventricular Arrhythmia Detection Algorithms (AAMI Recommended Practice)	AAMI ECAR-1987
Vascular Graft Prostheses (American National Standard)	ANSI/AAMI VP20-1986

Electrical Safety

Safe Current Limits for Electromedical Apparatus (American National Standard)	ANSI/AAMI ES1-1985

Maintenance and Design

Establishing and Administering Medical Instrumentation Maintenance Programs, Guideline for (AAMI Recommended Practice)	AAMI MIM-3/84
Human Factors Engineering Guidelines and Preferred Practices for the Design of Medical Devices (AAMI Recommended Practice)	AAMI HE-1988

General Surgical Devices

Autotransfusion Devices (American National Standard)	ANSI/AAMI AT6-1981
Blood Transfusion Micro-Filters (American National Standard)	ANSI/AAMI BF7-1982
Electrosurgical Devices (American National Standard)	ANSI/AAMI HF18-1986

Nephrology

First Use Hemodialyzers (American National Standard)	ANSI/AAMI RD16-1984
Hemodialysis Systems (American National Standard)	ANSI/AAMI RD5-1981

Hemodialyzer Blood Tubing (American National Standard)	ANSI/AAMI RD17-1984
Reuse of Hemodialyzers (AAMI Recommended Practice)	AAMI ROH-1986

Neurosurgery

Implantable Peripheral Nerve Stimulators (American National Standard)	ANSI/AAMI NS15-1984
Implantable Spinal Cord Stimulators (American National Standard)	ANSI/AAMI NS14-1984
Transcutaneous Electrical Nerve Stimulators (American National Standard)	ANSI/AAMI NS4-1985

Sterilization

Automatic, General-Purpose Ethylene Oxide Sterilizers and Sterilant Sources for Use in Health Care Facilities (American National Standard)	ANSI/AAMI ST24-1987
BIER/EO Gas Vessels (AAMI Standard)	AAMI BEOV-3/82
BIER/Steam Vessels (AAMI Standard)	AAMI BSV-3/81
Biological Indicators for Ethylene Oxide Sterilization Processes in Health Care Facilities (American National Standard)	ANSI/AAMI ST21-1986
Biological Indicators for Saturated Steam Sterilization Processes in Health Care Facilities (American National Standard)	ANSI/AAMI ST19-1985
Determining Residual Ethylene Oxide in Medical Devices (AAMI Recommended Practice)	AAMI EOR-1986
Ethylene Oxide Gas—Ventilation Recommendations and Safe Use, Good Hospital Practice (AAMI Recommended Practice)	AAMI GVR-1987
Hospital Steam Sterilizers (American National Standard)	ANSI/AAMI ST8-1989
Industrial Ethylene Oxide Sterilization of Medical Devices, Guideline for (American National Standard)	ANSI/AAMI ST27-1988
Industrial Moist Heat Sterilization of Medical Products, Guideline for (American National Standard)	ANSI/AAMI ST25-1987

Performance Evaluation of Ethylene Oxide Sterilizers—EO Test Packs, Good Hospital Practice (AAMI Recommended Practice)	AAMI EOTP-2/85
Process Control Guidelines for Gamma Radiation Sterilization of Medical Devices (AAMI Recommended Practice)	AAMI RS-3/84
Steam Sterilization and Sterility Assurance, Good Hospital Practice (AAMI Recommended Practice)	AAMI SSSA-1988
Steam Sterilization Using the Unwrapped Method (Flash Sterilization), Good Hospital Practice (AAMI Recommended Practice)	AAMI SSUM-9/85

13

Role of ECRI

JOEL J. NOBEL *Emergency Care Research Institute, Plymouth Meeting, Pennsylvania*

I. INTRODUCTION

ECRI is an independent nonprofit agency, established in 1955, committed to the improvement of health care technology. (Previously known as the Emergency Care Research Institute, ECRI's role has grown over the years and it has adapted its acronym as its full name.) Since 1968, it has operated a broad range of programs dedicated to upgrading the safety, efficacy, reliability, and cost effectiveness of medical devices, health care facilities, and the related skills of health professionals. Many of its activities have a significant impact on medical device manufacturers, and most companies have developed policies and responses to ECRI that usually redound to their benefit. Some companies, however, are not fully aware of ECRI's range of programs, and some do not understand how to relate to ECRI in an effective manner, consistent with their own self-interest. Almost two decades of experience with ECRI-industry relations have clearly demonstrated that the long-term interests of a manufacturer are best served when it recognizes that ECRI has a well-accepted role within the health community and among national medical device regulatory authorities worldwide and also has a significant impact on the selection, procurement, maintenance, and operation of products by customers. ECRI also affects broader aspects of technology policy within individual health care institutions and at regional and national levels.

ECRI's underpinnings are science based, stressing multidisciplinary analysis rather than a legalistic, regulatory, or market-driven approach to making judgments about medical devices. It has served, for two decades, as the fourth force in the medical device marketplace, alongside buyers, sellers, and regulatory agencies, redressing the traditional power imbalance between buyers and sellers of medical devices. It does so by providing information and guidance, widely regarded as objective, highly relevant, and practical for hospitals, health professionals, and regulatory agencies and industry as well. At the same time, the interdisciplinary nature of ECRI's staff, which includes attorneys, risk managers, cost analysts, and social scientists, as well as health professionals, engineers, life scientists, and computer specialists, gives it a far broader view of technology than would be assumed of a simple testing laboratory.

ECRI's contributions in the area of technology assessment spring from its basic philosophy. ECRI believes that an educated and informed marketplace, which consciously selects, purchases, and uses superior technologies in preference to suboptimal products, encourages and even forces manufacturers to respond by producing better products. Individuals who use ECRI's information must make day-to-day decisions in hospitals, nursing homes, home health care programs, and clinics. These individuals require firm action recommendations. They are forced to make operational decisions. They cannot study a problem indefinitely and can ill afford, given the pressures of liability, malpractice attorneys, and patients, to make incorrect decisions. An informed marketplace discourages marginal technologies, designs, and producers—but contrasts with a centralized or regulatory approach to technology policymaking. ECRI helps achieve rational and widespread results through a mechanism that is, according to the conventional wisdom, upside down. By influencing thousands of practical day-to-day decisions, it changes aggregate behavior, making technology policy, in a sense, from the bottom up rather than the top down. It uses the power of the marketplace.

While ECRI operates at arm's length from the medical device industry, it supports that industry in many ways that are often not apparent to manufacturers. At the same time, however, one of ECRI's fundamental roles is to make judgments about the relative worth of competing products in its publications, information services, or consultation activities for its member hospitals. Depending on the results of tests and studies, its judgments inevitably will please, disappoint, or anger a manufacturer.

Almost two decades of experience have shown, however, that ECRI does make careful, independent judgments based on reliable analyses. Its technical methods and administrative procedures, its review processes to assure the integrity and quality of its work, and—most important—its results have all withstood the test of time. That experience has also demonstrated that ECRI exercises and carries out its mission with a strong sense of responsibility. Manufacturers should recognize, further, that ECRI's policies, procedures, evaluations, and ratings of individual products undergo a stringent review process directed toward achieving technical validity and a fair and balanced position.

This chapter describes those major ECRI programs that affect medical device manufacturers, its policies for relating to manufacturers, and how a manufacturer's relationship with ECRI can be managed constructively.

II. ECRI POLICIES AND THE MANUFACTURING COMMUNITY

Both ECRI and medical device manufacturers share the same customers, providing services or products, respectively, that sometimes conflict, but are not competitive. The manufacturer sells the health care facility or health care professional products and associated services. ECRI sells that same customer information, data, and judgment about those very manufacturer-supplied products and services. ECRI influences the buying practices of the manufacturers' customers frequently, intensively, and on a widespread basis. Hospitals representing 70% of the acute-care beds in North America are members of ECRI's information and consultation systems. ECRI also wields significant influence in many other nations. Its services are sufficiently valued that its membership renewal rates have exceeded 95%, year after year, for almost two decades.

ECRI's services are targeted to the entire range of decision makers in the health community, ranging from trustees, chief executive officers, and chief financial officers to materials managers, directors of environmental services, biomedical and clinical engineers, physicians, nurses, and technicians. The periodicals, journals, and newsletters produced by ECRI, directed toward different segments of the manufacturer's potential customer base, have extended ECRI's influence broadly.

Some manufacturers have learned that, in a sense, ECRI is simply another of *their* customers and that it deserves the same or even greater attention if the manufacturer wishes to do better in the marketplace. While ECRI's criteria for determining the value of a product are clearly more stringent than those of most customers, it is, inevitably, persuaded by demonstrable technical, clinical, and economic facts. No other approach to influencing ECRI's views and judgment works.

Neither ECRI nor its staff accept financial support from the medical device industry or undertake work commissioned by industry. No ECRI employee may consult for or own stock in a medical device company, and each employee's federal income tax return is examined each year to assure conformance to this rule. While ECRI staff frequently visit manufacturing plants, they do so only at ECRI's own expense.

While, clearly, ECRI operates at arm's length from manufacturers, it is not hostile to industry and shuns the typical antibusiness mentality of consumer advocates and their hunger for publicity. It does not publish its hazard reports in the public press to gain attention or support. Furthermore, its staff understands that it is easier to criticize rather than create and manufacture products, that there are legitimate trade-offs between time, money, marketplace realities, and the technical perfection of products, and that perfection itself is elusive and, simply, often unnecessary.

While ECRI obviously places great value on superior products (considering such attributes as safety, performance, reliability, serviceability, and cost effectiveness), it recognizes that these attributes evolve not simply as the outcome of product design and development, but during the product life cycle in the marketplace as well. It therefore values continuing communication with manufacturers and is, itself, open to revising its judgments. It does so quickly when presented with appropriate technical data or new test results. With rare exception, ECRI refrains from criticism of manufacturers, focusing almost exclusively on technical examination and criticism of products per se. Broader criticism of manufacturers themselves, in ECRI's publications or consulting reports, is rare and is invoked if a manufacturer violates ECRI's no-commercialization policy on the use of its monthly journal *Health Devices*, or if a manufacturer, in the presence of an obviously defective product, refuses to communicate and stonewalls against reasonable efforts to examine the problem, or in the exceedingly rare instance of frank dishonesty.

While it would be unrealistic to expect industry to regard ECRI as a congenial partner in improving the quality of health care technology and to applaud its efforts, it is also unrealistic to regard ECRI as an antagonist to industry generally or toward specific companies. Medical device manufacturers are diverse in nature, leadership, policies, and attributes. ECRI is a single, consistent entity and has constructive relations with many companies. When difficult relations exist, it is usually a short-term issue involving legitimate technical disagreements about a product or the significance of a deficiency or, alternatively, a general problem resulting from company policy or attitudes. Sometimes, it is simply a matter of a difficult-to-get-along-with personality.

Should a manufacturer feel that an individual ECRI employee with whom it is dealing is closed-minded, unreasonable, or simply difficult to work with, it may simply request the intervention of that employee's supervisor or, if necessary, consult with a member of ECRI's executive staff to seek resolution of a conflict. While ECRI listens carefully to technical arguments related to a product, threats of legal action by a manufacturer have, historically, consistently proven counterproductive. It is simply the wrong arena in which to contest ECRI's position. Truth is the ultimate defense of a technical position as well as a legal one.

There are many safeguards built into ECRI's internal review processes to assure reasonable technical judgment and fairness in dealing with manufacturers. Unresolved controversies with companies are routinely reviewed by ECRI's executive and internal legal staffs when brought to that level of attention either by an ECRI employee or by the manufacturer. While these mechanisms exist, it is important to keep matters in perspective. In two decades, no manufacturer has ever brought legal action against ECRI, nor has ECRI ever instituted formal legal action against a manufacturer. Even the specter of legal action resulting from a difference in technical views rarely arises, due, in large part, to the common sense of that vast majority of manufacturers that do understand that differences of technical opinions are most effectively and quickly resolved at the lowest possible technical level in the two organizations.

As might be expected, ECRI is privy to many trade secrets and confidential company information, much of it provided directly by manufacturers during the course of a product evaluation or the investigation of a hazard. ECRI's policies protect that information, and it will not voluntarily be communicated to third parties. ECRI will, further, resist its discovery during medical device related litigation.

While often asked for insights into companies and products by investment firms, financial analysts, product developers, medical device companies, or entrepreneurs, ECRI will not express views on investment, profit-making, or marketplace potential of a company or a proposed or existing product to such parties. ECRI is also frequently asked to examine or test product concepts, working models, prototypes, or equipment that is not yet being sold. Doing so would place ECRI in the position of participating in product development where its knowledge might give unfair competitive advantage to one party, inadvertently transfer know-how from one company to another, or make it difficult to evaluate the product objectively in the future. Such requests are consistently declined.

III. ECRI AND MANUFACTURERS: THE WORKING INTERFACE

Manufacturers may expect to encounter and interact with ECRI in the following general areas:

Product evaluation and testing
Problem investigation
Publications and information services
Consultation services
Forensic engineering and litigation

A. Product Evaluation and Testing

ECRI's traditional technical strength derives from its extensive two-decade-old product testing activities. The primary purpose of its product testing is to develop objective data to differentiate competing products in order to support selection and procurement decisions by health care institutions and professionals. The information derived from product testing is published in ECRI's monthly journal *Health Devices* and in derivative newsletters for medical and surgical specialists. Knowledge derived from testing also contributes to other ECRI publications related to risk management, quality of care, technology assessment, and similar areas. It also provides a strong knowledge base to support ECRI's comprehensive consultation services provided to health care facilities and agencies.

ECRI decides which classes of medical devices it will test and evaluate based on the needs of its member hospitals, as expressed through questionnaires, telephone interviews, focus groups, and the day-to-day experience of its staff. An evaluation project always encompasses a single class of devices, such as portable electrocardiographs, disposable surgical drapes, mammographic units, blood glucose monitors, or anesthesia machines.

ECRI has, historically, evaluated a wide range of capital equipment and single-use products employed in health care facilities. It has not, to date, undertaken formal full-scale evaluations of implantable devices, such as heart valves, intraocular lenses, or hip prostheses. It does examine implants involved in hazard reports and forensic investigations and maintains an extensive database related to implants.

An evaluation is typically assigned to a single project officer, usually a biomedical engineer or physicist, with experience related to the type of device. Following a market analysis to determine what products and manufacturers exist; a literature review to develop knowledge about clinical needs, applications, and utility; and a database search to examine reported problems, a study protocol is created by the project officer assigned responsibility for the evaluation. The protocol builds on experience, existing standards, and ECRI's own evaluation criteria, and standard test methods created by ECRI or others. Unlike accepted standards that tend to memorialize a decade-old state of generally accepted knowledge, ECRI's evaluation criteria push the state of the art, encouraging better technology. Usually, new types of test methods must also be developed and, frequently, specialized test fixtures, equipment, or systems as well. The evaluation criteria and test methods undergo internal review, refinement, and additional review as necessary. The study protocol is then reviewed by outside specialists in clinical and engineering disciplines. It is also, together with a letter requesting participation, sent to all manufacturers of these specific devices. The manufacturers are invited to review and comment on the test protocol and to provide test samples. ECRI's operating assumption is that at the beginning of any new study, manufacturers know far more about the technology than does the project officer and have a significant contribution to make that will help assure objective testing procedures and results. Some manufacturers offer to provide plans for specialized test fixtures or instruments and, on occasion, to loan test instruments. The perspectives, criticisms, and suggestions by manufacturers and outside clinical and engineering consultants are combined with the originally proposed study protocol, again reviewed internally, and refined into a working document.

Manufacturers will generally be asked to provide test samples. The vast majority of manufacturers do so, despite the obvious burden and expense of having equipment taken from inventory for some months. Manufacturers sometimes decline to provide test samples or participate in the study, generally for the following reasons:

They are in the middle of model changeovers.
There is simply no equipment available because of limited production or delivery commitments.
They do not believe that their equipment will fare well and believe that being omitted will prove less damaging to sales.
They dominate the marketplace for this particular device category and believe that participating in the study cannot help them gain market share and may harm them.

ECRI's response to the failure of a manufacturer to participate depends on the type of device and that manufacturer's market share. If ECRI believes that a particular manufacturer's device is likely to have special virtues or commands significant market share, it will try to persuade the manufacturer to provide a test sample and, failing to do so, will often conduct tests on recently delivered equipment in a member hospital.

Obviously, there is a higher probability that a manufacturer's products will be tested and prove favorable when it fully participates in the process by providing input during development of the test protocol, provides a new test sample and in-service training, and, finally, carefully reviews the study results and prepublication draft of the full evaluation.

The ECRI project officer may contact a manufacturer from time to time during the typical 6-month evaluation cycle. The project officer may have unresolved questions or may unexpectedly encounter a hazard that should be communicated to the manufacturer immediately, or the equipment may fail, requiring repair or replacement. A manufacturer should feel free to communicate with the project officer at any time.

Following completion of product testing, the project officer drafts the evaluation article as it is likely to appear in print. That full document undergoes intensive internal review by an ECRI staff committee in a process that is similar to a Ph.D. thesis defense. After internal review, the project officer may undertake additional tests, repeat certain tests, or otherwise refine the study or document. The second draft will then undergo another internal review and, frequently, a third or even fourth cycle of review if improvement of the report is required. The internal review process, at all levels, typically involves five to seven senior staff members at ECRI representing engineering, clinical, and sometimes other disciplines. The review process is comprehensive and frequently includes examination of the data audit trail through laboratory notebooks, computer records, and the provisional report. It is, beyond question, far more thorough than the peer review process typical of academic journals.

When ECRI's staff is satisfied that the report deserves external review, it is sent to outside clinical and engineering consultants and all manufacturers whose products were evaluated. Generally, the external review cycle must be completed and telephone or written comments returned to ECRI within 2 weeks. This is a critical time for manufacturers during which they have an opportunity and obligation, to themselves, to thoroughly review the test results, recommendations, and product ratings for technical validity, fairness, and balance. It is also the time when interaction between ECRI and manufacturers is most intensive.

Inevitably, some products will fare better than others. External review may lead ECRI to recognize a need for additional testing, a change in emphasis, and even revision of product rankings. While ECRI listens carefully to all of the reviewers' comments, and this activity is monitored by the project officer's supervisor and others at ECRI, in the ultimate analysis, ECRI must take responsibility for the report contents based on the underlying technical evidence and the judgment of its staff. The final document, as published, does not, therefore, necessarily represent a diplomatic compromise between ECRI's views and that of its external reviewers, manufacturers included, but represents a statement of technical truths and judgments refined by taking into account all views and evidence. The final statement represents the most fair and balanced one that ECRI can make at the time. The article is then quickly published. If products change or new perspectives are developed, manufacturers should communicate such information to the project officer as quickly as possible since, with the *Health Devices* monthly publication schedule, ECRI can quickly bring additional information to its electorate.

B. Problem Investigation

ECRI has operated an international medical device problem reporting network since 1971, predating the U.S. Food and Drug Administration (FDA) Problem Reporting Program (PRP) and Medical Device Reporting (MDR) systems (which, together, comprise the Device Experience Network). The data provided to ECRI by health care institutions and professionals tend to be of higher quality, better documented, and in lower volume than those reported to the FDA. Incoming reports are logged in a computer register and assigned to specialists for assignment of priority and investigation. About 10% of all problems reported result in publication as significant hazards.

Hazard reports are published when there is significant danger of injury, death, or environmental damage due to a defect in a specific model of device, a general problem associated with a class or type of device, or where operator error in using particular brands and models or types or classes of devices seems both likely and clinically significant. ECRI hazard reports contain specific action recommendations to be implemented by the hospital.

Those problem reports that have a low probability of significant injury and little probability of causing death or significant environmental damage but that create significant inconvenience, an economic burden, or maintenance problem for the owner are sometimes published as User Experience Reports in *Health Devices* and derivative publications. The format of these reports is rather different from a hazard report. It consists of a statement by the reporting institution, a statement by the manufacturer, and, finally, an interpretation or statement by ECRI. Approximately 20% of all problem reports received are published as user experience reports. About 70% of all problem reports received and investigated do not warrant publication either as a hazard or user experience report.

ECRI's first act is to inform a manufacturer that such a problem report has been received and to seek the manufacturer's cooperation in investigating and resolving the issue. The response of manufacturers to communications from ECRI when a problem report has been received about a product varies significantly. Most companies have systems in place to deal quickly and effectively with such information and to cooperate in resolution of issues. Some have tried to suppress such information, and a very few simply refuse to communicate at all. Since ECRI's obligation is to inform its member hospitals and health professionals worldwide of significant problems or hazards, it will do so, based on its best judgment, resulting from careful investigation, with or without the cooperation of the manufacturer.

While engaged in extensive publishing, ECRI does not lack material for publication, and manufacturers should not believe that ECRI's pressing to resolve a reported problem or hazard is motivated by a need for copy. ECRI holds itself out to its member hospitals as a resolver of such problems, and that is precisely what it needs to do—resolve the problem. When the problem does not deserve broader attention, ECRI notifies the reporting institution.

Sometimes, although a problem with a specific brand and model device has been reported, investigation demonstrates that it is typical of a whole class of devices, and other brands and models may have the same limitation. In such cases, a generic (not brand- or model-specific) hazard report may be published, although the specific examples reported to ECRI may be cited. Frequently, operator error is the cause of the problem. In such cases, a hazard report addressing operator error, sometimes using the specific device as an example, may be published. Since ECRI is as committed to elimination of operator error as it is to product improvement, it frequently publishes articles in association with evaluations, or generic hazard reports, directed toward improving user competence. It has had a significant favorable impact in reducing user-caused patient injuries and deaths across a wide spectrum of devices over two decades.

Publication of hazard reports, like formal FDA product recalls, has mixed effects on manufacturers. The ultimate economic and psychological impact is often determined by how effectively the company handles communications, as much as by the nature of the technical problem and its resolution or even the harm to patients. Clearly, while manufacturers are not enthusiastic about a telephone call or letter from ECRI notifying them of problems, they can do much to assure a more favorable outcome. The real measure of a manufacturer's virtue as perceived both by ECRI and the manufacturer's customers is less a zero-defects record than a competent, ethical, and rapid response to a reported problem.

Perfection is impossible and problems and hazards are inevitable. The quality of response to newly discovered problems differentiates manufacturers. Open communication and cooperation in resolution of a problem, an appropriate technical fix (whether it be a retrofit, change in labeling or instructions, major modification, or other solution), and prompt communication of this to users are the critical factors. Exercised well, what appears to be an unfavorable crisis can redound to the advantage of the manufacturer. There are other paradoxes. It is often the largest or most competent manufacturers, typically those with the greatest market shares, that have the most frequent problems. That should not be a surprise. A greater number and variety of their devices are in use.

ECRI's position is that product defects are often unpredictable and often fail to show up in premarket testing; sometimes, no amount of design validation uncovers them. Sometimes a problem or accident-or even an injury or death—must occur to create the very first awareness of a problem. No manufacturer deserves condemnation for producing a defective product unless it is clearly the result of intent or extraordinary carelessness. A manufacturer is, however, unworthy when it fails to respond constructively and quickly to news of a problem. What counts is that the problem or hazard be resolved in a constructive manner that is good for patients, the health care facility, and professionals who care for them, and that, ultimately, does best for the manufacturer who supplies the device.

C. Publications and Information Services

ECRI produces a series of 30 primary and derivative publications. Product evaluations, hazard reports, and User Experience Network™ reports are routinely published each month in *Health Devices*. In addition, ECRI publishes *Health Devices Alerts*, the *Health Devices Sourcebook*, *Product Comparison Systems*, and *Health Technology Trends*. *Health Devices* is technically based and leads to judgments about products. The *Health Devices Sourcebook* and *Product Comparison Systems* simply disseminate information about products without making judgments, and inclusion of information about all manufacturers' products is clearly and unquestionably in their self-interest. *Health Technology Trends*, to be discussed below, frequently makes judgments and recommendations about technology but not about specific products.

The *Health Devices Sourcebook* is the premier purchasing directory for medical equipment. Utilizing ECRI's standard nomenclature and coding system to designate each product category, it lists all North American manufacturers, their products, and basic information about each company. It is used worldwide by materials managers, procurement officers, and others to identify sources of supply and to organize information about medical products. It is provided, in addition, to over 200 U.S. embassies, consulates, and commercial missions without charge. While the *Sourcebook* contains a number of sections, such as product categories, device lists by medical and surgical specialties, manufacturers' product lines, third-party service firms, and other suppliers to health facilities, virtually all of the data in it derive from answers provided by each company on one or two relatively simple short forms each year. A few manufacturers have to be coaxed to respond by repeated letters, telegrams, and telephone calls, usually because the manufacturer simply forgets to respond to the questionnaire rather than intending not to respond. Not having one's company or products listed in a vital, commonly used purchasing guide is clearly counterproductive, and manufacturers are urged to respond completely and quickly.

ECRI produces several multivolume loose-leaf *Product Comparison Systems* covering a wide range of medical equipment, including that used for acute care, in clinical laboratories, diagnostic imaging, and surgery. The *Product Comparison Systems* are neutral and nonjudgmental about the value of specific products. Each Product Comparison contains a brief general description of a generic technology, how it works, its principles of operation, reported problems and risks, and any specialized vocabulary necessary to understand it, and lists all relevant manufacturers. It then provides specifications for all brands and models of that particular product category in a uniform format. This procurement guide is used extensively throughout North America and Western Europe and in some other countries. It brings to the attention of potential customers all those brands and models worthy of consideration. The telephone hotline for users of the *Product Comparison Systems* (*PCS*) does not provide advice on purchasing specific models. Manufacturers are frequently asked to provide or update product information for *PCS* and to verify specifications.

Health Technology Trends, a monthly newsletter, is directed toward hospital chief executive officers, chief financial officers, trustees, clinical department heads, and high-level purchasing executives, as well as health care planners and marketing specialists outside the hospital. Its purpose is to alert key decision makers and help them come to judgments about establishing, upgrading, or continuing technology-based services in their institutions. It examines the entire range of critical factors: patient demographics, maturity and cost of the technology, clinical utility, ethics, law, risk, and other issues. It also contains briefs that describe new or changing technologies and their sources.

D. Consultation Services

ECRI has an extensive consultation support program that provides a wide range of services to health care facilities and professionals. ECRI's SELECT™ Program and other activities that involve product specification and procurement intimately involve manufacturers.

SELECT is a customized product selection program in which ECRI makes specific recommendations to hospitals to buy or reject one brand or model over another. It is based on all information available at ECRI and on information provided by the manufacturer. The hospital selects a limited number of brands and models that it wishes to consider, and ECRI examines their relative merits. While formal test results may be available to support some of ECRI's conclusions in these custom consulting reports, recommendations are made not simply on the technical merits of a given brand or model, but are also based on user experience with clinical and handling characteristics, service support, and also on a life-cycle cost analysis undertaken by ECRI.

User experience data are derived from questionnaires, telephone interviews, and focus groups. Manufacturers will be asked not simply for technical literature and data, but for the names of customers from whom user experience may be gathered. The larger the list of users provided, generally, the more fair and balanced the interviews and report.

SELECT's life-cycle cost (LCC) analysis comparing the economic value of competing brands and models is based on a computer model and is probably the most sensitive area for manufacturers. A manufacturer will be asked by ECRI for pricing data for a specific brand, model, or system configuration. If list price is used by one manufacturer and discounted prices by others, clearly the former is at a disadvantage. The result of a life-cycle cost analysis is often dominated by the price of disposable products associated with the technology, such as infusion pump cassettes. A manufacturer may be confronted by the dilemma that if it only provides the list price, its products may turn out relatively poorly in that economic analysis. If, on the other hand, it provides its lowest selling price, it is disclosing that information to a potential customer up front rather than during negotiations. Nonetheless,

ECRI believes that the use of a life-cycle cost analysis is more realistic and fair than basing a decision exclusively on technical issues and first or acquisition price. The acquisition price often becomes relatively unimportant as a differentiating factor, compared with ongoing service or supply costs. Even if list price is used, however, the analysis helps hospitals decide which cost elements are most important. Manufacturers should obviously seek—not simply in dealing with ECRI and its member hospitals, but with the marketplace generally—to minimize the life-cycle cost of their products if they are to succeed in a highly cost-sensitive era.

For over two decades, ECRI has specified equipment for health care facilities. Manufacturers will be asked for information about products and pricing and to participate in equipment demonstrations and other activities that give them an opportunity on bidded or negotiated projects. This activity is distinct from ECRI's SELECT Program inasmuch as it is conducted on site and is frequently supported by extensive studies of patient demographics, computer utilization models, the hospital's ability to assimilate the technology, and other factors unique to the institution. ECRI also has a formal program of strategic technology planning for hospitals, which sets up 5- or 10-year plans for technology acquisition and replacement, together with budgeting and marketing programs consistent with the hospital's strategic plan. Some of these programs can result in turnover of 30-50% of a hospital's technology base over a decade.

Working closely with ECRI's consulting staff is clearly in the interest of any manufacturer. Some of ECRI's consulting activities are offered through major hospital purchasing groups. In Canada, some of ECRI's services are provided in conjunction with the Canadian Hospital Association. In Australia, Sweden, Benelux, and other countries, some of ECRI's services are provided through other nonprofit laboratories or institutions.

Information and judgment gained from these consulting activities contributes significantly to ECRI publications for the benefit of the entire health care community. This linkage between its consulting activities and open contribution of knowledge is a critical element in ECRI's mission.

While ECRI maintains an actual price paid database for capital equipment, drawn from thousands of transactions, it is used to support price negotiations and financial models. It is *not* published and cannot be directly accessed by hospital clients. Because of the many variations in capital equipment transactions (e.g., extended warranties, spare parts, nonprice concessions, freight), making raw data broadly available would often prove misleading. ECRI's perspective is that the most useful information to support price negotiation is not specific transaction data, but the best percentage discount from list price for a given manufacturer's specific product line. ECRI also protects these data because it recognizes that there are ethical and legal considerations ranging from the privacy of a transaction between two parties to potential antitrust and price-fixing implications.

E. Forensic Engineering and Litigation Support

While forensic engineering, accident investigation, and litigation support provide a relatively small proportion of ECRI's revenues, they are important functions because these activities provide important information on injuries and deaths associated with medical technology that would otherwise be unavailable not simply to ECRI, but to the health community generally. ECRI is typically engaged by a hospital, health professional, insurance carrier, or attorney. The insurance carrier or attorney may represent any of the preceding parties or a patient or manufacturer. Manufacturers are sometimes angered by ECRI's participation in a product liability suit. They are generally unaware that in most such engagements, ECRI concludes that operator error was the primary cause or a major significant contributing factor. There are quite a few cases in which an investigation undertaken by ECRI on behalf of a patient, hospital, or insurance carrier rules out the device and leads to an out-of-court settlement by the hospital or physician, and the manufacturer is never even aware of the process. Sometimes, negligence is on the part of service personnel who may or may not be associated with a manufacturer, and sometimes, of course, assignment of cause, on purely a technical basis, implicates the manufacturer.

ECRI has very specific rules of engagement for participating in forensic and accident investigations and litigation support. Generally, clients are served on a first-come basis. First, however, ECRI undertakes a case merit review to determine if there is cause for action, and attorneys are often advised that there is none. Recent review of 500 cases handled by ECRI showed that rapid assignment of technical cause resulted in out-of-court settlement in over 95% of the cases, with manufacturers held responsible in full or in part in about half. That same historical review showed, startlingly, that ECRI clients prevailed in all but a handful of the cases. This probably results from careful case selection and a reluctance to be involved in litigation support unless it either produces new information to improve patient care, supports ECRI's overall objectives, or is clearly meritorious—three interrelated, compatible factors.

F. Referrals for Manufacturers

Medical device manufacturers, developers, entrepreneurs, or others seeking premarket testing of medical devices, consultation support, or other services, which ECRI will not provide because of its conflict-of-interest rules, are routinely referred to individuals, health care institutions, and laboratories that ECRI believes are competent to provide such assistance. ECRI accepts no commissions or finder's fees for such referrals.

IV. MANUFACTURERS' USE OF ECRI INFORMATION

Medical device manufacturers may access ECRI's public databases or subscribe to any ECRI publications without limitation. They may purchase ECRI publications for education and training of their employees. They may purchase bulk quantities of certain ECRI publications for distribution, subject, however, to ECRI's formal and consistent no-commercialization policy. If a manufacturer wishes to distribute any ECRI publication to parties outside its own organization (i.e., within the health community), it must submit in advance any letter or copy intended to accompany that distribution.

In order to maintain its credibility as an objective source of product information, ECRI will not permit manufacturers or distributors of medical devices to exploit its product evaluations published in *Health Devices* to support marketing and sales. Generally, *Health Devices* product ratings are specific and highly qualified. Selective quotation or reproduction is usually misleading. Further, even distribution of an entire issue of *Health Devices* by a manufacturer (which would eliminate the issue of distortion) may violate the policy if accompanied by statements implying endorsement of the product or the existence of some special arrangement between ECRI and the manufacturer. Quoting *Health Devices*, implying product endorsements by *Health Devices* or ECRI, in advertising or promotion, is also contrary to that policy. Most manufacturers respect ECRI's traditional policy, and transgressions are few and far between. ECRI will take appropriate legal action to protect its work product and reputation. ECRI's hospital members, the manufacturer's very customers, have been protective of ECRI's policy and have strongly expressed disapproval of manufacturers who have violated it in the past.

Disapproval of this practice, uncommon as it is, is also expressed in *Health Devices* editorials detailing the violation. Manufacturers should understand that under no circumstances does one of their competitors obtain permission from ECRI to exploit our studies or publications in this manner, and manufacturers are invited to draw such policy violations to our attention.

Consistent with its mission of protecting the public, however, ECRI will provide bulk quantities of a hazard report to the manufacturer of the subject product for broad distribution.

V. RELATIONS WITH REGULATORY AGENCIES

ECRI routinely exchanges certain types of information with some regulatory agencies and responds to some special requests for information. ECRI does not usually make information from ongoing hazard investigations available to regulatory agencies before they are internally approved for publication. In practice, regulatory agencies subscribing to ECRI publications receive information on hazards at the same time as does the health community. On occasion, the nature of a specific hazard investigation justifies exchanging information at an earlier time. ECRI will sometimes ask the FDA to consider action where there is risk of death or significant injury and a manufacturer refuses to communicate about or resolve the issue. The FDA will, of course, follow its own policies and procedures and make an independent decision about its own response.

When there is specific technical expertise within a regulatory agency that is likely to contribute to an ongoing study of any type, the agency may be asked to review a study protocol, test method, or similar document. ECRI never involves regulatory agencies in judging the merits of products, rankings, or recommendations.

ECRI maintains very close working relationships with regulatory agencies in Australia, Canada, and many European countries. For some such agencies, it provides computer software and databases that underpin or contribute to their regulatory programs. ECRI often acts as their agent in investigating problem reports involving equipment of U.S. origin.

VI. INTERNATIONAL PROGRAMS

ECRI has a long history of acting on behalf of patients and health systems throughout the world. One of its earliest efforts was directed toward restricting export of medical devices manufactured in the United States that were defective. Television clips of the 1969 assassination of Tom Mboya, a major political figure in what was then Kenya, showed futile attempts to resuscitate him with a U.S.-made defective resuscitator. This led to ECRI's testimony in the United States Congress in 1973 about the need to prohibit export of recalled medical devices. ECRI's interest in patient welfare has led it to adopt a free trade, open marketplace international philosophy. It is opposed to the use of standards to suppress imports and has an active, organized effort to promote free trade of quality goods.

ECRI's international programs affecting manufacturers include the following additional activities or relationships:

World Health Organization (WHO) Collaborating Center
International Working Group for Medical Device Testing
Information systems
Fellowships
Referrals
CITECH

A. WHO Collaborating Center Status

ECRI is a Collaborating Center of the World Health Organization (WHO) for health care technology. It is designated as the worldwide information-gathering point for problems and hazard reports related to medical devices and as the disseminator of such information. WHO itself does not collect or disseminate this data, and ECRI alone is responsible for its accuracy. It is also responsible for promulgating the Universal Medical Device Nomenclature and Coding System to facilitate international communication and exchange of databases and, generally, serves as a key source of information on health care technology. It produces, on WHO's behalf, a newsletter entitled *Health Equipment Management*, which is focused on regulation, procurement, service, and user training. Its purpose is to upgrade knowledge and competence in these areas among member nations. This newsletter is also the medium of communication for the Medical Applications Section of the International Atomic Energy Agency and other organizations.

B. International Working Group for Medical Device Testing

In 1985, ECRI invited representatives from the European Community and six nations that operate medical device testing laboratories to meet and discuss the potential value of cooperative activities. That group, now expanded to include seven additional countries, shares information on planned projects and technical and organizational methods through periodic meetings, exchange of documents, and personnel, and undertakes joint studies. A cooperative evaluation of insulin pumps carried out by four laboratories was published by ECRI in November 1987, and a joint study of mammographic units by ECRI and Sprima (Swedish Testing Institute for Medical Supplies) was published in 1989. The participating laboratories are shown in Table 1.

C. Information Systems

ECRI has developed many of the fundamental information tools required by national medical device regulatory authorities, hospitals, health professionals, and other key decision makers and has created a solid foundation for inter-

TABLE 1 International Working Group for Medical Device Testing

Organization	Type	Nation
APSF, Australian Patient Safety Foundation	Independent, nonprofit	Australia
FDA	Government	Canada
VTT, Technical Research Centre of Finland	Quasi-governmental	Finland
CNEH, Centre National de l'Equipement	Government	France
EMTEC	Independent, nonprofit	Federal Republic of Germany
ORKI, National Institute for Hospital and Medical Engineering	Government	Hungary
ISS, Instituto Superiore Di Sanita	Government	Italy
TNO, Netherlands Organization for Applied Scientific Research	Quasi-governmental	Netherlands
RTF, Center for Industrial Research	Government	Norway
Sprima, Swedish Testing Institute for Medical Supplies	Quasi-governmental	Sweden
DHSS, Department of Health and Social Security	Government	U.K.
ECRI	Independent, nonprofit	U.S.A.
VNIIIMT, All-Union Scientific Research Institute for Medical Engineering	Government	U.S.S.R.

national exchange of information, knowledge, and judgment related to medical devices.

The key elements of the ECRI system, some of which are already widely adopted around the world, serve as basic building blocks for national medical device regulatory authorities:

1. An international medical device problem reporting and hazard resolution network, described in Section III.B (introduced in 1971).
2. A Universal Medical Device Nomenclature and Coding System. The purpose of this system is to facilitate identifying, processing, filing, storing, retrieving, transferring, and communicating data and information about medical devices. Thousands of hospitals, government agencies, regulatory agencies, and other organizations presently employ the system in applications ranging from hospital inventory and work-order controls to national regulatory systems. The nomenclature and related numerical codes are incorporated in both print products and computer software provided by ECRI and other sources.

The system is available for use on a license basis without charge to certain public agencies and organizations, to hospitals that are members of the ECRI Health Devices System, and to publishers, information systems providers, software firms, and others by special arrangement. Manufacturers are encouraged to use ECRI nomenclature and codes to identify themselves and their products in product literature and packing.

The ECRI English-language standard nomenclature has been translated into other languages and a matched medical device dictionary is in development to facilitate such translations. Regardless of language barriers, each distinct medical device is designated by a unique five-digit computer code, which permits identifying and exchanging data on specific devices despite language barriers. Translations have been made into Chinese, Dutch, French, German, Norwegian, Portuguese, Russian, Spanish, and Turkish. Other translations are in progress.

Nomenclature systems are dynamic and must meet changing needs. The ECRI system, is, therefore, a fully supported system with frequent updates and immediate assistance available to system users for assignment of new nomenclature and codes; it is available on paper and magnetic media. International assistance on nomenclature and codes is available from the system's Gatekeeper immediately via telex, fax, telephone, or letter.

3. A standard, noncomputerized equipment control program (introduced in 1971, expanded in 1979). The concept of a hospital equipment control program was first enunciated in *Health Devices* in 1971. Over the next decade, this model was adopted as a basic systems concept as well as in operational detail and was implemented in thousands of hospitals throughout the world. It proved to be the conceptual link that caused the interest and priority formerly given to biomedical engineering, with its research emphasis, to shift to clinical engineering and its support of the practical day-to-day operating requirements of hospitals.

   The primary thrust of the original system and its variants implemented in many hospitals was to:

   Improve safety.
   Control assets.
   Help rationalize the selection and purchasing of medical equipment.
   Provide a system of management and accountability for technology that is properly documented in order to meet the tests of liability and accreditation.

   While many institutions introduced minor variations to the basic concept, the fundamental idea and most of the originally enunciated details for implementing it have held up well over the years. The equipment control program was further amplified and updated in *Health Devices* in 1979. Protocols, procedures, and forms have been translated into many other languages and put to use in many countries from the Netherlands to Turkey and from Saudi Arabia to Japan.

4. A computerized Hospital Equipment Control System™ (HECS™) (introduced in 1985). The manual equipment control program provided the experience and need to develop a computerized version. HECS is far more than an automated version of the traditional paper system. Its special advantages and new capabilities have become quickly apparent. HECS does the following:

Serves as the primary building block of a new type of technology management system that can provide information that has not been previously available, in part because dedicating the labor to extract such information from paper records was impractical.

Provides new types of objective data based on an adequate statistical foundation, which can be obtained only by combining similar data from a number of institutions.

Provides new types of economic, financial, cost-benefit, and productivity data within each hospital to help optimize the cost-effectiveness of the hospital's technology base as well as of the clinical engineering and materials management department operations.

Provides more detailed and relevant financial data to support cost-center accounting and pinpoint potential savings by detailing costs associated with hospital locations, departments, vendor support, and specific technologies, brands, and models.

Provides more efficient support and analysis of clinical engineering operations through detailed work scheduling for inspection, preventive maintenance and repair, generation of work orders, and analysis of productivity of personnel by both department and individual.

While the basic types of data entered into a computerized equipment control program are not significantly different in type or quantity from those typically required for a traditional paper system, the system's information output extends well beyond inspection, maintenance, performance, safety, and basic financial data. It can provide new information support, including:

Data on the relative reliability of competing brands and models of clinical equipment to support purchasing decisions or consideration of service alternatives.

Repair/replace decision support based on reliable historical data and simple analytical models.

Work norms and, therefore, relative labor requirements and costs for inspection, preventive maintenance, and repair of competing equipment to support future equipment selection and evaluation of departmental and individual productivity.

Reliable data on the cost of service, inspection, and preventive maintenance to underpin equipment life-cycle cost models, in order to further optimize buying decisions.

Better support for technology-related risk management—for example, rapid inventory database searches for equipment involved in life-threatening hazard reports and recalls.

HECS is currently being evaluated for possible use as a national standard in several countries, and some nations have translated HECS to facilitate wider use.

5. A computerized universal medical device registration and regulatory management system (in development). This is a universal microcomputer-based stand-alone system. It will facilitate exchange of information and electronic databases, tracking products for recall, safety, and reliability statistics, inventory control, and myriad other purposes.
6. The world's most comprehensive medical device databases on medical device problems, hazards, recalls, evaluations, and technology assessments, including all U.S. FDA, Canadian, Australian, and other national report-

ing systems (e.g., FDA, DEN, Canadian MDA). Hard-copy updates are issued weekly and electronic tape updates are available monthly.
7. Training fellowships available by special arrangement, often with international agency or foundation support.

D. Fellowships

ECRI provides opportunities for continuing education and training to officials of Ministries of Health and National Medical Device Regulatory Authorities, procurement officials, physicians, engineers, health service administrators, and others involved with medical equipment and supplies. Application for fellowships and financial support should be made through normal government channels to UNESCO, WHO, or other supporting agencies if not organizationally funded.

Fellowships are typically awarded for periods of 1 month to 1 year, and programs are tailored to meet the specific needs of the sponsoring agency or the individual.

Previous fellowships have been awarded to deputy ministerial-level personnel, academics, physicians, engineers, and others from Brazil, Finland, France, Hong Kong, India, Japan, Malaysia, the Netherlands, People's Republic of China, Saudi Arabia, Spain, Pakistan, the Philippines, the Soviet Union, and Turkey.

E. Referrals

ECRI frequently receives requests from outside North America for assistance in establishing relationships with U.S. agencies and manufacturers. Requests may be simply to arrange a visit to a manufacturing plant or to explore a commercial connection for importing, exporting, sales, or service of medical devices or offshore or coproduction arrangements. Manufacturers will occasionally receive calls or correspondence from ECRI staff seeking to make such introductions.

F. CITECH

CITECH (Center for Information on Technology for Health Care) is a nonprofit agency chartered in Sweden in 1987 to improve the quality and availability of medical devices and the competence of device users. ECRI has supported the concept and early development of CITECH, as have TNO (the Netherlands Organization for Applied Scientific Research) in the Netherlands and Sprima in Sweden. CITECH will have a significant impact on the medical device industry through its rapidly developing international medical device certification system.

Concern over growing trade restrictions, based on disparate and conflicting regulations and standards adopted by various nations, led to a suggestion, by TNO, that the first focus of CITECH be on development of a medical device certification system directed toward assuring universal recognition of quality products. Through agreements with national governments, no restrictions would be placed on free import and export of such certified products. CITECH would both accept test results based on existing standards and develop measures of product quality, giving special attention to the relationship between human factors design and operator error and to issues of performance and efficacy. Most commercial product-testing standards and listing agencies focus on safety certification and have traditionally expressed a lack of interest in the issues of clinical efficacy and the equipment-user interface. This has sometimes led to listed or certified products that were "intrinsically safe" but ineffective. Obviously, ineffective medical devices are often dangerous to patients.

CITECH has signed agreements with ECRI, Sprima, and TNO for technical support. ECRI's traditional ground rules against accepting support from the medical device industry constrain ECRI's role in CITECH operations to providing a physical home and laboratory environment in which CITECH, rather than ECRI, employees carry out product certification activities. ECRI's agreement with CITECH also specifies that no ECRI employee may serve as an employee or trustee of CITECH with financial compensation.

CITECH, working in consultation with industry, had, by the spring of 1988, developed its Manufacturer's Manual and began trial operations to test and evaluate its procedures. It also established industry, user, scientific, regulatory agency, and policy advisory councils and has begun negotiations with various national governments.

Industry's reaction to CITECH is mixed. Some companies welcome it as a mechanism to diminish trade restrictions and the problem of dealing with disparate national standards. Some companies are strongly opposed to the concept, in part because they object philosophically to any initiatives to rationalize international medical device activities that are not under industry control. Others, lacking firsthand knowledge of the facts, incorrectly perceive CITECH as a possible United Nations agency-related activity or supranational regulatory agency. It is not. Most trade organizations are taking a wait-and-see attitude. Many manufacturers have already submitted products for certification and are strongly supportive. Some see it as yet another bureaucratic hurdle despite the fact that it accepts test results from other qualified laboratories and has significant added value elements.

VII. ECRI MEDICAL DEVICE DATABASES

ECRI produces a number of databases on medical devices. Some are for internal research or management use only. Others are published as print products or made available electronically via DIALOG or through Compuserve or custom database searches. Still other databases are under development. All ECRI internal and external databases uniformly employ the standard ECRI-developed medical device nomenclature and coding system that has been broadly accepted and implemented internationally.

Health Devices Alerts (HDA): Government and manufacturers' product recall data and new product approvals; abstracts of the English-language engineering, medical, and legal literature on medical device problems, evaluations, hazards, and health care technology assessment studies. Approximately 2200 abstracts are added annually to a 10-year-old database. HDA is issued as a weekly print product and is available electronically via DIALOG or via custom database searches.

Health Devices Alerts Action Items: A selective database drawn from *Health Devices Alerts*, dealing with those medical device recalls and problems for which hospitals and health professionals can and should take immediate action to protect patients and staff from injury and death, or equipment and environment from damage. Published weekly in hard copy and available electronically via DIALOG.

User Experience Network: An umbrella database encompassing information from ECRI's International Medical Device Hazard Reporting Network, problem reporting network, and medical equipment user experience derived from questionnaries, interviews, and other information-gathering methods. Information from this database is incorporated in various ECRI print and electronic information products and is available via custom database searches. As of 1989, approximately 130,000 reports were on file.

Problem Reporting Program (PRP): ECRI's restructured and enhanced PRP database consists of short reports replicating the entire FDA DEN database, in which all voluntary user reports to the FDA have been assigned proper nomenclature, coded, and reformatted to facilitate database searches. This database is available electronically via DIALOG and Compuserve or via custom database searches. About 30,000 reports are on file; new reports are added monthly.

Medical Device Reporting (MDR): This is an ECRI restructured and enhanced database derived from the FDA's mandatory Medical Device Reporting (MDR) system. It has been reorganized and reformatted with universal nomenclature and codes assigned to facilitate searching. Between 1000 and 2000 manufacturer reports are added per month, a significant number of which convey information on injuries or deaths. Information from this database is incorporated in various ECRI print and electronic information products and is available electronically via DIALOG and Compuserve and via custom database searches.

Health Devices Sourcebook: This database lists all types of medical devices produced throughout the world; all North American manufacturers, their product lines, and key contacts; in which clinical specialties the devices are used; and manufacturers' addresses and toll-free and standard telephone numbers. It includes third-party service and parts vendors and many other features. Meticulously organized and cross-referenced, with a broadly accepted standard nomenclature and coding system, it is updated monthly, published annually as the *Health Devices Sourcebook*, and available electronically via DIALOG.

Actual Price Paid Database: This database contains prices paid for a wide variety of brands and models of medical equipment by U.S. hospitals, based on examination of invoice data from hundreds of institutions and thousands of purchases. It is employed internally to produce ECRI's SELECT and similar specialized custom consulting reports for hospitals and life-cycle cost studies, which compare the relative economic advantages of one brand and model of medical equipment over others. It is not available electronically, as a print product, or via custom database searches, but only on a selective basis for specific client institutions.

Medical Device Manufacturers Database: This database contains basic business information about each North American medical device manufacturer and is employed to make judgments about the economic stability and developmental capabilities of companies. It is employed as an internal database to support custom consulting reports such as SELECT or other procurement consulting activities and is not publicly available.

VIII. STAFF AND FACILITIES

ECRI has a full-time interdisciplinary staff of approximately 200 that provides a broad range of analytical, clinical, management, and technical expertise. ECRI staff members include certified clinical engineers, chemists, facilities engineers, nurses, physicians, physicists, physiologists, planners, and registered professional engineers and architects.

Additional full-time personnel include attorneys; biomedical, electrical, electronic, management, materials, and medical engineers; and specialists in clinical technology, hospital administration, law, financial and policy analysis, risk management, information, and computer science. Research and editorial advisory boards add the knowledge and perspectives of hundreds of additional experienced individuals. ECRI's staff members frequently publish in medical, hospital, and technical literature and participate in professional conferences and symposia.

ECRI owns a 12-acre research campus in suburban Philadelphia and a modern 120,000 square foot facility with offices and extensively instrumented laboratories for electronic, environmental, mechanical, and physical testing, calibration, chemistry, forensic studies, microbiology, photography, and surgical physiology. In addition, there are specialized laboratories for examining cardiovascular, imaging, laser, rehabilitation, and respiratory technologies. A television studio and printing and publication, conference, and classroom facilities provide supporting services. It has a powerful IBM computer system with sophisticated database management software and more than 120 terminals to provide on-line retrieval of information and text editing and typesetting capabilities. A variety of minicomputer and microprocessor systems facilitates laboratory and field data collection and processing.

IX. CONCLUSION

As ECRI enters its third decade of evaluating medical products and assessing health care technology, it must, just as most manufacturers and regulatory agencies, rise to meet tough new technical and economic challenges and opportunities. Among them are:

Validating software associated with the microprocessors embedded in a growing number of devices and systems.

Determining the relative long-term costs and benefits associated with competing products.

Coping with internationalization of the medical device industry and regulatory issues.

Dealing with the serious and continuing problem of operator error and the failure of medical and nursing schools to prepare health professionals adequately to deal with technology.

Achieving a reasonable balance between an injured patient's right to economic succor and the innovation-suppressing impact of today's product liability environment.

Showing hospitals how to minimize maintenance and service costs associated with technology.

Determining the value of linkages between specific medical technologies and general health care facility information systems.

Preparing for the impact of biotechnology and replacement of many current diagnostic and therapeutic modalities.
Helping its member institutions to cope with a growing flood of often conflicting information about medical devices and systems.

ECRI will respond to these and many other challenges constructively and will maintain its traditional arm's-length and neutral position vis-à-vis industry. It will also continue, however, to seek out new areas of cooperation with industry that will benefit the health care community and its patients that, in the ultimate analysis, both industry and ECRI serve.

14

Role of ASTM in the Voluntary Standards Process

KENNETH PEARSON *American Society for Testing and Materials, Philadelphia, Pennsylvania*

I. INTRODUCTION TO ASTM

A. General Information

The American Society for Testing and Materials (ASTM), a nonprofit corporation organized in 1898, is one of the world's leaders in the development of voluntary consensus standards for materials, products, systems, and services. It provides a legal, administrative, and publications forum within which producers, users, ultimate consumers, and those representing the general interest (representatives of government and academia) can meet on a common ground to write standards that best meet the needs of all concerned.

ASTM's membership of 32,000, of which 4000 are international members, is organized into 134 different main technical committees that do the actual work of writing standards. Committee members voluntarily contribute their time and efforts.

ASTM publishes over 8500 standards in such diverse technical fields as ferrous and nonferrous metals, energy, environmental, coatings, constructions, petroleum, textiles, plastics, medical devices and services, consumer products, biotechnology, and computerization. ASTM is unique in that its management system can be used to address virtually any standards activity.

ASTM also publishes special technical publications (STPs), which are collections of peer-reviewed technical papers reflecting the state of the art in subject areas spanning the scope of ASTM activities. Most STPs are based on sympsia sponsored by ASTM technical committees. Other publications include compilations of ASTM standards, reference radiographs, standards adjuncts, and data series, and ASTM Info Briefs, quarterly updates of new and revised ASTM standards. The Society also publishes a monthly magazine, *Standardization News*, and five journals: *Journal of Testing and Evlauation; Cement, Concrete, and Aggregates; Geotechnical Testing Journal; Journal of Composites Technology and Research;* and the *Journal of Forensic Sciences*.

B. Principles of the Society

Timely and adequate notice of a proposed standard undertaking to all persons known to the Society to be likely to be materially affected by it.

Opportunity for all affected interests to participate in the deliberations, discussions, and decisions concerned both with procedural and substantive matters affecting the proposed standard.

Maintenance of adequate records of discussions, decisions, and technical data accumulated in standards development.

Timely publication and distribution of minutes of meetings of main committees and subcommittees.

Adequate notice of proposed actions.

Distribution of letter ballots to those eligible to vote.

Timely and full reports and results of balloting.

Careful attention to minority opinions throughout the process.

Maintenance of records of drafts of a proposed standard, proposed amendments, action on amendments, and final promulgation of the standard.

II. DEVELOPMENT, APPROVAL, AND MAINTENANCE OF ASTM STANDARDS

A. Definition and Types of Standards

ASTM defines a standard as "a rule for an orderly approach to a specific activity, formulated and applied for the benefit and with the cooperation of all concerned." An ASTM standard is an accepted rule of behavior developed by democratic procedures.

ASTM develops six different types of full consensus standards. They are:

1. Standard test method—a definitive procedure for the identification, measurement, and evaluation of one or more qualities, characteristics, or properties of a material, product, system, or service that produces a test result.
2. Standard specification—a precise statement of a set of requirements to be satisfied by a material, product, system, or service that also indicates the procedures for determining whether each of the requirements is satisfied.
3. Standard practice—a definitive procedure for performing one or more specific operations or functions that does not produce a test result.
4. Standard terminology—a definition or description of terms, or explanation of symbols, abbreviations, or acronyms.
5. Standard guide—offers a series of options or instructions, but does not recommend a specific course of action.
6. Standard classification—a systematic arrangement or division of materials, products, systems, or services into groups based on similar characteristics such as origin, composition, properties, or use.

B. Committee Structure

The ASTM committee structure is comprised of a main committee, subcommittees, task groups, and an executive subcommittee.

Each main committee has a clearly stated scope of activities, which defines the technical area in which it may work to develop standards. The main committee is comprised of the entire committee membership. Its officers are a chairman, a vice-chairman (perhaps more than one), a recording secretary, and a membership secretary.

Main committees are further divided into subcommittees, which are responsible for developing standards in specific areas relating to the main committee's scope. Officers include a chairman, a vice-chairman when the subcommittee is large enough, and a secretary.

Subcommittees are further divided into task groups. The function of the task group is to initiate draft revisions or new standards. This is where the standards development work actually begins. The task group membership can include individuals who are not ASTM members but who are willing to provide input and expertise.

The executive subcommittee manages the administrative matters of the main committee and the subcommittees. It is usually comprised of the main committee officers, the subcommittee chairmen, and frequently some members at large.

C. Balloting Procedures

One of the most important phases of committee activity is balloting on proposed new standards and revisions to existing standards. The integrity of ASTM standards is based largely on the fact that several procedural requirements govern the balloting at each level. First, all ballots must be conducted by mail. This ensures that every member has an opportunity to vote. Second, all ballots must be issued for a period of at least 30 days. This allows members adequate time to critically review the document. And third, whenever a ballot leads to substantive changes in a document, the revised version must be reballoted at all levels in order to ensure that technical changes to a document are made on a consensus basis. The "Regulations Governing ASTM Technical Committees" detail the requirements that the ASTM Committees follow in the development of full consensus standards.

D. Subcommittee, Main Committee, and Society Ballots

After the task group has prepared a draft document, it is then ready for the first level of review, the subcommittee ballot. If the document is approved by two-thirds of those returning ballots (a minimum of 60% of the voting interest must return ballots), the document proceeds to a main committee ballot. Here, 90% of those returning ballots (again a 60% return is required) must approve the document. It then goes to a Society ballot located in the ASTM monthly magazine *Standardization News* and is distributed to ASTM's 32,000 members as a final public review.

E. Handling of Negative Votes

All negative votes received on subcommittee, main committee, and Society ballots must be considered by the subcommittee from which a document originated before it can progress to the next ballot level or be published as a standard. To be valid, a negative vote must be accompanied by a written explanation of what the writer considers improper technical or procedural consideration.

It is not uncommon for a negative's reasoning to be persuasive. Few, if any, ASTM standards are developed without the help of persuasive negative votes. When a negative vote is persuasive, the document must be rewritten by the subcommittee to incorporate the reasoning supporting the negative and then reballoted. If a negative vote is judged not persuasive, the document proceeds to the next balloting level along with the record of the ballot item.

F. Committee on Standards

Once the document has received approval via the ASTM Society ballot, the ballot summary and the documentation of the handling of negatives votes by the main committee and subcommittee shall be submitted to the Committee on Standards, a review committee appointed by ASTM's Board of Directors. The Committee on Standards shall determine, on behalf of the Society, whether the procedural requirements of the Society have been met. The negative voter(s) and a representative of the committee are invited to be present and participate in the deliberation. If the Committee on Standards determines that the requirements of the Society have been followed, the standard is approved and published as an official ASTM standard.

G. Review and Maintenance

ASTM standards maintain their high level of technical credibility due to the Society procedures, which assure review and action on a timely and continuous basis. Once published, a standard should be reviewed yearly by the sponsoring subcommittee and recommended for revision when appropriate. Most ASTM committees meet twice a year; thus there is ample opportunity for frequent review of standards both at the meetings and in between via the mail. Subcommittees and task groups can hold additional meetings if necessary to expedite action on a standard. It is important ot note that revisions to standards may be initiated by any interested party at any time. In addition, each ASTM standard contains a caveat encouraging the users of the document to contact ASTM Headquarters with any suggested revisions. As a minimum, standards that have been published for 4 years without technical revisions must be reviewed by the responsible committee and balloted through the full ASTM process as a reapproval, revision, or withdrawal.

III. MEMBERSHIP, CLASSIFICATION, BALANCE OF INTEREST

ASTM membership is open to all interested parties. Anyone who is qualified or knowledgeable in the area of a committee's scope may become a member. Individuals may participate on task groups without being members of ASTM; however, ASTM membership is required to vote on subcommittees and main committees standards actions. ASTM members pay an $50.00 Society membership fee. ASTM does not charge any other fees for its services. Members are classified as producer, user, consumer, or general interest. This classification is assigned with regard to the scope fo the committee. Balance of interest on a main committee or subcommittee requires that the number of voting producer members cannot exceed the combined number of voting user, consumer, and general interest members. In addition, the chairman of an ASTM Main Committee can not be from the producer classification. Further, ASTM ensures that no one group or organization can dominate an activity by permitting only one official vote per voting interest on a main committee or subcommittee.

IV. ASTM HEADQUARTERS STAFF

The ASTM staff of over 200, located at Society Headquarters in Philadelphia, exists to serve the needs of the Society and provide extensive management and administrative assistance to the ASTM committee officers and members. ASTM also maintains an office in Washington, DC, to facilitate liaison with government agencies and regulatory bodies in areas that may affect technical committee activities.

ASTM Headquarters: 1916 Race Street, Philadelphia, PA 19103. Telephone 215-299-5400; telex 710-670-1037; fac 215-977-9679.

Washington Office: 300 Metropolitan Square, 655 Fifteenth St., NW, Washington, DC 20005. Telephone 202-639-4025; fax 202-347-6109.

V. ASTM ACTIVITIES IN THE MEDICAL AREA

Within ASTM's structure of 134 technical committees, Committees F04 on Medical and Surgical Materials and Devices and F-29 on Anesthetic and Respiratory Equipment are the two primary activities that address standards in the medical device area. There are other committees that have specific subcommittees addressing medical devices and equipment. The following are brief descriptions of the committees.

A. Committee F04 on Medical and Surgical Materials and Devices

Committee F04, organized in 1962, has 450 members, jurisdiction over 118 standards, and has published several special technical publications.

The scope of the committee is the development of terminology and nomenclature, test methods, specifications, and performance standards for medical and surgical materials and devices. The committee will encourage research in this field and will promote liaison with other ASTM committees and outside organizations with mutual interests.

Subcommittee and Task Group Structure

F04.01 Administrative
F04.02 Resources
- F04.02.01 Polymeric Materials
- F04.02.02 Metallurgical Materials
- F04.02.03 Ceramic Materials
- F04.02.04 Composite Materials
- F04.02.05 Test Methods
- F04.02.06 Biocompatibility
- F04.02.08 Device Retrieval Analysis

F04.02.09 Corrosion of Implant Materials
F04.02.11 Precision and Accuracy
F04.03 Orthopaedics
F04.03.01 Osteosynthesis
F04.03.02 Arthroplasty
F04.03.03 Soft Tissue Replacement
F04.03.04 Surgical Instruments
F04.03.06 Packaging and Labeling
F04.03.07 Orthopaedic Application of Electrical Stimulation
F04.03.09 Performance Considerations
F04.03.10 Biodegradable Orthopaedic Implants
F04.04 Cardiovascular
F04.04.01 Nomenclature
F04.04.02 Materials
F04.04.04 Implementation
F04.04.04 Retrieval
F04.04.05 Regulatory Compliance
F04.04.06 Legal and Forensic Affairs
F04.05 Neurosurgical
F04.05.01 Materials
F04.05.02 Aneurysm Clips and Clip Appliers
F04.05.04 Carotid Clamps
F04.05.05 Shunts
F04.05.06 Aneurysmorrhaphy and Tissue Adhesives
F04.05.07 Electrodes
F04.05.09 Cranial Tongs for Skeletal Traction
F04.05.10 Stereotaxic Instruments
F04.05.11 Operating Microscopes
F04.06 Plastic and Reconstructive Surgery
F04.06.01 Mammary Implants
F04.06.02 Non-Mammary Materials and Devices
F04.06.03 Surgical Instruments
F04.07 Spinal Devices
F04.07.01 Terminology
F04.07.02 Component Testing
F04.07.03 Component-Component Interfaces
F04.07.04 Bone Implant Interfaces
F04.07.05 Constructs
F04.07.06 Classification
F04.08 Medical/Surgical Instruments
F04.12 Urological Materials and Devices
F04.19 Orthopaedics, External Prosthetics, and Mobility Aids
F04.19.10 Materials
F04.19.20 Prosthetics
F04.19.30 Orthopaedics
F04.19.40 Mobility Aids
F04.19.91 Nomenclature and Definitions
F04.19.92 Research
F04.19.93 ISO/TC 168
F04.20 Reprocessing of Medical Devices
F04.92 Planning

STP 898, Vascular Graft Update: Safety and Performance.
STP 953, Quantitative Characterization and Performance of Porous Implants for Hard Tissue Applications (in cooperation with American Academy of Orthopedic Surgery's committee on Biomedical Engineering).
STP 859, Corrosion and Degradation of Implant Materials.
STP 800, Medical Devices: Measurements, Quality Assurance, and Standards.
STP 810, Cell-Culture Test Methods.
STP 796, Titanium Alloys in Surgical Implants.
STP 684, Corrosion and Degradation of Implant Materials.
STP 386, Plastics in Surgical Implants.

Standards Developed by Committee F04

A list of F04 standards is included as Appendix 1.

B. Committee F-29 on Anesthetic and Respiratory Equipment

Committee F-29, organized in 1983, has 158 members and jurisdiction over 15 standards.

The scope of the committee is the development of terminology, definitions, units of measure, identification, dimensions, tolerances, performance, and methods of test of anesthetic and respiratory equipment and systems for function and safety. The work of this committee will be coordinated with other ASTM committees and other organizations having mutual interest.

Subcommittees of Committee F-29

F29.01 Division One on Anesthesia Gas Machine
- F29.01.01 Continuous Flow Anesthesia Gas Machine
- F29.01.02 Anesthesia Breathing Systems/Performance
- F29.01.03 15/22 mm Connectors and Adapters
- F29.01.04 Breathing Systems/Fittings
- F29.01.05 Pollution Control

F29.02 Division Two-Working Group
- F29.02.01 Tracheal Tubes and Tracheal Tubes for Long-Term Use
- F29.02.02 Naso/Oropharyngeal Air Ways
- F29.02.03 Breathing Tubes/Bags
- F29.02.04 Tracheal Tube Connectors
- F29.02.05 Tracheostomy Tubes-Adult
- F29.02.06 Tracheostomy Tubes-Pediatric
- F29.02.07 Laryngoscopes-Bulbs/Handles/Blades (Rigid)
- F29.02.08 Laryngoscopes and Bronchoscopes (Flexible)
- F29.02.09 Bronchoscopes (Rigid)

F29.03 Division Three on Ventilators and Ancillary Devices
- F29.03.01 Lung Ventilators (Other than for Anesthetic Use)
- F29.03.02 Lung Ventilators for Use in Anesthesia
- F29.03.03 Resuscitators
- F29.03.04 Harmonization of Alarms
- F29.03.05 Blood Gas Monitoring
- F29.03.06 Cutaneous Gas Monitoring
- F29.03.07 Hunidifiers
- F29.03.08 Oxygen Analyzers

F29.03.02 Lung Ventilators for Use in Anesthesia
F29.03.03 Resuscitators
F29.03.04 Harmonization of Alarms
F29.03.06 Cutaneous Gas Monitoring
F29.03.07 Humidifiers
F29.03.08 Oxygen Analyzers
F29.03.09 Home Care Ventilators
F29.03.10 Pulse Oximiters
F29.03.11 Capnometry
F29.04 Division Four on Terminology
F29.04.01 Terminology and Nomenclature
F29.04.02 Equipment Life Expectancy (Inactive)
F29.06 Division Six on Medical Gas Supply Systems
F29.06.01 Non-Flammable Medical Gas Supply Systems
F29.06.02 Oxygen Concentrators
F29.07 Division Seven on Suction and Drainage
F29.07.01 Medical-Surgical Suction Systems
F29.07.02 Suction Catheters

Standards Developed by Committee F-29

A list of F-29 standards is included as Appendix 2.

C. Committee F-30 on Emergency Medical Services

Committee F-30, organized in 1984, has 500 members and jurisdiction over ten approved standards. F-30 has over 40 standards in the balloting phase of standards development.

The scope of the committee is the promotion of knowledge, stimulation of research, and the development of standards (classifications, guides, practices, specifications, test methods, and terminology) for quality emergency medical services. Standards for clinical medical practice shall be excluded from the work of this committee. Coordinate these activities with other organizations having mutual interests.

Subcommittee Structure

F30.01 EMS Equipment
F30.02 Personnel, Training and Education
F30.03 Organization/Management
F30.04 Communications
F30.05 EM Care Facilities
F30.06 Terminology

Subcommittee F30.01 on EMS Equipment

F30.01 is the subcommittee within Committee F-30 that deals specifically with the medical equipment area. Scope of the subcommittee is the development of standards for air, water, and gorund vehicles, support devices, and other related EMS equipment. The following are the task group structure and approved and draft standards.

T.G. F30.01.01 on Ground Vehicles: F 1230-89, Standard Specification for Minimum Performance Requirements for EMS Ground Vehicles.

T.G. F30.01.02 on Water and Air Vehicles:

F XXXX-XX, New Standard Specification for Basic Life Support (BLS) Rotary Wing Transport Units
F XXXX-XX, New Standard Specification for Advance Life Support (ALS) Rotary Wing Transport Units
F XXXX-XX, New Standard Specification for Specialized Medical Transport Units
F XXXX-XX, New Standard Practice for National Air Medical Transport Units Resources and Specification Catalogs
F XXXX-XX, New Standard Specification for Basic Life Support (BLS) for Fixed Wing Transport Units
F XXXX-XX, New Standard Specification for Advanced Life Support (ALS) for Fixed Wing Transport Units
F XXXX-XX, New Standard Specification for Specialized Medical Transport Units
F XXXX-XX, New Standard Practice for National Air Medical Transport Units Resources and Specification Catalogs

T.G. F30.01.03 on Airway Management Equipment: F XXXX-XX, New Standard Guidelines for the Manufacture and Use of Oxygen Administration Equipment for Use in Pre-Hospital Emergency Medical Care

T.G. F30.01.04 on Circulatory Support Devices: F XXXX-XX, New Standard Guidelines for the Performance Evaluation of Defibrilators

T.G. F30.01.05 on Skeletal Devices and Transportation: F XXXX-XX, New Standard Specification for Skeletal Support Devices

T.G. F30.01.06 on Rescue Equipment: F XXXX-XX, New Standard Guide for Equipping and Training EMS Personnel Who Provide Ambulance-Based Rescue Services

T.G. F30.01.08 on Hazardous Environmental Incidents Equipment

T.G. F30.01.09 on Medical Equipment for EMS Vehicles:

F XXXX-XX, New Standard Specification for Basic Life Support (BLS) Equipment and Supplies
F XXXX-XX, New Standard Specification for Advanced Life Support (ALS) Equipment and Supplies

D. Committee F-31 on Health Care Services

Committee F-31 was organized in 1987 and has 110 members. Its scope is to develop needed standards (classificaitons, guides, practices, specifications, terminology, and test methods) for the delivery of cost effective quality services within a coordinated health care system through the voluntary cooperation of agencies, organizations, industry and consumer groups. The work of this committee will be coordinated with other ASTM committees and outside organizations working in the field to avoid duplication of effort.

Subcommittees of Committee F-31

F31.01 In-Home Services
F31.02 Patient Characterization
F31.03 Institutional Long Term Care
F31.04 Coordination of Health Care Services

Subcommittee F31.01 is currently developing a standard for respiratory care services in the home and is exploring the need for standards for products used by consumers at home as part of health care.

E. Subcommittee E20.08 on Medical Thermometry

E20.08, a subcommittee of E-20 on Temperature Measurement, has 40 members and jurisdiction over six standards. Its scope is to develop nomenclature and definitions, methods of testing, and specifications, for clinical thermometers and also to promote the knowledge of the theory, use, and applications of clinical thermometers through sponsoring symposia, manuals, or other publications.

Standards Developed by E20.08

E 667, Standard Specification for Clinical Thermometers (Maximum Self-Registering, Mercury-in-Glass)
E 825, Standard Specification for Thermistor Sensors for Clinical Laboratory Temperature Measurement
E 1061 Standard Specification for Direct Reading Liquid Crystal Forehead Thermometers
E 1104-86, Specification for Clinical Thermometers, Probe Covers, and Sheaths
E 1112-86, Specification for Electronic Thermometers for Intermittent Determination of Patient Temperature
E 1299-89, Specification for Reusable Phase Change-Type Fever Thermometer for Intermittent Determination of Human Temperature

VI. CONCLUSIONS

ASTM develops full consensus standards in the belief that with a broad input into the standard from the beginning of its development, the result will be technically competent and will have the highest credibility when critically examined and used as the basis for commercial or regulatory actions.

ASTM standards are used by thousands of individuals, companies, and agencies involved in the entire range of technical endeavor. Purchasers and sellers write them into contracts, scientists and engineers use them in their laboratories, architects and designers use them in their plans, government agencies reference them in codes, regulations, and laws, and all manner of technical people refer to them for guidance.

ASTM's management system assures all interested parties a voice in the development of standards that will affect their organization and industry. It exempts them from any personal liability in the development of standards. And it ensures their right to due process when they dissent.

ASTM standards are written by volunteer members who serve on technical committees. Anyone who is qualified or knowledgeable in the area of a committee's scope may become a committee member. If you have an interest in the activities of an ASTM Committee, you are welcome and encouraged to participate in the process. You need only to contact ASTM Headquarters and request information.

APPENDIX 1: STANDARDS OF COMMITTEE F04 ON MEDICAL AND SURGICAL MATERIALS AND DEVICES

Subcommittee F04.02 on Resources

F 55-82	Specification for Stainless Steel Bar and Wire for Surgical Implants
F 56-82	Specification for Stainless Steel Sheet and Strip for Surgical Implants
F67-88	Specification for Unalloyed Titanium for Surgical Implant Applications
F 75-87	Specification for Cast Cobalt-Chromium-Molybdenum Alloy for Surgical Implant Applications
F 86-84	Recommended Practice for Surface Preparation and Marking of Metallic Surgical Implants
F 90-87	Specification for Wrought Cobalt-Chromium-Tungsten-Nickel Alloy for Surgical Implants
F 136-84	Specification for Wrought Titanium 6AI-4V Eli Alloy for Surgical Implant Applications
F 138-86	Specification for Stainless Steel Bars and Wire for Surgical Implants (Special Quality)
F 139-86	Specification for Stainless Steel and Strip for Surgical Implants (Special Quality)
F 361-80	Replaced by F 981, Practice for Assessment of Compatibility of Metallic Materials for Surgical Implants with Respect to Effect of Materials in Muscle and Bone
F 382-86	Test Method for Static Bending Properties of Metallic Bone Plates
F 383-73(1989)	Recommended Practice for Static Bend and Torsion Testing of Intramedullary Rods
F 384-73(1986)	Recommended Practice for Static Bend Testing of Nail Plates
F 451-86	Specification for Acrylic Bond Cements
F 560-86	Specification for Unalloyed Tantalum for Surgical Implant Applications
F 561-87	Practice for Analysis of Retrieved Orthopedic Implants
F 562-84	Specification for Wrought Cobalt-Nickel-Chromium-Molybdenum Alloy for Surgical Implants
F 563-88	Specification for Wrought Cobalt-Nickel-Chromium
F 601-86	Practice for Fluorescent Penetrant Inspection of Metallic Surgical Implants
F 602-87	Criteria for Implantable Thermoset Epoxy Plastics
F 603-83	Specification for High-Purity Dense Aluminum Oxide for Surgical Implant Application
F 604-87	Classification for Silicone Elastomers Used in Medical Applications
F 619-79(1986)	Practice for Extraction of Medical Plastics
F 620-87	Specification for Titanium 6AI-4V Eli Alloy Forgings for Surgical Implants
F 621-86	Specification for Stainless Steel Forgings for Surgical Implants

F 624-81	Guide for Evaluation of Thermoplastic Polyurethane Solids and Solutions for Biomedical Applications
F 629-86	Practice for Radiography of Cast Metallic Surgical Implants
F 639-79(1985)	Specification for Polyethylene Plastics for Medical Applications
F 640-79(1987)	Test Methods for Radiopacity of Plastics for Medical Use
F 641-86	Specification for Implantable Epoxy Electronic Encapsulants
F 648-84	Specification for Ultra-High-Weight Polyethylene Powder and Fabricated Form for Surgical
F 665-80(1986)	Classification for Vinyl Chloride Plastics Used in Biomedical Application
F 668-88	Specification for Wrought Cobalt-Nickel-Chromium-Molybdenum Alloy Plate, Sheet
F 702-81(1985)	Specification for Resin for Medical Applications
F 719-81(1986)	Practice for Testing Biomaterials in Rabbits for Primary Skin Irritation
F 720-81(1986)	Practice for Testing Guinea Pigs for Contact Allergens: Guinea Pig Maximization Test
F 732-82	Practice for Reciprocating Pin-On-Flat Evaluation of Friction and Wear Properties of Polymeric Materials for Use in Total Joint Prosthesis
F 745-81(1988)	Specification for Stainless Steel for Cast and Solution Annealed Surgical Implant Applications
F 746-87	Test Method for Pitting or Crevice Corrosion of Metallic Surgical Implant Materials
F 748-87	Practice for Selecting Generic Biological Test Methods for Materials and Devices
F 749-87	Practice for Evaluating Material Extracts by Intracutaneous Injection in the Rabbit
F 750-87	Practice for Evaluating Materials Extracts by Systemic Injection in the Mouse
F 754-88	Specification for Implantable Polytetrafluoroethylene (PTFE) Polymer Fabricated in Sheet, Tube, and Rod
F 755-87	Specification for Selection of Porous Polyethylene for Use in Surgical Implants
F 756-87	Practice for Assessments of the Hemolytic Properties of Materials
F 763-87	Practice for Short-Term Screening of Implant Materials
F 786-82	Specification for Metallic Bone Plates
F 799-87	Specification for Thermomechanically Processed Cobalt-Chromium-Molybdenum Alloy for Surgical Implants
F 813-83(1988)	Practice for Direct Contact Cell Culture Evaluation of Materials for Medical Devices
F 882-84	Performance and Safety Specification for Cryosurgical Medical Instruments
F 895-84	Test Method for Agar Diffusion Cell Culture Screening for Cytotoxicity
F 897-84	Test Method for Measuring Fretting Corrosion of Osteosynthesis Plates and Screws
F 961-85	Specification for Cobalt-Nickel-Chromium-Molybdenum Alloy Forgings for Surgical Implant Applications
F 981-87	Practice for Assessment of Compatibility of Biomaterials (Non-Porous) for Surgical Implants with Respect to Effect of Materials in Muscle and Bone

F 982-86	Specification for the Disclosure of Characteristics of Surgically Implanted Clamps for Carotid Occlusion
F 988-86	Guide for Specifying Carbon-Fiber Randomly Reinforced Ultra-High-Molecular-Weight Polyethylene for Medical Devices
F 997-86	Specification for Polycarbonate Resin for Medical Applications
F 1027-86	Practice for Assessment of Tissue and Cell Compatibility of Orofacial Prosthetic Materials and Devices
F 1044-87	Test Method for Shear Testing of Porous Metal Coatings
F 1088-87	Specification for Beta-Tricalcium Phosphate for Surgical Implantation
F 1108-88	Specification for TI6A14V Alloy Castings for Surgical Implants
F 1109-87	Specification for Porous Composites of Polytetrafluoroethylene and Carbon for Surgical
F 1147-88	Test Method for Tension Testing of Porous Metal Coatings
F 1160	Practice for Constant Stress Amplitude Fatigue Testing of Porous Metal-Coated Metallic Materials
F 1185-80	Specification for Ceramic Hydroxylapatite for Surgical Implants
New Standard	Methods for the Determination of Total Knee Replacement Classification
F 1223-89	New Standard Test Method for Determination of Total Knee Replacement Constraint
New Standard	Practice for Friction and Wear Evaluation of Materials and Prosthetic Hip Design Combinations in Hip
New Standard	Test Method for Determining Strength and Setting Time of Synthetic Water Activated Polyurethane Orthopedic
New Standard	Terminology Relating to Polymeric Biomaterials in Medical and Surgical Devices

Subcommittee F04.03 on Orthopedics

F 116-85	Specification for Medical Screwdriver Bits
F 117-79(1985)	Test Method for Driving Torque of Self-Tapping Medical Bone Screws
F 339-71(1985)	Specification for Cloverleaf Intramedullary Pins
F 340-71(1987)	Specification for Pilot-Type Medical Countersinks
F 342-71(1987)	Specification for Counterlocking Screws (Thornton Type)
F 343-71(1987)	Specification for Counterlocking Screws (McLaughlin Type)
F 344-71(1987)	Specification for Conterbored Hex-Head Bolts (Thornton Type) Made from Stainless Steel
F 345-71(1987)	Specification for Counterbored Hex-Head Bolts (McLaughlin Type) Made from Stainless Steel
F 346-71(1987)	Specification for Cannulated Hex-Head Bolts (Thornton Type) Made from Cobalt-Chromium Alloy
F 347-71(1987)	Specification for Cannulated Hex-Head Bolts (McLaughlin Type) Made from Cobalt-Chromium Alloy
F 348-71(1987)	Specification for Hex-Head Bolts (Thornton Type) Made from Cobalt-Chromium Alloy
F 349-71(1987)	Specification for Hex-Head Bolts (McLaughlin Type) Made from Cobalt-Chromium Alloy
F 350-71(1987)	Specification for Hex-Head Bolts (Thornton Type) and Lock Washers Made from Stainless Steel

351-71(1987)	Specification for Hex-Head Bolts (McLaughlin Type) and Lock Washers Made form Stainless Steel
F 352-71(1987)	Specification for Lock Washers Made from Stainless Steel
F 353-71(1987)	Specification for Tri-Fin Nails Made from Cobalt-Chromium Alloy
F 354-71(1987)	Specification for Tri-Fin Nails Made from Stainless Steel
F 366-82(1987)	Specification for Fixation Pins and Wires
F 367-81(1985)	Specification for Holes and Slots for Inch Cortical Bone Screws
F 367M-82	Specification for Holes and Slots with Spherical Contour for Metric Cortical Bone Screws [Metric]
F 370-73	Specification for Proximal Femoral Prosthesis
F 453-76(1986)	Specification for Hooked Intramedullary Pins
F 454-76(1986)	Specification for Intramedullary Pins
F 455-78(1986)	Specification for Solid Cross-Section Intramedullary Nails
F 543-82	Specification for Cortical Bone Screws
F 544-77(1986)	Reference Chart for Pictorial Bone Screw Classification
F 564-85	Specification for Bone Staples
F 565-85	Practice for Care and Handling of Orthopedic Implants and Instruments
F 642-79(1984)	Specification for Stainless Steel Flexible Wire for Surgical Fixation for Soft Tissue
F 643-79(1984)	Specification for Wrought Cobalt-Chromium Alloy Flexible Wire for Surgical Fixations for Soft Tissue
F 644-79(1984)	Specification for Wrought Cobalt-Chromium Alloy Flexible Wire for Surgical Fixations for Bone
F 666-80	Specification for Stainless Steel Flexible Wire for Surgical Fixation for Bone
F 787-82(1986)	Specification for Metallic Nail Plate Appliances
F 983-86	Practice for Permanent Marking of Orthopedic Implant Components
F 1028-86	Specification for Tri-Nails
New Standard	Specification for Cementable Total Hip Prostheses with Femoral Stems
New Standard	Specification for Cementable Total Knee Prostheses
New Standard	Specification for Intramedullary Fixation Devices

Subcommittee F04.05 on Neurosurgical

F 452-76(1986)	Specification for Preformed Cranioplasty Plates
F 500-77	Specification for Self-Curing Acrylic Resins Used in Neurosurgery
F 622-79(1986)	Specification for Preformed Cranioplasty Plates That Can Be Altered
F 647-85	Practice for Evaluating and Specifying Implantable Shunt Assemblies for Neurosurgical Application
F 700-81(1986)	Practice for Care and Handling of Intracranial Aneurysm Clips and Instruments
F 701-81(1986)	Practice for Care and Handling of Neurosurgical Implants and Instruments
F 1043	Specification for Requirements and Disclosure of Aneurysm Self-Closing Clip Appliers
New Standard	Specification for the Requirements and Disclosure of Self-Closing Aneurysm Clips

New Standard — Performance Requirements for Cerebral Stereotactic Instruments

Subcommittee F04.06 on Plastic and Reconstructive Surgery

F 703-81(1986) — Specification for Implantable Breast Prostheses
F 881-84 — Specification for Silicone Gel and Silicone Solid (Non-Porous) Facial Implants

Subcommittee F04.08 on Medical/Surgical Instruments

F 899-84 — Specification for Stainless Steel Billet, Bar, and Wire for Surgical Instruments
F 921-85 — Definitions of Terms Relating to Hemostatic Forcers
F 1026-86 — Specification for General Workmanship and Performance Measurements of Hemostatic Forcers
F 1078-87 — Terminology for Surgical Scissors—Inserted Blades
F 1079-87 — Specification for Inserted and Noninserted Surgical Scissors
F 1089-87 — Test Method for Corrosion of Surgical Instruments

Subcommittee F04.12 on Urological Materials and Devices

F 623-81 — Performance Specification for Foley Catheter

APPENDIX 2: STANDARDS FOR COMMITTEE F29 ON ANESTHETIC AND RESPIRATORY EQUIPMENT

Subcommittee F29.01 on Division One on Anesthesia Apparatus

F0984-86 — Specification for Cutaneous Gas Monitoring Devices for Oxygen and Carbon Dixoide

Subcommittee F29.01.01 on Anesthesia Gas Machine

F 1161-88 — Specification for Minimum Performance and Safety Requirements for Components and Systems of Anesthesia

Subcommittee F29.01.02 on Breathing Systems/Performance

F 1208 — Specification for Minimum Performance and Safety Requirements for Anesthesia Breathing Systems

Subcommittee F29.01.03 on 15/22 mm Connectors and Adapters

F 1054-87 — Specification for Conical Fittings of 15 mm and 22 mm Sizes

Subcommittee F29.02 on Division Two on Endoscopes and Airways

New Standard — Specification for Cupped and Uncupped Tracheal Tubes

Subcommittee F29.02.03 on Breathing Tubes/Bags

F 1204 — Specification for Anesthesia Reservoir Bags

F 1205 Specification for Anesthesia Breathing Tubes

Subcommittee F29.02.04 on Tracheal Tube Connectors

New Standard Specification for Tracheal Tube Connectors

Subcommittee F29.02.06 on Tracheostomy Tubes—Pediatric

F 927-86 Specification for Pediatric Trachostomy Tubes

Subcommittee F29.02.07 on Laryngoscopes-Bulbs/Handles/Blades (Rigid)

F 965-85 Specification for Rigid Laryngoscopes for Tracheal Intubation-Hook-On Fittings for Laryngoscope Handles

F 1195 Specification for Rigid Laryngoscopes for Tracheal Intubation

Subcommittee F29.02.09 on Bronchoscopes (Rigid)

New Standard Specification for Bronchoscopes (Rigid)

Subcommittee on F29.03.01 on Lung Ventilators (Other Than Anesthetic Use)

F 1100 Specification for Ventilators Intended for Use in Critical Care

Subcommittee F29.03.02 on Lung Ventilators for Use in Anesthesia

F 1101 Specification for Ventilators Intended for Use During Anesthesia

Subcommittee F29.03.03 on Resuscitators

F 920-85 Specification for Minimum Performance and Safety Requirements for Resuscitators Intended for Use With Humans

Subcommittee F29.03.09 on Home Care Ventilators

New Standard Specification for Electrically-Powered Home Care Ventilators Part I—Positive Pressure Ventilators

Subcommittee F29.07.01 on Medical and Surgical Suction Systems

F 960-86 Specification for Medical and Surgical Suction and Drainage Systems

15

Role of USP

ARTHUR HULL HAYES *EM Pharmaceuticals, Inc., Hawthorne, New York*

JOSEPH G. VALENTINO *United States Pharmacopeial Convention, Inc., Rockville, Maryland*

I. DEVICE STANDARDIZATION

A. Early Activities

Unlike that of many other bodies, the standards-setting process of the *United States Pharmacopeia* (USP) is *not* based on consensus. Rather, the USP Convention utilizes a "republic-type" system in which scientific decisions are voted on by experts elected by individuals representing the various organizations in the pharmaceutical, medical, and other health-related fields. To appreciate fully the operation and role of the USP Convention, one must know something of its history.

The United States Pharmacopeia was founded in 1820 by physicians concerned with the proper identification and uniformity of the medicines used in their practice. The approach taken by these founding physicians (and later, pharmacists) was to standardize articles utilized in their practice that they already believed to be safe and effective. They standardized them through the provision of formulations by which they could be consistently compounded. They gave the preparations definite names so as to prevent confusion. In the late 1800s, the formula for compounding gave way to the standardization of products by end-product characterization, utilizing the analytical procedures of the developing sciences of chemistry, physics, and microbiology.

In sum, the USP standards and methods of analysis were not designed to measure the performance of the articles in humans, but were designed to identify or characterize an acceptable product that would reasonably be expected to produce the desired result or therapeutic effect in humans.

Because of its widespread acceptance, the USP became the "official" compendium of medicine and pharmacy—USP quality was expected and became the rule. A "not USP" product was an exception. Thus articles bearing the title or name used by USP were considered by health professionals to be of USP quality, whether or not the initials "USP" were present on the label. State food and drug laws and, later, the first federal food and drug act simply incorporated this theory and provided that a drug sold under a name recognized in the Pharmacopeia shall be deemed to be adulterated unless it met the standards of strength, quality, and purity contained in the Pharmacopeia. This section of the act allowed products to deviate from the standards, but only if the difference was clearly stated on the label so as not to mislead the professions and the public.

At the meeting of the USP Convention held in 1910, it was recommended that the next Pharmacopeia include reagents, with standards for their strength and purity as needed for the proper execution of tests used in diagnosis. The ninth revision of the Pharmacopeia, published in 1916, contained an extensive chapter providing formulas and methods of preparation for diagnostic reagents and chemical tests. This chapter did not appear in USP X, but appeared in the *National Formulary* (NF), a sister publication to the USP. (The NF and USP are now both revised by the same General Committee of Revision and are published together in one book with common supplements.) Consistent with compendial procedures, the reagents were standardized by specification of their formulations.

The ninth revision of USP also introduced a chapter on sterilization, not only for medicinal substances but covering such items as surgical dressings (cotton, bandages, gauzes, ligatures, etc.), glass, and metal utensils.

This was not the first attempt to include standards for a device. Cotton entered the USP as a monograph in 1880: first on a "drug basis" as an article needed for the preparation of pyroxylin, but later as an absorbent.

A monograph on surgical gut (chorda chirurgicalis) first appeared in the Second Supplement to USP XI published in 1939. Entry into the "surgical arena" expanded in USP XII, which contained monographs on absorbent gauze, adhesive absorbent gauze, sterile absorbent gauze, gauze bandage, adhesive plaster, sterile adhesive plaster, silk sutures, and sterile surgical silk, in addition to the monograph on surgical gut. Monographs on such surgical items continue to appear in the Pharmacopeia.

In order to establish uniform international standards for sutures an *ad hoc* Advisory Committee on Sutures was established in 1972. From this committee came an Industry (Revisions) Committee and later the U.S. Suture Industry Advisory Group. Since 1975 the group has acted as a technical advisor to the USP in the establishment of uniform suture standards and test methods. Group membership is open to all U.S. companies manufacturing sutures or prime components of sutures.

In order to help ensure the safety and efficacy of large-volume intravenously administered pharmacopeial preparations, USP XV introduced a chapter containing tests for transfusion and infusion assemblies.

USP XVI contained a separate chapter with monographs to cover "Adjuncts and Clinical Reagents." The introduction to those special monographs justified their presence as follows: "Pharmacopeial standards are provided in this section for several adjuncts and reagents used in medicine which, although not used on the person of the patient, are of such nature as to be closely associated with drugs so used." The chapter also declared that the standards should have the "same force and effect as those of the monographs section." Included were monographs for such items as dusting powder, blood grouping and typing serums, filtering media, and carbon dioxide absorbants.

Thus it seems that although historically, at times, USP attempted to distinguish those articles it did not consider to be drugs and took a somewhat limited approach, it was willing to provide standards when deemed necessary for articles utilized in medical practice, no matter how classified by regulatory authorities.

B. Current Standards

The 1938 Food, Drug, and Cosmetic Act recognized the *United States Pharmacopeia* as an "official compendium" and defined the term "drug" as an article recognized in it. No such recognition was extended in the definition for devices until the 1976 Medical Device Amendments. Although these amendments recognized articles in the USP as devices, there is, for medical devices, no adulteration provision comparable to that for drugs. Section 502(e) of the act requires a device to bear its established name if it has one. That section recognizes the official title contained in the USP or the NF as the established name, in the absence of any designation by the FDA. Section 508 of the act requires that two or more names may not be applied to a single device or to two or more devices that are substantially equivalent in design and purpose.

Thus it seems that the act anticipates some differentiation of medical devices by the USP or NF that will be governed by the misbranding provisions of the act.

In 1980, USP XX, in the General Notices, adopted new terminology for distinguishing articles for which monographs were provided. An "official substance" could also be a component of a finished device. A dosage form or a "finished device" was defined as the finished preparation or product of one or more official substances formulated for use on or for the patient.

At its meeting in 1975 the USP Convention had passed a resolution to expand the coverage of medical devices and diagnostic reagents.

Sentiment for standardizing devices was again expressed at the 1985 meeting, when the Convention delegates requested, by resolution, that the Committee of Revision identify those devices for which it could appropriately set standards.

The 1985 USP-NF (USPXXI-NFXVI) contains over 30 monographs for medical devices, including three newly developed monographs with widespread impact on the industry—biological indicators for sterilization cycles. Currently official devices and diagnostic agents include:

Adhesive Bandage
Gauze Bandage
Barium Hydroxide Lime
Biological Indicator for Dry-Heat Sterilization, Paper Strip
Biological Indicator for Ethylene Oxide Sterilization, Paper Strip
Biological Indicator for Steam Sterilization, Paper Strip
Anti-A Blood Grouping Serum
Anti-B Blood Grouping Serum
Blood Grouping Serums
Blood Grouping Serums Anti-D, Anti-C, Anti-E, Anti-c, Anti-e
Leukocyte Typing Serum
Oxidized Cellulose
Oxidized Regenerated Cellulose
Purified Cotton
Absorbable Dusting Powder
Erythrosine Sodium Topical Solution

Erythrosine Sodium Soluble Tablets
Fluorescein Sodium Ophthalmic Solution
Fluorescein Sodium Ophthalmic Strips
Absorbent Gauze
Petrolatum Gauze
Absorbable Gelatin Film
Absorbable Gelatin Sponge
Glucose Copper-Reduction Reagent Tablets
Glucose Copper-Reduction Reagent Tablets
Glucose Enzymatic Reagent Tablets
Glucose Oxidase and Sodium Nitroprusside Test Strip
Heparin Lock Flush Solution
Purified Rayon
Soda Lime
Sodium Nitroprusside Test Strip
Sodium Nitroprusside Reagent Tablet
Absorbable Surgical Suture
Nonabsorbable Surgical Suture
Adhesive Tape

There are also chapters dealing with devices including transfusion and infusion assemblies and elastomeric closures. In addition to monographs or chapters covering specifically "devices," the USP General Chapters have been adopted or adapted by purchasers, manufacturers, and the government and applied to medical devices, whether or not the device itself was recognized in the USP. For example, USP chapters on sterilization, sterility, safety testing, pyrogens, the bacterial endotoxins test, and plastics have had far-reaching impact.

II. THE ORGANIZATION

A. The Convention

The United States Pharmacopeial (USP) Convention is an organization composed mostly of individual members appointed as delegates from approximately 275 organizations in pharmacy and medicine. Organizations entitled to appoint delegates include each accredited college of medicine and pharmacy in the United States, each state medical and pharmacy society, 23 national organizations, and nine departments of the federal government (Fig. 1).

In addition, up to 30 individuals can be appointed as members at large by the USPC Board of Trustees because of their special scientific or administrative expertise, or to represent the public, or from foreign nations that recognize USP in their laws.

The United States Pharmacopeial Convention first met in 1820 and was incorporated in 1900 as a nonprofit organization.

The members of the USP Convention meet every 5 years, in Washington, D.C. At these meetings the officers and two "governing" bodies are elected by the members:

1. A Board of Trustees, which oversees the business management of the Convention.

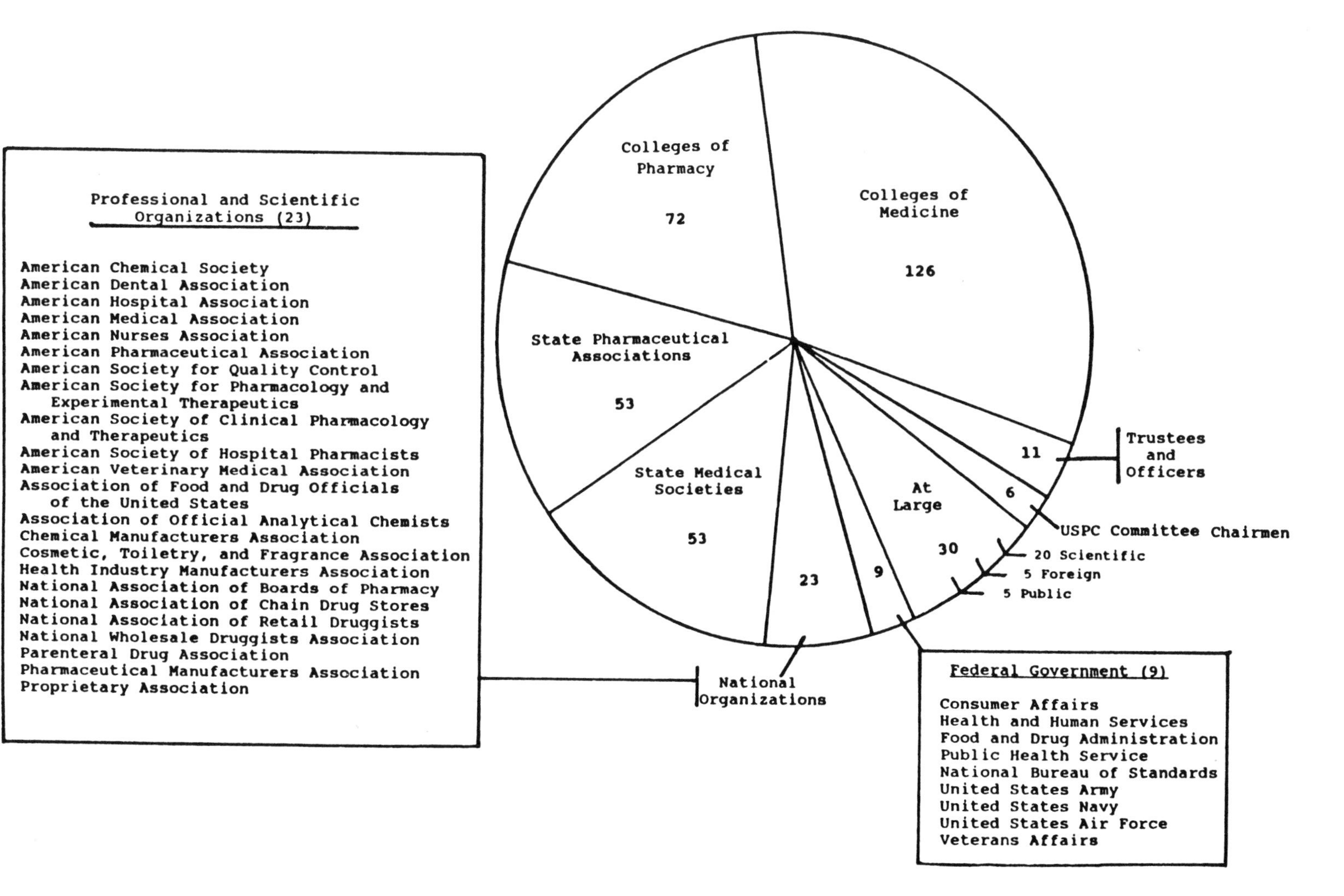

FIGURE 1 Composition of U.S.P. Convention.

2. The General Committee of Revision, which is responsible for the scientific revision of the various texts and data bases.

At this quinquennial meeting the delegates pass resolutions to give guidance to these two bodies for the ensuing 5-year period.

B. The USPC Board of Trustees

The 11-member Board of Trustees is composed of the officers plus seven elected members: two representatives of the medical sciences, two representatives of the pharmaceutical sciences, two at-large, and one public interest member. The Board meets at least annually.

C. The General Committee of Revision

The Secretary to the Board is also the Chairman of the Committee of Revision and Secretary of the Convention.

The General Committee of Revision for 1985-1990 is composed of 95 individuals. The total number can be expanded or contracted during a revision cycle depending on perceived need. Policy for the Committee is determined by the Executive Committee of Revision, which also serves as an appeals body.

The Committee of Revision is divided into two divisions: (a) the Drug Standards Division and (b) the Drug Information Division. Individuals are elected to the Committee of Revision as individual scientists, not as representatives of a particular manufacturer, government agency, university, or organization. Approximately 1 year before the quinquennial meeting, the classifications of service and the number of persons needed in each classification are determined. A call for the names of potential candidates is widely distributed throughout the industrial, academic, association, and government world. From the names submitted, a nominating committee nominates at least two individuals for each position. A background statement on each candidate is prepared and submitted to the Convention members prior to the election at the meeting of the Convention.

D. The Drug Standards Division

The Committee of Revision members elected to classifications of service needed in the revision of USP and NF are organized into subcommittees (17, in the 1985-1990 revision period). Each subcommittee is responsible for a particular assigned portion (Table 1), and individual monographs may be assigned to an individual member within the subcommittee. The chairmen of the subcommittees and the Executive Director constitute the Drug Standards Division Executive Committee. Expert advisory panels are appointed when needed to assist Subcommittees in the consideration of specialized subject areas (Table 2). Proposed revisions of the USP and NF are published in a bimonthly journal, entitled *Pharmacopeial Forum*, which is available by subscription. Supplements to the USP and NF are published every 6 months. Interim Revision Announcements are published as needed.

The revision process is continuous, and requests for revision of published standards may be made at any time. An appeals process is built into the system, but appeals are usually not necessary since the regular revision process can always reconsider a previous decision (see Fig. 2).

TABLE 1 Drug Standards Division Subcommittees

ANT.	Antibiotics
B&M.	Biochemistry and Microbiology
CH1.	Steroids
CH2.	Inorganics
CH3.	Chemistry
CH4.	Natural Products
CH5.	Chemistry
C&M.	Containers and Materials
GCR.	General Chapters and Reagents
HHC.	Home Health Care
IVT.	In-Vitro Toxicity
M&S.	Medical and Surgical Products
PAR.	Parenteral Products
PH1.	Dosage Forms and Excipients
PH2.	Biopharmaceutics
PUR.	Chemical Purity
RAD.	Radiopharmaceuticals

Documents under which the USP Convention operates are published in the USP and its Supplements. These include its Articles of Incorporation, Constitution and Bylaws, Rules and Procedures of the Committee of Revision, the USP Communications Policy, and the Document Disclosure Policy.

E. The Drug Information Division

The Committee of Revision members elected to classifications of service needed in the revision of USP DI (Dispensing Information) serve in the Drug Informa-

TABLE 2 Appointed Panels

Panel on Biological Tests and Assays
Panel on Bulk Packaging
Panel on In-Vitro Toxicity Testing
Panel on Moisture Specifications
Panel on Radiopharmaceuticals
Panel on Sterility and Microbial Attributes
Panel on Sterilization Indicators

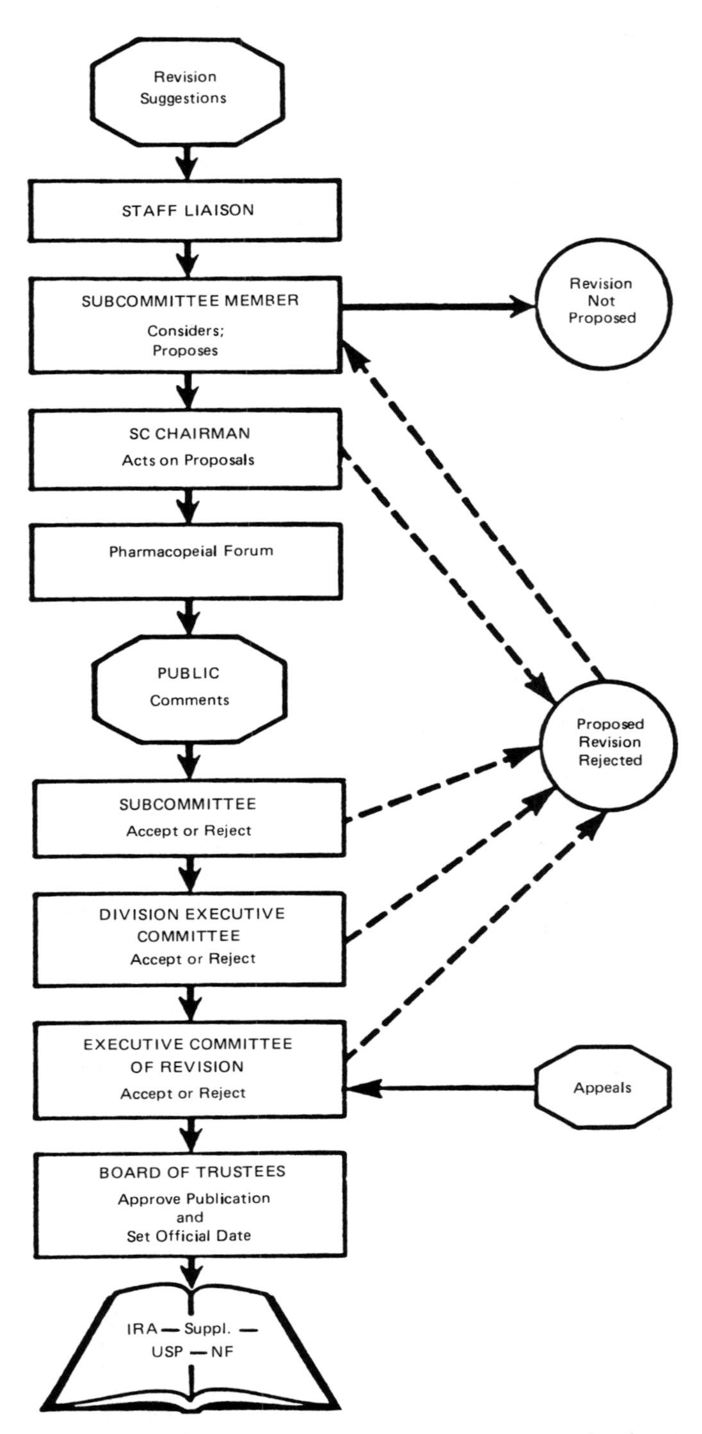

FIGURE 2 Revision process for USP-NF standards.

tion Division. USP DI is published annually in three volumes: Volume I, *Drug Information for the Health Care Professional* (two books); Volume II, *Advice for the Patient* (one book); and Volume III, *Approved Drug Products and Legal Requirements*. As is indicated by their titles, the USP DI volumes are concerned primarily with information about drugs. Information about medical devices has not yet been developed.

The Drug Information Division has a structure different from the Drug Standards Division. Only one representative from each medical specialty is actually elected to the Committee of Revision. That member then assembles a nationwide panel of experts in his field to assist in the work. These panels are primarily responsible for revising the text of USP DI, but they also provide their medical and professional views on standards-setting questions that arise in the course of USP-NF revision.

III. CURRENT ACTIVITIES FOR FUTURE STANDARDS

In the drug field, the *United States Pharmacopeia* is the preeminent pharmacopeia worldwide and is widely followed. About a fourth of the USP chemical reference standard materials, for instance, are sent outside the United States.

In considering the role of the USP in setting standards for devices, it must be remembered that the Pharmacopeia is not published in response to a statute or a regulation, but in anticipation of statute and regulation. The scope of compendial standards-setting depends on the desires of the professions. Once developed, the standards can be expected eventually to have far-reaching legal significance and consequences, either because of special recognition by law or through ordinary commercial contract requirements.

Not bound by the constraints of legal definitions and regulatory distinctions between devices and drugs found in statutes, the USP will continue to provide public standards for these articles based on perceived need by the members of its Convention and its Committee of Revision.

It is likely that Convention activities in the future will continue to focus on those medical devices either associated with medication delivery and utilization (e.g., syringes, drug delivery systems, etc.) or associated with physicians' decisions in the selection of treatment based on chemical testing (e.g., *in vitro* biological or physicochemical diagnostic tests).

A recent survey of complaints relative to *in vitro* diagnostics has indicated to the USP Committee of Revision that a need for public standards

TABLE 3 Primary Publications of the USP Convention

The United States Pharmacopeia[a]
The National Formulary[a]
The Pharmacopeial Forum
USP DI
USP DI Review
USAN and the USP *Dictionary of Drug Names*

[a]Published as one volume.

exists in the areas of *in vitro* diagnostic products frequently used by consumers at home, by physicians in their offices, by laboratories, and by hospitals. The Medical and Surgical Products Subcommittee has initiated the development of monographs for *in vitro* diagnostics for the determinationa of pregnancy and of fecal occult blood, and it has been decided to appoint a special advisory panel to provide expert advice and assistance in the development of public standards for pregnancy tests.

The new initiative of USP in the development of monographs for biotechnology-derived products will impact on the development of public standards for *in vitro* diagnostics, since a number of these products contain monoclonal antibodies derived via biotechnological processes.

Other activities in medical and surgical products include the development and publication of a proposed monograph on sterile disposable syringes and cartridges [*Pharmacopeial Forum*, *14*(5), 1988]. This proposed monograph contains requirements for particulate matter, sterility, bacterial endotoxins, and accuracy of delivery. Notable also is the proposal of requirements that the plastics and other polymers used in these syringes be tested by the biological reactivity tests. The latter were proposed following the USP Open Conference on Alternatives to Animal Testing—The In-Vitro Option. That conference, held in January 1988, set the direction of USP in its use of alternatives to *in vivo* tests for biological reactivity (toxicity) for elastomers, plastics, and other polymers, as well as the exploration of replacing identity, potency, and purity tests that use animals by *in vitro* biological and physicochemical tests. For example, it is proposed to expand the transfusion and infusion assemblies chapter by applying the *in vitro* testing requirements to all devices that are in direct contact with the cardiovascular system and other soft body tissues. Included are intravenous solution administration sets, blood administration sets, intravenous catheters, heart valves, vascular grafts, dialysers, and dialysis tubing and accessories.

The USP revision process is an open, scientific process, and industry is afforded ample opportunity to participate in the process. Employees from industry sit as individual scientists on USP Committee of Revision subcommittees and panels. Manufacturers interested in keeping abreast of USP activities should subscribe to the *Pharmacopeial Forum*. When exploring new areas, USP will often have open conferences designed to solicit opinions of experts in the area. At the least, manufacturers should inform USP of their interest in a particular device or subject area so that the USP can notify them in the event of a proposal or conference which may affect their product.

Proposals are published early in the revision process in *Pharmacopeial Forum* (PF) and are subject to comment and suggestions from all interested persons. USP recently instituted the PF "open house" concept to allow for free and open discussion at USP headquarters of any matter appearing in the most recently released PF.

USP standards and recognition may make marketing differences. When there is an issue of differing quality of products, the initials USP on the label of an article help to give assurance to health professionals that the article meets acceptable standards. Thus, USP requirements are often included in purchase specifications by hospitals and other large purchasers.

Pharmacopeial standards and specifications devised from this open process have benefited the public as well as purchasers and sellers of drugs and medical devices. The USP Convention is a unique organization, which can offer the professions and manufacturers an alternative to government regulation based solely on government standardization.

part four

Responding as Companies to the Regulatory Environment

16

How to Submit a Successful 510(k)

HOWARD M. HOLSTEIN *Dickstein, Shapiro & Morin, Washington, D.C.*

I. INTRODUCTION AND OVERVIEW OF THE 510(k) PROCESS

The Food and Drug Administration (FDA) oversees the investigation, manufacture, and marketing of medical devices and diagnostic products under the authority of the 1976 Medical Device Amendments (the Amendments) to the Federal Food, Drug, and Cosmetic Act (FDCA).

All medical devices[1] marketed by a manufacturer for the first time after the enactment of the Amendments on May 28, 1976, must undergo some form of FDA review before they can be sold. This is true whether the device is being marketed for the first time (a "new" device) or whether it is an "old" or "pre-Amendments" device that has undergone a significant change.

---

[1] A medical device is defined as "an instrument, apparatus, implement, machine, contrivance, implant, in vitro reagent, or other similar or related article, including any component, part, or accessory, which is—

(1) recognized in the official National Formulary, or the United States Pharmacopeia, or any supplement to them,
(2) intended for use in the diagnosis of disease of other conditions, or in the cure, mitigation, treatment, or prevention of disease, in man or other animals, or
(3) intended to affect the structure or any function of the body of man or other animals, and

which does not achieve any of its principal intended purposes through chemical action within or on the body of man or other animals and which is not dependent upon being metabolized for the achievement of any of its principal intended purposes" [21 U.S.C. 321(h)].

FDA's review of products sought to be marketed falls into one of two categories. For those post-Amendments devices that are truly new, for pre-Amendments devices that have truly new uses, and for certain pre-Amendments devices designated by FDA, a manufacturer must provide FDA with evidence of the product's *safety and effectiveness*. This is done by filing a premarket approval application (PMAA). For all other devices (e.g., newly introduced or changed post-Amendments devices that are similar to pre-Amendments devices) a manufacturer need only provide FDA with information to demonstrate the new or changed product's *substantial equivalence* to a pre-Amendments or "predicate" device. This information is provided to FDA in a premarket notification or 510(k).

In some instances where a device has the same intended use as a preamendments device, but incorporates a different technology, the FDA might require test data in what has become known as a "hybrid 510(k) notice" [1]. To many companies, the most important distinction between a hybrid 510(k) notice and a PMAA is that the 510(k) notice is processed more quickly than PMAAs. FDA's latest reports show that the average review itme for 510(k) notices was 69 days from the date of filing, while the average review time for PMAAs was 337 days from the date of filing. Furthermore, 510(k) notices have a greater chance of approval than PMAAs, and require less, if any, data. Many device experts who have been through the device review process have observed that hybrid 510(k) notices undergo less rigorous statistical analysis than PMAAs. Also, 510(k) notices can be approved by the Division Director without review by the Director of the Office of Device Evaluation (ODE). Also significant is the fact that advisory panel review is not required for acceptance of 510(k) notices, while all PMAAs require some form of panel review [2].

Since the 510(k) process applies to so many devices, its importance cannot be overemphasized.[2] It is the purpose of this chapter to provide the reader with an understanding of the substantive and procedural requirements that must be met in order to file a successful 510(k).

II. STATUTORY SCHEME

A. Classification

The FDCA specifies that a post-Amendments device (a medical device marketed after May 28, 1976) or a significantly changed pre-Amendments device (e.g., a new intended use for an existing device) is automatically placed in Class III. Before such a product can be marketed, it must undergo premarket approval or it must be reclassified into Class I or Class II, *unless* (1) it can be shown to be substantially equivalent (SE) to a device that was in commercial distribution prior to May 28, 1976, or (2) it can be shown to be substantially equivalent to a device that was introduced after May 28, 1976, when that device has been reclassified into Class I or II.

A post-Amendments device that has been found to be SE will be placed in the same class, and will be subject to the same requirements, as the device

[2]An average of nearly 5000 510(k)s has been filed each year since 1976.

to which it is SE.[3] In effect, this is a "grandfather" provision to ensure equal treatment for similar devices regardless of whether they were first sold before or after May 28, 1976.

B. Section 510(k)

In addition to establishing these broad product categories, Congress also determined that FDA should be provided with advance notice before a manufacturer initially markets a device. Section 510(k) of the Act, therefore, provides that:

> [E]ach person who is required to register under this Section and who proposes to begin the introduction or delivery for introduction into interstate commerce for commercial distribution of a device intended for human use shall, at least ninety days before making such introduction or delivery, report to the Secretary (in such form and manner as the Secretary shall by regulation prescribe)—
>
> (1) the class in which the device is classified under section 513 or if such person determines that the device is not classified under such section, a statement of that determination and the basis for such person's determination that the device is or is not so classified; and
>
> (2) action taken by such person to comply with requirements under section 514 [standards] or 515 [PMA] which are applicable to the device.

Thus, if read literally, Section 510(k) merely requires a person who plans to introduce a new or changed device into commercial distribution to notify FDA at least 90 days in advance of the first product sale.

---

[3]The actual statutory language setting forth this requirement is contained in Section 513(f)(1) of the FDCA. This section states as follows:

> [A]ny device intended for human use which was not introduced or delivered for introduction into interstate commerce for commercial distribution before the date of the enactment of this section is classified in class III unless—
>
> (A) the device—
>
> (i) is within a type of device (I) which was introduced or delivered for introduction into interstate commerce for commercial distribution before such date and which is to be classified pursuant to subsection (b), or (II) which was not so introduced or delivered before such date and has been classified in class I or II, and
>
> (ii) is substantially equivalent to another device within such type, or
>
> (B) the Secretary in response to a petition submitted under paragraph (2) has classified such device in class I or II.
>
> A device classified in class III under this paragraph shall be classified in that class until the effective date of an order of the Secretary under paragraph (2) classifying the device in class I or II.

In practice, however, FDA often uses 510(k) notifications to grant permission to market a device based on a determination of whether a new device is, or is not, SE to either a pre-Amendments device or a reclassified post-Amendments device. If a device is found to be not substantially equivalent (NSE) a PMA must be obtained before the device can be sold.[4] An understanding of the criteria used by FDA to determine "substantial equivalence" is of great importance to a manufacturer seeking market entry via the 510(k) route.

Unfortunately, in enacting Section 510(k) of the law, Congress provided no guidance on the appropriate scope of premarket notifications. What little guidance was given to help FDA and manufacturers relates solely to the term "substantial equivalence."[5] FDA has issued its own guidance that outlines what factors the agency considers in making a determination that a device is, or is not, substantially equivalent.[6] In FDA's view, substantial equivalence must not be too liberally nor too conservatively interpreted. If this sounds confusing, in practice it often is, especially with regard to changed products. However, the best way to try to understand how a product is determined to be substantially equivalent is to begin by examining the requirements of the 510(k) regulation.

III. 510(k) REGULATORY GUIDANCE

A. FDA's Regulatory Manual

Since 1977 when the final 510(k) regulation was published, FDA has issued two guidance documents on the 510(k) program. Initial FDA pronouncements attempting to provide some guidance on 510(k) submissions were contained in the preamble to the final 510(k) regulation as discussed previously. FDA next issued a manual used in workshops conducted by its Division of Small Manufacturers Assistance [3]. This manual has been updated several times since 1977, the most recent being in 1985. It is, however, fairly general in scope. The most useful information contained in the 510(k) chapter consists of several examples of simple 510(k)s.

---

[4]As will be described more fully within, a 510(k) generally is a relatively simple application, it often requires little or no data to demonstrate substantial equivalence, and it must be reviewed within 90 days. In contrast, a PMA is far more complex and lengthy. It must prove safety and effectiveness based on scientific evidence. Generally it requires clinical data for support, and the statutorily prescribed FDA review time is 180 days.

[5]The pertinent House report, in describing when a device will be substantially equivalent to another for purposes of section 510(k) clearance, states that "differences between 'new' and marketed devices in materials, design, or energy source, for example, would have a bearing on the adequacy of information as to a new device's safety and effectiveness" and that "devices whose variations are immaterial to safety and effectiveness" would not necessarily require notification [H.R. Rep. No. 853, 94th Cong., 2d Sess. 36-37 (1976)]. This language seems to indicate that Congress intended to include some, but not all, device changes in the premarket notification requirement.

[6]This "Guidance on the Center for Devices and Radiological Health's Premarket Notification Review Program" ("Guidance") is discussed in greater detail in Section III.

B. FDA's Guidance Documents—Initial Submissions[7]

1. Background

The most important, and the most useful, guidance was issued by FDA on June 30, 1986. Entitled "Guidance on the Center for Devices and Radiological Health's Premarket Notification Review Program" (Guidance), this document is intended to outline FDA's internal requirements for 510(k) review and to explain what the Center for Devices and Radiological Health (CDRH) considers when making a substantial equivalence determination.

2. Content

The Guidance specifies that all new devices and newly changed devices that are marketed under a 510(k) must be shown to be substantially equivalent to a predicate device. A predicate device is defined as one that

> was in distribution prior to enactment of the Amendments, or is a post-Amendments device that was subject to reclassification from Class III to Class I or Class II. [4]

Significantly, the guidance says that FDA does not routinely require manufacturers to determine specific predicate devices that were available in 1976, nor does FDA even routinely require that all 510(k)s initially provide information on a predicate device. Rather, CDRH will require 510(k) submitters to provide information that compares the device sought to be marketed to a device of a similar type that is presently on the market, regardless of whether this marketed device was marketed before or after enactment of the Amendments, or before or after this type of post-Amendments device was reclassified.

This means that a manufacturer can claim substantial equivalence to a post-1976 device so long as the predicate product itself was found to be substantially equivalent to a pre-1976 device. This is sometimes referred to as "piggybacking."

This is an extremely significant concept since it provides FDA and manufacturers with a reasonable amount of flexibility to make changes in devices and to keep up with medical and technological advances, while still being able to take advantage of the more simple 510(k) mechanism. In this manner, manufacturers may ultimately market devices that are different from a pre-1976 model without having to file a PMA.[8]

Even though the law speaks in terms of "substantial equivalence," not safety, the Guidance makes it clear that FDA will be more likely to find a device to be substantially equivalent if FDA is given assurances that the prod-

---

[7]The guidance document was written by FDA to assist with the submission of 510(k)s for new products. However, some of the information it provides is equally applicable as well to 510(k)s for changed devices, as will be discussed in this chapter.

[8]The Guidance goes on to say that the ultimate burden of demonstrating the substantial equivalence of a new device to a predicate device remains with the 510(k) submitter, and on those occasions when CDRH is unfamiliar with certain aspects of the predicate device, the submitter will be required to provide information that substantiates a claim of substantial equivalence.

uct is in fact safe and effective and that the changes that have been made will not raise public health concerns.

The remainder of the Guidance document describes in detail what CDRH considers during a 510(k) review to justify claims of substantial equivalence.

CDRH, in examining a 510(k), will ask the following key questions:

1. Does the new device have the same intended use as a predicate device?
2. Does the new device have the same technological characteristics, that is, the same materials, design, energy source, etc.?
3. If the new device has new technological characteristics, could they affect safety or effectiveness?

A new device generally will be considered to have the same intended use if it has the same labeled indications, but even a device with different indications may have the same intended use if the differences do not alter the intended therapeutic effect. The key is whether the new indications raise questions about safety or effectiveness different form those raised by the marketed device. For example, a dialyzer for a heart-lung machine, to be used after surgery, has been held to have the same intended use as a dialyzer for an artificial kidney system for patients with renal failure, as both are intended to remove excess water from the vascular system.

FDA says that if the new device has a new intended use (a critically important concept, discussed next), it is automatically considered not substantially equivalent. In contrast, if the new device has the same intended use as a predicate device and the same technological characteristics regarding safety and effectiveness, it is considered substantially equivalent. If, however, the new device has the same intended use as a predicate device but new technological features that could affect safety or effectiveness, then FDA asks the following additional questions:

1. Do the new technological features pose the same types of questions about safety or effectiveness as are raised by the predicate device with the same intended use? If not,
2. Are there accepted scientific methods for evaluating whether safety or effectiveness has been adversely affected as a result of the use of new technological characteristics? If so,
3. Is there information to demonstrate that the new technological features have not diminished safety or effectiveness?

If the reviewer finds the intended use of the new device to be equivalent to that of a predicate device, FDA's next step is to determine whether the new device has novel technological characteristics that would raise new types of safety or effectiveness questions. If so, the device is not substantially equivalent. If the safety and effectiveness issues are familiar, the substantial equivalence of a device with new technology that could significantly affect safety or effectiveness is demonstrated by submitting to the agency data to demonstrate that the new device is similar to a marketed device in terms of safety and effectiveness.

Manufacturers submitting 510(k)s for new devices that may raise safety or effectiveness questions, or where changes from the pre-1976 product are significant, should be certain to include answers to all of these questions in a 510(k) submission.

The Guidance amplifies these points and essentially establishes a road map to follow to make an SE determination. First, FDA suggests asking whether the new or changed device has the same intended use.

In order to have the same intended use, FDA recognizes that new or changed devices do not need to be labeled with precise therapeutic or diagnostic statements that are identical to those that appear on predicate device labeling. CDRH says that it will rely on its own scientific expertise to exercise "considerable discretion in construing intended uses in labeling and promotional materials for predicate and new devices" [4, p. 5].[9] As long as the label indications do not raise questions about safety or effectiveness different from those posed by the predicate device, a new device may be substantially equivalent even if its labeling differs from the predicate device—in other words, it does not have the precise intended use that the predicate device has.

CDRH will assess differences in intended use or in labeled indications by looking at the safety and effectiveness questions that may be raised. In doing this, CDRH will consider such matters as physiological purpose, the condition or disease to be treated or diagnosed, whether the product is for professional or lay use, and other similar matters. As long as FDA finds that the device has the same general intended use, the device can be found to be substantially equivalent.[10]

Different intended uses automatically rule out substantial equivalence determinations. Thus, as noted above, it is critical to carefully determine on which products a claim of substantial equivalence will be based and to favorably compare the intended uses of these products with the device that is sought to be marketed. There is no reason why substantial equivalence cannot be claimed to two or more products, borrowing something from each and combining these in one device.

When a new device seeks substantial equivalence to types of devices from different classes, FDA will examine each of the features and relate them to predicate devices from those classes. When such a new device is found to be substantially equivalent, FDA will place it in the regulatory category of the highest predicate device (e.g., Class III rather than Class II).

---

[9] As a general rule, intended use will be determined by reference to product labels, labeling, or promotional claims. FDA defines labels and labeling as follows:

> The term "label" means a display of written, printed, or graphic matter upon the immediate container of any article; and a requirement made by or under authority of this Act that any word, statement, or other information appear on the label shall not be considered to be complied with unless such word, statement, or other information also appears on the outside container or wrapper, if any there be, of the retail package of such article, or is easily legible through the outside container or wrapper.
>
> The term "labeling" means all labels and other written, printed, or graphic matter (1) upon any article or any of its containers or wrappers, or (2) accompanying such article. [21 U.S.C. 321(k) and (m)]

[10] Pages 6-7 of FDA's 510(k) Guidance lists five product examples to illustrate the concept of intended use. Safety concerns clearly affect FDA's intended use interpretations.

Once FDA has determined that the new or changed device has the same intended use as a predicate device, it next is necessary to determine if there are differences in technological features. If not, the device will be substantially equivalent. If there are different technological features but these could not affect safety or effectiveness, the device also will be substantially equivalent. Finally, if there are new technological features and these could affect safety or effectiveness, the device may or may not be substantially equivalent. The answer to these questions depends on whether, even if the technological differences may affect safety or effectiveness, there is sufficient information available to assure FDA that safety and effectiveness have not been diminished.

FDA primarily is interested in differences that are medically and scientifically significant. Thus, FDA will look at how the product is to be used, what FDA knows about the material or the changes from other products, and who will be using the product.[11]

An additional area covered by the Guidance concerns the often-asked questions of what type and what amount of data are required to support a 510(k). FDA says that in order to appropriately determine whether or not a product is substantially equivalent, it is necessary to always include some descriptive data that help to explain a new or changed device's use, physical composition, method of operation, specifications, performance claims, etc.

FDA also may require performance testing information in certain circumstances. Performance testing information includes such things as bench data, animal tests, or clinical studies. These data are supplied to demonstrate that the product performs according to its description. Those 510(k)s that contain such test data are sometimes referred to as "hybrid 510(k)s."

A 510(k) is not intended to demonstrate a product's safety or effectiveness. This is the purpose of a PMAA. Therefore, data must support comparability of a new device to a predicate product, not absolute safety or effectiveness. While this may be a less rigorous standard, the data supplied to FDA still must be valid, on point, and supportive of the equivalence claim.

FDA generally does not require performance test data if a new device does not have an important descriptive or functional difference and its descriptive characteristics are sufficiently precise to ensure that comparability will exist. In contrast, performance data generally will be required if a new or changed device has important descriptive differences from a predicate product and it is not clear that the new or changed device has an intended use or a technological change that makes it not substantially equivalent. Performance test data also may be required if the new or changed device has descriptive characteristics that are too imprecise for FDA to evaluate performance comparability.

The Guidance document described above applies to all product areas. Additionally, FDA, at times, will issue specific guidance relating to individual product types. Specific guidance often will describe required labeling, data,

---

[11]Pages 8-11 of the Guidance list examples of devices that have undergone changes in technological features. The Guidance analyzes each of these examples to illustrate FDA's concerns and describes how the CDRH internal review process reached a substantially equivalent/not substantially equivalent decision with respect to each product.

performance characteristics, or warnings or contraindications.[12] Meeting this specific guidance may be a prerequisite to the filing of a successful 510(k). Unfortunately, there presently is no efficient mechanism for determining the existence of specific guidance documents within all of FDA's reviewing divisions. It is therefore a good idea to check with FDA or with FDA-knowledgeable persons before filing a 510(k) to determine whether any guidance exists.

C. FDA's Guidance Documents—Modified Devices

1. Background

In establishing regulations for premarket notification, FDA has indeed taken the position that certain modifications require premarket notification. According to the regulations, a premarket notification submission ("510(k) notice" or "510(k)") is required when "the device is one that the person currently has in commercial distribution . . . but that is about to be significantly changed or modified in design, components, method of manufacture, or intended use" [21 C.F.R. 807.81(a)(3) (1988)]. Thus, only significant changes require new notifications.

The regulations define as significant changes those that "could significantly affect the safety or effectiveness of the device." Examples provided include significant changes in design, material, chemical composition, energy source, or manufacturing process [21 C.F.R. 807.81(a)(3)(i)]. In addition, a major change in intended use would require a new notice [21 C.F.R. 807.81-(a)(3)(ii)]. By implication, then, a modification that would not significantly affect safety or effectiveness or a minor change in the intended use may not require a new 510(k) notice.

Obviously, these regulations add very little of practical use to the guidance in the Act and its legislative history. The regulations are explained to some extent in their preamble, which was published in the *Federal Register* [5]. Even the preamble, however, does not provide much substantive help in determining when changes may be considered significant. The preamble states that the agency does not intend a company to submit a premarket notification for every change in design, material, chemical composition, energy source, or manufacturing process. However, at least in theory, a notice is required for a significant change even if the change actually increases the safety of effectiveness of the device. FDA states that the burden of determining whether a premarket notification should be submitted falls initially on the manufacturer, and adds that FDA will inform the manufacturer if a notice that has been submitted was not needed [6].

FDA has for years promised the industry a guidance document to describe specifically when a 510(k) is required for a change to a medical device, but none has been forthcoming. Accordingly, in deciding when to file a 510(k) notification, the industry has relied on various informal documents issued by the agency and on knowledge gained from experience in negotiating 510(k) clearances.

---

[12] Examples of specific guidance include guidance for ultrasound products and for TENS devices. FDA also has issued generic guidance to its reviewers regarding software review as part of the 510(k) process: "Reviewer Guidance for Computer-Controlled Medical Devices" (July 15, 1988).

Primary among the existing FDA guidance documents regarding determinations of substantial equivalence is a June 30, 1986, memorandum from Dr. Kshitij Mohan, discussed above. While this memorandum does not specifically address the issue of when to file 510(k)s for changes in medical devices, it does describe the factors a reviewer will take into consideration in determining substantial equivalence. By inference, these factors may be useful to determing whether a change in a device is sufficiently significant to require a premarket notification.

2. Content

As noted, the June 30, 1986 guidance does not specifically address the issue of when modifications to a marketed device rise to a level requiring a new Section 510(k) notification. There are, however, two other documents that shed some additional light on the agency's views regarding modifications to a marketed device. First, FDA wrote in 1984 to Rodney R. Munsey, Esq., purporting to describe the criteria that the agency applies internally to determine the kinds of changes that might be exempted form Section 510(k) filing. Mostly, the letter reiterates FDA's regulations in stating, first, that changes in design, energy source, materials or components, manufacturing process or method, or operational principle may be significant enough to require a 510(k) filing, and second, that the decision is up to the company. However, the letter also provides examples of changes that would require filing:

Design: a change or modification in design that could significantly affect the safety and effectiveness of the device (e.g., a proposed change that may result in increased or decreased output from the device, or physical breakage).

Energy source: a change or modification in energy source that could significantly affect the safety and effectiveness of the device (e.g., a change from a mechanically powered to an AC-powered device).

Materials or components: a change or modificaiton in a material or component that could significantly affect the safety or effectiveness of the device (e.g., a change to a material that is not biocompatible, or to a component that lacks sufficient strength).

Manufacturing process or method: a change or modification in a method or process that could significantly affect the safety or effectiveness of the device (e.g., a change from conventional methods of antibody production to the use of monoclonal antibody technology).

Operational principle: a change or modification in an operational principle that could significantly affect the safety or effectiveness of the device (e.g., for a diagnostic device, a change in the formula used to calculate the test results, the interpretation of results, or changes in the sequence of procedures performed during the determination).

The second document, an undated FDA draft entitled "Process Changes," contains guidance concerning when a 510(k) filing is necessary for changes to sterilization processes. A sterilization process falls within the "manufacturing process" category, one of the areas singled out as possibly having a significant effect on safety or effectiveness. FDA's guidance regarding process changes states that sterilization processes present certain assurances of safety and effectiveness in that they are subject to GMP requirements for process validation and change control. Moreover, this area is inspected by

FDA biennially. Accordingly, the guideline states that "some manufacturers with highly qualified personnel and substantial experience may feel confident in determining that a particular change in either the process or finished-device release mechanism will not significantly affect the safety, effectiveness or quality of the device." According to the document, assurance that a change will not significantly affect safety and effectiveness can be supplied through change control procedures, equipment qualification and calibration, process validation, personnel training, and development of routine sterilization procedures. FDA advises that if a manufacturer remains uncertain as to whether the change could affect safety or effectiveness, then a 510(k) should be submitted so that the agency can identify undesirable aspects of the proposed changes if they exist. A company should keep on file the data supporting a decision not to file a 510(k) notice.

IV. 510(k) DECISION MAKING

A. Presubmission Considerations

The first question to answer is whether it is necessary to file a 510(k). In some respects, this may be both the simplest and the most difficult part of the 510(k) process. It is simple to determine whether a 510(k) is necessary if you have never marketed the product before. In this case a 510(k) always must be filed. It is more difficult to determine whether to file a 510(k) when you are making a product change. As discussed above, in this situation you must determine whether the change could have a significant effect on safety or effectiveness. Although this determination should be based on objective facts, it often leads to a largely subjective evaluation.

Once you decide to file a 510(k) it then is necessary to determine who actually will prepare the submission. In many companies, the marketing department may be the preparer of a 510(k); in other companies it may be regulatory affairs, outside legal consultants, or some combination of the above.

While there is no perfect formula for every company, it generally is a good idea to have a representative from all of the above groups at least involved in a review of the 510(k) before it is submitted. This is because each group brings to the review process expertise that is both different and valuable. Regulatory affairs will help make certain that all FDA requirements are met, marketing will ensure that sales and marketing needs are reflected in the 510(k), and legal will help ensure that FDA and other statutory requirements (such as patents and trademarks) and product liability implications are considered.

Once it is determined who will be responsible for drafting and for reviewing the 510(k) the next decision to make is what products will serve as predicate devices. Clearly, if you are marketing a virtual copy of an existing device, the decision is not difficult. However, when you are marketing new products or new technologies, some creativity may be required to be certain that sufficient information will be given to FDA to permit CDRH to find your product to be substantially equivalent. This may necessitate using two or more devices as predicate products, or conducting bench, *in vitro*, or clinical tests to support certain product applications.[13]

---

[13] It may be necessary to obtain an investigational device exemption (IDE) from FDA in order to clinically study a product for which a 510(k) is intended to be filed.

An additional presubmission consideration is to be certain that you have carefully and completely defined the device for which you intend to submit a premarket notification. Lack of an appropriate definition can be extremely problematic. For example, defining the intended use of a product in an overly broad manner may take it outside the realm of predicate devices and necessitate the filing of a PMAA. More narrowly defining intended use, even if the product may be used more broadly, may permit you to obtain a 510(k). The broader use can then be the subject of a later application, or of no application at all as long as no labeling claim for the "unapproved" use is made.

One final point with respect to presubmission considerations concerns timing. By law, FDA has 90 days to review a 510(k). Generally, CDRH responds to 510(k)s within this 90-day period. However, not all 510(k)s receive responses within 90 days.[14] While the vast majority of products go through the 510(k) process within the 90-day period, the more difficult 510(k)s, or those involving more sophisticated products or products that have undergone significant changes, often take longer. This is because FDA scrutinizes these applications much more closely and generally adopts a cautious approach to their review. Additionally, FDA may raise questions during the review process. If these questions are lengthy or require a significant amount of time to prepare a response, FDA may restart the 90-day review clock. FDA also can treat the submission of new information as starting the process over and thereby turn the clock back to day one. This allows an additional 90 days to review the 510(k). Thus, particularly for complex 510(k)s, the actual review period, from submission to clearance, may exceed 90 days.

Many companies do not begin to consider 510(k) implications until changes either have been implemented or are close to being finalized, or until new products are nearing the stage at which they can be introduced to the market. Making these 510(k) decisions at the last minute is a poor strategy. It puts pressure on the company to decide whether to file a 510(k) based on marketing rather than legal and regulatory concerns, and it can result in a hurried submission or a poor application. All of these factors can result in significant delays in entering the market. Companies should therefore include time considerations in their decision whether to file a 510(k), and should permit time for preparation and review in their overall marketing strategy. This means that marketing, sales, engineering, and other departments all may need to be aware of 510(k) requirements. All these groups also may need to be involved in the development of new or changed products to facilitate the timely submission of a 510(k).

B. When Is a 510(k) Not Necessary?

1. Regulatory Exemptions

There are certain situations in which it is clear that 510(k)s need not be filed. First, a 510(k) is not required if the potential submitter is not subject to the FDA's registration requirements described earlier in the Guidance manual. Second, a 510(k) is not required for products that are exempt from the need to file a premarket notification. FDA has published a list of predominately

[14] If you do not receive any word from FDA within 90 days of submitting a 510(k), technically you are free to begin marketing the product. The subject is discussed further in Section V,C.

Class I devices that have been exempted from 510(k) requirements. Additionally, an even longer list of devices proposed for exemptions also has been published. Third, it is not necessary to file a 510(k) for products that are marketed by another company under its own 510(k) where you are merely serving as the distributor or repacker of the product and where you do not make labeling or other changes. Fourth, it is not necessary to submit a 510-(k) when you are selling a custom device.[15]

2. *Nonsignificant Changes*

The above exemptions from 510(k) are relatively limited. The broadest potential exemption, yet one that is not defined in the regulations with any specificity, is for product modifications that do not significantly affect safety or effectiveness. As discussed above, both the 510(k) regulation and the FDA's Guidance provide some broad suggestions on what may or may not be a change that significantly affects safety or effectiveness. The regulation refers to a significant change or modification as being one that affects "design, material, chemical composition, energy source, or manufacturing process" [21 C.F.R. 807.81(a)(3)(i)].

In sum, the decision whether to file a 510(k) for a changed product can be a difficult one. The decision, however, rests with the company marketing the product. It is up to the manufacturer to determine whether a change could significantly affect safety or effectiveness and therefore require a 510(k).

The company should not make this decision lightly. Whatever decisions are made must be well supported. It is a good idea in making a determination whether or not to file a 510(k) to follow the path set forth in FDA's Guidance. If the decision is made that the change is not significant then no 510(k) needs to be filed. On the other hand, if a change is significant but falls within the FDA's Guidance, then a 510(k) should be filed and the company should expect to receive a substantial equivalence determination from FDA.

FDA, of course, is free to disagree with a manufacturer's determination that it is not necessary to file a 510(k). Generally, FDA will not be aware of a company's decision. However, if FDA becomes aware of a change and disagrees with the company, there are several steps that a company can take to protect itself. Decisions to file or not file a 510(k) always should be based on facts and on legal/regulatory requirements. Once a decision is made not to file a 510(k), it is a good idea to prepare a file memorandum documenting the reasons for the decision. FDA may question the decision, but if it does, a contemporaneous memorandum prepared by the company will be of great assistance in demonstrating the basis for and the correctness of the company's decision. The rationale will not sound as strong if it is made "after the fact"—in other words, after FDA has objected.

---

[15] A custom device is one that is not generally available in finished form for purchase and is not offered through labeling or advertising by the manufacturer, importer, or distributor for commercial distribution and the device meets one of the following conditions: (1) it is intended for use by a patient named in the order of a physician or other health care professional; and (2) it is intended solely for use by a physician or other health care professional and is not generally available to, or generally used by, other physicians or health care professionals.

Just because FDA disagrees with a 510(k) decision does not mean that a 510(k) must be filed. There are opportunities to discuss this with the FDA reviewing division in the Office of Device Evaluation and with more senior FDA management as well. Here, too, a memorandum in support of your position can help convince FDA that your decision was carefully thought out and based on a supportable rationale. Even if FDA continues to disagree and asks for a 510(k), unless there are serious public health concerns associated with use of the product without agency clearance, CDRH may not request that you stop marketing the device. Ultimately, if a company and FDA are not able to reach agreement on whether a 510(k) is necessary, appeal may be had to the courts.[16]

V. HOW TO FILE A 510(k)

A. Types of 510(k)s

1. *Simple Submissions for Identical Devices*

The 510(k)s for simple changes, or for identical or "me-too" devices, should be kept simple and straightforward. The submission should refer to one or more predicate devices, it should contain samples of labeling, it should have a brief statement of equivalence, and it may be useful to include a chart listing similarities and differences.

2. *Submissions for Equivalent but Not Identical Devices*[17]

This type of 510(k) should contain all of the information listed above. It also should contain sufficient data to demonstrate why the differing characteristics or functions do not affect safety or effectiveness. Submission of some functional data may be necessary. It should not be necessary, however, to include clinical data—bench or preclinical testing results should be sufficient. Preparing a comparative chart showing differences and similarities with predicate devices can be particularly helpful to the success of this type of application.

3. *Complex Submissions for Complex Devices or for Major Differences in Technological Characteristics*

This is the most difficult type of submission, since it begins to approach the point at which FDA will need to consider whether a 510(k) is sufficient or whether a PMAA must be submitted. The key is to demonstrate that the new features or the new uses do not diminish safety or effectiveness and that there are no significant new risks posed by the device. In addition to submission of the types of information described in the above examples, this type of submission will almost always require submission of some data, possibly including clinical data.

---

[16] This is not a course to be undertaken lightly. Litigation with FDA can be expensive, time-consuming, and the courts generally favor FDA's scientific determinations.

[17] This category includes combination devices where the characteristics or functions of more than one predicate device are relied on to support an SE determination.

As a general rule, it often is a good idea to meet with FDA to explain why the product is substantially equivalent, to discuss the data that will be submitted in support of a claim of substantial equivalence, and to learn FDA's concerns and questions so that these may be addressed in the 510(k) submission. Also, the FDA's Guidance document can be of greatest use in preparing this type of submission. The Guidance can suggest the type of questions FDA is likely to ask and can provide clues to the type of information that will satisfy the agency's concerns.

This is the type of 510(k) that may, from the date of initial submission until final clearance, exceed the 90-day review period, since FDA may raise questions during the review process that restart the regulatory clock.

4. *Software-Controlled Devices*

FDA recently has expressed greater interest and concern about software-controlled devices. This concern is illustrated by FDA's recently released "Reviewer Guidance for Computer-Controlled Medical Devices" (revised July 25, 1988).

This guidance applies to the software aspects of 510(k) submissions for medical devices. "Software" includes programs and/or data that pertain to the operation of a computer-controlled system, whether they are contained on floppy disks, hard disks, magnetic tapes, "laser" disks, or embedded in the hardware of a device, usually referred to as "firmware."

The depth of review by FDA is determined by the "level of concern" for the device and the role that the software plays in the functioning of the device. Levels of concern are minor (for example, a biofeedback analyzer), moderate, and major (for example, a software-controlled linear accelerator therapy device).

In reviewing such 510(k) submissions, FDA maintains that end-product testing may not be sufficient to establish that the device is substantially equivalent to the predicate devices; therefore, a firm's software development process and/or documentation should be examined for reasonable assurance of safety and effectiveness of the software-controlled device functions, including incorporated safeguards. This guidance document covers FDA's expectations for (1) the development and documentation of software-controlled devices, (2) the applications for premarket review, and (3) the approach to reviewing computer-controlled devices.

With respect to 510(k) review, reviewers are told they should be able to assess the following questions:

1. How thoroughly has the manufacturer analyzed the safety of critical functions of the system and software? If testing has not been completed, how thoroughly will these functions be analyzed?
2. How well has the manufacturer established the appropriateness of the function(s), algorithm(s), and/or knowledge on which the system and software are based?
3. How carefully has the manufacturer implemented the safety and performance requirements in the system and software? If testing has not been completed, will the test plan adequately demonstrate that the manufacturer implemented the safety and performance requirements?
4. Based on this information about the software component of a medical device and review of the other aspects of the device, is the device substantially equivalent to the predicate device?

The 510(k)s that are heavily software dependent will receive greater FDA scrutiny, and the questions posed above must be satisfactorily addressed. Prior to submitting such 510(k)s companies should be familiar with FDA's guidance and should ensure that the submission addresses all appropriate facets of the guidance.

B. 510(k) Format

The actual 510(k) submission will vary in complexity and length according to the type of device or product change for which an SE is sought. Generally, however, a 510(k) submission should contain the following:

1. Device name: classification name, common/usual name, trade/proprietary name
2. Establishment information: name, address, establishment registration number (if any)
3. Classification information: classification (final or proposed), classification panel (if known)
4. Performance standards: list applicable performance standards developed by FDA (if any) and the action(s) taken to comply
5. Labeling Information[18]
6. Substantial equivalence information: device description,[18] predicate device,[18] statements of similarities and differences with a predicate device[18]
7. Confidentiality statement
8. Possible appendices
   a. Device description
   b. Device label and labeling (e.g., users, installation or service manual)
   c. Predicate device labeling
   d. Similarities and differences regarding the predicate device
   e. Data, if appropriate: bench, preclinical, or clinical study results; tables; charts; graphs

C. The 510(k) Administrative Process

A 510(k) should be accompanied by a brief cover letter that clearly identifies the submission as a 510(k) premarket notification. To facilitate prompt routing of the submission to the correct reviewing division within FDA, the letter can mention the generic category of the product and its intended use.

Upon receipt by FDA of two copies, the 510(k) will be date and time stamped. The 90-day clock is then running. The document also will be given a unique "K number," also called a document control (or D.C.) number, for identification purposes. (All written communications to FDA following the initial submission must bear the K number.)

FDA will send the submitter an acknowledgment letter stating when the 510(k) was received and listing the assigned K number.

During the 90-day review period, the FDA reviewer may have questions to ask the submitter. While each reviewing division, and in fact each reviewer, operates somewhat differently, some generalizaitons can be made:

---

[18]For these sections of the 510(k) submission, appendices may be appropriate.

Reviewers often will ask simple questions by telephone. If the answers are straightforward, no delay in the process will arise.

More complex questions generally are asked in writing, although the telephone is sometimes used (with or without written follow-up). FDA often will put the review on hold for 30 days pending receipt of answers. This tolls, but does not restart, the 90-day clock.

Complex questions will be asked in writing. The submitter is given 90 days to respond and the 90-day clock is restarted on receipt by FDA of the response.

If a device is required to have a 510(k), it is not permissible to sell such a device until FDA has issued the 510(k). This provision of the law raises questions with respect to advertising, displaying, or taking orders for devices for which 510(k)s have not yet been cleared. FDA takes the position that while it may be permissible to advertise or display (e.g., at a trade show) such devices, it is *not* permissible to sell, take orders, or even be prepared to take orders, unless use of the device is limited to research or investigational purposes (and even then sale is strictly limited). This is a complex and potentially risky area, with serious compliance ramifications. Companies should consult counsel expert in device matters before proceeding to undertake any of these steps.

One final issue frequently raised is whether a device may be sold once the 90-day review period has elapsed if FDA has not yet responded to the 510(k). The technical answer to this question is yes. However, the practical answer is not as simple. First of all, if a company has not heard from FDA within 90 days, the submitter should contact the agency to determine why no response was received.[19] Second, a company must recognize the potential problems with marketing a device in advance of receiving an SE letter:

The product could be found not substantially equivalent, in which case it will not be on the market legally.

FDA could ask for labeling changes, in which case the product's labeling is subject to recall.

FDA could ask that product sales stop, or that the product itself be recalled.

While the burden of removing the product from the market may shift to FDA, the agency is not likely to ignore unapproved marketing. Additionally, marketing without FDA clearance can raise significant product liability problems. Finally, for companies that are concerned about their image with FDA, marketing without clearance is not likely to foster good relations with the agency. The best approach when a company has not heard form FDA within the 90-day time period is to discuss the matter with the agency before beginning to market.

As with the question of product sales, questions involving a decision to market a device for which no 510(k) response has been received are best answered after consultation with legal counsel.

---

[19] It is a good idea if there has been no contact from FDA to not wait the full 90 days before checking. A status call at the 60- to 70-day mark may be useful.

VI. CONCLUSION

Depending on the nature of the product for which a 510(k) is sought, the process can range from extremely simple to extremely complex. Nevertheless, even for a complex submission, the 510(k) mechanism provides device manufacturers with an efficient and timely procedure to obtain FDA clearance for marketing new or changed products. Understanding the system, and being familiar with the scope, intent, and purpose of the 510(k) regulation, will facilitate the submission process and enhance the prospects of a successful 510(k).

REFERENCES

1. J. S. Kahan, Special FDA Premarket Clearance Problems of the Start-up Medical Device Company, *Clinica 241*, March 27, 1987, p. 14.
2. J. S. Kahan, Medical Device Clinical Studies, *Clinica 311*, August 3, 1988, p. 13.
3. Regulatory Requirements for Medical Devices—A Workshop Manual, HHS Publication FDA 85-4165, U.S. Government Printing Office, Washington, D.C., or National Technical Information Service, Springfield, Va., 1985.
4. "Guidance on the Center for Devices and Radiological Health's Premarket Notification Review Program," FDA, Washington, D.C., 1986, p. 3.
5. 42 *Fed. Reg.* 42520 (August 23, 1977).
6. 42 *Fed. Reg.* 42520, 42522-42523 (August 23, 1977) (preamble to the final rule).

17

Compiling a Successful PMA Application

EDWARD M. BASILE and ALEXIS J. PREASE *King & Spalding, Washington, D.C.*

I. INTRODUCTION

As the importance of medical technology to the health care system has increased, so have the complexity and sophistication of that technology. It is the most sophisticated and complex medical devices that generally require review under FDA's premarket approval application (PMAA) process. Less sophisticated devices are also required to undergo PMAA review if they are represented for use in supporting or sustaining life, or are of substantial importance in preventing impairment of human health, or present a potential unreasonable risk of illness or injury.

The Federal Food, Drug, and Cosmetic Act (the Act) requires that any Class III medical device intended for human use, which was not commercially distributed before May 28, 1976, and which is not substantially equivalent to a device distributed before that date, be approved by FDA before it can be legally distributed. In addition, section 515(a) of the Act requires that Class III devices that were marketed before May 28, 1976 (preamendment devices) be approved when FDA promulgates a regulation under section 515(b) of the Act.

Although the statute requires FDA to complete its review of each PMA application within 6 months of filing, in 1987 the average FDA review time was over 11 months. These delays in product review time are not only significant to the medical device industry and FDA, but also to the public, which benefits from improved health care resulting from new medical device technology.

To FDA's credit, there have been improvements in the FDA review time for PMAAs. The average review time has steadily decreased from 13 months in fiscal year 1986, to 11 months in fiscal year 1987, to 9 months for the first half of fiscal year 1988. These improvements have resulted from improved management, a number of regulatory reforms, and increased Congressional funding dedicated specifically to the hiring of new reviewers.

Among the regulatory reforms aimed at increasing efficiency in the PMAA review process was the promulgation of the PMA regulations, which became effective on November 19, 1986, and can be found at 21 C.F.R. Part 814. The regulations are intended to inform PMA sponsors of precisely what FDA requires for inclusion in a PMAA, the procedures FDA will follow in reviewing a PMAA, and the time frames to which both FDA and manufacturers submitting PMAAs must adhere.

The PMA regulations were reviewed by the Office of Management and Budget under the Paperwork Reduction Act of 1980 (44 U.S.C. Chapter 35). Where a requirement was narrowed as a result of the OMB review, that limitation is noted.

There are two reasons a manufacturer should make every effort to comply with the provisions of the PMA regulations. First, a manufacturer must satisfy FDA in order to gain FDA's approval of a device. Second, it is possible that a third party, such as a competitor, could cause troublesome delays in marketing by challenging FDA's approval of another company's device on the grounds that FDA failed to comply with the PMA regulations. A PMA applicant should therefore spend the time necessary to make sure its PMAA meets the criteria of the regulations.

II. OVERVIEW OF PMAA TIME FRAMES

Before discussing the specifics of the regulation, we will briefly review the overall process and the important milestones. The general time frames of the premarket approval process are shown in Fig. 1.

The first step in the PMAA process is the filing of the investigational device exemption (IDE) application (for significant-risk devices). The statutory time frame for IDE review by FDA is 30 days; in practice, the average time required for approval of an IDE is about 37 days. Once an IDE is accepted, the sponsor can proceed with clinical trials. Figure 1 is obviously not to scale—the performance of clinicals can take years. The 180-day "clock" (i.e., the statutory time limit for review of a PMAA) begins to run at the time the PMAA is filed by FDA. The filing date is the date on which an acceptable PMAA was received by FDA. About 45 days after submission, FDA makes a decision on whether or not to "file" the PMAA. If the PMAA is not filable, the manufacturer must start all over again. If the PMAA is filable or any deficiencies are correctable, then substantive review of the submission can begin.

Section 814.20(e) of the PMA regulations provides that the first periodic report is due 90 days after the PMAA is submitted. As will be discussed more fully later, the periodic report is not an automatic filing; it is required only if certain information comes to the attention of the applicant. Following advisory committee review of the PMAA, the applicant will receive either an approvable letter or a not approvable letter or simply an approval or denial of the PMAA. As discussed above, the average review time is 267 days. This is an improvement over the average time required a year ago, but still exceeds the 6 months that is required by the Act.

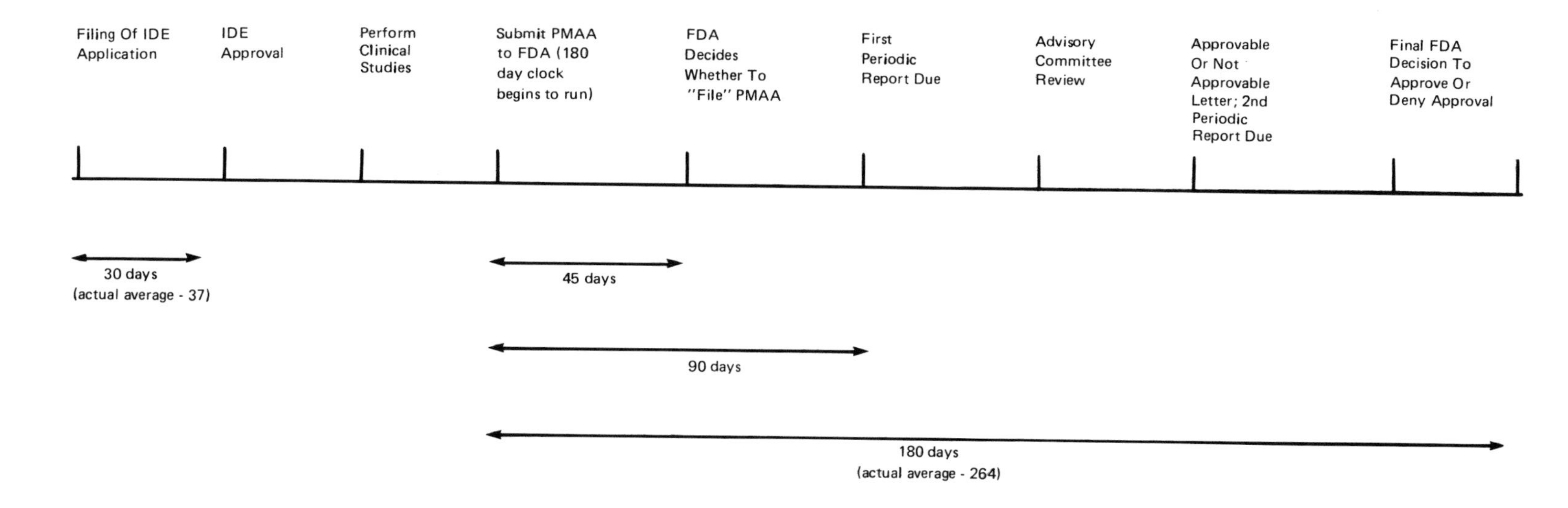

FIGURE 1 PMAA time frames.

III. CONTENTS OF A PMAA (21 C.F.R. SECTION 814.20)

Section 814.20 of the PMA regulations defines what must be included in an application. An overview of the required contents is set forth below and, where the requirement is not self-explanatory, an explanation is given.

Name and address, 21 C.F.R. 814.20(b)(1).
Application procedures and table of contents, 21 C.F.R. 814.20(b)(2).
Summary, 21 C.F.R. 814.20(b)(3).
Complete device description, 21 C.F.R. 814.20(b)(4).
Reference to performance standards, 21 C.F.R. 814.20(b)(5).
Nonclinical and clinical investigations, 21 C.F.R. 814.20(b)(6).
Justification for single investigator, 21 C.F.R. 814.20(b)(7).
Bibliography, 21 C.F.R. 814.20(b)(8).
Sample of device, 21 C.F.R. 814.20(b)(9).
Proposed labeling, 21 C.F.R. 814.20(b)(10).
Environmental assessment, 21 C.F.R. 814.20(b)(11).
Other information requested by FDA, 21 C.F.R. 814.20(b)(12).

A. Application Procedures and Table of Contents [21 C.F.R. 814.20(b)(2)]

The PMAA must be submitted in six copies to the PMAA Document Mail Center (HFZ-401), Center for Devices and Radiological Health, FDA, 1390 Piccard Drive, Room 282, Rockville, MD 20850. The six copies should be bound in one or more numbered volumes. Each volume should be less than two inches thick. In at least one copy, trade secret or confidential commercial or financial information, which is included in all six copies, should be identified.

The table of contents must include separate sections on nonclinical laboratory studies and clinical investigations involving human subjects.

B. Summary [21 C.F.R. 814.20(b)(3)]

The summary should include indications for use, a device description, a description of alternative practices and procedures, a brief description of the marketing history, and a summary of studies. This summary should be in sufficient detail to enable the reader to gain a general understanding of the application.

The PMAA must include the PMA applicant's foreign and domestic marketing history, as well as any marketing history of a third party marketing the same product [21 C.F.R. 814.20(b)(3)(iv)]. The regulations qualify this requirement with the phrase "if known," but it is not clear from the regulation how much of a burden is on the applicant to find out information about the experience of another company marketing a similar device. (This requirement was narrowed as a result of a review of the PMA regulation by OMB under the Paperwork Reduction Act of 1980; see 51 *Fed. Reg.* 43343.)

C. Description of Device [21 C.F.R. 814.20(b)(4)]

This item should include a complete description of the device, including pictorial representations. Each of the functional components or ingredients should be described, as well as the properties of the device relevant to the diagnosis, treatment, prevention, cure, or mitigation of a disease or condition. The principles of the device's operation should also be explained.

Information regarding the methods used in, and the facilities and controls used for, the manufacture, processing, packing, storage, and installation of the device should be explained in sufficient detail so that a person generally familiar with current good manufacturing practices can make a knowledgeable judgment about the quality control used in the manufacture of the device. Manufacturing information should be submitted to FDA in a separate volume.

D. Reference to Performance Standards [21 C.F.R. 814.20(b)(5)]

The PMAA must also include a reference to any performance standard for the device under section 514 of the Act or the Radiation Control for Health and Safety Act of 1968 that is in effect or proposed at the time of the submission of the PMAA. The PMAA must also include a reference to any voluntary standard, whether domestic or international, that is relevant to the safety or effectiveness of the device about which the applicant is aware or should reasonably be aware (see 51 *Fed. Reg.* 43343).

To determine which standards must be addressed, PMA applicants may ask members of the appropriate reviewing division of the Office of Device Evaluation (ODE) or consult FDA's list of relevant voluntary standards, the Medical Device Standards Activities Report (HHS Publication-FDA 87-4219), which can be obtained from National Technical Information Services, 5285 Port Royal Road, Springfield, VA 22161 (telephone 703-487-4650).

In addition to a reference to the FDA performance standards, the PMAA should contain information demonstrating how the device meets these performance standards. The PMAA should also contain a "justification" for any deviation from an FDA performance standard and an "explanation" for any deviation from a voluntary standard.

E. Nonclinical and Clinical Investigations [21 C.F.R. 814.20(b)(6)]

The nonclinical laboratory studies section of the PMAA should include microbiological, toxicological, immunological, biocompatibility, stress, wear, shelf life, and other laboratory or animal tests, as appropriate. This section should also include a statement that each study was conducted in compliance with the requirements of 21 C.F.R. Part 58, the good laboratory practice (GLP) regulations. If the studies were not conducted in compliance with GLPs, a brief statement of the reason for noncompliance should be given.

The heart of a PMAA submission is the clinical trials supporting the application [21 C.F.R. 814.20(b)(6)(ii)]. Prior to preparing PMAAs, therefore, PMA applicants should ensure that the clinical trials on which the PMAA will be based are designed and implemented properly so that they provide full support for the claims to be made in the PMAA and the labeling of the product.

In order to ensure proper design of a clinical trial, the indications for a product should first be determined. After determining the device's indications, the entrance criteria for clinical trials should be broad enough to include all categories of patients likely to be treated with the device. The reason for having these broad entrance criteria is that FDA will examine the patient population in the clinical trials to determine whether that population is representative of the patient population that will use the device.

In order to assure proper implementation of its clinical trials, sponsors of clinical trails should closely monitor the clinical trials to ensure that the protocol is being followed and to keep accurate records of the reasons patients did not complete the study. Keeping accurate records of patients who do not complete the study is particularly critical where a patient stopped participating because of a serious injury caused by the device. Where serious or minor adverse reactions occur, FDA will want to know all of the details. In addition, applicants should be prepared to defend the statistical treatment of those patients who were dropped from a study.

The PMAA must provide a statement of compliance with the informed consent regulations in 21 C.F.R. Part 50 and with the Institutional Review Board (IRB) regulations in 21 C.F.R. Part 56 [21 C.F.R. 814.20(b)(6)(ii)(A)]. If a PMAA includes published studies or studies conducted by a third party, it may be impossible for the applicant to make this certification. The regulations do not provide any guidance on this. The applicant should contact FDA if questions arise regarding the certification in these circumstances.

F. Justification for Single Investigator [21 C.F.R. 814.20(b)(7)]

Section 814.20(b)(7) requires a justification if clinical data from only one investigator support the PMAA. A PMAA can therefore be based on data from a single investigator, but the regulation implies that data from a single investigator are generally not sufficient. Consequently, the applicant must explain why data from only one investigator were submitted and must demonstrate that the data are sufficient to demonstrate safety and effectiveness.

G. Bibliography [21 C.F.R. 814.20(b)(8)]

A bibliography of all published reports that were not submitted under 21 C.F.R. 814.20(b)(6) must be submitted under this section, whether those reports are adverse or supportive. The bibliography must include all published reports that relate to the safety or effectiveness of the device that are known or should reasonably be known to the PMA applicant.

It is not advisable for an applicant to make a judgment that a particular study is so poorly done that it is unworthy of inclusion. An FDA reviewer may disagree with that assessment, and the company may later be charged with having withheld relevant information. The better practice is to include the study and explain to FDA why it is invalid.

H. Sample of Device [21 C.F.R. 814.20(b)(9)]

One or more samples of the device must be supplied if requested by FDA. If supplying a sample of the device is impractical, the applicant should specify the location at which FDA may examine and test the device.

I. Proposed Labeling [21 C.F.R. 814.20(b)(10)]

The PMAA must include copies of all proposed labeling and available advertising that constitutes labeling (see 51 *Fed. Reg.* 43343). Section 201(m) of the Act defines labeling as all labels and other written, printed, or graphic matter upon any article or its containers or wrappers, or accompanying the article. The word "accompany" has been interpreted by the courts very broadly. Under the case law, literature shipped to consumers at a different time and separate from shipment of the product promoted by the literature has been considered labeling [1]. For example, print advertising sent to prospective purchasers as part of an initial promotion along with samples or prototypes of a product, constitutes labeling.

J. Environmental Assessment [21 C.F.R. 814.20(b)(11)]

The PMAA should include an environmental assessment under 21 C.F.R. 25.22 (a)(18). In lieu of the environmental assessment, the applicant can claim a categorical exclusion under 21 C.F.R. 25.24(e)(4) or (5). If a categorical exclusion is claimed, the PMAA should contain sufficient information to establish to FDA's satisfaction that the requested action is included within the categorical exclusion. Before a PMA applicant prepares an environmental assessment, FDA's PMA staff should be consulted regarding the need for the assessment.

K. Other Information Requested by FDA [21 C.F.R. 814.20(b)(12)]

Despite all the information that a company is required to submit in the original PMAA, FDA can ask additional questions and request more information. The FDA need not go to an advisory committee for authorization if the additional information requested falls into any of the following categories: safety and effectiveness investigations, device description, manufacturing description, relevant mandatory performance standards, samples of the device, or labeling [21 U.S.C. 360e(c)(1)(A)-(F)]. The concurrence of an advisory committee is, however, required for information other than that listed in section 515(c)(1)(G) of the Act, but there is very little that does not fall into one of these categories.

The additional questions asked by FDA may seem trivial and the additional information requested may seem irrelevant. In the long run, however, it is usually better for the PMA applicant to respond to FDA's requests as fully and as quickly as possible rather than insisting that FDA obtain the concurrence of an advisory committee, which could cause a delay of several months.

If FDA makes a request for additional information, the applicant has 180 days in which to respond. If the applicant does not respond within 180 days or request an extension with an appropriate justification, FDA may consider the PMAA to be voluntarily withdrawn [21 C.F.R. 814.44(g)(1)]. If FDA subsequently receives a claim from another party that the applicant has or appears to have abandoned the PMAA, FDA will notify the applicant. The applicant then has 7 days to voluntarily withdraw the PMAA. If the applicant does not respond within 7 days, FDA will declare the PMAA abandoned [21 C.F.R. 814-9(g)(1)(i)(A) and (1)(ii)]. Whether the PMAA is voluntarily withdrawn or declared by FDA to be abandoned, all safety and effectiveness data and other information not previously disclosed to the public, and that do not constitute trade secret or confidential commercial or financial information, are available for public disclosure. (See the following discussion of 21 C.F.R. 814.9, which deals with the confidentiality provisions of the PMA regulations.)

IV. OTHER IMPORTANT ISSUES

A. Periodic Updates [21 C.F.R. 814.20(e)]

Periodic updates are a new requirement for PMAAs. (This requirement was narrowed as a result of a review of the PMA regulation by OMB under the Paperwork Reduction Act of 1980; see 51 *Fed. Reg.* 43343.) The filing of a periodic report is not automatic. The regulations provide that certain narrowly defined information be submitted (1) 3 months after a PMAA is filed, (2) upon receipt of the approvable letter, and (3) as otherwise requested by FDA.

The only information required to be reported is new, relevant data from ongoing, completed, or discontinued studies that may reasonably affect an evaluation of the safety or effectiveness of the device or that may reasonably affect the statement of contraindications, warnings, precautions, and adverse reactions in the labeling for the device. In some cases, for example, data from an ongoing study may suggest the need for changes in the contraindications for the device or the precautions for use, or the data may alter some of the statistics submitted in the PMAA, which would give a different picture of device adverse reactions. In these situations, a periodic update will have to be filed. The preamble suggests that the applicant consult with FDA about the format and content of this submission (51 *Fed. Reg.* 26351). A periodic update report will be considered a PMAA amendment.

B. Master Files [21 C.F.R. 814.3(d)]

The so-called "master file" is basically a reference source submitted to FDA to provide certain background information about a company or device. A master file may contain information on a manufacturing facility, a manufacturing process or methodology, or a component of a device. FDA ordinarily accepts device master files from firms other than the PMA applicant. The master file system is a means of facilitating the filing of information with FDA where the information is common to more than one PMA applicant or is confidential to the owner of the master file, but is essential to FDA's review of the PMAA. Device master files should not be submitted to provide information that is otherwise required to be included in the PMAA.

When a company files a PMAA, a particular master file will be referenced, as, for example, the source of information about the manufacturing facility that will be used in the production of the device. [FDA will only accept device master files from those persons who have not submitted or will not directly submit to FDA the same information in a PMAA, IDE, 510(k), or other device-related submission. FDA will not permit PMA applicants to submit information in DMFs that should be included in the PMAA. A PMAA may, however, include by reference applicable information submitted by the applicant in previous PMAAs or device-related submissions.] The master file is thus incorporated by reference into the PMAA, and once incorporated it is given the same confidentiality afforded the PMAA. If a master file is not referenced within 5 years, FDA will return it to the submitter (51 *Fed. Reg.* 26343-44). If it is referenced even once during the 5-year period, FDA will maintain the master file indefinitely.

When a PMAA contains information that is later required for a subsequent PMAA filed by the same sponsor, it is sufficient for the subsequent PMAA to refer to the earlier PMAA.

C. Foreign Research (21 C.F.R. 814.15)

In many respects, the criteria for acceptance of foreign data for premarket approval of new devices resemble those that apply to new drugs. Indeed, the PMAA and New Drug Application regulations regarding foreign research are identical in many respects.

In the PMA regulations, foreign research is divided into two basic categories: research that is conducted under an Investigational Device exemption (IDE) [section 520(f) of the Act], and research that is not conducted under an IDE. IDEs are not required for foreign clinical studies. FDA may, however, require the submission of nonclinical information (similar to that required in an IDE) in order to approve the export of a device for investigational use under section 801(d)(2) of the Act.

Research conducted under an IDE must comply with 21 C.F.R. Part 812 unless there is a waiver under section 812.10 of the IDE regulations [21 C.F.R. 814.15(a)]. For research not conducted under an IDE, compliance with part 812 is not required, but there are other criteria that must be met, depending on whether the research was begun after November 19, 1986 (the effective date of the PMA regulations), or before November 19.

For research that is begun after November 19, 1986, the criteria are set forth in section 814.15(b). First, the data must be "valid." What "valid" means is not clear because the term is not defined in the regulations or in the preamble. FDA can be expected to apply the same standards applied to studies conducted in the United States. The second criterion is that clinical studies must be performed in conformance with either the Declaration of Helsinki or the laws and regulations of the foreign country where the research is performed, whichever affords more protection to the subjects.

This criterion poses a contradiction. Under section 801(d) of the Act, a manufacturer can export and market a device in a foreign country if the device conforms to the laws of the foreign country and if certain other conditions are met. According to the PMA regulations, however, if the Declaration of Helsinki provides more protection than the laws of a particular foreign country, a company can export to that country and market in that country, but information obtained from a study conducted in that country may not be acceptable for a PMAA. This apparent contradiction has not yet been resolved.

If the research is begun before November 19, 1986, a different standard applies [21 C.F.R. 814.15(c)]. First, the data must be "scientifically valid." The difference between "valid" data and "scientifically valid" data is not obvious. The preamble to the regulations states that to be "scientifically valid," clinical studies must be well designed, conducted, and performed (51 *Fed. Reg.* 26346). The second criterion for studies begun before November 19, 1986, is that there must be no violation of the "rights, safety, and welfare" of human subjects. The preamble also states that the Declaration of Helsinki and the laws of the foreign country will be a reference point for determining violations of the rights, safety, or welfare of human subjects. Even for research initiated before the effective date of the regulations, the manufacturer should therefore be prepared to explain whether that research conforms to the Declaration of Helsinki or the laws of the foreign country. This is not an absolute requirement, though, so the manufacturer has some leeway provided that it can show that the subjects in the foreign study were not at undue risk.

Like a New Drug Application, a PMAA can be approved based solely on foreign data. Section 814.15(d) of the PMA regulations provides that a PMAA based solely on foreign data is acceptable if:

The data are applicable to the U.S. population and medical practice;
The studies are performed by clinical investigators of recognized competence; and
The data are considered valid without need for on-site inspection, or FDA can validate the data through on-site inspection or other appropriate means.

In what is probably good advice, FDA suggests in the preamble that companies with PMAAs based only on foreign data seek presubmission meetings with agency staff.

D. Filing of PMAs (21 C.F.R. 814.42)

Although the 180-day time clock for FDA review of a PMAA begins to run immediately after a PMAA is submitted to FDA, the PMAA is not technically "filed" until approximately 45 days after submission of the PMAA. After the PMAA is submitted, it is assigned to the appropriate reviewing division of the ODE, and to the Statistics Branch of the Office of Science and Technology. The Statistics Branch performs an in-depth review of the design of, and the data derived from, clinical studies.

During the 45 days after submission, FDA decides whether the application is acceptable for filing. An acceptable application is "filed" as of the date it was first received by FDA. In general, FDA will try to determine in the initial review whether the application is sufficient on its face and whether it contains the information required by the statute and regulations or, if information is omitted, that the omission is justified [21 C.F.R. 814.42(a)]. If the application is acceptable for filing, the applicant will be so notified by FDA, and substantive review of the PMAA will begin. At this time, an assignment is usually made to the Office of Compliance for a GMP inspection of the manufacturing facilities unless FDA decides a GMP inspection is not required.

Review of the PMAA will be delayed if a notification by the applicant under section 510(k) of the Act is pending for the device. The agency suggests that an applicant consult with FDA before filing both a PMAA and a 510(k) [21 C.F.R. 814.42(e)(3)]. Additionally, if the PMAA contains a false statement of material fact, it will not be accepted for filing [21 C.F.R. 814.42(e)(4)].

If FDA refuses to file a PMAA, the applicant will receive a letter enumerating the deficiencies in the PMAA [21 C.F.R. 814.42(d)]. The applicant then has a number of options. First, the PMAA may be amended with the additional information, in which case the filing date is the date FDA receives the amended PMAA [21 C.F.R. 814.42(d)(1)]. Alternatively, the applicant may request in writing an informal conference with the Director of ODE [21 C.F.R. 814.42(d)(2)]. A meeting will be held within 10 days. The purpose of this meeting is to provide the applicant an opportunity to demonstrate that its PMAA is complete. If the applicant agrees with FDA that its PMAA is incomplete and the applicant plans to amend it, the refusal to file the PMAA should not be appealed. Instead, the applicant should schedule a meeting with the appropriate reviewing division of ODE and negotiate the contents of the PMAA with them. The agency is required to make a decision within 5 days of the meeting between ODE and the PMA applicant. If the PMAA is accepted for filing, the application is filed as of the date of acceptance, not as of the date the PMAA was first submitted to FDA.

If the Director of ODE decides that the application should not be filed, a request for reconsideration can be made to the Director of the Center for

Devices and Radiological Health. The Center Director's decision is the final decision for purposes of judicial review. If an appeal to the Center Director is denied, the only remaining recourse is judicial review.

E. Substantive PMAA Review and Approval (21 C.F.R. 814.44)

The PMA regulations also describe the procedures for substantive FDA review of a PMAA. After a PMAA is filed, ODE performs an in-depth review of the PMAA and prepares a "pre-panel summary." The PMAA is then sent to the appropriate panel of experts for review, as required by section 515(c)(2) of the Act.

The panel may elect to hold an open meeting at which anyone may address the pending matter. If the panel holds an open meeting, FDA has a right to present its findings and the PMA applicant has a right to make a presentation. It is often counterproductive, however, for the PMA applicant to attempt to use the panel meeting as a forum to review issues that are adequately covered in its PMA submission.

Following the open meeting, the panel decides whether to recommend approval and states the basis for its decision on the record. After the panel makes its recommendation, the reviewing division of ODE completes its review and makes its decision whether to issue a not approvable or an approvable letter.

If a not approvable letter is issued, the firm has four options. The firm may (1) voluntarily withdraw the PMAA; (2) amend the PMAA to cure the deficiencies within 180 days; (3) request that FDA grant an extension of time in which to amend the PMAA; or (4) treat the not approvable letter as a denial and request that FDA initiate proceedings for a hearing on the denial.

If an approvable letter is to be issued, ODE drafts an approval package. The draft approval package is then reviewed by the PMA staff and, sometimes, by the Office of General Counsel (OGC). Comments from the PMA staff and OGC are considered and, when the PMA staff, OGC, and ODE reach a common agreement, the approval package is sent to the Director of ODE for concurrence and issuance of an approvable letter.

The approvable letter may only list postapproval requirements to be imposed as a condition of approval. The applicant will be asked to submit an amendment agreeing with, disagreeing with, or proposing alternatives to, these postapproval requirements. If a company has not had a recent FDA inspection that was satisfactory, the approvable letter will require that an FDA inspection be performed to determine whether the manufacturer makes the device according to current good manufacturing practices. The approvable letter will often require that minor labeling questions be resolved.

In most cases, the prudent response to an approvable letter will be to give FDA the information requested, in which case, presumably, the PMAA will be approved. Alternatively, if the applicant disagrees with the agency, the applicant can consider the approvable letter to be a denial of approval and file a petition for reconsideration [21 C.F.R. 814.44(e)(2)(ii)]. Before an applicant decides to consider the approvable letter to be a denial of approval, it is generally prudent to discuss this matter with FDA. Finally, the applicant can simply decide to withdraw a PMAA in response to an approvable letter [21 C.F.R. 814.44(e)(2)(iii)].

When FDA is satisfied with the company's response to the approvable letter, an approval order is issued. The agency may issue an approval order based on draft labeling. If a company receives an approval order based on draft labeling, the company can begin marketing before FDA sees the final printed labeling.

For every device for which FDA approves a PMAA, seciton 520 of the Act requires FDA to prepare a detailed summary of safety and effectiveness data. The summary is made available to the public and for purposes of judicial review would be the document a court would examine to determine the rationale on which the agency relied to approve a company's PMAA.

Although the summary of safety and effectiveness data is prepared by FDA, an applicant can expedite its PMA review by preparing a draft summary for FDA. The best way to write a summary of safety and effectiveness data is to refer to the format and content requirements of 21 C.F.R. 814.20(b)(3) or to find a model from a recent approval for the same or a similar device and follow it. There simply are no points to be gained in this process for being creative. More likely than not, trying to be different will only create problems.

Most important, do not try to make claims in the summary of safety and effectiveness data that your device is better than another. Such claims will be questioned by FDA and will delay the approval process. The summary should detail the scientific literature completely and accurately. Applicants should also be sure the data referenced in the summary are sufficient to meet FDA's requirements for approval of a PMAA and that the data are correct.

If questions about the indications statement in the device's labeling have not been raised by the time FDA begins preparation or review of the summary of safety and effectiveness data, FDA is likely to raise them at that time. For orthopedic devices, FDA has limited indication statements to a particular surgical method that was used in the clinical trials for the device even though there were four or five other surgical methods that were appropriate for the device. As discussed earlier, clinical trials should be designed to assure the inclusion of as broad a patient population as possible so as to be representative of the intended patient population.

F. PMAA Amendments (21 C.F.R. 814.37)

A PMAA amendment is information submitted to FDA to modify a pending PMAA or PMAA supplement. It can be submitted either on the applicant's own initiative or at the request of FDA. The review time for a PMAA can be extended up to 180 days in the case of a "major" amendment, that is, the submission of significant new data from a previously unreported study, significant updated data from a previously reported study, detailed new analyses of previously submitted data, or significant required information previously omitted [21 C.F.R. 814.37(c)(1)]. If the applicant refuses to submit a major amendment requested by FDA, FDA's review time is extended by the time elapsed between FDA's original request for the major amendment and the applicant's refusal to respond [21 C.F.R. 814.37(c)(2)].

G. Withdrawal and Resubmission of PMAAs [21 C.F.R. 814.37(d)]

When a request for a PMAA amendment is received, an applicant may simply decide to withdraw the PMAA. Thus, the provision of the regulation pertaining to amendments also covers withdrawal and resubmission of a PMAA. In addition to voluntary withdrawal, a PMAA will be deemed to be withdrawn if there is no response to an FDA request for a PMAA amendment within 180 days. There will be no notice to the applicant prior to the expiration of 180 days. The only notice will be in the letter from FDA requesting the amendment. The letter will state that if there is no response within 180 days, the PMAA will be considered to be voluntarily withdrawn. There will be no notice thereafter. Within this 180-day period, the PMA applicant may request an extension of time in which to submit an amendment, along with an appropriate justification and a projected date for submission of the amendment. FDA will notify the applicant whether the extension is granted.

The regulation specifically provides that an applicant can resubmit a withdrawn PMAA [21 C.F.R. 814.37(e)]. Of course, the 180-day clock will begin to run all over again in that event.

H. Denial of Approval (21 C.F.R. 814.45)

The PMA regulations describe the grounds for denial of approval of a PMAA. Approval may be denied for failure to comply with the statute, with labeling regulations (sections 801 and 809), or with the requirements for valid scientific evidence (section 860.70) [21 C.F.R. 814.45(a)(2)].

The agency may also decide to deny approval of a PMAA if it contains a false statement of material fact [21 C.F.R. 814.45(a)(1)]. It is important to note that an omission of information can constitute a false statement of material fact [21. C.F.R. 814.3(i)]. For example, if the applicant fails to include a study that FDA considers to be material, FDA can deny approval of the PMAA because it contains a false statement of material fact. The omission or other false statement need not be intentional. It can simply be inadvertent. Even if an applicant has simply forgotten to include some relevant data, FDA can deny the PMAA. Obviously, the agency has some discretion here, but it is clear that it is not necessary for FDA to establish that a false statement was intentionally made.

The other grounds for denying approval of a PMAA are the failure of the manufacturer to permit an FDA inspection, the failure of "essential" nonclinical studies to conform to GLP regulations (21 C.F.R. Part 58), and the failure of any clinical study to conform to the informed consent or the IRB regulations [21 C.F.R. 814.45(a)(3), (4), and (5)]. The requirement that all clinical studies conform to the informed consent or IRB regulations sometimes presents a problem. For example, the manufacturer may not have conducted all the clinical studies or a particular clinical study may not be very important, but was included for completeness. The agency will not necessarily penalize an applicant in these circumstances, but a strict interpretation of the regulation would allow a denial of the PMAA.

According to the procedures for denial of PMAA approval, the first step is the issuance of a not approvable letter [21 C.F.R. 814.44(f)]. The not approvable letter will indicate that there are major deficiencies in the application and will state what the manufacturer must do to bring the PMA into an approvable state.

In response to a not approvable letter, the applicant may do one of three things. The first option is to amend the PMAA as requested in the not approvable letter. Such an amendment will be considered a major amendment under 21 C.F.R. 814.37(c)(1), and FDA may then extend the review period up to

180 days. The second option is to consider the not approvable letter to be a denial of approval and petition for review under section 515(d)(3) of the Act by filing a petition for reconsideration under 21 C.F.R. 10.33. The final option is to withdraw the PMAA.

If the applicant chooses to amend the PMAA as requested and a ground for denying approval still applies or if the applicant notifies FDA in writing that the requested amendment will not be submitted and files a petition for reconsideration, FDA will issue an order denying approval. FDA will also issue an order denying approval if the applicant petitions for review [21 C.F.R. 814.45(e)].

FDA's order of denial will state the deficiencies in the PMAA, which should be the same as those stated in the not approvable letter, and the grounds for denial under the statute and the regulations. The order will include a notice of an opportunity to request review under section 515(d)(3) of the Act.

The FDA will also give public notice of an order denying approval of a PMAA by publishing a notice in the *Federal Register* [21 C.F.R. 814.45(d)]. When such a notice is published, data and information in the PMAA file will be made publicly available under 21 C.F.R. 814.9. The *Federal Register* notice will announce the availability of the safety and effectiveness data that formed the basis of FDA's denial action.

If an applicant requests review under section 515(d)(3) of the Act, he may petition for a formal evidentiary hearing before an administrative law judge. The petition will be granted unless it is "without good cause," that is, there is no issue of material fact in dispute to be tried. A formal evidentiary hearing normally takes at least a year. Alternatively, the PMA applicant may request a review of the record by an advisory committee of experts.

Either the administrative law judge following a hearing, or the advisory committee following its review of the record, will submit a proposed decision to the Commissioner of Food and Drugs, who then issues his decision, which is final agency action. Typically the Commissioner's office, rather than simply adopting the administrative law judge's decision, writes its own decision—a decision that is longer and more fully researched in order to withstand judicial review. If the Commissioner's decision is adverse, it can be appealed to the U.S. Court of Appeals for the District of Columbia.

I. Withdrawal of Approval of PMAA (21 C.F.R. 814.46)

The grounds for withdrawal of approval of a PMAA are those set forth in section 515(e)(1) of the Act and the failure of the applicant to comply with any postapproval requirements, failure of any "essential" nonclinical study to comply with the GLPs, and failure of any clinical study to comply with informed consent or IRB regulations.

In deciding whether to withdraw approval of a PMAA, FDA may consult with an advisory committee or use information other than that contained in the PMAA file [21 C.F.R. 814.46(b)]. Under the procedures for withdrawal of approval, the manufacturer will first receive a notice of an opportunity for an informal hearing [21 C.F.R. 814.46(c)]. This notice is followed by an order of withdrawal, which is published in the *Federal Register* and which states FDA's reasons for withdrawing approval [21 C.F.R. 814.46(d)]. Under 21 C.F.R. 814.9, all safety and effectiveness data in the PMAA file "and other information not previously disclosed to the public" will be available for public disclosure after an order of withdrawal of the PMAA is published in the *Federal Register*.

J. Postapproval Requirements (21 C.F.R. 814.82)

The FDA may impose postapproval requirements in a PMAA approval order. These postapproval requirements include:

Restrictions on sale, distribution, or use
Continuing education and periodic reporting on safety, effectiveness, and reliability
Prominent display of warnings in labeling and advertising (for restricted devices)
Identification codes on the device or, if the device is an implant, in cards given to patients
Maintenance of records
Periodic reporting
Batch testing
Patient follow-up

Postapproval requirements can be negotiated with FDA prior to the issuance of the approval letter. In most cases, FDA will issue an approvable letter that will contain FDA's proposed postapproval requirements. PMA applicants should discuss these matters with their clinical investigators to determine whether any of the requirements can be eliminated or reduced. Before FDA issues the approval letter, the applicant should suggest to FDA alternative means to achieve FDA's objectives. Some time spent at this stage can substantially reduce the amount of work that will be required after the product is on the market. FDA cannot issue the order of approval unless the applicant concurs in FDA's postapproval requirements or the applicant and FDA reach an agrement on revised requirements.

The preamble to the PMA regulations clearly states that postapproval requirements are not to include new clinical studies (51 *Fed. Reg.* 26359). FDA may, however, require additional information, which can be obtained by expanding a study to include more patients or more variables. To avoid the necessity of starting a new study when FDA requires additional information, FDA recommends that PMA applicants not terminate an existing study before or shortly after the PMAA is filed. FDA also suggests that PMA applicants contact the approriate reviewing division of ODE before discontinuing an existing study.

As a postapproval condition, FDA can require that identification codes be assigned to devices or, for implants, that cards be given to patients [21 C.F.R. 814.82(a)(4)]. The regulation does not seem to recognize that it may be difficult for manufacturers to obtain this sort of information because the manufacturers will have to rely on physicians—who may or may not be cooperative—for the return of the necessary data. (This requirement was narrowed as a result of a review of the PMA regulation by OMB under the Paperwork Reduction Act of 1980; see 51 *Fed. Reg.* 43343.)

K. PMAA Supplements (21 C.F.R. 814.39)

The PMA regulations require PMAA supplements for changes "affecting the safety or effectiveness of the device" [21 C.F.R. 814.39(a)]. The regulation gives examples of the sorts of changes that would require supplements: new indications for use of the device; other labeling changes; changes in manufacturing facilities, methods, or quality control procedures; changes in sterilization methods; changes in packaging; changes in performance or design specifications, circuits, components, ingredients, principles of operation, or physical layout of the device; or an extension of the expiration date unless the extension is made in accordance with a previously approved protocol.

It is important to note that the regulations require supplements for these types of changes only "if they affect the safety or effectiveness of the device" [21 C.F.R. 814.39(a)]. FDA has taken the position in a number of cases, however, that a supplement is required for any change that has the potential to affect the safety or effectiveness of the device. If the manufacturer tests a device to determine whether a change affects the device's safety or effectiveness, FDA is likely to take the view that a supplement is required, regardless of what the testing shows. FDA will argue that the decision to test the product demonstrates that the change has a potential effect on safety or effectiveness. This standard is so broad as to require supplements for many minor changes that should not require a PMAA supplement.

Not on the list of changes that require supplements is a change in supplier, subcontractor, or distributor in those instances when the component, ingredient, or subassembly for the new subcontractor, supplier, or distributor meets the specifications approved in the PMAA (51 *Fed. Reg.* 26353). Generally, a supplement need not be filed for this type of change. Where a change does not necessitate a PMAA supplement, FDA will nevertheless receive the information because the manufacturer must report the change in its periodic update [21 C.F.R. 814.39(b)].

The regulations establish a procedure for FDA to provide alternative methods for making changes that would otherwise require prior FDA approval of a regular PMAA supplement before implementation [21 C.F.R. 814.39(e)]. For generic devices (e.g., pacemakers in general), a formal request for an advisory opinion must be made. For a specific device, the PMAA holder can simply write a letter to FDA and ask whether a PMAA supplement is necessary for a specific change. FDA may respond and state that no supplement is needed. Alternatively, the regulation also authorizes FDA to require that the change be described in the periodic report to FDA filed under section 814.84 or that the manufacturer file a 30-day supplement, rather than a regular supplement [21 C.F.R. 814.39(e)(1) and (2)]. In the latter case, the supplement becomes approved automatically if the manufacturer does not receive an adverse response from the agency within 30 days of the filing. Thus, there are some ways to reduce the burden of PMAA supplements.

The regulations identify some changes that can be put into effect immediately before FDA approval [21 C.F.R. 814.39(d)]. These changes are basically those that enhance safety or effectiveness, such as labeling changes that add or strengthen a contraindication, warning, or precaution, or that add information about an adverse reaction; labeling changes that add or strengthen an instruction that is intended to enhance the safe use of the device; labeling changes that delete misleading, false, or unsupported indications; and changes in quality control manufacturing procedures that add a new specification or test method, or otherwise provide additional assurance of the purity, identity, strength, or reliability of the device.

For changes that do not require FDA's preapproval, the supplement and its mailing cover must be plainly marked "Special PMA Supplement—Changes Being Effected." The supplement must also identify the date the changes are being made and a full explanation of the basis for the changes.

L. Confidentiality (21 C.F.R. 814.9)

The confidentiality providions of the PMA regulations are a PMA applicant's only protection against a competitor obtaining trade secret and confidential commercial information from a PMAA. The disclosure of information held by the government is regulated by the Freedom of Information (FOI) Act (5 U.S.C. 552). In general, the FOI Act encourages the disclosure of all documents to the public. There are a few exemptions, however, and the one that is most important to device manufacturers is the exemption for trade secret or confidential commercial information [5 U.S.C. 552(b)(4)].

Also, under the Food, Drug, and Cosmetic Act, it is a crime for FDA to disclose trade secret information [21 U.S.C. 331(j)]. Although the PMA regulations do provide for the disclosure of certain information in a PMAA, trade secret or confidential commercial information is never disclosable. Moreover, the PMAA regulations specify that its confidentiality provisions apply to all information in a PMAA file. This would include an IDE if it is incorporated by reference and any PMA amendments, PMAA supplements, or reports that may be filed [21 C.F.R. 814.9(a)].

Before a PMAA is approved or denied, its existence is not publicly disclosed unless the applicant or a third party discloses its existence [21 C.F.R. 814.9(b)]. Once such a disclosure is made—even if it is through no fault of the PMA applicant that information on the existence of a PMAA is disclosed by a third party, and even if the disclosure is illegal—FDA will consider the existence of the PMAA to be public information and will disclose its existence to anyone who requests it.

There are circumstances in which important public interests may dictate that the agency disclose information about a PMAA before it is approved or denied. Section 814.9(d) of the regulations therefore authorizes FDA to disclose a summary of the safety and effectiveness information in a PMAA before an approval or denial is issued. If a PMAA involves an important issue for public consideration, the agency may decide to hold an advisory committee meeting in public session.

After approval or denial of a PMAA, the existence of a PMAA file will be disclosed along with the routine disclosure of the summary of safety and effectiveness information in the application [21 C.F.R. 814.9(e)]. In addition, the following types of information are disclosable after approval or denial of a PMAA: information previously disclosed; protocols, assays, or analytical methods; adverse reaction reports and customer complaints; and any correspondence or memos of telephone conversations between FDA staff and the sponsor [21 C.F.R. 814.9(f)]. Again, FDA cannot disclose trade secret or confidential commercial information.

In addition to the approval or denial of a PMAA, there are other triggers for disclosure that are important. First, disclosure will be triggered if a PMAA is "abandoned" [21 C.F.R. 814.9(g)(1)]. This is a concept that runs throughout the regulations. There are basically two situations in which a PMAA will be considered abandoned:

1. If FDA makes a request for additional information, the applicant has 180 days in which to respond. Absent a response to such a request within 180 days, FDA will initially consider the PMAA to be voluntarily withdrawn [21 C.F.R. 814.9(g)(1)(i)(A)]. If FDA subsequently receives a claim from another party that the applicant has or appears to have abandoned the PMAA, FDA will then notify the applicant that such appears to be the case. The applicant then has 7 days to voluntarily withdraw the PMAA before FDA will declare the PMAA abandoned (see 51 *Fed. Reg.* 26345).
2. If FDA has reason to believe that the manufacturer is no longer working on the application because of the manufacturer's lack of attention to the PMAA, FDA will send a letter to the applicant asking whether the PMAA is still active or the intent is to abandon it. If the applicant fails to respond to that letter within 7 days, the PMAA will be considered abandoned [21 C.F.R. 814.9(g)(1)(i)(B)].

The second situation discussed above has not yet occurred.

In either of the situations described above, section 814.9(g) of the regulations suggests that the ensuing disclosure may be braoder than simply the summary of safety and effectiveness information made publicly available upon approval or denial of a PMAA. The abandonment of a PMAA may trigger the disclosure of all information in the PMAA that is not trade secret or confidential commercial information.

Other triggers for disclosure are the withdrawal of approval of a PMAA, the reclassification of a device that is the subject of a PMAA into Class I or Class II, or a finding that a device that is the subject of a PMAA is substantially equivalent to a Class I or Class II device [21 C.F.R. 814.9(g)(2)-(5)]. Finally, the voluntary withdrawal of a PMAA by the applicant will trigger disclosure [21 C.F.R. 814.9(g)(6)]. This aspect of the regulation may present difficulties to companies. For example, this may cause problems when a manufacturer has filed a PMAA and is pursuing it, but then runs into problems. Perhaps FDA requests information and the applicant realizes that it will be necessary to conduct a study that will take a year. The applicant may conclude that the best course of action is to withdraw the PMAA, conduct the study, and resubmit the PMAA at a later date.

Unfortunately, if the applicant voluntarily withdraws the PMAA, its existence becomes public and, if requested, FDA will disclose safety and effectiveness information. As a result, there is pressure on a company not to voluntarily withdraw a PMAA when this might actually be the most sensible thing to do. One option available to the applicant is to request an extension of time to submit the additional information. This request must be submitted before the 180-day response period expires. By following this procedure, an applicant may be able to temporarily avoid the voluntary withdrawal of the PMAA if the request for an extension contains (1) an appropriate justification for the extension and (2) a realistic estimation as to when the requested information will be submitted. FDA may also require periodic reports regarding the additional studies being conducted by the applicant to determine whether the applicant is seriously pursuing approval of the PMAA.

Information that FDA usually will not disclose includes manufacturing methods or processes; production, sales, or distribution data; and quantitative or semiquantitative formulas [21 C.F.R. 814.9(h)]. These data generally encompass the trade secret or confidential commercial information likely to be in a PMA.

There are a number of things that a manufacturer can do to help ensure confidentiality. First, under seciton 20.44 of FDA's FOI regulations [2], the manufacturer can obtain a presubmission review of a confidentiality request. [Section 20.44(h) of 21 C.F.R. provides that a request for a presubmission review of confidentiality will not be accepted if the status of the records involved is already determined by section 20.111 or any other regulation in Part 20. Section 20.111 sets forth the type of information in a PMAA that FDA will disclose. PMA applicants should consult section 20.111 of FDA's regulations and other applicable regulations before submitting to FDA a request for presubmission review of the confidentiality of information in a PMAA.] The presubmission review allows the applicant to review the PMAA with FDA and obtain a commitment from the agency on what information will be afforded confidential status. Second, the manufacturer should clearly identify confidential information in the submitted PMAA. Section 814.20(b)(2) of the PMA regulations requires that confidential information be included in all copies of the PMA and identified in at least one copy. Third, if there is an FOI request for information in a PMAA file and the agency is uncertain about whether the information is confidential, section 20.45 of the FOI regulations requires FDA to consult with the applicant.

Here the manufacturer has another opportunity to persuade the agency that the information is confidential. Finally, the agency may agree with an applicant that certain information is confidential, but FDA will not defend an FOI lawsuit filed by a third party (21 C.F.R. 20.53). The agency will instead contact the company and require the company to defend the suit. Defending an FOI lawsuit is a very burdensome, time-consuming process, but it can be done successfully. The manufacturer must make a judgment on whether the confidentiality of the information is worth the time and financial resources necessary to undertake the suit.

V. CONCLUSION

While the PMA regulations may seem arcane and the PMA process burdensome, the important thing to remember is that there is no way around complying with those regulations if the medical device a company wishes to market is one that requires an approved PMAA. In planning the market launch for a product, a crucial aspect of any company's plans to market a medical device for which a PMAA is required is to assure that the clinical trials on which the PMAA will be based are designed and implemented properly so that they provide full support for the claims to be made in the PMAA and the labeling of the product.

Once the clinical trials are completed, the PMAA should be compiled systematically so as to make the information as accessible as possible to the FDA. After submission of the PMAA to FDA, applicants should respond to FDA's questions and concerns as fully and as quickly as possible in order to avoid causing unnecessary delays.

When FDA has indicated that the agency is about to issue an approval letter that contains postapproval requirements, those postapproval requirements should be negotiated with FDA. Once those requirements are in place, the manufacturer must, however, adhere to the requirements or risk withdrawal of the approval.

After marketing has begun, some changes are likely to be made in the device itself or in its labeling or manufacture. Manufacturers should carefully monitor the regulations and supplement the PMAA application when required by the regulations.

REFERENCES

1. *United States v. Kordel*, 335 U.S. 345 (1948).
2. 21 C.F.R. Part 20.

18

Optimal Use of FDA Advisory Panels

AMIRAM DANIEL *Health Industry Manufacturers Association, Washington, D.C.*

I. INTRODUCTION

The product approval process for Class III medical devices dictates that a firm wishing to market a new product must request permission from the Center for Devices and Radiological Health (CDRH) to do so. As part of the approval process, a company's application is presented to an FDA advisory panel composed of clinical scientists in the medical specialty in question plus representatives of both industry and consumers.

The advisory panel reviews the application, which is composed of research and development data, manufacturing and quality control information, results from clinical trials, and proposed labeling for the device. Based on a company's presentation, the answers provided to the panel's questions, and the panel's expertise in the device field in question, the panel recommends to the agency either to accept, reject, or conditionally accept the application.

The review process can have profound effects on the business plan of the company. Firms have been thrown into financial disarray because their devices failed to gain timely FDA marketing approval.

In contrast, situations exist where companies gain prominence in certain device fields because of timely approval. The panel plays a central role in the outcome.

The company, however, is not without a friend. The Industry Representative (IR) is an invaluable resource to every company whenever its product comes before the panel. In the following sections, some of the ways in which the IR can assist companies in their quest for product approval will be described. The IR is very knowledgeable about panel members, the approval process, FDA regulations, and procedures applicable to Class III Premarket Approval Applications (PMAA). He or she is central to the panel's review process. How a company can best prepare for securing a positive approval recommendation from the advisory panel will also be discussed.

II. THE ADVISORY PANELS

A. Legal Basis for Advisory Panels

FDA's advisory panels are charged with assisting the Center for Devices and Radiological Health (CDRH) in making decisions concerning new devices and diagnostic products. The panels' input into the product-approval process is technical in nature, advising in such areas as product methodology, product safety, medical use, adequacy of clinical trials, and labeling efficacy. The advisory panels deal with all medical devices on the market today and are composed of medical experts in numerous fields of diagnosis and treatment.

Occasionally, the panels are called on to assist CDRH in classifying products that were not on the market prior to May 28, 1976 (the date the Medical Devices Amendments were signed into law) or to provide advisory opinions to the agency regarding petitions for reclassification of devices (usually from regulatory Class III to Class II). The panels are also consulted periodically by the agency about matters on which the agency is seeking advice—from the labeling content for home-use IVD devices to guidelines for premarket approval applications (PMAAs) for artificial ligaments. A list of FDA Advisory Panels appears in Table 1.

The panels are convened by the CDRH staff on a periodic basis as the need dictates.* When CDRH decides that a PMAA is complete it will ask the

TABLE 1 FDA Devices Advisory Panels

1.	General and Plastic Surgery
2.	Ear, Nose and Throat Devices
3.	Orthopedic and Rehabilitation Devices
4.	Gastroenterology-Urology Devices
5.	General Hospital and Personal Use Devices
6.	Radiological Devices
7.	Neurological Devices
8.	Immunology Devices
9.	Circulatory Systems Devices
10.	Hematology and Pathology Devices
11.	Clinical Chemistry and Clinical Toxicology Devices
12.	Microbiology Devices
13.	Anesthesiology and Respiratory Therapy Devices
14.	Obstetrics-Gynecology Devices
15.	Dental Devices
16.	Ophthalmic Devices
17.	Blood and Blood Products
18.	Good Manufacturing Practices Committee

*Panel 17 Blood and Blood Products panel is administered by Center of Biologics Evaluation and Research.

commercial sponsor to present it to the panel for its scientific review and recommendation(s). The sponsor has the legal right to demand a panel's review of its application at any time during the application review process, but chances for a positive recommendation from the panel in face of FDA's opposition are very minute.

FDA's authority to operate such panels is contained in the Federal Food, Drug, and Cosmetic Act (FD&C Act), Sections 513(b)(1) and (c)(1). These sections of the law specify how the panels should be convened, administered, and utilized by the FDA. The law specifies that each panel shall be composed of scientists/physicians who are experts on the subject(s) before the panel as well as one representative each of the consumer and industry interests. The panelists are appointed for a 3-year duration. The panels are also subject to the provisions of the Federal Advisory Committee Act, which specifies the administrative procedures that apply to all federal advisory committees to ensure adequate and unbiased discussion of the issues. Since the device field is a rapidly developing one, the CDRH staff makes periodic assessments of the nature of the issues likely to be brought before the panel and recruits the experts to address them. In this respect, the scope of the work of the panels can and does change from time to time.

There are two additional advisory committees that deal with device matters. The CDRH has empaneled an advisory committee that assists the agency with matters relating to the Good Manufacturing Practices regulations (GMPs) and associated guidelines, guidance documents and points to consider, and the composition of the Critical Devices List. This committee differs from the product approval panels in that it has two representatives each for the industry and consumer's interests. The Center for Biologics Research and Review employs a Blood and Blood Products Advisory Committee, which, among other things, deals with materials classified as biologicals and used in *in vitro* treatment of blood.

B. Historical Role of the Panels

FDA has employed device advisory panels for a long time. As early as 1973, FDA's Associate Commissioner for Medical Affairs convened a panel of experts in clinical chemistry to advise the agency on labeling for *in vitro* diagnostic devices. The adoption of the Medical Device Amendments in 1976 authorized the Health and Human Services Secretary to establish advisory panels and described their composition and method of operation. Initially, the panels served as classification panels. They evaluated the total universe of devices on the market at that time, and classified them into one of three regulatory categories based on the degree of risk to the patient associated with the use of each device. The advisory panels thus classified the entire world of medical devices based on use information provided by FDA staff and the panelists own experience. Most of these classification lists have now been published in the Code of Federal Regulations, Vol. 21 Part 800.

C. Changing Nature of the Panels

In the last few years, the role of the advisory panels has changed. The panels have turned their attention primarily to premarket approval applications and to petitions for product reclassification. Since review of these items requires less time to accomplish, the panels tend to meet less frequently. In addition, the FDA has periodically convened special ad hoc advisory panels to deal with specific generic issues, such as home use device labeling and proposed guidelines for biotechnology-derived products. These special panels are usually staffed by selected members of existing panels, including the consumer representative and the IR. The FDA, to facilitate the process of product approval, has published numerous product-specific guideline documents to provide the sponsors with insights into the conduct of clinical trials or the PMAA process itself. The FDA has requested the advisory panels to review these documents and advise the agency about their scientific and clinical accuracy.

III. THE REVIEW PROCESS

As indicated above, FDA has the regulatory responsibility to review the safety and efficacy of all new medical devices about to be introduced. The level of review depends on the degree of risk to the user that the device poses in case of failure or misuse. For this reason, the agency has classified all devices into one of three regulatory categories depending on the degree of risk. For more complete details on this subject consult other chapters in this book or contact the Division of Small Manufacturers Assistance, CDRH, 1-800/638-2041. In addition, consult the 1989 Regional Seminars Binders, Health Industry Manufacturers Association (HIMA 202/452-8240), for additional information on 510(k) notifications. HIMA also publishes a Premarket Approval (PMAA) Manual detailing the PMAA process.

In this section, however, I would like to briefly describe the PMAA path through the agency and the review process itself. All incoming PMAAs are logged in the CDRH/ODE (Office of Device Evaluation) Document Control Center. They are given a number and then routed to the appropriate division for review. The Document Center also tracks the various submissions to ascertain that legally mandated deadlines are met. In addition, the Document Center also is the corresponding entity for ODE, notifying submittors concerning questions raised during the review process.

Once the application has arrived at the appropriate division, it is reviewed from numerous points of views. Is the application complete? That is, have all the appropriate sections (such as Phase I, II, and III) been addressed? For specific products, there may exist FDA guidelines highlighting numerous questions of safety and efficacy that the agency wants addressed. The reviewer will evaluate the application for answers to such questions, and will also review the results of the preclinical and clinical studies. If at any time the reviewer feels that clarification of existing data or additional data are needed, the agency will ask the firm to supply the information. The PMAA will be reviewed by the agency's biostatistical group whenever appropriate. The agency's compliance group gets into the act through its review of the proposed manufacturing protocols and the conduct of prelicense inspection. The agency may employ a consultant to review aspects of the PMAA in those areas where the FDA staff may disagree on the merits of the submitted material or where no expertise exist in-house for the device or technology under consideration.

When the FDA has finished its review of the PMAA, and all questions have been satisfactorily answered, the agency will arrange for an Advisory Panel to review the application as well. The agency is obligated by law to announce, in the *Federal Register*, all panel meetings at least 2 months before the proposed date. This deadline, however, may slip from time to time. The Executive Secretary to the Panel will distribute to the panel members all relevant materials, the meeting agenda, and copies of FDA and other prepared presentations. The applicant should make every effort to obtain copies of all relevant communications between the Executive Secretary and the panel. Once the panel has recommended approval and summaries of safety and efficacy have been written (either by the company, FDA, or both), the application is sent to the FDA General Counsel for his concurrence. FDA recently announced that the agency reserves the right not to send me-too PMAAs (PMAAs from different manufacturers that are identical to an already approved product) to the General Counsel. The last step in the process is for the agency to communicate its decision to the firm in question.

On occasion, the agency will request the company, as part of the approval, to perform postmarketing surveillance of the product. The agency may impose a deadline for such data collection and will request the sponsor to present those data to the Advisory Panel for review.

IV. THE IR ROLE IN THE PANEL REVIEW PROCESS

The IR is considered an integral part of the PMAA review process. He/she must participate in all panel deliberations (see Panel Executive Handbook, CDRH ODE publication, 1987) and is charged with the task of bringing industry's views before the panel. If a company wants to be successful in its presentation before the panel, it should be acquainted with the panel and its members, heed FDA's advice regarding product approval (as disseminated through official publications and in private meetings), and consult with the IR. It is also desirable to meet with the review staff prior to the submittal of an Investigational Device Exemption (IDE), the submittal of the PMAA, or before the meeting of the panel.

The panel meeting is the ultimate test for the applicant and the work embodied within the application. It is very unusual for an application to receive panel approval and still be rejected by the FDA staff. The presentation of the PMAA before the panel takes on, therefore, added importance. To be successful, an applicant should have a coherent and well-organized presentation. Such a presentation should be well rehearsed in advance. The applicant should collect as much information as possible about panel members and their areas of expertise, past panel decisions on the subject, FDA's evaluation of the application, and the current state-of-the-art for this and similar devices. The IR can play an important role in this process, by acting as a consultant to the applicant in terms of how to structure the PMAA presentation. The IR can assist the manufacturer in the following ways:

1. While the Federal Advisory Committee Act prohibits discussion of substantive issues at premeetings, it is appropriate for a panel to have a premeeting to discuss the procedures to be followed at a panel meeting. The IR will participate in all panel premeetings that may be called by the panel chairman to discuss the specific PMAA.

   In some instances, the IR may not be able to participate in a given panel meeting. In such cases, the IR can and should nominate another industry representative as a substitute (on a one-time basis). The IR must secure verbal approval from the Committee Management Staff at

CDRH for such a change before the substitute can participate in the panel's deliberations. IRs can request HIMA's assistance in securing a suitable replacement.

2. The professional qualifications of the panelists, the consultants, and the consumer representative(s) should be communicated to the applicant. It will assist the PMAA sponsor in structuring his presentations to the panel. For this reason, consultations with the IR, although they may pose some competitive difficulties, are probably more advantageous than detrimental.
3. The IR can alert interested companies, through HIMA and/or other appropriate groups, to forthcoming panel meetings, agenda items, and other issues that may affect the review process.

   In essence, the IR serves as the industry's eyes and ears on matters brought before the panel. By providing companies with pertinent information (through HIMA), the IR ensures the dissemination of such information as quickly as possible to all affected companies. The IR may also utilize other means at his/her disposal as appropriate, for the dissemination of such information.
4. The IR may convene, through HIMA when necessary, an industry briefing to help prepare either the industry in general or the sponsor in particular, for panel discussions and actions on upcoming generic or specific issues that may affect the particular PMAA review. The IR may also request a meeting of all affected companies to brief them or seek their views on various issues. It is the IR's responsibility to present properly the industry's views to the panel in a manner consistent with his or her own scientific perspectives. The IR may also be consulted by HIMA or other organizations on matters to be discussed between HIMA and the FDA and which may affect the panel. Such issues, when relating to a PMAA and its review, would be of a technical or an administrative nature.
5. The IR represents the whole industry to the panel. As such, he or she may be able to assist interested companies in preparing a presentation to the panel. Such assistance may take the form of actually helping in structuring the presentation or by offering helpful comments. Such assistance should be provided voluntarily but only at the request of the presenting company.
6. The IR serves as a resource to the panel in understanding the design, testing, and manufacturing processes for medical devices. Companies should consider using this neutral avenue when bringing information to the attention of the panel. The IR stands ready to assist the sponsors in bringing before the panel design, manufacturing, marketing, clinical, and associated regulatory issues specific to the PMAA. Such assistance may take the form of seminar presentations to the panel, tutorials, plant tours, and informal discussions, and can even be made during the panel meeting itself. For this reason, companies should consider including the IR as part of their delegation during closed panel meetings when confidential information is discussed. The IR is otherwise excluded from closed panel meetings.
7. The IR should, whenever possible, summarize the panel meeting developments as soon as it is over, highlighting all significant discussions and actions taken by the panels as well as all significant administrative decisions rendered by the FDA. Written summaries can be very helpful to other IRs and industry. Such summaries are also helpful to future applicants, and they should be forwarded to HIMA for distribution to all other

IRs and interested companies. The IR should exercise judgment as to what constitutes significant discussion and action(s) that necessitate the writing of summary minutes. Companies should be aware that review of minutes from past panel meetings is an important source of information that can be of great help in future PMAAs.

8. Industry periodically presents oral and/or written comments to the FDA or the panels concerning generic or product specific issues. The IR may assist HIMA, other groups or individual firms in the formulation of such presentations (either written or oral) and should be consulted on such issues before presentation to the panel.

V. HIMA'S ROLE IN PRODUCT APPROVAL

The industry associations such as HIMA, also have a role to play by assisting companies in optimizing their presentations to the advisory panels. This role may take several forms:

1. When an issue to be presented before the panel has industry-wide implications, HIMA can assist the presentor in all aspects of the preparation of the presentation. If the situation warrants, HIMA may make the presentation on behalf of its member companies.
2. HIMA alerts interested companies to new panel developments through newsletters, telephone calls, and/or special mailings or meetings. HIMA's role in this process is to ensure that all information received is quickly disseminated to all interested parties by one or more of the various communication modalities available to the association.
3. Periodically, FDA announces a vacancy for an IR on one of its panels. HIMA has taken on the role of soliciting qualified nominations from industry for presentation to the FDA. HIMA, in response to the FDA vacancy announcements, coordinates the nominations of qualified individuals to serve as IRs. The IR can also submit to HIMA a nomination for his/her successor on the panel. HIMA may also recommend to the FDA potential candidates to serve as panel members or consultants as the need arises.
4. When significant issues are to be presented before the advisory panel, either the IR or interested companies may wish to convene a meeting of the IR and the industry segment to discuss the issues and brief the IR on industry's position. HIMA's role is to facilitate the pre- or postpanel meeting of companies and/or the IR. The IR or member(s) may request a post panel meeting to discuss issues that arose during the panel meeting. The IR or the firm(s) may also wish to take the opportunity to alert HIMA to various scientific and administrative issues that are of concern and request the association to take appropriate action. HIMA serves as the focal point for such requests and subsequent actions.

VI. CONCLUSION

The essential elements of the product approval process and the central role of the Advisory Panel in this scheme have been described. Naturally, every company that has one of its products before the panel wants to be successful and to secure unrestricted marketing entry for its device. The IR represents a valuable free resource for the company. It is this author's hope that the entire industry makes use of this resource that has been legally granted to the industry to represent its rights before the FDA and its advisory panels.

19

Complying with the IDE Regulation

EARLE L. CANTY *Ventritex, Inc., Sunnyvale, California*

I. INTRODUCTION

While drafting the Medical Device Amendments of 1976, Congress realized that it was necessary to make provisions for the clinical evaluation of new technologies and new life-supporting or life-sustaining devices. They chartered FDA with the responsibility for developing appropriate regulations for investigational devices. The Investigational Device Exemption (IDE) regulation defined the mechanism for conducting the clinical investigation of a new device.

While a large company has significant financial and human resources that can be devoted to clinical investigation of new devices, the IDE regulation does not represent an insurmountable challenge for a small company. My objective in this chapter is to provide some insight into compliance with the regulation and some suggestions for efficiently and effectively navigating through the IDE process.

The Investigational Device Exemption regulation provides the mechanism for accomplishing three activities. First, the regulation exempts a medical device undergoing clinical evaluation (for the purpose of establishing safety and efficacy) from certain regulatory requirements, including compliance with the Good Manufacturing Practices (GMP) regulation. Second, it establishes the requirements for conducting a clinical investigation of a medical device. Third, the regulation defines the means by which investigational devices can be imported and/or exported.

A substantial amount of information regarding compliance with the IDE regulation can, and should, be obtained directly from FDA. The Division of Small Manufacturers' Assistance (DSMA) serves the device industry, and particularly the small manufacturers that comprise a majority of the companies in the industry, as an information source within FDA. To encourage utilization of that resource, DSMA has a toll-free telephone number, 800-638-2041.

There are a number of publications that can be obtained from DSMA as aids for complying with the IDE regulation. The *Federal Food, Drug, and*

Cosmetic Act, as Amended, and Related Laws provides the specific language of the legislation that serves as FDA's charter for regulating medical devices and enacting the IDE regulation [1]. The *Code of Federal Regulations, Title 21, Parts 800 to 1299* contains the Investigational Device Exemption regulation [2]. Part 812 is applicable to all devices except intraocular lenses (governed by Part 813). The regulations regarding Informed Consent and those regarding Institutional Review Boards are 21 CFR Parts 50 and 56, respectively, and can be obtained individually or in *Code of Federal Regulations, Title 21, Parts 1 to 99* [3]. A manual regarding the regulation, entitled "Investigational Device Regulation—Regulatory Requirements for Medical Devices FDA 46-4159," is also available from DSMA [4].

Within the Office of Device Evaluation, a checklist is used when reviewing IDEs. This "Original IDE Review Form" identifies the essential components of an IDE and serves as a tool during the review process [5]. It can be obtained by contacting the IDE Staff office. While an IDE that addresses all elements of the form does not guarantee approval, it eliminates all nontechnical issues as causes for disapproval and focuses the discussions on the technical issues.

II. PLANNING/SCHEDULING

A. Preparation

Most companies underestimate the time required to design, develop, and test a device, and to secure FDA approval for commercial distribution. For small, privately held and privately financed companies, this can be very damaging. Securing additional funding when schedules have slipped and milestones have been missed can be a very time-consuming, energy draining, and expensive process. During the planning phase, it is important to prepare a reasonable schedule for accomplishing each task. When presenting the plan to the financial community, the tasks that are most likely to incur delays and the scenario for dealing with those delays should be discussed. This candidness minimizes the damage to credibility associated with unpleasant surprises.

Thoughtful planning is the key to preparing an IDE. "Reinventing the wheel" is a time-consuming and expensive task that should be avoided whenever possible. The Freedom of Information (FOI) Act provides industry with a legitimate and valuable means for obtaining some understanding of FDA's expectations regarding the information necessary to support a notification or application for commercial distribution of a medical device. Early in the planning stages for an IDE, FOI requests should be made for the 510(k)s or PMAs filed by competitors for devices similar to the device that is to be the subject of the IDE. FOI requests can be made directly to FDA, indirectly through services specializing in FOI requests, or through consultants or legal counsel. The request should ask for all releaseable information in the file and should include the device name, the model number, and the manufacturer's name.

B. Significant Risk or Nonsignificant Risk

The question of whether the device represents a significant or nonsignificant risk to the user or subject has important implications. A nonsignificant risk device does not require FDA approval for an IDE, whereas an IDE for a significant risk device does require FDA approval.

"Significant risk device" is defined by FDA [2] as follows:

1. Is intended as an implant and presents a potential for serious risk to the health, safety, or welfare of a subject.
2. Is purported or represented to be for a use in supporting or sustaining human life and presents a potential for serious risk to the health, safety, or welfare of a subject.
3. Is for a use of substantial importance in diagnosing, curing, mitigating, or treating disease, or otherwise preventing impairment of human health, and presents a potential for serious risk to the health, safety, or welfare of a subject; or
4. Otherwise presents a potential for serious risk to the health, safety, or welfare of a subject.

The sponsor should develop (and document in the case of a nonsignificant risk device) a position regarding the significance of risk of the device by reasonably interpreting the definition and maintaining objectivity. When the device clearly appears to be a nonsignificant risk device, the device should be represented as such to a participating Institutional Review Board (IRB). IRB agreement constitutes approval of the IDE, and FDA approval need not be obtained.

If the device is quite possibly, though not clearly, a nonsignificant risk device, the device should still be represented as a nonsignificant risk device to a participating Institutional Review Board (IRB). Again, IRB agreement constitutes approval of the IDE and obviates the need to file an IDE application with FDA. Should the IRB conclude that the device presents a significant risk to a subject, an IDE application must be filed with and approved by FDA. Throughout an investigation, should any reviewing IRB determine that a device, considered by other IRBs to be nonsignificant risk, is significant risk, all participating IRBs must be notified and an IDE must be obtained from FDA.

C. Selecting Investigators

Choosing investigators for a clinical investigation can be a contentious issue. The functional groups within each company have their own vested interests in the investigation of a new device and in the selection of clinical investigators. It is in the company's best interest to put aside special agendas and to make these important decisions for the right reasons. Enticing a sales target by offering to make that individual an investigator is not an appropriate practice for selecting investigators.

Although often an unpopular position with management, the best investigations of most devices are kept small and focused. A study with a few investigators at a small number of institutions is easier to control and easier to monitor. As the number of institutions increases, dependency on travel or on-site monitors grows in order to maintain a presence at each site. The clinical training and interface can be better focused when the institution base is small, and the desired quality and consistency of the data can be maintained more easily.

There are several types of investigators, each of whom has strengths and weaknesses. Although the good scientist and clinician is the most desirable from the scientific standpoint, the real world is filled with practitioners that represent a broad spectrum of scientific and clinical ability. A study comprised solely of good scientists and clinicians might yield biased results not indicative of the entire practitioner community.

The size of the relevant subject population seen by an investigator can significantly impact the time required to collect the necessary amount of clinical experience with a device. FDA has guidelines regarding the number of subjects that should be studied in order to establish the safety and efficacy of some devices. FDA also has guidelines regarding the number of investigators using some devices at certain cumulative experience stages of a study. In addition to those constraints, for some devices, follow-up data at some specified period (long-term) subsequent to the use or implantation of a device are necessary to support approval. An investigator with access to a large subject population can be very valuable, particularly during the early stages of a study for which long-term follow-up data is required. Typically, the negative aspect of working with such investigators is that data collection becomes more difficult. The investigator with a large subject population is often very busy and struggles with finding time to complete the data collection process.

Another category consists of clinicians who are influential within their peer group. Though this criterion should never be the sole basis for selecting an investigator, stature and influence can be valuable when combined with one or both of the previously mentioned attributes. Such combinations exist, and can substantially increase the credibility of a good device within the user community.

The inventor investigator can provide real dilemmas. Interest and enthusiasm are high, and the motivation to develop the device is often due to a large subject population seen by the investigator. This same personal involvement in the concept can create division that impedes progress.

III. STUDY DESIGN

There are several key elements that must be considered when designing a clinical investigation. The objective, the protocol, the informed consent, the case report form, and the statistical analysis can significantly influence the success of the study and the time frame required to collect the data needed to support approval for commercial distribution.

The importance of clearly defining the study objective cannot be overemphasized. In some cases, the specific indications for a device may be limited and known at the outset. In other cases, the device might be appropriate for several indications and the purpose of the study may be to determine the most appropriate of those indications. The first type of study is a classic clinical investigation that should generate safety and efficacy data regarding the device. The second type of study is more of a feasibility study that should generate data indicating which applications of the device appear to be the most promising. While both types of studies are appropriate, the methods and magnitudes of the studies differ and should be viewed differently.

Of equal importance is the need to ensure that the objective is consistent with the proposed indications for use. Unless the study demonstrates in a scientifically valid way that the device is safe and effective for the proposed indications for use, the sponsor will have a difficult time convincing FDA to approve commercial distribution of the device.

Investigator participation during the study design process is desirable. The investigator has the most comprehensive understanding of the clinical environment in which the device is to be used. That perspective can help the sponsor to develop a meaningful yet streamlined protocol that is free of activities lacking significant clinical relevance. Protocol deviations, particularly consistent protocol deviations, are less likely to occur under those circumstances.

IV. INFORMED CONSENT

The informed consent document should describe the procedure to prospective subjects in terms and concepts that are understandable to a lay person. It must also present the risks, benefits, and available alternatives to the subject, as well as communicate who is financially responsible for treatment necessitated by problems arising during the study. IRBs and FDA look carefully at this document to make certain that the rights of prospective subjects are protected.

Development of the informed consent document is another study element that can benefit from investigator participation. Though accustomed to using medical terminology, they are also good at describing a complex medical procedure in understandable terms. They are also best equipped to analyze the risks and benefits of participation and to evaluate the likelihood of complications arising during participation or during any of the available alternative procedures.

In some cases, the informed consent document included in the investigational plan will be used at an institution after adding any institution-specific language. Compliance with the essential elements of informed consent for these documents will rarely be a problem because the sponsor will have incorporated those elements during drafting of the document. For all other informed consent documents, the sponsor should verify that all mandatory elements are included. If elements are missing, the sponsor should request that the document be appropriately modified.

V. CASE REPORT FORMS

Well-designed case report forms can facilitate data collection. A good case report form should be easy to complete and unambiguous. If the form allows for multiple interpretations or presentations of some information, instructions should be provided to guide the personnel completing the form. The form should provide for recording all relevant information and little, if any, information that is not likely to be relevant. Though that sounds obvious, in practice it can be difficult to achieve when a study is just beginning and much is yet unknown. The case report form should have the minimum number of pages, though this objective must be tempered by the constraints of the data collection and processing systems.

Three actions during case report form development can help optimize the form. Because the investigators or their assistants will often be completing the forms, their participation in the development process should be sought.

It is also wise to attempt to complete some hypothetical case report forms. Although much time may be devoted to considering all the possible permutations and combinations, it is likely that one or two relatively frequently occurring situations have been overlooked during form development. Data handling and data processing should also be important considerations during case report form development. A multiple-phase study will require multiple pages and may require that some data be recorded on more than one page. Encoding data digitally will facilitate statistical analysis. Since the investigator, the sponsor, and FDA will want a copy of the case report form, there must be some mechanism for generating multiple copies. Multicopy forms, carbon paper, and photocopying are all viable alternatives and all have positive and negative points. The sequence of events during a study should also be considered during form development to facilitate completion.

VI. IRB APPROVAL

IRB approval is not typically difficult to obtain. In most cases, the hardest part of the process is to get the investigator to prepare and submit the necessary information to the IRB. Sponsors should try to assist the investigator, whenever possible, by preparing paperwork or attending the IRB meeting(s). Most importantly, the sponsor must be doggedly, though tactfully, persistent in communicating with the investigator regarding the status of the process.

VII. PRE-IDE REVIEW MEETING WITH FDA

In some cases, the sponsor may have a clear understanding of what will be necessary in order to secure FDA approval of an IDE. In other cases, that will be less clear. For either case, a pre-IDE meeting with FDA should be sought. For the sponsor with a clear understanding, the meeting affords the opportunity to introduce the company to FDA and to let the key players for the sponsor and FDA meet face to face. When the sponsor is less certain about FDA's requirements, the meeting enables the company to probe FDA's position, as well as to introduce the company and key players to one another.

There are several keys to making these meetings happen and making them successful. A rapport must first be developed with the personnel within the appropriate reviewing branch. New technologies can use education as a premise for the meeting. When there is no precedent with a device, it is to FDA's benefit to get a preview of what is coming. When the device is not the first of its kind, guidelines may have been developed by the reviewers to assist with reviews of subsequent devices. In some cases, these guidelines can be obtained. A phone call regarding the availability of guidelines can be an excellent way to begin the rapport development process.

Any meeting should be scheduled well in advance. An agenda should be developed and sent to FDA just prior to the meeting. Selection of the sponsor's attendees should be done carefully. Choose good advocates who are also good listeners. Good preparation, highly focused discussions, and practice will maximize the probability of accomplishing the objectives of the session. The sponsor should seek FDA's opinion of the sponsor's plan, rather than asking FDA to define a plan for the sponsor. At the end of the meeting, the sponsor should summarize the discussions and any action plans emanating from those discussions. A letter containing the same information should be sent to FDA as soon as possible so that the discussions are documented.

VIII. CHARGES FOR INVESTIGATIONAL DEVICES

The Investigational Device Exemption regulation contains language regarding charges for investigational devices. A sponsor is allowed to charge for an investigational device as long as the revenue received does not constitute commercial distribution of the device. In the application for the IDE, the sponsor must convince FDA that the price to be charged satisfies this criteria. For most studies, it is easy to make the argument that the charges do not constitute commercial distribution of the device. Reasonable changes (i.e., a price comparable to the price of a similar device) will rarely be perceived by FDA as commercial distribution of the device.

IX. INVESTIGATOR SOLICITATION

Several years ago, abuses by sponsors of clinical investigations prompted FDA to issue guidelines regarding the solicitation of investigators [6]. The nine points of the guideline specified the manner and the forum(s) that FDA considered appropriate for communicating the fact that a device is available for clinical evaluation. The points are:

1. Availability should only be announced in medical or scientific publications or conferences focused on experts best qualified to investigate the safety and effectiveness of the device.
2. Clearly state that the purpose is only to obtain investigators.
3. Limit the information to that needed by a qualified expert considering participation in the study.
4. Use only limited mailings directed at qualified experts.
5. Include the investigational device caution statement printed in the largest type size used for other information in the presentation (or clearly state in any oral presentation that the device is under clinical investigation and only available for investigational use).
6. Make only objective statements concerning the physical nature of the device.
7. Make no claims that state or imply, directly or indirectly, that the device is reliable, durable, dependable, safe, or effective for the purpose under investigation, and make no claims that the device is in any way superior to any other device.
8. Make no comparative descriptions of the device to any other device.
9. Do not include any pricing information.

These guidelines are available from the IDE office or from DSMA.

X. MONITORING

Effective monitoring of a clinical study involves two related tasks, monitoring compliance by the investigators with the study requirements and handling of investigators not complying with those requirements. Who will serve as monitor for the study should be dictated by (1) the complexity of the device and the procedure for which it is intended, (2) the magnitude of the study, and (3) the proximity of the investigational sites to the sponsor's home base.

Successful introduction of complex devices and procedures often necessitates that the sponsor provide clinical support to help the investigator learn to use the device properly. It is generally desirable for the clinical support personnel to be employees of the sponsor. In this case, a reasonable alternative is for one or more of the clinical support personnel to serve as the monitor(s).

A very focused study is often best handled by personnel within the company. For focused studies, the proximity issue ceases to be significant because it is not that difficult to cover a small number of study centers (five or less) with personnel from the home office.

For a large study to be conducted all over the country, the sponsor must decide whether to establish clinical support "offices" at key places throughout the country, whether to contract with monitors residing near the investigational sites, or whether to have personnel within the company travelling from the home office. Each of these alternatives has advantages and disadvantages relative to the following factors: (1) employee versus contractor; (2) practical frequency of visits versus distance from site; (3) need for clinical support during and after the investigational phase; (4) company personnel addressing compliance problems versus independent party addressing those problems. The best monitor(s) for a study will be selected by considering these and any other factors that are relevant to the study.

XI. IMPORT AND EXPORT

Investigational devices can be imported and exported. An IDE is required to import an investigational device into the United States. The importing agent must act as the sponsor or must ensure that there is another person acting as the agent and sponsor. While an IDE is not mandatory for exportation of an investigational device, it makes the process of securing an 801(d) letter much easier.

Foreign countries can be an important marketplace for devices that are investigational in the United States. However, the process requires 90-120 days, minimum, and grows more complicated each year as foreign governments enact device regulations. Successfully executing a foreign distribution plan for an investigational device requires action well in advance of the anticipated time for commencing shipments. Fortunately, the tasks that need to be accomplished are, in most cases, straightforward.

The letter to the foreign government should concisely introduce the company, describe the device, and state that the purpose of the letter is to secure that government's approval to import the device. Another FDA publication available from DSMA, "Import/Export—Regulatory Requirements for Medical Devices FDA 85-4160," provides a listing of the health ministries of most foreign countries and the appropriate mailing addresses [7].

After receiving a letter from a foreign government that indicates that importation of the device is not contrary to the laws of that country, a copy of that letter should be forwarded to FDA with a letter requesting approval to export the device to that country. If the letter from the foreign government is not written in English, a certified translation must be obtained and sent with the copy of the letter from the foreign government and the letter requesting approval to export the device to that country. Export requests are reviewed by the Division of Compliance. When the device is the subject of an IDE, Compliance checks with the Office of Device Evaluation to learn whether there are any reports of problems that would make it advisable to deny the request. Export approvals are typically issued within 45 to 75 days, the time required being a function of the workload in Compliance.

XII. CONCLUSION

Obtaining an IDE need not be viewed as an overwhelming task that requires tremendous financial and personnel resources. A sponsor can present an investigational device to the medical community and can generate revenue with the device (from domestic as well as foreign users). There are several useful publications readily available from FDA that present the appropriate regulations and guidelines for obtaining an IDE.

Good preparation will facilitate good planning and scheduling. Careful thought regarding the selection of investigators, the study design, the informed consent document, and the case report form will make it possible to conduct a scientifically valid study and to maintain compliance with the IDE regulations. These are the keys (in addition to the safety and efficacy of the device itself) to a successful clinical evaluation of a medical device.

REFERENCES

1. Federal Food, Drug, and Cosmetic Act, as Amended, and Related Laws. HHS publication no. (FDA) 86-1051, U.S. Government Printing Office, Washington, D.C., 1986.
2. Code of Federal Regulations, Title 21, Parts 800-1299. U.S. Government Printing Office, Washington, D.C., 1987.
3. Code of Federal Regulations, Title 21, Parts 1-99. U.S. Government Printing Office, Washington, D.C., 1987.
4. Investigational Device Regulation—Regulatory Requirements for Medical Devices FDA 46-4159. Division of Small Manufacturers' Assistance, Center for Devices and Radiological Health, U.S. Food and Drug Administration, Rockville, MD.
5. Original IDE Review Form. Investigational Device Exemption Staff, Center for Devices and Radiological Health, U.S. Food and Drug Admin-Rockville, Spring, MD.
6. Guidelines for Preparing Notices of Investigational Medical Devices. Center for Devices and Radiological Health, U.S. Food and Drug Administration, Rockville, MD.
7. Import/Export—Regulatory Requirements for Medical Devices FDA 85-416. Division of Small Manufacturers' Assistance, Center for Devices and Radiological Health, U.S. Food and Drug Administration, Rockville, MD.

20

Clinical Trials and Labeling Claims

ROBERT A. DORMER *Hyman, Phelps & McNamara, Washington, D.C.*

I. INTRODUCTION

In a 1973 decision upholding the Food and Drug Administration's (FDA's) regulations establishing minimum standards for adequate and well-controlled clinical investigations for drugs, the Supreme Court stated that the "impressions or beliefs of physicians [about whether a drug is effective], no matter how fervently held, are treacherous" [1]. This skepticism about what FDA often refers to disparagingly as "anecdotal evidence" underlies the FDA approach to data of all kinds. Unless data are generated from adequate and well-controlled studies, FDA will place little credence in them. This mindset has a controlling effect on what a manufacturer can and cannot say about its product and frequently on whether a product can be marketed at all.

II. BASIC CONCEPTS

All devices are classified by FDA into one of three classes. Class I devices present the least risk to patients; Class II devices present an intermediate risk; and Class III devices present the greatest risk [2]. Under the Federal Food, Drug, and Cosmetic Act (FDC Act), Class III devices are those that are "purported or represented to be for a use in supporting or sustaining human life or for a use which is of substantial importance in preventing impairment of human health, or . . . [which] presents a potential unreasonable risk of illness or injury" [3]. Only Class III devices require a prior determination by FDA that the product is safe and effective. A manufacturer must submit a premarket approval application (PMAA) before a Class III device can be marketed. Class I and Class II devices, on the other hand, can be marketed more easily by submitting a premarket notification of "510(k)" to FDA 90 days prior to marketing the product [4].

Devices can be marketed under the 510(k) procedure provided they are "substantially equivalent" to a device that was on the market prior to passage of the Medical Device Amendments of 1976 and that is in Class I or Class II [5]. Generally, clinical investigations are not necessary to demonstrate "substantial equivalence." In some circumstances, however, such as where a device has different technological features than the predicate device, FDA will request manufacturers to submit clinical data to demonstrate that, in fact, the device works as well as the earlier product [6].

The 510(k) process is, however, neither an "approval" process nor an agreement by FDA that a device is either safe or effective for its labeled indications [7]. Although modifications can be made to a marketed product or to a product's labeling, FDA requires the submission of a new 510(k) if the change could "significantly affect safety or effectiveness" [8]. Among the types of changes that require a new 510(k) is a "major change or modification in the intended use of the device" [9]. In fact, some changes may be so significant that "substantial equivalence" to a predicate device is no longer possible and the manufacturer must instead submit a PMAA for the new indication.

Obtaining approval of a PMAA is a considerably longer and more expensive process than the 510(k) procedure.[1] The FDC Act provides that a PMAA is generally required to be supported by "well-controlled investigations, including clinical investigations where appropriate" [10]. Although the Act does allow a PMAA to be approved on the basis of "other" valid scientific evidence [11], as a practical matter FDA always requires clinical investigations to support a PMAA. Those clinical investigations must be well-controlled and must show that "the device will have the effect it purports or is represented to have under the conditions of use prescribed, recommended, or suggested in the labeling of the device" [10]. Among other requirements, a PMAA must contain "full reports of all information published or known to or which should reasonably be known to the applicant, concerning investigations which have been made to show whether or not such device is safe and effective" [12].

III. LABELING

"Labeling" is the primary means by which a manufacturer communicates to health care providers, third-party payers, and patients the uses for which a device is intended. When one designs a clinical trial to establish the scientific support for a labeling claim, it is essential that one have a complete understanding of the relationship between labeling and science. This is true irrespective of whether the clinical study is intended to support initial approval of a PMAA, to support a "substantial equivalence" determination for a premarket notification under section 510(k) of the FDC Act, or to support new claims for a product already on the market.

For purposes of this chapter, the discussion of claims for a product will not be limited solely to those materials prepared by or for a manufacturer that constitute what one would traditionally consider to be labeling—for example, labels, user's manuals, service manuals, package inserts—but rather will include any materials intended by the manufacturer to describe how the

[1] Although FDA has 180 days to review a PMA under the FDC Act [FDC Act 515(d)(1), 21 U.S.C. 360e(d)(1)], it rarely meets that deadline.

product is to be used, and/or to promote sales of the product. Therefore much of the discussion will focus on materials that companies typically consider to be advertising rather than labeling.

The term "labeling" is defined in the FDC Act as "all labels and other written, printed, or graphic matter (1) upon any article or any of its containers or wrappers, or (2) accompanying such article" [13]. A label is "a display of written, printed, or graphic matter upon the immediate container" of a device [14]. For a document to "accompany" a device, it is not necessary that it physically accompany the product. Provided the document and the device have a common origin, a common destination, and are related textually, a document can be considered "labeling" even if it is shipped at a different time than the device [15]. For example, if a manufacturer disseminates copies of an article from a medical journal that suggests that the manufacturer's device is safe or effective for an unlabeled use, FDA will consider that article to be labeling for the device. Because the device does not bear adequate directions for this use, FDA will consider the device to be misbranded. Similarly, advertising such as print ads in medical journals can under certain circumstances also be deemed to be labeling.

The FDC Act requires that devices bear labeling that contains "adequate directions for use" [16]. FDA has long interpreted the term "adequate directions for use" as directions under which "the layman can use a device safely and for the purposes for which it is intended" [17]. Therefore, under the regulatory scheme that FDA has constructed, if a device cannot bear adequate directions for lay use, that is, because it can be used safely only by, or under the supervision of, a physician, the device is deemed "misbranded" unless FDA has issued a regulation exempting the device from the requirement to bear adequate directions for lay use and the device complies with the conditions for exemption.

The two most important exemptions for purposes of this discussion are for prescription devices and for *in vitro* diagnostic products [18]. A prescription device is one that "because of any potentiality for harmful effect, or the method of its use, or the collateral measures necessary to its use, is not safe except under the supervision of a practicioner licensed by law to direct the use of such device" [19]. Among other requirements pertaining to prescription devices, the device's labeling is required to bear "information for use, including indications, effects, routes, methods, and frequency and duration of administration, and any relevant hazards, contraindications, side effects, and precautions under which practitioners licensed by law to administer the device can use the device safely and for the purpose for which it is intended, *including all purposes for which it is advertised* or represented" [20]. *In vitro* diagnostics are similarly required to bear information about how to use and store the product [21]. These provisions state what is required to appear in labeling; they do not describe how one goes about developing data to support claims that a company wishes to make.

Although FDA does not directly regulate advertising for most devices,[2] the agency uses the concept of "intended use" to exercise considerable con-

[2]FDA does have authority over the advertising of "restricted" devices. Restricted devices are those which FDA has, *by regulation*, limited to prescription status [FDC Act 520(e), 21 U.S.C. 360j(e)]. The only restricted devices are PMAA products. FDA issues an order or regulation approving the product and restricting it to use by or on the order of a practitioner.

trol over advertising. FDA defines the term "intended use" as the "objective intent of the persons legally responsible for the labeling of devices" [22]. This intent may be shown by "labeling claims, *advertising matter*, or oral or written statements by such persons or their representatives" [22, emphasis added]. If a product is advertised for a use for which it is not labeled, FDA will assert that it is misbranded because it fails to bear adequate directions for its intended use. In this way FDA is able to take regulatory action if the advertising for a device is, in the agency's view, false or misleading.

The Federal Trade Commission (FTC) is the primary federal agency that regulates advertising [23]. It is a violation of federal law to disseminate any false advertisement [24].

The basic rule of advertising from the perspective of the FTC is that advertisers must have a reasonable basis for the claims they make [25]. The level of substantiation required will depend on a number of factors, including whether the claims are express or implied. It is important to recognize that the FTC requires an advertiser to have substantiation at the time it makes the claims [25]. Postclaim substantiation can be used in certain cases to support a claim, but to comply with the FTC's requirements there must be substantiation when the claims are made [25].

The federal Lanham Act gives private parties the right to sue and recover damages from and obtain injunctive relief against a company that makes false representations about its own products [26]. Although false advertising litigation has generally involved widely advertised consumer goods [27], it is not surprising that such litigation is also being used to stop false advertising directed to physicians and other practitioners. One observer points out that such suits have been brought against the sellers of products such as "home pregnancy test kits advertised in medical journals, fluoride gel products sold to dentists, pregnancy test kits sold to doctors, surgical needles, specialty bedding products sold to health care facilities, and theophylline, a prescription drug used for the treatment of asthma" [28].

Although there may be some variations among the different statutory provisions regarding the level of proof necessary to show that an advertisement or labeling claim is true (or false, depending on who has the burden of proof), in no case will a manufacturer's claims be able to pass muster unless they are supported by valid scientific evidence.

IV. CLINICAL TESTING

The most important question that a company needs to ask itself before it embarks upon a clinical testing program is: What is it that the clinical study is intended to accomplish? Only when that question is answered can a company begin to determine whether the design of the study is adequate for the purpose for which it is intended. There are, for example, a variety of clinical end points that can theoretically be measured in a study but that may support dramatically different labeling claims. It is also important to recognize that the principles for designing scientifically acceptable clinical studies were developed in the drug area. Drug studies tend to be very similar in design. Because of the wide variability in the types of devices on the market and under investigation, it is simply not feasible to have as much uniformity in device studies as one does in the drug area.

For example, if a company is testing a plasmapheresis machine for the purpose of demonstrating that the machine is capable of separating plasma from the other blood components and returning the remaining components to a patient safely, the test will be relatively simple. If, on the other hand, a manufacturer wants to show that this procedure has some therapeutic effect in preventing or treating disease, the clinical evaluation becomes much more complicated. Then the company needs to assess how best to measure the therapeutic effect that is being studied. Similarly, the ability to measure or detect the presence of various components or substances in blood or other body fluids will require a different type of test than one intended to show that the particular measurement being made is predictive of the patient developing a disease.

FDA obviously prefers that data be obtained from well-controlled studies,[3] regardless of whether the studies are conducted under an investigational device exemption (IDE). But is a double-blind study necessary to demonstrate effectiveness? Is such a study technically feasible? Is it ethical? These are just some of the questions that need to be answered in designing a clinical study.

To demonstrate that a device is effective, FDA's regulations provide specifically that a manufacturer shall submit evidence derived from "well-controlled investigations . . . unless the Commissioner authorizes reliance upon other valid scientific evidence which the Commissioner has determined is sufficient evidence from which to determine the effectiveness of a device, even in the absence of well-controlled investigations" [29]. The FDA commissioner recently stated that "to justify the use of information drawn from a source other than a well-controlled clinical investigation in humans, a person seeking approval to market a device under seciton 515 of the act must show that there is something about the type of device or the mode of operation of the device that would prevent the development of data in such a study" [30].

The elements of a well-controlled clinical investigation include a protocol that contains a clear statement of the objectives of the study, a method of selection of the subjects that provides assurance that the subjects are suitable for the purposes of the study, and a method for assigning subjects to test groups so as to minimize any possible bias [31]. In addition, the protocol should assure comparability between test groups and any control groups with regard to variables such as sex, severity or duration of the disease, and use of therapy other than the test device. There should also be an explanation of the methods of observation to be used in the study and an explanation of how the results of the study will be recorded, including the variables to be measured, quantitation, assessment of any subjects' response, and steps taken to minimize possible bias of both subjects and observers. Finally, the protocol should contain a method for comparing the results of treatment or diagnosis with a control [32].

---

[3] In order to conduct clinical testing of a device a manufacturer must obtain from FDA approval of an IDE, be exempt from the IDE requirements, or satisfy the abbreviated requirements for non-significant risk devices (21 C.F.R. Part 812). As with FDA's classification regulations, the IDE regulations impose differing levels of FDA control and approval of investigational studies depending on the risk the device presents.

Generally FDA recognizes four types of valid comparisons that can be made in a clinical study: no treatments, placebo control, active treatment control, and historical control [33]. It is obviously important in designing a study to determine which method of comparison is the most appropriate for the device being studied and the conditions for which it is intended to be used. In some cases, it will be unethical to use a placebo control. A study to test a barrier contraceptive for the prevention of AIDS obviously could not employ a placebo control. In other instances, a placebo control is the only way to eliminate the possibility of a placebo response to the device. In a pain study, for example, it is well known that there is a high placebo response, and therefore double-blinded studies with large numbers of patients are often required, from both a regulatory and a statistical standpoint, to demonstrate effectiveness.

A report of a clinical investigation needs to include a summary of the methods of analysis and an evaluation of the data derived from the study, including any appropriate statistical methods that are utilized. The study should use a test device that is standardized in its composition or design and performance. This last point is particularly important. Often, a manufacturer will begin a clinical investigaiton with a prototype device and change the device during the course of the study. Although this is certainly permissible, it may generate a requirement that additional data be obtained before approval of the product. Unless the manufacturer can show that the changes were not likely to have a significant effect on safety or effectiveness, FDA may not permit data from two different versions of the product to be pooled for demonstrating effectiveness. The information may, however, be used to support safety.

With these protocol concerns firmly in mind, the next issue is how to go about selecting investigators and monitoring the study [34]. The issue of investigators raises another question about where, that is, in what country, the investigation is to take place. FDA allows a company to submit non-U.S. data as the sole support for a PMAA [35]. Further, it is not required that foreign investigations be conducted under an IDE.[4] If, however, a company intends to use foreign investigators, it is important that (1) the investigators are well-recognized clinicians, (2) there are no serious medical or racial differences between a comparable test population in the United States and the test group outside the United States, and (3) FDA must have the ability to inspect the investigational site. FDA has approved several PMAAs solely on the basis of foreign clinical data.

In addition to selecting investigators, the importance of monitoring a study cannot be overemphasized. In analyzing data, FDA will treat study drop-outs or others who are lost to followup as treatment failures. In assessing the efficacy of the study, the denominator (i.e., the total number of patients in the study) will thus include all those who completed treatment, as well as those who dropped out of the study or were lost. Therefore, measuring the effectiveness of a device will be made considerably more difficult if a significant number of study participants have failed to complete the study and such failures cannot be adequately explained.

---

[4] If a non-U.S. study is not conducted under an IDE, a manufacturer must obtain FDA approval of an export request to ship an unapproved device outside the country. [FDC Act 801(d)(2), 21 U.S.C. 382(d)(2)].

Outside advisory panels also play an important role in the approval process. All PMAAs (and significant PMAA supplements) are required to be reviewed by an appropriate advisory panel of outside experts [36]. Before a company submits its own PMAA, it is worthwhile to attend advisory committee meetings where similar products are being reviewed. In this way the company can determine what concerns advisory panel members have had so that these concerns can be addressed either through clinical investigations, in the PMAA, or through other appropriate means. In addition, a manufacturer is well advised to follow the research done by companies that have already obtained approval for similar devices. FDA is required to publish a summary of safety and effectiveness data for each device approved under a PMAA [37]. This summary will provide detailed information about the type of studies conducted by the applicant. Because FDA is required to treat applicants equally, it stands to reason that if a company was able to obtain approval of a product by studying a patient population of 200 people, a subsequent manufacturer should be able to do the same, unless new safety or effectiveness issues have arisen that require a larger patient population to address. It is not necessary for each manufacturer to develop a novel study methodology for its clinical trials. Following work that was done before by other companies on similar products is an extremely useful guide to obtaining approval.

It is equally important that a company consult early with FDA reviewers during the preparation of an investigational study. FDA will say in its correspondence approving IDEs that it does not guarantee that if the investigational plan is followed, the studies will generate sufficient data to support approval of a product as safe and effective. Nevertheless, the likelihood that a study will generate such data is increased if FDA has had an opportunity to participate in the design of the study and offer suggestions and criticisms for improving study design and implementation. In this regard, FDA in some areas has issued guidelines about what information is required to obtain approval of particular products. These guidelines, although not legally binding, are important and should be followed unless there is a valid scientific reason for not doing so.

V. CONCLUSION

The purpose of doing clinical investigations of medical devices is twofold. First, from a scientific standpoint (as well as from a product liability perspective), a company must assure itself and its customers that a product is indeed safe and effective. Second, the company must assure the FDA that a product meets the legal requirements for marketing under the FDC Act. If a company bears in mind that its science is intended for both medical and regulatory purposes, it will develop the necessary data and write the appropriate marketing application in such a way that FDA will be able to stand behind the data from a legal standpoint. FDA has to be convinced legally that a device meets the requirements of the statute. FDA cannot rely on data in another company's application, for example, to approve a competitor's product. FDA has to know that a company's application contains sufficient data to stand by itself and that the FDA can approve the product without fear that it will be attacked because it had violated the statute. If these dual purposes are properly addressed, a company will assist itself and thereby obtain early approval of its product. By addressing forthrightly the issues that FDA will want to see addressed, the company will enhance its credibility and provide assurance to FDA that the data are acceptable. Trying to take shortcuts is often counterproductive and, rather than leading to prompt approval, may result in protracted delays while the company attempts to patch together an application that should never have been submitted in the first place. As in other areas of life, doing it right the first time is the best advice.

REFERENCES

1. *Weinberger v. Hynson, Westcott and Dunning, Inc.*, 412 U.S. 609, 619 (1973) (footnote omitted).
2. See generally FDC Act 513, 21 U.S.C. 360c.
3. FDC Act 513(a)(1)(C)(ii), 21 U.S.C. 360c(a)(1)(C)(ii).
4. This submission is called a "510(k)" because the procedure is set forth in section 510(k) of the FDC Act, 21 U.S.C. 360(k).
5. FDC Act 513(f)(1)(A), 21 U.S.C. 360c(f)(1)(A).
6. See R. M. Cooper, Clinical Data under Section 510(k), *Food Drug Cosmet. L.J. 42*:192 (1987).
7. 21 C.F.R. 807.97 [FDA acceptance of a 510(k) "does not in any way denote official approval of the device"].
8. 21 C.F.R. 807.81(a)(3).
9. 21 C.F.R. 807.81(a)(3)(ii).
10. FDC Act 513(a)(3)(A), 21 U.S.C. 360c(a)(3)(A).
11. FDC Act 513(a)(3)(B), 21 U.S.C. 360c(a)(3)(B).
12. FDC Act 515(c)(1)(A), 21 U.S.C. 360e(c)(1)(A). FDA's regulations governing the requirements for a PMAA are at 21 C.F.R. Part 814.
13. FDC Act 201(m), 21 U.S.C. 321(m).
14. FDC Act 201(k), 21 U.S.C. 321 (k). The term "immediate container" does not include package liners [FDC Act 201(1), 21 U.S.C. 321(1)].
15. *United States v. Kordel*, 397 U.S. 1 (1970).
16. FDC Act 502(f)(1), 21 U.S.C. 352(f)(1).
17. 21 C.F.R. 801.5.
18. See 21 C.F.R. 801.109, 801.119, and 809.10.
19. 21 C.F.R. 801.109.
20. 21 C.F.R. 801.109(c) (emphasis added).
21. 21 C.F.R. 809.10.
22. 21 C.F.R. 801.4.
23. 15 U.S.C. 52.
24. 15 U.S.C. 52(a).
25. See 49 *Fed. Reg.* 30999 (August 2, 1984).
26. 15 U.S.C. 1125.
27. See, for example, *McNeilab, Inc. v. American Home Prods. Corp.*, 848 F.2d 34 (2d Cir. 1988) (litigation involving over-the-counter pain relievers).
28. Morrison, T. C., The False Advertising of Specialty Medical Products under the Lanham Act, *Food Drug Cosmet. L. J. 44*:265, 266 (1989).
29. 21 C.F.R. 860.7(e)(2).
30. In the Matter of Howmedica Simplex P Antibiotic Bone Cement, Docket No. 84P-0346 Med. Device Rep. (CCH) paragraph 15,101 at 14,669 (March 28, 1988).
31. For a good discussion of how to plan for FDA approval of a drug product, see F. L. Hurley, Planning Research and Development of New

Drugs to Assure Regulatory Approval, *Food Drug Cosmet. L. J. 39*:312 (1984).

32. 21 C.F.R. 860.7(f).
33. 21 C.F.R. 860.7(f)(1)(iv).
34. The sponsor's obligations to monitor an investigational study are set forth at 21 C.F.R. 812.43 (selecting investigators and monitors) and 812.46 (monitoring investigations).
35. 21 C.F.R. 814.15.
36. FDC Act 515(c)(2), 21 U.S.C. 360e(c)(2).
37. FDC Act 520(h), 21 U.S.C. 360j(h).

21

Establishing a Training Program for Plant Personnel

NOEL L. BUTERBAUGH and LEIF E. OLSEN *Whitaker Bioproducts, Inc., Walkerville, Maryland*

I. INTRODUCTION

A. Objective of Chapter

There can be little doubt of the need for training in industry in the United States. The need to be competitive in the domestic marketplaces as well as in the international arena coupled with the increasing concern over the nation's illiteracy problem would seem to be reasons enough for companies to establish training programs. In the medical and health care industries this need goes beyond business considerations such as competitiveness and productivity. It is a U.S. government requirement. This requirement is specified by the Food and Drug Administration (FDA), in the Code of Federal Regulations (CFR).

The purpose of this chapter is to provide guidance in setting up a comprehensive training program that will meet government requirements and establish guidelines for employee development. In addition to training in the performance of assigned responsibilities, documentation and auditing of the program will be addressed.

To supplement the discussion on how to establish in-house training programs, an effort has been made to provide information on outside sources for training to include audiovisual training aids and off-site training programs. Although many small companies simply cannot afford expensive in-house staff and programs, it should be recognized that there are numerous training programs and seminars available. Further, state funds are often available to supplement training expenses.

Finally, another objective of this chapter is to go beyond the need to meet FDA requirements and to bring out the overall importance of training in achieving high quality and productivity. If a company's goal is merely to satisfy FDA requirements, the real value of a training program will not be realized. Although there may be considerable rhetoric on the importance of training, often the need for a strong commitment on the part of top management is overlooked. A successful training program is usually the result of a company's philosophy that recognizes the need for training in achieving high quality.

B. The Need for Training

The implementation of new manufacturing technology cannot be successful without training. In an industry where technology is dynamic, such as the health care industry, companies must continually train and retrain their employees in new skills required to meet new technology demands. This training is essential in order to compete in the business arena.

This situation is further exacerbated by the pressure of a reduction in the size of the available work force. A declining younger population heavily weighted with educationally disadvantaged and untrained people is expected to result in a severe employment shortage by the early 1990s. However, with technological advances creating new jobs and the need for new skills, "the private-sector employment demand is expected to reach 156.5 million jobs by 1990. That total, nearly twice the demand of 10 years ago, will mean a shortage of 23 million persons in the work force" [1].

The need for increased emphasis on training is leading to dramatically increased corporate spending. Corporate spending to educate the work force is running approximately $40 billion annually, more than half the $60 billion spent by the university system in the United States.

Preparing individuals to handle different jobs and new technology through training can be termed "employee development." However, the need for training encompasses more than development. This need parallels the need for product and service quality. There is probably no one factor that has contributed to a decline in the United States leadership position in the world market more than inferior quality. This position as a world leader can be regained only through a renewed emphasis on quality. Workers must recognize the need and value of quality and be thoroughly trained in their jobs to achieve this goal.

In the medical device industry, product quality can be a key factor in product safety. Through the 1980s FDA spokesmen have placed increased emphasis on the importance of quality and product performance. Further, the FDA has identified training in the use of medical devices as a primary goal over the coming years.

This concern for training can be found in the CFR. Under 21 CFR Part 820.25, the requirement for training is clearly stated:

> (a) *Personnel Training.* All personnel shall have the necessary training to perform their assigned responsibilities adequately. Where training programs are necessary to assure that personnel have a thorough understanding of their jobs, such programs shall be conducted and documented.

This knowledge of FDA requirements, combined with the economic advantages of training, leaves little doubt that a comprehensive training program must be an integral part of a company's business structure. This applies to companies of any size and should be given a high priority regardless of financial strength. More than any other factor, the lack of training can lead to failure in achieving goals.

An objective of this chapter is to provide help in setting up training programs regardless of company size.

II. ESTABLISHING THE PROGRAM

A. Overview

No matter how large or small a company is, it is necessary to have a well-defined training program. For those companies that do not have a formal program, the following offers useful information to help establish one. If your company does have a program, this information may be useful as a comparison.

A company should develop a written policy for training. A written program will provide consistency in employee development. It will also be used as the basis for periodic review and revision, in order to keep the program current and to improve the quality of the program as new training techniques are introduced.

There does not appear to be much written on how to establish a company-wide training program. However, there are numerous resources available that can be used to address specific training needs, such as worker safety, GMP compliance, or management development. These will be discussed later in this section.

How then do companies establish an in-house program to meet their specific needs? There are four basic approaches: (1) hire a full-time training professional to develop and maintain a program, (2) hire consultants on an as-needed basis, (3) send employees to workshops and seminars specific to their job needs, and (4) combine any two or all of the above. Whatever the approach, there should be one individual responsible for coordinating the program and maintaining necessary documentation of employee training.

One approach to establishing a successful program is outlined below. This represents a program for an average-size medical device manufacturing company that has a variety of professional, technical, and other skilled employees. This approach is basic and can be used as the foundation to build upon.

B. The Written Program

The written program should define the company's training philosophy. This program can be divided into two documents: (1) company policy statement and (2) training procedure. In most cases these are developed by either the Director of Personnel or a training specialist. Once written, the policy should be reviewed and approved by the company's upper management. At this time the program can be finalized and put into effect with full management support.

A basic company policy should include:

a. *Purpose*: The basic goal of the training program is to improve quality and productivity through improved job performance. This program should include, but should not be limited to, new employee orientation and continuing employee development.
b. *Authority and responsibility*: The Director of Personnel should have the authority to initiate and approve all training topics. The Director of Regulatory Affairs should review the training program on a periodic basis to ensure compliance with local/state/federal government regulations. The training specialist should be responsible for scheduling, preparing

curriculum, coordinating or administering training or both, providing cost/benefit information to management and record keeping of attendance.

c. *Documentation*: A written record of training sessions should be maintained.

The written training procedure should include:

a. *Purpose*: This section should be consistent with the overall company policy statement and could simply restate it.
b. *Authority and responsibility*: This section should specify who has authority and responsibility in carrying out the various training functions.
c. *Scheduling*: The trainer should coordinate all training sessions with department managers and provide a schedule listing the training subject and attendees.
d. *Resources*: This section should outline the various training aids available (e.g., videos, slides, films, manuals, overheads). Any reference to consultants and local colleges and universities could be included in this section.
e. *Subjects*: This section should make reference to the specific training that is available, such as good manufacturing practices (GMP), hazard communication program, safety, supervisor development, and specific technical training. This list will grow as the company program develops.
f. *Documentation*: A written record of all training should be maintained. This information should include the subject, a brief description of the course, date of training, and the names of attendees. In a regulated industry, documentation is essential in order to have evidence that the firm is complying with the various regulations. Software programs are available to automate the documentation system. See references in Section II.E on training and management programs.

C. The Trainer

Historically, trainers have emerged in industry from the ranks of the personnel department. With the changing emphasis of technical and regulatory requirements in the medical device industry, skills in addition to human resource management appear to be necessary for a trainer to be effective.

The training professional must have knowledge of the scientific basis of the operation, statistical methods for use in needs assessment and evaluation, and regulatory issues impacting on the industry.

The training professional also should have skills in management and communication. Technical skills or knowledge in any of the subjects being taught would be valuable.

Texas A&M researchers examined "the specific tasks the entering training/human resource development professional is expected to perform and . . . the frequency and level of competence expected in the performance of each task" [2]. According to their findings, tasks most frequently performed by this individual included the following:

Working effectively with both individuals and groups
Applying interpersonal relations skills to on-the-job situations
Establishing good relations with management
Using methods of adult education
Keeping abreast of training and development concepts

Assisting in group and organization development
Conducting classroom training

How does one go about finding a person with these skills and abilities? "Within the industrial training profession there is no certification or degree program that prepares practitioners to enter it. University programs to prepare persons specifically for the private sector industrial training profession are only a recent development" [3]. The company should review its training needs and determine the type of individual it requires. It is unlikely that advertising will produce a large pool of applicants for these positions.

One resource for trainer development is the pool of public high school teachers with scientific backgrounds. These individuals already have the necessary information and practice in learning theory and instructional implementation. A detailed orientation program geared to knowledge of the particular company—its products and operation—could produce an effective training professional in a relatively short period of time.

Another resource is from within the company. Since the training function is directly affected by developments in industry, it is possible to groom employees already in the company. Identifying technically oriented employees with good communication skills can lead to developing the trainer of the future.

D. Consultants

Companies without technical trainers or trainers with emerging skills could enlist the services of consultants. Even in companies with qualified trainers, a staff of one or two persons cannot provide all the necessary training.

Often consultants can be found who specialize in a particular subject. Qualified consultants can save a company time and money by providing excellent training in a timely manner.

Before hiring a consultant the company should have a clear picture of what it wants to accomplish. The company should develop written objectives, scope, and expectations of the training to be performed.

On the initial contact with a consultant, the company should be specific about its needs and discuss its written objectives with the consultant. Before deciding on a particular consultant, it may be beneficial to see the candidate in action. You may want the consultant to present a proposal developed in response to your training needs. Also, track records of consultants can be checked by contacting former clients. Polling their past contracts can give a general idea of their effectiveness.

Two areas of commitment are needed to form a strong working relationship with a consultant. The first is the commitment of time. The company needs to provide the consultant with staff assistance in terms of its in-house professionals or experts or both. Consultants must learn as much as possible about the company and gain an understanding of the industry. This will take time and effort from the company's employees.

The second area of commitment is money. Consultants can be expensive, and some of them are very expensive. A price range should be established before embarking on a consultant search. When a particular consultant has been chosen, a contract should be developed.

Information on consultant pools for regulatory and technical training can be obtained from the American Society for Training and Development [4]. Additionally, contact with adult education departments of community colleges and universities can provide information of local sources to fill this need. In fact, some community colleges and universities provide consulting services. The departments of continuing education can supply this information.

Do not overlook "free" consultant help in the form of government service agencies. State agencies responsible for job training programs often have resources which can be borrowed or shared. Industrial training programs in various state capitals can provide current information concerning contacts for consulting.

Other sources for consultant contact include the Health Industry Manufacturers Association, Division of Small Manufacturers Assistance at the FDA, Better Business Bureau, the local chamber of commerce, and the industrial engineering department at a state university.

E. Resources

There is a variety of sources available for specific information needed to supplement a company's training program. The following is an outline by topic of specific organizations available for assistance.

The topics listed are divided into three basic subjects: (1) technical, (2) regulatory, and (3) management. The organizations listed under each category will provide various types of assistance from workshops and seminars to written publications and other training materials. These organizations are listed with their addresses and telephone numbers for your assistance.

1. *Technical Training*

The following organizations offer assistance from specific seminars to training aids such as technical guides and audiovisuals.

	Organization	Type of assistance
a.	Center for Professional Advancement P. O. Box 964 East Brunswick, NJ 08816-0964 Telephone: (201)613-4500	Seminars
b.	STAT-A-MATRIX Institute 2124 Oak Tree Road Edison, NJ 08820 Telephone: (201)548-0600	Seminars
c.	Health Industry Manufacturers Association 1030 Fifteenth Street, N.W. Washington, DC 20005 Telephone: (202)452-8240	1. Technical monographs 2. Task forces on specific issues
d.	International Center for Technology Transfer, Inc. 13014 N. Dale Mabry Avenue, Suite 142 Tampa, FL 33618 Telephone: (800)654-1147	1. Workshops 2. Reference materials
e.	Interpharm P.O. Box 530 Prairie View, IL 60069 Telex: 499-5880 INTPHRM Telefax: (312)459-6644	1. Seminars 2. Workshop manuals

f. Pharmaceutical Manufacturers Association 1100 15th Street, N.W. Washington, DC 20005 Telephone: (202)835-3552	Training resource booklet
g. American Society for Quality Control 310 W. Wisconsin Avenue Milwaukee, WI 53203 Telephone: (414)272-8575	1. Seminars on quality topics

2. Regulations

The following organizations offer assistance such as regional seminars, guidance manuals, and audiovisuals.

Organization	Type of assistance
a. Health Industry Manufacturers Association (see address above)	1. Regional seminars 2. Guidance manuals 3. Audiovisuals
b. Food and Drug Administration Division of Small Manufacturers Assistance 5600 Fishers Lane (HFZ-220) Rockville, MD 20857 Telephone: (301)443-6597	1. Regional seminars 2. Guidance manuals 3. Audiovisuals
c. Interpharm (see address above)	Seminars
d. International Center for Technology Transfer, Inc. (see address above)	Seminars
e. Center for Professional Advancement (see address above)	Seminars
f. STAT-A-MATRIX Institute (see address above)	Seminars
g. Washington Business Information, Inc. 1117 North 19th Street, Suite 200 Arlington, VA 22209 Telephone: (703)247-3434	Trade magazines
h. Executive Enterprises, Inc. 22 West 21st Street New York, NY 10010-6904 Telephone: (212)645-7880	Workshops
i. GMP Institute 2823 Pacific Avenue Cincinnati, OH 45212 Telephone: (513)631-3600	1. Workshops 2. Guidance manuals
j. Regulatory Affairs Professional Society 1010 Wisconsin Ave., N.W., Suite 630 Washington, DC 20007 Telephone: (202)857-1148	Seminars

3. Training and Management Programs

The following organizations offer training assistance such as specific topics for management development and training guides.

	Organization	Type of assistance
a.	The American Society for Training and Development P.O. Box 1443, 1613 Duke Street Alexandria, VA 22313 Telephone: (703)683-8124	Contact for scientific manager development programs
b.	Training Media Distributors Association 198 Thomas Johnson Drive, Suite 206 Frederick, MD 21701 Telephone: (301)662-2727	Catalog of training aids
c.	Training Magazine Lakewood Building 50 South Ninth Street Minneapolis, MN 55402 Telephone: (612)333-0471	Training research and development information
d.	Roundtable Film and Video 113 North San Vicente Boulevard Beverly Hills, CA 90211 Telephone: (800)332-4444	Audiovisual aids
e.	American Management Association 135 West 50 Street New York, NY 10020 Telephone: (518)891-5510	Management seminars
f.	Blanchard Training and Development, Inc. 125 State Place Escondido, CA 92025 Telephone: (800)821-5332	Catalog of training aids
g.	Dun and Bradstreet Business Education Services P.O. Box 3734, Church Street Station New York, NY 10008-3734 Telephone: (212)312-6880	Management seminars
h.	Addison-Wesley Training Systems Rt. 128 Reading, MA 01867 Telephone: (617)944-3700	Management seminars
i.	Association for Manufacturing Excellence, Inc. 380 West Palatine Road Wheeling, IL 60090 Telephone: (312)520-3282	1. Workshops 2. Conferences
j.	National Society for Performance and Instruction (NSPI) 1126 Sixteenth Street, N.W. Suite 214 Washington, DC 20036 Telephone: (202)861-0777	1. Journal 2. Conferences 3. Task forces

III. MAINTAINING THE PROGRAM

A. Overview

Once a training program has been established, the second and equally important phase is maintenance. In order for a training program to be successful it must be maintained properly. This will involve the oversight of the trainer or individual assigned to coordinate the training program.

B. New Employees

As new employees come into the company they will need to be trained. Depending on the background of the individual, the extent of training needs may vary. However, all new employees will need basic information. This will include general company policies, company history, structure, and safety program. This type of training is usually performed on the employee's first day on the job and can be viewed as a orientation session.

The first few days on the job are an important time for the new employee. It is a time when the employee beings to more fully understand the company, its goals, how he or she fits into the organization, and the importance of one's job relative to the rest of the company.

C. Cycling Employees

For those employees who have already been through the various training sessions required for their job, there must be a mechanism to keep them current with up-to-date information.

This is an ongoing process that should be closely monitored by the trainer. Periodic refresher courses should be scheduled at least once each year. Specific training topics can be planned well enough in advance to determine which individuals should be scheduled.

D. Documentation

Earlier in this chapter reference was made specifically to FDA's regulations requiring a training program. These regulations also require documentation of the training that was performed. Other regulatory agencies, such as the Occupational Safety and Health Administration (OSHA), also require training documentation. Regulatory agencies may take the position that if you cannot prove training occurred then it never happened.

Not only is documentation important for regulatory reasons, but it makes good business sense to maintain a record of what was taught, who learned it, and when. Documentation can be used as a tool for determining who requires new information on a specific subject in order for the employee to maintain quality performance.

Documentation can be found in many forms, that is, the overall company policy statement, the general company-wide training procedure, specific training programs, and records of specific training sessions. Some of these have already been discussed. The documentation that is completed and filed after each training session is important to demonstrate regulatory compliance.

Every training session should be documented with a dated attendance sheet and include a course outline and copies of any handouts, overhead transparencies, and quiz sheets [5]. If a written program on the subject is available it should be referenced.

A central file should be maintained for all records of employee training. The file should be updated as training sessions are held in order to have a current record available at all times.

E. Keeping Current

One of the more important aspects of a good training program is keeping the various subjects current with the most recent and up-to-date information. If the company has a full-time trainer on staff, this should be part of his or her responsibility. The number of training programs a company has will determine the amount of work necessary to maintain current information.

Maintaining an active membership in a trade organization is one means of keeping current with various subjects. For example, the Health Industry Manufacturers Association (HIMA) (see Section II.E for address) can provide current regulatory information. Also, through HIMA's science and technology section, a company can participate in any one of a number of committees and task forces dealing with technology issues relative to its products or processes.

Organizations such as HIMA sponsor seminars and workshops throughout the year to present recent developments on a wide variety of topics. By attending these meetings one also has an excellent opportunity to meet and exchange information with others in the industry. Through this networking, valuable information can be obtained that can be used to enhance a company's training program. Other organizations are listed in Section II for additional resources.

F. Scheduling

The timing when training topics are presented is another important part of the overall training program, for both new and long-term employees.

It is important for new employees to be trained immediately upon entry into the organization. As presented earlier in this chapter, training of new employees begins with an orientation program. Due to time constraints, the information presented is usually basic and general. Scheduling for job-specific training should be made by the training coordinator as soon as possible, preferably within several days.

All employees should receive periodic training as refresher courses and to update prior training. These training sessions can present a problem due to routine work schedules. However, if these sessions are properly organized this disruption can be minimized.

Another aspect of scheduling is the length of time for the session. Thirty- to sixty-minute time period frames are recommended in order to maintain interest and provide enough time to highlight the important points of the topic.

The training cycle (frequency of retraining) will vary due to the complexity and dynamics of a given job or routine, however, it is recommended that refresher courses be carried out annually as a minimum.

G. Auditing

Since training employees is a regulatory requirement, it is important that a company review its program periodically for compliance to the regulations. This review can be part of the internal GMP audit program and should be conducted at least once each year.

Observations made during the audit should be for conformance to the company's written training program. Examples of some of the questions that need to be answered during the audit include: Are the people who are responsible for training following their procedures? Are employees receiving the training required? Is the documentation up to date?

An important part of the audit is the follow-up. If there are deficiencies noted in the audit report, these should be addressed and corrected as soon as possible. It is better for you to find problems with the program than the regulatory agencies.

IV. EVALUATING THE PROGRAM

A successful training program must be part of the company's culture. An ongoing program requires a system for evaluating the results of the program.

Training should be documented from the point of employee entry and the orientation program through training in technology skills. Large companies may even provide training in basic reading, writing, and mathematics; employees are tested and must receive a passing grade before going on to the next unit of training.

The basis for an evaluation system should include a review of the written program and the standard operating procedures (SOPs). These SOPs should be well-written documents explaining how various tasks are performed. Since they are the written guides for employees to follow, they should not leave anything to the imagination. To a large extent the success of a training program depends on the support of a company's top management team. This support continues to be an important factor in the evaluation of the training program from the standpoint of product quality and profitability.

A correlation should be made between training and profits, "the bottom line." An effective training program can be expected to impact favorably on product quality, productivity, employee morale, and compliance with government regulations. These factors in turn can be expected to have a favorable impact on sales and profits. However, there are many other factors that impact sales and profits, so it is difficult, if not impossible, to determine a dollar return on the cost of a training program. On the other hand, there are measurements that should be used in evaluating training programs. For example:

Did the participants (trainees) enjoy the program and feel that it met a need?
Were the goals of the program met with respect to whether or not the trainees learned what was expected?
Do the trainees apply and realize the benefit of skills learned through the program?

These seem like simple enough questions, which could be answered inexpensively and could go a long way in evaluating a training program. Surprisingly, many companies do not even go this far in evaluating their training programs. According to one study, less than 12% of 285 companies studied evaluated the results of supervisory training programs in management [6]. Trainers and management should ensure that training programs are evaluated and that these programs are meeting the needs of the company. Effective evaluation programs must show that the training programs get results. Proving that training programs get results need not be expensive or complicated. A simple inexpensive program for a small company could involve management walking through the work area and surveying the trainees. Asking questions about how the trainees felt about the program, what they feel they got out of the program, and whether they are using what they learned can be very effective. This face-to-face exchange will not only provide an audit mechanism, it will also serve to improve communications in the company.

The next level of evaluation for a company with available time and resources would be the use of a questionnaire. A questionnaire evaluation may range from a simple form sent to trainees upon completion of a program to a more comprehensive program where pre- and posttraining questionnaires, which address all the areas the course was designed to affect, are sent to trainees [7].

Finally, a combination of face-to-face interviews and written questionnaires can be used. Whatever system is used, management and the training staff should be aware that the evaluation program is not merely an academic exercise. The evaluation program should be planned in advance to ensure that questions are formulated that determine whether the company's training goals are met and to provide information to improve future programs.

An example of a large company with a training evaluation program is IBM. IBM's training and development program was shown to be the most admired in America in *Training's* 1984 readers' poll. IBM uses four levels of evaluation:

1. *Reaction*: Asks trainees how valuable they found the course.
2. *Testing*: Pre- and posttests that measure the trainees' knowledge and skills before and after training.
3. *Application*: Shows whether or not the trainees are using the new skills.
4. *Business results*: What the organization gets out of the training.

It should be noted that Jack Bowser, the Director of Education for IBM, observes that it is possible to carry an evaluation to level 4 (dollar impact), but only in certain cases, and most of those cases involve technical training, such as training customer engineers to repair equipment [8].

For the reader who wishes to use a comprehensive evaluation program, an excellent source for further information and evaluation forms is the *Training and Development Handbook*, sponsored by the American Society for Training and Development [9].

In summary, an evaluation system should be established with the objective of determining whether or not the company's training goals are being met. This can be done on a simple, inexpensive "verbal survey" basis, or by using comprehensive written questionnaires. In either case the system should be documented and provide management with a level of confidence that the training program is being monitored and that the money spent on such a program is meeting the company's needs.

V. SUMMARY

The importance of training programs, both as a business strategy and as compliance with FDA and other agencies, has been discussed in this chapter. The information presented outlines suggestions for establishing, executing, documenting, maintaining, and finally evaluating the training program.

In addition to the discussion of in-house programs, an effort has been made to provide sources for information on the subject. A listing of organizations that can provide training and information on various subjects is included.

The first steps in establishing an effective training program are management commitment and recognition of the need for a well trained work force.

The information presented in this chapter should serve as a useful guide in achieving this goal.

REFERENCES

1. *Workplace Regulatory Report*, Volume II, No. 2, February 1988.
2. So You Want To Be A Trainer?, *Training, 2*:85 (1984).
3. R. A. Swanson, Industrial Training, in *Encyclopedia of Educational Research*, 5th ed., Harold E. Mitzel, ed., Collier MacMillan Publishing Co., London, 1982, p. 864.
4. American Society for Training and Development, P.O. Box 1443, 1613 Duke Street, Alexandria, VA.
5. C. M. Orelli, Good Manufacturing Practices Training, *BioPharm*, April, p. 38-41 (1988).
6. L. R. Smeltzer, Do You Really Evaluate or Just Talk About It?, *Training*, August: 6-8 (1979).
7. S. K. Quinn and S. Karp, Developing an Objective Evaluation Tool, *Training and Development Journal*, May (1986).
8. J. Gordon, Romancing the Bottom Line, *Training*, June: 31-42 (1987).
9. R. L. Craig, ed., *Training and Development Handbook, A Guide to Human Resource Development*, 3rd ed.

22

Complying with GMPs: Government Perspective

W. FRED HOOTEN *Center for Devices and Radiological Health, Food and Drug Administration, Rockville, Maryland*

I. INTRODUCTION

The medical device good manufacturing practice (GMP) regulation was authorized by the Medical Device Amendments of 1976 and was written to ensure that proper controls are used for the manufacture, packing, storing, installation, and distribution of medical devices intended for human use. Since December 1978 all manufacturers of medical devices have been required to comply with the applicable parts of the regulation unless specifically exempted. The complete text of the regulation is contained in the Code of Federal Regulations (CFR), 21 CFR 820.

The GMP is referred to as an "umbrella" GMP, because its requirements are intended to cover the manufacture of all medical devices. In order to achieve this coverage, the GMP provides only the general framework within which each manufacturer is expected to develop and implement a quality assurance (QA) program appropriate for its' product and manufacturing process. The requirements that make up the framework merely identify the controls that must be established without specifying methods for meeting the requirements. Because of this, some new manufacturers, especially small manufacturers, have had difficulty interpreting application of the GMP requirements and determining what is necessary to meet the requirements. Some manufacturers have not done enough while others develop controls to meet the GMP requirement when they are not always needed. While the GMP is a mandatory regulation with which all manufacturers must comply unless exempted, FDA has generally been flexible in applying the regulation when specific requirements provide little or no contribution to preventing the production of nonconforming devices. Therefore, manufacturers should use good judgment when attempting to comply with the GMP, and consult with FDA whenever questions arise.

This chapter is intended to assist manufacturers in complying with the GMP by providing an overview of the requirements and activities associated with the GMP. Additional guidance is provided in supplemental documents that have been developed since 1978. These are referenced throughout this chapter and can be obtained by calling the Division of Small Manufacturers' Assistance (DSMA), toll free, at 800-638-2041.

II. PREPRODUCTION CONSIDERATIONS

Before an effective GMP program can be established, certain preproduction considerations should be made. The preamble to the GMP states that it is vitally important that devices be manufactured in accordance with QA principles that help *prevent* (my emphasis) the production of defective products that can endanger consumers [1]. Section 820.1, Scope, of the GMP states that the regulation is intended to *assure* (my emphasis again) that devices will be safe and effective. It is clear that the GMP is intended to prevent the distribution of unsafe and ineffective devices. Anyone familiar with the scope of the GMP, however, knows that compliance with the GMP regulation, alone, will not assure a device is safe and effective. A device's safety, effectiveness, and reliability are characteristics established during the design stage. The GMP's contribution to assuring device safety and effectiveness is to assure that the device conforms to the accepted design specifications, that is, that the design is faithfully and consistently reproduced. Therefore, a good GMP program can assure a safe and effective device is produced only if the design represents a safe and effective device.

Of course, once a device is distributed, certain GMP requirements, if properly implemented, can be used to detect unsafe and ineffective devices. For example, manufacturers should quickly become aware of safety and effectiveness problems involving their devices if complaints are processed in a timely manner. However, this approach to assuring quality does not prevent the production of defective devices and can be costly to both the manufacturer and user. Costs to the manufacturer may include customer notification, product retrieval, redesign, rework, scrap, documentation changes, retraining, and, of course, product liability costs, if the device(s) causes death or injury. There may also be an increase in insurance costs and loss of customer confidence and market share. For the customer, the cost could be more severe if injury or death occurs due to the malfunction of a device.

The cost to a manufacturer to correct a design defect, or error, increases with each progressive stage of the life cycle of the device. Therefore, the earlier a design or process error or defect can be detected, the less costly it will be to correct. For example, it is much less costly to correct a design deficiency or error during the early stages of design than it is after the device is released and in production. The old TV advertisement for motor oil "You can pay now, or you can pay later" certainly applies here. It's going to cost a lot more later. An investment in assuring product quality has the most cost-saving impact when it is made during the preproduction phase.

Therefore, it is crucial that proper controls are in place during the design phase to assure that those device characteristics necessary for establishing the proper degree of safety, effectiveness, and reliability are defined and achieved, before the design is released to production.

A. Preproduction Quality Assurance Recommendations

In 1989, FDA finalized recommendations for establishing and implementing a preproduction quality assurance program that can be used by manufacturers to establish a program for the orderly review of the device design to assure it has the proper degree of safety, effectiveness and reliability needed prior to releasing the design to production [2]. Because these are only recommendations, manufacturers have no obligation to comply with the document. However, all manufacturers of medical devices should have a documented program for assuring the adequacy of the device design with respect to safety, effectiveness, and reliability before releasing the design to routine production. The threat of product liability alone warrants such a program. To be truly effective in assuring only safe and effective devices are produced, each medical device manufacturer's preproduction quality assurance program should be integrated with an effective process validation program.

B. Process Validation

While compliance with the GMP alone will not assure a device is safe and effective, failure to have acceptable manufacturing processes and procedures can prevent the achievement of these goals. Therefore, assuring the adequacy of the manufacturing processes used to reproduce the design is also important for assuring the acceptability of the device design. Adequate controls should be in place to assure that the manufacturing and quality assurance processes and methods, as designed or selected, will be successful in achieving conformance to device specifications, *before* the production process begins—that is, the process should be validated.

Process validation is considered a design transfer control and is interpreted by FDA to be a GMP requirement. It is most effective when conducted as a preproduction activity. However, FDA may insist that existing processes be validated when they are found to be ineffective or if effectiveness is questionable. The methods and procedures used to achieve process validation are left to the discretion of the manufacturer. For those seeking guidance, FDA has provided a validation guideline [3]. This guideline contains general principles that apply to the validation of manufacturing processes that are acceptable to FDA, but they are not mandatory for manufacturers. Manufacturers may conform to the guidelines with assurance that they are acceptable to FDA, or may choose other methods to prove the effectiveness of a process as long as it can be shown that the methods are scientifically sound and adequate for their intended purpose.

However, unlike the previously mentioned recommendations, which are advisory, guidelines identify areas in which manufacturers are expected to have controls [4]. Therefore, manufacturers are advised that during routine GMP inspections, FDA investigators may determine if manufacturers are complying with the FDA validation guideline. Those manufacturers who have chosen alternate methods to validate processes, should be prepared to provide documented evidence of the validity of the alternate method. It is not always necessary to apply the guideline to all manufacturing processes. The degree of evidence or assurance necessary to show that a process is adequate will vary with the significance and complexity of the manufacturing process. If a manufacturer can show through other means, such as inspection and/or test, that a process is effective, validation by methods suggested in the guideline may not be necessary. For example, a punching or cutting operation can in many cases be validated by first piece and last piece inspection. A typical EtO sterilization process, on the other hand, usually requires validation using the principles described in the guideline. It is left to each manufacturer to judge the need for applying the guideline, and FDA will evaluate these decisions during GMP inspections. Manufacturers should be prepared to provide the rationalization used for not applying the guideline when it is not used.

III. THE GMP

A. General

While FDA has interpreted the GMP to require process validation, and some degree of process validation is considered a necessary element in a manufacturer's GMP program, the preceding discussion of preproduction considerations (with the exception of validation and design transfer controls) are currently considered non-GMP obligations and considerations that should be made before a device is released to production for routine manufacturing. The remainder of this chapter will address a medical device manufacturer's GMP obligations, in addition to process validation.

1. Scope

The GMP applies to "finished devices" intended for human use as defined by 21 CFR 820.3(j). Finished devices are devices that appear suitable for use, whether or not sterilized, packaged, or labeled, and include accessories. The GMP is not intended to apply to component manufacturers [5]. Instead, the burden is placed on the finished device manufacturer who uses the components to assure components are acceptable for their intended application. FDA does conduct GMP inspections of firms that, to some, may appear to be component manufacturers. For example, FDA conducts GMP inspections of manufacturers of bulk sutures, breathing components, etc., when they are labeled and sold for health care purposes. Also, manufacturers of components used in medical devices should keep in mind that, by definition, components of medical devices are "devices" and component manufacturers may be visited by FDA if problems arise [6]. Additional guidance on who is subject to the GMP can be found in FDA publications written for FDA investigators and industry [7].

2. Critical Devices

The GMP regulation is referred to as a "two-tier" regulation. This means there are general requirements that apply to all medical devices and then there are additional requirements that apply only to "critical devices." Critical devices are defined in 21 CFR 820.3(f). The "two-tier" scheme was devised as a means to prevent overregulation of medical devices. The need for additional GMP requirements for critical devices is based on the additional risk to the user that a critical device presents should it fail. Critical devices are identified by FDA after consulting with the applicable device classification panel and the device GMP Advisory Committee, which will be described later in this chapter. A list of the devices FDA is currently regulating as critical devices is available from the Division of Small Manufacturers Assistance (DSMA) [8]. The list is updated periodically as additional critical devices are identified.

B. Organization

The GMP does not dictate any particular organizational structure for medical device manufacturers, and manufacturers have the discretion to organize their operations as they see fit. The only stipulation is that they formalize the organization and organize in a manner that will assure applicable GMP requirements are met. This means that the activities, responsibilities, and authorities necessary to establish and implement the QA program and verify compliance with QA program goals and objectives must be defined, documented, and approved by management.

1. *Quality Assurance Program*

As defined in the GMP [21 CFR 820.3(n)], "quality assurance" may appear to be limtied in scope to only the production process, i.e., "all activities necessary to verify confidence in the quality of the process used to manufacture a finished device." However, quality assurance also includes other activities, such as the compilation and evaluation of product experience data that may be contained in complaints, service requests, etc. This information provides feedback to the QA program which should be used to verify the adequacy of the device design and the manufacturing process. The extent of the QA program will vary from manufacturer to manufacturer, depending on the nature of the product and manufacturing processes. For example, some manufacturers will require more controls simply because their operations are larger and more complex.

FDA inspectional findings indicate that some small manufacturers believe that inspection and test are sufficient to meet the GMP quality assurance requirements. Inspection and test, traditionally considered quality control (QC) functions, are "sorting" activities used to separate the nonconforming from the conforming and do not *prevent* the production of nonconforming devices. As stated previously, the GMP is intended to be preventive, and while QC is an essential element of a good QA program, it is seldom the entire program.

Regardless of the QA program established, the program must have the support of upper management if it is to be viable and successful. Once an organization is in place and employees are properly trained and provided the proper tools, its primarily management's attitude that will determine whether or not a quality product will be produced. FDA has found more than one manufacturer who set up a comprehensive QA program, only to see it decay with time due to lack of management support. Upper management can actively participate in the quality assurance program by establishing quality policies and goals. Oversight of the quality program can be realized through requiring and reviewing quality audits. An essential part of management's role should be periodic feedback to employees to demonstrate an interest in the success of the QA program. Employees then tend to be more attentive to assuring that QA goals and objectives are met.

2. *Quality Audits*

There is some misconception on the part of some manufacturers as to what constitutes a GMP quality audit. The quality audit required by the GMP is not a product audit as some believe, although it may include a product audit. The audit required by the GMP is a comprehensive review and evaluation of all phases of the quality assurance and production program to assure that adequate operating procedures and policies are in place, and that they are being followed. Where possible, quality audits should also include suppliers and contractors. As mentioned earlier, the quality audit should also be used as a management tool, to provide upper management with information on the state of control of the manufacturing process, the quality assurance program, and compliance with regulations.

Inspectional records show that medical device manufacturers with less than 50 employees are the worst offenders of the quality audit requirements. The most common reason provided by the managers of these operations is that they did not believe an audit was needed because they were directly involved in day-to-day operations. This argument may be valid in very simple manufacturing operations, but generally cannot be supported. Realistically, a person cannot make an objective overall evaluation of operations while trying to meet daily production demands. Even the operator of a one-person firm should step back periodically and assess overall operations for compliance with established procedures and the continued adequacy of these procedures. FDA's experience with medical device manufacturers indicates that those firms found not complying with their own established procedures are usually those that do not have adequate quality audit programs. To FDA, the absence of an active quality audit is an indicator of an inadequate quality assurance program and a significant GMP deviation. Without an audit, there is no assurance that the QA program is being properly followed or is adequate, and the QA program becomes an open-loop system.

C. Buildings and Equipment

The GMP "buildings and equipment" requirements are basic to the manufacture of any product, encompassing the overall establishment of facilities and operations, a cleaning and sanitation program, environmental controls, and maintenance and calibration of equipment. Because these are basic requirements, noncompliance is easily detected by the FDA investigator.

The GMP stresses the need to design and arrange facilities and manufacturing operations so that the physical layout of the operations and the condition of the facilities will not adversely effect the device, such as contamination through unscreened windows, general traffic flow through restricted areas such as clean areas, or inadequate storage. One GMP concern is the prevention of mixups of components, materials, finished de-vices, and labeling. A common cause of mixups is inadequate space to prop-erly segregate acceptable from rejected components and work in process from finished product. FDA has monitored a number of recalls that were determined by the manufacturers to be due to mix-up of components, such as rejected components mixed with good components, use of wrong materials, etc. [9]. These recalls were directly due to the failure to adequately segregate and identify acceptable and rejected components.

Contamination due to inadequate sanitation and environmental controls can have a major impact on the performance of certain devices such as sterile devices, some *in vitro* diagnostics, raw materials, etc. Environmental controls must be considered when contamination generated by facilities, operations, or other sources, could adversely affect the device. Whenever environmental controls are determined to be necessary, specifications for the environment and controls must be developed, documented, and included in the device master record. Failure to include environmental specifications (where applicable) in the device master record has been a frequent GMP deviation.

Failure to properly maintain production equipment has resulted in recalls of defective devices due to inadequate package seals, molds for injection molding, etc. [9]. Therefore manufacturers should make sure that adequate equipment maintenance controls are in place and that the applicable GMP requirements are met.

Production and QA equipment must be maintained and calibrated to the extent necessary for it to perform as intended. The GMP does not specify the kind of maintenance or calibration necessary. This should be determined by the manufacturer based on the equipment manufacturer's recommendations and experience with the equipment or similar equipment. The GMP stresses the need for formalizing the maintenance and calibration program and the use of recognized calibration standards. Guidance on establishing calibration and maintenance programs can be found in the DSMA GMP Workshop Manuals [10].

D. Components Controls

A major GMP-related cause of device failures and subsequent recalls is the use of components that do not have the specified physical or performance characteristics (e.g., strength, capacity, stability, etc.) [9]. The GMP component acceptance requirements are intended to assure that components conform to design specifications, and if these requirements are met, recalls due to nonconforming components should be minimized. GMP component controls include assuring the acceptance status of components, establishing controls to prevent mixups and use of unacceptable components, and the control of component storage to prevent degradation and use of the wrong components.

Manufacturers can assure that components meet specifications by establishing an inspeciton/test program in house, or having acceptance procedures performed by a contractor or the component manufacturer or supplier. Some device manufacturers find it more cost-effective to have the component manufacturer conduct acceptance testing and supply the components with test data and certificates of conformance. This approach is acceptable if the finished device manufacturer can assure the supplier has the capability to consistently provide acceptable components and is providing valid test data and certificates.

Manufacturers typically spend lots of money and time during research and development assuring the adequacy of components, and the safety, effectiveness, and reliability of the finished device. Yet some put little effort in assuring the suppliers they select will provide quality components, although a device will be no more reliable than its least reliable component. The selection of suppliers who will consistently provide acceptable components is crucial to the success of a manufacturer's program to produce only quality devices.

Variations in the quality of supplied parts and materials can be minimized by providing specifications with sufficient detail to identify the needed physical and performance characteristics. Many times, suppliers provide inadequate components because the finished device manufacturer did not provide component specifications in sufficient detail to describe the physical and performance requirements needed.

Noncritical device manufacturers have the discretion to decide the level of control needed for their component acceptance program. However, critical device manufacturers must identify those critical device components that qualify as "critical components" and apply specific controls, defined by the GMP, to assure these components comply with specifications. Additional component controls are intended to assure traceability in case nonconforming critical components are inadvertently used that result in the distribtion of ineffective or unsafe devices. Critical components are defined in 21 CFR 820.3(e), and guidance for identifying critical components can be found in the DSMA GMP Workshop Manual [10].

E. Production and Process Controls

This section of the GMP is intended to ensure that manufacturers have sufficient and suitable production procedures and specifications. The most common observation made by FDA during GMP inspections is the failure to have adequate processing, inspection, and test procedures [9]. Each manufacturer should carefully plan its documentation program to assure that sufficient written manufacturing procedures and specifications are available to adequately define what is to be produced, how it is to be produced (including methods and equipment), how successful completion of processes and operations will be verified, and how to verify that finished devices meet specifications.

Throughout the GMP, there are specific requirements for written procedures. These procedures are intended to provide direction and ensure uniformity by describing the controls and operations necessary to produce devices. By defining the operations, the written procedures also provide a means by which changes to the operations can be managed.

FDA has been generally flexible in enforcing the GMP written procedure requirements, and manufacturers should not develop written procedures that will not contribute to ensuring the finished device meets acceptable specifications, e.g., that an activity is carried out properly. Written procedures are usually necessary for production line operations when unskilled workers are employed, for complex and critical operations, etc. Conversely, written procedures may not be needed in skilled activities such as machining, hand soldering, etc. In these situations other methods, such as training, workmanship standards, set-up procedures, and drawings, may be used in lieu of written operational procedures. Each manufacturer should carefully plan the documentation needed, to assure that no more is generated than can be managed, that is, kept current. However, when manufacturers have not established written procedures where required by the GMP, they should be prepared to provide logical reasons for not doing so.

1. Design Transfer

Part 21 CFR 820.100(a)(1) requires all manufacturers to establish controls that will ensure that the specifications prepared for producing the device, its components, software, etc., properly reflect the device design. This is the design transfer phase in the life cycle of a device, and the place where application of the GMP begins. Controls must be in place to ensure the design is properly translated into manufacturing procedures and specifications. The controls that are applicable include documentation review and process validation.

2. Change Control

Once released for use in production, all documentation must be placed under a formal documentation control program that will ensure changes are properly documented and evaluated before implementation. Once the device specifications are released for production, a manufacturer can no longer take the position that the GMP does not apply to design. Any changes to the device, its software, components, packaging, or labeling are subject to the GMP change control requirements which include qualification/validation of the change before implementation.

3. Process Control

One of the most common causes of recalls is failure to have adequate process controls [9]. The GMP requires manufacturers to establish process controls as necessary to ensure processes are properly carried out. In many cases recalls have occurred because process controls were not properly considered during process design and validated. Other process control failures were due to employee errors, indicating a need for better employee training programs, or better written operating procedures.

4. Critical Operations

Manufacturers of critical devices are required to identify those processes that qualify as "critical operations." They must assure that these operations are performed by qualified personnel and that the satisfactory completion of the operation is documented. The definition of a "critical operation" is found in 21 CFR 820.3(g). Even though by definition it may appear that all processes meet the critical operation definition, manufacturers can use judgment in identifying these operations. Guidance in this area can be found in the DSMA GMP Workshop Manual [10].

5. Reprocessing

During production, manufacturers may find it necessary to repeat processes or operations because of errors, defective components, etc. When it is necessary to repeat steps in the manufacturing process, manufacturers must assure that the reprocessing is done under GMP controls and the reprocessed device meets the original or approved, modified specifications. If the reprocessed device is a critical device or component, the GMP details the controls that must be in place. These include written reprocessing procedures, identification and segregation of reprocessed devices, and written testing procedures solely for reprocessing. If the reprocessing involves the same operation(s) as the original operations, the procedures used for the original operations can sometimes be used. If there are only slight differences, these can be written into the regular production procedures and identified as applying only to reprocessing. Where processes are routeinly repeated as a part of production, the reprocessing may actually be part of production and can be written into the routine production procedures. For example, orthopedic implant manufacturers may recycle product to the polishing operation at various stages in the process to improve cosmetic appearance. Polishing in this example possibly could be considered part of routine manufacturing.

The reprocessing requirements were not intended to apply to devices that are returned from distribution and reworked, modified, etc., and resold. Instead these activities may be considered manufacturing depending on the circumstances involved. If the devices are reprocessed and returned to the user, these activities may be considered repair or service.

F. Packaging and Labeling Control

1. *Labeling*

The GMP labeling control requirements are basic, but failure to have such controls has resulted in a number of recalls. A mislabeled device is usually a misbranded device subject to seizure by FDA. A significant number of recalls monitored by FDA are due to labeling mixups or incorrect labeling content [9]. Labeling includes not only the labels attached to the device but operations manuals, maintenance manuals, and any other written promotion material that accompanies the device. Labeling may also include software that is used to provide operation instructions.

The GMP labeling requirements address both controls to prevent labeling mixups and controls to assure labeling content meets specifications before use. There is also a requirement that the labeling attached to the device must stay attached and legible for the useful life of the device. Section 21 CFR 820.120(a) of the GMP requires that labeling be proofread before release to inventory. This requirement was intended to apply to purchased labeling which is received and then placed into inventory. Today, many manufacturers generate labeling by computer as it is needed. In such cases, the proofreading requirement prior to inventory may not be practical. Instead the controls that should be considered by the manufacturer may include validation or checking of the software and the examination of the labeling for content accuracy before use.

One of the more common causes of labeling mixups involves labeling that is similar except for subtle differences. These types of problems can be avoided through such methods as color coding or segregation of labeling.

2. *Packaging*

The packaging section of the GMP is one of the few GMP areas that addresses design. These requirements are intended to insure that packaging is adequately designed to protect the device from adverse effects during shipping and storage. Packing design is more commonly a problem when sterile devices are involved. For example, the packaging for sterile devices must be designed to maintain sterility after the packaged device is sterilized. For other devices, where it may be impractical to design packaging that can protect the device, labeling warnings may be sufficient. For example, if a device can be affected by extremes of temperature, it may be labeled with warnings against exposure to temperature extremes or list the acceptable storage temperature ranges for the device.

G. Holding, Distribution, and Installation

1. Distribution Controls

The GMP does not require noncritical device manufacturers to maintain distribution records, although to properly conduct business all manufacturers probably maintain distribution records. Distribution records are mandated only for critical devices. This sometimes presents a problem for FDA because, while all manufacturers maintain distribution records, FDA has access to only those distribution records required by the GMP. Occasionally this has severely hampered FDA efforts to follow-up on recalls of noncritical devices, and FDA usually depends on the cooperation of manufacturers in providing noncritical device distribution records.

All manufacturers are required by the GMP to have written procedures for warehouse and distribution control that will ensure that only approved devices are released for distribution. However, although this requirement is mandatory, there are situations where a written procedure may not be needed to meet the intent of the GMP, and in these cases FDA usually does not enforce this requirement. For example, firms that ship only small numbers of devices each year may not need written procedures describing how devices are to be held in the warehouse and released.

2. Installation

The GMP installation requirements apply when a device or device system, such as diagnostic X-ray, computerized axial tomography, etc, is installed by a manufacturer or their representatives. Basically, the requirements are intended to ensure that the device meets specifications after installation. Manufacturers of these systems must either assure that the system is properly installed through inspection or provide customers with adequate written instructions for installation.

H. Device Evaluation

1. Finished Device Inspection

All finished device(s) must be examined to some degree to ensure they meet specifications before they are released to the warehouse or distribution. Testing/inspection may be on a sampling basis or 100%. When sampling is used, the sampling plan should be statistically valid, with an acceptable risk based on the significance of the device. One approach to meeting this requirement is to use existing recognized sampling plans such as MIL-STD-105 with an AQL appropriate to the significance of the device. When manufacturers develop their own plans evidence must be available to show they are valid.

Test and inspection methods are left to the discretion of the manufacturer, but as a minimum should verify the acceptance of those characteristics important for acceptable device performance, (e.g., growth characteristics of an IVD medium, electrical current characteristics of an electrical nerve stimulator, etc.). Manufacturers should ensure that adequate written inspection and test procedures are established. Failure to have adequate inspection/test procedures is one of the more common GMP deviations [9].

One of the more important requirements in this section of the GMP is the "testing under simulated use conditions" requirement. This requirement states that "Where practical, a device shall be selected from a production run, lot or batch and tested under simulated use conditions." If those manufacturers who have experienced recall due to defective devices had fully complied with this requirement, a large portion of the recalls could have been avoided [9].

2. Failure Investigation

Whenever a manufacturer learns that a distributed device has failed to meet performance specifications, the manufacturer has an obligation to investigate the failure, determine why the device failed, and take appropriate corrective action. Of course the investigation and findings, including corrections made, must be documented to provide objective evidence that an investigation was satisfactorily conducted. Critical device manufacturers must also follow this procedure for devices that fail in house (21 CFR 820.161). Guidance on conducting failure investigations is provided in references 12 and 13.

I. Records

1. Authority to Review and Copy Records

FDA has the authority to review and copy all records required by the GMP. This authority is provided by section 519 of the Food, Drug and Cosmetic Act.

2. Confidentiality

While FDA has the authority to copy records, manufacturers have a say in whether the records can be released outside FDA. The GMP recommends that manufacturers mark "confidential" all records that a manufacturer believes are confidential to assist FDA in deciding what records, collected by FDA or submitted by the manufacturer, are releasable under the Freedom of Information Act (FOI). A word of caution: do not stamp every document "confidential." If every document is stamped confidential, it may appear that a manufacturer is not sincere or objective in identifying confidential documents, and FDA may not honor all of the manufacturer's selection.

3. Record Retention

All records required by the GMP, or maintained by a manufacturer to comply with the GMP, must be retained as long as the devices are expected to be in use, but in no case for less than 2 years from the time the devices are released for use. Usually, FDA allows manufacturers to use their own judgment in determining how long to maintain records after the 2 year minimum.

4. Device Master Record

The Device Master Record (DMR) required by the GMP consists of one or more files containing, or referencing the location of, all the documents that a manufacturer uses to produce finished devices. The required types of documentation is specified in the GMP, parts 820.181 and 820.182.

FDA investigators are instructed to thoroughly review the DMR during comprehensive GMP inspections. During this process, the investigator will check to assure that documentation required by the GMP is complete and current. A review of inspection results indicates that failure to have adequate documentation or a complete absence of certain required documentation (e.g., processing procedures, inspection/test procedures, and specifications for the device and its components) was a common observation made by FDA investigators [9]. Once the investigator has reviewed the DMR, he or she will typically go to the production area to verify that the DMR documentation properly reflects what is actually being done during production. Some investigators may do this in reverse—that is, they may first observe production before reviewing the DMR.

Although the absence of required procedures appears to be a major GMP deviation, manufacturers should not react by overdocumenting simply to satisfy a GMP requirement. Manufacturers should have sufficient documentation to assure that device(s) can be produced to meet design specifications and that the applicable controls required by the GMP are established. However, the documentation must be carefully planned so that it can be properly managed and maintained.

5. *Device History Record*

The Device History Record (DHR) is intended to provide objective evidence that all things required by the Device Master Record were completed. The DHR is also sometimes crucial to conducting failure investigations. A significant number of manufacturers were found to have inadequate DHRs, that is, the DHR was incomplete or absent [9]. Guidance on how to minimize the effort in developing an adequate DHR can be found in the DSMA GMP Workshop Manual [10].

6. *Complaint Files*

The GMP requires all manufacturers of medical devices to maintain a complaint file and to review, evaluate, and investigate complaints. No medical device manufacturer is exempt from the GMP complaint file requirements, not even those exempted by the classification regulations or by petition. The definition of a complaint is basically that described in the GMP requirement. Guidance has also been written by FDA in this area [13]. When FDA conducts a GMP inspection, one of the first areas that the investigator will review is the manufacturer's complaint-handling procedures. The investigator will typically review complaints and Medical Device Reporting (MDR) submissions contained in the district files pertaining to a particular manufacturer prior to the inspection. During review of a manufacturer's complaint files, the investigator will not only evaluate compliance with the GMP complaint handling requirements, but will also be looking for information that may indicate design or manufacturing problems. The investigator will also be looking for complaints that meet the MDR reporting criteria that have not been reported by the manufacturer. Therefore, to minimize the chance of the investigator making adverse observations relative to the complaint handling and MDR program, manufacturers must consider both their GMP and MDR obligations when setting up complaint handling procedures.

The MDR regulation requires manufacturers to report to FDA when they become aware of information that their devices may have been involved in a death or serious injury, or may have malfunctioned and, if the malfunction recurs, is likely to cause or contribute to a death or serious injury [11]. Just about all MDR reportable events can be considered complaints, but not all complaints meet the MDR reporting criteria. To assure that all reports are properly considered as possible complaint candidates and MDR reportable events, the two programs should be managed closely together. Additional information can be obtained from FDA's MDR Question and Answer Booklet [12].

While the GMP does not require manufacturers to establish written procedures for processing complaints, manufacturers can usually substantially improve their chances of meeting GMP requirements if a written procedure is prepared specifying the criteria to be used in identifying complaints, the responsibilities for the program, and the procedures to be followed in receiving and processing complaints. While the manufacturer has no obligation to provide this procedure to the FDA, a written procedure can be used as evidence that a proper complaint handling program is in place and that the manufacturer is attempting to comply with the regulation.

One of the more common deficiencies that FDA finds in reviewing complaint handling programs is the manufacturer's failure to follow up properly on complaints. As mentioned previously, the FDA investigator will be looking for indications of design, manufacturing, and other problems when reviewing complaints. If problems are found, the investigator will determine what kind of follow-up action the manufacturer has taken to rectify the problem.

One of the points of contention between FDA and some members of the medical device industry is whether the GMP requires trend analysis. Trend analysis is a necessary element of each manufacturer's QA program, crucial to assuring the safety and effectiveness of medical devices. Actually, most manufacturers *do* conduct trend analyses as it is relatively easy to do with the use of computers. The point of contention appears to be whether or not FDA should have access to the analysis. Many FDA investigators are now trained and equipped with portable computers, have authority to review all complaints, and can readily prepare a trend analysis on site. It would appear to be to the advantage of each manufacturer to prepare his or her own trend analyses and make it available to the FDA.

FDA often finds that complaints are maintained in legal files, failure investigation files, or other interim files. All complaints must be maintatined in the GMP complaint file. The GMP does not make provisions for maintaining complaints in files other than the GMP complaint file, which must be open to review by FDA. However, this does not preclude manufacturers from maintaining *copies* of complaints in other files.

IV. GMP ADVISORY COMMITTEE

During our GMP workshops, we have found that few manufacturers are aware of the GMP Advisory Committee (GMPAC). The GMPAC is required by the Medical Device Amendments of 1976 and was established to advise FDA on, and make recommendations concerning, proposed regulations regarding methods, facilities, and controls used in the manufacture, packing, storage, and installation of medical devices. The Committee may, upon request, also make recommendations with respect to approval or disapproval of petitions for exemption or variance from the GMP regulation. The Committee consists of nine members: three representatives of federal (not FDA), state, or local government, two health care representatives, two industry representatives, and two consumer representatives. Meetings of the Committee are held when appropriate topics or issues arise.

Announcements of meetings are published in the *Federal Register* and the CDRH's Medical Device Bulletin. The meetings are open to the public and anyone may make presentations at the meetings, as long as the FDA is provided with advance notice of the proposed presentations so that time can be planned. Typical items addressed by the Committee have included additions to the critical device list, process validation, and *in vitro* diagnostic guidelines.

V. PETITIONS FOR EXEMPTION OR VARIANCE FROM THE GMP

The Medical Device Amendments of 1976 contain provisions whereby manufacturers, organizations, or other interested persons may seek exemption or variance from those parts or all of the GMP that, in their judgment, are not appropriate to their manufacturing operations [14]. Guidance for the preparation and submission of GMP petitions can be found in reference 15.

Manufacturers should realize that FDA will not process a petition for exemption or variance while an inspection or investigation of a manufacturer's quality assurance program is ongoing [2].

Exemptions or variances may be obtained on either an individual or product class basis. In some cases FDA may decide to initiate action in the absence of a peition, if it is in the public interest to do so.

In all cases, manufacturers should consult with the FDA before filing an exemption or variance petition. Because of the flexibility of the application of the GMP, we have found that petitions are seldom necessary. It is not the Agency's intention to insist upon compliance where such compliance will have no demonstrable benefit on the safety, effectiveness, or quality of the device(s). Therefore, in many cases FDA can make interpretations of application of the GMP to a certain industry that will serve the purpose of a petition, and relieve manufacturers from petitioning.

VI. GMP INSPECTIONS

The Food, Drug, and Cosmetic (FD&C) Act requires FDA to inspect manufacturers of Class II and III devices at least once every 2 years. In 1979 the Agency decided that this mandatory requirement could only be satisfied by conducting GMP inspections. The number of manufacturers of Class II and III devices has increased at an annual rate of approximately 15% since 1979. Since 1982, FDA's field force has been reduced by approximately 15%. To conserve resources and better focus resources on problem areas, the FDA adopted in FY'86 a two-track inspectional strategy. As the name implies, the strategy allows the FDA investigator to follow one of two tracks depending upon the quality history of the firm. A Track I inspection is a limited inspection consisting of a review of documents that would typically contain evidence of device problems (i.e., complaints, repair/service records, failure investigation records, etc.). A Track II inspection is a comprehensive inspection that is intended to determine compliance with all or selected portions of the GMP. A more detailed description of the inspectional process can be found in the GMP Compliance Program, CP 7382.83. Information on inspections of sterile device manufacturers and contract sterilizers is contained in CP 7382.830A and 7382.830B, respectively.

VII. CONCLUSION

The GMP provides a framework within which each manufacturer of medical devices is expected to develop a quality assurance program that will assure that finished devices conform to specifications. The actual methods and procedures used to achieve this goal are left to the discretion of each manufacturer. However, the procedures and methods selected must be defined and translated into written specifications, SOPs, etc. The extent and comprehensiveness of each manufacturer's program should be determined by a number of factors, including the significance of the device, the complexity of the manufacturing processes, and the controls necessary to prevent adverse effects caused by the manufacturing processes and environment. In order to minimize the cost and maximize the benefits of the GMP program, manufacturers must understand their operations and the intent of each GMP requirement in sufficient detail to assure the appropriate controls are selected. Manufacturers may find that they need more or less controls than are required by the GMP regulation to achieve a state-of-control. FDA is flexible in applying the GMP regulation, and manufacturers can deviate from the specific GMP requirements as long as they can provide sound logical rationale for doing so. FDA will evaluate the adequacy of each manufacturer's GMP program with respect to meeting the intent of the GMP requirements during GMP inspections.

While the GMP outlines a basic quality assurance program that can be used to prevent the production and distribution of devices that fail to conform to design specifications, application of the GMP alone will not assure the design specifications reflect a safe and effective device. The safety and effectiveness of a device are established during the design phase. In order to assure that appropriate safety and effectiveness goals are met, the GMP must be integrated with an acceptable preproduction quality assurance program to produce a total quality systems program that extends over the complete life cycle of the device. A total quality program has many advantages for the medical device manufacturer. Such a program can assure that devices consistently meet customer expectations and are produced at optimum cost by reducing the cost of repair, rework, scrap, reinspection, retest, market retrievals, etc. As a result, productivity is improved, products become more attractive to customers, more products can be sold, and market share and profits can be increased. Implementation of an effective quality systems program that includes design controls can also reduce a manufacturer's product liability exposure. Manufacturers who hope to sell medical devices within the European Economic Community (EEC) if, and when harmonization is realized, will also be substantially in compliance with the EEC GMP, which is proposed to include design control requirements.

REFERENCES

1. Food and Drug Administration, Center for Devices and Radiological Health, "Regulations Establishing Good Manufacturing Practices for the Manufacture, Packing, Storage and Installation of Medical Devices," Preamble 43 FR 31508, July 21, 1978.
2. Food and Drug Administration, Center for Devices and Radiological Health, "Preproduction Quality Assurance Planning; Recommendation

for Medical Device Manufacturers," September 1989.
3. Food and Drug Administration, Center for Drugs and Biologics and Center for Devices and Radiological Health, "Guideline on General Principles of Process Validation," May 1987.
4. 21 CFR 10.90, Food and Drug Administration regulations, guidelines, recommendations, and agreements.
5. 21 CFR 820, Good Manufacturing Practice for Medical Devices, Subpart A—General Provisions, 820.1 Scope.
6. Food, Drug, and Cosmetic Act, Chapter II—Definitions, Section 301(h).
7. Food and Drug Administration, Center for Devices and Radiological Health, "Medical Device GMP Guidance for FDA Investigators," April 1984.
8. Food and Drug Administration, "Advisory List of Critical Devices—1988," 53 FR 8854, March 17, 1988.
9. Food and Drug Administration, Center for Devices and Radiological Health, "Device Recalls and Inspection Results: A Study of Quality Problems," 1989.
10. Food and Drug Administration, Center for Devices and Radiological Health, "Device Good Manufacturing Practices Manual," 4th ed., November 1987.
11. 21 CFR 803, Medical Device Reporting.
12. Food and Drug Administration, Center for Devices and Radiological Health, Compliance Guidance Series, "Medical Device Reporting Questions and Answers," February 1988.
13. W. Fred Hooten and Richard I. Bimonte, "GMP Complaint Files: How They Relate to Reports Required under MDR," *Medical Device and Diagnostic Industry*, 7(5):57-63 (1985).
14. Food, Drug, and Cosmetic Act, Section 520(f)(2)(A).
15. Food and Drug Administration, Center for Devices and Radiological Health, "Petition Guidance on Exemption or Variance from the Device GMP Regulation," September 1979.

23

Complying with GMPs: Industry Perspective

WILLIAM H. DUFFELL *Syntex (U.S.A.) Inc., Palo Alto, California*

I. INTRODUCTION

The purpose of this chapter is to offer practical advice for complying with GMPs in order to meet regulatory obligations. Each section is written independently and can be read and used as a reference for the specific subject covered within the section. However, all of the sections are interrelated and some repetition is desirable.

Each section begins with a brief summary statement of the principal point(s) within the section.

II. GMPS ARE GOOD MANAGEMENT PRACTICES

Medical device companies can comply with the FDA's GMP regulations by implementing good management practices. FDA investigators think of GMPs (good *manufacturing* practices) as requirements codified in Part 820 of Title 21 of the Code of Federal Regulations (21 CFR 820). Industry executives should think of GMPs as good *management* practices. Fortunately, the implementation of good management practices will generally result in compliance with the FDA's good manufacturing practice regulations.

III. GMPS ARE AN UMBRELLA REGUATION

There are numerous references to terms such as "periodic," "adequate," "appropriate," "responsible," and "necessary" in the FDA's GMP regulations. Each FDA investigator and each quality assurance professional often applies his or her own definition to these terms. Since little guidance exists for defining many GMP terms, medical device manufacturers have a challenge to determine what is required of them by the regulations. Sometimes it has been necessary for the courts to interpret the regulatory definition of some of these terms. Nevertheless, it is preferable in the long run to have a "general" regulation rather than a "cookbook" regulation.

Management must develop *specific* procedures in accordance with the intent of the regulations. Although the medical device GMPs are intended as a general umbrella regulation, there are specifics within the regulation that require industry management to establish and implement manufacturing procedures that do not vary too greatly from the written regulation. The degree of industry compliance with the FDA's GMP requirements is often related to the size of the device manufacturer. A critical point to keep in mind when establishing good management practices for production of medical devices is the fact that the FDA will often determine a manufacturer's conformance to its own documented procedures. When the FDA observes a deviation between what the manufacturer has documented in a written procedure and what it actually does in practice, there is much grief. It is my opinion that when a manufacturer is allowed to develop its own manufacturing standards, it will be much harder on itself than the FDA would have been under similar circumstances. It is for this reason that many of the vague terms such as "adequate" and "appropriate" offer challenges in complying with the FDA's regulations. Each device manufacturer is responsible for determining whether its procedure is "adequate" and "appropriate."

IV. INDUSTRY'S CONCERNS WITH GMPS

Management is criminally liable for failure to comply with the FDA's GMP regulations. There are two significant court cases of importance to management for consideration when discussing the development of good management practices for the manufacture of medical devices. These are the Dotterweich [1] and the Park [2] cases. These two court cases helped establish the premise that management, as the most responsible individuals of the firm, can be subject to criminal prosecution for violations of the FDC Act, even though the individuals themselves had no direct knowledge or awareness of wrongdoing by their employees. The basic principle that the judge emphasized in the Park case is restated, "the FDC Act imposes not only a positive duty to seek out and remedy violations when they occur, but also, and primarily, a duty to implement measures that will insure that violations will not occur." Management's best defense to criminal charges is to require internal audits to help assure regulatory compliance with the FDA-mandated GMPs. Management must review audit reports and assure itself that noted deficiencies are corrected. The reliance of management solely on the FDA's inspections to assure its compliance status with good manufacturing practices is not prudent. Some device managers have a tendency to boast that their firms have never been cited for any GMP violations by the FDA as evidence of their companys' compliance. Reliance exclusively on such factors is based on a false sense of security.

V. THE INTENT OF GMPS

Management's challenge is to implement practices that comply with the *intent* of the FDA's regulations. Since FDA investigators are not experts in the manufacture of all types of medical devices, it is not reasonable to expect that an individual FDA investigator should know the best procedures that can assure compliance of medical devices with the *intent* of the regulations. Each device manufacturer is its own best expert for the development of manufacturing practices that will result in the production of competitively priced, safe, and effective medical devices. There are generally several alternative procedures and methods of documentation that will comply with the intent of the FDA's good manufacturing practices. The challenge lies in convincing an FDA investigator that a unique practice that is different from one with which the FDA investigator is familiar is adequate to assure the production of safe and reliable devices.

VI. ORGANIZING FOR GMP COMPLIANCE

Compliance with the FDA's GMPs depends on the company's organization, with knowledgeable professionals responsible for quality and regulatory functions. Medical device manufacturers should approach the task of complying with the FDA's regulations by establishing authoritative positions high enough within the company's organizational structure to implement good management practices for the produciton of medical devices. Such individuals must have a good understanding of business management, in addition to a good understanding of the FDA's regulations. The first objective of these business-oriented regulatory and quality professionals should be the review and evaluation of all regulatory/quality procedures. It is possible that procedures implemented *specifically* to assure compliance with the FDA's regulations may serve merely to accommodate the FDA investigators in their audits.

VII. GMPS ARE NOT STATIC

The FDA's interpretations of the GMP requirements are constantly changing. Each FDA investigator has his own idea of what constitutes a *current* good manufacturing practice [3]. FDA investigators, having seen a practice that they believe to be a good manufacturing practice at one device manufacturer, will consider that practice to be a GMP requirement for the next device manufacturer that they visit. These differing experiences lead to diversity in the expectations of FDA investigators and to a leapfrog development of what constitute *current* good manufacturing practices.

VIII. PREPRODUCTION MANAGEMENT PRACTICES

Good management practices that address only the manufacturing processes cannot provide a high degree of confidence that medical devices are reliable, safe, and effective unless there is also a program that describes preproduction practices applicable to the design of new medical devices and changes in existing designs. The *inherent* safety, efficacy, and reliability of medical devices are established during the design phase of a device life cycle.

The specifications for a device that are developed during the design phase are a major contributor to the manufacturing procedures that follow. The specifications define the design concept in terms of measurable characteristics based on the intended use of the device. The original specifications and subsequent changes made to the specifications should be documented and evaluated for effects on reliability, safety, and efficacy. Several independent design reviews should occur at transitions to succeeding states of the device's design phase. A formal review of subsystems, software, packaging, labeling, drawings, etc., should be documented and approved by representatives of quality, research and development, regulatory, and engineering departments.

One or more design reviews should include a failure mode and effects analysis (FMEA) to identify potential defects, the probability of occurrence, and an assessment of their effect on safety and performance. The FMEA should include an analysis of potential device user-induced failures or defects. The design reviews will also assess reliability of the device and establish acceptable failure rates.

Guidelines issued by the FDA offer good advice, but can be easily mistaken for legal requirements. It appears that the FDA has been placing more and more reliance on informal guidelines. Informal FDA guidelines have been issued on process validation and preproduction quality assurance. These guidelines state procedures or standards of general applicability that are not legal requirements but are acceptable to the FDA. Generally, a device manufacturer may rely on a guideline with the assurance that it is acceptable to the FDA, but guidelines cannot be used in administrative or court proceedings as a legal requirement.

IX. CRITICAL AND NONCRITICAL DEVICE GMPS

The FDA has defined some medical devices as "critical" and has imposed stricter GMP requirements for these. The different FDA requirements for critical and noncritical devices are found in the good manufacturing practices regulation codified as 21 CFR 820 [4]. Good *management* practices do not necessarily need to distinguish between critical and noncritical devices. If the procedure does not make good management sense, it is likely that it is not necessary for either critical or noncritical devices. This distinction of critical and noncritical devices is an area within the FDA's good manufacturing practices in which the development of good *management* practices offers management many opportunities for creativity. The more stringent requirements of the FDA's regulations for critical devices often add considerable requirements for additional documentation.

X. GMPS EQUALLY APPLICABLE TO ALL DEVICE CLASSES

Device classes are not to be confused with the critical and noncritical designations used by the FDA. The FDA's good manufacturing practice regulations apply equally to Class I (general controls), Class II (performance standards), and Class III (premarket approval) devices, unless specific Class I devices have been exempted from one or more of the GMP requirements. Common sense and good management practices dictate the development of appropriate manufacturing procedures applicable to all three classes of devices. Whether a device must comply with a performance standard or whether it must be approved prior to marketing is not related to the development of good management practices for manufacturing that medical device.

XI. QUALITY CONTROL VERSUS QUALITY ASSURANCE

Do not confuse quality control with quality assurance.

Quality control is responsible for day-to-day activities associated with device production and component procurement. Quality control personnel typically report to the manufacturing organization, whereas quality assurance is usually a very small staff typically reporting to the president or general manager. The quality assurance staff conducts periodic audits of the quality control function to assure that the policies and procedures developed by quality assurance are appropriately implemented by the quality control function.

The FDA's good manufacturing practices for medical devices describe very broad regulatory requirements for a quality assurance program. The points to be emphasized in management's development of a quality assurance program involve primarily organization and responsibility. The scope of the quality assurance activity should include involvement in the early design of the product, product reliability, and direct overview of the quality control functions of the organizaiton. Notice the distinction between quality control and quality assurance.

Long-range quality planning and quality cost-effectiveness are responsibilities of the quality assurance activity. One-year and 5-year quality plans are typical responsibilities of the quality assurance staff. The quality assurance staff is intimately involved in new product designs and changes in products. Quality assurance is also invovled in facility and equipment planning.

XII. DOCUMENTATION REQUIRED BY GMPS

The FDA's GMP regulations require a large amount of documentaiton. There are 33 references in 21 CFR 820 to "written procedures," 53 citations to "record requirements," and 30 requirements for "written specifications."

Management's primary defense for allegations of noncompliance with the FDA's GMP regulations is dependent on appropriate documentation. The degree of documentation that is developed is usually directly proportional to the size of the manufacturer. In small companies, management is usually available to communicate orally information to a relatively few workers. When procedures in small companies are written, they are usually more concise than those generated by "procedure writers" in large manufacturing operations. The most useful procedures usually begin with an explanation of the need for the procedure, followed by examples where known problems result from failures to follow the procedure described. The procedure should differentiate between major and minor requirements. In general, a well-written procedure will be brief and concise, thoroughly objective, and written in simple language. The basic function of a procedure is to communicate management's requirements to workers and others responsible for their implementation.

XIII. PROCUREMENT OF COMPONENTS UNDER GMPS

The quality of the components largely determines the quality of the finished device. Procurement of components is centered around documentation of supplier qualification, supplier process control, and audits of suppliers. Occasional verification of supplier performance by testing and examination of components is necessary, and the frequency is dependent on the history and experience with each supplier. Certifications of actual tests and examination results should be required from each supplier with each shipment of components.

The certifications should *not* be simply statements by the supplier that all required tests and examinations have been performed and found to be acceptable. Rather, the certifications should contain details of the mutually agreed upon sampling plans, actual measurements and observations, and explanations of the rejects noted—even though the quantity of rejects is within the mutually agreed upon average quality level (AQL). The certifications combined with supplier audit visits (frequency depending on supplier history and performance) and prior supplier qualification are sufficient to allow shipment of components directly to stock without performing duplicate testing/examination upon receipt at the device manufacturer.

Good management practices involve the selection of qualified, ethical suppliers who are capable and can be depended on to manufacture and supply quality components. The premium prices paid to qualified suppliers are appropriate compensation for the quality of components required by the device manufacturer. It is not necessary to retest and reexamine each shipment of components from properly qualified and controlled suppliers. The qualification of suppliers is a time-consuming task and often invovles a great deal of training and education of the supplier. Agreements that suppliers will not make significant changes without notifying the device manufacturer should be documented [5].

Good management practices require that finished products be competitive in cost as well as quality. The quality of device components is the primary concern of the FDA. It is preferable that the number of suppliers of medical device components be kept to a minimum to assure uniformity and quality of components. However, there are several important reasons (such as competitive pricing and consistent, on-time deliveries) for not having a single source for any component. Quality control departments should evaluate potential suppliers prior to consummating contracts for components. The quality control management of both the finished device manufacturer and the device component manufacturer must be in total agreement as to the required quality of the components. This includes agreement as to the required methods of manufacture of components and any required control of the component manufacturer's processes.

Only after this understanding has been established can the finished device manufacturer's purchasing representatives negotiate price with the component manufacturer. The quality control department can be assisted in its evaluation of device components and potential component suppliers by members of the device manufacturer's research and development and engineering functions. This assistance may include joint site visits to component manufacturers, in addition to the joint development of specifications.

When initially qualifying component suppliers, medical device manufacturers will sometimes need to test "first article" components upon receipt to verify compliance with specifications. Medical device manufacturers can cooperate with the component supplier to assure that the component supplier will ship only components that have been found to meet mutually agreed upon specifications so that components can be received directly into stock for assembly into finished medical devices without quality control having to duplicate the tests and examinations which were performed by the supplier prior to shipment. The basis for this procedure is a thorough understanding through open communication between the component supplier and the device manufacturer as to what is expected and what will be acceptable. Periodic audits of the component supplier and certifications of each shipment, accompanied by occasional examinations of components upon receipt, can assure receipt of acceptable components.

XIV. COMPONENT AND PRODUCT TRACEABILITY UNDER GMPS

It is not necessary or practical to trace each component used in the manufacture of every medical device. Critical components of critical devices offer the biggest challenge. Management may decide to institute control procedures that bracket manufacturing or receiving dates rather than providing traceability of each component. There is no problem with this practice for noncritical devices except that more customers than necessary may have to be contacted during a product recall. This practice represents a tradeoff because management has decided to forego the cost of individual component traceability during manufacture and has opted to assume the risks associated with a larger than necessary product recall. Component traceability must include final product traceability through distribution and sales records.

XV. PROCESS VALIDATION

End testing of devices by itself cannot assure compliance with the FDA's GMP regulations. A good management practice that will help assure compliance with the FDA's regulations is the validation of processes used in the manufacture of medical devices. Once a process has been validated, products produced by that process will conform to the same specifications as long as no changes are made in the process *and* the quality of the components does not change. Various tools, such as statistical process controls (SPC) and process capability studies, can be used to assist in controlling processes. Prospective validation of equipment, processes, and products, and retrospective validation of processes are all acceptable for process validation. Routine end-product testing by itself is insufficient for assuring product quality. Each step of the manufacturing process must be controlled to maximize the probability that medical devices will meet all quality and design specifications.

XVI. CUSTOMER COMPLAINT HANDLING UNDER GMPS

A practice that the FDA frequently finds deficient during its inspections is management's failure to properly handle customer complaints in a manner that corrects the reported problems. A complaint represents a customer's dissatisfaction with a product. All complaints need not be handled in the same way. The receipt of two devices instead of only the one device that the customer ordered is a type of customer complaint that will not result in patient injury. Another type of complaint is that of the customer who received a device that failed to perform as labeled and resulted in an injury to the user. A suggested procedure for handling customer complaints is to provide for representatives to receive telephone reports 24 hours a day, 365 days a year via toll-free telephone numbers from users or health-care professionals. Each customer contact should be documented as completely as possible at the time of the original customer contact, and *each* report should be reviewed and evaluated by a second individual who is responsible for determining an appropriate follow-up action.

The person originally receiving a customer report may make a preliminary determination that the report does not involve a misuse or malfunction of the device. The person may also decide not to consider a particular customer contact to be a complaint. All customer contacts should be documented, whether or not the report is thought to represent a complaint. The person receiving the customer contact may determine that the reported problem was due to misuse of the device by the customer and does not, in his opinion, require further evaluation or investigation. Such a decision may constitute an error because labeling for the product may need to be revised or salesmen and customers may need additional training or instructions on proper use of the device.

If more than one person has responsibility for evaluating customer contact reports, precautions should be taken to assure that there will not be differences of opinions between the reviewers as to whether further follow-up is required for similar complaints. In addition, trends may be missed because the same product problems amy be documented differently by the different reviewers. This can result in a delay in recognizing problems.

It is important to determine the identification of each specific device involved in a complaint by a lot number or serial number. If the investigation of a complaint reveals that the cause of the problem is related to a specific process or component, other affected customers may be contacted and advised as to proper procedures to prevent additional problems. Proper distribution records must be maintained so that affected customers can be quickly identified and contacted.

Good management handling of complaints should include a follow-up contact with customers after the evaluation of each complaint has been completed. An appropriate procedure would acknowledge receipt of each complaint to the customer with a description of the complaint as it was recorded and an assurance to the customer that the device manufacturer has identified and corrected the problem. It is not necessary to disclose to the complainant the *details* of exactly what was found on investigation and what was done as a result of the investigation.

XVII. INTERNAL AUDITS ASSURE GMP COMPLIANCE

The *internal* audit is management's most important tool for assuring compliance with the FDA's regulations. Management is often hard pressed to keep informed of the status of manufacturing procedures under its cognizance. The concept of the internal audit came about as a response to the need of management to inform itself of the status of compliance with policies and other requirements. The FDA has stated that it will only seek access to internal audit reports during litigation under applicable procedural rules and will not *routinely* ask to review the manufacturer's internal audit reports during FDA inspections [6]. However, there is no assurance that internal audit reports will not become accessible to the FDA at some future time. It would be prudent for device manufacturers to correct all observed deficiencies as expeditiously as possible. The success of the internal audit depends on the corrective actions instituted as a result of the internal audit. The failure to perform internal audits is a frequently cited GMP deficiency by the FDA investigators.

The development of a management-oriented internal audit program will assist management in the identification and correction of deficiencies before they adversely affect products and the company's reputation for quality. During an internal audit, the interaction of people with systems, processes, procedures, and operations should be evaluated to assure uniformity of products that comply with the customer's needs and expectations. The internal audits should also evaluate the company's compliance with the FDA's regulations.

The internal audit should concentrate on documentation and should review production records (particularly device master records and device history records), purchase orders and vendor agreements, product specifications, design drawings, software, and process validation results. The audit should result in a written report, which should be reviewed during the next audit to ascertain that the necessary corrective actions have taken place and are effective. The purpose of the audit should be to either (1) provide management assurance that there are no problems in the areas audited or (2) detect and correct problems at the earliest practical stage.

The FDA uses its inspections in the same way that management uses internal audits. Both audit procedures stress the in-depth review of records and documentation. The primary difference between the two functions is the involvement of the auditors in the corrective actions that follow an audit. When an FDA investigator conducts an inspection, he often leaves a list of objectionable conditions (FD-483). Those items represent that individual investigator's observations as to conditions that, *in his opinion*, could result in a violation of the FDC Act. The manufacturer's corrective actions that follow an FDA inspection are usually directed toward correcting the specific items noted by the FDA investigator.

The FDA investigator usually comes to a facility unannounced and leaves a list of objectionable conditions without suggestions as to possible corrective actions. To be most effective, an audit would not be a "surprise" visit.

The persons and areas to be audited need to have time to prepare for an audit. An announced audit represents the best use of everyone's time. *Significant* deficiencies generally cannot be corrected on short notice by advance preparation for an audit. Management's internal audits should be announced and planned in advance. The internal audit reports should include recommendations for correction of deficiencies, and the internal auditors should be responsible for assisting the managers of the audited areas in the development and implementation of the corrective actions.

XVIII. CONCLUSION

Each company has its own philosophy for complying with GMP regulations. The advice given in the preceding sections is based on the experiences of one person who initially worked with the Food and Drug Administration and subsequently has held increasingly responsible quality and regulatory positions with both large and small device-manufacturing companies.

An attempt was made to offer practical advice on subjects involving controversial issues. Industry and the FDA continue to work together through ad hoc committees in conjunction with trade associations such as the Health Industry Manufacturers Association (HIMA) and the Pharmaceutical Manufacturers Association (PMA) to resolve the differences of opinions that arise from time to time. The FDA's GMP advisory committee has also been helpful in providing recommendations to the FDA concerning proposed GMP regulations.

REFERENCES

1. *U.S. v. Dotterweich* (U.S. Sup. Ct. 1943), 320 U.S. 277, 277, 64 S. Ct. 134.
2. *U.S. v. Park* (U.S. Sup. Ct. 1975), 421 U.S. 658, 95 SUP. Ct. 1903.
3. 21 CFR 820.1.
4. 43 *Fed. Reg.*, 31508, 7/21/78, 78.
5. 21 CFR 820.3(f),(m).
6. 43 *Fed. Reg.*, 31508, 7/21/78, 37.

24

Developing a Recall Program

DONALD F. GRABARZ *DFG & Associates, Inc., Salt Lake City, Utah*

MICHAEL F. COLE *Laxalt, Washington, Perito & Dubuc, Washington, D.C.*

I. INTRODUCTION

The process of recalling a product presents a traumatic and potentially costly experience for companies confronted with a situation causing them to conduct a recall. Over the past several decades, it would be safe to say that the majority of manufacturers of medical devices and diagnostic products have, at least one time in their history, been involved with the recall decision process. The fact of recall is a serious matter, since it deals with the basic issues of safety and effectiveness of medical products. It should be kept in mind that there are tens of thousands of medical devices used safely in multimillions of procedures each year. The number of situations giving rise to a recall is extremely small and, with few exceptions, manufacturers of medical devices and diagnostic products have acted in both an effective and responsible manner.

In this chapter, we will discuss in detail the laws, regulations, and policies governed by the Food and Drug Administration as they relate to recalls. We will provide practical considerations, the do's and don't's, along with ramifications and consequences. We will suggest procedures to be followed that are both known and proven, and will outline some recent statistics that will hopefully focus thought toward prevention.

Many manufacturers of medical devices have the perception that a recall is the physical removal or retrieval of a product from the marketplace. This only represents one aspect of what may be considered in recall. It should be understood that a recall covers a number of actions taken by a company. In addition to the physical recovery of a product, it can involve field actions, such as repair and service, retrofit, and labeling, among others. Not to be confused with what is termed a market withdrawal, as defined below, a recall is the physical removal of a product from the marketplace, the repair and/or retrofit of a product in the field, or the changing of a product's labeling, each in order to correct that which is considered a violative situation by FDA.

The distinction between a recall and a market withdrawal will be discussed further in this chapter.

In order to approach the recall procedure, responsible individuals within any company need understand the factors, as well as certain definitional terms that can lead to the recall process.

As previously discussed, the term recall, from a legal and regulatory perspective, suggests a product to be violative within a statutory definition. Accordingly, there may be reasons causing a recall to take place that are beyond any actual occurrence of a safety or effectiveness issue. Certain administrative situations, such as failure to have an effective premarket approval application or an effective investigational device exemption application, or failure to comply with good manufacturing practices (GMP) or an administrative detention order, or a device deemed to be banned by FDA can cause a product to be recalled. While some of these administrative situations may have caused a manufacture to recall a product, the occurrence is rare, with the exception of GMP violations. However, while all these factors need be considered, since the effect is the same in administering the process, company credibility, reputation, and *cost* are at stake. Loss of any of the previously mentioned can portray the demise of many companies.

The most prevalent reason for recalling a product focuses on issues of safety and effectiveness, rather than administration. The clear majority of recalls over the past years have been related to GMP violations, rivaled by those for lack of proper preproduction analysis and evaluation in the design phase. This can be seen from the charts presented in Figs. 1, 2, and 3, compiled by the Food and Drug Administration. Reviewing numbers compiled by

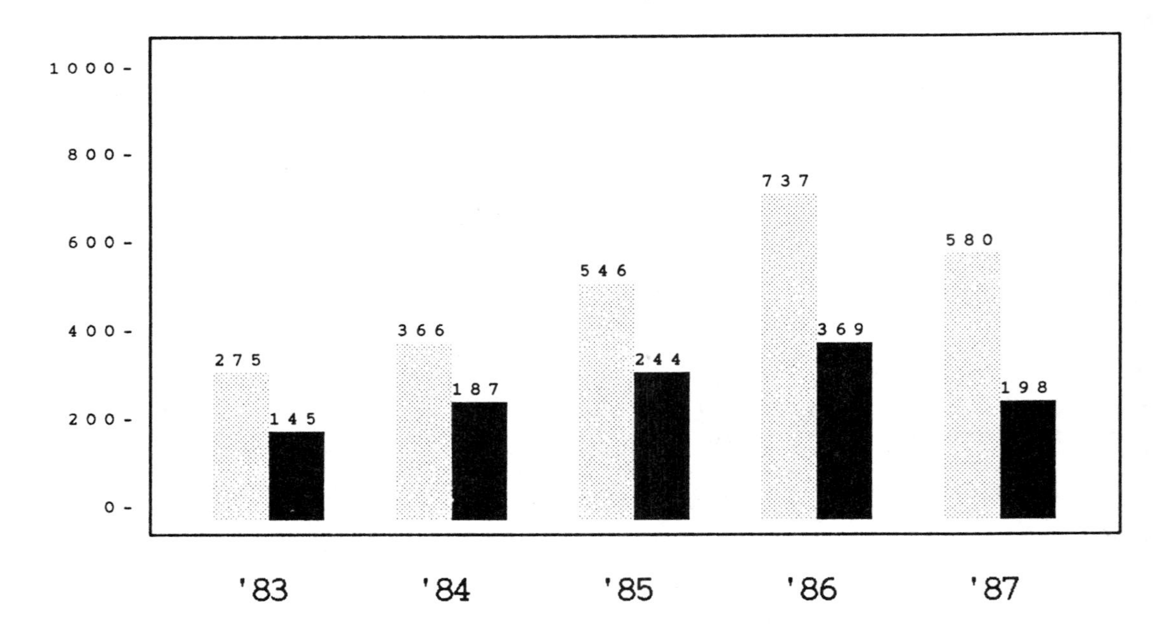

FIGURE 1

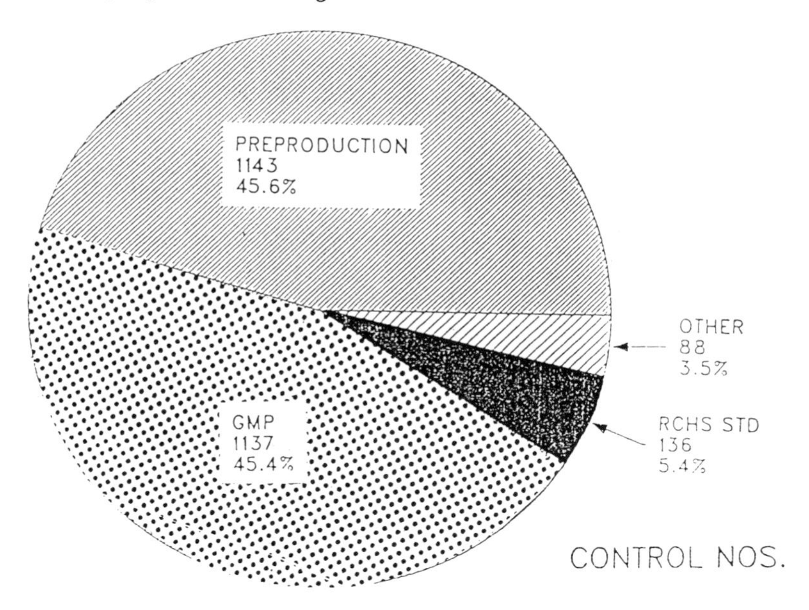

FIGURE 2

the FDA from FY83 through FY87, there have been over 2500 recalls. Ninety percent of these have been related to causes resulting from good manufacturing situations, or from preproduction design product problems. Figure 3 depicts the distribution of preproduction causes from product recalls. This number has been on the increase in FY88 and FY89. Lack of attention in strategic planning in the design and preproduction phases, as well as loose manufacturing controls, is suggestive of the major reasons causing products to be recalled.

II. THE LAW AND REGULATIONS

The term "recall" does not appear in the federal Food, Drug, and Cosmetic Act, as amended. Instead, the recall mechanism has developed over the past decades as an informal way of resolving substantial problems faced by both FDA and the health-care industry. These problems and issues have arisen when violative products appear in the marketplace.

Congress did give FDA a variety of enforcement options and tools to use to get violative medical devices out of commerce. The remedies are potent and include criminal prosecution, fines, imprisonment, seizures, injunctions, and repair, replacement, or refund of the purchase price of defective devices (the 3R's). While this is an impressive arsenal, FDA cannot always use the remedies efficiently to deal with problems. The remedies each involve judicial proceedings that are cumbersome, costly, and time-consuming. In choosing one or more of these remedies as a course of action, FDA must very carefully develop a full and complete evidentiary case. Even when FDA prevails with one of these remedies, the net result may not be altogether useful. For ex-

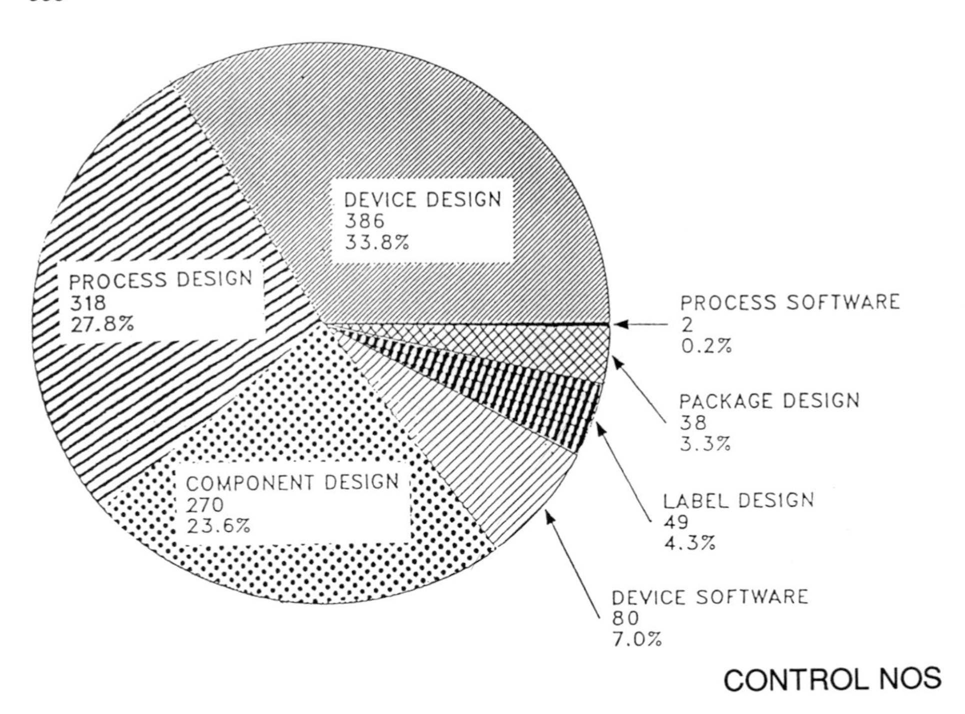

FIGURE 3

ample, FDA can initiate an action to seize noncomplying goods. If it prevails, FDA must then physically locate the goods in order to seize them. In addition, a separate seizure order normally must be filed for each location where specified goods are actually found at the time the order is served. Other devices of the same type at different locations may still be sold. This process of ferreting out the relevant devices makes it virtually impossible to use a seizure course of action as a quick way of addressing a product problem, unless the problem is very limited in scope. FDA may also seek to enjoin further production and/or distribution of violative goods. This action again means careful preparation of a case, and even if FDA succeeds, the goods already in distribution are not usually subject to the order, particularly if already sold to a consignee. The FDA can also institute criminal action against firms and/or individuals and has done so. However, this again requires substantial evidentiary finding, is time-consuming and costly, and may involve intermediate administrative hearings. Attempting to get fines levied against a company may also be as laborious, preventing rapid correction of violative conditions. The reality, therefore, is that FDA seldom relies on its statutory authority and powers, given resources and the need to react quickly. Thus, as discussed in Section III of this chapter, FDA predominately relies on regulatory policy, on the threat of statutory remedies, and on effectuating a recall for voluntary compliance with companies to correct problem situations.

The device industry capitulates so as to avoid the impact of formal FDA action. While several of the remedies may be imperfect from an enforcement point of view, they are very serious matters for a manufacturer and its responsible individuals. No one relishes facing a criminal indictment and the potential penalties that may be imposed. The use of seizure or injunctive powers by FDA can cause a firm serious financial loss not only in direct dollars such as may be associated with the administrative legal fees and retrieval of product but in market share for future sales, as well as the cost related to imposition of fines. Therefore, manufacturers of medical devices much prefer the practice of a "voluntary" recall in lieu of having to be confronted with those legal remedies that can be imposed by FDA. It provides a more palliative alternative toward correcting problems in the marketplace. As such, recall has evolved as the single most useful tool to accomplish what both FDA and industry need. From FDA's point of view, violative product is more rapidly and efficiently removed from commerce, or modified so that it conforms to regulatory requirements. From the industry point of view, recall forestalls more serious legal action on the part of FDA. So recall becomes the single most important concept for dispute resolution on a product problem. It is therefore necessary to thoroughly understand the rules governing this process, as well as the law and regulatory options.

FDA does not contend that it has the authority to order a recall; however, it does contend that it has the power to determine whether the action taken is appropriate and complete. As authortiy, FDA relies on its general powers to promulgate rules for the implementation of the Food, Drug and Cosmetic Act, as amended. FDA also points out that if it does not agree with the proposed action to remedy a defect, it would have the power to take further action. This leaves the question as to whether a manufacturer contemplating or conducting a recall should notify FDA of its action. A company may wish to conduct a recall without ever notifying FDA. The reason is that some firms feel that involving FDA only unnecessarily complicates the conduct of a recall: time must be spent in educating investigators as to the problem that has arisen, leading to a diversion of resources in answering questions often misdirected by FDA. In addition, since FDA publishes the fact of a recall in its Enforcement Reports and, as a result, information may become more readily available through the Freedom of Information Act, companies become sensitive to both public image and competitive pressures. Finally, there is a sensitivity toward the occasional inaccuracy of information published by FDA concerning a recall, often confusing consignees and causing unnecessary follow-up communications.

Under the law and regulations, can a manufacturer legally not notify the FDA of a recall action? While the law does not specify any requirement for a manufacturer to report a recall action, it does provide FDA with authority under certain implementing regulations to do so. FDA has the authority to require a manufacturer of medical devices to report field corrective actions under conditions of approval upon issuance of a premarket approval letter as per 21 CFR 814.84 *et seq*. In addition, the premarket approval regulations specifically reference adherence to the medical device reporting regulation, 21 CFR, Part 803 (MDR). With respect to the latter, questions arise concerning the requirement to inform FDA of a recall on any device. By virtue of 21 CFR 803.24(e)(8)-(9), FDA can compel submission of information about a proposed remedial "recall" action as part of the follow-up to an initial MDR report. Many manufacturers have come to view this section of the MDR regulations as a requirement to report remedial "recall" actions.

On the other hand, if FDA inquires as a result of an initial MDR filing, any remedial action must be discussed. However, two other situations exist.

First, the manufacturer may be contemplating a recall in a situation where there does not exist an order under a premarket approval application to provide a report or where an MDR report is not required. Second, the manufacturer may have filed a report under MDR or as may be required under a PMA order, without having received any request from FDA for "recall-type" information. Should the firm notify FDA of its intentions to recall the Product? In the opinion of the authors, while such circumstance does not legally require a firm to provide notification, in consideration and practice for continued credible relations with FDA, we would urge notification and agency involvement.

Irrespective of the law and regulations, FDA strongly advises manufacturers to notify the agency when they contemplate conducting a recall. In many situations, FDA will become aware of a recall being conducted, independent of any formal notification from the company.

It is particularly instructive to review the reasons why FDA contends it should be notified of a proposed recall. FDA states:

> The Commissioner advises that persons involved in the distribution of products regulated by FDA have responsibilities and duties of an unusually high order. The public relies every day on the quality, purity, and safety of these products. The Commissioner, therefore, strongly believes that it is essential that FDA be notified of a firm-initiated recall: (1) So that the agency can assist, where necessary, in making certain that all units that need to be recalled are in fact recalled, e.g., by helping to locate the product subject to a recall; by notifying the public, if necessary, through warnings so that undetected units may not be used by possibly serious public detriment; (2) so that FDA can assist in locating the cause of the problem and its possible solution in order to prevent future injury or damage to the public; (3) to enable FDA to check similar firms and/or similar products to determine whether like problems (detected or undetected) are being encountered and, if so, to take steps 1 and 2 with respect to such other firms in order to prevent or stop injury or damage to the public; and finally (4) so that FDA with this accumulated knowledge can in the future recognize the same or similar problems as they are developing or even before they actually appear, so to prevent or mitigate injury or damage more readily and efficiently and be more helpful to industry in diagnosing and solving problems that may, if undetected or unsolved, cause injury or damage to the public.
>
> The Commissioner agrees, however, with the comments that § 7.46(a) should indicate that FDA notification of a recall is not a requirement under these guidelines; and appropriate revisions have been made. The Commissioner points out, however, that this revision does not alter the agency's position that firms are expected voluntarily to notify FDA whenever they decide to initiate a recall and where otherwise required by other FDA regulations. [1]

For the above reasons, a firm facing a recall should consider consulting with FDA in an attempt to conclude corrective action, minimizing later distractions and additional costs and the potential of having to administratively reconduct a recall action.

Repair, replacement, and refund provisions of the Food, Drug and Cosmetic Act, as amended [2], as they relate to recall are important to understand. This section of the act empowers FDA with significant authority. Beginning with this section of the act, Notification, the Director of the Center for Devices and Radiological Health (CDRH) can, if he deems that a device represents an unreasonable risk of substantial harm and that notification is necessary because no other means are available, order a manufacturer to notify all health-care professionals who use or prescribe the device, as well as others, in order to eliminate the risk. The only requirement needed by FDA to invoke this section is consultation with the manufacturer. There is no judicial hearing.

If such notification will not eliminate risk to health, and certain conditions are met, then the Director can order a manufacturer to replace the device, repair it, or refund the purchase price.

On its face, Section 518 looks very much like a statutory recall provision. Indeed, when the recall guidelines were first proposed in 1976, the preamble contained a statement saying that implementing rules under Section 518 would work to eventually exclude devices from the guidelines. However, in the final policy, FDA recanted, and the agency's reasoning was instructive in determining the future relationship of recalls and notifications:

> Notification and remedy under section 518 are limited to situations where the agency determines that a commercially distributed device intended for human use presents an unreasonable risk of substantial harm to the public health.
>
> Furthermore, a remedy other than notification cannot be ordered in such situation unless FDA also determines that there are reasonable grounds to believe that: (1) The device was not properly designed and manufactured with reference to the state of art as it existed at the time of its design and manufacture and (2) the unreasonable risk was not caused by the negligence of a person other than a manufacturer, importer, distributor, or retailer of a device. If a situation with a device does meet these prerequisites, the agency will consider using its authority under section 518.
>
> The Commissioner notes, however, that the authority in section 518 does not apply to the following situations: Violative veterinary devices and investigational devices; devices that present some risk to health, but do not present a risk notwithstanding the fact that the device was properly designed and manufactured with reference to the state of art at that time; or devices which were rendered dangerous because of the negligence of persons other than a manufacturer, importer, distributor, or retailer of a device. In each of these situations, action cannot be taken under section 518, although the agency does have the authority to initiate seizure of the device under section 304 of the act in order to protect the public from misbranded or adulterated medical devices. Since a recall is an alternative to seizure, these guidelines should be also applicable to violative medical devices, notwithstanding the authority in section 518. [3]

Therefore, recall and notification coexist, with notification being utilized when certain conditions and criteria are met. The statutory provisions of section 518 have two undeniably appealing features from FDA's point of view. First, FDA can order an action rather than just encouraging one. Second,

FDA can dictate specific forms of economic relief, something that cannot be done in the "voluntary" recall situation. The economic relief that FDA can order involves:

a. Repair of the device to remove a defect;
b. Replacement of the device with one which conforms to the Act; or
c. Refund of the price of the device, less an allowance, if the product has been in use for more than 1 year.

FDA can order one of the above after an informal hearing is offered to a manufacturer. However, FDA must be able to demonstrate that an unreasonable risk exists, that notification alone will not eliminate the defect, and that the risk exists because of a defect and not just due to obsolescence of misuse of the product. This, as stated earlier, requires strong evidentiary findings on the part of FDA. Inasmuch, if FDA prevails, the manufacturer must submit a plan of action effecting relief, which FDA can either accept or change.

III. REGULATORY POLICY

The Food and Drug Administration has at its disposal several remedies, as previously discussed. While these remedies involve certain statutory and regulatory tools that can be invoked against a manufacturer, FDA has most often supported the voluntary route, and accordingly has a policy that outlines its expectations.

Generally, it is only when a manufacturer acts irresponsibly or resists taking appropriate action with respect to a violative product that FDA will commence more formal legal remedies. In the worst-case situation, such remedies can not only result in a product being removed from the marketplace, but also involve civil and/or criminal penalties.

The Food and Drug Aministration within its policy guide covers a number of definitional terms and procedural matters [4]. First and foremost is that which FDA considers a recall as different from other field corrective actions a manufacturer may take. According to FDA's policy, the definition of a recall is a firm's removal or correction of a marketed product that is considered to be in violation of the laws administered by the agency, for which the agency would otherwise initiate legal actions.

Other actions, such as a market withdrawal or stock recovery, are not considered to be a recall. Accordingly, the distinction between these latter terms and that of recall revolve around a number of interpretive issues. Market withdrawal is considered by the agency to be an action taken by a manufacturer that invovles a very minor violation or no violation of the law. Such actions would involve normal stock rotation, routine maintenance and repair, and the like. Stock recovery would involve retrieval of a product that has not left the direct control of a manufacturer. In stock recovery, the product in question will not have been placed in commercial distribution, such as to an institution or distributor.

Typically, a stock recovery action is focused on product maintained within a manufacturer's own distribution network. Two major questions that are posed with respect to market withdrawal and stock recovery concern interpretation. First, what is a minor violation of law, and, second, to what extent is "control" with respect to a distribution network?

What should be considered in this process is the distinction between what is considered a recall and that which is considered a market withdrawal/stock recovery. Simply, these represent regulatory definitions having a number of practical implications. In regulatory terms, a recall constitutes removal or field correction of a violative product, within FDA terminology, where the product is out of direct control of the manufacturer. A market withdrawal/stock recovery constitutes removal or correction of a product that is basically not in any violation under the law. The latter, market withdrawal/stock recovery, is often highly interpretive as to legal and regulatory terminology.

The FDA will consider a number of factors in a recall situation. The primary one deals with the health hazard evaluation [5]. The conclusion in this evaluation will dictate many, if not all, aspects surrounding the conduct of a recall. These include the classification of a given recall, the depth of action, and the level of effectiveness checks. An important consideration is the health hazard evaluation, which takes into account six major factors. These include:

a. Death, diseases, injuries, or other adverse reactions that have already occurred from the use of the product.
b. Existing conditions that could contribute to a clinical situation that could expose individuals to a health hazard.
c. Assessment of the population who may be at risk.
d. Assessment of the degree of seriousness of a health hazard to a population at risk.
e. Assessment of the probability of occurrence.
f. Assessment of the long-range and/or immediate consequences of any health hazard.

Given the above factors, FDA will classify a recall into one of three classes, with Class I being the most serious, Class II less serious, and Class III the least serious. FDA provides definitional terms for each as follows:

Class I: a situation where there is reasonable probability that the use of, or exposure to, a violative product will cause serious, adverse health consequences or death.
Class II: a situation in which the use of, or exposure to, a volative product may cause temporary or medically reversible adverse health consequences or where the probability of serious health consequences is remote.
Class III: a situation in which the use of, or exposure to, a violative product is not likely to cause adverse health consequences.

The determinations and findings described above will establish the framework and actions that need to be taken in conducting a recall, including the development of an overall recall strategy. The depth of a recall, the level of effectiveness checks, and the use of publicity will be mostly determined by the health hazard evaluation and the classification decision.

The Food and Drug Administration defines effectiveness checks within its policy. Effectiveness checks will dictate the follow-up a manufacturer must conduct during the course of a recall in order to assure competent administration of the process. According to FDA policy there are five levels of effectiveness checks. These are defined as follows:

Level A: 100% of the total number of consignees to be contacted (at the specified recall depth).
Level B: some percentage of the total number of consignees to be contacted, which percentage is to be determined on a case-by-case basis, but is greater than 10% and less than 100% of the total number of consignees to be contacted (at the specified recall depth).
Level C: 10% of the total number of consignees to be contacted (at the specified recall depth).
Level D: 2% of the total number of consignees to be contacted (at the specified recall depth).
Level E: no effectiveness checks.

IV. INTERNAL POLICY

Thus far, significant aspects of the law, regulations, and policies administered by the Food and Drug Administration have been outlined. The next two sections are intended to focus on the procedural and practical considerations in dealing with a recall.

The first and perhaps foremost consideration is planning, recall preparedness. That is, contingency planning through internal procedures and policies should be in place before being confronted with having to conduct any recall. Management awareness and understanding of the process and financial reserves should be established before a recall becomes a reality. As stated in the introduction to this chapter, being confronted with a recall is a traumatic and costly experience. Cost of product, loss in sales, administrative time, market share, and potential product liability are but a few factors that can affect the image and financial stability of a company in a recall situation. Proper planning and anticipation can go a long way in softening this difficult experience.

A. Management Awareness

Management awareness is perhaps one of the key elements that can provide sanity and calm in this process. It should be understood that virtually every discipline within a company needs to be involved in the process. However, the ultimate and final decision as to whether or not to recall a product rests with the chief executive office (CEO) of a company. This is not to say that other disciplines are unimportant. To the contrary, they comprise a multifunctional team in assessing, evaluating, and recommending a course of action or nonaction. A core group of responsible individuals, with focused roles, should be charged with the responsibility of developing and presenting a recommendation to the CEO. Most typically, such individuals will be comprised of key personnel from quality assurance, regulatory affairs, medical, research and development, manufacturing, marketing, and law departments.

In order to manage this core team, one individual should be designated to chair this activity. Experience has demonstrated that the most appropriate individual to chair this activity is from either regulatory affairs or quality assurance, with other disciplines providing significant input to the decision process.

Medical affairs needs to play a key role in the health hazard evaluation. Manufacturing and quality assurance are important in assessing the extent of a potential problem. Research and development most typically will evaluate design changes to correct product problems encountered. Marketing should be prepared to deal with customer relations, and the law department will assess any legal ramifications, including potential product liability issues. Understanding their individual roles and participating as team members are important in making the whole process work in an organized, efficient, and cost-effective manner.

B. Internal Policies and Procedures

Given the fact that management understands the process, internal policies and procedures should be reduced to writing. Doing so helps to assure continuity and preserves a singular approach tailored to the style and philosophy of the company. However, the establishment of policies and procedures should only be a guide, with flexibility in dealing with specific situations. In developing a policy and a procedure, it is best to keep such simple and to the point. The overall purpose, responsibilities, definitions, and procedures to be followed should be not only documented but also agreed to by those who may ultimately be involved in the recall process.

More often than not, considerations to recall a product are precipitated through information derived from users, such as product complaints. In addition, information may arise from internal findings or scientific publications that suggest a product problem. Regardless of the source, when there is an indication that a problem may exist, a thorough investigation must be conducted. Should such investigation confirm the fact that there is in fact a problem with the product in question, and a corrective action plan needs to be developed. One aspect of the corrective action plan is the recall consideration. Here is where the decision process starts. Throughout this activity, diligence to documentation is extremely important. Attention should be paid to assure that related internal procedures and policies are followed and adhered to. If, as a result of the decision analysis, a conclusion is reached that no field action (recall) is necessary, then clear and concise documentation leading to this decision must be maintained. An important part of such documentation is the rationale for such conclusion, which must include a health hazard evaluation.

C. The "Recall" Decision

There are no magic numbers that trigger a recall decision. Each situation must be weighed on its own. Obviously, the seriousness of the health hazard or the potential frequency for product failure, coupled with the size of the exposed population, enters the equation in any recall decision. Another factor that enters this equation is the potential product liability exposure.

Given the foregoing, let us say that a decision has been made to conduct a recall. Administrative and practical matters need to be considered and placed in motion. First, one individual should be assigned as recall coordinator with responsibility for assuring the proper conduct of the recall plan and strategy. This individual should be the focal point for all information concerning a recall and for directing associated activities, including interface with FDA.

While the basic nature of each recall plan have many similarities, the differences will most often revolve around the depth of the recall as well as the physical action to be conducted. Depending on the product and the severity of the problem, recalls can range from the consumer level to physicians, hospitals, nursing homes, and/or distributors. Field actions can involve retrieval of a product, repair, labeling changes, warnings, or other related instructions. Undoubtedly, the most difficult recall to conduct is that on a consumer level. Fortunately, the majority of recalls on medical devices have been focused at the level of institutions and professionals.

It is extremely important that all the potentially affected product be thoroughly identified. This must include each catalog number or product code, every lot or serial number, with the total quantities of each produced. The quantities should be shown as what has been distributed versus what remains in the direct control of the company. Each consignee must be identified and, with respect to quantities of product, it is important to associate all product distributed to each consignee. The greatest major reason to be diligent in this process is to provide the maximum possible assurance that all product in question has been identified. The last thing a company needs is to have to repeat the recall process for lack of having identified all affected product. To this end, many companies will consider including surrounding lots of the product subject to recall. A manufacturer's ability to adequately and effectively conduct this important aspect of the recall process relies on the quality of the traceability system.

A number of other steps and decisions need be taken. Many of these can and should be concurrent with the above. Such steps should include, but are not necessarily limited to, the depth at which the recall will be conducted and the methods of communication, that is, whether personal visit, telephone, telex, and/or mail should be used. In addition, decisions need be made with respect to repair, replacement, or refund. Clear and concise directions have to be developed, instructing consignees with the method and means by which repairs will be made, or how and where to return product. Regarding the latter, a company will have to consider such matters as whether all consignees should be instructed to return product directly to the company or through an intermediary, such as a distributor, especially where there is multilevel distribution. Decisions should also take into account the policy concerning administrative reimbursement to consignees for their participation.

While all of the elements discussed above are individually important, they must be bundled into a single recall plan for implementation. In addition to these administrative considerations, the plan should include a health hazard evaluation, a company assessment as to the class of recall, and the level and method of effective checks. An estimated time of completion for returned or repaired product disposition and a description of the corrective action to prevent recurrence should be included in this plan. With regard to product to be returned, the plan should also include a description of any rework or destruction.

Recall communication is important to success in correcting a field problem in an effective manner. Depending on the circumstances dictating the nature of a recall, the appropriate method should be chosen. Regardless of the method, written communication to the consignee should always be made. If, as discussed earlier, personal visits or telephone is the primary method of choice, a written script should be developed and adhered to, followed by written communication. Clear and complete documentation of all means of communication must be maintained.

V. PRACTICAL CONSIDERATIONS

From a practical point of view, the most frequent medical device and diagnostic product recalls involve a network, working from manufacturers through distributors, to hospitals and to health-care professionals. While there are situations that may differ, such as for regional or limited recalls, recalls involving international distribution or those warranting publicity targeted at the consumer level can be profiled by the following typical scenario.

The facts have been sorted, the decision has been made, your company is about to conduct a recall, and the recall plan has been established. The product is in national distribution. The product is intended for intravascular access. It is single-use and sold as sterile. You have received reports of open packages and through investigation have confirmed the fact of this possibility in your packaging process, where such has gone undetected for a period of time. The health hazard evaluation suggests that a product could in fact be nonsterile at the time of use which could pose a serious, but yet reversible, chance of infection. Because of the circumstances, you have decided to recall the product and have assessed the situation as being a Class II recall.

Given this situation, in deciding to recall the quantity of distributed product involved, you are confronted with thousands, perhaps tens of thousands, of product to recover. This volume of product, having been marketed through your wide distribution network, may have found its way to its ultimate destination in any number of hospitals throughout the country and/or internationally. The practical aspects of your recall plan need take into consideration the "how." First, your records have identified all distributors to whom you have sold the product. You know that their redistribution network is extensive, involving hospitals in every state. In order to maintain focused control, your plan is designed to contact both distributors and hospitals. You decide distributor contact will be by both telephone and follow-up correspondence. Hospital contact will be initial letter, but you need to develop a list.

With regard to distributors, it is important to inform them not only of your intention to recall a product (with the according details) but also that you will be making direct contact with those hospitals in their area regarding the same. Considering the number of hospitals that may be involved in this scenario, you would best be advised to use one of several mailing services that maintain up-to-date lists of institutions nationally.

Examples of letters that can be used are portrayed in Figs. 4 and 5. It should be recognized that these letters are typical examples that need to be tailored to each individual situation. These examples reflect the basic information needed. Such letters should not incorporate marketing ploys. Phrases such as "we will replace your returned product with our new and improved model" or the incentive "two for one offer" should be avoided.

A return reply card should be incorporated in all initial recall correspondence. Figure 6 shows a typical return reply card, which should be enclosed with each letter. Many companies will incorporate the consignee mailing label on the return reply card, using a windowed envelope. The advantage this has is to make it easier to keep track of those replies received in associating each with the consignees. In addition, the use of a return reply carries the possibility of satisfying the need for potential effectiveness checks.

Proper administration of a recall will speak to its success. It will expedite the execution and certainly its completion. It will go a long way toward keep-

IMPORTANT PRODUCT RECALL

Distributor Name
Address

Dear . . .:

Confirming our conversation of (date), we are recalling those products as listed below:

Product name
Catalog numbers
Lot/serial numbers

We ask that you discontinue any further distribution of the above products and arrange for the immediate return of any inventory which you may have in your possession.

The reason for this recall is based on potential package failure which may compromise sterility. Such failure could cause infection if used.

Our records show that we have shipped (XXX) units of this product to you. We ask that you confirm the quantity currently on hand and complete the enclosed reply card, returning it to our attention as soon as possible. Full credit and/or replacement for all product returned with regard to this recall will be made.

You should be advised that the Food and Drug Administration has been notified of this action.

Product to be returned should be forwarded to the following:

(Name—Address—Attention)

Should you have any questions concerning this matter please contact (name) at (telephone number).

Sincerely,

FIGURE 4 Distributor letter.

ing costs at a minimum, maintaining customer relationship and company image, and satisfying regulatory concerns, thus minimizing FDA invovlement. It involves the designation of a recall coordinator and the choice of appropriate individual disciplines that need to direct and manage the process. The considerations for involvement deal with internal company communications, FDA relationships and communications, establishment and adherence to company policies and procedures, control of recall product disposition and resolution, and public relations, as may be necessary.

IMPORTANT PRODUCT RECALL

Hospital Administrator
Hospital Name
Address

Product name
Product code(s)
Lot/serial numbers

Dear Administrator:

Our company is initiating an immediate recall on all subject products which may be in your possession. The reason for this action is the potential compromise in sterility due to a packaging problem. As a result some of the product may be nonsterile, and if used could cause systemic infection.

We ask that you inventory each area where this product may be located, placing it on hold and arranging for return as instructed below. Further, with respect to of your inventory, we ask that the enclosed return reply card be completed and mailed to us, whether you have any of the subject within your institution or not.

Arrangements will be made to replace all returned product.

Please be advised that the Food and Drug Administration has been informed of this action.

Product returns should be made freight collect to:

(Name—Address—Attention)

We apologize for any inconvenience this may cause. Should there be any questions concerning this matter please contact (name) at (telephone number).

Sincerely,

FIGURE 5 Hospital letter.

Experience has demonstrated that administration of a recall that relies on sales and marketing personnel, versus a nonmarketing recall coordinator, is more time-consuming and less effective. While these disciplines, sales and marketing, should be informed, it is suggested that they not have direct involvement in the administration of the process. In the situation where a company is seeking to retrieve product, not only do warehousing arragnements to receive and quarantine product need to be made, but also a system to expedite elements such as freight collect charges should be set up. A public relations scenario should be established so as to have a uniform response to potential questions that may arise from the media as well as stockholders.

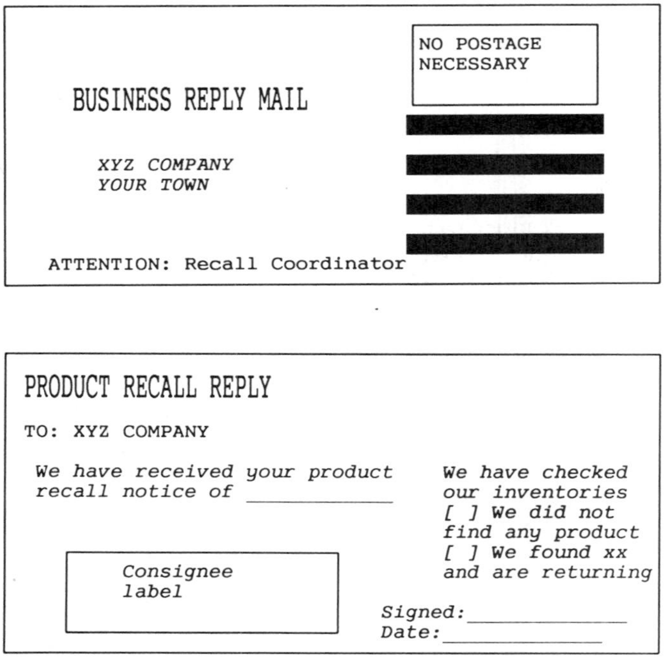

BUSINESS REPLY MAIL

NO POSTAGE
NECESSARY

XYZ COMPANY
YOUR TOWN

ATTENTION: Recall Coordinator

PRODUCT RECALL REPLY

TO: XYZ COMPANY

We have received your product recall notice of ____________

Consignee label

We have checked our inventories
[ ] We did not find any product
[ ] We found xx and are returning

Signed:____________
Date:____________

FIGURE 6 Return reply card.

A. Reporting Requirements

Proper administration of a recall plan must also take into account reporting requirements to the Food and Drug Administration, and FDA's involvement. First, once a decision has been reached to conduct a recall, a report should be submitted under provisions of the medical devices reporting rule. Second, the district office local to the company should be notified. These actions will precipitate a visit by FDA personnel to investigate those circumstances leading to the recall, as well as the recall action itself. Important to this event is anticipation of questions FDA may pose. This is where the documented recall plan provides value equal to that necessary for internal administrative requirements. A written package containing all the basic elements of the recall plan, including the events and findings leading to the final decisions, should be made available to the FDA investigator. This informational package, if organized to answer appropriate FDA concerns, may well satisfy the investigation, and thus minimize potential enforcement action on the part of FDA.

In consideration of FDA requirements and involvement, periodic status reports will need to be provided. Depending on the circumstances and reasons for the recall, such reports may need to be weekly or monthly. Given the conduct of the recall, it is advisable for the company to provide information to the FDA along with a final report, including an estimation of completion time. FDA may, at this point, conduct a follow-up inspection, which could include the witnessing of recalled product destruction, and/or rework, as may be appropriate.

VI. CONCLUSION

The process of a product recall is a trying, dramatic, and costly experience for any company. While the simplistic answer is prevention, and most certainly a goal to be strived for, the realistic fact of life is that at any given time, your company may be confronted with the prospect of having to recall one of its products. Errors will occur, whether in product design or in manufacturing practices. Anticipation through situation and potential problem analysis with planning and administration can ease the recall process. The apparent negatives for a company in having to conduct a recall can also present a potential opportunity. Seizing an apparent negative, correcting an errant situation, can ultimately turn into a positive event, improving or enhancing a company image, as well as providing a better product within the health-care system. Acting in a responsible and prudent manner can stave off user, competitive, and regulatory criticism and gain long-term respect. Should your company find itself in a situation where a recall of one of its products may be imminent, prudent and timely action will become important, maintaining the delicate balance between haste and procrastination. Remember, each situation warrants independent review.

While we have attempted to present the basic information that needs to be considered from a legal, regulatory, and practical point of view, there are many scenarios that will differ from the one described. We trust that this chapter will give pause for thought on this subject and that you will preplan the eventuality of having to conduct a recall.

Should you or your company find itself in the unfortunate situation of having to even consider recalling one of your products, we strongly urge that you seek professional expertise to assist in the deliberations and to help in any strategic planning. Time spent in this early phase has more often proved to be most cost-effective in the long run, although preventing ever having to be confronted with a recall decision is of course the least costly route.

REFERENCES

1. Preamble to the Enforcement Policy, 43 *Fed. Reg.* 26212, June 16, 1978.
2. 21 USC 518, Federal Food, Drug and Cosmetic Act.
3. 21 CFR Part 7, Subpart C; 42 *Fed. Reg.* 15567, March 22, 1977; 43 *Fed. Reg.* 26218, June 16, 1978.
4. 21 CFR Part 7, Subpart C.
5. 21 CFR 7.41, Health hazard evaluation and recall classification.

25

Complying with the Medical Device Reporting Regulation

STEPHEN D. TERMAN *Olssen, Frank and Weeda, Washington, D.C.*

I. INTRODUCTION

The Medical Device Reporting (MDR) regulation took effect in December 1984. It is probably second only to the Good Manufacturing Practice regulation in its significance to medical device companies.

The MDR regulation requires device manufacturers and importers ("companies") to report to the Food and Drug Administration (FDA) whenever they have reason to believe that one of their marketed devices may have caused or contributed to a death or serious injury or has malfunctioned and the malfunction would be likely to cause or contribute to a death or serious injury if the malfunction were to recur.

Under the regulation, companies have to report irrespective of the cause of the event—whether the fault of the product, or even if caused by product misuse.

Reporting must occur quickly (five calendar days for death or serious injury), so companies must react immediately when information is received that may be reportable. Also, since many of the key definitions, such as serious injury and malfunction, are so broad, a lot of the information companies receive may well be reportable. Further, according to FDA, if a health-care professional attributes a death or serious injury to the product, it is "per se" a reportable event. Failure to report can subject a company to a variety of FDA regulatory and legal actions. The key point to remember is that FDA and Congress take medical device problem reporting very seriously and are watching companies very carefully—so compliance is essential.

The intent of the MDR regulation is to assure that FDA is informed promply of all serious problems or potentially serious problems associated with marketed medical devices. Authority for the regulation stems from Section 519 of the Federal Food, Drug, and Cosmetic Act (FD&C Act). Section 519, which was added to the FD&C Act as part of the Medical Device Amendments of 1976, gives FDA the authority to require persons who manufacture, import, or distribute medical devices intended for human use to maintain such records, make such reports, and provide such information as FDA may, by regulation, reasonably require to assure that such devices are not adulterated or misbranded and to otherwise assure their safety and effectiveness.

In 1980, FDA published the Mandatory Experience Reporting rule. The proposal was criticized by the medical device industry as being overly broad. FDA, after reviewing the public comments submitted in response to the proposed rule, placed the proposal in abeyance in November 1981 pending further review. In July 1982 Congressman John Dingell, Chairman of the subcommittee on Oversight and Investigations of the House Committee on Energy and Commerce, held oversight hearings criticizing FDA for not implementing medical device reporting regulations. In May 1983, FDA issued a reproposed MDR rule. Although the industry disagreed with several portions of the final rule, it supported the concept of a reporting rule and did not object to its full and appropriate implementation.

In September 1984 the General Accounting Office (GAO) issued a report entitled "Federal Regulation of Medical Devices: Problems Still to Overcome." This study faulted FDA for failing to develop adequate reporting requirements for medical device problems. Congressional hearings citing FDA's failure to promulgate an MDR regulation were again held in February 1984 by Congressman Henry Waxman, Chairman of the House Committee on Health and the Environment, and in March 1984 by Congressman Dingell. The final MDR regulation was published on September 14, 1984.

The MDR regulation has been in effect for 5 years and is still being criticized for both its substance and implementation by FDA. The industry has criticized the regulation as being overly broad and ambiguous in certain areas; Congress, GAO, and consumer groups have criticized it for not being comprehensive enough and have taken FDA to task for not adequately enforcing the regulation. Legislation was introduced into the House of Representatives in August 1989 (H.R. 3095) by Representatives Waxman and Dingell calling for an extension of the MDR requirements to hospitals, nursing homes and ambulatory surgical centers. Whether new legislation is passed or not, FDA is clearly under pressure to utilize its current MDR authority to compel reporting of adverse incidents to a greater extent; that may mean more MDR inspections of medical device facilities, stricter enforcement of the MDR regulation, and more regulatory and enforcement actions brought for noncompliance.

II. THE MEDICAL DEVICE REPORTING REGULATION

The final MDR regulation was published in the *Federal Register* on September 14, 1984 (49 *Fed. Reg.* 36326), and became effective December 13, 1984. The regulation, 21 C.F.R. 803, requires a device manufacturer or importer to report to FDA whenever it becomes aware of information that reasonably suggests that one of its marketed devices (1) may have caused or contributed to a death or serious injury, or (2) has malfunctioned and that the device or any other device marketed by the company would be likely to cause or contribute to a death or serious injury if the malfunction were to recur. In order for a company to comply with the MDR regulation, it is essential that its employees understand the requirements of the regulation. However, the MDR regulation is no different from other government regulations in that many of its terms and requirements are undefined, broad, and open to different interpretations. It is therefore necessary for companies to also understand FDA's interpretation of the regulation. Armed with this knowledge, a company will be able to both comply with the regulation and avoid confrontations with FDA. To that end, the MDR regulation may be viewed as a set of individual concepts each of which must be defined in general and specific terms. The regulation can be broken down into the following concepts:

A. Who must report?
B. When does a company "become aware of" MDR reportable information?
C. What constitutes "information that reasonably suggests" a reportable event has occurred?
D. Reportable events: What do the terms death, serious injury, and malfunction mean?
E. When and how must reports be made to FDA?
F. What information must be reported to FDA?
G. When are reports not required?
H. What records must the company keep and for how long?

A. Who Must Report?

Any manufacturer or importer who is required to register under 21 C.F.R. 807 is subject to the MDR regulation. According to FDA the following entities are also required to report when they receive MDR reportable information:

Original equipment manufacturers.
Relabelers, repackers, and refurbishers.
Specification developers that are required to register.
Foreign-owned U.S. distributors.
Sole U.S. distributors for foreign devices.
Both a kit manufacturer and the manufacturer of a component of the kit (if the component is labeled by the original manufacturer) must report if they individually receive reportable information from anyone other than each other.

B. When Does a Company "Become Aware of" MDR Reportable Information?

Information concerning reportable events may come from *any source*, including:

1. Individuals or institutions—for example, physicians, nurses, hospital administrators, patients, consumers, company technicians or sales representatives, lawyers, newspapers or magazines, legal documents, hospitals, nursing homes;
2. Medical or scientific literature; or
3. Through the company's own research, testing, evaluation, servicing, or maintenance of its medical devices.

Obviously, the importance and validity of information differs depending upon the source of the information. The source of the information, however, plays no role in determining if and when a company "becomes aware of" MDR reportable information.

Information of a "reportable" event may come from any source. The source of the information is only important in answering the next question.

C. What Constitutes "Information That Reasonably Suggests" a Reportable Event Has Occurred?

A reportable event (i.e., death, serious injury, or malfunction) must be based on "information that reasonably suggests" that the company's medical device may have caused or contributed to a death or serious injury, or has malfunctioned and such a malfunction is likely to cause or contribute to a death or a serious injury. The MDR regulation defines the concept of "information that reasonably suggests" a reportable event by dividing all information into two types. The first is information (such as professional, medical, or scientific facts or opinions) from which a reasonable person would reach a conclusion that a reportable event has occurred. For this type of information, which does not come from a health-care professional, the regulation establishes a reasonable person standard, based upon professional, scientific, or medical facts or opinions. Mere allegations are insufficient. The company must use its judgment when receiving such information to determine whether the information received by them is "information that reasonably suggests" that a reportable event has occurred.

The second type of information is a statement to a company by a health-care professional reaching the conclusion that a reportable event has occurred. The regulation presumes that when a company receives such information, it must treat the statement as fact and the event must be reported to FDA (the "per se" rule). According to the regulation and current FDA interpretation, a health-care professional is a person who has medical or special education, training, experience or knowledge. A health-care professional includes, but is not limited to, physicians (including doctors of osteopathy, dental surgery, podiatry, chiropractic), optometrists, pharmacists, dentists, nurses, respiratory therapists, biomedical engineers, and hospital administrators.

D. Reportable Events: What Do the Terms "Death," "Serious Injury," and "Malfunction" Mean?

1. Death

Death is defined in the dictionary as the act or fact of dying; the permanent ending of all life in a person. Although the question of what constitutes "death" has undergone intense medical, legal, and moral discussion over the last twenty years, it is immaterial to whether a company should report such an incident since death and serious injuries are reported in the same manner under the MDR regulation.

2. Serious Injury

The regulation defines serious injury as an injury that:

a. Is life threatening, or
b. Results in permanent impairment of a body function or permanent damage to body structure, or
c. Necessitates medical or surgical intervention by a health-care professional to (i) prevent permanent impairment or damage, or (ii) relieve unanticipated temporary impairment or damage.

Temporary impairment or damage is unanticipated if reference to such impairment or damage is not made in the device labeling or is occurring with greater frequency or severity than expected, irrespective of whether reference to such impairment or damage is on the labeling.

The regulation's definition of "serious injury" has been criticized by the medical device industry as being too vague, and inadequate to distinguish between minor injuries and truly serious injuries. However, until the regulation is changed or successfully challenged in court, it is FDA's interpretation that must be understood.

When a health-care professional provides information indicating that a serious injury has occurred and uses the word "serious injury" or equivalent language, it is "per se" a reportable event and the company must report the event to FDA. If the health-care professional describes an event with no mention of a death or serious injury, the company may use the serious injury definition and the results of its own investigation to determine if a serious injury has in fact occurred. There are other situations where reports of a serious injury submitted to a company by a health-care professional do not mean that an MDR report is required under the regulation. The health-care professional must indicate that the company's device may have caused or contributed to the serious injury. Keep in mind, however, that the report does not have to make the actual statement, and that if the report reasonably suggests or implies that the company's device may have been the actual or suspected cause, that would be sufficient to require an MDR report.

With regard to the definition of "serious injury," the first two provisions, life threatening and permanent impairment or permanent damage to the body, are fairly straightforward. It is the third part of the definition that has been difficult for companies to understand and apply consistently. According to FDA, to be a reportable event, medical intervention must take place to prevent permanent impairment or damage or relieve unanticipated temporary impairment or damage. There are, however, situations where medical intervention is necessary but the injury is not serious. Consequently, not every situation involving medical intervention is required to be reported as a serious injury.

For example, according to FDA, surgical intervention is likely to be reportable when it is taken to prevent or relieve a condition. However, FDA does not consider the explant of a device that has reached its labeled, anticipated end of life to be a reportable event, if that is why it is being explanted. In addition, any event that requires medical or surgical intervention to relieve temporary impairment of a body function or temporary damage to body structure is not reportable if (a) it is referenced in the labeling or (b) it is not occurring more frequently or with greater severity than stated in the labeling or than is usual for the device if no statement is in the labeling. Unfortunately, FDA has provided little other guidance in this area. Experience and/or direct advice from FDA are the best ways to understand what FDA requires.

3. *Malfunction*

The malfunction definition has also come under criticism from the medical device industry as being overbroad and often unworkable. The regulation states that a malfunction is reportable if the device or any other device marketed by the manufacturer would be likely to cause or contribute to a death or serious injury if the malfunction were to recur.

A malfunction is defined as a failure of a device to meet any of its performance specifications (which include all claims made in the device labeling) or othrwise to perform as intended. The intended performance of a device takes into account labeling claims, advertising matter, oral or written statements by the manufacturer and its representatives, and known uses of the device to perform a function for which it is neither labeled nor advertised.

The "per se" rule applies to malfunctions as it does to reports of deaths and serious injuries. Whenever a health-care professional provides the company with information indicating that a malfunction has occurred and that it would be likely to cause or contribute to a death or serious injury if the malfunction were to recur, the company must report the event to FDA. If the health-care professional describes a malfunction with no mention of the possibility of a death or serious injury, the company must use the malfunction definition and the results of its own investigation to determine if a reportable malfunction has occurred.

As defined in the regulation, a malfunction occurs whenever a device fails to meet any of its performance specifications or otherwise fails to perform as intended. This not only includes written performance specifications, but also labeling claims, advertisements, and statements made by the manufacturer and its representatives. After it has been determined that a malfunction has occurred, the company must determine if the malfunction would likely cause or contribute to a death or serious injury if the malfunction were to recur. According to FDA, "likely" means more probable than not. This is not a statistical probability. The standard to be used should be both a quantitative and qualitative evaluation of the likelihood that a recurrence of the malfunction will cause or contribute to a death or serious injury.

According to FDA, the regulation presumes that a malfunction will recur, and thus a company must determine if it is more probable than not that a death or serious injury will result from any subsequent malfunction. FDA provides the following guidance in making such a determination:

1. Fail-safe systems. Does the device have a properly functioning fail-safe system that will prevent a death or serious injury if the malfunction were to recur? If so, the malfunction is not reportable.
2. Device performance. Does the device history provide documented evidence, available to FDA, that the malfunction in question has never resulted in a death or serious injury, or that the injury potential for this malfunction does not meet the test of "more probable than not" because of the low injury rate associated with the event and product? If so, the malfunction is not reportable.
3. Anticipated malfunctions. Any malfunction that is anticipated (i.e., its severity known and its rate of occurrence statistically predictable) is not reportable if the event is adequately described in the labeling and it is occurring at or below the rate established or usual for the device, as described in 21 C.F.R. 803.24(d)(iii).
4. Service, repair, and maintenance. Defects and malfunctions discovered during scheduled service, repair, and maintenance procedures are not reportable as malfunctions if the rate of failure and/or severity is both anticipated and usual for the device as indicated by 21 C.F.R. 803.24(d)(3)(iii). Requests for service, repair, and maintenance due to a malfunction are subject to the "likely to cause" conditions previously described.

5. Trends. Firms may detect a trend—that is, a pattern of failures that indicates a problem is recurring. Once a trend is identified, the event is subject to reporting as a malfunction if it is likely to cause death or serious injury. Trend analysis is an integral aspect of complaint review and essential to document a firm's rationale for not reporting selected events.

The MDR regulation provides for an exception to reporting a "reportable" malfunction if all of the following conditions are met:

1. A death or serious injury has not occurred.
2. The device's labeling sets forth information concerning the potential for death or the type of serious injury to which the malfunction may cause or contribute.
3. The device's labeling describes the malfunction and the routine service, repair, or maintenance instructions to correct the malfunction.
4. The malfunction has occurred or is occurring at or below the frequency and severity stated in the labeling for the device or, if there is not any pertinent statement in the labeling, at or below the frequency and severity that are usual for the device.
5. The malfunction does not lead the manufacturer or importer to undertake remedial action involving any device other than the device product in which the malfunction occurred.

E. When and How Must Reports Be Made to FDA?

The time periods for making MDR reports are for the most part straightforward. Deaths and serious injuries must be initially reported within 5 calendar days and malfunctions within 15 working days. Specifically, whenever a company receives or otherwise becomes aware of information that reasonably suggests that one of its marketed devices may have caused or contributed to a death or serious injury, the company must report the event to FDA by telephone within 5 calendar days of initial receipt of the information. The company must follow up the telephone call with a written report within 15 working days of initial receipt of the information.

Reportable malfunctions must be reported to FDA within 15 working days of initial receipt of the information. Telephone reports of malfunctions are not required, but may be made prior to submitting the written report.

In calculating time frames, the clock begins to run the day the company initially receives information that reasonably suggests a reportable event. Telephone reports of deaths and serious injuries are due by the fifth calendar day after the company receives the information. No allowances are made for weekends or holidays. Written reports must be forwarded to FDA (or post-marked) by the fifteenth working day after the company receives the information. Saturdays and Sundays, or any other two days that a company is routinely closed for business in place of Saturday and Sunday, and scheduled federal holidays are considered nonworking days.

One problem many companies have is deciding "when" they have enough information to decide a report needs to be filed and the time to file should begin. FDA advises that in its telephone or written report, the company should include its evaluation of the validity of the information or identify additional information that is needed or will be submitted concerning the reported event. It should be noted that an MDR report is required even if the company believes that the reportable event is due to user error, failure to service or maintain the device properly, or use of the device beyond its labeled useful life.

F. What Information Has to Be Reported to FDA?

The MDR regulation requires seven items of data to be submitted with every telephone or written report. FDA has issued the following guidance on the information it requires to be submitted in an MDR report:

1. Identify the device by providing the following information:

   Complete brand name of the device as it appears on the product's label.

   The common name (a generic description of the device) or FDA's classification name or code that is used when a firm lists its devices may be used.

   Model, lot, serial, catalog, PMAA (premarket approval application), and the firm's reference numbers are always useful and sometimes essential in determining the significance of an event. Firms should make every attempt to determine the model, lot, serial, catalog, or any other number that can be used to identify a device. If this information is not available when the initial telephone report is made, it must be provided to the extent known in the written report. Since MDR applies to devices with approved PMAAs, firms also need to provide FDA with the PMAA number when applicable.

   The MDR system was designed to allow a firm to provide information on its internal control number. A firm, however, is expected to use the number that FDA assigns to each MDR submission rather than its internal control number. The MDR number is assigned when a death or serious injury report is submitted by phone. For malfunction reports submitted by mail, firms are provided with an access number via an FDA acknowledgment letter.

   If the device has been the subject of a 510(k) submission, it would also be helpful to FDA if this fact and the 510(k) number were provided (this is not required by the MDR regulation).
2. Identify the manufacturer, or in the case of an imported device, identify the importer.

   Domestic firms should always provide the address of the site where the device involved is manufactured. While this may not be the same firm location from which MDR reports are transmitted to FDA, the identification of the manufacturing site is needed in the event further followup is required.

   In the case of an importer who is identifying a foreign firm, the importer should ensure that the identification is complete, including the date that the report was forwarded to the foreign manufacturer for evaluation, and that a copy of the foreign firm's evaluation will be returned to the importer. In addition, the importer should indicate whether it intends to forward a copy of the evaluation to FDA.
3. Identify by name, address, and telephone number the individual making the report to FDA.

This person is also known as the "MDR contact person." If the address and telephone number are the same as those of the manufacturing site discussed in item 2 above, please inform the person taking your call.

Individuals submitting reports should indicate any changes in their firm's MDR contact person, address, telephone number, or content of a "disclaimer" if one is used.

4. Describe, to the extent known, the event giving rise to the information received by the manufacturer or importer, including (a) whether any deaths or serious injuries have occurred and (b) the number of persons who died or were seriously injured. If the report is required by reason of a medical or scientific journal publication, the journal reference is to be given in the phone call and the written report is to include a copy of the article.

   Firms are required to indicate if the event is being reported as a death, serious injury, or malfunction. The date that the event occurred will be requested by FDA as "additional information" authorized under 21 C.F.R. 803.24(e), if the date is not reported in the 5-day or 15-day report.

   The type of injury is also a necessary item of information. The type of injury must be determined by the firm as part of its decision that an event is reportable under MDR. "Type" refers to the "serious injury" definition and, for convenience, has been broken down into the following levels for clarity:

   a. Death (use only for malfunctions that could lead to death)
   b. Life-threatening
   c. Permanent impairment of body function
   d. Permanent damage to body structure
   e. Requires medical intervention to preclude permanent impairment of body function
   f. Requires intervention to preclude damage to body structure
   g. Temporary impairment of body function
   h. Temporary damage to body structure

   Firms should attempt to obtain as much information as possible about the device and the circumstances involved in the alleged incident. In responding to the requirement to "describe to the extent known, the event giving rise to the information," this can include age, sex, weight of the patient, a complete description of the event, the language used by the reporter to describe the occurrence, and the like. In addition, the determination of an incident's reportability under MDR requires an evaluation of the medical significance, the health of the patient at the time of the incident, and the additional treatment required; this information is useful for both the firm and FDA.

A Government Accounting Office report indicated that 83% of all device problem-related reports are transmitted verbally between medical facilities and the manufacturer or distributor. Firms that receive most of their adverse reaction data by telephone often provide their staff with forms and/or a list of questions to minimize the loss of data associated with "word of mouth" reports. FDA believes that this is an excellent practice and that it aids the firm in determining the significance of incidents involving their devices, both to determine need for action on their own part and also in making a determination of the event's reportability under MDR. It is also important for the firm to have developed criteria that aid in making accurate and consistent decisions that a reportable death, serious injury, or malfunction has occurred based on the information gathered.

Some of the questions that can provide useful information for both firms and FDA include the following: age of the device, last service date, how the device was being used when the problem was observed or the event occurred, and any special or unusual environmental conditions surrounding the use of the device.

If any of the aforementioned (or any other necessary) information is not available when a report is first made to a firm, it should be obtained during the firm's investigation of the report and added to the 15-day written report or sent to FDA as a separate item.

5. Identify by name and address the person submitting the information to the manufacturer or importer.
6. State whether the manufacturer or importer intends to submit additional information and, if so, when such information will be submitted.

This allows a firm the opportunity to indicate if (a) any information that has not been provided under questions 1-5 will be submitted at a later date and (b) the firm intends to provide any additional data, beyond that required by MDR, to explain the possible cause or nature of the reported event. The provision of information at a later date gives firms an opportunity to clarify a reported event by means of their evaluation/findings. Additional data submitted by firms are evaluated by FDA analysts and added to the MDR text in the data base. This provides those requesting information under the Freedom of Information Act and others with the opportunity to read the firm's analysis/evaluation of the reported events and may mitigate any misunderstandings of the data.

Firms must indicate the nature of the information to be submitted—that is, lab results, failure analysis, lot numbers, etc.—and the approximate date that the data will be sent to FDA.

7. State whether the reported event has occurred or is occurring more frequently or with greater severity than is stated on the labeling for the device or, if there is no pertinent statement in the labeling, than is usual for the device, if such information is available.

This question should be approached in the following manner:

(a) If a firm has no information in its product labeling about either the frequency or severity of the event being reported, the answer to the first part of the question is "no," and the answer to the second part of the question is "no" if the firm does not maintain or have any knowledge about the frequency or severity of the reported event. It is unlikely, however, that firms do not have some information about what is "usual" for many of the events that are reported under MDR. At a minimum, many firms now have over 4 years of experience with MDR reports. This experience should enable them to determine if an event is occurring more or less frequently than "usual" and/or with more or less severity than "usual."

(b) If a firm has stated in its labeling the expected frequency (*X* percent) of an event (e.g., valve failure) and/or its severity (e.g., embolism), the firm is required to determine if the rate and/or severity is greater than listed on the label. This determination is based upon the firm's calculations of reported events and respective severity. The firm must also be prepared to explain to FDA how it arrives at its frequency and severity rates and how it determines what is acceptable.

(c) If a firm does not provide in its labeling any frequency and/or severity data, but does maintain information on or is aware of the frequency/severity of the reported event, then the second part of question 7 is answered in the same manner as described in part (b) above.

G. When Are Reports Not Required?

The MDR regulation and its preamble contain seven types of events that are not MDR reportable.

Certain events do not have to be reported. In no particular order, they are:

1. If a company receives independent reports of the same event involving the same patient, only the first event reported to the company needs to be reported to FDA. Subsequent reports do not have to be reported. However, similar events involving the same patient and identical events involving different patients must be reported.
2. Events involving general purpose articles not specifically labeled or promoted for medical use do not have to be reported.
3. Reports dealing solely with device effectiveness (that do not relate to death, serious injury, or malfunction) are not required to be reported.
4. Reports concerning devices manufactured and sold outside the United States are not reportable. This applies even if the foreign manufacturer is owned by a U.S. company. However, reports involving a device manufactured in the United States and only sold abroad and a foreign-manufactured device sold in the United States must be reported.
5. Reports made to distributors that are not a wholly owned subsidiary of the manufacturer need not be reported unless the information is conveyed to the manufacturer.
6. FDA may notify a company in writing that MDR reports of a particular type of event are no longer required.
7. A report is not required if before the expiration of the required time period, either 5 or 15 days, the company can determine that (a) a death or serious injury did not occur, or (b) the device that is involved in the death or serious injury was manufactured or imported by another company, or (c) the criteria for not reporting a malfunction referred to in Section II,C is met.

H. What Records Must the Company Keep and for How Long?

The regulation requires companies to retain copies of records of all information received by the company concerning an MDR reportable event. This information includes written or oral communications received by the company. The company must also retain a copy of any MDR report submitted to FDA. This includes:

The initial death or serious injury report submitted by telephone to FDA within 5 calendar days after receipt of information.

The written death or serious injury report submitted within 15 working days after receipt of information.

The written malfunction report submitted within 15 working days after receipt of information.

All subsequent follow-up data submitted to FDA voluntarily or as requested by FDA.

Any failure investigation report prepared pursuant to the good manufacturing practice (GMP) regulations.

The company must retain these records for a period of 2 years from the date that the MDR report or additional information was submitted to FDA or the period of time equivalent to the design and expected life of the device, whichever is greater. These records may be maintained as part of the company's GMP complaint file.

III. CONCLUSION

The MDR regulation is a complicated federal regulation. Companies required to comply with it need to understand the intent, the requirements, and FDA's interpretation of the regulation. Once the regulation is understood, a company should develop an MDR policy/procedures manual that provides guidance to those persons in the company responsible for submitting MDR reports to FDA. A typical MDR procedure manual contains such topics as the company's general MDR policy including an appropriate disclaimer to be included on every MDR report, a decision tree or chart to facilitate appropriate reporting decisions, directives to company personnel on who, what, when, and how to report to the designated MDR responsible person, the responsibilities of the company's MDR responsible person, and procedures on how to evaluate and follow up on MDR reports. Once this document has been developed, it should constantly be discussed and reviewed with appropriate personnel. As a company gains experience in complying with the MDR regulation, the procedure manual should be modified and updated to incorporate the information learned from that experience.

Medical device problem reporting is important—for the company and for FDA. Companies should not view the MDR regulation exclusively as an inconvenience forced upon them by FDA. Companies can benefit from the information gathered from their efforts in complying with the MDR regulation by using it to improve product quality and streamline customer service. Viewed in a constructive manner, compliance with the MDR regulation will benefit both the company and FDA.

ACKNOWLEDGMENTS

I would like to thank Edward Basile, Thomas DiPasquale, Melvin Drozen, Stanley Garber, Michael Gropp, Howard Holstein, and James Hulse for their ideas and suggestions.

BIBLIOGRAPHY

The following materials were used in preparation of this article:

Federal Food, Drug, and Cosmetic Act, 21 C.F.R. 301 et. seq.
Code of Federal Regulations, 21 C.F.R. Part 803.
48 *Fed. Reg.* 24014 (May 27, 1983).
49 *Fed. Reg.* 36326 (September 14, 1984).
Medical Device Reporting Manual, Vol. I, HIMA Report No. 84-6 (1984).
Medical Device Reporting Manual, Vol. II, HIMA Report No. 86.1 (1986).
Medical Device Reporting Questions and Answers, HHS Publication FDA 88-4226 (February 1988).

26

FDA/GMP/MDR Inspections: Obligations and Rights

RODNEY R. MUNSEY and HOWARD M. HOLSTEIN *Hogan & Hartson, Washington, D.C.*

I. INTRODUCTION

Each year, the Food and Drug Administration (FDA) inspects establishments involved in the manufacture, processing, and distribution of foods, drugs, and medical devices. The purpose of these inspections is to protect the public health and safety by assuring the quality of products that fall under FDA's jurisdiction, including the safety and effectiveness of medical devices. Although extensive device regulation is a recent addition to FDA's bailiwick, device inspections are similar in many respects to other inspections performed by FDA. In general, manufacturers of Class II and III devices can expect to encounter an FDA investigator at least once every 2 years, and more often if problems arise with the company or the products it manufacturers.

This chapter will provide some basic information about the rights and obligations of medical device companies during and following an FDA inspection. First, this chapter will briefly describe FDA's authority to conduct inspections. FDA's field force, their powers, and the guidelines investigators follow during inspections will be addressed in the sections that follow. Finally, this chapter will focus on the methods that companies use to coordinate, manage, and respond to FDA inspections.

II. FDA'S AUTHORITY TO INSPECT

The FDA's power to inspect originates in Section 704 of the Federal Food, Drug, and Cosmetic Act (the FDC Act). This provision allows FDA officials to inspect "any factory, warehouse, or establishment in which . . . devices . . . are manufactured, processed, packed or held, for introduction into interstate commerce or after such introduction." In addition to the "establishments" specification in Section 704, FDA is permitted to enter any vehicle used to transport or hold regulated products for export or in interstate commerce.

This inspection power is specifically extended to medical device manufacturers by Sections 519 and 520 of the FDC Act.

In addition to the foregoing inspectional authorities, Section 519 also grants to the Secretary of Health and Human Services (who has delegated this authority to FDA) the power to require reports from manufacturers, importers, or distributors of medical devices. This provision, which supported the promulgation of the medical device reporting (MDR) regulation, was designed to help assure the safety and effectiveness of devices. By providing authority to require reports from device companies, Congress provided that FDA would be alerted to any significant or potential problems that could arise with respect to devices. Thus, with early notice, FDA could trigger an inspection to detect and prevent further risks or harm to the public health.

Finally, Section 520(f) of the FDC Act grants FDA the power to promulgate regulations "requiring that the methods used in, the facilities and controls used for, the manufacture, packing, storage, and installation of a device conform to current good manufacturing practice." Because compliance with good manufacturing practices (GMPs) and medical device reporting are important issues arising in the context of medical device inspections, this chapter will focus on these areas with respect to FDA inspections.

A. The Statutory Basis for Inspections

Every FDA investigator is authorized by law to inspect all equipment that is used in the manufacturing process. Furthermore, investigators may examine finished and unfinished devices and device components, containers and labeling for regulated products, and all documents that are required to be kept by the regulations (e.g., device master records and device history records, MDR reports).

For restricted devices[1] the inspectional authority is broader than for all other devices. Because restricted devices inherently have a greater potential for "harmful effect," FDA believes that they pose a more significant risk to the public health than other devices and that greater inspectional powers are justified. This power, under Section 704 of the FDC Act, allows FDA investigators to inspect anything within a "factory, warehouse, establishment, or consulting laboratory" where restricted devices are "manufactured, processed, packed, or held." The inspectional authority extends over "records, files, papers, processes, controls, and facilities" which "may have bearing on whether the restricted devices are adulterated or misbranded," or otherwise in violation of the FDC Act.

Despite the broad inspectional authority over restricted devices, the statute provides that regardless of the device's restricted status, certain information is excluded from FDA's inspectional ambit. The kind of information to which FDA does not have access includes financial data, sales data (other than shipping data that provide the destination of devices), pricing data, personnel data (except information on the qualifications of technical and professional personnel performing job functions that are subject to FDA's jurisdiction), and research data (excluding data relating to new drugs, antibiotic drugs, and devices subject to reporting and inspection under certain provi-

[1]Restricted devices can be distributed and sold only upon the written or oral authorization of a licensed practitioner or upon any other conditions prescribed by FDA regulations. See 21 U.S.C. 360j(e).

sions of the FDC Act). Thus, FDA may request to inspect research data that are a part of or form the basis for a submission to FDA, such as an application for investigational device exemption (IDE) or a premarket approval application (PMAA).

As previously noted, Section 703 of the FDC Act does permit FDA to obtain copies of records that show the shipment of devices in interstate commerce. For medical devices, however, this provision is less important than for other FDA-regulated products. The Medical Device Amendments of 1976 (the Amendments) presume interstate shipment and although the presumption is rebuttable, it is both factually unlikely and extremely difficult to demonstrate that a device is outside FDA's jurisdiction because it is manufactured and distributed wholly intrastate.

Section 703 also permits FDA to gain access to and obtain copies of records showing the movement of products from persons carrying, holding, and receiving devices in interstate commerce. These people cannot lawfully refuse a written request for such records. The FDA, however, offers some protection by prohibiting these records from being used as evidence in a criminal prosecution against the individual who complied with the FDA request.

Every device establishment that is subject to the registration requirement (Section 510) is subject to inspection under Section 704. This includes establishments, not otherwise exempt, that are engaged in the "manufacture, preparation, propagation, compounding, assembly, or processing of a device intended for human use." Registration requirements also apply to any person who initiates or develops specifications for a device to be manufactured by a third party, manufactures a device for distribution by another person or company, repackages or relabels a device, initially distributes a device imported into the United States, or manufactures components or accessories that are ready to be used for any intended health-related purpose and that are packaged or labeled for commercial distribution for such a purpose or that must be further manufactured by a licensed practitioner to meet the needs of a particular patient (e.g., contact lens "blanks").

FDA does routinely inspect most manufacturing facilities every 2 years and may inspect more often if there is cause to suspect problems. At present, for device companies, FDA has established a two-track inspection program that provides for less in-depth inspections of firms whose routine inspections did not reveal any serious compliance problems. This program provides for in-depth GMP inspection once every 4 years.

B. The Reasonableness Standard

Although FDA's inspectional authority is broad, the FDC Act does impose certain limitations on the conduct of inspections under Section 704. FDA investigators may inspect only within "reasonable" limits and may enter facilities only at "reasonable" times. Inspections are to be started and completed with "reasonable" promptness. Since there is no clear definition of reasonableness for the purposes of the Act, courts have been willing to provide FDA with a great deal of latitude in making decisions about the reasonableness of inspections. The FDA Inspection Operations Manual (IOM),[2] for example, states that

---

[2]The IOM is the investigator's "bible," containing general information on FDA inspectional policies, procedures, and specific information on various areas of concern to FDA, for example, contract sterilization. It is available to the public.

although a court may ultimately determine what is reasonable, FDA "assumes this authority extends to what is reasonably necessary to achieve the objective of the investigation."

Generally, the "reasonableness" language of the statute has been interpreted as permitting FDA to inspect a facility during normal operating hours; this may mean that FDA can inspect companies that operate around the clock in the middle of the night. As part of the reasonableness standard FDA inspectors are also directed to adhere to the same personnel practices as those followed by company employees. Inspectors routinely don the same protective clothing that employees in a manufacturing area wear and, in fact, should not be permitted to enter an area if not properly attired.

It is also incumbent upon the inspector not to interfere with the normal production process. It is improper for inspectors to ask that production be halted to allow them to take samples or inspect equipment that is in operation. Companies are likewise not expected to run a manufacturing process or operate specific equipment if they would not ordinarily do so. In most circumstances, FDA investigators will neither ask for nor expect manufacturers to deviate from their normal operating procedures. However, in the event of such a request, it is reasonable for the manufacturer to contact the local district office for guidance.[3]

Most FDA regulatory counsel and experienced regulatory affairs personnel recognize the wisdom of attempting to cooperate to the greatest degree possible with FDA investigators. This is the most practical approach. Manufacturers must remember that FDA does have a great deal of inspectional authority and that it is seldom wise to "step on the tail of the lion." It is important, however, for companies to remember that FDA investigators are, in essence, law enforcers whose job it is to find violations and, if possible, prosecute the offenders. A company must know its legal rights and obligations with respect to FDA inspection and should have established procedures for handling investigators and the inspection process.

III. FDA'S FIELD FORCE

FDA inspections are usually conducted by consumer safety officers (CSOs) or investigators from one of FDA's district offices or resident inspection posts. As of August 1988, FDA had six regional offices, 21 district offices, and 130 resident posts in the United States.[4] The district offices are headed by a district director who is responsible for operations from that office. Generally, district offices consist of a compliance branch and administrative branch, among others. District personnel may also include experts or specialists in certain specific areas, as well as the field investigators who perform inspections.

FDA inspection personnel are not limited to CSOs from district offices or resident posts. Personnel from other divisions of the Center for Devices and Radiological Health (CDRH) may also be involved in inspections for

---

[3]See Section IV.D regarding the appropriate times for contacts with the district and headquarters during FDA inspections.

[4]Due to the closing of several district offices this number has decreased slightly.

a number of reasons. These individuals may be in training, or may be assigned to deal with specific technical or scientific issues. In addition, FDA has engineers on staff who may accompany investigators on inspections of new facilities or may be called in to provide input on facility design or other technical areas in cases where design or other problems are believed to be an issue.

FDA investigators have a wide range of backgrounds and varying expertise. All investigators must have a college degree, and many have degrees in scientific or technical areas. Even the newest investigators have gone through an FDA training program and will complete a certain number of inspections under the supervision of an experienced investigator. All FDA investigators receive periodic retraining in inspectional techniques. Further, FDA retraining sessions may focus on specific areas of current interest to FDA, such as sterilization processes. In some districts, FDA investigators are selected for inspections based on their particular area of specialization; for example, some investigators are experts in the field of inspections of in vitro diagnostic manufacturers, while others have received specialized training in computerized manufacturing processes and computerized devices. In other districts, most investigators are generalists and will find themselves equally at home in a food processing plant or a device facility. If the investigator is willing to discuss his or her background it is always useful to be aware of this information. It is often possible to tactfully direct the conversation during the initial discussion, to enable the company to provide some background information to an investigator who is not familiar with a particular type of device or facility. Provided it is handled properly, many investigators will appreciate a company's efforts to familiarize them with a new area.

The FDA field force that is responsible for performing inspections of foreign device manufacturers is called the Division of Field Investigations of the International Programs and Technical Support Branch and is located within the Office of the Associate Commissioner for Regulatory Affairs. The Division, with assistance from selected members of the district offices, performs this task worldwide. Because of limitations on available staff, inspections of foreign manufacturers are much less frequent than inspections of U.S. companies and are generally performed after advance notice of impending inspection. However, investigators who inspect foreign establishments must meet the same requirements as U.S. investigators. In addition, these individuals must have the ability to work in unfamiliar situations and deal with foreign nationals. Although foreign inspections will not be covered here in great detail, the inspection process abroad is unquestionably somewhat less stringent.

IV. BEGINNING THE INSPECTION

The company's ability to handle an inspection from the moment an FDA investigator enters the front door of the facility is key to FDA's perception of the company. This section provides some practical advice as to approaches that have been successful in demonstrating to investigators that the company is operating in a state of control. Naturally, the following is not the only possible approach. If your company's procedures have worked well in the past, they may in fact be the most appropriate for the district and the type of inspection involved.

A. The Inspection Policy and Manual

It is important for companies that are subject to FDA inspection to implement written policies for these occasions. The inspection policy should designate the individual who is responsible for meeting the investigator and coordinating the inspection. Further, the policy should provide a list of the other individuals in the company who should be notified when an investigator enters the facility and the individual who should notify them. The policy should explicitly identify the individual who is authorized to receive the Notice of Inspection and the individual who will receive the FD-483 containing the list of observations that is issued at the end of almost every inspection. The policy should also address the company's position with respect to photographs, affidavits, copying documents, and product sampling. By expressing company policy ahead of time, employee confusion will be eliminated and the inspection will go more smoothly.

Another useful aid to have on hand for inspections is a notebook or folder that contains specific information that most investigators will ask for in their preliminary discussions with a company. This folder should include a history of the business, a copy of the organizational chart for the operating entity and the parent company (if any), a list of products or a current catalogue, and any other general background information that may be of value. This information should be kept, optimally in duplicate, in the reception area of the facility and another copy should be filed by the inspection coordinator.

The selection of an inspection coordinator is a very important one. The inspection coordinator should be an individual who is well versed in FDA regulations and who is familiar with the most recent developments in the compliance area. Training is available, through trade and professional associations, for individuals who will act as inspection coordinators, and periodic updates provide current information on evolving issues. It also is important for the inspection coordinator to be quite knowledgeable about the company's facilities, the devices manufactured, and the manufacturing process itself. With this kind of background, the individual will be able to either answer the investigator's questions directly or refer them to the best-qualified individual. Beyond these capabilities, however, the inspection coordinator must be a personable individual who is not afraid to set out the ground rules for the inspection and will refuse the investigator access to records not subject to inspection in a firm yet nonconfrontational manner. Perhaps the most important attribute of an inspection coordinator is the ability to "sit down and shut up." It is important to answer the investigator's questions fully and in a manner that projects that the company has nothing to hide. It is equally important, however, to stop talking when the answer is complete. Never volunteer unasked-for information to an FDA investigator.

B. The Notice of Inspection

Companies are not usually given advance notice of inspections, except in the case of foreign inspections and, on occasion, when FDA is involved in a fact-finding mission. There are, however, certain situations in which FDA inspection can be anticipated, even if the precise time cannot be predicted. If a company has filed an MDR report relating to a death or serious injury or has notified FDA of the recall of a product due to serious adulteration or misbranding issues, FDA often will inspect, and the company should be prepared and waiting. Companies are often surprised, in fact, at the length of time that elapses between a report to FDA that such a problem exists and the appearance of an investigator at the door. Because the field force is often understaffed and frequently has a number of top priority problems to handle, inspectors may not arrive for several months, or even for more than a year after the problem is reported. However, regardless of whether the inspection is routine or for cause, it is inevitable.

Companies should be aware that, although the inspectional process is usually a surprise, it should not be a cause for panic. Inspection days are not the time to try out new processes or procedures. Companies should not implement programs that are not in regular use, but are designed and intended to impress an investigator. In keeping with the laws of Murphy, everything that can go wrong will. Naturally, the wrong impression will be created. Rather, it is best to emphasize to all personnel the importance of continuing with the usual routine during the course of an inspection.

At the beginning of the inspection, the inspector will present his or her credentials and ask to meet with the owner, operator, or the most responsible person on the premises. It is often desirable to have a high-ranking or senior person greet the inspector. This gives the investigator the impression that management is concerned and involved with FDA inspectional activities and is prepared to cooperate with FDA. It is not necessary to have this individual present during inspection, although he or she may return for the exit interview. In fact, it is best to turn the inspection over to the inspection coordinator at the earliest possible time.

In addition to making a record of the investigator's name and badge number, and obtaining a copy of the notice of inspection (FD 482), the inspection coordinator should attempt to ascertain the purpose of the inspection as quickly as possible. By determining the purpose of the inspection, the coordinator can establish an agenda for the inspection that will make it possible for the investigator to accomplish his or her goal in the shortest possible period of time. Thus, the inspection coordinator should assure that appropriate individuals will be available as needed, that documents will be efficiently retrieved, and that the processes the investigator is interested in reviewing are, in fact, taking place.

It is appropriate for the company to suggest a sequence for the inspection. Some companies, for example, will suggest that the investigator follow a product from raw materials through finished product testing and distribution. It is usually wise to suggest that the investigator perform the walk-through portion of the inspection first. Then, later, the inspector can review the relevant paperwork in a conference room or other area that is removed from the actual manufacturing process. The inspector, of course, has the authority to make the final decision regarding the course of the inspection and may, if he wishes, choose to spend his entire visit on the floor of the production area. Many investigators are, however, amenable to gentle direction from the inspection coordinator.

C. Warrants, Photographs, and Tape Recordings

As most companies are aware, FDA is not required to obtain a warrant prior to initiating inspection of a regulated facility. In contrast to other types of administration inspections (such as those involving housing code violations, for example), inspections of industries that are "pervasively regulated" can be performed without a warrant. Within the scope of the warrantless inspection, FDA has the right to observe manufacturing operations and to obtain copies of documents that are required by the statute and regulations to be kept. Several situations, described below, illustrate the circumstances under which FDA will seek a warrant to permit a portion of an inspection.

One of the most significant issues today regarding FDA's inspectional scope is whether an investigator has the authority to take photographs during the course of an inspection. FDA's position is that photographs are within the definition of the "reasonable" inspection permitted by Section 704. Furthermore, FDA states that there is substantial judicial precedent to permit FDA to request and receive permission to take photographs.

Conversely, most companies have established a policy that photographs are not permitted by the FDC Act. In recent years, the U.S. Department of Agriculture (USDA) has received explicit legislative authorization to take photographs during inspections. FDA, however, has not received similar authorization. Companies have a number of reasons for prohibiting FDA from taking photographs in production facilities. Aside from the obvious objection to permitting FDA to graphically document violative conditions for later display in court, many companies have a blanket policy preventing photographs as a means to protect proprietary manufacturing processes and to prevent the release of information related to new, developmental products. Moreover, many companies fear that the photographs, once taken, may be disclosed to a company's competitors under the Freedom of Information Act (FOIA).

Regardless of the basis for the company's objection, a recently decided Supreme Court case is often cited by FDA as the basis of its authority. In *Dow Chemical Company v. United States* [1] despite the company's refusal to permit on-site inspection, the Supreme Court upheld the authority of the Environmental Protection Agency to fly over and photograph a chemical plant. In this case, the Court held that aerial photography did not constitute an illegal search prohibited by the Fourth Amendment. It is not clear, however, whether taking aerial photographs when on-site inspection was refused is analogous to taking photographs within a facility during the course of a factory inpsection. Some critics maintain that there is a different expectation of privacy within a building than that which exists with respect to an outside area that is open to public view.

Prior to *Dow*, FDA investigators cited as authority for taking photographs *U.S. v. Acri Wholesale Grocery Co.* [2]. In *Acri*, a U.S. District Court ruled that FDA could use as evidence photographs taken during inspection of a food warehouse. The company had, however, given the FDA investigator permission to take photographs, but later asserted the photographs could not be used in evidence. Rather than providing authority for FDA to take photographs without the concurrence of management, *Acri* merely stood for the proposition that, once taken, photographs could be used as evidence.

In accordance with its own interpretation of the *Dow* case, FDA has revised the IOM to instruct investigators that they should ask for permission to take photographs and, if refused, should explain that photographs are an integral part of inspection and permitted by *Dow*. If the company continues to refuse permission to take photographs, the IOM instructs investigators to contact the district office for a decision on whether a warrant should be sought to compel permission to take photographs. FDA's pre-*Dow* position held that if a company refused permission to take photographs, the investigator would advise management that this would be noted in the establishment inspection report (EIR) as a refusal to permit a portion of the inspection.

However, at that point, the matter would be dropped and the inspection would be completed without photographs. This change in FDA's policy signals that FDA is now willing to pursue the matter of the agency's authority to require companies to permit photographs to be taken during inspections.

In light of FDA's policy change, our advice to companies with respect to the issue of photographs must likewise be modified. Although previously, regulatory counsel and regulatory affairs personnel often recommended blanket refusal of permission to take photographs, the same advice msut now be given with a caveat. It is unlikely that FDA will seek a warrant or take action against every company that refuses photographs. This is particularly true for companies that are substantially in compliance with FDA regulations, because in those cases the purpose of the photographs would less clearly be related to FDA efforts to preserve evidence for later litigation. It is equally true, however, that FDA will unquestionably be looking for test cases. Thus, if a particular company is guilty of significant GMP violations, a refusal to permit photographs may result in further action by FDA.

As previously noted, FDA may seek a warrant to permit the taking of photographs by the same procedure the agency uses to seek a warrant if entry to a device facility is refused. In addition, FDA can also seek prosecution for refusal to allow inspection if photographs are not permitted. In the latter case, it is not clear whether refusal to allow photographs is a valid basis for prosecution under the Act.

With respect to warrants, FDA can seek either an administrative warrant or a "search and seizure" warrant from a district court judge or magistrate. The administrative warrant is used by FDA in routine cases. It is based upon a showing that the company is regulated by FDA, devices in the facility are held, stored, or manufactured for interstate commerce, FDA has reason to inspect, and the company refused FDA entry. A search and seizure warrant, on the other hand, is not grounded on Section 704 and will permit inspection far in excess of that permitted by Section 704 or that usually performed by FDA. Such a warrant, however, does require a showing of probable cause to believe evidence of a crime will be found.

Companies should be aware that a significant legal controversy exists over the authority of FDA investigators to take photographs. In addition, a company's refusal to allow photographs may cause the investigator concern that the company is attempting to conceal information or is trying to obstruct the investigator in the performance of his duties. The decision on whether to allow an FDA investigator to take photographs must be made carefully and, as previously noted, should be made before an investigator enters the facility. If a company decides to permit photographs, management should be made aware before an investigator enters the facility. If a company decides to permit photographs, management should be sure to claim confidentiality for the contents of the photographs in order to try to prevent their release under the FOIA. This does not, however, provide absolute assurance that FDA will not release the photographs.

FDA investigators, in contrast, seldom request permission to tape record any aspects of an inspection, since the IOM directs investigators not to request permission to record inspection proceedings. Likewise, it is the policy of most companies that neither the company nor FDA should be permitted to tape inspections. From the company's perspective, tapes, like photographs, can be used as evidence against a company in court.

Despite the fact that photographs and tape recordings may be refused, FDA investigators are free to record all of their inspectional observations, including, if they wish, verbatim records of conversations or word-for-word copies of confidential documents. Investigators typically carry a bound notebook for this purpose, and it is from these notes that the observations appearing on the FD 483 and the narrative EIR are prepared. Companies do not have access to investigators' notebooks and have no right, under the usual circumstance, to be made aware of the contents.

The inspection coordinator should also take notes for the company's benefit during the course of the inspection. These notes, which will be discussed in greater detail below, will help company management recap the inspection at the end of each day, and will serve as a reminder of anything the inspector requested but did not receive, as well as helping to prepare for the next day's inspection and the exit interview. The company notes should consist of all questions asked by the inspector, the company's responses, any observations that the inspector makes, and a list and copies of all documents reviewed by or copied and given to the inspector, as well as information on any areas or operations viewed by the investigator.

V. MANAGING AN INSPECTION

A. Inspection Responsibilities

As discussed, it is the responsibility of each company to select an inspection coordinator and several alternates, whose duties include accompanying the FDA investigator at all times, providing documents the investigator requests when appropriate, and responding to the investigator's questions or directing the questions to the most knowledgable person in the firm. The inspection coordinator also will usually make decisions about immediately correcting conditions that the investigator considers deviations from GMPs and will hold a daily meeting with management during the course of an inspection to go over the activities of the day, to delegate responsibilities for corrective action, and to plan for the remainder of the inspection. Finally, it is the inspection coordinator's responsibility to properly document the inspection and to prepare the company response to the FD 483.

In order to most effectively "cover" an inspection, it is valuable to have, in addition to the inspection coordinator, another individual present whose job it is to take notes, to obtain documents and other materials requested by the investigator when appropriate and assure that copies are made and filed of each document reviewed or copied by the investigator, and to generally assist as a link with the "outside world." This individual need not be a decision maker; in fact, anyone with basic clerical abilities can perform this task with proper instruction.

B. Inspection Venue

The optimal way to manage an inspection is to direct the investigator immediately upon his entry into the facility to a conference room or other nonproduction area, which serves as a base for the inspection. If possible, with the exception of the walk through, the remainder of the inspection can best be handled if the inspection coordinator directs the investigator back to the base and allows the investigator to review documents there. As questions arise, individuals with the most expertise in that particular area can be called to the base to respond to the investigator's questions. The less an investigator is permitted to wander around a facility, the greater the likelihood that the company will be able to continue normal operations without concerns that GMP deviations will be detected. Although, as previously noted, it is impossible to confine an unwilling investigator to a conference room, many investigators are quite responsive to having inspections handled in this fashion, since it is easier to review documentation in this manner. As long as the investigator does not feel that the company is attempting to keep him or her out of a given area or to conceal conditions or information, few difficulties are encountered with this approach.

C. Responding to Inquiries and Requests for Documents

A key quality that is desirable in each individual who will have contact with the FDA investigator during the course of an inspection is the ability to understand the questions that are posed, respond to them directly without providing extraneous information, be willing to admit when the answer to a question is not known, and otherwise remain quiet. If a question that the investigator poses is not clear, it is preferable to ask for an explanation, rather than to attempt to answer. One of the possible ill effects of trying to answer a question that one does not truly understand is that the answer may be wrong or it may give the investigator the impression that the company is trying to mislead the investigator. The consequences of generating mistrust can range from an inspection that lasts two or three times as long as normal to possible criminal prosecution.

The heart of most FDA device inspections is the investigator's review of the documentation generated by a company, specifically those policies, procedures, and records that are required as part of GMPs. It is wise for a company to know and to set forth in its inspection policy those records to which an FDA investigator has access, as well as those that can and sometimes should be refused. An FDA investigator is generally entitled to inspect and copy all records required to be kept under the GMP regulations, including:

1. Documentation of the quality assurance program
2. Audit procedure
3. Records that audits have been performed
4. Records of personnel training
5. Documentation of inspections of any required environmental control systems
6. Procedures, schedules, and documentation for facility cleaning and sanitation
7. Procedures for contamination control
8. Procedures and schedules for maintenance, adjustment, and cleaning of equipment
9. Records of equipment maintenance activities
10. Records of equipment inspections
11. Limitations or ranges for equipment operating parameters
12. Procedures for use and removal of manufacturing material
13. Procedures for calibration, inspection, and checking of measurement equipment

14. Records of calibration, inspection, and checking of measuring equipment
15. Documentation of testing of computer software programs for calibrating, inspecting, and checking measurement equipment
16. Procedure for approval of changes to software programs
17. Procedures and limits for accuracy and precision in calibration of measurement equipment
18. Records documenting traceability of calibration standards
19. Procedures for acceptance of components
20. Records of acceptance or rejection of components
21. Records of the disposition of obsolete, rejected, or deteriorated components
22. Procedures for accepting, sampling, testing, and inspecting lots of critical components
23. Records of percentage of defective components for each lot and the percentage of lots rejected.
24. Records of agreements with suppliers of critical components
25. Procedures and specifications for manufacturing processes
26. Procedures for specification control measures to assure that device, components, and packaging design result in approved specifications
27. Documentation of specification changes
28. Procedure for approval of change of manufacturing process
29. Record of critical operations in device history records by responsible individual
30. Procedure for reprocessing of devices or components; procedures for reprocessing of critical device or component
31. Documentation of the effect of reprocessing in the constant reprocessing of a critical device or component
32. Procedures for approval to institute or alter any reprocessing procedure for critical devices or components
33. Procedures for testing and sampling during reprocessing and control of critical devices and components
34. Record of examination of labeling materials
35. Record including signature and date for proofreading of labels or other labeling of critical devices
36. Procedures for warehouse control and distribution of finished devices
37. Records of distribution of critical devices
38. Procedures including adequate directions for installation of critical devices
39. Procedures for finished device inspection and sampling plans
40. Record of investigation of critical device or component that does not meet specifications
41. Record of check of acceptance records and documentation for critical devices and components
42. Records of investigation of failure of device or components after release for distribution
43. Device master record (DMR)
44. Device history record (DHR)
45. Records of written and oral complaints
46. Record of complaint investigation or reason no investigation was performed
47. Complaints pertaining to injury, death, or any hazard to safety
48. Reports filed to comply with the MDR regulation

As previously noted, FDA is also entitled to inspect documentation required by the IDE regulations, including:

1. Correspondence relating to the study
2. Records of device shipment and disposition
3. Signed investigator agreements
4. Records concerning adverse effects
5. Any other records FDA requires to be maintained, including, for non-significant risk studies, a brief explanation of why the device does not present a significant risk, the name and address of each investigator and IRB, and a statement of the extent to which GMPs will be followed

Following is a list of items that FDA is not permitted to inspect:

1. Financial data
2. Sales data
3. Pricing data
4. Personnel data, except qualifications of technical and professional personnel
5. Research data
6. Internal audits, reports, and memoranda on inspections, except those relating to a device defect or failure
7. Distribution records
8. Shipment records for materials received, unless the investigator offers a written request

As previously noted, investigators are allowed to take copies of any document they are allowed to inspect. In the interest of efficiency and orderliness, it is easiest to make all copies at one time, such as at the end of the day. The inspection coordinator or assistant should assure that two copies of each document are made, one for the investigator and one for the company's records. Before the copies are turned over to the investigator, any proprietary or confidential information should be marked.[5] Some investigators will initial the original of each document they review or copy on the reverse side, while other investigators will not do so. It goes without saying, of course, that prior to producing a document for review or copying, the inspection coordinator should assure that it is the most recent revision of the document, that it contains approval signatures and effective date, and that it does not contain extraneous notes or other markings on it.

With respect to an investigator's request to review complaint files, it will be in the company's best interest if the inspection coordinator attempts to avoid a fishing expedition. The investigator should be asked to name the products for which files are desired or the period of time the records span. It is important that the company's complaint policy contain a clear definition of complaint that is in accord with the meaning ascribed in the GMP regulations.

If the company maintains complaint files at a site other than the manufacturing facility—for example, at a distribution center—a duplicate copy of the

---

[5]Companies should be judicious in marking materials they consider confidential, since FDA makes the final decision as to what should be released. Unnecessarily broad claims by a company are more likely to be disregarded by FDA.

record of the investigation of any complaint must be transmitted to and maintained at the manufacturing site in a designated file. It is most important that complaint files be well organized and up-to-date and that appropriate records of complaint investigations be available and be performed in timely fashion.

Many companies have no policy or are otherwise unclear as to how they should respond if the investigator asks to see documents that the company is not obligated to provide. In many situations, it may not be wise to deny access outright, particularly if the information is already in the public domain. Unreasonable refusals to provide information, even if the information is not required by law to be provided, will be documented by the investigator as refusal of part of the inspection. Particularly with respect to sensitive areas, companies may wish to confer with their regulatory counsel prior to refusing to produce documents. If it is decided to refuse the investigator access to a particular type of document, it may be wise to advise that the company's legal counsel has made the decision that the document should not be provided. This tactic may help to prevent the investigator from developing animosity toward the inspection coordinator or other company officials.

D. Samples and Affidavits

FDA investigators are empowered to request and receive samples of raw materials, in-process materials, finished products, and labeling. Product samples may be used by FDA as evidence of adulteration, misbranding, insanitary conditions, or other violations of law; nevertheless, FDA does have an absolute right to obtain such samples. The inspection coordinator should take a duplicate sample for company testing purposes or simply to retain until FDA test results, if any, are obtained.

For each sample that is taken, the investigator must provide the company a receipt for samples, form FD 484. The sample product code, lot number, and number of units taken should be shown on the receipt. Although some companies will sign a "sample affidavit" indicating that a sample has been obtained, the company is not required to do so. It is also important to note that companies are permitted to charge for samples, and many companies do so if the amount of sample is great or the cost is high. If the samples involve reasonable quantities, many companies provide them to FDA free of charge.

If FDA takes samples for testing, the agency is often willing to provide the company with the results of any testing that it performs. A written request to the district office usually suffices to obtain test results. Alternatively, the company may file an FOI request for test results. Any company whose products are sampled by FDA should definitely try to obtain a copy of test results.

In addition to the "sample affidavit," FDA investigators will sometimes prepare affidavits relating to observations made during the inspection and will ask the inspection coordinator or other company official to sign. These affidavits usually relate to possible violations of the law that have been identified during the course of the inspection. Such an affidavit can be used later in legal proceedings against the company and will be admissible as an admission of wrongdoing by the company.

There is no legal obligation for a company to sign any affidavit under any circumstances, and it is the wise company that makes it a part of its inspection policy to refuse to sign any and all affidavits. If an investigator is persistent in asking for a signature or initials on an affidavit, the inspection coordinator may defuse the situation by asking that a copy, unsigned, be provided so that the company's legal counsel can review it before signing. At this point, most investigators will cease insisting.

Not only should a company official refuse to sign an affidavit, but some companies instruct employees to refuse to read an affidavit or to permit the affidavit to be read to them by the investigator. The reason for this stringent policy is that if the official indicates by word or action that he or she agrees with the contents of the affidavit, the investigator may note that response as an affirmation of the information contained thereon. Thus, the consequences, for legal purposes, may be the same as a signature. The only document that the inspection coordinator should sign during an inspection is a certification that quality assurance audits are being performed.

E. Controversies Arising During Inspection

Although it is not particularly common during FDA inspections, on occasion some controversy will arise between the investigator and the company as to what GMP compliance actually entails or what FDA's policy is with respect to a given area. In such a case, a company's first response may be to consult outside experts in the area or its own legal counsel for an interpretation. A question arises, however, as to the course the company should puruse if its judgement is confirmed, but the investigator cannot be persuaded to retreat from his or her position. In such a case, companies often wonder if it is appropriate to contact the district office or CDRH.

As a general rule, unless a controversy cannot be resolved after considerable discussion, the district office and CDRH should not be brought in. If, however, the company believes the investigator's conduct or position is totally unreasonable or unfounded, the inspection coordinator should inform the investigator that the company plans to contact the district. This is not a threat that should be employed lightly, and it must be used with the understanding that the district may well back its investigator, leaving the company with no immediate recourse. In certain situations, however, where an investigator is clearly out of line, a call to the district may rapidly resolve the problem.

Contact with CDRH is appropriate only for matters of policy and should be undertaken during inspection only with the express permission of the investigator and often the district office. Without express permission, CDRH will not be willing to talk with a company while an inspection is taking place. One example of a case when contact with CDRH was approved involved an inspection where an issue of sterility levels was raised by the investigator but could not be resolved. In this case, the investigator contacted CDRH and asked the resident expert there to speak to directly with company officials. In the usual case, however, a company should neither ask for nor expect to receive permission to contact CDRH during inspection. Policy issues that require CDRH expertise should be resolved either prior to or following the actual inspection.

F. Recall and MDR Inspections

A company faces a slightly different situation when it is the target of a directed inspection rather than the subject of a routine GMP inspection. In many cases, if FDA becomes aware of a product recall or MDR report related to a device, the agency will follow up at the time of the next GMP inspection. It is also possible, particularly in a situation where multiple recalls or MDR reports are involved and the products at issue have caused serious injuries or death, that an FDA investigator will pay a special visit to follow up. In this situation, the investigator does not have to look for problems; the problems are already apparent. How the company handles the inspection or this portion of the inspection, however, will be most important in the sense of damage control.

When a company has experienced a problem with one of its products that is serious enough to merit recall or MDR reporting, it should prepare adequate documentation at once to be provided to the investigator. This information should be gathered in a file folder or notebook, properly labeled and stored until needed. Included in this information will be a description of the problem, the cause of the problem if known, any investigation performed into the source of the problem, and the results of that investigation. A list of possible corrective actions to prevent recurrence of the problem should be compiled and, if appropriate, a solution should be implemented and documented. The records should also contain copies of any customer complaints related to the product, the total number of products in distribution, the account numbers or other identifiers of customers who received the product, and any letters or documentation of contacts with customers in which the company has been involved.

If recall of the product is appropriate, the file should contain records showing field correction or return of defective product and disposition of returned product. Also, for recall, the company should complete a health hazard evaluation that FDA can use in evaluating the class of the recall and determining what further action, if any, the agency needs to take. The company should use its technical expertise to make a convincing argument to FDA that little or no health hazard exists and that as a result of the actions the company took no further action is required. It is not good strategy to allow FDA on its own to make such determinations, although if the agency wishes to disagree with the company's assessment, it can, of course, do so. Again, it goes without saying that in a case where a company anticipates an FDA inspection due to recalls or MDR reports or other problems that are likely to come to the attention of the agency, such as an irate customer calling a pal at FDA to complain, the company should immediately prepare comprehensively for inspection, not only by documenting that the specific problem is solved, but also by performing an internal audit and correcting, to the extent possible, any GMP deficiencies.

VI. THE EXIT INTERVIEW

FDA investigators generally give a company advance notice of the end of an inspection, for purposes of scheduling the exit interview. Some investigators will finish the inspection and leave the premises for a day or more to prepare the FD 483, particularly where the list of observations is extensive. In any case, the inspection coordinator should assure that company management with sufficient authority to provide for corrective actions is at the exit interview.

Prior to the exit interview, the inspection coordinator, as mentioned previously, should be holding daily inspection recap meetings. Based upon the notes taken during the inspection, it will usually be possible for the company to accurately predict the observations that will appear on the FD 483 and to prepare its response. Many companies, during the course of an inspection, will make any corrections that they deem appropriate and will provide documentation to the investigator either at once or at the exit interview. There are two important reasons to take immediate corrective actions. First, it may help to shorten the list of observations on the FD 483. Second, correction of operations will help to demonstrate the company's good faith and intent to comply with GMPs.

If the company has any inkling that the investigator may have received inaccurate or misleading information, corrections can be made at the exit interview, if not before. Corrective actions during inspection may include the generation of documents that did not previously exist, such as an audit procedure or complaint investigation, but care should always be taken that the dates on such documents accurately reflect the actual time of preparation.

The investigator usually begins the exit interview by providing a copy of the FD 483 for the company's review and by making a standard statement that the observations on the FD 483 represent conditions that the investigator believes represent violations of the law and against which FDA may decide to take legal action. Many investigators will then read through the FD 483 item by item, giving the company an opportunity to ask questions or engage in discussion after each item. Sometimes, if the FD 483 is extensive, the investigator may simply present it for review and offer to discuss any items that the company wishes. The most responsible individual in the exit interview should be given the responsibility of promising correction where appropriate, but anyone present who is unclear as to the meaning of an observation should be permitted to ask questions that will help to clarify it. If observations that have been corrected appear, the company should ask to have them deleted, and if the investigator refuses to do so, ask that the observation be qualified with a "prior to" statement, for example, "Prior to [date of correction], the company failed to have a written procedure for calibration of measurement equipment used in the quality control lab." In this way, it will be clear to the district compliance branch that the company has remedied the problem.

Certain possible violations of the FDC Act and regulations will not generally appear on the FD 483. For example, violations of the MDR regulations may not appear on the FD 483. Likewise, if the investigator believes there are problems with device labeling, samples will be taken and provided to FDA's compliance branch for review. The investigator may or may not mention these issues at an exit interview.

VII. POSTINSPECTIONAL ACTIVITIES

Once the investigator has left the premises, the inspection coordinator should meet with all responsible parties to prepare a timetable for corrective action. Each item on the FD 483 should be delegated to the appropriate individual for response. Many companies schedule a weekly meeting to review the status of corrective actions. In the interim, the inspection coordinator will begin to prepare a response to the FD 483, to be sent to the district. It is usually wise to plan to submit a written response within 30 days of the receipt of the FD 483, unless only a very few minor deviations were noted. The 30-

day time frame may prevent the district issuing a notice of adverse finding (NAF) letter or regulatory letter, which requires a response and may, in the latter case, give rise to a 6-month reinspection.

It is important that the inspection response letter provide a reasonable time frame for corrective actions, in the event that a reinspection is scheduled to confirm corrective actions. It is best for the company to be conservative in estimating the amount of time it will take to achieve compliance. If the observations on the FD 483 were of such significance that further regulatory action by FDA is anticipated, the company should make an effort to make as many corrections as possible within 30 days and may want to provide FDA with documentation of those corrective actions. The response letter should also advise the district that the company would like to have FDA confirm that its response is acceptable and, that if no written response is received from the district, a company official will phone the district after 30 or 45 days. At the same time, the company should submit an FOI request for the establishment inspection report (EIR) related to its most recent inspection.[6] A review of the EIR may make a company aware of potential compliance problems that did not appear on the FD 483, and it may also give the company an opportunity to review the EIR for any confidential information that the company would like to have stricken from the report.

If the observations on the FD 483 are sufficiently serious, in addition to preparing a written response, company management may want to schedule a meeting with the district office to discuss corrective actions. This is also wise if the EIR discloses additional problems to which the company did not already respond. The purpose of such a visit is the same as the purpose of the response letter: to assure FDA that the company is willing to make the corrections needed to attain GMP compliance. Such an attitude on the part of a company is of great benefit not only in averting regulatory action, but also in maintaining a good working relationship with FDA.

VIII. CONCLUSION

Unquestionably FDA inspections can be among the more traumatic experiences that any company faces, but it is clear that several factors can contribute to successful inspection results. First, it is important for a company to know its rights and obligations with respect to FDA inspections. Second, clearly defined policies and procedures should be in place for the handling of inspections. The individuals who will be involved in inspections should be well trained and adequately prepared for inspection at any time. Internal audits should be conducted, as they are another valuable aid for preparing for inspections. Not only do they help to keep a company apprised of potential inspectional problems, but they also serve as a motivator to encourage personnel to correct deficiencies on an ongoing basis.

Following an inspection, it is important to take prompt and appropriate corrective actions and, if indicated, to keep FDA informed of the company's progress. Should enforcement action follow an inspection, timely action and a good relationship with the district may help to mitigate unpleasant results.

---

[6] The company can receive an unpurged copy of its own EIR, usually soon after the inspection is concluded. A purged copy can be obtained by a third-party FOI request.

REFERENCES

1. 106 S. Ct. 1819 (1986).
2. 409 F. Supp. 273 (S.D. Iowa 1976).

27

Product Liability Implications of Regulatory Compliance or Noncompliance

THOMAS A. BOARDMAN and THOMAS DiPASQUALE
3M Company, St. Paul, Minnesota

I. INTRODUCTION

In many industries regulatory compliance is an overriding consideration in the business decision-making process. For manufacturers in these industries the primacy of these regulatory considerations is essential because the overall health of the company (indeed, in some cases, its very existence) may depend on regulatory compliance. However, in the past few years, and particularly in the medical device and diagnostic industry, product liability concerns have taken a more prominent position due to the increase in product liability litigation and the insurance "crisis" of 1985-1986. Risk assessment that places a disproportionate emphasis on either the regulatory or the product liability implications of desired conduct will result in a distorted analysis that may have serious consequences for the long-term financial viability of a medical device or diagnostic manufacturer. Further, a lack of understanding of the relationship between a firm's regulatory posture and its product liability risk is an additional deficiency to be avoided in the risk assessment process. Our goal in this chapter is to assist the lay business person in risk assessment decision making by discussing the product liability implications of both compliance and noncompliance with certain Food and Drug Administration (FDA) statutes and regulations that govern the medical device and diagnostic industry.

II. COMMON-LAW PRODUCT LIABILITY CAUSES OF ACTION

Before discussing specific regulatory requirements and their product liability implications, the reader should understand the basic theories of product liability law today. Our purpose is to provide background information only, and no attempt is made to provide anything more than a brief overview of the subject.

The three most common theories of liability for which a manufacturer may be held liable for personal injury caused by its product are negligence, strict liability, and breach of warranty. These are referred to as common-law causes of action, which are distinct from causes of action based on federal or state statutory law. Although within the last decade federal legislative action that would create a uniform federal product liability law has been proposed and debated, no such law exists today. Thus, such litigation is governed by the laws of each state. There are three basic liability theories that, with few exceptions, are recognized in all 50 states.

A. Negligence

The focus in any negligence suit is on the conduct of the defendant. This focus generally is broken down into two components: first, the existence of a duty running from the defendant to the plaintiff, and second, the reasonableness of the defendant's conduct in meeting that duty. The alleged negligent conduct can occur at any point in a product's life cycle. Thus, the plaintiff may allege that the defendant was negligent in its product design, testing, manufacture, marketing, or labeling, or negligence may be alleged with respect to all of these items. In addition, the plaintiff must establish that some actual harm (usually personal injury) resulted as a consequence of the manufacturer's negligence. Thus, the plaintiff can prevail at trial by establishing all the elements of the negligence claim: (1) a duty; (2) a breach of that duty; (3) actual harm; and (4) proximate cause between the breach and the harm.

B. Strict Liability

Unlike the negligence suit, in which the focus is on the defendant's conduct, in a strict liability suit the focus is on the product itself. The doctrine of strict liability has become almost universally accepted in the United States. Most states have adopted the formulation of strict product liability as described in the Restatement (Second) of Torts. Section 402A of the Restatement provides:

> (1) One who sells any product in a defective condition unreasonably dangerous to the user or consumer or to his property is subject to liability for physical harm thereby caused to the ultimate user or consumer, or to his property if:
>
> (a) the seller is engaged in the business of selling such a product, and
> (b) it is expected to and does reach the user or consumer without substantial change to the condition in which it is sold.

Therefore, the critical focus in a strict liability case is on whether the product is "defective" and "unreasonably dangerous." A common standard applied in medical device cases to reach that determination is the risk/benefit analysis—that is, whether the benefits of the device outweigh the risks attendant with its use.

Because of the special nature of the risk/benefit analysis as applied to medical products cases, the Restatement provides a unique defense for drug, biologic, and medical device manufacturers. Known as the "Comment k" defense because of the Restatement section from which it comes, the defense applies to products that, because of the present state of human knowledge, are incapable of being made completely safe for their intended use and therefore carry unavoidable risks [1]. In cases involving these kinds of products, strict liability is inappropriate if the product is "properly prepared, and accompanied by proper directions and warnings" [1]. The shift in focus to the propriety of preparation, directions, and warnings changes such a case into a negligence case. The ambiguous language of Comment k, and the restrictive manner in which courts have interpreted and applied it, have resulted in an erosion of what appeared to be a reasonable approach to this issue and produced a rather weak defense for device manufacturers, which should not be mistaken as a safe harbor from liability.

C. Breach of Warranty

A third cause of action that may be asserted by a plaintiff is breach of warranty. There are three types of breaches of warranty that may be alleged: (1) breach of the implied warranty of merchantability; (2) breach of the implied warranty of fitness for a particular purpose; and (3) breach of an express warranty. These warranty causes of action do not offer any advantages for the injured plaintiff that cannot be obtained by resort to negligence and strict liability claims and, in fact, pose greater hurdles to recovery. Thus, although a breach of warranty claim is often pled in the plaintiff's complaint, it is seldom relied on at trial as the basis for recovery.

III. THE PRODUCT LIABILITY-REGULATORY INTERFACE

The federal Food, Drug, and Cosmetic Act (the Act) [2] provides the basis for FDA's extensive authority over the medical device and diagnostic industry. This authority is supplemented by extensive regulations issued pursuant to the Act and that legally have the same binding effect as legislation passed by Congress. The scope of FDA's regulatory power extends to virtually every phase of a product's life cycle, including clinical testing, clearance for marketing, labeling, manufacturing, and monitoring of clinical use. This extensive federal oversight can strongly influence the outcome of a product liability suit. Plaintiffs will often cite violations of FDA regulations as evidence that a product was defective or that a company was negligent. Manufacturers, on the other hand, may argue that compliance with these regulations is evidence of their due care and that the product is safe and effective for its intended applications. Similarly, both sides can cite FDA regulations as establishing behavioral and product standards. Invariably, courts view these regulatory requirements as setting minimal standards of conduct to which adherence will not immunize a manufacturer from liability when a reasonable manufacturer would be expected to take additional precautions [3]. Therefore regulatory compliance is no guarantee against product liability, and as we shall see, the fact of compliance itself can have direct liability implications by, for example, creating evidence that can be used against the manufacturer.

If legislative and regulatory compliance can sometimes carry unexpectedly negative consequences in product cases, noncompliance with those laws and regulations can have potentially devastating effects on a medical device or diagnostic manufacturer. In addition to the regulatory enforcement activity that noncompliance with FDA regulations may trigger, the risk that a manufacturer will incur liability in a products case is increased substantially. This is so, in part, because of the current status of the law relating to government regulations and their evidentiary value in civil lawsuits. Courts have repeatedly held that violation of government regulations is negligence *per se* [4]. The practical effect of the negligence *per se* doctrine is to reduce the plaintiff's burden of proof by allowing the jury to presume the defendant manufacturer was negligent, thus leaving whether the presumed negligence was a cause of the plaintiff's injuries as the only substantive issue for the plaintiff to prove.

A second and perhaps far more serious consequence of noncompliance with FDA regulations is the increased exposure to punitive damages. Punitive damages are intended to serve two primary functions—to punish the wrongdoer and to deter similar wrongdoing. Punitive damages need bear little relationship to actual damages, and they frequently do not. To recover for punitive damages, the plaintiff must generally prove that the defendant acted with some form of malice or with intentional or reckless disregard of the plaintiff's rights or safety. Plaintiffs may attempt to meet this standard by arguing, for example, that the manufacturer did not recall a product when it would have been prudent to do so, or that the manufacturer's failure to comply fully with FDA record-keeping and reporting requirements was an attempt to mislead or deceive the government and consumers on matters relating to a product's safety.

The appropriateness of punitive damages in these situations may be open to debate, but one must nevertheless recognize the serious consequences to a business against which punitive damages are sought. Those consequences are at least fourfold. First, an allegation of punitive damages will inevitably increase the settlement value of a case, since the risk of a "runaway" verdict is enhanced. Second, punitive damage allegations increase the risk of adverse publicity because complaints containing multimillion-dollar damage claims tend to attract media attention. The final two consequences flow from a successful claim for punitive damages—the potential for subsequent claims against the same product as a direct outgrowth of the prior litigation and, finally, the financial consequences to the business itself, which often cannot insure against punitive damage liability. This rather unhappy state of affairs can often be traced to a firm's failure to comply with FDA regulations but may also be the result of compliance efforts. Therefore, in an effort to provide some guidance to the unwary, we will explore some of the many FDA regulatory requirements for which both compliance and noncompliance can have serious product liability implications down the road.

A. Noncompliance with Preapproval Regulations

No new medical device or diagnostic product can enter the market without some form of FDA approval [5]. FDA is also specifically charged by Congress with determining whether a clinical trial for a new product should be undertaken [5]. To fulfill that task, the FDA has promulgated regulations that require detailed submissions to the agency and establish internal mechanisms for reviewing whether the proposed investigation presents an unreasonable risk to patients [6]. The FDA approval process for medical devices is described elsewhere in this text and will not be discussed in any detail here. However, it should be noted that noncompliance with those regulatory approval processes can have a significant impact on the outcome of product liability suits.

One of the most obvious implications of noncompliance in this area is that the offending firm is faced with the twin evils of the negligence *per se* doctrine and the increased risk of punitive damages. Let us suppose a firm has fraudulently obtained FDA approval for its product by failing to report or fraudulently reporting data to the FDA in a premarket approval application or in subsequent reports summarizing clinical experience. It is this type of willful disregard for public safety that punitive damages were intended to reach. Of course, it would be left to the court and fact-finder to determine whether the defendant firm's conduct was in fact an intentional and fraudulent circumvention of the approval regulations, or merely an oversight attributable to negligent conduct and poor internal procedures. In either case, when asked to determine on whom the financial loss resulting from the firm's noncompliance should properly be placed—the injured claimant or the defendant firm—most juries will place the loss on the "deep pocket" defendant.

Beyond complicating the defense to a product liability claim, noncompliance with preapproval regulations may even form the basis for the claim itself. For example, firms that sponsor medical device investigations are "responsible for selecting qualified investigators" [7]. A sponsor who chooses a researcher or physician who lacks adequate training or is otherwise unsuitable might be held to have violated this regulation. Furthermore, sponsors are charged with monitoring the progress of the study [8]. If the device sponsor discovers that the investigator is not complying with the study protocol or FDA regulations, compliance must be promptly obtained or shipments of the device discontinued [9]. Therefore, a clinical subject who was injured by a device because an investigator had deviated from the clinical protocol might have a valid negligence claim against the sponsoring firm if the subject can show that the injury may have been prevented by better monitoring.

While noncompliance with preapproval regulatory requirements is likely to trigger enforcement activity by the FDA, there is a relative paucity of case law in which liability is premised on this sort of conduct. Far more prevalent are product liability suits in which the alleged product defect (be it a manufacturing defect, design defect, or labeling defect) or the alleged negligent conduct arose subsequent to FDA approval for commercial distribution. We therefore turn to postapproval regulatory requirements and their impact on product liability litigation.

B. Good Manufacturing Practices and FDA Inspections

Proof of a specific manufacturing defect has typically posed a significant burden for a plaintiff seeking to recover on such a theory. However, that burden was made much easier when FDA adopted Good Manufacturing Practice regulations (GMPs) for medical devices in 1978. These regulations impose extensive record keeping requirements on the device manufacturer, covering everything from personnel and equipment to product complaints [10]. Although most health care firms have developed comprehensive GMP programs, compliance remains problematic for many companies. Approximately half of all medical device recalls occur because of GMP deficiencies [11].

Noncompliance with these regulations can be critical evidence in product liability cases for several reasons. First, failure to adhere to GMPs renders a device adulterated, which itself could trigger FDA enforcement action. The enforcement action, in turn, can be used against the firm at a later trial. Even in the absence of FDA action, violation of the GMPs, like the violation of other regulations designed for the public safety, would probably be considered negligence *per se* in most jurisdictions.

Moreover, it has been suggested that the GMPs provide an objective standard of what constitutes proper manufacturing procedures [12]. The GMPs define minimum acceptable standards for a manufacturer in a variety of phases in the manufacturing operation. Failure to meet these minimum standards creates an impression of unreasonable behavior, or a slipshod operation, that can be difficult to overcome even if the regulatory shortcoming did not affect product quality.

The GMP regulations also require certain affirmative compliance action on the part of manufacturers. For example, master files, which include device specifications, quality assurance procedures, and packaging and labeling specifications, must be compiled for each device. The manufacturer must then perform the tests prescribed in the device master record and document that it has done so [13]. These records must be maintained for at least two years, and longer in most cases [14]. Accordingly, an injured plaintiff is able to reconstruct the manufacturing history of a particular device to determine if the company adhered to its own specifications. Failure to maintain or retain these records not only serves as evidence of negligence, but also effectively shifts the "burden of explanation" to the defendant as to why such records are not available. Departure from specifications is often too great a burden for the defendant to overcome, especially in those instances where the particular specification is arguably related to the cause of the injury.

Yet another area of concern from a compliance standpoint relates to GMP inspections. In this context the reference to GMP inspections includes both the firm's own internal audit mandated by the GMP regulation [15] and the FDA authorized inspection [16].

By its very nature the internal GMP audit is designed to point out those areas that are in need of attention because of noncompliance with GMP regulations. The regulatory provision that mandates such audits also requires that the results be documented in written reports and that corrective action be taken where indicated [17]. These written audit reports are akin to road maps, which could lead the reader directly to regulatory deficiencies in the manufacturing process. In many instances, the audits will create an impression of a process not in compliance or "out of control," which, in turn, may have a significant enough general impact to lead to an adverse jury verdict or to prompt a substantial settlement.

Finally, as mentioned above, the FDA is required to inspect periodically firms operating under its jurisdiction. Frequently the focus of these inspections is compliance with GMP regulations, but the inspection may also be focused on a particular product or a particular complaint. These inspections can provide additional evidence for the plaintiff at a later product liability trial. A manufacturer can be required in an inspection to produce records and product samples. In addition, in 1985, FDA instructed its investigators to be more aggressive with respect to taking photographs of a manufacturing facility [18]. Further, employees may provide statements to the FDA inspector during the course of the inspection. From this activity the inspector will prepare an Establishment Inspection Report that details the findings. If conditions are observed that violate the GMPs or other legal requirements, a form FD-483 may be issued, and if the violations are deemed serious enough, a regulatory letter or notice of adverse findings may also be issued. These documents can later be obtained through a Freedom of Information Act request and, in many cases, provide useful evidence of shortcomings in the defendant's manufacturing operations.

Finally, internal memoranda as well as any company response commenting on either the inspection or FDA's formal documentation are subject to discovery. This documentation may prove to be fertile ground for a plaintiff's attorney looking for damaging admissions or circumstantial evidence of product defects. It is very easy for a manufacturer confronted with an FDA inspection to focus on the inspection process alone. It is also very short sighted. As we have hoped to impress upon the reader, noncompliance in the area of GMP regulations can have a ripple effect that extends many years after the violation, and can take on much greater significance than the original violative act itself.

C. Labeling Requirements [19]

The FDA reviews device product labeling in the preapproval process, but for the most part manufacturers are given wide discretion to determine the specific contents of their package inserts, promotional literature and other forms of labeling. However, there are general labeling requirements that require device labeling to state all known precautions, warnings, contraindications, and directions for use. The wording of the package insert and other product labeling often forms the principal battleground on which the issue of liability is fought. Generally speaking, if the literature provides adequate warnings and directions for use, the manufacturer will prevail on the labeling issue. If there are shortcomings in the product literature that played a contributory role in bringing about the plaintiff's injuries, then the failure to warn theory is almost a certain vehicle to the plaintiff's recovery. It must be noted that the area of adequate warnings has become predominantly a creature of judicial case law, to the extent that regulatory compliance alone has virtually no salutory affect at trial and regulatory noncompliance is seldom the focus of the plaintiff's argument. It is therefore advisable for a manufacturer to seek legal counsel for an interpretation of the common law in the area of what constitutes an adequate warning for a particular product.

Nevertheless, we can focus on some examples of regulatory noncompliance with labeling requirements that could affect the product liability defense. One situation that complicates the defense enormously is where the package insert is claimed to have not been included in the product packaging. This, of course, constitutes misbranding, and could provide highly persuasive evidence at trial of negligence in the manufacturing process. Quality control procedures should be such that this does not occur.

Many other aspects of product labeling could give rise to both allegations of misbranding and increased product liability exposure. As stated before, the labeling must set out the information needed to use the product safely, such as contraindications, warnings, and adverse reactions. New findings in the medical literature or in the manufacturer's own marketing experience that call the accuracy of these statements into question may require revision of the labeling. Failure to do so could result in product misbranding. The need to modify product labeling may also be triggered by knowledge that the product is being used for other than indicated or approved uses. In this situation, FDA regulations require the manufacturer to "provide adequate labeling . . . which accords with such other uses" [20].

In summary, it is fair to say that while the issue of the adequacy of labeling looms large in most product liability trials, the status of the defendant's regulatory compliance or noncompliance seldom plays a predominant role in the resolution of that issue.

D. Additional Record Keeping and Reporting Requirements

We have emphasized the important role that documents play in product liability cases. They can be used to show that the manufacturer exercised reasonable care in the design, manufacture, and sale of its product, or they can be used as an effective tool against the manufacturer in proving liability. There are numerous record-keeping and reporting regulations for medical devices, some of which have been alluded to earlier. Failure to comply with these requirements, in and of itself, can support a finding of liability against a manufacturer. For example, in *Stanton v. Astra Pharmaceutical Products, Inc.* [21], the court held that the defendant's failure to file adverse drug experience reports regarding Xylocaine, as required by federal regulation, was negligence *per se*. The court also affirmed the jury's finding that failure to file the reports was a contributing cause of the plaintiff's injury. This finding was based on the belief that the FDA would have intervened by means of regulatory enforcement activity designed to protect the public safety had it been properly informed of the reportable incidents.

The *Stanton* case highlights the significance of FDA reporting requirements. Although the case dealt specifically with adverse drug experience reporting, its teaching applies equally to the more recently enacted Medical Device Reporting regulations [22]. The MDR regulations require device manufacturers to report to the FDA any deaths or serious injuries associated with the use of a firm's device as well as malfunctions that could lead to a death or serious injury. Regulatory and legal risk assessment in the area of MDR reporting should be decidedly conservative. It is our view that underreporting of these incidents presents greater risk to the company than overreporting, if for no other reason than that failure to report brings into play the prospect of punitive damages based on a theory of "concealment" from the government. It must also be recognized that neat lines between compliance and noncompliance are especially difficult to draw when discussing the MDR regulations because they are so poorly drafted that much is left to the reader's interpretation of the regulatory language. Again, this is a reason for conservative decision making to avoid even the appearance of noncompliance in a product liability trial.

On the other hand, it must be recognized that the rule of thumb to "report when in doubt" carries its own negative consequences at trial. For example, MDRs submitted to the FDA are discoverable both under the Freedom of Infor-mation Act from the government and through civil discovery from the company. These records can be dispositive on one of the most strongly contested issues at any product liability trial—that is, when a manufacturer knew or should have known of a product risk [23]. MDRs can provide the plaintiff with evidence that fixes the notice period more narrowly than almost any other available evidence. Moreover, it can be a frightful sight for a company executive to see MDRs stacked several feet high in front of the plaintiff's counsel, especially when the jury is made fully aware of what information those documents contain.

Concern over these issues was raised by the medical device industry when the MDR regulation was proposed. FDA responded by including language in the regulation that an MDR report "does not necessarily reflect a conclusion by the manufacturer . . . that the report or information constitutes an admission that the device caused or contributed to a death or serious injury or malfunctioned" [24]. The regulation further states that the manufacturer may specifically deny that the report constitutes an admission [24]. This language opens the door to the use of disclaimers on all MDR reports, which may provide some help in the litigation context. At a minimum such a disclaimer should state that the information reported to the manufacturer may not be verified or verifiable and may be incomplete or inaccurate as reported. Further, it should state that the information in the MDR is provided in compliance with the MDR regulations and does not constitute an admission that the device has malfunctioned or that a connection exists between the product and the reported death or injury. On balance, however, it is felt that strict compliance with the MDR regulations presents the path of least risk in product liability litigation.

Since the Good Manufacturing Practice regulations were first promulgated in 1978, manufacturers of medical device and diagnostic products have been required to investigate failed devices and maintain complaint files [25]. These provisions of the GMP regulation also require the generation of a written record documenting the findings and the retention of such records for 2 years or more [25]. The plaintiff in a product liability action can demand that a defendant manufacturer produce for inspection all GMP complaint files and failure investigation files that relate to the product in question or similar products. These files can be a virtual cornucopia of information for the plaintiff with respect to potential design defects, problems with the product, and notice to the manufacturer of such problems. Even though many such complaints are not directly relevant to the issue being litigated, the mere volume of complaints can create an impression of a "problem product." An experienced attorney for the plaintiff can take one look at a GMP complaint file and structure an entire case around the information gleaned from the file.

E. Recalls

The FDA's regulations governing recalls can have a substantial impact on product liability litigation. A recall is required if a product is in violation of FDA laws and the FDA would initiate legal action (seizure, injunction) if the product is not recalled by the manufacturer [26]. Recalls are classified based on the level of health risk presented by the product.

Courts have taken different views as to whether evidence of a recall is admissible at trial. Some courts have refused to allow introduction of evidence concerning recalls because the evidence is deemed either irrelevant or its probative value is outweighed by its tendency to prejudice the jury [27]. In any event, the decision whether to recall an FDA-regulated product is dictated largely by a firm's interpretation of the recall regulation, and it is not our purpose to discuss the factors that go into making that decision. Instead, we emphasize the substantial benefits that can be derived from responsible corporate action when confronted with a recall situation, and suffice it to say that incalcuable damage can be wrought by ignoring the responsibility to recall when it is otherwise indicated by regulation or sound business judgment.

The consequences of a recall can usually be dealt with at trial. For example, it can often be argued that the defect that led to the recall was not present in the device that is the subject of the litigation. Further, it can be argued that even if the defect did manifest itself in the product, the particular defect did not cause or contribute to the plaintiff's injuries. But there is not room for argument by the defense when users of a device are left to their own risk against a product that should have been recalled. Our only counsel in this situation is that the case should not be tried.

IV. CONCLUSION

In today's society, the specter of product liability hangs over all manufacturers. In the health care industry the risk of product liability has a special presence. This is due in part to the fact that significant technological advances have allowed for the treatment of patients who are seriously ill on an increasing basis. Thus, the "medical miracle" of 20 years ago is no longer a miracle, but rather a routine expectation. To this extent, the medical device and diagnostic industry is a "victim" of its own success. Nevertheless, as we have discussed, there is much that manufacturers can do to minimize the risk of product liability.

In this chapter we have pointed out that there are product liability implications to both compliance and noncompliance with FDA regulations. Manufacturers of medical device and diagnostic products need to employ a vision in their business decision-making process that goes beyond the issue of regulatory compliance/noncompliance. They need to consider *all* of the implications of the decisions made throughout a product's life cycle and, where appropriate, seek professional counsel.

REFERENCES

1. Restatement (second) of Torts 402A, Comment k (1965).
2. 21 U.S.C. 321 et seq. (1988).
3. In the medical device and diagnostic industry a strong argument can be made that legislative and regulatory requirements are so pervasive that they should constitute the standard of conduct and not just minimum requirements. This argument is the strongest with respect to design and labeling issues either where FDA has reviewed and approved the design or label or where FDA has mandated design changes or labeling text. Such is not infrequently the case with devices subject to the premarket approval process. This argument forms the basis for the position taken by tort reformers advocating the elimination of punitive damage liability for products that are in compliance with FDA rules and regulations.
4. For example, *Brooks v. Astra Pharmaceutical Prods.*, 718 F.2d 553 (3rd Cir. 1983); *Palmer v. A. H. Robins Co.*, 684 P. 2d 187 (Colo. 1984).
5. 21 U.S.C. 360e (1988).
6. 21 CFR 812.1-.150 (1988).
7. 21 CFR 812.40 (1988).
8. 21 CFR 812.43(d) (1988).
9. 21 CFR 812.46(a) (1988).
10. 21 CFR 820.1-.198 (1988).
11. McDonnel, An Analysis of the Assignable Causes of Device Recalls, Speech to American Society for Quality Control (April 1, 1986).

12. Gibbs, J. N. and Mackler, B. F., Food and Drug Administration Regulation and Products Liability: Strong Sword, Weak Shield, *Tort and Insurance Law Journal* (1987).
13. 21 CFR 820.184 (1988).
14. 21 CFR 820.180(b) (1988).
15. 21 CFR 820.20(b) (1988).
16. 21 U.S.C. 360(h) (1988).
17. 21 CFR 820.20(b) (1988).
18. Morin and Yeager, Responding to FDA's Insistence on Taking Photo graphs During an Administrative Inspection, *Food, Drug and Cosmetic Law Journal, 42*:485-499 (1987).
19. 21 CFR 801.1-.430 (1988).
20. 21 CFR 801.4 (1988).
21. 718 F.2d 553 (3d. Cir. 1983).
22. 21 CFR 803.1-.36 (1988).
23. FDA is obligated to delete from MDR reports information that constitutes trade secret or confidential commercial or financial information as defined in 21 CFR 20.61 (1988). Note, however, that simply claiming that information is confidential does not make it so. (*See* 21 CFR 20.27 (1988). The best advice would seem to be when in doubt, claim confidentiality.
24. 21 CFR 803.24(f) (1988).
25. 21 CFR 820.162 and 820.198 (1988).
26. 21 CFR 7.3(g) (1988).
27. See *Smyth v. Upjohn Co.*, 529 F.2d 803 (2nd Cir. 1975).

28

Developing a Labeling Compliance Program

MAX SHERMAN *Zimmer, Inc., Bristol-Myers Company, Warsaw, Indiana*

I. INTRODUCTION

The Medical Device Amendments of 1976 and subsequent regulations provided the Food and Drug Administration (FDA) with new authority to ensure the safety and effectiveness of medical devices. Manufacturers were then and are now faced with responsibility for complying with a host of new requirements, including premarket notification; rules governing advertising for restricted devices; classification procedures to determine controls; performance standards; good manufacturing practices; banned-device regulations; premarket approval applications; investigational device exemptions; exporting; medical device reporting; and additional misbranding requirements. While these responsibilities seem diverse, all share the important characteristic of specific labeling requirements. This should come as no surprise, because through its entire history the FDA has had one essential assignment: to assure that the products it regulates are safe and truthfully labeled. The determination of proper labeling is a pivotal concern and, in fact, the very essence of the medical device amendments.

II. LABELING OBLIGATIONS

To illustrate this point, each of the above obligations will be examined separately.

A. Premarket Notification Procedures

The Code of Federal Regulations, 21 CFR 807.87(e), stipulates that proposed labels, labeling, and advertisements sufficient to describe the device, its intended use, and directions for use should be part of the information submitted in a premarket notification submission [510(k)].

B. Advertising for Restricted Devices

Section 502(r) of the Federal Food, Drug, and Cosmetic Act declares a restricted device misbranded unless the advertising (labeling) includes a brief statement of that device's intended uses, along with a delineation of relevant warnings, precautions, side effects, and contraindications.

A proposed rule for restricted devices was published in the *Federal Register* on October 30, 1980. In that notice, the agency determined that restricted devices, because of their potential for harmful effects, or the collateral measures necessary for their use, can be used safely and effectively only by, on the order of, or under the supervision of a health professional. The agency also noted that so-called "prescription" devices require the special training, knowledge, and experience of a health professional to be used safely and effectively. By their nature, these devices require surgery or are used in the diagnosis and treatment of diseases that can be properly recognized and treated only by persons who have special training and education. Admitting that the current determination that a device is a "prescription" device (21 CFR 801.109) is quite subjective, the agency proposed that the following types of devices be restricted:

1. An invasive device that is intended to penetrate or pierce the skin or mucous membranes of the body, the ocular cavity, the urethra, the ear beyond the external auditory canal, the nose beyond the nares, the mouth beyond the pharynx, the anal canal beyond the rectum, or the vagina beyond the cervix or;
2. A device intended to be implanted into the body or one that replaces or corrects a body part or function;
3. A device intended to introduce energy on or into the body;
4. A device intended to deliver medicinal gas to the body;
5. A device, other than an *in vitro* diagnostic product, intended for use in the diagnosis of diseases or other conditions, or for monitoring physiological functions; or
6. An *in vitro* diagnostic product, as defined in 21 CFR 809.3(a), that is intended to be used, or to provide information that will be used, interpreted, or analyzed by a health professional.

The proposed rule was then withdrawn on November 24, 1981, because of a number of comments and Executive Order 1229, which requires federal agencies to base administrative decisions on adequate information concerning the need for, and consequences of, the decisions and to undertake regulatory action only when the potential benefits of the action outweigh the potential costs. Manufacturers are thus left with the "subjective" nature of the prescription device regulations to make the proper determination, while FDA relies on the inspectional authority available under U.S.C. 374(c) for inspection of records required under the good manufacturing practices regulation (21 CFR 820) as well as the dispensing and labeling requirements of the present prescription device regulation and other applicable regulations, in regulating medical devices. This subject will be discussed later.

C. Classification Procedures

Regulation 21 CFR 860.7(b)(2) dictates that, in determining a device's safety and efficacy for purposes of classification, or in establishing performance standards for Class II devices or premarket approval for Class III devices, the FDA commissioner and the classification panels will consider, among other relevant factors, "the conditions of use for the device, including conditions . . . prescribed, recommended, or suggested in the labeling or advertising of the device."

D. Performance Standards

Section 514(a)(i) of the Act provides that FDA "may" by regulation establish a performance standard for a Class II device. The establishment of a performance standard is required for a Class II device in Section 513(a)(1)(B). Although none have been written to date, the agency has invited offers to submit or to develop standards for vascular graft prostheses, central nervous system fluid shunts, breathing frequency monitors, and continuous ventilators. Among other requirements, the standard [514(a)(2)(c)] shall, where appropriate, require the use and prescribe the form and content of labeling for the proper installation, maintenance, operation, and use of the device.

E. Good Manufacturing Practices

Regulations 21 CFR 820.120 and 820.121 delineate what constitutes adequate controls for maintaining labeling integrity and for preventing labeling mixups. Regulation 21 CFR 820.182(b), which concerns critical devices and device master files, dictates that manufacturers keep complete records of labeling procedures for specific devices, as well as copies of all approved labels and other labeling. Additionally, a petition for exemption or variance from specific GMP requirements cannot be obtained unless a complete record of labeling, including package inserts and promotional material, has first been submitted (as described in the sections on administrative practices and procedures, 21 CFR 10.30 and 10.20).

F. Banned Devices

Regulation 21 CFR 895.25(a)-(d) permits the FDA commissioner to require that a manufacturer, a distributor, or an importer—or any individual(s) responsible for the labeling or advertising of a device—revise a device's labeling or advertising if such revision is necessary to eliminate or reduce risks or dangers to health. The commissioner may also require that additional information pertaining to these hazards be listed for a specified period of time. This warning or notice should appear in the manner or form prescribed by the commissioner.

G. Premarket Approval Applications

In 21 CFR 814.20(b)(10), there is a requirement that the application for a premarket approval application (PMA) contain copies of all proposed labeling for the device. Such labeling may include, for instance, instructions for installation and any information, literature, or advertising that constitutes labeling under Section 201(m) of the Food and Drug Act. This section defines the term "labeling" to mean all labels and other written, printed, or graphic matter (1) upon any article or any of its containers or wrappers or (2) accompanying such article. It is important to know that approval of an application can be withdrawn [per Section 515(e)(i)] if the Secretary of Health and Human Services finds that a device is unsafe or ineffective under the conditions of use that have been prescribed, recommended, or suggested in the labeling, or if the labeling is false or misleading.

Other labeling obligations for holders of PMAs are found in Conditions of Approval that accompany the approval letter. An advertisement for the device, if it is limited to prescription use, must include a brief statement of the intended use, relevant warnings, precautions, side effects, and contraindications. A mixup of the device or its labeling with another article must be reported within 10 days as a Device Defect Report. Additionally, any adverse reactions, side effects, injury, toxicity, or sensitivity reaction attributed to the device that has not been addressed in the labeling or that has been addressed in the labeling but is occurring with unexpected severity or frequency must also be reported in 10 days as an Adverse Reaction Report.

H. Investigational Device Exemptions

Regulation 21 CFR 812.5(a) stipulates that an investigational device bear the manufacturer's, packer's, or distributor's name and place of business as well as the quantity of contents. Also required is the statement "CAUTION—Investigational device. Limited by federal [or United States] law to investigational use." Regulation 21 CFR 812.5 dictates that the labeling must not be false or misleading and must not imply that the device is safe or effective or both when used for the purpose(s) for which it is being investigated. The labeling must also include a description of any relevant hazards, contraindications, adverse effects, interfering substances or devices, and precautions that prior investigation or experience with the investigational device (or any related device) has revealed. A copy of all labeling is required by 21 CFR 812.2(b)(10). An investigational device shipped solely for research on or with laboratory animals must bear on its label the following statement: "CAUTION—Device for investigational use in laboratory animals or other tests that do not involve human subjects."

I. Exporting

Section 801(d) states that a device intended for export shall not be deemed to be adulterated or misbranded under the act if it (a) accords to the specifications of the foreign purchaser, (b) is not in conflict with the laws of the country to which it is intended for export, (c) is *labeled* on the outside of the shipping package that it is intended for export, and (d) is not sold or offered in domestic commerce.

J. Medical Device Reporting

In some cases the malfunction of a medical device does not have to be reported. The malfunction, however, must meet each of the following under 21 CFR 803.24(a)(3)(iii): (a) a death or serious injury has not occurred; (b) the device's labeling sets forth the information concerning the potential for death or the type of serious injury that the device may cause or contribute to; (c) the device's labeling describes the malfunction and the routine service repair or maintenance instructions to correct the malfunction; (d) the malfunction has occurred or is occurring at or below the frequency and severity stated in the labeling for the device or, if there is not any pertinent statement in the labeling, at or below the frequency and severity that are usual for the device; and (e) the malfunction does not lead the manufacturer or importer to undertake a remedial action involving any device other than the device product in which the malfunction occurred.

K. Other Misbranding Requirements

Section 502 of the Federal Food, Drug, and Cosmetic Act, in addition to stating concerns for full-disclosure information with regard to restricted devices, requires that a device be considered misbranded (1) if its labeling is false or misleading; (2) if the label does not contain the name and address of the manufacturer, packer, or distributor, as well as an accurate statement of the quantity of contents; (3) if the information required on the label is not prominent, legible, and written in English; (4) if the established name of the device does not appear in type that is at least half as large as the proprietary name used; (5) if labeling does not bear adequate directions for use and adequate warnings against unsafe use; (6) if the device is dangerous to health; (7) if the device is restricted and is not sold, distributed, or used in compliance with regulations; (8) if the device does not comply with an applicable performance-standard labeling requirement; and (9) if there was a failure to comply with a requirement under section 518 (notification and other remedies) or section 519 (records and reports) of the Federal Food, Drug, and Cosmetic Act.

There are several exceptions from the section in the Act that refer to the requirement for adequate instructions for use. These are listed in 21 CFR 801. A device intended for processing, repacking, or use in manufacturing another drug or device is exempt providing its label bears the statement "CAUTION—for manufacturing, processing or repacking."

Devices are also exempt if they are shipped or sold to, or in the possession of, persons regularly and lawfully engaged in instruction in pharmacy, chemistry, or medicine, not involving clinical use, or engaged in law enforcement, or research not involving clinical use, or in chemical analysis, or physical testing, and are to be used only for such instruction, law enforcement, research, analysis, or testing. As a general rule in these cases the manufacturer should attach a label that reads, "For testing, research, or analysis."

Specific labeling provisions for medical devices are contained in (1) the Federal Food, Drug, and Cosmetic Act, (2) the Fair Packaging and Labeling Act, (3) the Radiation Control for Health and Safety Act, and (4) Title 21 of the U.S. Code of Federal Regulations, Part 801, for general devices and Part 809 for *in vitro* diagnostic products.

III. ANALYSIS OF LABELING COMPLIANCE

Compliance with labeling requirements actually comprises two separate issues: label control and labeling content. Label control implies label integrity—a GMP requirement that was introduced to minimize labeling mistakes, problems that have resulted in many device recalls. Probable causes identified with product mislabeling include (1) label mixups, (2) product mixups, (3) label printing errors, (4) label versus container errors, and (5) wrong inserts or outserts. The issue of labeling content concerns the definition of permissible copy and proper format. The following section will focus on device labeling content and will discuss the similarity, in this context, of device labeling to drug labeling. Label control methods will be discussed later.

IV. FULL DISCLOSURE

The reference to intended uses, relevant warnings, precautions, side effects, and contraindications in section 502(r) of the Federal Food, Drug, and Cosmetic Act constitutes what is generally called *full disclosure*. With regard to prescription drugs, this term is defined in 21 CFR 201.100(c)(1), published in the *Federal Register* on September 6, 1961. The section has since been recodified and now states, "Labeling on or within the package from which the drug is to be dispensed bears adequate information for its use, including indications, effects, dosages, routes, methods, and frequency and duration of administration, and any relevant hazards, contraindications, side effects, and precautions under which practitioners licensed by law to administer the drug can use the drug safely and for the purposes for which it is intended."

This requirement for full-disclosure labeling, fulfilled through the use of a package insert, also applies to devices. In fact, the 1962 Kefauver-Harris amendments stated that the manufacturer or distributor of a device is responsible for providing adequate information under which practitioners licensed by law to employ that device can do so safely and for the purposes for which it is intended. Such information was to include a device's indications; its effects; its routes, methods, frequency, and duration of administration; and all relevant hazards, contraindications, side effects, and precautions.

The first regulation pertaining to the format of a device package insert appeared in the *Federal Register* on May 10, 1977 (vol. 42, no. 90). This regulation, 21 CFR 801.427, dictated the format for professional and patient labeling of intrauterine contraceptive devices (IUDs). The format for the professional labeling of IUDs was as follows: (1) description; (2) mode of action; (3) indications and usage; (4) contraindications; (5) warnings; (6) precautions; (7) adverse reactions; (8) directions for use; and (9) clinical studies. It is safe to assume that subsequent regulations for specific devices will remain the same.

These headings are quite similar to those published in the *Federal Register* for the labeling of human prescription drugs (see vol. 44, no. 124, June 26, 1979).

If a format for the labeling of devices is to be established, a similar regulation must be promulgated for restricted devices. Until this is done (and it may be some time, since the prescription-drug format remained a proposal for more than 4 years), it is prudent for manufacturers to emulate the format for IUD labeling.

V. ADDITIONAL CONSIDERATIONS

A. Product Liability

In a sense, the entire philosophy underlying device labeling has changed since the implementation of the Medical Device Amendments, in as much as this regulatory constraint gave manufacturers the impetus to take a closer look at labeling. However, an equally significant factor affecting device labeling has been the recent increase in product liability cases. It is recognized, for example, that liability can be established even without fault on the part of the manufacturer—and even if no other reasonable alternatives were available to the manufacturer at the time of a product's development.

There are three major methods of recovery that can be employed by injured persons in product liability suits: negligence, breach of warranty, and strict liability. In each of these instances, the plaintiff must demonstrate that the defendant breached a legal duty owed to the plaintiff, thereby causing injury.

From a labeling standpoint, negligence and strict liability are areas in which the device manufacturer is particularly vulnerable. Negligence is a violation of the manufacturer's duty to use due care with respect to a person to whom such duty is owed. A manufacturer must exercise reasonable care in product testing and design, and in providing adequate warnings. In cases involving strict liability, a manufacturer incurs liability by marketing a product in a "defective" condition, thus causing injury to the patient. In this context, there are three principal ways in which a product may be judged defective: improper manufacturing, improper design, or failure to warn of product dangers. To reduce liability exposure, manufacturers must ensure that products are properly labeled. Clearly worded warnings can be an effective means of reducing liability exposure.

B. Interrogatories

As a consequence of these product liability cases, manufacturers are being subjected to more and more interrogatories—that is, formal, written questions presented by the plaintiff's attorneys. In most such cases, attorneys are requesting the submission of product labeling to help determine if the manufacturer acted negligently by not providing adequate warnings and instructions. In addition, an interrogatory may delve into a manufacturer's earlier labeling practices in attempts to determine why various changes were made. For this reason, it is essential that the manufacturer maintain an accurate historical file and document each labeling change that is implemented. For litigious products, the manufacturer may find it advisable to keep a notebook containing actual label specimens, along with backup reference materials to substantiate each revision.

C. Advertising

The label compliance manager must remind other individuals in the company that labeling encompasses not only the material directly on the product or the shipping container, but also items such as advertisements, circulars, audio, slide, or video programs, and other written or oral statements respecting the product. As stated previously, inadequate labeling can seriously increase liability exposure. Manufacturers may be held liable for failing to (1) give adequate warnings about known risks, thus encouraging or causing misuse, or misleading users who might have acted differently had they known of the actual risks, and (2) warn against improper uses, thus tacitly encouraging such misuse.

D. Demands From Health Care Personnel

Another stimulus for labeling compliance has come from nurses and other health care personnel. Hospital personnel are now demanding that manufacturers provide detailed labeling information pertaining to a device's maintenance, uses, and hazards. In light of the increasing threat of malpractice litigation, many hospital staff members are more alert to the necessity of following labeling instructions closely.

All of the above considerations—governmental, ethical, medicolegal, and economic—serve to underscore the significance of adequate product labeling.

VI. DEVELOPING A DEVICE PACKAGE INSERT

The general principles of labeling development are the same for all medical devices. In all cases, the manufacturer's goal is to provide lucid, accurate, and complete product information that fulfills full disclosure requirements. The formulation of device labeling is a creative, challenging, and interesting task. Unlike drug labeling, which often relies heavily on guidelines delineated in the *Federal Register*, device labeling requires a considerable amount of research and documentation—and the writer and reviewers must supply the actual text. The following hierarchy for prescription drugs applies equally to medical devices and should be carefully considered when providing labeling: what the product is, what its benefits and risks are, how it should be used, what risks in use there are and how these risks can be avoided, and what should be done if the product is not used properly.

A. Regulations to Consider

Part 801 of Title 21 of the Code of Federal Regulations outlines the general labeling provisions for medical devices. These provisions constitute the guidelines for labeling compliance. However, much of this section is interpretative and subjective (a salient exception is subpart H, which provides special requirements for specific devices).

B. Restricted (Prescription) Devices

One of the first things that the manufacturer must define before providing labeling is the term *restricted* device. Under the authority given FDA in section 520(e) of the Federal Food, Drug, and Cosmetic Act, a device's sale, distribution, or use can be restricted to prescription dispension only if there is potential for harm or if collateral measures are necessary for the proper use of the device. When the medical device amendments were implemented, FDA determined that the category of restricted devices included all prescription devices defined in 21 CFR 801.109. This interpretation has since been challenged in the courts. The U.S. Court of Appeals has ruled that FDA must publish a formal regulation before a device can be deemed restricted. At the present time, therefore, the manufacturer must decide what constitutes a restricted device. Regulation 21 CFR 801.109 describes a prescription device as one which, "because of any potentiality for harmful effect, or the method of use, or the collateral measures necessary to its use, is not safe except under the supervision of a practitioner licensed by law to direct the use of such device, and hence for which adequate directions for use cannot be prepared." This section goes on to say that such a device "shall be exempt from Section 502(f)(1) of the [Federal Food, Drug, and Cosmetic] Act." It is obvious that this is a gray area for physical medicine or orthopaedic device manufacturers with relation to such products as postoperative knee dressings, shoulder immobilizers, certain types of arm slings, knee braces, cervical collars, sacrolumbar supports, and wrist/ankle immobilizers. If these or similar products are deemed "restricted" or "prescription," they might conceivably be exempt from stipulations in section 502(f) of the Federal Food, Drug, and Cosmetic Act regarding adequate usage directions for the layman.

Even if these products were exempt, however, they would still require labeling, as defined in 21 CFR 801.109(c), that "bears information for use, including indications, effects, routes, methods, and frequency and duration of administration, [as well as] any relevant hazards, contraindications, side effects, and precautions under which practitioners can use the device safely and for the purpose for which it is intended." The next decision that must be made by the manufacturer follows from the above in that this information can be omitted if it is common knowledge among practitioners.

C. Exemptions

The last part requiring interpretation is 21 CFR 801.110, which states that a device can be exempt from 21 CFR 801.109 "if, at the time of delivery to the ultimate user or purchaser by the licensed practitioner, in the course of his professional practice, or upon prescription or other order lawfully issued in the course of his professional practice [it is provided] with labeling bearing the name and address of the licensed practitioner and the directions for use and cautionary statements, if any, contained in such order." To further complicate matters, 21 CFR 801.116 states that a device is exempt from section 502(f)(1) of the Federal Food, Drug, and Cosmetic Act insofar as adequate directions for common uses thereof are known to the ordinary individual. Initially, then, the manufacturer must determine whether the product is a "prescription" device. If adequate usage directions for the layman can, in the manufacturer's estimation, overcome a device's potential for harm, that product should not be designated a prescription or a restricted device.

VII. PATIENT LABELING

Because of the vague and perplexing nature of the regulatory requirements for "ambivalent" products and the recent emphasis on patient labeling, it may be prudent to provide combination labeling that is directed both to the physician and to the patient. This combination package insert can be perforated; thus, the physician can provide his patient with information to reinforce or supplement any oral instructions that have been given. Patient labeling is modeled after physician labeling requirements, but it includes only information that is appropriate to the patient in laymen's language. Such information is advantageous in that the patient often plays a major role in determining the ultimate success or failure of a device; this extends not only to the gray area of soft goods but to implanted devices as well.

FDA has announced (*Federal Register*, vol. 44, no. 172, September 4, 1979) that it may require the development and dissemination of patient information on the uses, benefits, and risks of medical devices, as is currently required for IUDs and hearing aids. The agency believes that patient information is needed (1) if there are alternatives that the patient should know of; (2) if there are substantial risks or discomforts associated with the product's use; (3) if the cost of the product is significant; (4) if the patient must adhere strictly to a specific treatment regimen; and (5) if there is substantial public or professional controversy regarding the device or its related procedures.

VIII. LABELING DEVELOPMENT

A. Literature Review

The procedures that are used to develop product labeling are basically the same whether a product is old or new. For these pruposes, find a library that affords access to MEDLARS (Medical Literature Analysis and Retrieval System) and Lockheed DIALOG data bases. For all new labeling, a literature search should be conducted before each manuscript is developed. By this means, pertinent articles can be retrieved to serve as documentation for label claims and to substantiate hazards of use. These articles are retained in product files that are maintained by labeling compliance personnel.

B. Format

In developing labeling, employ a format similar to that used for IUDs. If the product is sold unsterile, it may be appropriate to add a section heading entitled "sterility," and describe the recommended method of sterilization. Where applicable, resterilization information can also be provided if there is adequate documentation.

Briefly, the sections used for devices and a description of their contents are as follows:

Description: Includes the product's dimensions, materials, design, intent, and sterility status; compares the product to similar devices.

Indications and usage: Outlines the device's clinical uses and limitations; describes adjunctive or primary treatment and length of treatment. Some indications and usage information can be obtained from the final regulations, published by the classification panels in the *Federal Register*.

Contraindications: Lists situations in which the device should not be used because its risks outweigh its benefits.

Warnings: Includes potential adverse effects as well as any effects that might result from the use of components that have not been manufactured to required specifications, that is, using components from another manufacturer. Generally, a warning should include the probable result of noncompliance.

Precautions: Delineates any precautionary measures that might improve the prospect of beneficial uses; and includes information for the patient.

Adverse effects: Describes side effects that can be reasonably associated with the use of the device. They are generally derived from the scientific literature, MDRs, complaint files, or from reports during a clinical trial.

Directions for use: References an appropriate surgical technique or provides comprehensive instructions; illustrations may be required to clarify directions.

C. Determining Adverse Effects

One difficult problem that the label writer must resolve concerns the discussion of adverse effects. As stated so eloquently by Ray W. Gifford, M.D. [Primum Non Nocere, *Journal of the American Medical Association, 238*(7), August 1977], this discussion constitutes perhaps the most burdensome and unauthenticated portion of an insert. Dr. Gifford comments that "rigorous scientific evidence of efficacy is required before an indication can be included in the [drug] package insert, [and yet] poorly documented, isolated, unpublished, and uncritical observations are listed under adverse reactions." Dr. Gifford's comments on pharmaceutical labeling requirements are particularly appropriate and equally relevant to medical devices.

The writer of a medical device package insert must also face the task of determining what has caused an adverse effect. With implantable devices, for example, many adverse effects are induced at surgery and thus bear no direct relation to the actual risks of the device itself. One can remedy this problem by issuing a disclaimer on the insert stating, "The following effects have been noted: they may or may not be device related." One may also list adverse effects on the insert and explain their possible causes.

D. Indications for Use

If a product is undergoing a premarket approval process, the indications for that product's use can be determined by means of adequate and well-controlled investigations, and additional information can be provided in the package insert. If this is not the case, then indications must be derived from the scientific literature or from personal communication with physicians. Information thus derived must be analyzed and monitored continually, both in the preparation and in the revision of package inserts. In order to supplement this process, it is helpful to have a consulting physician review product labeling when necessary.

E. Two Important Considerations

At a minimum, the following two questions must be carefully evaluated on a periodic basis and considered with respect to all labeling: (1) Does it contain specific instructions that are necessary to foster proper device use, and (2) do the written instructional materials accompanying the product set forth how the device is to be used, when it should not be used, additional precautions to observe, potential adverse effects, and what to do in the event of a failure or other problem? For example, where a device must be set up or assembled by the user or by someone who takes it apart for cleaning, the manufacturer must seek ways to make misassembly impossible or unlikely. This can be accomplished by including pictorial or written warnings or instructions on or with the product so they will be seen every time. The other alternative is to provide the instructions in a manual, recognizing that it will not be read by all users or assemblers.

IX. LABEL CONTENT APPROVAL

The labeling compliance manager should work in concert with packaging development. Personnel in this department must determine the size of the insert or manual, the type of paper to be used, and the number of folds that will be necessary for the package that will be employed. Some type of label content approval form should be used, and the draft manuscript of the insert should be circulated to other departments, such as marketing, legal, medical, and engineering. After this operation, the manuscript can be returned to the labeling compliance manager for final review and editing. It can then be returned to packaging development for typesetting or sent to an outside printing service. Before final printing, a silverprint of the copy should be inspected by the labeling compliance manager or designee to ensure that no sections have been inadvertently deleted and that pages are in the proper order. Each package insert must list the date of issue or date of revision for historical purposes. (Prescription drug labeling regulations require that the date of issue or revision be prominently placed after the last section of labeling.)

X. THE LABEL

A. Definition

Section 201(k) of the Act defines the term "label" as a "display of written, printed, or graphic matter upon the immediate container of any article."

B. OTC Devices

Label contents differ for over-the-counter (OTC) and prescription devices. For example, OTC device labeling must comply with the principal display panel requirements listed in 21 CFR 801.60 and with the statement of identity and intended action section (801.61). There is also the requirement under 801.62 for the declaration and location for the net quantity of contents. All of the above are requirements to enforce the Fair Packaging and Labeling Act (see 21 CFR Part 1) to enable the consumer to obtain accurate information as to quantity of contents to facilitate "value" comparisons. OTC devices are required to be labeled with adequate directions for use either on the label or in the packaging enclosure (801.5).

C. OTC and Prescription Devices

Both OTC and prescription devices must include the name of the company, street address, city, state, and zip code on the label. The street address, however, may be omitted if it is shown in a current city or telephone directory. If a company manufacturers, packs, or distributes a device at a place other than his principal place of business, the label may state the principal place of business in lieu of the actual place where such device was manufactured or packed or is to be distributed, unless such statement would be misleading (see 801.1). Additionally, when a device is not manufactured by the company whose name appears on the label, the name shall be qualified by a phrase that reveals the connection the company has with such device, such as "manufactured for" or "distributed by" or any other wording that expresses the fact.

Prescription and OTC device labels also should contain a lot or control number (a necessity in the event of recall, repairs, modifications, or a failure investigation, although such numbers are only required for "critical" devices; see 820.121 and 820.185).

D. Prescription Devices

In addition, a prescription device label would include a catalog number, the quantity of contents, the established name of the device, or, if none, the compendial name, or, if none, the common or usual name, and the prescription device legend (Caution: Federal law restricts this device to sale by or on the order of a physician, dentist, or veterinarian, or the descriptive designation of any other practitioner licensed by the law of the State in which he practices to use or order the use of a device). There must also be a statement regarding sterility where applicable. In the case of a single-use sterile device, many manufacturers include labeling to advise against resterilization or reuse. The label for multidevice kits or packages containing a combination of sterile and nonsterile products must not imply that all contents are sterile. If there is a need for users to have instructions on how to open a sterile device package to avoid contamination, these instructions should be included on the label.

E. Other Considerations

1. *Trademarks*

A trademark is a word, name, symbol, or a combination used by a manufacturer to identify its goods and distinguish them from others. Trademark rights continue indefinitely as long as the mark is not abandoned and is properly used. On the other hand, a federal trademark registration is maintained by filing a declaration of use during the sixth year after its registration, and by renewal every 20 years as long as the mark is still in use. The federal law provides that non-use of a mark for 2 consecutive years is ordinarily considered abandonment, and the first subsequent user of the mark can claim exclusive trademark rights. Trademarks, therefore, must be protected or they will be lost. They must be distinguished in print from other words and must appear in a distinctive manner. Trademarks should be followed by a notice of their status. If it has been registered in the U.S. Patent Office, the registration notice ® or Reg. U.S. Pat. OFF. should be used. Neither should be used, however, if the trademark has not been registered but the superscripted letters TM should follow the mark, or an asterisk can be used to refer to a footnote stating, "A trademark of - - -." The label compliance manager should remember that trademarks are proper adjectives and must be accompanied by the generic name for the product they identify, i.e., TIVANIUM® Tl=6Al-4V alloy. Trademarks are not to be used as possessives, nor in the plural form (because a trademark is not a noun).

2. *Patents*

A patent, granted only by the federal government, constitutes the right to exclude others from making, using, or selling an invention. After a patent application has been filed, the informal legend "Patent Applied For" or "Patent Pending" frequently is used on patented articles or processes in an advertisement for them. After a patent has been issued, the following patent notice may be used: the word "patent" or "Pat." and the number of the patent. (For a patent issued in the United States, it may be more proper to state "U.S. Patent" or "U.S. Pat.") This notice is not mandatory but may be necessary to obtain damages from an infringer. On the other hand, use of an improper patent notice is punishable by a fine. A utility patent or plant patent lasts

17 years from the date of issuance of the patent. A design patent lasts either $3\frac{1}{2}$, 7, or 14 years from the date of issuance, depending on the fee the inventor chooses to pay. A patent cannot be renewed, and since it is publicly disclosed when issued, anyone can make, use, or sell in the U.S. devices embodying the formerly patented invention after the patent has expired or been declared invalid.

3. *Marking Country of Origin*

It is the general rule of the United States customs laws that each imported article produced abroad, which is not merely in transit through the United States, under a bond to insure its exportation, or otherwise specifically exempted from marketing requirements, must be legibly marked in a conspicuous place with the name of the country of origin in English in order to show the ultimate purchaser in the United States the name of the country in which the article was manufactured or produced. There are certain classes of articles that are exempted from marking to indicate the country of their origin: (1) articles that are incapable of being marked, (2) articles that cannot be marked prior to shipment to the United States without injury, or (3) articles that cannot be marked prior to shipment to the United States except at an expense economically prohibitive of their importation. Although such articles are exempted from marking to indicate the country of their origin, the outermost containers in which such articles will ordinarily reach the ultimate consumer in the United States must be marked to show the country of origin or such articles. When the marking of the container of an article will reasonably indicate the country of origin of the article, the article itself is exempt from such marking. If the article is a surgical instrument, the marking should be done in a manner that will not impair the mechanical properties or the corrosion resistance. The marking should be located at a point of low stress.

4. *Copyrights*

It is prudent to have manuals, brochures, package inserts, or other comprehensive labeling copyrighted. A copyright protects the writings of an author against copying. Copyrights are registered in the copyright office in the Library of Congress and identified with a © and the year.

5. *Hazardous Materials*

The label compliance manager should be familiar with, and have available for reference, packaging, labeling, and marketing regulations for hazardous materials. Hazardous material is a substance or material including a hazardous substance that has been determined by the Secretary of Transportation to be capable of posing an unreasonable risk to health, safety, and property when transported in commerce and that has been so designated (49 CFR 171.8). The manager should also become familiar with 49 CFR 100-199, particularly part 172, which contains the marking, labeling, and placarding sections.

6. *Overlabeling*

One question frequently asked by quality assurance personnel concerns overlabeling. There may be occasions when manufacturing may want to use a new label on an already labeled device to preclude destroying a product that has been mislabeled or where the label may be unacceptable for other reasons and the label cannot be removed. While overlabeling of an old label is not encouraged, it is acceptable as long as the new label is designed, printed, and applied so as to remain legible and in place during the customary conditions of processing, storage, handling, distribution, and use [21 CFR 820.120(a)]. This may not be true, however, for marking products destined for the military. The label compliance manager should check Mil-Standard 129H and the contract to see whether or not overlabeling is permitted for government orders.

7. American National Standards Institute (ANSI)

There are currently four ANSI standards now under ballot respecting color codes, facilities and safety signs, criteria for safety symbols and product safety signs and labels. The latter (ANSI 534.4) is particularly important as it suggests that a warning for a product contain three elements: (1) a signal word to clearly identify the nature and extent of the danger; (2) a symbol or pictorial to convey the hazard and/or the consequences of not avoiding the hazard; and (3) clear, concise, and understandable words to describe the hazard and how it may be avoided. Copies of the proposed standards are available by contacting the National Institute for Standards and Technology, Gaithersburg, MD.

XI. LABEL CONTROL

Label control is actually a quality assurance function. The GMPs require that adequate controls be instituted to maintain labeling integrity. Most companies assign a quality control inspector to the function of label control. This individual proofreads labels for accuracy, both during the printing operation and before issuance of the labels. The label control areas should be restricted to authorized personnel. All preprinted labels and labeling should be maintained and stored in a manner that will ensure proper identification and prevent mixups. This can be enhanced by using shrink wrapping and special well-marked bins. All labels, once they are issued, should be double counted, and a sample from each lot should be kept for historical purposes. The label control manager should obtain a copy of the quality audit program for industry, available from the Food and Drug Administration. In this document, under subpart G, there is a section devoted to packaging and labeling control.

The checklist for area inspections, storage and labeling materials includes the following:

Prior to the implementation of any labeling and packaging operation, there is an inspection of the area where the operation is to occur by a designated individual to assure that devices and labeling materials from prior operations do not remain in the labeling or packaging area.

Any such items found are destroyed, disposed of, or returned to storage prior to the onset of a new or different labeling or packaging operation.

Labels and labeling are stored and maintained in a manner that provides proper identification and that is designed to prevent mixups.

Labeling materials issued for devices are examined for *identity* and *accuracy*.

A record of such examination, including the date and person performing the examination, is maintained in the device history record.

Critical devices have more stringent requirements, and those listed in the quality audit are:

Labels issued for critical devices contain a control number.
The signature of the individual who proofreads the label and other labeling and the date of the proofreading are recorded.
Access to labels and labeling is restricted to authorized personnel.

XII. CONCLUSION

The Medical Device Amendments of 1976 have had a profound impact on the manufacture of medical devices. Nowhere is this impact more evident than in the area of labeling. The law impinges on an actual content of labeling, as well as on its control. Countless product recalls have resulted from mislabeling, from labeling that has failed to include adequate instructions for the device's use, or from labeling that is false or misleading. For these reasons, all device manufacturers should concentrate on, and train specialists in, the area of labeling compliance. (Additional information on this subject can be obtained at workshops sponsored by professional organizations such as the Regulatory Affairs Professional Society, HIMA, or the Food and Drug Law Institute.) Manufacturers should also hold periodic meetings for label control personnel; at such meetings, individuals can be made aware of defects or errors they are likely to encounter as part of their quality assurance function. The major negative factors found to increase the possibility of labeling mixups are as follows:

1. Use of cut labels.
2. Use of labels similar in size, color, and shape.
3. Evidence of poor label control practices.

Each of these is a primary cause of label mixup recalls. It should be noted that most packaging and labeling mistakes can be assumed to be human errors. Although the FDA has identified some common factors, both positive and negative, that can be taken into consideration to approve labeling operations, in the final analysis, the most important factor is human. Packaging line and label control employees who have not been trained, who do not understand the potentially serious consequences of labeling errors, who are taught that high production is more important than extreme care, will make mistakes. As with all critical operations in the medical device industry, assuring the proper qualification, experience, and training of the personnel involved should take the highest priority.

Labels, warnings, and use instructions should be part of the checklist and discussed during formal design review to help assure that safety, efficacy, and reliability are optimally enhanced prior to releasing a medical device to final production and distribution. The evaluation of the effectiveness of the warnings and instructions does not end, however, once the product is marketed. The manufacturer must continue its evaluation by monitoring field performance of the product. Field performance may indicate that changes are needed or that additional post-sale warnings must be given.

Remember that words that are lucid, precise, conspicuous, and neither overstated nor understated, whether on the label or in literature furnished with the product, will be most effective.

For writing labeling, the best rule is the old rule: Keep it simple, make it clear.

SUGGESTED READING

1. Hadden, S., *Read the Label*, Westview Press, Boulder, CO, 1986.
2. King, L. S., *Why Not Say it Clearly*, Little, Brown, Boston, 1978.
3. U.S. Department of Commerce, Patent and Trademark Office, *General Information Concerning Patents*, Washington, D.C.
4. Department of Treasury, U.S. Customs Service, *Importing into the United States*, Washington, D.C.
5. *Trademark Management, A Guide for Executives*, United States Trademark Association, Clark Boardman Company Ltd., New York, 1981.
6. U.S. Department of Commerce, *Metric Laws and Practices in International Trade*, Washington, D.C.
7. Food and Drug Administration, *Import/Export of Medical Devices, A Workshop Manual*, Rockville, MD, September 1983.
8. Food and Drug Administration, *Device Good Manufacturing Practices*, 4th ed., Rockville, MD, November 1987.
9. Food and Drug Administration, *Sterile Medical Devices, A GMP Workshop Manual*, 4th ed., Rockville, MD. January 1985.
10. Food and Drug Administration, *Labeling—Regulatory Requirements for Medical Devices*, Rockville, MD, February 1986.
11. Strunk, W., and White, E.B., *The Elements of Style*, 3rd ed., Macmillan, New York, 1979.
12. Viscusi, W. K., and Magut, W. A., *Learning about Risk*, Harvard University Press, Cambridge, MA, 1987.
13. Flynn, G. W., Product Liability—Warnings, Labels, and Instructions, 1989 ASQC Preproduction Quality Assurance and Liability Prevention Conference Proceedings. March 1989, Bethesda, Maryland.

29

Home Use Diagnostics: Special Considerations

STEPHEN C. KOLAKOWSKY *Carter Products, Division of Carter-Wallace, Inc., Cranbury, New Jersey*

I. INTRODUCTION

A. Why Home Use Diagnostics?

The past decade has seen a number of technological advances in the area of *in vitro* diagnostic (IVD) test kits for the laboratory. Principal among these advances has been the development of monoclonal antibodies. The incorporation of monoclonal antibodies into IVD test kits has allowed the production of highly specific tests, which in turn provided the opportunity for greater test sensitivity. Increased specificity and sensitivity has yielded IVD tests with greater overall accuracy. This greater accuracy together with the development of less complex test procedures have provided an opportunity for development of IVD tests that can be used by individuals other than the traditional clinical laboratory worker—the home user.

The idea of home use IVD tests is not new. Urine glucose tests have been available for home use since before the 1950s. Some of the first IVD tests developed for pregnancy detection were originally intended for home use. However, the U.S. Food and Drug Administration (FDA) was not ready to accept home use pregnancy tests in the 1960s any more than FDA is willing to accept some tests for the home use market today.

Home use IVD tests are only one part of at-home self care. For centuries people have been self-diagnosing their ailments and performing self-treatment. Many of the methods of self-diagnosis have lacked a sound scientific and medical basis, and some have been nothing more than superstitions or "old wives" tales. Self-treatment, although sometimes suffering from the same deficiencies as the use of some home remedies, has been more successful through the use of over-the-counter (OTC) drugs. But now, with the advent of highly accurate and simple medical devices and IVD test kits designed for home use, consumers will eventually be able to supplement self-treatment with more accurate self-diagnosis.

Self-care is not new, but the degree to which people are willing to take greater responsibility for their own bodies and their health is new. Besides the technological advancements of the last decade in IVD testing, there has been a greater awareness of health consciouness, especially preventative health care, and physical fitness among the public in general, particularly among the "baby boomer" generation. This increased awareness and concern with health, the convenience and privacy offered by home use IVD tests, and increases in health care costs have created a demand by consumers for home use test kits. Manufacturers recognizing this demand are seeking to fulfill it.

Home use IVD test kits provide manufacturers with a new market area into which to expand the use of their existing line of professional use products. It may not be obvious except to those intimately involved in the IVD industry, but many of the "new" home use IVD tests up to the mid 1980s have been nothing more than a repackaging and relabeling for at-home use of "old," existing laboratory test kits. Some of the more recently established biotechnology companies have seen the home use market as an opportunity to enter a market with their newly developed methods and reagents with less competition than the professional laboratory market. It is important, however, to recognize the differences in the needs of the professional and the home users. It is important to recognize that there might be some special considerations required for a home use IVD test that do not present a problem for a professional use IVD.

B. Sepcial Considerations

The home user, unlike the professional laboratory worker, is very probably untrained, inexperienced, and unskilled in the use of IVD tests. In addition, the home user very probably is lacking in scientific and medical education and may be lacking in education and reading ability in general. The home environment is also quite different from a clinical laboratory; therefore, the home use IVD test must be designed for the home. In addition to the user and the product itself and its labeling, the manufacturer considering introducing a home use IVD test must take special consideration of the regulatory requirements that FDA has and is developing, and the special marketing considerations for a home use IVD test.

II. USER CONSIDERATIONS

A. Characteristics of the User

The first and primary consideration for a company desiring to market a home use IVD test kit must be the user.

It must be recognized that in the vast majority of cases, if not almost all cases, the home user will be unskilled and inexperienced in laboratory methods. The most that may be expected is that the user may have performed a simple blood typing test in a high school general biology class 15 or 20 years ago, and maybe the user has performed one or two other home use tests.

The educational level of the home consumer is another factor that must be given consideration. Unlike the professional user, the consumer of a home use IVD will not always be a Ph.D. pathologist, biochemist, medical technologist, or someone else with scientific or medical training or experience with laboratory tests. The home consumer may not have completed high school.

education. Even if the home user is one with scientific or medical training, the user may not be familiar with laboratory tests, particularly a new home use IVD test kit. The company marketing a home use IVD cannot even assume the home consumer will have a good knowledge of the English language or speak or read English at all.

In addition to skill, education, and language barriers, the user characteristics of vision and dexterity must be considered. If the home use IVD test kit is likely to be used by the elderly, visual acuity could present a problem for the user. Small type size, fine measurement lines, or fine discrimination can present problems in the use of the test kit. Not only the elderly could have a problem, poor vision could afflict any age user. The dexterity required to manipulate test tubes or measuring devices or to push small buttons on an instrument is also likely to present a problem for the elderly home user, as well as for the user with arthritis, or even a big man with large hands and bulky fingers.

The product itself must be considered in light of the home user. Is the test procedure so complex that even a good, careful reader of the directions will have difficulty performing the test? Is the intended use of the test so broad and nonspecific that the home user will not understand for what disease or condition the test is intended? Will the user be able to understand the consequences of the results?

B. Needs of the User

Does the home consumer really need the home use IVD proposed to be marketed? Some consumers may not want to know what a diagnostic test might tell them. (No news is good news for some people.) Even if the consumer wants to know what is happening with his or her body, IVD tests requiring certain types of specimens would prevent the home consumer from using a home use IVD. For example, a spinal tap is out of the question; a venipuncture, absent some future technology development for obtaining the specimen, is also very much an unlikely specimen to be obtained by the ordinary home consumer; the ordinary home consumer of a screening test may also be reluctant to perform a finger prick to obtain a peripheral blood speciman.

Some individuals at the FDA and within the medical community argue that there is no need for consumers to perform diagnostic tests at home, that is why there are physicians and clinical laboratories. A physician can perform a physical examination and order the necessary IVD tests and other diagnostic procedures. These would be performed by professionals trained and experienced in performing such tests and procedures. The argument continues that a physician is trained to interpret test results and will take the necessary and appropriate follow-up action. But, barring those individuals to which no news is good news, there is likely to be a need for a home use IVD test by the home consumer.

Individuals, since the 1960s, have become more "self" oriented and more health conscious. They want to know what is happening to their bodies and want to take care of themselves. They feel they have a right to know what is happening with their bodies without having to go to someone else to tell them. But even before the 1960s, consumers have had a do-it-yourself attitude for one reason or another, usually as a means to save money. People have been diagnosing themselves and treating themselves for years with either home remedies or OTC medications, or a combination of both. The sequence of events often followed is to (1) do nothing and wait and see what develops, maybe it will go away, but if it does not then (2) some home remedy may be used. If the condition continues or worsens, (3) an OTC remedy might be used; and (4) finally, as a last resort, when all else fails, an appointment is made with a physician. The proper use of a home remedy or an OTC drug product depends on being able to properly diagnose the ailment. For nonserious conditions, an at-home use IVD test can help provide important information to the consumer.

Today's consumer, including the health care consumer, wants more information. As in other product and service areas, the informed consumer is a good consumer. A home use IVD test offers consumers privacy and the right to decide for themselves what they will do with their bodies. Some individuals may feel intimidated by a doctor and feel they have no choice or voice in their treatment. The opinions of physicians are no longer accepted dogmatically as they once were by many people. The 1960s resulted in the diminishment of recognized "heros," a lack of confidence and lack of acceptance of authority. These attitudes, right or wrong, on the part of consumers have contributed to the consumers' demand, the need, to have more information about whatever may be happening within their bodies.

What the consumer will do with the result from his or her home use test is an important consideration. Even if an individual feels intimidated by his or her physician, it is most often best for the individual to follow the physician's directions, especially when a serious disease or condition is involved. Likewise, it may be extremely important to emphasize strongly in the product labeling that the consumer performing a home use test follow directions exactly and seek a physician's consultation when a particular result is obtained, or sometimes even regardless of the result. The latter situation raises the question of the need for a particular home use IVD test in the first place. If the consumer performing a home use IVD test must seek a physician's consultation regardless of the test result, what utility is there in the home test? The answer to this question rests in the individual's right to know. The consumer's need for home use IVD tests also is justified by the sometimes too frequent situation where a patient is unable to obtain a doctor's appointment for several days or where the individual does not have a regular physician. Such a person is likely to postpone seeking medical attention until the condition worsens. A home use IVD test would also serve the need of the individual who has a "wait and see" attitude. This person does not want to incur the cost of a visit to the doctor unless something is "really" wrong. A home use IVD test may aid in this type of person seeking medical treatment sooner.

Consideration of the user's need for an at-home use diagnostic test also depends on the type of test. Not all IVD tests yield a "diagnostic" result. An IVD test may be for "screening," "diagnosing," or "monitoring" purposes.

A "screening" test is one used when signs and symptoms are not precise or are absent entirely. The test result only suggests a particular condition may exist. A "screening" test usually will have a high level of sensitivity but lack specificity in some degree. Such a test would be used to screen out negatives, those patients who most likely do not have the disease or condition in question.

A "diagnostic" test is used when specific signs and symptoms suggest a particular disease or condition. The test provides additional information for reaching a conclusive result confirming the existance of the disease or condition. A "diagnostic" test requires a high degree of both sensitivity and specificity; overall, it should be highly accurate, approaching 100%. FDA, at this time, will not allow a home use test to be considered "diagnostic," and manufacturers are unwilling to assume the liability risk of calling their produce a "diagnostic" test. Therefore, many "screening" home use IVD tests are actually highly accurate "diagnostic" tests with high sensitivity and specificity, for example, most home use pregnancy tests. The fact is that many true "screening" tests would not be acceptable to FDA because of the concern FDA has over false results. Also, since the user of a home use IVD test is unlikely to have the training and education to make a presumptive diagnosis based on signs and symptoms, a "diagnostic" test should not exist by FDA's definition. (This should be contrasted with FDA's prerequisite for OTC drugs that the home consumer be able to make a presumptive diagnosis in order to properly select the correct OTC drug.) The situation that exists, therefore, is a confusing one. FDA is defining IVD tests for "screening" and "diagnosing" when in reality neither a true "screening" test nor a true "diagnostic" test is likely to be acceptable to FDA. In reality, a "diagnostic" test is considered a "screening" test when it comes to home use.

The final category, a "monitoring" test, is one defined for use when a particular disease or condition has already been diagnosed. Such a test is used to monitor the disease or condition or the treatment, such as therapeutic drug monitoring or monitoring of glucose levels. A "monitoring" test is probably the most acceptable to FDA and the medical community because a "monitoring" test serves to harmonize professional care with self-care. A medical professional makes the diagnosis and decides on the appropriate treatment regiment for the patient. The patient then uses a home use IVD test to monitor the condition and treatment. Another factor in favor of "monitoring" tests is that the home user is usually instructed in its proper use and interpretation by a health care professional.

C. Effect of the Result on the User

Whether a screening test, a diagnostic test, or a monitoring test, the result of that test will have an effect on the home user. For a screening or diagnostic test, a positive test could indicate the possible existence of a serious disease or condition. A negative test would provide some assurance to the home user that the disease or condition does not exist. But if the disease or condition does or does not exist, then what is the home user to do? In the case of a positive result, hopefully the home user will be motivated to do something—seek the advice of a physician if the disease or condition is serious, or for less serious situations, to adjust their diet, life style, or physical conditioning, or initiate the proper regiment of treatment using OTC medications. In the case of a negative result, either no action is needed on the part of the home user or it might be necessary to perform a different home use IVD test or seek the consultation of a physician. Regardless of the result, therefore, the full range of possibilities must be fully explained to the home user in the product labeling.

In the case of a serious disease or condition, a positive result can create a great deal of anxiety for the home consumer. In the case of a false positive result, this would be unnecessary anxiety. A false positive result could also cause the consumer to go to the unnecessary expense of a visit to his or her doctor and possibly the expense of additional tests, possibly putting the patient at unnecessary risk. The counterargument is that the home consumer must have suspected some problem, and therefore, even if a false positive result served as the impetus for the consumer to seek medical attention, the test served a purpose. The concern is, however, that with some potential home use tests, such as a test for AIDS or cancer, the home user may not seek the proper medical care but instead may decide to take some other drastic or ill-conceived action, physically or financially, that could result in irreparable harm.

A false negative result could also have a profound effect on the home user. A false negative result could give a false sense of security or confidence to the home user. The false security could lead the home user to postpone seeking necessary medical care. This could allow a serious condition to worsen or the patient's condition to deteriorate to an incurable state or to a state that requires more extensive and longer treatment at additional expense. The false sense of confidence that may be created could lead to the spread of a serious infectious disease, such as a sexually transmitted one.

Whatever the result of a home use VID test kit, positive or negative, true or false, it must be recognized that the result will have an effect on the user. The home consumer will use the test result to make a decision regarding the state of his or her health and the care needed.

III. PRODUCT CONSIDERATIONS

A. Technology

Hot water bottles, fever thermometers, steam vaporizers, and other medical devices have been used by home users for decades. More recently, blood pressure monitors, urine and blood glucose tests, pregnancy tests, ovulation tests, and colorectal (occult fecal blood) tests have entered the home use market. A large proportion of these home use medical devices and IVD tests have been old professional use devices and laboratory tests repackaged into "new" home use products. Now, due to the technological advances made during the past decade, many more improved old and turly new medical care devices and IVD tests for home use are expected to enter the market.

For home use IVD test products, the most significant technological advance of the past decade has been the development of monoclonal antibodies. Monoclonal antibodies have made possible highly accurate IVD test kits. Because of the exceptionally high specificity of the monoclonal antibodies, the sensitivity of the tests has been able to be increased to greater levels without increasing the level of false positive results. [The sensitivity of a test is its ability to detect the substrate of interest (an antigen in an immunoassay); a lack of sensitivity leads to false negative results. The specificity of a test is its ability to discriminate between the substrate of interest and other similar molecules that may be present in the specimen sample; a lack of specificity leads to false positive results.] Another important factor is that monoclonal antibodies can be produced in relatively large quantities, relatively inexpensively. The increased accuracy afforded by the use of monoclonal antibodies has and will make possible the development of home use IVD test kits that before now lacked the reliability needed for the home user.

A second important technological development has been the enzyme immunoassay or EIA. This methodology is also known by several other names, most commonly ELISA for enzyme-linked immunosorbent assay. Most immunochemical methods utilize some kind of "tag" to allow the user to visualize the reaction of antigen and antibody. This so called tag can be a latex particle, as in the case of a latex particle agglutination assay (LPA); a red blood cell, as in the case of a hemagglutination assay (HAA); a radioactive molecule, as in the case of a radioimmunoassay (RIA) or radioreceptor assay (RRA); or a color-change reagent activated by an enzyme, as in an enzyme immunoassay (EIA). Except for EIA, the other immunoassays involve relatively complex procedures prone to user-induced errors or the use of radioactive materials, unsuitable for home use, in complex procedures. The EIA provides a simple procedure with a simple color-change reaction marking the end point, simple enough for home use. Together with monoclonal antibodies, EIA provides the means for the development of IVD tests for home use with the high accuracy and simplicity necessary for home use.

A third technological advance has been in the area of electronics and computerization. Advances in electronics and computerization have made once large, bulky, and complex laboratory instruments into small, low-cost, low-maintenance, simply operated chemical and biological sensors. Applications of this technology have already provided automatic blood pressure monitoring meters, electronic thermometers, and greatly improved blood glucose test meters. Further applications and developments will lead to further improvements in currently available metering devices and make possible the development of new home use IVD test kit products.

But even given these technological advances and improvements, a home use IVD test product (a reagent test kit or instrument or both) must have product characteristics designed with the home user in mind. Monoclonal antibodies, enzyme immunoassays, and computerized electronic instrumentation may be inherently simple and easy to use, but the actual product developed using these technologies must still be designed with characteristics that are user friendly for the home user.

B. The "Lab"

One must remember that the laboratory is in the home. The "lab" is not going to be a place with all the ancillary reagents, equipment, and other resources usually found in a clinical laboratory. The "lab" is going to be a kitchen, a bathroom, a bedroom, a den, or some other room in the home—it could also be the men's room or ladies' room at work or the office desk. Some special considerations must be given to this "lab" environment, which may be user friently but may not be product friendly.

The test procedure should not be too lengthy. Interuptions could impact on the home user's ability to perform the test. The home use IVD product's characteristics must take into consideration the disruptions that are likely to occur in the home "lab." The phone or door bell may ring, but can be ignored; children and other household members may not be as easily ignored. Young children especially, when ignored, can create more disruption. Even though the home use IVD test may offer the consumer privacy, the home "lab" may not offer the same privacy.

These disruptions of course cannot be controlled by the manufacturer. When designing the home use IVD test, however, the manufacturer should consider these likely disruptive factors. The test should be designed so that its procedure is rapid and flexible. A rapid test, by its short time required to perform the procedure, minimizes the probability of distruption. The test procedure should be designed to allow the disrupted user to delay the next step for a few minutes without causing the test to be invalidated.

In all likelihood the home "lab" will not have the space a commercial laboratory would have to set up and spread test reagents, directions, etc., over several square feet of laboratory work bench. The lab bench may be a small section of kitchen counter, a small area along the side of the sink in the bathroom or the top of the toilet tank, or the office desk. Therefore, space requirements are important for the home use IVD test. A home use IVD test kit should be compact and require little space for it to be performed. This is especially important for those tests that are likely to be carried away from home, to work or on a trip or vacation. For example, a glucose monitoring kit could be used every day and often several times a day; therefore, it would be necessary to be able to carry it easily or to pack it in a suitcase.

C. Safety

Safety considerations of IVD test reagents are important in the professional laboratory. In the home use environment, safety requires special consideration. Unlike the professional user of IVD tests and their reagents, the home user should be considered unfamiliar with the safe handling of hazardous reagents. Even IVD reagents handled routinely in the professional clinical laboratory by trained and experienced technologists may present a hazard to the untrained and inexperienced home user. If possible, hazardous reagents, such as strong acids and bases, live microorganisms, highly toxic poisons, etc., should be avoided in home use IVD test products.

The possibility of accidental ingestion and inhalation of hazardous reagents by the home user must be considered. Spills can create a potentially serious hazard for the user. When hazardous reagents are included in a home use IVD test kit, special cautions should be provided to the home user. Besides the home IVD test reagents, the test specimen may present a potential hazard, especially if the IVD test is intended for an infectious disease or a condition that may be associated with an infectious disease.

Routine laboratory procedures for the professional laboratory worker may present a hazard to the untrained, inexperienced, and uneducated home IVD test user. Consideration could be given to eliminating from home use IVD test kits glass test tubes, pipettes, ampules, etc. The application of heat to test reagents is a potential hazard that should also be avoided, if possible. in home use IVD tests.

Specimen collection techniques can present a hazard to the home user. Besides the potential for contamination of the user and others by the specimen, the collection technique itself may present a hazard. The user should be instructed to use special care when performing an invasive technique to obtain a specimen. A finger prick with a lancet can present a hazard if the user allows the lancet to become contaminated or somehow results in creating a wound much larger than intended for a finger prick. Even obtaining a simple throat swab or a vaginal swab could present a hazard without clear and simple directions to the home user.

Special consideration should be given to the disposal of the specimen and the used and remaining unused test reagents. Appropriate disposal instructions should be provided to the home user. Instructions might be given, for example, to wrap the specimen container and reagent vials in paper toweling or other adsorbent material and then place them into a palstic bag that can be sealed or tied closed before disposing in the trash—outside, if appropriate.

Special safety consideration should be given to children and animals. Home users should be clearly instructed to store the test kit where children and family pets will not be able to get to it. Disposal instructions should likewise provide cautions to keep items out of reach of children and family pets.

The one last special safety consideration a manufacturer of home use IVD test kits should be sure to address is the potential for a liability suit. Failure to address the safety characteristics of the IVD test product pertinent to its entry into the home environment can easily result in a single suit that, if successful, could completely wipe out any profits from the sale of the home use IVD test.

IV. LABELING CONSIDERATIONS

A. Readability

Many of the items discussed so far in this chapter regarding special considerations for home use IVD tests are associated with providing instructions and information to the home user of a home use OTC IVD test. These instructions and information find their place in the labeling of the home use IVD test product. Just as these items require special consideration in and of themselves, so does the labeling of a home use IVD test require special consideration.

The primary consideration for any instructions and information provided to anyone is that the instructions and information be provided in clear, concise, and understandable language appropriate to the audience to whom they are directed. For the user of a home use IVD test, the "readability" of these instructions and information is of paramount importance.

Because of the inherent technical nature of a home use IVD test product and the scientific and medical information that must be provided to the home user in the labeling of the product, the readability of the labeling can present a particularly formidable challenge to the manufacturer. If the labeling of a home use IVD test lacks readability, the home user may not be able to perform the test or interpret the results correctly. The home user may be led to reach an incorrect decision regarding the state of his or her health. This of course could then lead to other problems for the manufacturer of the home use IVD test kit.

The determination of the readability of the labeling of a home use IVD test can be approached in a systematic manner. The first step in preparing labeling for any IVD test, whether for professional use or home use, is to have the individual or individuals most familiar with the test write the directions for use. This individual is usually someone from the research department who has helped develop the product. The next step would be to have this research draft reviewed by the regulatory affairs department for compliance with applicable regulations and guidelines (as discussed below). The regulatory affairs professional should be able to "clean up" the first draft, adding any appropriate caution and warning statements, simplifying the language, and assuring the instructions and information presentation are clear, not misleading, informative, and appropriate to the home user and the IVD test in question. At this time, the package label and any other labeling that might accompany the product should be developed and reviewed.

The next step in the labeling development process is to attempt to assure the understandability or readability of the labeling. There are several readability formulas that have been developed just for this purpose [1-3]. By applying the particular formula selected, a reading grade level is determined, that is, the grade level in school the reader would have had to attain in order to be able to accurately read and understand the text. A reading grade level of about seventh or eighth grade is generally recommended to assure that most consumers can comprehend what is being said in the labeling of an OTC product [4]. Due to the technical nature of a home use IVD test, many scientific and medical terms will unavoidably find their way into the labeling and cause the reading level of some sections of labeling to be higher. As a guide, no section of the labeling should be beyond a tenth grade reading level.

After the labeling has been written to the appropriate reading grade level, the next step is to test the readability of the labeling by actually having consumers read the labeling. Several approaches could be taken for a consumer test of the labeling. One such approach might be as follows: A panel of consumers could be asked to read the product labeling, then complete a questionnaire and answer questions about the product and labeling. Such a consumer reading test could not only serve the purpose of determining if actual consuemrs can read and understand the labeling, the intended use of the product, and the need for appropriate follow-up, but also such a consumer test could serve a marketing purpose to determine probable likes and dislikes about the product, the consumer's feelings about testing for the disease or condition, the product's potential for consumer sales, etc. As with the reading grade level test, adjustments to the labeling can be made based on the results of the consumer reading test.

Following the consumer reading test, a simulated use test with another panel of consuemrs could be performed. In a simulated consumer use test, each panel member would be given a test kit and a simulated specimen. The simulated specimen panel should be composed of negative and various levels of positive samples for a qualitative test format. Each simulated specimen should be prepared in duplicate and the duplicate tested by a laboratory technician. Each consumer panel member would then be asked to read the instructions and perform the test, without any assistance from the test monitors or other panel members. Again, each consumer panel member should be asked to complete a questionnaire and to answer appropriate questions regarding the labeling and the product.

Following the above labeling development process should assure the readability of the product labeling. However, even though these steps are followed, there can be no guarantee that every home user of an OTC IVD test kit will be able to read and understand the instructions and information presented in the product labeling—additional customer support will be necessary (see Section VI). In addition to customer support, the labeling of a home use IVD test can be enhanced by other techniques to aid the consumer that still might have difficulty reading and understanding the labeling even at a seventh or eighth grade level.

The use of illustrations in the product labeling can enhance understanding, especially in the procedure section of the package insert. To aid the home user, each instruction step in the test procedure should be accompanied by an illustration of what the consumer is expected to do. An illustration of the test kit should also be used to aid the consumer in identifying each of the components. The illustration of the results that are expected, including inconclusive results, should be considered to aid the consumer in reading the test result. The use of headings and subheadings, bold face or other type faces, color, boxed text, and other typographic emphasis for warnings and precautions, important points, and other information requiring special attention by the consumer can all enhance the readability of a home use IVD test product.

A final consideration in providing readability in product labeling language is the use of a language other than English. If the home use IVD test product is expected to appeal to a segment of the population that may have difficulty reading and understanding English or that would be distributed in areas of the country where the population typically speaks another language, the use of supplementary labeling printed in another language, such as Spanish, should be considered [5].

B. Content

The information required to be on the package label and in the package insert labeling of a home use IVD product is provided in FDA regulations that can be found in Title 21 of the Code of Federal Regulations (21 CFR), Parts 801 and 809. FDA has also drafted a "points to consider" document [6] for home use IVD test products, which provides guidance to manufacturers (and FDA reviewers!). A manufacturer considering the marketing of a home use IVD test should obtain copies of these regulations and points to consider. For this reason and because these documents are always subject to revision, no further discussion of these FDA requirements will be provided in this chapter. The National Committee for Clinical Laaboratory Standards (NCCLS) is also preparing a guideline for the labeling of home use IVD test products [7].

V. REGULATORY CONSIDERATIONS

A. 510(k) versus PMAA

A difference between a 510(k) Premarket Notification and a PMAA (premarket approval application) is in the time and money that must be invested to get a product to market. A home use IVD test introduced to the market via the 510(k) route can be there with substantially less time and less money expended than with a PMAA. A 510(k) can usually clear the FDA within 90 days, and even when a "maxi-510(k)/mini-PMAA" is required, FDA clearance can be attained in substantially less time than a PMAA. The PMAA route to market requires extensive, time-consuming, and costly clinical studies, and an extensive amount of information about the product, the methods of its manufacturer, quality control, packaging and stability, and the facility where it will be made, including a FDA inspection for compliance with GMPs. Once all this information and data are put together and submitted to FDA, it will take FDA about a year to review it—longer if additional information is required by FDA.

It would appear then that the preferred route to FDA clearance to market a home use IVD test product would be by a 510(k) notificaiton. The 510(k) route is the faster and less expensive route to market, but a PMAA also offers some advantage. Because of the higher costs and greater amount of time required, not to mention the higher level of regulatory expertise required, to conduct the required studies, prepare the submission, and guide it through the FDA review process, few manufacturers will want to take the PMAA route to market, and many will forego the introduction of a product requiring a PMAA. Therefore, a product cleared by the PMAA route offers some extent of exclusivity to the manufacturer who is the first to introduce a new IVD test product. FDA, however, may not provide the choice of a 510(k) but may instead require a PMAA.

Most manufacturers presented with the choice would opt for the easier 510(k) route to the marketing of a home use IVD test. FDA, however, can place some formidable roadblocks in the way of getting a home use IVD test to market. Even with a 510(k), an FDA reviewer can make requests for additional information that turn the simple notification into a "mini-PMMA." Therefore, a small, inexperienced company introducing its first home use IVD product should consider retaining a regulatory affairs consultant with the institutional expertise of the workings of FDA to prepare and manage the 510(k) through the FDA review process.

Whether a home use IVD test product can proceed by the 510(k) route or must travel the PMAA route depends on the classification of the product. A new IVD test never marketed before April 28, 1976, must be reviewed through the PMAA process. An IVD test product that is substantially equivalent to another IVD test product that has been on the market may be required to have a 510(k) or a PMAA submitted, depending on the product's classification. (See Chapter 3 for an explanation of the three classes of medical devices and their regulatory requirements.) Separate regulations, including classifications, do not exist for home use IVD tests and professional IVD test products. No differentiation is made in the Federal Food, Drug, and Cosmetic Act or in the classification regulations between a home use IVD test and a professional use IVD test. Therefore, a Class I, Class II, or Class III *in vitro* diagnostic test is in the same class whether it is a home use IVD test or a professional use IVD test.

B. The 510(k) Submission

Other chapters provide detailed information regarding the preparation, content, and format of a 510(k) Premarket Notification. Those chapters should be referred to for general guidance. Here, only the special considerations that pertain to a 510(k) for a home use OTC-IVD test will be discussed. FDA's Center for Devices and Radiological Health (CDRH) is preparing a points-to-consider document that provides detailed guidance on the type of information and data that should be contained in a 510(k) for a home use IVD test product [6]. Any manufacturer considering the marketing of a home use IVD test should obtain a copy of this document from CDRH.

The 510(k) notification for a home use IVD test is basically the same as one for a professional use IVD test. They both require the same basic information required by 21 CFR 807.87—product name, company name, address, and registration number, product classification, applicable 514 standard, and labeling. The statement of substantial equivalence sections of the 510(k)s are also similar. However, for the home use IVD test, the 510(k) must contain an additional element, the consumer use study.

C. Performance Considerations

The performance characteristics of a home use IVD test when performed by the consumer must be comparable to those of the test when performed by a trained and experienced laboratory professional. This additional element of performance criteria is what makes a home use IVD test 510(k) different from that of a professional use IVD test.

An IVD test should provide accurate and reliable results whether it is intended for professional use or for home use. The first step once an IVD test has been developed is, therefore, to evaluate its performance in appropriate studies. The studies should determine the analytical sensitivity and specificity of the test, that is, the minimum level of analyte that can be detected and the degree to which the test is subject to interfering substances, respectively. Other technical performance parameters, such as the precision, accuracy, and reproducibility of the test, should also be determined. Clinical sensitivity and specificity should be determined via appropriate clinical studies or laboratory studies. Once the technical performance of the IVD test product has been defined, if the product is intended for home use, a consumer use performance study must be conducted. In the consumer use performance study, the ability of the typical home user to perform the test is compared to the professional laboratory worker's ability to perform the test using duplicate specimens.

Two consumer use studies of different design are useful. The first is a study conducted in a controlled situation where the consumers participating in the study can be monitored. For this type of study, actual specimens with previously determined values or specimens prepared ("spiked") with various levels of analyte are double blinded to the consumer participants and the professional laboratory worker(s) performing the test. A special consideration that should be given to conducting a consumer use performance study such as this might be the specimens to be used. With the possibility of exposure to HIV (the AIDS virus), special consideration should be given to treating the specimens to inactivate any virus that may be present or to using artificial specimens prepared to look like real specimens but without the potential for causing infection by HIV or any other microorganism.

The second type of consumer use performance study would involve an actual use situation. In this type of study, a panel of consumers is selected and asked to take the test kit home to use it. The benefit of this study design is that it evaluates the product in the actual use environment. This type of study, however, may present formidable obstacles. It may be difficult or impossible to impanel a sufficient number of consumers with a particular disease or condition at the time of the study. The use of the study subject's own specimen also creates the need to consider the implications of the Investigational Device Exemption (IDE) regulations. Most studies of IVD products are exempted from the IDEs because the experimental test result is not used to make a diagnosis or another procedure or method is used in parallel [see 21 CFR 812.2(c)(3)]. When consumer participants in a study utilize their own specimen, it must be made perfectly clear that the result obtained by use of the test under study is not to be used to make a decision regarding the state of their health. A home study also is impossible to monitor. A study monitor cannot be placed in each subject's home. Therefore, the information from the visual observation of the subject's performance of the test is lost in this type of study.

Once the technical performance study data and the consumer use study data are compiled, analyzed, and summarized, it may be incorporated into the 510(k) or PMAA and submitted to the FDA.

D. FDA

After the 510(k) or PMAA has been submitted, an FDA reviewer will first perform an administrative review of the submission. This is not a substantive review but a review to be sure that all the necessary pieces of information are there. If any part of the required information is missing, the sponsor of the submission will usually be notified within 30 days that the submission is incomplete and additional information is required.

Usually, if the sponsor of a submission does not hear from the FDA within the first 30 days after submitting a 510(k) or PMAA, it means the submission has at least passed the preliminary review and the submission has at least been compiled correctly. Now the in-depth scientific review beings. This phase usually takes up to a total of 90 days from the date of receipt of a 510(k) by FDA. With a PMAA, FDA has 180 days to perform its review according to statute. But, as with a 510Ik), the "clock" is restarted each time additional information is required to be submitted. Therefore, it not unusual for a 510(k) to take 4-6 months or a PMAA to take 12 or 18 months or longer to attain FDA approval.

FDA has drafted a points-to-consider document [8] to aid manufacturers in preparing 510(k) and PMAA submissions for home use IVD test products. The points-to-consider document also serves as a guideline for FDA reviewers. Therefore, this document is a double-edged sword. While helping the manufacturer to understand what FDA would like to see in a submission for a home use IVD test product, the points-to-consider document also requires the sponsor to justify why certain points are not applicable to the product. And, although this document only provides points to consider and is not a regulatory guideline as defined in FDA's regulations, some FDA reviewers may apply the points as definitive requirements necessary to attain FDA clearance to market. (This has been the case in many situations even though the points-to-consider document was not officially issued at the time of the writing of this chapter.) Therefore, although FDA may be making things easier for sponsors of home use IVD test product submissions, FDA is also making things tougher by delineating for FDA reviewers the "requirements" for clearance to market a home use IVD test product.

VI. MARKETING CONSIDERATIONS

A. Consumer Need?

One of the primary questions FDA asks about any home use IVD test product is whether the product provides a clinical utility. In FDA terms, does the product offer a medical benefit? In marketing terms, the question is whether the product meets a consumer need. Hopefully, the marketing perception of consumer need and FDA's perception of clinical utility and medical benefit are one and the same.

In the introduction to this chapter, the background elements of consumer demand for home use IVD test products was discussed. The question often raised by FDA and consumer organizations, however, is whether a new home use IVD test product is intended to meet an existing consumer need or is intended to meet a consumer demand created by marketing and advertising techniques. Advertising does not create a need, but it can create demand for a particular product—which is its intent. However, if there is not an underlying need, no amount of advertising will create sufficient consumer demand to support the marketing of any product, no less a home use IVD test. The costs associated with the research, development, and marketing of a home use IVD test product can be very considerable. The manufacturer of a proposed new home use IVD test product would be well advised to consider available market research information before making such a large investment and would be ill-advised to make such an investment where no real need exists.

The successful marketing of a home use IVD test product requires a disease or condition with a high prevalence in the general population or, in the case of a monitoring type IVD, a high amount of repeat and continuous usage. Market research should be able to identify potential diseases or conditions that might be appropriate to screen for or monitor with a home use IVD test. The market research should also consider the probability of a home consumer's willingness to perform the test at home. The prevalence of a disease or a condition does not equate with consumer acceptance for a do-it-yourself IVD test.

Once a consumer need for a home use IVD test has been identified, and even though there exists consumer acceptance for such a test, advertising and marketing techniques may still be necessary to make the public aware of the need for testing. (This is what FDA and consumer groups may actually be objecting to when they say advertising is creating a need for a product.) Consumers may not, however, be aware of their susceptibility to or the potential for contracting a particular disease or condition, or, if aware of the disease or condition, are not aware of the consequences of the disease and the benefit of early detection. People in general do not understand the risks certain diseases or conditions may present to their health and often underestimate these risks—the "it can't happen to me" syndrome. The marketing of a home use IVD test product thus must take on an educational role and, through advertising and other marketing media, educate consumers about the need for testing. Of course at the same time, consumers will be made aware of the home use IVD test kit from a particular manufacturer.

B. Educating the Consumer

Build a better mouse trap and the world will beat a path to your door. The introduction of a new home use IVD test kit, even a better one, is a different situation. Consumers know what a mouse trap is and how to use it. With a new home use IVD test, consumers will need to be led down the path through education. They will need to have many questions answered before making a purchase.

"What is the test for?" "What disease? I never heard of it!" "Someone at work had that, but I won't get it." "How does it work?" Questions and attitudes must be addressed to help educate consumers about the need to perform a screening test or monitoring test. How to educate consumers must be given considerable attention by the manufacturer. Not only must a manufacturer educate the consumer about the new home use IVD test, the disease, and its consequences, but also the manufacturer must reeducate consumers who have misinformation and must attempt to deal with uncontrollable information the consumer may receive form other family members, friends, and acquaintances.

The role of education is also to avoid early misuse of the new test before it has had a chance to establish its benefit to the public health. Education also can avoid misuse of the test after it is established in the marketplace and its benefits are accepted by the professional health care system. The pharmacist's and physician's role in helping to educate consumers should not be overlooked. In addition to pharmacists and physicians, consumer health agencies may exist that specialize in providing education and general support to consumers with particular diseases. These agencies, along with pharmacists and physicians, should be considered as possible recipients of educational materials directed to the consumer and solicited and encouraged to aid in educating and training home users.

In preparing educational materials for the home use IVD test consumer, the manufacturer should keep in mind the same elements required for the product's labeling. Like the product labeling, the educational materials must be clear and understandable to the average consumer. The same techniques used in preparing labeling should be used in preparing the educational materials (see Section IV).

C. Customer Support

The process of educating the consumer of a home use IVD test product does not end with the printing and distribution of some educational brochures or sponsoring the publication of an article or two in a magazine. The consumer education process is a continuous process. As long as the home use IVD test product continues to be actively marketed, new consumers will continue to enter the market and will have a need to be educated about the test: what it is for and how it is to be properly used in the overall scheme of personal health care. Existing home users will also be in need of continual reeducation to provide reinforcement of previous messages and to bring them new information. In addition, the home consumer will be in need of other support mechanisms, which may not be able to be provided through brochures and magazine and newspaper articles or even the product labeling. The home user may also not fully comprehend the message or directions.

One of the best customer support mechanisms, but also probably one of the most costly, is providing an 800 phone number. This allows the home user to call with questions or just for reassurance. An 800 number system can be established in-house with an appropriate support staff utilizing a medical laboratory technologist or a nurse to answer questions. Some companies may also consider having a physician or a lawyer available to answer some questions. An 800 number system can also be established through a telemarketing service company that supplies telephone operators that have been trained to field consumer questions about the particular product. Such a commercial 800 number system should have a mechanism preestablished to allow one of the manufacturer's representatives to contact the consumer in a timely manner should the commercial service's operator not be able to answer the question for the consumer.

In addition to an 800 number, a manufacturer may consider using supplemental educational materials that go into greater detail than the generally distributed materials. The availability of these supplemental educational materials can be noted in the labeling of the product or in the generally distributed materials or product advertising (a tear-out business reply card or a "circle the number" service can be utilized). As noted in the previous section, the pharmacist, physician, and consumer health agencies can serve as sources of customer support. These additional sources other than the manufacturer should also be noted in the product labeling and educational materials.

VII. CONCLUSION

This chapter has attempted to provide the reader with some insight into the special considerations that should be given to home use IVD test products. The most significant consideration for the manufacturer of a home use IVD test is that regarding the consumer—the home user, the user's needs, expectations, and education. The home consumer's need for home use IVD tests may be questioned, but, as with most new concepts, home testing will eventually be generally accepted. The home consumer may not fully comprehend the scientific and technological basis of a home use IVD test. Yet, the home user is able to benefit from the privacy, cost savings, early detection, and additional time to consider treatment alternatives made possible by the at-home OTC IVD test. The user of a home use IVD test product must be an informed consumer, and it is the manufacturer's responsibility, in addition to that of the FDA, the federal, state, and local public health services, and consumer health organizations, to make a significant contribution to this educational process. FDA has needs also. FDA must fulfill its statutory mandate to protect the public health. To this end FDA, must assure itself that the proposed new home use IVD test provides a clinical utility and will benefit the home user without exposing the home user to unreasonable risks. The high cost of research and development and marketing of a home use IVD test make it necessary for the manufacturer to give such a proposal special considerations.

REFERENCES

1. McLaughlin, G. H., SMOG grading—a new readability formula, *J. Reading 12*:639-646 (1969).
2. Flesh, R. F., A New Readability Yardstick, *J. Appl. Psychol., 32*:221-223 (1948).
3. Kincaid, J. P., Fishburne, R. P., Rogers, R. L., and Chissom, B.S., Derivation of New Readability Formulas (Automated Readability Index, Fog Count and Flesh Reading Ease Formula) for Navy Enlisted Personnel, National Technical Information Service, U.S. Department of Commerce (AD-D006-655), Feb. 1975.
4. According to U.S. Bureau of the Census unpublished data as of March 1985 (from *The World Almanac and Book of Facts 1987*, World Almanac, 1986, New York, p. 212), 25.6% and 26.5% of male and female population of the United States 25 years and over completed less than 4 years of high school.
5. According to the U.S. Bureau of the Census, in 1980, 19.4% of the U.S. population spoke English "not well" or "not at all" (*The World Almanac and Book of Facts*, 1987, p. 213).
6. Assessing the Safety and Effectiveness of Home-Use In Vitro Diagnostic Devices (IVDs): Points to Consider Regarding Labeling and Premarket Submissions, Center for Devices and Radiological Health, FDA, January 1988 (Draft).
7. Labeling of Home Use Diagnostic Products (Proposed Guideline, GP14-P), NCCLS, Villanova, PA, June 30, 1987 (Draft).
8. Assessing the Safety and Effectiveness of Home-Use In Vitro Diagnostic Devices (IVDs): Points to Consider Regarding Labeling and Premarket Submissions, CDRH, FDA, January 1988 (Draft).

30

Sterility and Bioburden: An Issue for *in vitro* Diagnostic Products

GLEN PAUL FREIBERG *Boehringer Mannheim Diagnostics, Division of Boehringer Mannheim Corporation, Indianapolis, Indiana*

I. INTRODUCTION

A. Background

Contamination has always been an issue for products under the Food and Drug Administration's (FDA's) jurisdiction. One of the most notable attempts to create a sound federal policy stemmed from the "Delaney Clause" [1] of the federal Food, Drug, and Cosmetic Act. This effort to remove cancer-causing substances from the food supply was not as simple as the original idea professed. Technology evolved, and measurement techniques found ever smaller quantities of forbidden substances in consumables.

The government, industry, and consumers have wrestled with the Delaney Clause's related contamination issue quite publicly in cases such as saccharin and color additives. As a result the general public has become much more aware of FDA in general, that this regulatory group has varied success in its efforts, and that "safe" products are not a simple cut-and-dried issue.

With this apparent tolerance of an imperfect food supply has also evolved an understanding by the public that drugs are unequally effective for all people, and in some cases not effective at all. Even safety is understood to be relative, since the average consumer understands allergic reactions. While the consuming public has "grown up" with the "food" and "drug" issues of the Food and Drug Administration, similar issues of relative safety and efficacy of *in vitro* diagnostics (IVDs) have only recently been subject to the public's emerging learning curve. Therefore, extremely high expectations remain, as if diagnostic issues are as simple as the original protective goals of the Delaney Clause.

B. Education

As an emerging challenge, sterility and microbial load ("bioburden") in *in vitro* diagnostic products will not necessarily benefit from the learning curves of other regulated products. The diagnostic industry instead will have to undergo its own learning curve with the areas of FDA it interacts with, mainly submission reviews, investigations, and district offices. To prepare for these interactions, each company's regulatory office must participate in sponsored meetings and review continuously the relevant government and industry publications, newsletters, and regulatory actions. Such reviews are the only way for an individual company to have an industry-wide perspective of "current good manufacturing practices" (CGMPs).

II. GOVERNMENT ACTIVITIES

In the last several years diagnostics have been the subject of many news items and publications. Unfortunately, the diagnostic companies who did not keep up with the developing issues are either gone or under regulatory sanction. In some of these cases, such an adverse result was not necessarily the only logical outcome.

Obviously, keeping current and attempting to address evolving topics will not be sufficient protection in any individual inspection because a firm's views of current GMPs may not coincide with those of an FDA investigator, or that investigator's district office. Media manufacturers were recently surprised to find that they, as a group, did not understand CGMPs for their products in the same way as FDA. This issue of sterility of products not labeled as such was approached with numerous regulatory letters, voluntary closings, and severe financial losses in the mid 1980s. FDA had summarily decided that media manufacturers must treat their products as sterile unless "bioburden" data on physical and performance specifications could be produced to the FDA's satisfaction. An example from one (8-page) regulatory letter [2] summarized some of these issues as follows, "If you do not believe that your product must be sterile to achieve acceptable performance, you must establish appropriate specifications for your product in terms of both physical and performance characteristics including the type and quantity of allowable contamination, both viable and nonviable, if such contamination could adversely affect product performance. You must demonstrate that the allowed contamination will not adversely affect product performance and stability." These two sentences by then acting FDA Boston District Director Lee DeBell summarized FDA's fundamental position.

While FDA investigators throughout the country applied this "fundamental" concept to microbiological media manufacturers in particular, one small firm decided that compliance was not appropriate. Two highly publicized FDA-initiated court cases resulted from this resistance, and while BioClinical Systems fought the two consecutive cases against them, other firms continued to suffer from FDA's overzealous scrutiny of a group of products not in contact with consumers, but for use by health professionals.

The cases were very costly to BioClinical Systems, but may have been more costly to those companies who did not take a stand. To produce the subject product lines to the FDA's determined "sterility assurance limit" (SAL) was not, in fact, "current" manufacturing practice, and the court agreed. In both the government's preliminary injunction and permanent injunction filings, U.S. District Court Judge J. Frederick Motz stated that the FDA had exceeded its statutory authority.

These cases are considered landmark losses for FDA by most observers. Nevertheless the general concepts promoted by the FDA will continue through the 1990s. In the same month (March 1988) that Judge Motz denied FDA's permanent injunction against BioClinical Systems, the FDA issued the second draft of their IVD guideline. Once again the FDA attempted to enhance the GMPs. One example is the reference to 21 CFR 809 (labeling regulations) for inspectional coverage. This is beyond the documented scope of the agency's authority [3], and the draft guideline once again blended the microbiological load or bioburden issue among all types of IVDs.

A hearing was held at FDA headquarters on May 4 and 5, 1988, on this second edition of the draft guideline. The FDA selected advisory panel agreed with the industry testimony that clarification was needed to delineate the bioburden issues on different product groups, for example, labeled as sterile, microbiological load sensitive, and other. (Neither the FDA response to the hearing nor an updated draft was available when this text went to press.) Barring congressional action, this evolving guideline and the GMP advisory committee may have the greatest impact on the readers of this chapter.

The FDA's approach to "supplemental GMPs" in the future will continue to be more subtle and better organized; therefore, individual manufacturers must review and carefully analyze each publication as to its applicability to their processes. In areas of discrepant conclusions between the methods in a guideline and needs of an individual company, process, or products, it is highly recommended that the manufacturers finding such a discrepancy do not "go it alone."

III. UNDERSTANDING THE LABELING ISSUES

Some diagnostic reagents not labeled as sterile may support bacterial or mold growth and still meet performance requirements. But how much growth? Has the microbial load reduction of terminal filtration been measured? Does it need to be measured? Are "current" GMPs changing without some diagnostic manufacturers paying enough attention?

Paying attention to the issues is essential. When the FDA produced the 1985 draft IVD guideline [4], industry was told clearly where government thinking was going. Ignoring this government thinking because the majority of diagnostics are not produced aseptically, not sold as "sterile" or claimed to be sterile, proved to be disaster for many companies, as described earlier. Of course, the answer was not to begin producing sterile IVDs, but *at least* to evaluate FDA's position, comment on the position directly, through appropriate trade associations or independently or both, and consider positions for the eventual inspection. Consultation can be appropriate, but the consultants will not be the target of FDA sanctions. When the FDA indicates that an issue such as sterility or microbial load will be a focus of its attention, the company position must be thought out and organized in advance. In addition, postinspection actions and options must be analyzed so that simple "reactions" to an investigator's observations, also known as "voluntary compliance," are only undertaken with the knowledge of *realistic* current good manufacturing practices.

Ongoing education for regulatory and quality departments cannot be overemphasized. All manufacturers, and especially new manufacturers, must review and consider the publications available. Many other documents, in addition to FDA's now famous and evolving draft IVD guideline, are available. A reasonable starting place is FDA's Division of Small Manufacturers' Assistance (DSMA). While the responses from this FDA office will always be somewhat cautious, this is an excellent source of "free" advice, publications, and training. The DSMA periodically provides training courses around the United States, including Puerto Rico, which are from the FDA perspective and are excellent starter courses in several subjects. But while these courses are excellent for general subjects, they will not deal with the specifics individual companies will need to cover for the emerging issues of sterility and microbial load in IVDs (see Chap. 8 for more about how DSMA can help you).

Many specific issues will often fall under the umbrella of "validation." How does one prove and document the effectiveness of the cleaning process? Is there a seasonal effect of humidity on the production line? The lesson of manufacturing in a regulated environment is that the more you learn, the more you find out there is to learn. Validation, product challenges, and aiming for release specifications that truly are equal to rejection criteria are ways to continue to develop this learning curve.

IV. RECOMMENDATIONS

Unfortunately, there are many diagnostic manufacturers who simply are not interested in continued development of their learning curves. In fact, products and processes may have worked well for years without understanding or knowing if specifications were accurate, valid, or reasonable. For this group of manufacturers, retrospective validation is a key concept. Microbial load may never have been measured, the products may never have been challenged, but you have all the data you need to document that you need not undertake such work. It is only a matter of organizing the data as a retrospective validation report to document historically how the process works.

Using FDA guidelines (specifically the Process Validation Guideline) and understanding the intent of 21 CFR 10.90(b) can be invaluable to a long-established firm recognizing microbial load as an emerging issue that could cost a great deal of money, time, and effort to document cohesively what the company "knows" to be an effective process. FDA's Guideline on Process Validation [5] should be evaluated and addressed on a case-by-case basis not only for microbial load and sterility, but for cleaning, environmental control, filter sterilization, and any associated process. There is a separate guideline for the subject of aseptic fill [6].

Although this guideline is intended for drugs, individual investigators will most likely broaden its scope. If this guideline evaluation is not undertaken before FDA arrives, the time and effort postinspection could be much greater. If such work is not undertaken and a recall ensues, the additional costs will be quite startling. Specifically, the guideline states that the aseptic process has more variability than terminally sterilized products; therefore FDA believes it is "more difficult to attain a high degree of assurance that the end product will be sterile." Topics such as air quality monitoring and testing may be obvious to manufacturers, but the guideline system does do as the regulations suggests: it sets the ground rules for data organization and needs so that the inspection may be coordinated between the FDA investigator and the manufacturing provider.

Manufacturers should be aware of, and read, each of the other FDA-generated documents relating to their manufacturing processes. They include the Inspector's Operations Manual (IOM), Compliance Programs, and the various informational guidelines and "points-to-consider" documents. The IOM is general and rarely has substantive updates. Compliance programs are issued yearly, may change somewhat, and often include specialty attachments. One such attachment is the checklist for Inspecting Sterilization Processes in the IVD compliance program [7].

"Points-to-consider" documents are usually reserved for developing technologies, where the approaches to control, inspection, and evaluation are in early stages. By far the documents with the most impact on emerging issues are "guidelines" as discussed previously. In each of these, the FDA presents an issue to be addressed and suggestions for dealing with the issue. Remember that if these suggestions are followed, FDA is not supposed to object. However, if the suggestions are not followed, the FDA investigator will expect to review alternate methods and data that deal with the topic. Recently, the FDA has begun to announce the availability of guidelines in the *Federal Register*. For industry regulated by the Center for Devices and Radiological Health (CDRH), guidelines may be requested from the Divison of Small Manufacturers' Assistance (DSMA) at 800-638-2041.

As discussed earlier, the regulatory world has improved its networking ability tremendously in the last 10 years, and this network should be used to discuss the issues! In addition, trade groups, legal consultation in advance of a crisis, and a fresh look at the process should be considered. When a critical event, such as an FDA GMP advisory committee meeting occurs, it is essential that the regulatory department in each company monitor such an event, network in the field, provide management with both results and their impact, and (where possible) participate in such meetings.

Since FDA processes, procedures, and enforcement practices normally evolve slowly, a knowledge of the adverse regulatory history of specific types of processes or products is a valuable educational tool. Under the provisions of the Freedom of Information Act, the public may request nonconfidential information from the government. One such document available from the FDA is the "regulatory letter." Issuance of these letters is recorded in the trade press and in the FDA's monthly "Enforcement Report." Each regulatory letter will document specifically alleged deviations from the regulations, provide citations, and state that FDA is prepared to take immediate legal action. A review of available relevant regulatory letters is a valuable training tool and supplement when conducting internal audits.

Manufacturers must watch for new questions (do you have preservative test validation data generated in the last month of a product's labeled shelf life?). Determine what such new or previously unaddressed questions mean to you, review the references in all the available publications, consult as needed, and have a position (and sometimes data) ready before the next inspection. Most important, remember that depth and degree of each inspection will be variable and entirely dependent on individual investigator's background, ability, interest, and motivation. Do not let an "easy" inspection delay data development or participation and preparation for emerging issues.

V. CONCLUSION

The Food and Drug Administration will continue in a step-by-step fashion to demand more data from manufacturers. In the areas of sterility and reduced microbial load, the government activity level is high.

REFERENCES

1. 21 USC 348 (1958).
2. Regulatory letter addressed to Scott Laboratories, W. Warwick, RI, November 19, 1986.
3. 21 CFR 809.20 General requirements for manufacturers and producers of in vitro diagnostic products. (a) [Reserved]. (b) Compliance with good manufacturing practices. In vitro diagnostic products shall be manufactured in accordance with the good manufacturing practices requirements found in Part 820 of this chapter.
4. In Vitro Diagnostic Devices Inspectional Guidelines (DRAFT), Division of Compliance Programs, CDRH, and Baltimore District Office and Division of Clinical Laboratory Devices, ODE, CDRH.
5. Guideline on General Principles of Process Validation, May 1987, Center for Devices and Radiological Health, FDA.
6. Guideline on Sterile Drug Products Produced by Aseptic Processing, June 1987, Center for Drugs and Biologics and Office of Regulatory Affairs, FDA.
7. CP 7382.830A, October 1985.

part five

Current and Emerging Issues in the Regulatory Environment

31

Process/Design/Facility Validation

EUGENE H. LaBREC *E. H. LaBrec & Associates, Rockville, Maryland*

LEIF E. OLSEN *Whittaker Bioproducts, Inc., Walkersville, Maryland*

I. INTRODUCTION

A. Need for Validation

Manufacturers of medical devices and diagnostic products are challenged today with increasing domestic and international competition. The challenge in today's market is to produce the highest quality product possible at the lowest cost.

In order for companies to maintain a competitive edge, many factors come into play, such as basic business practices, high quality standards for raw materials and finished products, and well-defined and controlled manufacturing practices. This chapter presents information that should assist large and small companies alike in better understanding the current and future challenges of manufacturing practices as they relate to process, facility, and product validation.

Not only does it make good business sense to be able to control manufacturing processes through well-documented validation studies, but it is also a government requirement. Validation is a requirement of the Good Manufacturing Practice (GMP) Regulations for Medical Devices, 21 CFR Part 820 [1].

Robert Kieffer, in an introductory chapter to a validation handbook for the pharmaceutical industry, stated three reasons why this industry is concered that its processes perform as consistently as expected: (1) government regulations, (2) assurance of quality, and (3) cost reduction [2]. A well-trained and organized validation team within a company can produce results that will affect, in a positive way, all three of these concerns.

B. Scope of Chapter

This chapter provides general guidance to the medical device and diagnostic industry on the basic components of validation. The subject matter has been divided into three areas:

1. Process validation
2. Facility design and validation
3. Product design and validation

In May 1987, the United States Food and Drug Administration (FDA) released its final document entitled, "Guideline on General Principles of Process Validation" [3]. This document was prepared jointly by the Center for Drugs and Biologics and the Center for Devices and Radiological Health, and is intended as a guideline for the manufacture of pharmaceuticals and medical devices. This document states principles and practices of general applicability that are acceptable to the FDA; it is recommended as a basic reference in establishing or maintaining a company's validation program.

The facility that contains all the processes associated with the manufacturing of the product is the second area of special concern. Whether it is a new or old facility, it must be designed and operated in such a way as to comply with GMP requirements. Manufacturers today are faced with constantly changing technologies, which may impact on decisions whether to renovate an existing facility or build a new one. Recommendations and suggested sources of information are provided in this chapter.

Another document that the FDA introduced in May 1987 is titled "Preproduction Quality Assurance Planning: Recommendations for Medical Device Manufacturers" [4]. This draft document is intended to assist manufacturers in planning and implementing a preproduction quality assurance (PQA) program that addresses the product design concern. The PQA program is intended to provide a high degree of confidence that medical device designs are proven reliable, safe, and effective prior to the initiation of manufacturing.

The subject matter that this chapter covers is so broad that books have been written on the specific topics. The purpose of this chapter, therefore, is to introduce the reader to validation and to highlight significant issues. For further reading and sources of information, Appendices A and B have been included to assist in better understanding specific topics.

II. DEFINITIONS

There are several key terms used in connection with validation. Several of the references made throughout this chapter use and often define many of these terms. Listed below are those most frequently used in describing validation activities. An understanding of these definitions will greatly help in building a comprehensive approach to process validation.

1. *Validation*: "Establishing documented evidence which provides a high degree of assurance that a specific process will consistently produce a product meeting its predetermined specifications and quality attributes" [3].
2. *Process validation*: "Establishing documented evidence that a process does what it purports to do" [5].
3. *Prospective validation*: "Establishing documented evidence that a system does what it purports to do based on a preplanned protocol" [5].
4. *Retrospective validation*: "Establishing documented evidence that a system does what it purports to do based on review and analysis of historic information" [5].

5. *Concurrent process validation*: "Establishing documented evidence that a system does what it purports to do based on information generated during actual implementation of the process" [5].
6. *Validation protocol*: A written procedure stating how the validation study will be conducted. It should include test parameters, product attributes, product equipment, and decision points on what constitutes acceptable test results. Also, the protocol will identify individuals responsible for various tasks.
7. *Validation committee*: Representatives from various disciplines within the company who are responsible for reviewing and approving validation protocols and final results. Typically these individuals are from quality control, quality assurance, regulatory affairs, manufacturing, research and development, and engineering departments.
8. *Validation report*: A documentation package summarizing the results of the validation study, including the validation protocol, quality control/assurance test results, and a validation committee review/approval sign-off form.
9. *Validation change control*: "A formal monitoring system by which qualified representatives of appropriate disciplines review proposed or actual changes that might affect validated status and cause corrective action to be taken that will ensure that the system retains its validated state of control" [5].
10. *Revalidation*: "Repetition of the validation process or a specific portion of it" [5].
11. *Worst case*: "A set of conditions encompassing upper and lower processing limits and circumstances, including those within standard operating procedures, which pose the greatest chance of process failure when compared to ideal conditions. Such conditions do not necessarily induce product or process failure" [3].
12. *Critical process parameters*: "Those process parameters which are deemed important to product fitness for use" [6].
13. *Installation qualification*: "Establishing confidence that process equipment and ancillary systems are capable of consistently operating within established limits and tolerances" [3].
14. *Process performance qualification*: "Establishing confidence that the process is effective and reproducible" [3].
15. *Product performance qualification*: "Establishing confidence through appropriate testing that the finished product produced by a specified process meets all release requirements for functionality and safety" [3].
16. *Calibration*: "Demonstrating that a measuring device produces results within specified limits of those produced by a reference standard device over an appropriate range of measurements. This process results in corrections that may be applied if maximum accuracy is required" [5].
17. *Certification*: "Documented testimony by qualified authorities that a system qualification, calibration, validation or revalidation has been performed appropriately and that the results are acceptable" [5].
18. *State of control*: "A condition in which all operating variables that can affect performance remain within such ranges that the system or process performs consistently and as intended" [5].
19. *Proven acceptable range*: "All values of a given control parameter that fall between proven high and low worst-case conditions" [5].

III. PROCESS VALIDATION

A. General

In 1983, the FDA introduced the concept of process validation in a draft guideline titled "Guideline on General Principles of Process Validation." After several interpretations of the draft, the final guideline was published in May 1987.

The FDA defines process validation as:

> . . . establishing documented evidence which provides a high degree of assurance that a specific process will consistently produce a product meeting its predetermined specification and quality characteristics [3].

In order to establish an effective company-wide validation program, upper management must understand the importance of validation and commit the necessary resources. It is recommended that a validation committee be formed to guide the various validation projects. It is recommended that the committee draft a company policy that would state the goals, objectives, responsibilities, and other important aspects of the company-wide validation program. A well-trained and organized validation committee will greatly assist in maintaining a company's compliance to the GMPs.

Members of the validation committee usually are representatives from various disciplines who are responsible for reviewing and approving validation procedures and final results.

Maintaining the program is important because of the various processes that require validation and the need for revalidation. In other words, the validation program is ongoing and requires a mechanism to maintain its effectiveness. Committee members are encouraged to keep informed of changes in technology as well as the regulations. There are many sources of information that are available, and several of them have been listed in Appendix A.

In the list of definitions of process validation, the one concept that has met with the most opposition by manufacturers is the concept and definition of "worst case." Principally, opposition is based on what to use as a definition of worst case and how to use it. The earliest versions (1983 and 1984) of the draft, "Guideline for Process Validation," defined worst case as the conditions in which the process would fail. In reviewing process validation documentation for equipment such as ovens and autoclaves, FDA uses the term worst case to define the stress limits required for validation of equipment: that is, load configurations for an autoclave or oven in which the required parameters of sterilization are most difficult to achieve. Other definitions have followed, such as "Edge of Failure: A control parameter value that, if exceeded, means adverse effect on state of control and/or fitness for use of the product" [7]. Worst case may be an appropriate definition for validations which involve something like load configurations for an autoclave. However, it should not be the operative phrase for defining validations of processes. Chapman [8] described a new concept in process validation called "process acceptable range" (PAR). In the product development process, most firms set narrow operating parameters and then set somewhat less stringent control parameters. In monitoring, the process must be within the control parameters. When the process exceeds the control parameter, action takes place to bring the process back to the original operating range. He states that the function of the PAR is to embrace and validate all corresponding operating and control ranges. The conditions at the "edge of failure" are outside the parameters of the process acceptable range. Critical to the concept of the PAR approach is the documentation system. The PAR documentation system has three components. Although developed for use in the pharmaceutical industry, it has application also in the device industry. One component, the dosage reference file (DRF), is a compilation of validation studies and references pertaining to validation of the process. A second component is a process control spreadsheet. The third component is the process validation certificate. Prospective validation activities must be an integral part of the product development process. When appropriate documentation of process validation is kept for all steps in the manufacture of a device or operating process, it is possible to include or update validation with data gathered by retrospective and concurrent validation.

B. Aseptic Process Validation

There are certain medical devices and diagnostic products that require a sterility assurance level (SAL) in regard to the design of the facility and process validation. The SAL is based on the product's intended use, risk to the patient, and current industry practice. Products of this nature require special attention.

These products generally fall into two categories: (1) products that are terminally sterilized and (2) products that cannot be terminally sterilized but are produced by aseptic processes. Both processes require similar methods in order to kill contaminating microorganisms and control the work environment where the products are handled. The following highlights are the major components involved in aseptic process validation.

1. Facility Design

The design of the facility is a major component in assuring that aseptic conditions are met at various stages throughout manufacturing. Whether a manufacturer is planning a new facility or renovating an existing building, validation begins with the design phase. For the manufacturing of sterile products, special materials, equipment, and systems must be evaluated and selected. Ovens, autoclaves, filtration systems, lyophilizers, steam generators, and other equipment used for sterilization processes should be selected with a view to proper validation. The materials used in the construction of, for example, clean-room walls, floors, and ceiling should be durable enough to withstand repeated cleaning and disinfection over time.

Room layout should be carefully planned to optimize productivity and contamination control. The area in which aseptic processing occurs should be located so as to prevent access by unauthorized personnel. Air locks are routinely used for personnel change rooms as well as material pass-through. Sterilizers should be located adjacent to the aseptic processing area in order to maintain the sterile integrity of the glassware, equipment, and other components used in production and final product assembly. If a product is to be lyophilized, the lyophilization equipment should be placed either within the aseptic processing area or in an area adjacent to it. If transportation of sterile components, equipment, or product is required during the process prior to final product assembly, that process must be validated to demonstrate sterility assurance.

This type of room layout can be visualized as an archery target with its concentric circles. The bull's-eye is the sterile core or aseptic processing area. The next zone is the area for gowning, materials pass-through, and pass-through sterilizers and lyophilizers. The next zone is the area in which component preparation and final packaging take place. The support laboratories, utilities, warehouses, and offices are located in the outer zone.

For assistance in designing a new or renovated facility, it may be prudent to seek the advice of consultants. Consultants can help assure that the design is correct the first time, which saves time and money. (Choosing the right consultant is important, so reference checks are recommended.) It may also be useful to review the plans with FDA. Agency officials can be very helpful in uncovering potential problems and sharing their experiences.

2. *Process Equipment and Instrumentation*

The equipment used in sterilization or aseptic processing will need to be selected based on its compatibility with the product and its ability to be qualified for its intended use. For instance, certain types of autoclaves and dry-heat ovens cannot be adequately validated following FDA, Parenteral Drug Association (PDA), or Health Industry Manufacturers Association (HIMA) guidelines.

Medical device manufacturers who already have their equipment installed and operating but have not yet performed qualification validation will need to design appropriate procedures and complete these studies. For assistance in equipment installation, qualification, and validation, see Appendix A.

The instrumentation identified with the process equipment will need to be calibrated on a routine schedule. When new equipment is installed and qualified for use, the associated instruments are calibrated at that time as a prerequisite for validation studies. However, in order to maintain the confidence of the validated piece of equipment, the associated instruments will need to be checked periodically, whether it be on an each-use, daily, monthly, or yearly basis. The firm will need to maintain a comprehensive calibration program for this to be effective.

3. *Environmental Control Systems*

The environmental control system is ultimately tied into facility design and operation, but it is a system in and of itself and should be carefully designed and evaluated. The heating, ventilation, and air conditioning (HVAC) that make up this system will control the quality of air for particulates, airborne viables, temperature, humidity, air flow direction, and air pressure. Special high-efficiency particulate air (HEPA) filters are part of this system that controls the air quality for particulates.

To verify that preestablished specifications are being met, the qualification of the system should be performed prior to validation. For example, the HEPA filters must be certified by dioctylphthalate (DOP) testing for Class 100 or lower specifications. Also, it must be verified that the air flow pattern directly over the product exposure area is laminar, not turbulent, and that the pressure differential is greater from the aseptic processing area than from adjacent, less controlled areas.

To assist manufacturers in validation of their environment in regard to air quality, they should become familiar with Federal Standard 209B for clean rooms and work stations. Recently Revision 209C, and now 209D, of the standard was approved by the General Services Administration, U.S. Government.

Revision D provides better definitions of cleanliness classes and methods for measuring and monitoring these areas.

4. *The Sterilization Process*

The process by which components, containers, equipment, and so forth are sterilized should be validated to assure a level of sterility appropriate for the products. The technical monographs published by the PDA (1978) and HIMA (1978) provide helpful guidance for validating autoclaves, dry-heat ovens, and sterilizable in-place systems.

Sterile filtration of liquid products should be validated to demonstrate that the end product is sterile prior to aseptic filling. The major filter manufacturers offer information on appropriate validation studies for their particular types of filters.

5. *Sanitization*

Another key component of assuring aseptic processing is cleaning and disinfecting the work area. The firm should develop an appropriate SOP to define qualified disinfectants to be used, the frequency of disinfection, and step-by-step procedures. Routine monitoring for surface viable microorganisms should be part of the quality assurance program to monitor the effectiveness of the process.

6. *Employee Gowning*

Employee gowning is another key factor in aseptic processing. Information provided by the PDA (1978) states that approximately 80% or more of environmental contaminants in an aseptic work area are due to inadequate employee gowning or poor work practices within the area. Thorough training of employees in proper gowning methods and practices while working in the aseptic processing area is of utmost importance. Before an employee is allowed to enter the area, his or her training should be validated.

7. *Aseptic Filling*

Finally, the process for aseptic filling must be validated. The validation study will tie together all of the above controls into an actual simulated product fill. In the procedure, a sterile liquid medium, such as trypticase soy broth or another appropriate growth medium, is passed through the entire aseptic filling setup, beginning with the filtration step. The liquid passes through the dispensing equipment and into the final product container, and the container closure is applied. The product is then incubated under controlled temperature conditions for a predetermined period of time and visually evaluated to determine a rate or percentage of contamination.

For IVD products labeled sterile, some manufacturers are using a 0.3 sterility assurance level (SAL) (that is, no more than 3 contaminated units in 1000 units). The number of units used in validation has been 3000 or more per run. If a normal production run consists of more than 3000 units, the manufacturer may want to consider doing a validation run using the highest number of production units; alternatively, the time period over which the validation run is performed can be extended to the time that would be required for the highest number of production units. This could be considered part of the "worse-case" rationale.

8. *Lyophilization*

If lyophilization is used as part of product manufacturing, the process of disinfecting or sterilizing the chamber prior to use, the loading process, and the lyophilization cycle should be validated to demonstrate contamination control. Bacteriological growth media can be used in product simulation runs for loading and cycle evaluation.

9. *Product Closure System*

The final product closure or packaging should be evaluated in order to verify the integrity of the container or package seal. Validation studies should be designed to stress whatever closure systems is being used. For further information see Appendix A.

IV. FACILITY DESIGN AND VALIDATION

Whether it is a new facility, or a renovation to a present facility for a new or established process, the design, construction, installation, and operation must be validated. Validation activities should not be limited to single processes, or units of process equipment; rather, one must take an integrated approach to validation. Only by a thoroughly integrated approach can a manufacturer significantly reduce the time required to bring a new product to market.

A. Facility Design Considerations

The single theme found in all references to planning and construction of facilities for manufacture of devices and drugs is: Know your product and develop a plan. The plan may be called a program of requirements [9], business plan, design package, or other name. Whatever the name, it must be an integrated plan for design, construction of facilities, installation of utilities, installation of equipment, and validation of equipment and product processes within the new or renovated facility. Although most published articles concerning planning, construction, and validation of facilities and processes refer to pharmaceutical facilities, much of the material can be used as a guide for device manufacturers. References concerning validation of biotechnology facilities and products are useful to both pharmaceutical and device manufacturers. Major issues that should be considered are listed below.

1. *Design and Validation Team*

At the start of the project, a design and validation group is assembled. Established firms will assemble a team from within that has expertise in facilities engineering, process engineering, quality assurance, quality control, and regulatory affairs. In addition to the expertise that the appointed group brings, it must be able to give adequate time to the project. If expertise is not available within the company or experts within the company cannot give the time required, outside consultants are hired for specific parts of the project. If the process involves manufacture of a complicated or new biotechnology product, the team should have a member familiar with state and local construction and utility codes in the area where the facility is to be built or renovated. Additionally, FDA regulations require, before approval of a new drug, biologic, and Class III medical device, that the manufacturer submit an environmental impact assessment (EIS) (21 CFR Part 25). If there are environmental impact considerations to construction of the facility, installation of utilities, or the product process, the firm should consider having a consultant on the team with environmental engineering experience.

Most start-up companies wishing to commercialize a new device or process technology lack the know-how to manage a building project. Such companies must rely on outside services to bring the device to market in a timely fashion. Whether the management team relies principally on in-house expertise or hires an outside firm to design, build, and validate the facility, it is imperative that they "manage" the project. Whether an established firm or a start-up, its expertise resides with its laboratory scientists or engineers; they know the product, and must be able to translate product specifications to the architectural/engineering (A/E) firm that will design and build the facility.

2. Define the Business Objectives

Will the facility be designed for a single process or product, or a multiplicity of products? If the design is for a single process or product, can the facility be designed in a manner that it can be put to other profitable uses? If the facility is to produce a variety of products, will the processes be similar or dissimilar? When products are produced by similar processes, there can be common areas to be used at different times for different products. If the processes are dissimilar or have different requirements for utilities, it will be difficult to design without expensive redundancies, and validation of the facility will be more difficult, costly, and time-consuming. For example, if the device utilizes monoclonal antibody preparations, it must be decided early whether production of the reagent will be by collection of ascites fluids from mice or by cell culture techniques, or both. The utility requirements for each are different, and without installation of special environmental controls and equipment, they could be incompatible in the same facility.

3. Define Manufacturing Capability

The manufacturing capacity must be defined. If the facility is designed at a predetermined level of production, what provisions will be made for expansion? Site selection is important. If capacity is to be expanded by multiple shifts, a number of factors must be considered in the design of the facility. The facility must be designed to handle extra personnel working additional shifts. Utility capacities may have to be increased. It may also be necessary to design into the facility redundancies of certain equipment and utilities to assure smooth product flow.

4. Facility Staffing

The number of workers required and the level of training must be defined. Founders of new companies employing new technologies often lack the necessary experience in scale-up of the process or a production line. A cadre of production, quality control, and facility engineering managers should be hired early in the design and construction cycle so the project can be properly managed to - assure a smooth turnover program (TOP), which includes validation of the facility and the processes.

5. *Environmental Controls*

Is the facility to be designed with specific requirements such as microbiological containment, aseptic processing, handling of hazardous agents such as radiochemicals, toxic chemicals, infectious agents, animals, etc.? Each of these issues must be addressed, because they have a major impact on facility design and installation of utilities.

6. *Regulations*

The design, construction, and validation of a facility and its operations must conform to existing federal, state, and local regulations, most important of which are GMP regulations for medical devices. There must be a thorough understanding of present regulatory requirements, and an awareness of possible new legislative and regulatory issues that could affect business objectives by the time the facility is completed or in the near future. For example, how will the facility be designed if the FDA adopts guidelines that suggest that certain bioprocesses should be sterile or have a defined bioburden? The initial construction and installation costs and product life-cycle costs will be much higher if an aseptic processing facility is required. If the issue is not settled before construction of a facility, the design should include provisions for installation of cleanroom modules.

Facilities built to manufacture the newer biotechnology products must be aware that many federal, state, and local agencies have regulatory agenda for these products. Importantly, the FDA says that it will regulate biotechnology products in the same manner as natural biological products; the agency does not foresee the need to promulgate further or new regulations. This may not be true of other agencies.

B. Process Flow Charts

In planning a new facility or a major renovation, process flow charts (PFC) should be developed. Well-constructed flow charts allow the design and validation team to better understand the concepts and the requirements for the facility. The process is described first, in general terms, for the concept and design of the process. The flow charts are then subdivided into modules that detail the requirements of the process, such as process equipment and instrumentation or environmental specification flow charts. The A/E firm then can develop the specification and drawings for construction of the facility. Once developed, these flow charts can be used in submissions to the FDA for a device premarket approval application (PMAA) or a biologics establishment license. The flow charts can also be included in the product device master record (DMR).

1. *General Layout Flow Charts*

The general layout PFC describes the different areas of use within the facility and general construction requirements. The product/process flow through the facility will be an important consideration in planning the general layout. To facilitate the manufacturing process, a linear arrangement is preferred to minimize material handling. However, it is important to maintain flexibility with efficient use of floor space. In planning product process flow, the placement of laboratory furniture and equipment should be carefully reviewed. Where similar processes can be assigned to an area, it should be determined if mobile process equipment can be used. When mobile equipment can be used, provisions should be made for easy access to utilities to provide for quick connect/disconnect. Strategic placement of central storage facilities near the processing area also enhances flexibility. When manufacturing raw materials and components are stored in one place, proper control and rotation of stocks are facilitated, assuring better compliance to GMPs. Processing and storage of glassware used in production and filling operations should be carefully planned to assure adequate space and proper environmental conditions for storage to minimize any bioburden that could be carried into the processing suite.

2. *Personnel Flow Charts*

Flow charts should also be developed to show the flow of personnel through the facility at different phases of the process. Considerations must be given to environmental and safety issues. Each step in the process must be carefully planned to assure that there is adequate protection of product and worker at the lowest cost.

3. *Manufacturing Process Flow Charts*

The product PFC illustrates all the steps in the process. If the product is a multicomponent device, flow charts are developed for each component. Each step in the process, from input of raw materials, intermediate process steps, through intermediates to finished product, is analyzed. The flow charts describe the space required, the equipment needed, the environmental conditions, the utilities required to meet the environmental conditions, etc. The PFC should describe where the process will require validation and how the process validation will be carried out. The actual process validation procedures must be developed, reviewed, and approved early in the planning and design phase. When validation of the product process is properly defined, problems in design, construction, and installation of equipment that could affect facility and process validation can be avoided. Failure to plan early for validation can be costly both in money and time lost in validating the facility.

4. *Process Equipment and Instrumentation Flow Charts*

Process equipment and instrumentation flow charts will illustrate placement of equipment, instrumentation, and utilities required to support the equipment. These flow charts should show where sampling might take place during validation of the manufacturing process. For example, if one is validating a fermentation process in a closed fermentation vessel, additional sampling ports in the stream might be required to validate the process. When these are provided for in the construction and installation, it simplifies process validation. If these ports have to be added after the equipment is installed in the facility, it will cause delay in installation qualification and validation of the fermentation process.

C. Validation of Systems

1. *General Environmental Specifications and Validation*

Perhaps one of the most important issues in developing a plan for a facility is developing a detailed specification for the plant/manufacturing environment. In developing process flow charts, one must take into account the environmental conditions required. From the area where raw materials are received to the reagent filling line or electronic assembly area for a critical device, the PFC must have detailed environmental specifications. A particular specification might be as simple as the requirement for air conditioning in a raw material and component holding area, to complex environmental specifications for aseptic processing of an *in vitro* diagnostic reagent.

Whether the device is an implant, an electromechanical diagnostic machine, or an *in vitro* diagnostic, validation of environmental controls is an essential part of GMP compliance. The heating, ventilating, and air conditioning system may be a central supply system or one in which specific areas are under control of individual units. Where environmental specifications call for several areas to be under individual environmental conditions, HVAC units specifically designed for each should be used. In this manner, the design and installation of units can be tailored to meet the specifications at the least cost.

HVAC requirements are often dictated by regulations adopted at the state and local level. Before any HVAC plan is approved, the air handling specifications for all areas must be reviewed to assure that usual criteria and special requirements for air handling are met and can be constructed in compliance with local regulations. Validation of the HVAC system will document that it complies with the regulations.

2. Water System

Water has many uses in manufacturing medical devices and *in vitro* diagnostics. It is used for general cleaning of equipment and laboratory glassware, for cleaning of product final containers, for preparing sanitization agents, and for product formulation. These varied demands require that there be written specificaitons for water quality and the treatment system that will be used to produce water of consistent quality according to specification. To demonstrate consistent control, the water system to be installed must be validated.

The key elements in a validation program include planning, qualification, validation, documentation, and finally maintaining control of the system. The validation committee will act as a steering committee to monitor the validation as it progresses and to help resolve problems as they arise.

The first step to be undertaken for installing a new or upgraded system will be the planning and designing phase. Factors to consider are incoming water quality, final use of the water, and treatment systems necessary to obtain the desired quality of water. Next, is the qualification phase, which is divided into two parts: installation and operation. During the installation qualification, documentation should be developed to show that the system is built according to specification. Once this is complete, the operation qualification begins. Here the system is operated to verify that each piece of equipment performs properly and that the total system will produce the intended water quality. Calibration of instrumentation and the creation of standard operating procedures (SOPs) for operation and maintenance should be developed during this time. The validation phase begins at the completion of system qualification activities. It is intended to demonstrate that the system will perform on a consistent basis over a preselected period of time. During this time, all sites monitored for water quality must meet the design criteria for the validation to be successful.

All of the documentation involved in the design, installation, qualification, and validation should be assembled into a validation notebook. A summary of the key elements of the validation should be written and this summary reviewed and approved by the validation committee members.

The final element of the validation program is maintaining a state of control over the validated water system. A routine monitoring program should be established to verify that the system is functioning as demonstrated during the validation. Routine maintenance and any alterations to the system should be evaluated for their impact on the system's performance. If necessary, the system may require revalidation to reestablish confidence that the system is continually producing the water quality required.

For assistance in establishing specifications, the *U.S. Pharmacopeia* [10] can be used as a reference. In establishing specifications the manufacturer should consider the intended use of the water in the manufacturing process and the ultimate intended use of the final product. It is not uncommon for a manufacturer to have more than one type of water used in its facility (i.e., potable, purified, and water for injection).

3. *Other Utilities*

Compressed Air Systems. In most medical device manufacturing facilities, compressed air is supplied by a central system. Air quality must be defined for each use. In general, various traps, filters, and desiccant dryers are used to assure air free of moisture, oils, and other contaminants. Where air pressure is used to operate process equipment, it is necessary to validate that the air supply is isolated from the product and will not contaminate it. Where air is supplied to the process, as in bioreactors, the air supply must be virtually free of all contaminants. The design, installation, and performance of the air system for bioprocesses are complex and require a program of validation from the initial design phase, installation, and operational qualification.

We have discussed in detail only a few of the systems for which it is important to develop design, construction, installation, and operations specifications with an eye to process validation.

4. *Computerized Systems*

The use of computers to operate various systems, processes, and equipment in the manufacturing facility is rapidly increasing. The need to validate the equipment and software is important to demonstrate control over these processes.

The validation of computers and computer software is addressed specifically in device GMPs, 21 CFR Part 820. Paragraph 820.61 states: "When computers are used as part of an automated production or quality assurance system, the computer software programs shall be validated by adequate and documented testing. All program changes shall be made by a designated individual(s) through a formal approval procedure." Paragraph 820.195, Critical Devices, Automated Processing, notes: "When automated data processing is used for manufacturing or quality assurance purposes, adequate checks shall be designed and implemented to prevent inaccurate data output, input, and programming errors."

Just as it has expressed concern over design and development of all medical devices and published a guideline on preproduction quality assurance, the FDA has expressed equal concern over the explosion of computerized products used in the health-care industry. FDA recently drafted a policy on regulation of computer products classified as medical devices. It is clear from the number of public policy statements written in industry journals and expressed in public forums, by officials in the FDA, that process validation, especially of computer software, is high on the regulatory agenda. An in-depth discussion of computer software regulation can be found elsewhere in this volume.

Today a large number of devices that were once electromechanical are now software driven, or the device or process is controlled by programmed computer chips, PROMs (programmable read-only memories). The policies and procedures established for documentation of development and manufacture of medical devices must be equally, or perhaps even more rigorously, applied to the development, installation, and validation of computer software. In the past few years, the Pharmaceutical Manufacturers Association (PMA) has sponsored several conferences in which significant time was given to discussions of software validation. The results of these round table discussions appeared in *Pharmaceutical Technology* in 1986 and 1987. In 1987, the FDA released a technical report titled "Software Development Activities, Reference Materials and Training Aids for Investigators." It is an excellent frame of reference for software development activities. The PMA series and the FDA technical report of 1987 make generous references to two standards organizations, the American National Standards Institute (ANSI) and the Institute of Electrical and Electronics Engineers (IEEE), who have collaborated to produce several guides to software development and validation activities.

The following is a general discussion of validation of computer- (microprocessor-) and software-driven medical devices, and automated equipment used to produce such devices. It borrows heavily from the above references.

While design, manufacture, and validation of computer hardware are relatively straightforward and are very much like validation of any other piece of equipment used in medical device manufacturing, the design and, more often than not, documentation of design, coding, and testing of software suffer from benign neglect. Yet documentation of the design and validation by appropriate testing of computer software are the most important activities of the project. Without complete documentation of design, development, and writing of code, it will be impossible at a later time to maintain the software and to make changes without a possible complete destruction of the program.

The phases of development of software parallel activities found in development of medical device hardware, or computer-assisted medical devices. A short discussion of each phase follows.

Requirements Phase or Product Concept Phase. The functional specifications for software or firmware are developed and documented at this stage. In this phase there is extensive discussion and finally documentation of the product requirements, specifications for operation, environmental operating conditions, requirements for integration with the hardware, requirements for user interfaces, and the testing to be conducted to validate the software.

Test planning, in this phase, translates function specifications into testable specifications. Test specifications are written in parallel with the product specifications. "Functional testing tests the system as specified with no knowledge of how the system is programmed (i.e., with no knowledge of the software)" [11]. This functional testing is also characterized as external or "black box" testing.

Software Design. The standards to be used in developing software are adopted. Most firms that have experience in development of software will have a published set of policies and procedures to be used in software design. A new project is defined following the standards in its policies and procedures. Such policies and procedures will contain programming constructs (algorithms), and how the code will be documented. The design of user and equipment interfacing is documented, and the tests to be performed to test the interfaces are written. File activities are defined in detail: the types of files, file access, reports to be generated from files, the number of files that will be accessed during operations, how they will be opened and closed, and the controls to be exercised by the software, the limits or parameters of control.

A system test plan is developed. The tests to be performed in developing software are often referred to as structural or "gray box" testing. The system test plan defines the tests to be performed as the software is developed, including bounds or boundary tests. Boundary tests are designed to determine if the software gives the correct answer: just below, at, and just above the control limits for input and/or output of data or initiation of mechanical and electrical function. The important branching routines in the program are defined, and tests are designed to assure that proper activity takes place for each branch.

Load or stress tests are designed. These tests determine if the software can handle increasing numbers of transactions without failure.

The structural test program, should plan for some regression testing. Regression tests are defined as repeat tests using data sets or constructs that were used in tests of other modules of the program.

Implementation or Programming Phase. The product requirements and design are written into a source code by the programmer(s) in the language(s) decided in the design phase. It is common for various modules of code to be written in different languages. Validation of the code is performed as the modules or structures are completed. Validation at this phase is internal testing of the different modules. This internal testing is often referred to as "white box" testing, and is not necessarily related to external or functional testing. The coded modules are tested with data sets described and documented in the design phase. In this phase, data sets are often constructed to show if the code and the machine design can be stressed to cause failure. Selected regression tests are performed. All of the test data, inputs, results, and all corrections to the code made to meet design and functional specifications are collected and documented in a data package during this phase. This package is reviewed and approved by members of management and the design team.

Installation and Checkout Phase. This is the final program validation phase. In this phase, the external function tests are performed to assure that the software design and the hardware interface will perform to the original requirements of specifications. In this, the "black box" test phase, the functional specifications are tested with data sets and other simulations developed in the requirements phase. Simulation tests are used to validate the operation of the software and the hardware. The simulation tests should include the "worst case" scenarios that could lead to failure of the software or instrumentation.

After functional tests have been completed in-house, the software and instrumentation should be tested outside by potential customers. This is the so-called beta test. Both in-house function testing and tests at beta sites will undoubtedly uncover bugs in the software. It is important when changes are made to the finished software that all changes be documented and appropriate regression tests be performed to assure that the changes have not affected other parts of the program. Changes in the finished product are equal to reprocessing. Under GMPs, reprocessed devices must be subjected to testing at least as stringent as the original testing. The documentation package developed for both internal structure testing and external function testing makes up the validation package. Validation activities for software, such as activities for manufacture of medical devices, should become part of the device master record (DMR).

Maintenance Phase. After changes have been made and documented, the next phase of the software life cycle is maintenance of the product. When all phases of the design and coding of the software package have been properly and well documented, maintenance of the product, even by programmers who did not take part in development, will be relatively easy.

In the early part of this phase, depending on the frequency with which bugs were discovered in the checkout phase, there will be further changes in the software. However, unlike mechanical and electrical or electronic instrumentation, which wears out, the quality of a software product improves with age. Later, there will be changes in user requirements or discoveries of ways to improve the performance of the software, or to add new features. Each of the proposed changes should be subjected to the same if somewhat less rigorous review of the changes. In the maintenance phase, it is important to perform selected regression testing to assure that the changes do not affect software performance.

D. Turn-Key Program

Finally, it is important that the design and validation team develop a "turnover program." This program, which can be graphically displayed with a process flow chart, describes the activities required for complete validation of each of the systems installed in the facility. Each system installed must have a validation program developed that (1) verifies correct installation, (2) verifies the operational characteristics and specifications, and (3) verifies product process performance.

If the facility is new, or the process is new, it is important that the team develop and validate certain operational procedures such as (1) SOPs for cleaning, testing, and validation of bioprocess and sterilizing filters, (2) SOPs for cleaning and sanitizing equipment, (3) SOPs for cleaning and sanitizing bioprocess areas, (4) SOPs for calibration of process and control instrumentation and measurement equipment, and (5) SOPs for preventative maintenance of equipment and uilities. Preventative maintenance SOPs are usually developed as part of the installation validation package.

V. PRODUCT DESIGN

We have previously discussed aspects of process and facility validation. In the discussion of facility validation, we stressed that to perform an adequate validation, the process for which a facility is being designed must be validated. In concert with validation of facility and manufacturing processes, the product design must be validated.

"In launching a major new product, the quality function is all pervasive; it involves all departments. . . . In addition, the growing emphasis on long life and reliability has demanded that the quality planning encompass the entire life cycle of the product, 'from cradle to grave'" [12]. Effective quality planning tools need to be employed to assure that a new product is brought to market in the shortest time possible at optimum development cost. New approaches in management policies for development and manufacture of products, such as quality function deployment rely on formal design review.

Why design review? Product design review, as a formal process, has evolved over the years because of demands of the customer for quality products, by the dramatic rise in product liability claims, and the ever-increasing product failures plaguing industries that manufacture complex products. During World War II and after, the U.S. defense establishment was plagued with hardware reliability problems. There evolved from these difficulties a system of quality referenced in such documents as MIL-Q-9858A [13].

The requirements of the space program and its demand for the ultimate in design reliability led to further refinements in the concept of design review. All NASA contractors now are required to implement design reviews of their products [14]. Today, most organizations perform some sort of design review during development of their products. How well the concept of design review is integrated into a management policy will signal how successful it will be.

The importance of design review cannot be overemphasized. Product reliability problems in the field follow this general pattern:

Responsibility for failure	Percent of total problems
Engineering design	40
Misuse or abuse	30
Manufacturing quality	20
Unclassified miscellaneous	10

In a recent article in *Quality Progress*, Gon-Fu Lin stated that research at a major aerospace firm disclosed that 50% of failures could be attributed to design failures [15]. In 1986, in a report from the Center for Devices and Radiological Health, FDA, it was estimated that approximately 48% of recalls stemmed from preproduction causes such as product or process design deficiencies. The concern FDA has for products that fail in the customer's hands led to the publication of a draft guideline on preproduction quality assurance [16].

The following is a discussion of design review programs as a tool to assure quality product design that will satisfy customer requirements and will meet the requirements of a GMP program.

A. Product Design Review

1. *Purpose*

The purpose of design review is to provide a matrix for coordination of all activities responsible for development, manufacture, quality control, marketing, distribution, and service to review and approve a product's design and development.

Examples of design review are found in all kinds of products. Each example is an individual fabric made up of common threads of different colors. The results are:

1. A product of mature concept that meets the requirements of the customer and the customer's perception of quality.
2. A product of high reliability and low liability.
3. A product of optimal quality and cost.
4. A product of good profitability.

2. *What Is Design Review (DR)?*

Design reviews are scheduled, systematic reviews and evaluation of a product design by persons not directly associated with its development, but who as a group are knowledgeable in and have responsibility for elements of the product through its life cycle.

The administrative function requires expenditure of both time and money. DR tasks are carried out in addition to the usual departmental responsibilities, including design, development, manufacture, packaging, distribution, and life-cycle maintenance. Each element has a quality component and objective of management. Those participating in DRs are responsible to management to maintain this objective. When these objectives are met, the result is a better product at lower cost earlier in the development cycle.

A program of formal DR costs more. However, the value returned will far outweigh the costs. In one case, cited by Daly and Ockerman of Westinghouse Electric Corp., the cost of a single design review ($1000) before installation of a piece of equipment at the customer's site saved the firm an estimated $100,000 [17].

An effective DR program starts with development of a strong design review policy. The policy will have the usual statements of scope/applicability, definitions, responsibilities of participants, and detailed procedures for conducting design reviews. But the most important statement in such a document is management's commitment to the concept of design review. Without it, the process and eventually the manufacturer are doomed to failure.

DR must be conducted throughout the cycle of development, manufacture and distribution. Too often, design problems faced and solved after a product design is frozen result in change decisions that are severely restricted in latitude and applicability. Therefore, it is important to look at design review activities as being much more than merely seeing that a design is ready to manufacture. Design review must have its influence early in the product planning stage when the wants and needs of the customer are being considered. Formal design reviews must be a part of the development cycle and are placed in the schedule, that is, milestone or PERT chart or other formal schedule of development activity.

Design reviews are typically conducted at the following phases in the product cycle: (1) product concept, (2) preliminary design, (3) intermediate, (4) pilot lot or preclinical, (5) preproduction, (6) final DR before product release, and (7) postproduction. Depending on the complexity of the product and whether it is a new concept or a product with a previous production history, the number and frequency of design reviews will vary.

The written DR policy addresses subjects for review at each phase of development. In developing the policy it is important to itemize what activities will be reviewed at each step, what will be the detail of review for each activity, and who will be principal reviewers. To allow for adequate discussion and review of product design, each member of the management matrix should have an agenda for each DR phase. The policy should address the development of checklists for each phase of design review. Without formalizing a method for review of each stage, issues will be overlooked with possible disastrous consequences. When a new product is being developed, the DR, at the concept stage, should address any changes in checklists required for monitoring development of the particular design.

Design review is conducted by a team of specialists not directly associated with development of the product, but who are responsible for some aspect of development, manufacture, quality control, marketing, distribution, and service of the product. These specialists review the design. Their expertise allows for optimization of the design. The designer must accept suggestions of the DR team for design changes. However, such input must be based on fact and not on hearsay. The designer has the option to accept or reject design modifications. Design review is, after all, a review of the design and not the designer.

Participants in the DR are responsible for the actions assigned to them during design review. Failure to prepare for and carry out the action items assigned and to adequately monitor the requirements of their functions will cause the DR system to fail.

3. *Participants and Responsibilities in Design Review*

Design Review Chairman. Design review requires a strong, responsible administrator with communication skills. The chairman of the DR committee administers the review process, avoiding any tendency toward an adversary relationship between the designer and the reviewers. The DR chairman, the representative of corporate management, is responsible to assure that activities and responsibilities of the reviewers are properly conducted, and that actions items resulting from design reviews are promptly executed.

Designer Engineer, Senior Scientist, Program Manager R&D. The process or product designer is responsible for developing the product within the constraints imposed by the product's design specifications agreed upon at the concept phase design review. In design reviews for each phase, the design engineer (DE) reviews the completed specifics of the product. The DE is responsible for test design and for all specifications for production, including raw materials, quality control of the process or final product, and the final product specification. The DE is responsible for all documentation produced throughout the development process. Documentation must be formal and all specifications written must be traceable to assure final approval within the company, and also by regulatory agencies that approve the product for safety and efficacy. The only product of research and development (R&D) is information. It must be accurate and committed to formal documentation.

Marketing Rerpesentative. The primary responsibility of the marketing representative is to assure that the marketing aims, first and foremost, represent customer desires and expectations. The representative has major responsibilities at the product concept stage.

Research and Development Engineer/Scientist. Research and development (R&D) also is represented by someone who is knowledgeable about the product or process being developed but who is not directly involved in the development.

Quality Assurance (QA) Representative. The QA representative is responsible for assuring that the device meets the quality aims and product specifications. This person is responsible for ensuring that process controls, inspection, and test equipment are appropriate, and also evaluates the processes and equipment required in order to make recommendations for appropriate validation activities. The QA representative is also concerned with development and validation of assay procedures used in quality control testing of intermediate processes and finished devices. Quality assurance develops a quality plan for assuring product quality throughout the product life cycle.

Process Engineer (PE). The PE is responsible for manufacture of the finished device and must be cognizant of process capabilities to ascertain manufacturability in present facilities or whether new facilities and/or equipment are needed to achieve success.

Distribution. Specialists in the area of product integrity should be part of design review to ascertain whether the product is subject to environmental damage or degradation during holding and distribution. Packaging consultants should be employed to assure product integrity.

Regulatory Affairs (RA) Representative. The regulatory affairs department is responsible for ensuring that the design meets regulatory requirements for all federal, state, and local agencies. The RA representative is responsible for developing strategies for approval of a device within the frame-work of the product specification. He or she must be cognizant in review of possible product liability problems. If the product or process design could present serious product liability problems, legal advice from an expert in product liability audits should be sought. "The products liability 'audit' is a method now being utilized by research intensive drug and device manufacturers in order to evaluate the legal risk factors at the earliest point in product development" [18].

These are some of the principal participants in the design review process. Where other specialties are required, such as general legal counsel, patent attorney, special medical consultants, or clinical trials specialists, they should participate at the appropriate phase of product development.

VI. CONCLUSION

Manufacturing processes, facility, and design validation are necessary for assuring product quality and compliance to GMP regulations. Process validation is most often associated with specific activities such as validation of process equipment, for example, dry heat or steam sterilizers. These individual activities are but a small part of the total concept of process validation. The FDA guidelines on the principles of process validation not only covers validation of individual pieces of equipment, but clearly addresses facility and product design validation as well. Product design validation extends to validation of software used in medical devices and used to manufacture medical devices.

We have stressed in this chapter that process validation is a coordinated activity, that facilities must be designed and constructed in a manner that assures that the facility, the equipment and utilities installed, and the process to be conducted within can be validated to meet requirements of the firm but also must satisfy regulatory requirements. We have stressed that each firm should approach process validation as a coordinated effort. Persons with expertise in the disciplines discussed in the chapter are responsible for bringing the facility and the process into compliance within the regulatory milieux. Process validation is not only the law, it is good business.

REFERENCES

1. Title 21, Code of Federal Regulations, Part 820.
2. F. J. Carleton, and J. P. Agalloco, Why Validation, *Validation of Aseptic Pharmaceutical Processes*, Marcel Dekker, New York, 1986.
3. FDA, "Guidelines on General Principles of Process Validation," Center for Drugs and Biologics, Center for Devices and Radiological Health, May 1987.
4. FDA draft, "Preproduction Quality Assurance Planning: Recommendations for Medical Device Manufacturers," Center for Devices and Radiological Health, May 1987.
5. K. G. Chapman, A Suggested Validation Lexicon, *Pharmaceutical Technology, 7*(8):51 (1983).
6. K. G. Chapman, et al., Process Validation Concepts for Drug Products, *Pharm. Technol., 9*(9):78, (1985).
7. M. H. Anisfield, "Validation Considerations in the Design of an Aseptic Processing Facility—An Overview," Second PMA Seminar on Validation Sterile Manufacturing Processes, Atlanta, GA, 1977.
8. K. G. Chapman, The PAR Approach to Process Validation, *Pharm. Technol., 8*(12):22 (1984).
9. G. B. Phillips, Development or a Program of Requirements, in *Design of Facilities for Manufacture of Medical Devices and Diagnostic Products: A Handbook*, Health Industry Manufacturers Association, Report No. 77-5, Washington, DC, 1977, p. 37.
10. *U.S. Pharmacopeia XXI*, Section 1231, January 1, 1985.
11. J. A. Campanizzi, Structured Software Testing, *Quality Prog., 17*(5): 14 (1984).
12. R. J. Pierce, Quality Planning, in *Quality Control Handbook*, eds. J. M. Juran, F. M. Gryna, Jr., and R. S. Bingham, McGraw-Hill, New York, 1974, pp. 6-7.
13. Department of Defense, "Quality Program Requirements," MIL-Q-9858A.
14. NASA, "Quality Program Provisions for Aeronautical and Space System Contractors," NHB 5300.4(1B).
15. G.-F. Lin, Designing a Solution, *Quality Prog., 21*(9):44 (1988).
16. FDA, "Preproduction Quality Assurance Planning: Recommendations for Medical Device Manufacturers," Division of Compliance Programs, CDRH, 1987.
17. T. A. Daly, and P. H. Ockerman, "Commercial Reliability Programs—A Good Investment," Westinghouse Electric Corp., Pittsburgh, PA.
18. K. A. Touby, Products Liability and the Device Industry, *Food Drug Cosmet. L. J., 43*:55 (1988).

APPENDIX A: BIBLIOGRAPHY

A. Process Validation

J. P. Agalloco, Practical Considerations in Retrospective Validation, *Pharm. Technol.*, *7*(6):88 (1983).

K. G. Chapman, A Suggested Validation Lexicon, *Pharm. Technol.*, *7*(8):51 1983).

K. G. Chapman, Chairman, PMA Validation Advisory Committee, Process Validation Concepts for Drug Products, *Pharm. Technol.*, *9*(9):78 (1985).

K. G. Chapman, The PAR Approach to Process Validation, *Pharm. Technol.*, *8*(1):22 (1984).

E. M. Fry, General Principles of Process Validation, *Pharm. Engr.*, *4*(3):33 (1984).

A. Hess, An Integrated Approach to Validation, *Biopharm. Manufact.*, *1*(3): 42 (1988).

R. F. Johnson, Process Validation: A Guide to Successful Application, *Med. Dev. Diag. Ind.*, *7*(1):73 (1985).

J. D. Nally, Validation Guidelines—Industry's Perspective, *Pharm. Engr.*, *4*(3):21 (1984).

T. H. Riggs, Principles of Process Validation, *Med. Dev. Diag. Ind.*, *3*(5):31 (1981).

M. Schlager, Pharmaceutical Process Validation: A Working Model Concept, *Pharm. Engr.*, *5*(4):15 (1985).

B. Sterilization Validation

1. General References

F. Carleton and J. Agalloco, *Validation of Aseptic Pharmaceutical Processes*, Marcel Dekker, New York, 1986.

FDA, *Sterile Medical Devices, A GMP Workshop Manual*, 4th Ed., HHS Publication (FDA) 84-4174, January 1985.

FDA, *Sterilization Questions and Answers*, Center for Devices and Radiological Health, January 1985.

HIMA, "Sterilization Cycle Development," Report No. 78-4.2, Health Industry Manufacturers Association, Washington, D.C., 1978.

HIMA, "Validation of Sterilization Systems," Report No. 78-4.1, Health Industry Manufacturers Association, Washington, D.C., 1978.

Pharmaceutical Manufacturers Association, "The Validation of Sterilization of Large Volume Parenterals—Current Concepts," Washington, D.C., 1979.

2. Steam Sterilization

R. DeRisio, Equipment Design: Moist Heat Sterilizer, *Pharm. Engr.*, *7*(6): 43 (1987).

Parenteral Drug Association (PDA), "Validation of Steam Sterilization Cycles," Technical Monograph No. 1, Philadelphia, PA, 1978.

J. J. Perkins, *Principles and Methods of Sterilization in Health Sciences*, Charles C. Thomas, Springfield, IL, 1973.

I. J. Pflug, *Syllabus for an Introductory Course in the Microbiology and Engineering of Sterilization Processes*, 4th Ed., Environmental Sterilization Services, St. Paul, MN, 1980.

I. J. Pflug, and M. R. Berry, Jr., Using Thermocouples to Measure Temperatures During Retort of Autoclave Validation, *J. Food Protect.*, *86*(4) (1987).
See List of General References.

3. Dry Heat Sterilization

PDA, "Validation of Dry Heat Processes Used for Sterilization and Depyrogenation," Technical Monograph No. 3, Philadelphia, PA, 1981.
R. T. Wood, *Parenteral Drug Association Short Course on Dry Heat Sterilization Validation and Monitoring*, Philadelphia, PA, 1982, pp. 12-19.
See List of General References.

4. Ethylene Oxide Sterilization

AAMI, *Standard for RIER/EO Gas Vessels*, Association for the Advancement of Medical Instrumentation, Arlington, VA, 1982.
See List of General References.

5. Cobalt-60 Radiation Sterilization

AAMI, "Process Control Guidelines for Radiation Sterilization of Medical Devices (Proposed)," Recommended Practice, RS-P, Association for the Advancement of Medical Instrumentation, Arlington, VA, 1982.
J. Masefield, Current North American Practices in Gamma Sterilization, in *Proceedings fo the International Symposium, Advances in Sterilizaiton of Medical Products*, University of New South Wales, Kensington, N.S.W., Australia, 1982.
See List of General References.

6. Aseptic Process Validation

H. L. Avallone, Control Aspects of Aseptically Produced Products, *J. Parenteral Sci. Technol.*, *39*(2):75 (1985).
F. Carleton, and J. Agalloco, *Validation of Aseptic Pharmaceutical Processes*, Marcel Dekker, New York, 1986.
J. W. Dirksen and R. V. Larson, Filling Vials Aseptically While Monitoring for Bacterial Contamination, *Am. J. Hosp. Pharm.*, *32*:1031 (1975).
FDA, *Guideline on Sterile Drug Products Produced by Aseptic Processing*, Center for Drugs and Biologics, June 1987.
E. M. Fry, FDA Update on Aseptic Processing Guidelines, *J. Parenteral Sci. Technol.*, *41*(2):56-60 (1987).
G. S. Gibbs and P. M. Rountree, A Laminar-Flow Sterile Products Department. An Example of Modern Biomedical Engineering, *Med. J. Aust.*, *1*:253 (1973).
HIMA, "Microbial Control in the Manufacturing Environment," Report No. 78-4.3, Health Industry Manufacturers Association, Washington, DC, 1978.
K. Z. McCullough, Environmental Factors Influencing Aseptic Areas, *Pharm. Engr.*, *7*(1):17-20 (1987).
PDA, PDA Response to FDA Guidelines on Sterile Drug Products Produced by Aseptic Processing, *J. Parenteral Sci. Technol.*, *42*(2):53 (1988).
PDA, "Validation of Aseptic Filling for Solution Drug Products," Technical Monograph No. 2, Philadelphia, PA, 1980.
H. L. Raiman, Panel Discussion: Environmental Sampling in an Aseptic En-

vironment, Microbiological Environmental Monitoring, *Parenteral Drug Assoc. Bull.*, *28*:253 (1974).
R. A. Tetzlaff, Aseptic Process Validation, *P and MC Ind.*, September-October: 25 (1983).
R. A. Tetzlaff, Regulatory Aspects of Aseptic Processing, *Pharm. Technol.*, *8*(11):38 (1984).
L. Vanell, Modernization of Existing Plant Air Purification Equipment, *Pharm. Engr.*, *3*(6):45 (1983).
J. Wasynczuk, Validation of Aseptic Filling Process, *Pharm. Technol.*, *10*(5): 36 (1986).

7. Calibration of Thermocouple Measuring Systems

I. J. Pflug, Temeprature Measurement Using Thermocouples, *Med. Dev. Diag. Ind.*, *3*(9): 23 (1981).

8. Depyrogenation Validation

PDA, "Depyrogenation," Technical Report No. 7, Philadelphia, PA, 1985.

9. Validation of Sanitization

S. R. Chesky, In Line Validation of Sanitizing Agents in Aseptic Processing Areas, *J. Parenteral Sci. Technol.*, *40*(4):169 (1986).
Parenteral Drug Association Task Force on Decontaminating Agents, Decontaminating Agents, *J. Parenteral Sci. Technol.*, *40*(3):104 (1986).

10. Validation of Filtration Process for Sterilization

S. H. Goldsmith and G. P. Grundelman, Validation of Pharmaceutical Filtration Products, *Pharm. Manufact.*, *9*(11):31 (1985).

11. Validation of Product Closure System

M. Bryant, Packaging Failures—Quality in Design Doesn't End with the Finished Product, *Med. Dev. Diag. Ind.*, *10*(8):30 (1988).
F. Carleton and J. Agalloco, *Validation of Aseptic Pharmaceutical Processes*, Marcel Dekker, New York, 1986, pp. 545-593.
FDA, "Sterile Medical Devices—A GMP Workshop Manual," 4th ed., No. FDA 84-4174, Reprinted January 1985.

C. Validation of Utilities

J. Y. Lee, Environmental Requirements for Clean Rooms, *BioPharm. Manufact.*, *1*(7):40 (1988).
NASA, "Standards for Clean Rooms and Work Stations for the Microbially Controlled Environment," NHB 5340.02, NASA, Washington, DC, 1967.

D. Validation of Lyophilizers

H. Avallone and A. Walk, Regulatory Aspects of Lyophilization, *J. Parenteral Sci. Technol.*, *40*(2):81 (1986).
FDA, "Lyophilization of Parenterals," Inspection Technical Guide, No. 43, April 18, 1986.

E. Water Systems

D. L. Jackman, Troubleshooting Your Pharmaceutical Water System, *Pharm. Engr.*, *8*(2):22 (1988).

PMA Deionized Water Committee (DIW), Validation and Control Concepts for Water Treatment Systems, *Pharm. Technol.*, *9*(11):50 (1985).

F. Software Validation

J. Agalloco, Validation of Existing Computer System, *Pharm. Technol.*, *11*(1): 38 (1987).

E. R. Atkinson, Understanding Software Validation in Automated Systems, *Med. Dev. Diag. Ind.*, *6*(9):10 (1984).

H. Bassen, J. Silverberg, F. Houston, W. Knight, C. Christman, and M. Greberman, *Computerized Medical Devices, Problems, and Safety Technology, IEEE/7th Annual Conf. of the Engineering in Medical and Biological Society*, 180, 1985.

J. A. Campanizzi, Structured Software Testing, *Quality Prog.*, *17*(5):14 (1984).

C. L. Carpenter and G. E. Murine, Measuring Software Product Quality, *Med. Dev. Diag. Ind.*, *6*(5):16 (1984).

A. S. Clark, Computer Systems Validation: An Investigator's View, *Pharm. Technol.*, *12*(1):60 (1988).

I. P. Cooper and B. F. Mackler, Regulation of Diagnostic Software: Preparative Steps, *Med. Dev. Diag. Ind.*, *7*(10):38 (1985).

K. Diggins, Development Standards for Software Used in Drug Production, *Pharm. Technol.*, *11*(9):56 (1987).

FDA, Associate Commissioner for Regulatory Affairs, Software Development Activities, Reference Materials and Training Aids for Investigators, Technical Report, Office of Regulatory Affairs, 1987.

R. M. Garwood, FDA's Viewpoint on Inspection of Computer Systems, FDA: Regulation of Medical Software Workshop—1983, 1983.

A. C. Grinath and P. H. Vess, Making SQA Work: The Development of a Software Quality System, *Quality Prog.*, *8*(7):18 (1983).

HIMA, Computer Technology in the Medical Device and Diagnostic Product Industry, Proceedings Computer Technology Session, HIMA Product Safety Seminar, October 6, 1982, Chicago, IL. Doc. No. 6, Vol. 4, 1982.

J. R. Harris, Chairman, PMA Computer Systems Validation Committee, Validation Concepts for Computer Systems Used in the Manufacture of Drug Products, *Pharm. Technol.*, *10*(5):24 (1986).

J. Jorgens, III, Computer Hardware and Software as Medical Devices, *Med. Dev. Diag. Ind.*, *8*(5):60 (1983).

J. S. Kahan, FDA Regulation of Computer Software as Medical Devices, *Med. Dev. Diag. Ind.*, *7*(3):51 (1985).

J. S. Kahan, Validating Computer Systems, *Med. Dev. Diag. Ind.*, *9*(3):48 (1987).

J. A. Keane, Computers and Quality Assurance in a Regulated Industry, *Med. Dev. Diag. Ind.*, *3*(10):45 (1981).

N. R. Kuzel, Fundamentals of Computer System Validation and Documentation in the Pharmaceutical Industry, *Pharm. Technol.*, *9*(9):60 (1985).

N. R. Kuzel, Quality Assurance Auditing of Computer Systems, *Pharm. Technol.*, *11*(2):34 (1987).

A. Lowery, Automated Production and QA Systems for Medical Devices, *Quality Prog.*, *17*(6):58 (1954).

G. Masters and P. Figarole, Validation Principles for Computer Systems—FDA's Perspective, *Pharm. Technol.*, *10*(11):44 (1986).
K. S. Mendis, Quantifying Software Quality, *Quality Prog.*, *15*(5):18 (1982).
P. J. Motise, What to Expect When FDA Audits Computer-Controlled Processes, *Pharm. Manufact.*, *8*(7):33 (1984).
PMA, Computer System Validation Committee, Validation Concepts for Computer Systems Used in the Manufacture of Drug Products, *Pharm. Technol.*, *10*(5):Reprint (1986).
D. S. Paulson and D. G. Vogel, Prepatory Software Documentation Validation of Manufacturing Operations, *Pharm. Manufact.*, *8*(5):37 (1984).
S. J. Shepherd, Manufacturing Software, *Manufact. Systems*, *3*(8):34 (1985).
N. Wong, Documenting Microprocessor-Based Medical Devices: An FDA Perspective, *Med. Dev. Diag. Ind.*, *7*(10):20 (1985).

G. Facility Design and Validation

G. Alperin, Validation Considerations in Pharmaceutical Process and Plant Design, *Pharm. Engr.*, *4*(3):15 (1984).
T. E. Byers, GMPs and Design for Quality, *J. Parenteral Drug Assoc.*, *32*: 22 (1987).
D. J. Cattaneo, Plant Validation Acceptance Criteria, *Pharm. Engr.*, *8*(4):9 (1988).
P. A. Cipriano, Designing Clean Rooms for FDA Process Validation, *Pharm. Technol.*, *7*(6):82 (1983).
J. C. Griffin and W. A. Pauli, Design Concepts for a Sterile Products Production Facility, *Parenteral Drug Assoc. Bull.*, *30*:293 (1976).
A. R. McGuire, Designing Clean Rooms for Anhydrous Product Manufacture, *Pharm. Manufact.*, March (1984).

APPENDIX B: STANDARDS/RESOURCE ORGANIZATIONS

ANSI—American National Standards Institute, 1430 Broadway, New York, NY 10018.
ANSI/IEEE Standard 729-1983, *Glossary of Software Engineering Technology*.
ANSI/IEEE Standard 730-1984, *Software Quality Assurance Plans*.
ANSI/IEEE Standard 828-1983, *Software Configuration Management Plans*.
ANSI/IEEE Standard 829-1983, *Software Test Documentation*.
ANSI/IEEE Standard 830-1984, *Software Requirements Specifications*.
ASHRAE—American Society of Heating, Refrigerating, and Air-Conditioning Engineers, 1791 Tullie Circle Northeast, Atlanta, GA 30329.
ASME—American Society of Medical Engineers, 345 East 47th Street, New York, NY 10017.
ASTM—American Society for Testing and Materials, 1916 Race Street, Philadelphia, PA 19103.
IES—Institute of Environmental Sciences, 940 East Northeast Highway, Mount Prospect, IL 60056.
Military Specification Mil-S-52779(AD), Software Quality Assurance Program Requirements, 1974.
NSF—National Sanitation Foundation, 3465 Plymouth Road, P.O. Box 1468, Ann Arbor, MI 48106.

32

Use of Risk Assessment Procedures for Evaluating Risks of Ethylene Oxide Residues in Medical Devices

STEPHEN L. BROWN *Environ Corporation, Arlington, Virginia*

I. INTRODUCTION

A. Ethylene Oxide Sterilization

Ethylene oxide (EtO) is the sterilant of choice for many medical devices, some of which cannot be effectively sterilized with radiation, steam, or any other alternative. Many types of disposable and reusable medical devices, ranging from the common adhesive bandage to packemaker systems and dialyzers, are sterilized with EtO by the manufacturer or custom sterilization companies. Table 1 gives a sample of device types potentially sterilized with EtO, drawn from a list of hundreds of device types, representing thousands of specific device designs and billions of annual units sold.

After sterilization, some of the EtO may remain adsorbed on the surface of the device or dissolved into its materials of construction. These residues dissipate with time at a rate depending on the materials of construction, the degree of ventilation of the product during storage, the type of product packaging, and other factors. During sterilization, some of the EtO may react with water vapor to form ethylene glycol (EG) or with chlorinated compounds to form ethylene chlorohydrin (ECH), either of which may also remain on the device. Manufacturers hold their products for a period prior to shipping to allow dissipation of EtO and, to some extent, the other two substances.

B. Potential Hazards of EtO

Each of these three compounds (ethylene glycol, ethylene chlorohydrin, EtO) has been shown to cause chronic toxicity or reproductive effects in laboratory animals if administered at relatively high doses. In addition, EtO has been reported to test positive in cancer bioassays with laboratory animals and has been associated with human cancers in epidemiological studies of uncertain significance.

TABLE 1 Examples of Medical Devices That May Be Sterilized with Ethylene Oxide

Bags, blood administration	Pacemakers and accessories
Bag, iv administration	Prostheses, breast
Bandages, adhesive	Reservoirs, cardiotomy
Catheters, intravascular	Sponges, gauze and cotton
Catheters, urological	Syringes, hypodermic
Dialyzers and accessories	Syringes, insulin
Filters, blood	Tips, surgical suction
Gloves, surgical	Tubes, infant feeding
Lenses, intraocular	Tubing, blood or iv
Needles, hypodermic	Valves, heart, prosthetic

Given these facts, it is reasonable to ask whether the residues of EtO on sterilized medical devices are dangerous for patients who use those devices. This chapter describes the results of a study of that question conducted by the author and his colleagues at ENVIRON Corporation for the Health Industry Manufacturers Association (HIMA). The full study, "Ethylene Oxide Residues on Sterilized Medical Devices," is available from HIMA as HIMA report no. 88-6. Our study examined only potential risks that might be incurred by patients or other end users of the devices, not to medical personnel or employees of manufacturers or sterilizing companies. Furthermore, we studied only the potential risks associated with medical devices delivered sterile to hospitals, physicians, or the patient, not the risks of devices that are sterilized or resterilized in the hospital or clinic.

C. Risk Assessment

Regulatory agencies and other organizations responsible for the protection of human health must justify their decisions regarding the levels of exposure to hazardous substances that can be tolerated without unacceptable risk. In the past, many decisions about the use of substances that might be toxic have been made through the use of qualitative arguments. Several examples of useful qualitative assessments, as well as assessments of different kinds of risks associated with medical devices, are discussed in the chapter by Alan Andersen (Chap. 33). Increasingly, however, decision makers have also begun to use *quantitative* risk assessment as an important input for decisions that entail consideration of economic, technologic, and political issues in addition to toxic hazards. This chapter is concerned principally with such quantitative assessments.

The National Academy of Sciences [1] has identified four steps in quantitative risk assessment: hazard identification, dose-response assessment, exposure assessment, and risk characterization. Hazard identification is concerned with the qualitative identification of health effects that can be caused by a substance such as EtO. Does-response assessment is the quantitative association of risks with specific levels of exposure to the substance. Exposure assessment quantitatively estimates the actual levels of the substance to which people are exposed. Risk characterization combines the results of the dose-response and exposure assessments into a quantitative portrayal of the magnitude of the public health problem that might arise from a continued or new pattern of use for the substance. The following secitons of this chapter follow this general outline.

II. HAZARD IDENTIFICATION

A. Ethylene Oxide

Identification of a hazard to human health is most positive when it is possible to demonstrate that a strong statistical association exists between human exposures to the substance in question and an excess of an adverse health effect *and* that a plausible biological mechanism exists that would explain the association. In the case of EtO, such human evidence is available for certain effects such as skin irritation or hypersensitivity reactions at very high exposures [2,3], but it is tenuous for the effects that might occur at the low exposures associated with the use of medical devices.

Proof of EtO's ability to cause mutations is available from experiments with bacteria and other organisms [4] and provides a plausible mechanism for the induction of mutations in cells that could lead eventually to cancers in humans. However, epidemiologic evidence of a statistical association is weak. For example, Hogsted et al. [5] have reported excesses of leukemias and other tumors in sterilizer operators, but exposures are not well characterized, other chemicals were present, and the statistical association was marginal. A comprehensive epidemiologic study of over 20,000 workers exposed to EtO is expected to be completed in 1989 by the National Institute for Occupational Safety and Health.

In the case of reproductive toxicity, Hemminki et al. [6] have reported an excess of "spontaneous" abortion in hospital sterilizing staff, but again exposure could not be characterized well and other substances were in use. Even less human evidence is available for other chronic health effects from exposure to EtO. A report of neurological symptoms in hospital workers [7] is sufficiently ambiguous that we did not consider it in our quantitative analysis.

Consequently, our best evidence of EtO hazards to human health comes from experiments in laboratory animals that are presumed by most toxicologists to be relevant for the identification of potential human hazards. The cancer-causing potential of EtO was demonstrated in studies with rats and mice that showed excesses of leukemias, brain tumors, and other cancers after exposure to air concentrations of 100 parts per million (ppm) or somewhat lower [8-10]. The fact that sites of cancer are similar in these species to those reported in the human studies lend credence to the assertion that EtO is carcinogenic in humans.

For noncancer effects, we have evidence of neurologic and hematologic effects in various laboratory species as well as damage to lungs, liver, kidney, and testes [4,8,11,12]. Moreover, a study by Snellings et al. [13] showed that EtO could cause longer gestation, decreased litter size, and fewer implantation sites in rats exposed to 100 ppm for 6 hours daily, but appeared to cause no adverse reproductive outcomes at exposure levels of 33 ppm.

B. Ethylene Glycol

No evidence is available to suggest that EG is carcinogenic. At very high oral doses, EG can lead to central nervous system dysfunction, kidney failure, or even death. The lowest doses known to cause effects in animals (deposition of calcium salts in the kidney) are in the vincinity of 40 mg EG per kilogram of body weight per day (mg/kg/day) [14]. Reproductive effects such as fetal death or malformation have also been observed at high doses [15], but are absent at 22 mg/kg/day [16].

C. Ethylene Chlorohydrin

Although some studies have suggested that ECH (2-chloroethanol) is carcinogenic, the most complete bioassay [17] was negative. Moderate doses of ECH in animals can retard weight gain in experimental animals, but 16 mg/kg/day leads to no apparent effects [18]. ECH can lead to fetal toxicity at doses above 60 mg/kg/day [19].

III. DOSE-RESPONSE ASSESSMENT

A. Principles of Quantitative Toxicology

Partly because of the nature of the relationships between exposure and effect and partly because of different traditions of analysis and use of toxicity information, the quantitative assessments of dose-response relationships differ for cancer and noncancer health effects. For substances such as EtO, which appear to cause cancer through inducing mutations in cells, it is usually assumed that any exposure to a substance *could* cause a cancer to develop, but that the probability decreases (more or less linearly) with dose for low doses, becoming negligibly small for very small doses. The risk of exposure can then be estimated from the slope of this linear relationship and the estimated doses to which people are exposed, and the risk can be compared with criteria of acceptability. For noncancer effects, by contrast, it is usually assumed that a critical nonzero degree of exposure is required before *any* clearly adverse effect can be observed. (Obviously, even small doses could cause subtle biochemical changes, but there is no reason to believe that they alone are detrimental to the health of the exposed individual.) For such effects, the objective is to identify a level of exposure below which adverse effects are extremely unlikely to occur. This dose can then be compared with estimated actual exposures to determine whether the exposures are of concern.

B. Quantitative Analysis

The carcinogenicity of EtO is described in terms of its potency or unit cancer risk (UCR), the slope of the dose-response relationship projected to hold at low doses. All of the noncancer effects are described by the no-observed-effect level (NOEL), the chronic daily dose below which no effects of consequence have been seen in any species studied, or the lowest-observed-effect level (LOEL), also from the most sensitive species, if no NOEL has been clearly identified.

The cancer potency of EtO is based on the above-mentioned studies in animals after exposure to air concentrations of 100 ppm or somewhat lower

[8-10]. These data were analyzed by the method of Gaylor and Kodell [20], which produces an upper confidence limit on the low-dose response to a carcinogen, assuming equal potency in humans and animals on the basis of daily dose per unit body weight. The UCR of EtO predicted by this method is 4.8×10^{-2} $(mg/kg/day)^{-1}$.

The NOELs for noncancer effects are obtained from other studies in laboratory animals. Snellings et al. [8] showed that no effects were observed in rats exposed to 10 ppm for 6 hours per day, 5 days per week. The corresponding NOEL in humans is calculated to be 2 mg/kg/day. The reproductive NOEL for EtO is estimated to be 9 mg/kg/day, based principally on a study by Snellings et al. [13], in which no adverse reproductive outcomes were seen in rats exposed to 33 ppm for 6 hours daily.

Neither EG nor ECH is assumed to be carcinogenic in humans, and no UCR is estimated for them. The LOEL for chronic toxicity of EG is taken to be 40 mg/kg/day [14], and its NOEL for reproductive effects is taken to be 22 mg/kg/day [16]. The NOEL for chronic toxicity of ECH is taken to be 16 mg/kg/day [18], and its LOEL for reproductive effects is taken to be 60 mg/kg/day [19].

A summary of the toxicity values for all three substances appears in Table 2.

IV. EXPOSURE ASSESSMENT

For many environmental pollutants—for example, chemicals discharged from industrial plants into air or water—exposure assessment is a reasonably straightforward activity that combines direct measurement of ambient concentrations with various well-developed mathematical models that represent the transport, transformation, and fate of pollutants. By contrast, methods for estimating exposure to EtO residues on sterilized medical devices have not been well developed. We know of only one other attempt to do so [21]. Novel methods were needed to estimate patient exposure to residues of EtO on medical devices.

Two distinct ways of estimating exposures were employed. One, the "composite individual" method, is designed to characterize exposure to the entire U.S. population for the purpose of assessing the overall carcinogenic risks of EtO. The other, the "specific procedures" method, is designed to estimate the risk of either cancer or other effects from exposure to medical devices in the course of specific medical procedures.

A. Composite Indivudal

A "composite individual" is a hypothetical person who uses exactly his or her share of all the medical devices sold in the United States every year. For example, if 6 billion adhesive bandages are sold annually in the United States and the U.S. population is 240 million, then the composite individual uses 6000/240 or 25 bandages per year. Over a 73-year lifetime, about 1800 bandages would be used. The corresponding numbers for infrequently used devices such as intraocular lenses are considerably lower; even when we multiply the annual use rate by an expected lifetime of 73 years, the composite individual may still "use" considerably less than one device over a whole lifetime. The composite individual thus represents a compromise or average between people who use one or more devices of a specific type over a lifetime

TABLE 2 Summary of Toxicity Values

Compound	Cancer	Chronic toxicity	Reproductive toxicity
EtO	UCR = 4.8×10^{-2}	NOEL = 2 mg/kg/day	NOEL = 9 mg/kg/day
EG		NOEL = 16 mg/kg/day	LOEL = 60 mg/kg/day
ECH		LOEL = 40 mg/kg/day	NOEL = 22 mg/kg/day

and those who use none at all. Table 3 shows a random sample of device use by the composite individual from a set of about 75 devices for which we have estimated exposures.

B. Device Use

To derive these estimates and ones like them, we needed to estimate the number of devices used in total across the United States in the course of a year. No comprehensive source of data on the use of medical devices in the United States came to our attention. We therefore estimated annual usage through a combination of complementary approaches.

For a very limited number of devices, the Census of Manufactures [22] tabulates data on number of units sold, total sales in dollars, or both. Some information on unit or dollar sales can also be found in the files of trade publications that serve the biomedical product industry. If prices of devices can be obtained from catalogs or industry sources, they can be divided into total dollar sales volume to estimate unit sales. Although not every unit sold is necessarily used on a patient, sales volume is a satisfactory surrogate for use volume for many types of devices. The principal exceptions involve items that enjoy widespread use for other than medical services, such as the use of hypodermic needles in biological and chemical research, the use of surgical-quality gloves in laboratories, or the use of absorbent towels in hospitals for general purposes instead of patient contact.

Device use can also be estimated (independently, in part) from statistics on disease rates and medical procedures. The National Center for Health Statistics [23] publishes data on the number of surgical, diagnostic, and therapeutic procedures undertaken in short-term care hospitals. Table 4 gives some examples. To obtain an estimate of the number of devices of a given type used per year, we can add the numbers estimated to be used in various procedures. For each procedure, that number is the product of the annual number of procedures performed and the number of devices of the given type that is used in each procedure. Estimates of device use per procedure are sometimes obvious (one pacemaker system per pacemaker implant operation) but are best provided by medical professionals. We asked active hospital nurses and nurse supervisors to estimate the average number of devices used in approximately 70 different procedures.

The NCHS does not report data separately on certain common procedures, such as anesthesia, that are considered a routine part of another procedure, or on procedures that are rarely performed inside the hospital, such as dialysis or insulin injections. The frequency of these kinds of procedures can sometimes be estimated from the prevalence in the population of persons with the corresponding disorders (kidney failure and diabetes, respectively).

TABLE 3 Examples of Use of Medical Devices by a Hypothetical Composite Individual[a]

Device type	Lifetime no. of uses
Bandages, adhesive	1800
Clamps, umbilical crod	1
Drapes, surgical	45
Gloves, surgical	19
Pacemakers and accessories	0.03
Sponges, surgical	142
Tubes, endotracheal	9
Tubing, blood or iv	15

[a]Note: In this and succeeding tables, the numerical values presented may appear precise, as they often are taken directly from computer printouts. The accuracy of the numbers usually would support no more than one *significant* figure, however.

C. Exposure per Use

Once we have estimates of the numbers of medical devices sterilized with EtO each year, we can combine that information with estimates of the exposures obtained from one use of each of the devices to estimate the aggregate exposure to the composite individual. The mass of EtO received over a lifetime is the product of the number of uses per lifetime and the mass absorbed into the body during each use. To match the information we have about toxicity, however, we need instead the dose (or more precisely, the dose rate) received in an average day. Dose is defined as the number of milligrams of EtO received per kilogram of body weight. The lifetime average daily dose (LADD) is just the total number of milligrams received over a lifetime by our hypothetical composite individual divided by 26,650 (the number of days in a 73-year lifetime) and by 60 kg (the assumed average weight of a person over a lifetime, including both men and women). With a little additional arithmetic, we can express the LADD as

$$\text{LADD} = \Sigma(D_d N_d)/(60 \times 365P) \qquad (1)$$

where D_d is the dose (mg) from one use of device d, N_d is the number of devices used annually, 60 is the weight of the composite individual (kg), 365 is the number of days per year, P is the U.S. population, taken to be 240 million, and the units of LADD are mg/kg/day.

The next step is to estimate the absorbed dose, in milligrams, per use of each device. The quantity of EtO *available* for exposure, in micrograms, is equal to the residue on the device (ppm by weight) times the weight of the device, in grams. The actual *absorbed* dose is estimated by applying a "reduction factor" that accounts for the tendency of EtO to escape to other

TABLE 4 Example Procedures Performed in Short-Term Care Hospitals

	Annual number of procedures (thousands)	Lifetime procedures per person	Dose (MDD) per procedure (mg/kg)
Appendectomy	294	0.09	0.11
Arthroscopy of knee	237	0.07	0.06
Bypass anastomosis for heart revascularization	202	0.06	0.53
Cesarean section and removal of fetus	814	0.24	0.14
Dilation and curettage of uterus	744	0.22	0.07
Excision or destruction of intervertebral disc	203	0.06	0.12
Insertion of cardiac pacemaker system	170	0.05	0.21
Left heart cardiac catheterization	220	0.07	0.07
Partial excision of large intestine	166	0.05	0.13
Repair and plastic operations on the nose	235	0.07	0.08
Total cholecystectomy	484	0.15	0.13
X-Ray of urinary system	480	0.14	0.05
IV administration	20,000[a]	6.00	0.09

[a]ENVIRON estimate.
Source: NCHS [23].

media (air or fluids leaving the body) or to remain on the device after use because the time of contact with the patient was insufficient for total absorption. In equation form, the absorbed dose is given by

$$D_d = F_r R_d m_d \quad (2)$$

where m_d is the weight of the device (g), R_d is the residue of EtO on the device (ppm), F_r is the reduction factor applicable to use of such devices, and the units of absorbed dose are μg.

No data base of residues on devices at the time of use is known to be available; although data are gathered about the residues on devices prior to release for sale, a comprehensive compilation of these levels has never been attempted, to our knowledge. Furthermore, little is known about the average shelf life of devices, and dissipation of ethylene oxide during storage is correspondingly difficult to estimate. We therefore made a strong (and probably substantially conservative) assumption about residue levels to arrive at a plausible risk assessment. We relied on the levels proposed by the FDA [24] as limits for residues. These proposals, which have never been finalized by the FDA, are shown in Table 5. Our adoption of these levels for the purposes of illustration does not necessarily imply our endorsement of either the values or the way in which they were expressed.

We were also unable to find any systematic compilation of device weights. For easily obtained devices, we weighed typical samples. We also obtained estimates from manufacturers or judged them from similarity to other devices for which weights were known.

Most uncertain of all are the values of the reduction factors. We were unable to find direct information on the transfer of EtO from sterilized medical devices to patients or even to experimental animals. Some measurements of "dissipation curves" have been made to understand how fast EtO escapes from sterilized devices as a function of storage conditions [25-30], and these

TABLE 5 FDA-Proposed Residue Limits

	Concentration limits (ppm)		
	EtO	ECH	EG
Implant			
Small (<10 g)	250	250	5000
Medium (10-100 g)	100	100	2000
Large (>100 g)	25	25	500
Intrauterine device	5	5	10
Intraocular lenses	25	25	500
Devices contacting mucosa	250	250	5000
Devices contacting blood ("ex vivo")	25	25	250
Devices contacting skin	250	250	5000
Surgical scrub sponges	25	25	250

may indicate the rapidity with which ethylene oxide would move into the air from devices applied to the skin. A very few experiments have been conducted to measure the transfer of EtO from devices to fluids [31], such as would occur in the course of transfusions, iv administration, or dialysis involving tubing sterilized with EtO.

The rates of transfer appear to vary significantly, depending on materials of construction and other factors. Many dissipation curves show that the rate of loss drops as the EtO source is depleted, as would be expected from the physical chemistry of the condition. This behavior can be represented in terms of "half-lives" ranging from a few hours or less to more than a week.

This variation is less important than it first may seem because those devices that transfer EtO rapidly to the body also rapidly lose EtO between release for sale and actual use. In the absence of measured values for the reduction factor, we assumed that the transfer would occur at a rate of about 3% per hour of contact, which is reasonably consistent with a half-life of 24 hours (see Fig. 1).

Times of contact of devices with patients range from a second or two for a gauze sponge to many years for implants. We assumed that any device remaining in place longer than 36 hours transfers all of its EtO residues, either to the patient or to external fluids (air, urine). Transfer for shorter periods of contact is assumed to be proportionally lower. We obtained estimates of time of contact from our nurse consultants, from industry sources, or from our own observations.

Devices applied to the skin or mucosa are assumed to transfer less EtO to the body than a device used invasively for the same period. Table 6 shows assumed reduction factors by type of exposure for several periods of exposure.

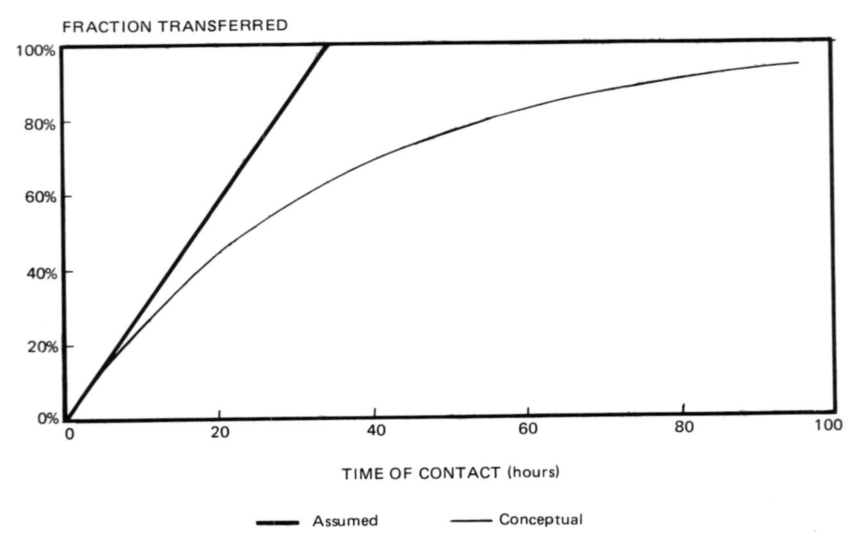

FIGURE 1 Assumed dependence of absorbed dose on time of exposure to EtO residues on medical devices.

TABLE 6 Assumed Transfer Reduction Factors

Time of contact	Devices contacting: Blood or internal organs	Mucosa or external conduits	Skin
1 s	7.0×10^{-6}	3.5×10^{-6}	2.0×10^{-6}
1 min	4.0×10^{-4}	2.0×10^{-4}	1.3×10^{-4}
1 h	2.5×10^{-2}	1.3×10^{-2}	8.0×10^{-3}
1 day	6.0×10^{-1}	3.0×10^{-1}	2.0×10^{-1}
Prolonged	1.0	5.0×10^{-1}	3.3×10^{-1}

When all of these data are combined, a data base of exposures per use of each device can be constructed. Table 7 gives a sample of estimated exposure per use from the data described above. In examining these estimates, one should be aware of the great uncertainty in the data used to derive them; most of the necessary assumptions are in the direction of overestimating the exposure per use. In particular, use of the FDA-proposed residue limits may have little relationship to the true values of residues on devices.

One can also compute the contribution to the LADD for the composite individual from these doses by multiplying by the number of exposures per lifetime and dividing by body weight and number of days per lifetime. These results are also displayed in Table 7 for the selected devices.

D. Specific Procedures

The appropriate measure of exposure for chronic toxic effects other than cancer is not the LADD but the daily dose over a shorter period. The appropriate duration depends on the specific type of effect under consideration. For example, in the case of birth defects (teratogenesis), the critical period may be measured in days; for most other chronic effects, it is much longer.

To be conservative, we assumed that all the dose from all the devices used in a medical procedure could be accrued in a relatively short time, generally one day unless we knew that several devices were used over a period of several days. The maximum daily dose, MDD_p, from a given procedure p is thus estimated by

$$MDD_p = \Sigma(n_{dp} D_d)/60 \qquad (3)$$

where n_{dp} is the number of devices of type d used per procedure p, D_d is the dose per use of device d, as defined earlier, and 60 is the weight of the average person (kg). The units of MDD are thus also mg/kg/day.

Discovering which types of devices are typically used in which procedures and how many of each are typically used was by no means a simple endeavor. Sometimes the answer is obvious, such as the association of an umbilical cord clamp with childbirth and the typical birth involving exactly one clamp. Most

TABLE 7 Selected Estimates of Exposure per Use and Corresponding Lifetime Average Daily Dose

Device type	Exposure per use (mg)	LADD[a] (ng/kg/day)
Bags, blood administration	0.042	0.19
Bandages, adhesive	0.008	7.82
Catheters, intravascular	0.125	2.34
Dressings, surgical	0.825	6.19
Gloves, examination	< 0.001	0.02
Lenses, intraocular	< 0.001	< 0.01
Oxygenators, blood	7.500	0.30
Prostheses, breast	12.500	0.07
Sponges, laparotomy	0.075	0.39
Tubes, infant feeding	0.008	0.32
Tubing, drainage	1.250	7.50

[a]LADDs for the composite individual are so small that they are best expressed in ng/kg/day (one millionth of a mg/kg/day).

are not obvious, however, because the rate of wastage is unclear and the rate of replacement is not obvious. In some cases, a rough idea of use can be obtained by counting the contents of a prepackaged sterilized kit destined for use in a specific type of procedure, but even here, the degree of redundancy is not obvious. Therefore, most of the identities and numbers of devices used per procedure were provided by our nurse consultants.

Using these estimates with the estimates of exposure for one use of each device, we can estimate the total exposure associated with all the devices used in a specific procedure and the corresponding maximum daily dose, MDD, by applying Equation (3). Selected results are shown in Table 4.

The information on exposures for specific procedures can also be used as a semiindependent chack on the exposure estimated for the composite individual. The number of procedures accomplished per year in the United States [23] is multiplied by the expected number of years in a lifetime (73) and divided by the population of the United States (240 million) to obtain an estimate of the expected number of procedures for the composite individual over a lifetime. See Table 3 for some examples. That estimate can then be multiplied by the exposure from all the devices used in the procedure to obtain an estimate of the lifetime exposure from the expected number of procedures. For example, about 120,000 mastectomies are performed annually; that projects to a lifetime risk of mastectomy of $0.121 \times 73/240$ or 0.04 (one in 25). The contribution of mastectomies to lifetime exposure for the composite individual is obtained by multiplying the exposure for one mastectomy by the factor 0.04.

These estimates can be summed over all the types of procedures considered to obtain a second estimate of lifetime dose for the composite individual. This estimate is not entirely independent of our earlier estimate, because the same estimates of exposure per use of one device are employed and because our estimates of the annual number of devices used were sometimes obtained from the annual number of procedures performed. The procedure method will tend to underestimate total exposure, because not all procedures can be included in a list of reasonable length and because uses of medical devices occur outside hospitals, especially in doctors' offices and homes. These omissions may result in underestimation by a factor of 2 to 3.

V. RISK CHARACTERIZATION

The nature of any risks posed by exposure to residues of EtO on medical devices can be understood by combining the results of the dose-response assessment and the exposure assessment. This process is conducted differently for the risks of cancer and of health effects other than cancer.

A. Cancer

The analysis of the risks of cancer from exposure to EtO residues assumes conservatively that some risk of cancer exists at any nonzero level of exposure. For low exposures, the lifetime risk is assumed to increase linearly with increasing total dose, expressed in daily terms as the LADD. [Lifetime risk is the overall chance of developing cancer; in the United States, this risk amounts to about one in four or five (20-25%) for all potentially fatal forms of cancer from all causes.]

For this assumed dose-response relationship, the constant of proportionality is the UCR or potency for EtO, which was given in Table 2 as 0.048 $(\text{mg/kg/day})^{-1}$. Under these assumptions, the lifetime risk R of developing cancer from EtO on medical devices is

$$R = 0.048(\text{LADD}) \tag{4}$$

For example, if the LADD turned out to be exactly 1 mg/kg/day, then the lifetime risk R would be 0.048, or a little under 5%, and would be a very significant contributor to the cancer burden in the United States, similar to that of smoking.

This equation can be evaluated either for all devices combined [LADD from Equation (1)] or for all the devices used in a specific procedure [dose from Equation (3) averaged over a lifetime (LADD = MDD/26,650)]. For the composite individual, the lifetime cancer risk computed by this procedure is about seven in a million (7×10^{-6}). Although this figure is above the criterion of one in a million sometimes quoted for environmental hazards, it applies to all medical devices sterilized with EtO; the risks from any one device or from all the devices used in any one procedure are much lower. For example, the cancer risk from a bypass operation, which is the procedure with the highest aggregate dose, is estimated as $0.048 \times 0.53/26{,}650$ or about one in a million. Furthermore, the benefits of a successful operation obviously outweigh this small risk by an enormous factor. And the recipients of these operations are ordinarily middle aged to elderly, which lessens the likelihood of cancer development in comparison to that from the same dose at an earlier age.

B. Chronic and Reproductive Toxicity

The risks of health effects other than cancer may be evaluated by comparing the MDD [Equation (3)] with some criterion of minimal toxic potential, in our case the NOEL (LOEL, if a NOEL is not available). The ratio of the NOEL or LOEL to the MDD is defined as the margin of safety, MOS:

$$MOS = NOEL/MDD \tag{5}$$

If the margin of safety drops below unity, concerns about possible health effects in humans should be raised. If the margin of safety exceeds the uncertainty factor usually applied to a NOEL derived from animal experiments to estimate a safe exposure in humans, typically a factor of 100, then essentially no concern at all is justified. In between, the acceptability of the margin of safety must be judged by the quality and quantity of data used to fix the NOEL, the likelihood that sensitive groups of humans may be more susceptible to the effects of EtO than any of the animal species and strains tested, and the degree of conservatism invested in the estimate of the MDD.

In our investigation of the risks of EtO, we found no instance of an MOS that fell below 1.0. In the aforementioned example of the bypass operation, the MDD was about 0.5 mg/kg, assumed to be delivered in 1 day. That dose is 4 times lower than the NOEL (2 mg/kg/day). Furthermore, the NOEL was for chronic exposure; a single day of dosing is not likely to be as dangerous.

The MOS is always larger for reproductive effects, because the NOEL is 9 mg/kg/day. In that the reproductive effects on which the NOEL was based were observed after exposure during pregnancy, the procedures of concern are those likely to be performed on pregnant women who are not terminating the pregnancy. These also tend to have much smaller MDDs than for a bypass operation.

Because children weigh less than 60 kg, the same exposure (if measured in mg) produces a larger MDD. The difference is greatest for the newborns, who weigh about 3 kg. For them, the exposures due to childbirth itself may be most important. Using extreme assumptions that are unlikely to prove true for most infants, we estimated the MDD to approach 2 mg/kg/day, the NOEL for chronic effects. Under our assumptions, however, these doses would persist no more than 3 days, and the application of a chronic NOEL would appear to be adequately protective. In this and other cases where the estimated MOS is less than 100, however, further research is indicated to reduce the current high level of uncertainty about the true exposure levels.

The reaction products of EtO—EG and ECH—are less toxic than EtO. By examining the relationship of the NOELs and LOELs in Table 2 with the residue limits in Table 3, one can determine that whenever the risks of EtO are acceptable, so too are those of the reaction products, with the possible exception of the reproductive effects of EG and the exposures to ECH via surgical scrub sponges. We have examined these cases as well and find similar or greater margins of safety.

VI. SUMMARY AND CONCLUSIONS

This chapter has described a quantitative risk assessment of the potential hazards of EtO residues on sterilized medical devices. The risks of both cancer and other chronic toxicity have been assessed, both for a hypothetical composite individual and for persons undergoing a variety of specific medical procedures. The cancer risk for the composite individual has been estimated conservatively to be seven in a million from exposure to EtO in all the medical devices studied. Furthermore, exposures from specific procedures have been estimated to be lower than the minimum levels causing chronic health effects, often substantially so.

Given the obvious health benefits of using a proven and cost-effective method for sterilizing medical devices, these risks are well within the range usually considered acceptable, especially in consideration of the fact that major uncertainties have been taken into account through the use of conservative assumptions.

The techniques described in this chapter were developed specifically to assess the risks of EtO and its reaction products. The basic concepts and many of the key data are also applicable, however, to other potentially toxic substances present in medical devices. The principal requirement would be to develop the concentrations of the substance in the various device types and the corresponding transfer reduction factors. In conclusion, quantitative risk assessment is a valuable tool in the evaluation of the potential hazards of medical devices.

ACKNOWLEDGMENTS

Thanks go to the Health Industry Manufacturers Association for supporting this work and to my colleagues at ENVIRON, especially Joseph Rodricks, Susan Rieth, Andrew Fragen, and Resha Putzrath, for their contributions.

REFERENCES

1. National Academy of Sciences/National Research Council, *Risk Assessment in the Federal Government: Managing the Process*, National Academy Press, Washington, D.C., 1983.
2. J. L. Shupack, S. R. Anderson, and S. J. Romano, Human Skin Reactions to Ethylene Oxide, *J. Lab. Clin. Med.*, *98*:723-729 (1981).
3. R. J. Caruana, R. W. Hamilton, and F. C. Pearson, Dialyzer Hypersensitivity Syndrome: Possible Role of Allergy to Ethylene Oxide, *Am. J. Nephrol.*, 5:271-274 (1985).
4. U.S. Environmental Protection Agency, Health Assessment Document for Ethylene Oxide, Office of Health and Environmental Assessment, Washington, D.C., EPA/600/8-84/009F, 1985.
5. C. Hogstedt, L. Aringer, and A. Gustavsson, Epidemiologic Support for Ethylene Oxide as a Cancer-Causing Agent, *J. Am. Med. Soc.*, *255*:1575-1578 (1986).
6. K. Hemminki, P. Mutanen, I. Saloniemi, M.-L. Niemi, and H. Vainio, Spontaneous Abortions in Hospital Staff Engaged in Sterilizing Instruments with Chemical Agents, *Br. Med. J.*, *258*:1461-1463 (1982).

7. W. J. Estrin, S. A. Cavalieri, P. Wald, C. E. Becker, J. R. Jones, and J. E. Cone, Evidence of Neurologic Dysfunction Related to Long-term Ethylene Oxide Exposure, *Arch. Neurol.*, *44*:1283-1286 (1987).
8. W. M. Snellings, C. S. Weil, and .R .R Maronpot, A Two-Year Inhalation Study of the Carcinogenic Potential of Ethylene Oxide in Fischer 344 Rats, *Toxicol. Appl. Pharmacol.*, *75*:105-117 (1984).
9. D. W. Lynch, T. R. Lewis, W. J. Moorman, J. R. Burg, D. H. Groth, A. Khan, L. J. Ackerman, and B. Y. Cockrell, Carcinogenic and Toxicological Effects of Inhaled Ethylene Oxide and Propylene Oxide, *Toxicol. Appl. Pharmacol.*, *76*:69-84 (1984).
10. National Toxicology Program, NTP Technical Report on the Toxicology and Carcinogenesis Studies of Ethylene Oxide (CAS No. 75-21-8) in B6C3F1 Mice (Inhalation Studies), Board draft, NTP TR 326, NIH publication no. 86-2582, 1986.
11. H. Sprinz, H. Matzke, and J. Carter, Neuropathological Evaluation of Monkeys Exposed to Ethylene and Propylene Oxide, Midwest Research Institute, Prepared for National Institute for Occupational Safety and Health, Cincinnati, OH, NIOSH contract no. 210-81-6004, MRI project no. 7222-B, PB83-134817, 1982.
12. W. M. Snellings, C. S. Weil, and R. R. Maronpot, A Subchronic Inhalation Study of the Toxicologic Potential of Ethylene Oxide in B6C3F1 Mice. *Toxicol. Appl. Pharmacol.*, *76*:510-518 (1984).
13. W. M. Snellings, J. P. Zelenak, and C. S. Weil, Effects on Reproduction in Fischer 344 Rats Exposed to Ethylene Oxide by Inhalation for One Generation, *Toxicol. Appl. Pharmacol.*, *63*:382-388 (1982).
14. Union Carbide, A Summary of Studies Conducted on the Long-Term Toxicology of Ethylene Glycol, Submitted in an internal Union Carbide memo from B. Ballantyne to V. H. Johnkoski, April 16, 1985.
15. J. C. Lamb, R. R. Maronpot, D. K. Gulati, V. S. Russell, L. Hommel-Barnes, and P. S. Sabharwal, Reproductive and Developmental Toxicity of Ethylene Glycol in the Mouse, *Toxicol. Appl. Pharmacol.*, *56*:16-22 (1985).
16. Union Carbide, Letter regarding inhalation teratology study of ethylene glycol from J. Browning, Union Carbide, to Document Control Officer, Information Management Division, Office of Toxic Substances, U.S. Environmental Protection Agency, Washington, D.C., October 7, 1985.
17. National Toxicology Program, NTP Technical Report on the Toxicology and Carcinogenesis Studies of 2-Chloro-ethanol (Ethylene Chlorohydrin) (CAS No. 107-07-3) in F344/N rats and Swiss CD-1 mice (Dermal Studies), NTP TR 275, NIH publication no. 86-2531, 1985.
18. M. K. Johnson, *Food Cosmet. Toxicol.*, *5*:449 (1967). Reported in Rowe, V. K., and McCollister, S. B., Alcohols. In *Patty's Industrial Hygiene and Toxicology*, G. D. Clayton and F. E. Clayton, eds., Wiley, New York, 1982, pp. 4675-4684.
19. C. Jones-Price, T. A. Marks, T. A. Ledoux, J. R. Reel, P. M. Fischer, L. Langhoff-Paschke, M. M. Marr, and C. A. Kimmel, Teratologic Evaluation of Ethylene Chlorohydrin in CD-1 Mice, Research Triangle Institute, Research Triangle Park, N.C., PB85-172104, 1985.
20. D. W. Gaylor and R. L. Kodell, Linear Interpolation Algorithm for Low Dose Risk Assessment of Toxic Substances, *J. Environ. Pathol. Toxicol.*, *4*:305-312 (1980).
21. Rijksinstituut voor Volksgezondheit en Milieuhygiëne, Ethylene Oxide

Residues in Medical Devices, National Institute for Public Health and Environmental Hygiene (Netherlands), n.d.
22. *Census of Manufactures*, 1982, Industry Series, MC82-1-38B: Medical Instruments; Ophthalmic Goods; Photographic Equipment; Clocks, Watches, and Watchcases, U.S. Bureau of the Census, U.S. Government Printing Office, Washington, D.C., 1985.
23. National Center for Health Statistics, Detailed Diagnoses and Procedures for Patients Discharged from Short-Stay Hospitals, United States, 1984. Series 13, No. 86, DHS publication no. (DHS)86-1747, U.S. Government Printing Office, Washington, D.C., 1986.
24. Food and Drug Administration, Ethylene Oxide, Ethylene Chlorohydrin, and Ethylene Glycol: Proposed Maximum Residue Limits and Maximum Levels of Exposure, *Fed. Reg.*, *43*:27474 (June 23, 1978).
25. P. Vink and K. Pleijsier, Aeration of Ethylene Oxide-Sterilized Polymers, *Biomaterials*, *7*:225-230 (1986).
26. S. R. Anderson, Ethylene Oxide Toxicity, *J. Lab. Clin. Med.*, *77*:346-356 (1971).
27. E. Baan, Ethylene Oxide Absorption and Desorption of Elastomers and Plastics, *Bull. Parenteral Drug Assoc.*, *30*:299-305 (1976).
28. W. Henne, W. Dietrich, M. Pelger, and G. v. Sengbusch, Residual Ethylene Oxide in Hollow-Fiber Dialyzers, *Artificial Organs*, *8*:306-309 (1984).
29. R. G. McGonnigle, J. A. Renner, S. J. Romano, and R. A. Abodeely, Jr., Residual Ethylene Oxide Levels in Medical Grade Tubing and Effects on an In Vitro Biologic System, *J. Biomed. Mater. Res.*, *5*:273-283 (1975).
30. R. C. Simpson and C. H. White, A Dissipation Study of Residual Ethylene Oxide from Various Medical Device Thermoplastics and Rubbers, Technical report no. 24, Plastics Division, The West Company, 1979.
31. R. Kroes, F. Bock, and L. Martis, Ethylene Oxide Extraction and Stability in Water and Blood, Travenol Laboratories, Morton Grove, IL, 1985.

33

Medical Device Risk Assessment

F. ALAN ANDERSON *Center for Devices and Radiological Health, Food and Drug Administration, Rockville, Maryland*

I. INTRODUCTION

A wide variety of medical devices is available to diagnose and treat illness or injury.

Device manufacturers who design and produce this wide variety of products and provide them to health care professionals or patients must deal with various federal and state regulatory agencies who have responsibility for assuring such things as safety, effectiveness, proper record keeping, and appropriate reimbursement. If health care professionals or patients find problems with devices, those same federal and state regulatory agencies take on a problem-solving responsibility to make certain that the manufacturer remedies the situation or to ensure the problem is resolved in some other way. One such federal agency is the Food and Drug Administration (FDA).

This chapter will look at medical device risk assessment as a tool FDA uses to carry out its responsibilities.

In approving new medical products for market, removing problem products from the market, or taking other public health actions, it is necessary for FDA to have a basis for decision-making. Legislative constraints influence such decisions. An example is the Delaney Amendments to the federal Food, Drug, and Cosmetic Act, which prohibit known carcinogens from certain uses.

The overall Medical Device Amendments to that Act (P.L. 94-295), however, are best characterized as risk-balancing legislation. That is, the medical benefit of a device is a factor, in addition to risk, that should be considered in determining the appropriate level of regulatory control.

Another legislative authority that is applied to medical devices is the Radiation Control for Health and Safety Act (P.L. 90-602), which provides authority to assure the public health and safety with respect to radiation emitting electronic products. This authority makes an important distinction between necessary radiation and unnecessary radiation. In effect, necessary radiation is handled by balancing risk with other factors, including benefit.

Risks from unnecessary radiation are generally considered undesirable, and technologically feasible product and/or user behavior changes are sought to reduce the risk to a reasonably achievable level.

In one form or other, the concept of risk appears in all decision making. This places medical device risk assessment at the root of problem-solving actions that FDA takes. Since the route to market is through FDA and since FDA takes problem-solving actions to remove products from the market or otherwise assure safety and effectiveness, the approaches used by FDA have an impact on the medical device industry.

Given this level of importance of medical device risk assessment, it is critical to understand the elements that are involved in that process. The major factors that are considered important to medical device risk assessment will be described, and examples will be cited that illustrate why they were important. The primary purpose is to provide basic information to the reader, especially those whose experience is mostly in chemical risk assessment. Any discussion of medical device risk assessment should be part of a continuing dialog between industry, health care professionals, and government. To that end, the author would appreciate any comments, suggestions, or additional information.

II. FACTORS IMPORTANT TO MEDICAL DEVICE RISK ASSESSMENT

FDA's experience with medical devices suggests that the factors that are important to medical device risk assessment relate to the device itself and to how it is used. Table 1 shows those categories and the major elements of each that will be discussed below. The FDA experience on which this discussion is based includes over 12 years of effort to implement medical device legislation and over 20 years of work in the medical radiation arena.

A. Device Design and Manufacture

A device can present a hazard if it is poorly manufactured or if insufficient attention is paid to design elements that influence performance. For example, the failure to design a structural component to resist the stress to which it will be subjected could lead to its fracture. Quality control or quality assurance during manufacture may not correct the problem. FDA's experience

TABLE 1 Major Factors Considered in Medical Device Risk Assessment

For the device	For device use
Design	Adequacy of instructions
Human factors engineering	Training of users
Manufacturing quality control/assurance	Interaction with other devices
Materials toxicity	Human factors engineering
Materials degradation	

suggests strongly that good manufacturing practices do not eliminate inherently bad designs, and could simply ensure that a "bad" design is faithfully produced. Alternatively, inattention to manufacturing quality assurance could unintentionally cause a device defect. For example, a manufacturing process that inappropriately stresses a component may lead to its weakening and its ultimate fracture during use. In the first case, the design was at fault, and in the second, the implementation of the design was flawed. The kind of hazard may not be different in either case, but in medical device risk assessment it is important to distinguish between the two, or determine if both are involved. This knowledge is needed because the approaches to risk management are different in each case.

Device hazards also can be considered in terms of whether the product does something that can cause harm, or whether the product fails to provide a benefit that it should. For example, patients and health professionals may be harmed by an electric shock from a supposedly insulated device. A missed diagnosis, on the other hand, could lead to inappropriate therapy or no therapy at all, each with dire consequences for the patient. It is important to the eventual management of risk to determine whether the hazard arises from omission or commission.

B. Device Materials

Materials from which devices are made are a part of the design and manufacturing process considerations discussed above. Proper design includes selection of appropriate materials. Good manufacturing practice includes process validation to ensure that the materials properties are not compromised. It is also important to consider materials separately when the device in question is applied to or implanted in the body. In this regard, both the effect of the material on the body and the effect of the body on the material are important.

In assessing the risk of a material used in an implant, its toxicity is a key element. The first question, given a particular device, relates to determining just what materials are in the device. Determining just what material is in a particular device can be difficult. Any one device can contain (1) the material making up the basic molecular structure, (2) catalytic material that was used in synthesis, (3) contaminant material, or (4) residual material to which the device was exposed during manufacture. Which is the toxic agent? How much of the toxic agent is in the device? Does the toxic agent leach from the device? Is the part of the body exposed sensitive to the injury produced by the toxic agent? The answers to each of these questions are basic to assessing the risk posed by a device.

The basic questions about materials toxicity are complicated by the fact that the part of the body exposed to the toxic agent may not be an independent variable. The tissue environment in which the material exists can influence the nature of the toxic agent, that is, through metabolic activation or metabolic inactivation. The tissue environment can also influence leachability. Finally, the tissue environment can affect whether the toxic agent remains localized or is distributed throughout the system.

The effect of the body on the material may not be limited to altered toxicity. The tissue environment can be hostile, resulting in promoting, initiating, or accelerating the degradation of the material. Whether a particular environment is "hostile" depends on the material and the length of time it resides in the environment. For example, certain polyurethanes degrade rapidly in a low pH solution. The stomach is a "hostile" environment for such a material. A risk assessment would have to include information on the total time the polyurethane resides in that hostile environment, and include a judgment made as to whether that total time was sufficient to result in compromising the performance of the material for that particular application.

C. Device Users

FDA's experience has repeatedly and consistently been that how a device is used is a significant part of the overall safety and effectiveness of that device. From the sample collection procedures that can influence the result obtained from an *in vitro* diagnostic device to the poor X-ray film processing techniques that lead to the need for a retake, experience clearly indicates the importance of the user.

The user of a medical device can be a health care professional, a patient, or a family member. There is a large variability in users' skills with devices, due to factors such as education and training, health status, environment, and motivation. For example, the visual acuity of some diabetic patients using a blood glucose monitor to determine blood sugar levels can be poor because of diabetic retinopathy, while others may have limited manual dexterity because of arthritis. Whatever the capability of a particular user, circumstances occur that can be a controlling influence, such as the inattention that can result when a tired individual tries to perform a repetitive task.

The user also represents the decision point on the application of medical technology. The decision may be wrong. Patients may inappropriately rely on over-the-counter diagnostic devices. A historical example of inappropriate use by health care professionals is the use of oxygen therapy for premature infants, which is now understood to damage the retina of the eye.

The kinds of hazards that may result from user error are not unusual in themselves. What is unusual is that the incidence of such hazards occurring can range from nonexistent to frequent depending on the influence of the user. Thus, the training of the user and the adequacy of the instructions provided with the device relate directly to assessing the above risks.

D. Human Factors

Human factors are those device design elements and use conditions that influence how the device and the user interact. This interaction places an additional complexity on the ability to identify the hazards that might be associated with a particular device. For example, infant apnea monitors are equipped with both an "off button" and a reset control. Early devices had the two adjacent to each other. Since false alarms do occur, it is expected that the units will be reset. The instructions are clear on how that is to be done, the reset control is identified differently from the "off button." In the middle of the night, however, a sleepy parent, after checking the infant being monitored, could easily press the "off button," which shuts off the alarm just as would the reset control. Inadequate attention to the human factor in early devices led to the problem. Without considering human factors as part of the risk assessment, it would have been difficult to consider the appropriate risk management approach.

E. Medical Devices Systems

The device-intensive nature of certain medical circumstances results in many devices used within close proximity to each other. Such configurations may have been considered by the manufacturer as part of the design, but the ingenuity of device users to devise new systems may outstrip the manufacturer's expectations. Depending on the configuration and number of devices involved, hazards to the patient can result from the interference of one device with another, for example, electromagnetic interference. Users may also attempt to interchange incompatible device components in an attempt at repair, etc., and create a risk to the patient. Even if the manufacturer intended for the device to be used in an environment with many others, patients can be at risk because of user behavior in dealing with the total amount of information (dials, charts, displays, alarms, etc.) that must be monitored. For example, incessant false alarms from different devices could lead to inattention to the one alarm that should be needed. Again, developing appropriate problem-solving approaches depends on accurate problem identification.

III. SPECIFIC DEVICE EXAMPLES

The factors discussed above are considered each time an assessment of risk associated with a particular device is done. Not all factors, however, play a pivotal role in every risk assessment. There may not be enough information to make a determination about the role of the user, for example, and a risk assessment is done without it. In other cases, material toxicity might turn out not to be critical, so the risk assessment is done without emphasis on that factor. The examples below are provided to show such practical applications of the principles of medical device risk assessment (see also Chap.   on Ethylene Oxide).

A. Vitamin B_{12} Assay Kits

Ten years ago a problem was found with vitamin B_{12} assay kits in that normal vitamin B_{12} levels could be reported for certain patients who actually have a deficiency that should be detected and treated. The only alternative to the use of the assay kits, however, was a costly, time-consuming biochemical assay procedure.

The problem with the assay kits was found in the material used to bind vitamin B_{12} as a step in its quantitative measurement. The binders were a mixture of a material called intrinsic factor, which is very specific for vitamin B_{12}, and glycoproteins (collectively known as R-proteins), which are not specific for vitamin B_{12} and will bind to vitamin B_{12} analogs. Blood may contain both vitamin B_{12} and vitamin B_{12} analogs. R-Proteins will bind to either, resulting in both being measured. The results obtained for a given patient, therefore, depended on the relative proportion of the two binders in a particular assay kit and on the amount of vitamin B_{12} and its analogs present in the patient's serum.

The risk assessment required information about the knowledge in the clinical community about the expected levels in blood of vitamin B_{12} relative to the levels of vitamin B_{12} analogs. Also important was the availability of alternative product designs that would eliminate, inactivate, or block R-proteins in an assay kit's binders. The risk management effort that followed included changes both in the test kit methodology and in the instructions for use. The benefit of the inexpensive, rapid diagnostic tool was retained and the risk was eliminated.

B. Anesthesia Systems

Over the past several years, evidence has accumulated indicating that surgery patients being given anesthesia gas and being supported by mechanical ventilation can be accidentally disconnected from the system. Such disconnections are so frequent that many anesthesiologists consider them routine. The connection is reestablished and the event unreported. Sometimes the error is not found and the patient dies.

The analysis of this risk indicated that the equipment generally is manufactured well, but that the design of the connectors could be improved and that device features could be added to provide for an indication when a disconnect has occurred. Independent of any such changes, it was clear that users would continue their practice of balancing the need for a tight connection against the value of a weak link. A weak connection is valued because it would separate before a "wrong move" actually pulled out the tracheal tube inserted into the patient.

Identification of such a strong influence of user behavior led in turn to a more careful look at how the user interacts with the device. This further risk assessment indicated that anesthesia incidents occur because of user error in setting up the equipment and because of faults present at setup that are not detected by the user. This understanding of the role of the user developed in the risk assessment has influenced the risk management efforts of FDA and others. Manufacturers, users, and FDA have combined to provide a checklist procedure that can be quickly run through and that detects most preexisting problems and assures attention to setup issues, including the need to avoid disconnects.

During this period in which activity focused on the role of the user, reports appeared of several patient deaths with one manufacturer's anesthesia device. The problem appeared to be failure to deliver oxygen with the anesthesia gas. Investigation revealed a rubber material in a valve that swelled when in contact with the lubricant and anesthesia gas. This swelling caused the valve to stick in a position that created the failure to deliver oxygen. Had users followed exactly all manufacturer-recommended procedures, anesthesia gas would not have contacted the valve. Actual use, however, was known not to be so rigorous. So, while the user aspect of the problem was added to the risk management effort described above, a specific problem-solving approach was directed to the one manufacturer's problem.

C. Heart Valves

In 1979, a prosthetic heart valve was introduced into general use that was considered a major advance in the state of the art based on the reduced frequency of clot formation around the valve. Such clots are the major life-threatening complication with artificial replacement heart valves, so the new valve was much heralded. Unfortunately, a rare but persistent pattern of mechanical failure began to emerge after the valves were marketed.

Postmarket surveillance data indicated that other valves did not exhibit the pattern of mechanical failure. It was not possible to compare complication rates across the industry, however, because the surveillance data did not include expected complications. Information from the clinical community suggested that the procedures for implanting the valves were not a likely source of the problem.

A review of the manufacturing process revealed several changes in procedure to eliminate repeated bending of one component that might be weakened by the bending. A review of the design specifications indicated that a particular mode of valve closure could exist that might result in unusual stress being placed on that same component. If the problem was in the manufacturing process, then the changes could have been sufficient to eliminate the mechanical failures. If the design was flawed, the changes in process would not necessarily solve the problem. The manufacturer's contention was that the absence of recent failures indicated that the change in the process had solved the problem. An actuarial life table analysis appropriate to evaluating such mechanical failures did not support that conclusion. In the final risk assessment, it was not possible to distinguish between the design and the manufacturing process, or both, as the root cause of the continuing problem.

Because there were data that compared the reduced complication rate of this valve with others made by the same company, a careful statistical review of that information was done. This analysis showed that no difference could be shown between the clot complication rate between the failing valve and earlier designs made by the same company that had not mechanically failed. With that information, it was not of immediate necessity to determine whether the manufacturing process or a design flaw had been the cause of the failures. The approach to risk management became one of eliminating a risk because there was no compensating benefit. Because of the unacceptable mechanical failure rate, the manufacturer removed the valve from the market.

IV. SUMMARY AND CONCLUSIONS

FDA uses primarily the risk-balancing and technological control approaches in its legislative authority to assure the safety and effectiveness of new devices and those on the market. FDA's experience at medical device problem solving demonstrates that risks can be managed when there is an understanding of why the risk exists in the first place. Thus, when problems are uncovered, FDA has found that a successful assessment of risk depends on considering aspects of both the device and its use. Before a risk management approach can be determined, the role of the device design and the manufacture of the device must be considered. At the same time, the role of the user in terms of training, available instructions, and human factors are evaluated. The specific cases show that information on each of these factors is not always available, but that sufficient understanding can be reached to support one or more approaches to problem-solving. The examples also support the view that risk assessment that ignores either the role of the device or the role of the user may lead to ineffective risk management.

34

The AIDS Epidemic: Its Impact on Manufacture and Use of Medical Devices

J. THOMAS LOWE *Office of Health Affairs, Food and Drug Administration, Rockville, Maryland*

LaWAYNE R. STROMBERG *Baxter Healthcare Corporation, Deerfield, Illinois*

NORMAN F. ESTRIN *Health Industry Manufacturers Association, Washington, D.C.*

I. INTRODUCTION

Since almost its inception the AIDS epidemic has catapulted the discipline of public health to a central role in our society and affected virtually every aspect of life in the United States. As we gain knowledge about the transmission of AIDS and how to reduce the risk of transmission, it became clear that medical devices would be essential in reducing the risk of transmission and could also be a mode of transmitting AIDS. This link between devices and AIDS is resulting in significant and far-reaching changes in the medical device industry, which this chapter will attempt to explore..

II. THE AIDS EPIDEMIC

As of the writing of this chapter in 1989, over 100,000 cases [1] of AIDS, acquired immunodeficiency syndrome, have been reported to the Centers for Disease Control of the U.S. Public Health Service. It is estimated that 1-1.5 million people in the U.S. are infected with human immunodeficiency virus (HIV), which causes AIDS [2].

HIV is transmitted through the exchange of bodily fluids in a manner similar to that of the hepatitis B virus. Primarily, transmission is through sexual intercourse and through blood-to-blood transmission. Such sexual practices as anal, oral, and vaginal intercourse, both homosexual and heterosexual, have resulted in HIV transmission. HIV is found in both the semen of a male and the vaginal fluid of a female. The presence of HIV in blood and blood products resulted in transfusion-related transmission prior to 1985,

---

The views presented in this chapter are those of the individual authors and do not necessarily represent the views of the Food and Drug Administration.

when measures were instituted to ensure the safety of the nation's blood supply. An ever-increasing number of cases of AIDS are the result of needle sharing by those who use illegal IV drugs. HIV may also be transmitted by human tissue products and from an infected mother to her fetus.

III. THE ROLE OF CONDOMS

The first wave of the AIDS epidemic in the United States was in the Gay community. With the identification of a virus as the causative agent and sexual intercourse as the mode of transmission came the first hint of the role of medical devices in the epidemic. The latex condom was a product that seemed almost an anachronism in America of the 1970s and early 1980s. With the entry of oral contraceptives into the U.S. market and the use of other methods of contraception such as diaphragms and spermicides, the condom seemed a product of those days prior to the sexual revolution. With the knowledge that HIV was being transmitted like other sexually transmitted diseases, those who were working to stop the spread of AIDS began to promote the use of latex condoms as a method of reducing one's risk of sexual exposure to the virus. Such efforts took place first in the Gay community. By 1986, with the issuance of the Surgeon General's report on AIDS, the effort to promote the latex condom as a method of AIDS prevention made those who manufacture or market condoms in the United States the first in the medical device industry to be affected by the AIDS epidemic.

In a very short time, condoms were a front-page topic along with AIDS. A product whose sales had been relatively constant for years saw a dramatic increase. Condom companies that had been fairly anonymous were contemplating ads on television. An industry that has been subject to a degree of federal regulation commensurate with the public view of the product was about to see increased FDA oversight. With the issuance of the Surgeon General's report and its endorsement of condoms, FDA instituted a series of changes in how it regulated condoms. First, FDA sent a letter to all of those who market condoms in the United States. This letter allowed those who market a latex condom to add to their labeling that their product is for the prevention of sexually transmitted diseases, including AIDS, without the submission of a premarket notification as would be normally required. In view of research that had indicated that natural membrane (skin) condoms might not be effective due to their natural porosity, this labeling change was not allowed for that product. Also, FDA instituted a sampling and testing program for latex condoms. This sampling and testing was performed on both domestic and imported condoms using a water leak test to assess the quality control measures employed by the companies.

The years since the establishment of this testing program find condom quality still a very popular topic in the press. The results of the FDA testing, the release of a condom rating from National Institutes of Health (NIH) funded research, and the publication of a cover story on condom quality by Consumer Reports have spawned countless articles on the subject of how good condoms are and which condoms are best. This higher visibility has resulted in both increased sales and increased headaches for condom companies, a mixed blessing.

AIDS has heightened the interest on the part of the public in any new method of barrier contraception that a company may be planning to market in the United States and has resulted in some questioning the appropriateness of market entry being dependent on a determination of substantial equivalence to a barrier contraceptive in the pre-AIDS era.

IV. GLOVES AND OTHER BARRIER DEVICES

As well as serving as a barrier to the sexual transmission of HIV, medical devices serve as a barrier to the transmission of HIV in a health care setting. The number of documented cases of HIV transmission to health care workers is quite small. About 25 documented cases were reported as having occurred in the health care setting at the 1989 Workshop on AIDS and Personal Protective Equipment [3]. Although this indicates that the level of risk is relatively low, those who work in a health care setting perceive that their risk is high. The highest risk faced by health care workers is from a parenteral exposure to HIV-infected blood, such as from a needle stick. In needle stick studies done at the National Institutes of Health and in San Francisco, the risk of becoming infected after an HIV contaminated needle stick is about 0.5% or 5/1000 [4]. Interestingly, in contrast with the 25 or so cases of HIV transmission that have occurred in health care settings to health care workers, as many as 200 persons die annually from hepatitis B acquired in a health care setting. Of the approximately 1,000,000 people in the United States infected with hepatitis B, about one-third will ultimately die of the results of the infection. Also, the risk of infection after a hepatitis B positive needle stick is as much as more than 10 times that of the risk of infection from an HIV-positive needle stick [4].

The risk from exposure to blood resulted in the issuance by the Centers for Disease Control of the Recommendations for the Prevention of HIV Transmission in Health Care Settings [5]. With an update issued in 1988 [1], the recommendations that apply to blood and other bodily fluids rely strongly upon medical gloves as a barrier to contact with blood and bodily fluids. The effect of the epidemic on the condom industry was but a hint of what was to come with devices to prevent the far less likely transmission in a clinical setting.

The demand for gloves and other methods of protection against HIV exposure in the clinical setting increased dramatically. At present, more medical gloves are sold per month in the United States than condoms in a year. As the demand for gloves skyrocketed, shortages occurred. Also, the number of firms trying to market gloves in the United States increased, as did the production of those already in the market, to attempt to take advantage of or meet the demand. As with condoms, gloves should be of a high quality to serve as an effective barrier. As the market boomed, so did statements on the part of users that poor-quality products were being encountered. On the regulatory front, FDA modified its approach to medical gloves. First, FDA rescinded the exemptions that had existed for examination gloves from the requirements for submission of a Premarket Notification and from the Good Manufacturing Practice guidelines. FDA required that those who were marketing or intending to market examination glove submit a Premarket Notification. To date FDA has received approximately 1500 such notifications. Those entities in this aspect of the industry were almost tenfold greater than some had speculated, their ranks swollen by late entries resulting from a burgeoning market generated by increases in the numbers of health care workers who were using gloves and from an increase in the numbers of procedures for which gloves were now routinely used. An additional shift in the regulation of medical gloves is about to be made by the FDA. FDA plans to publish a notice in the *Federal Register* that will establish a sampling and testing program for medical gloves, similar to that in place for condoms. As one can see, the increased opportunity for profit available to those who market a device that is AIDS protective has also brought to the marketers the cost of increased surveillance on the part of the FDA in assuring that the device is of sufficient quality to actually provide acceptable levels of risk reduction to the user.

V. RISK OF HIV TRANSMISSION FROM MEDICAL DEVICES

As well as serving as a barrier to the transmission of HIV, medical devices may act as a vehicle to transmit the virus. The most obvious case of this has been the transmission of the virus by syringes shared by illicit IV drug users. Needle sticks in clinical settings have also resulted in a small number of device-mediated transmissions of HIV. In response to these types of transmission, a number of syringe design modifications have been contemplated. The ultimate aim of such design modifications would be to produce a syringe that has a passive mechanism to prevent reuse and needle sticks. For those devices that come into contact with blood, other bodily fluids, or mucous membranes and that must be reused, the AIDS epidemic has heightened concerns about their disinfectability. Such a concern may warrant the assessment by manufacturers of the design of any reusable device to determine if the design may act as a barrier to the effective disinfection of the device. Additionally, a prudent company would also evaluate the labeling and instructions of any reusable device to assure perfect clarity in the statements about its reusability and the directions for assuring adequate disinfeciton of the device for reuse.

VI. CONCLUSION

Hospital-acquired infections affect at least 2 million hospitalized patients annually in the United States with a cost of over $3.5 billion. These infections result in about 20,000 deaths annually and contribute to another 60,000 deaths. With the AIDS epidemic has come an increased interest in hospital-acquired infections among both the clinical community and the general public, not to mention those with the responsibility for assuring the safety and effectiveness of those medical devices that may contribute to the transmission of a hospital-acquired infection. As the epidemic progresses, more and more health care will be provided to HIV-infected individuals. With this will come the potential for HIV transmission in a health care setting being positively linked to a specific medical device. It would be most unfortunate if such a case could have been prevented by a simple modification in the instructions or design of the device.

It may be that the tragedy of the AIDS epidemic offers a unique challenge and opportunity to the medical device industry as a result of the potential link between devices and AIDS. The challenge to the industry is to move beyond the old technology and the old assumptions. Many of the barrier products employed to reduce the risk of the sexual or clinical transmission of HIV or other infectious agents are not much different from the products of the past decades. Manufacturers of these barrier devices should reexamine the device in the light of currently available data on disease transmission and barrier technology to optimize the device's effectiveness in protecting the user. The users of these products, clinicians and the general public, are no longer willing to accept the level of risk they would accept in the pre-AIDS era. Although their actual risk from the failure of barrier products may not be significantly increased from that of the past, the users perceive that their risk is much greater. This perception will drive the market and increase user acceptance of new technology and paying for new technology. The potential, however, for losing the basic simplicity of a device through a technological advance has risks, as well.

For better or worse, AIDS has changed and will continue to change the medical device industry. The epidemic has changed practices of protection among the users of medical device. It has shifted markets and caused huge expansions in certain areas. It has created an atmosphere for the development of new technology and for the acceptance of that technology. It has also created a new degree of expectations on the part of the user of devices as to the quality, effectiveness, and safety of devices that may well spill over into the areas of litigation and legislation, as has already been the case in the area of federal regulation. Ignoring these changes can have tremendous costs for the industry. Responding to these changes in an appropriate fashion provides an opportunity to assist in the effort to control the epidemic as well as to make a responsible profit for that effort.

REFERENCES

1. CDC, *Morbidity and Mortality Weekly Report*, HIV Surveillance Report, Nov. 1989:1-16.
2. CDC, *MMWR Supplement*, AIDS and Human Immunodeficiency Virus in the United States, 1988 Update, 1989:38(No. S-4).
3. The Workshop on AIDS and Personal Protective Equipment, University of Washington, May 17-18, 1989.
4. CDC, *MMWR*, Guidelines for Prevention of Transmission of Human Immunodeficiency Virus and Hepatitis B Virus to Health Care and Public Safety Workers, 1989:38(No. S-6).
5. CDC, *MMWR Supplement*, Recommendations for Prevention of HIV Transmission in Health-Care Settings, August 21, 1987:36(No. S-2).
6. CDC, *MMWR*, Update: Universal Precautions for Prevention of Transmission of Human Immunodeficiency Virus, Hepatitis B Virus, and Other Bloodborne Pathogens in Health-Care Settings, June 24, 1988:37(No. 24).

35

The Animal Testing Issue

FRANKIE L. TRULL and BARBARA A. RICH *National Association for Biomedical Research, Washington, D.C.*

I. INTRODUCTION

The use of animal models in all research and product safety testing has become increasingly controversial as a result of growth in the modern animal rights movement. From corporate shareholder to town hall meetings and in all forms of the media, this always emotional, sometimes contentious, issue is the subject of debate.

The scientific community, including developers of medical devices, must keep abreast of the issue and be thoroughly familiar with the rules and regulations which govern animal research and testing. This chapter will provide a broad overview of the animal rights movement and its effects on research and development. Major emphasis is given to existing legal requirements and to current trends which will, no doubt, effect the way research and safety testing is conducted in the future.

A. Today's Animal Rights Movement

The animal welfare/animal rights movement in the United States evolved from 19th century animal protection activities in England and Europe. Building on an anticruelty law passed in the early 1800's, British antivivisectionists called for abolition of scientific research using animals. These efforts resulted in only partial success with the passage of the British Act of 1876 which amended earlier anticruelty laws and began the statutory regulation of research.

At about the same time, an animal welfare movement began in the United States. Starting with the formation of the American Society for the Prevention of Cruelty to Animals (ASPCA) in 1866, many humane societies formed around the country to promote the concept of humane care of animals. These animal welfare organizations expressed legitimate concern for animals, and addressed such issues as overpopulation, hunting and trapping, owner neglect and abuse and the humane care and treatment of laboratory animals. However, at the turn of the century, the U.S. Congress considered, but eventually rejected, legislation similar to the British Act of 1876 and designed to limit animal research.

During the past 100 years the issue has ebbed and flowed based primarily on the events of the day. During times of war or other adversity, society has had little interest in the antivivisectionist cause. Still, in times of peace and prosperity, social attention can and has been turned to issues such as laboratory animal usage.

Today, there are thousands of organizations concerned with animal protection and more than 400 organizations in the United States concerned specifically with animal rights. The organizations range from the traditional and more moderate local humane societies to the abolitionist organizations which, like their British antivivisection antecedents, demand an end to all use of animals in biomedical research, as food, in product safety testing and in education. Some factions of the movement even decry the ownership of animals as pets.

Moderate animal welfare organizations recognize the necessity to use animals for advancing knowledge about medical problems. Members of these organizations and research scientists usually find themselves in agreement over basic issues. They believe that the care and treatment of the animals used in research must be the best available without hampering the progress of science. They insist, and scientists agree, that for data to be valid animals must be well cared for and healthy. They call for the use of analgesics and anesthetics in surgical or other experiments where animals may experience pain. They want animal nutrition and exercise to be appropriately monitored and a routine part of the animals' veterinary care program.

Animal welfare moderates believe that laboratory animals need and deserve every attention to their physical welfare that humane and caring people are able to provide. Scientists actively promote these concepts in their institutions and consider themselves animal welfare supporters.

Nevertheless, in the last 10 years, there has been a dramatic shift in philosophy from the concept of animal welfare to the promotion of "animal rights." Most observers cite Australian philosopher Peter Singer as the originator of the modern American animal rights movement. In his book *Animal Liberation* [1], Singer maintains that to consider the rights of one species of animal—humans—over those of another makes one a "speciesist."

Adherents to the "animal rights" position believe animals, particularly vertebrates, are entitled to the same rights the human animal is accorded in our society. Regardless of the consequences to research and the future beneficiaries of that research, they believe animals have the "right" to be free from pain and to live their lives outside the environment of the research laboratory. To bring this viewpoint to public notice and to foster the impression of widespread support for this concept, they have initiated national grassroots campaigns focusing on specific research projects or facilities that are vulnerable to criticism by the lay public.

The new activist organizations have brought great sophistication to the movement. They understand and effectively use public and media relations techniques and are politically astute. They work toward their goal of stopping animal research by undermining public confidence in science, as well as through the promotion of restrictive legislation, making research more costly.

Extremists in the animal rights movement regard illegal activities as the most effective strategy for the accomplishment of their goals. In the last 5 years they have broken into laboratories to steal animals and destroy facilities, equipment and research data causing millions of dollars in damages. They have also held public demonstrations and practiced civil disobedience to draw attention to their cause. Such extremists pose a real threat to the continued use of animals in biomedical research and education, and their agenda must be viewed seriously.

B. Scientific Community Position

Scientific excellence and research ethics dictate that the users of laboratory animals adhere to the highest standards of humane care and treatment. Nothing less is acceptable. When the use of an animal model is necessary, the minimum number of animals required for valid conclusions should be used and any pain or distress these animals may experience should be minimized. Where adjunct methodologies will produce the necessary information to help answer a research question, then these methods should be developed and used. However, researchers insist that decisions about the use of animals must be based on the best available scientific evidence and not be arbitrarily forced by nonscientific pressures.

Beyond personal moral standards, scientists have adopted strict ethical codes for the care and use of laboratory animals. Professional societies—the Society for Neuroscience, the Federation of American Societies for Experimental Biology, the American Physiological Society and many others—have addressed the issue of the humane care and treatment of animals through official position statements. Some of these codes and policies were established early in this century and have served as guides for the system of government regulation which has since evolved. Scientific journals have also set standards for the acceptable use of animals and these standards govern the research on which they will report.

As part of its national movement to establish standards in laboratory animal care, the scientific community organized the American Association for the Accreditation of Laboratory Animal Care (AAALAC) in 1965. The purpose of this organization is to encourage "optimal care for laboratory animals by providing a mechanism for peer evaluation of animal care programs by the scientific community" (also see II.G).

Voluntary health organizations like the American Heart Association, the American Cancer Society, and the Juvenile Diabetes Foundation, among others, have taken a public stand on the need for animal research. Many have adopted written policies outlining acceptable standards for the care and use of laboratory animals. Research funded by these organizations must meet these criteria.

National coalitions of scientists, citizens, and industry have been formed to develop effective methods of communicating their position to the public and the media. Scientists are joining clinicians, patients and teachers in explaining biomedical research and its impact on the future health of our nation.

Despite scientists' efforts to date, the animal research/animal rights debate remains a public issue. Activities have been successful in capturing the public's imagination through media exposure of a few isolated cases of inadequate animal care. The nonscientists' perception of animal research is in danger of being shaped by these isolated, sensational incidents. Scientists, educators and their institutions are now faced with the difficult task of preventing this misrepresentation from becoming the standard by which future research is measured.

A thorough understanding of the regulatory framework governing research is essential to ensure that work with laboratory animals adheres to all professional and legal requirements. In addition, such comprehensive understanding is extremely helpful in public education efforts. A national survey conducted for the Foundation for Biomedical Research has indicated, while the vast majority of the public understands and supports the use of animals in research, very few know that these federal and state laws exist. Bringing this information to the attention of lay persons is an integral part of any discussion of research and testing involving animals. Following is the outline of existing laws about which all those involved in animal research and product safety testing must be knowledgeable.

II. LAWS AND REGULATIONS GOVERNING THE USE OF ANIMALS IN RESEARCH AND TESTING

It is imperative that all those working with and caring for laboratory animals are thoroughly aware of not only their ethical obligations to these animals, but also their legal responsibilities. Virtually every animal research facility, individual investigator and research or animal care staff member is governed by one or more federal laws. They may be subject to state and local laws as well.

The Animal Welfare Act [2] applies to all research facilities in the United States that use covered species of laboratory animals in basic and biomedical research, education and product safety testing, whether or not federal funding is involved. These facilities include postsecondary schools, research institutions and private industry. In addition, companies conducting animal studies to support applications for research or marketing permits issued by the Food and Drug Administration or the Environmental Protection Agency must comply with those agencies' *Good Laboratory Practice Standards* [3].

All vertebrate animals used in research, training, testing or related purposes supported by U.S. Public Health Service (PHS) agencies are protected by the *PHS Policy for the Humane Care and Use of Laboratory Animals for Awardee Institutions* [4] and must comply with the federal Animal Welfare Act and the *Guide for the Care and Use of Laboratory Animals* (the *Guide*) [5]. Other federal agencies, such as the National Science Foundation have animal research policies which apply to internal research and to grant or contract recipients. These policies are similar to those of the PHS, are based on the *Guide* and also require adherence to the Animal Welfare Act. The underlying tenets for the policies of each federal agency engaged in supporting external research and testing with laboratory animals are expressed in the *U.S. Government Principles for the Utilization and Care of Vertebrate Animals Used in Testing, Research and Training* [6].

In summary, the overwhelming majority of laboratory animal users in the U.S. are subject to at least one federal regulation. Such regulations embody strong, prudent and humane national standards for laboratory animal care and treatment and now also place animal care programs under the oversight of structured, local review committees which include representation from the general public. Compliance with these national standards requires commitment of considerable time, effort and resources. Still, scientists recognize this commitment is imperative for good science and for maintaining public confidence in the science. *The Biomedical Investigator's Handbook for Researchers Using Animal Models* [7], published by the Foundation for Biomedical Research, provides more detail regarding legal requirements as well as other information on laboratory animal care.

A. Animal Welfare Act [2]

All research facilities in the U.S. that use covered species of laboratory animals in basic and biomedical research, education and product safety testing must comply with the federal Animal Welfare Act (AWA) and register with the U.S. Department of Agriculture (USDA). For purposes of the Act, research facilities are defined as any individual institution, organization or postsecondary school that uses or intends to use live animals in research, tests or experiments and that purchases or transports live animals in commerce or receives Federal funds for research, tests, or experiments. At the present time, rats, mice, birds, and farm animals used for agriculture research are not covered by AWA regulations, so facilities using only those species do not have to be registered by the USDA. According to the latest USDA Annual Report, some 1,308 registered research facilities exist in the United States.

Originally passed in 1966 and subsequently amended in 1970, 1976, and 1985, the Act falls under the jurisdiction of the Secretary of Agriculture, with responsibility for administration delegated to the USDA Animal and Plant Health Inspection Service (APHIS). APHIS is responsible for regulating not only research facilities, but also animal dealers, exhibitors (including zoos, aquariums, circuses, etc.) and intermediate handlers of animals, including air and truck lines.

Current USDA/APHIS regulations [8] establish requirements for animal handling, housing, feeding, watering, sanitation, ventilation, shelter, and adequate veterinary care including the use of anesthetic, analgesic and tranquilizer drugs. Because one of the Act's original purposes was to prevent the sale of stolen or lost pets for research, the regulations also cover allowable means of acquiring dogs and cats, federal licensing of dealers, transportation, record-keeping and identification of animals.

Final regulations implementing the most recent amendments to the Act (1985) were promulgated, in part, in the *Federal Register* of August 31, 1989 These new regulations were effective on October 31, 1989, and additional changes in animal care standards contained in Part 3 of the USDA/APHIS regulations are anticipated in 1990. Following is an overview of the Animal Welfare Act as amended.

Standards—the Secretary of Agriculture is directed to establish standards including minimum requirements:

For animal care, treatment, and practices in experimental procedures to ensure that animal pain and distress are minimized, including adequate veterinary care with the appropriate use of anesthetic, analgesic, tranquilizer drugs, and euthanasia.

For exercise of dogs, as determined by an attending veterinarian in accordance with general standards promulgated by the Secretary, and for a physical environment adequate to promote the psychological well-being of primates.

That alternatives to painful procedures are considered.

That a doctor of veterinary medicine be consulted in any practice which could cause pain to animals.

That no animal be used in more than one major operative experiment from which it is allowed to recover except in cases of scientific necessity; or other special circumstances determined by the Secretary.

That exceptions to such standards may be made only when specified by research protocol and any such exception shall be detailed and explained in a report . . . filed with the Institutional Animal Committee.

Statutory language prohibits promulgation of rules or regulations with regard to "design, outlines, guidelines or performance of actual research" with exception of the standards described above.

Institutional Reporting Requirements—Registered research facilities will show on inspection and report annually that provisions of the Act are being followed and that professionally acceptable standards governing the care, treatment and use of animals are being followed during actual research and experimentation.

Institutional Animal Committees—Research facilities are required to have at least one Institutional Animal Committee appointed by the chief executive and composed of at least three members including a veterinarian and a member not affiliated with the facility "intended to provide representation for general community interests in the proper care and treatment of animals." Committees are to conduct semiannual inspections including review of practices involving pain to animals and the condition of animals and file certification reports with the research facility. Reports shall remain on file for 3 years and be available to APHIS inspectors. After a facility is notified of any deficiency, had an opportunity to correct it and fails to do so, the Committee shall report to APHIS and any federal funding agency.

Information Services—The Secretary is directed to establish an information service at the National Agriculture Library to work in cooperation with the National Library of Medicine to provide information on employee training, prevention of unintended duplication of research, and improved methods of reducing or replacing animal use and minimizing pain.

Training of Personnel—Research facilities shall make training available for scientists, technicians and others involved with animal care as required by the USDA Secretary.

Increased Penalties—Federal agencies will suspend or revoke funds when it has been determined that conditions of animal care, treatment or practice in a particular project are not in compliance despite notification and an opportunity for corrections. In addition, monetary fines for violations of the Animal Welfare Act are authorized: $2,500 for each day of noncompliance and an additional $1,500 per day for failure to obey a cease and desist order.

Trade Secrets—"Nothing in the Act shall be construed to require a research facility to disclose publicly or to the Institutional Animal Committee during its inspection, trade secrets or commercial or financial information which is privileged or confidential." Penalties for release of trade secrets by committee members are established.

Inspections—Under the Animal Welfare Act the Secretary is directed to conduct unannounced site visits to all registered research facilities at least once a year with follow-up inspection until any deficiency is corrected. However, the latest USDA Annual Report indicates that USDA-APHIS is making an average of two visits per institution per year.

B. The Public Health Service Policy on Humane Care and Use of Laboratory Animals [4]

Institutions conducting U.S. Public Health Service (PHS) supported animal research must comply with the *Policy on Humane Care and Use of Laboratory Animals*. The Policy, administered through the National Institutes of Health, Office of Protection from Research Risks (OPRR), has evolved over the last 20 years. The Health Research Extension Act of 1985, Section 495, Animals in Research, put a statutory foundation under the policy.

As revised to meet new legal requirements effective in September 1986, following are the key elements of the policy:

Assurances—Written documentation of the institution's commitment to animal welfare is required in the form of detailed information on the animal care and use program. Institutions are required to designate clear lines of authority and responsibility for those involved. Each institution must identify an official who is ultimately responsible for the institution's animal program and a veterinarian qualified in laboratory animal medicine who will participate in implementation. The program description must also give details concerning veterinary care, animal husbandry and all animal facilities. A synopsis of mandatory training or instruction in the humane practice of animal care and use, as well as the occupational health program for all personnel involved with animals, must also be included in the assurance statement. The research institution must ensure that the laboratory animal facilities meet the requirements outlined in the *Guide for the Care and Use of Laboratory Animals* [5].

The *Guide*, prepared under contract with the National Institutes of Health (NIH) by the Institute of Laboratory Animal Resources, Commission on Life Sciences, National Research Council of the National Academy of Sciences, is a companion piece to the PHS Policy and should be referred to regularly when developing grant proposals and research design, as well as in the actual care and treatment of research animals. The *Guide* specifically addresses institutional policies, laboratory animal husbandry, veterinary care and other information of which institutions need to be aware in order to insure compliance with the new PHS policy.

In addition to the *Guide*, institutions must assure compliance with applicable provisions of the Animal Welfare Act and other federal statutes and regulations relating to animals.

Institutional Animal Care and Use Committees—The centerpiece of each research facility's animal research program is the mandatory Institutional Animal Care and Use Committee (IACUC). The committee must be composed of at least five members including a veterinarian, a scientist experienced in animal research and an individual unaffiliated with the institution.

The IACUC must evaluate the institution's overall animal research program and every 6 months must inspect all the sites where laboratory animals are located; prepare a report on the evaluation and inspection and make recommendations to the appropriate institutional official; review grant applications and proposals in relation to animal care and use, and if necessary, suspend PHS-supported activities.

The IACUC in its submission of semiannual reports to the designated institutional official, must describe the institution's adherence to the *Guide*, including reasons for any departure, distinguishing significant deficiencies from minor deficiencies, including a plan and schedule to correct any deficiencies.

When reviewing research projects in relation to the care and use of the laboratory animals, the IACUC must consider the procedures to be utilized; the use of analgesics or anesthetics, the method of euthanasia, the medical care that will be provided and the qualifications of the personnel involved in the research project.

Requirements for Submission of Grants—When grant applications are submitted to PHS components, five points must be addressed:

Number and species of animals to be used
Rationale for the use of the animals
Description of use
Procedures to be used to minimize pain
Method of euthanasia

Record Keeping and Reporting—The research institution must keep on file the animal welfare assurance, the minutes of IACUC meetings, records of IACUC review of research projects, and semiannual IACUC reports and recommendations and records of determinations of accrediting bodies. Annual reporting requirements must include gain or loss of accreditation by the American Association for Accreditation of Laboratory Animal Care (AAALAC), program or facility changes, IACUC membership changes, the dates of the semiannual IACUC reports and any minority views. If there are no changes, a statement indicating such must be submitted.

If the IACUC suspends a research activity, the institutional official must review reasons for suspension, take action and report that action with full explanation to OPRR.

The PHS awarding unit must have an animal welfare assurance verification of IACUC review and approval on file for any project it conducts.

C. Good Laboratory Practices [3]

In 1978, the Food and Drug Administration (FDA) adopted Good Laboratory Practices (GLP) rules and implemented a laboratory audit and inspection program covering every regulated entity which conducts nonclinical laboratory studies for product safety and effectiveness. The FDA GLPs were amended in 1984. The Environmental Proteciton Agency (EPA), in 1983, issued its own GLP rules for its toxic substances control and pesticide programs.

Both GLP standards address all areas of laboratory operations including requirements for a Quality Assurance Unit to conduct periodic internal inspections and keep records for audit and reporting purposes; Standard Operating Procedures (SOPs) for all aspects of each study and for all phases of laboratory maintenance; a formal mechanism for evaluation and approval of study protocols and their amendments, and reports of data in sufficient detail to support conclusions drawn from them.

Provisions relating to care and housing of test animals are identical in both agencies' GLP rules. They specify SOPs for housing, feeding, handling and care with additional standards on separation, disease control and treatment, identification, sanitation, feed and water inspection, bedding and pest control.

The FDA inspection program includes GLP compliance, and a data audit to verify that information submitted to the agency accurately reflects the raw data. An inspection may also result from questionable data. Follow-up inspections are conducted for correction of previously discovered deficiencies.

D. Other Federal Departments and Agencies

The majority of federally supported research and testing involving laboratory animals is related to the U.S. Public Health Service. However, in total, seven federal departments (agriculture, commerce, defense, energy, health and human services, interior, and transportation) conduct animal experimentation within federal facilities, support some extramural research with animal models or both. Four federal agencies, the Consumer Product Safety Commission, Environmental Protection Agency, National Aeronautics and Space Administration and National Science Foundation, directly or indirectly are also involved in supporting research, testing, or training which relies on animal models.

Effective December 1986, each federal research facility using laboratory animals was required to establish an Animal Care and Use Committee with composition and function as described in the 1985 amendments to the Animal Welfare Act. Federal committees will report to the head of the federal entity conducting animal experimentation.

Each federal entity involved in animal research has endorsed the *U.S. Government Principles for the Utilization and Care of Vertebrate Animals* (see II,G). These principles underlie each department or agency's separate policies, procedures, and guidelines for animal research. Generally, all federal policies require adherence to the Animal Welfare Act and the *Guide for the Care and Use of Animals*. The various policies are similar to the *Public Health Service Policy on Humane Care and Use of Laboratory Animals by Awardee Institutions*.

E. U.S. Government Principels on the Utilization and Care of Vertebrate Animals [6]

Due to the number of rules and regulations governing research animal use mandated by various federal agencies and in an effort to insure complementing rather than competing regulations the Interagency Research Animal Committee (IRAC) was formed in 1983. The IRAC is comprised of representatives from 14 federal entities and in 1984 issued the *Principles for the Utilization and Care of Vertebrate Animals Used in Testing, Research and Training*. The purpose of the IRAC is to ensure humane and responsible use of animals in research both within and outside federal agencies.

The Principles state that "Whenever U.S. Government agencies develop requirements for testing, research or training procedures involving the use of vertebrate animals, the principles shall be considered; and whenever these agencies actually perform or sponsor such procedures the responsible institutional official shall ensure that these principles are adhered to:

I. The transportation, care, and use of animals should be in accordance with the Animal Welfare Act (7 U.S.C. 2131 et seq.) and other applicable federal laws, guidelines, and policies.

II. Procedures involving animals should be designed and performed with the due consideration of their relevance to human or animal health, the advancement of knowledge, or the good of society.

III. The animals selected for a procedure should be of an appropriate species and quality and the number required to obtain valid results. Methods such as mathematical models, computer simulation, and *in vitro* biological systems should be considered.

IV. Proper use of animals, including the avoidance of discomfort, distress, and pain when consistent with sound scientific practices, is imperative. Unless the contrary is established, investigators should consider that procedures that cause pain or distress in human beings may cause pain or distress in other animals.

VI. Procedures with animals that may cause more than momentary or slight pain and distress should be performed with appropriate sedation, analgesia, or anesthesia. Surgical or other painful procedures should not be performed on unanesthetized animals paralyzed by chemical agents.

VI. Animals that would otherwise suffer severe or chronic pain or distress that cannot be releived should be painlessly killed at the end of the procedure or, if appropriate, during the procedure.

VII. The living condition of animals should be appropriate for their species and contribute to their health and comfort. Normally, the housing, feeding, and care of all animals used for biomedical purposes must be directed by a veterinarian or other scientists trained and experienced in the proper care, handling, and use of the species being maintained or studied. In any case, veterinary care shall be provided as indicated.

VIII. Investigators and other personnel shall be appropriately qualified and experienced for conducting procedures on living animals. Adequate arrangements shall be made for their in-service training, including the proper and humane care and use of laboratory animals.

IX. Where exceptions are required in relation to the provisions of these principles, the decisions should not rest with the investigators directly concerned, but should be made with due regard to Principle II, by an appropriate review group such as an Institutional Animal Care and Use Committee. Such exceptions should not be made solely for the purposes of teaching or demonstration."

F. State and Local Laws

Many states have statutes and regulations in place relevant to laboratory animals. Typical state requirements fall into these broad categories.

Availability of Pound Animals for Research—Currently thirteen states restrict the use of animals obtained from pounds for research, testing and experimentation purposes. Twelve states prohibit pounds in that state from making animals available; they are: Connecticut, Delaware, Hawaii, Maine, New Hampshire, New Jersey, New York, Pennsylvania, Rhode Island, Vermont. and West Virginia. However, research facilities in these states may obtain pound animals outside the state through USDA licensed dealers. In Massachusetts, pound animals may not be used at all whether from within or outside the state.

Five states, Iowa, Minnesota, Oklahoma, South Dakota, Utah, and the District of Columbia, require pounds to release abandoned pound animals for research. Eight states allow research use including Arizona, California, Michigan, North Carolina, Ohio, Tennessee, Virginia, and Wisconsin. In these states where statutes explicitly or implicitly allow research use as well as in the twenty-three remaining states with no stated prohibition, county and local government often exercise jurisdiction.

Animal Cruelty—Long-standing laws against intentional cruelty to animals exist in every state. In twenty-two states and the District of Columbia, properly conducted research and testing is exempted.

Regulation of Research Facilities—Twenty states and the District of Columbia have laws concerning licensing of research facilities. For licensing purposes, state officials may set rules, regulations and standards of animal care and treatment and a number of states also provide for state inspection of facilities.

Please refer to the National Association for Biomedical Research publication, *State Laws Concerning the Use of Animals in Research*, for details and specific citations for individual state laws [9].

G. American Association for Accreditation of Laboratory Animal Care [10]

The American Association for Accreditation of Laboratory Animal Care (AAALAC) is an independent, non-government, non-profit association founded in 1965 for the purpose of accrediting laboratory animal programs on a voluntary basis. AAALAC is sponsored by 26 medical and biological research and education organizations. The *Guide for the Care and Use of Laboratory Animals* is the primary standard used by AAALAC in evaluating animal care programs.

The voluntary AAALAC accreditation program requires that an institution submit an application, which is reviewed for administrative detail by the office of the Executive Secretary. A site visit date is then established and site visitors are selected. The site visitors conduct the visit to collect information on the institution's entire animal care and use program. A complete report is written and submitted to the Executive Secretary. The report, plus the evaluation, is then reviewed by the Council at the next meeting and determination of accreditation status is made. Site visits and evaluation reports are repeated every three years for continuation of accreditation.

III. CURRENT CONCERNS AND FUTURE TRENDS

Throughout the history of the animal welfare/animal rights movement there have been recurring accusations that animals are used unnecessarily in research and safety testing because nonanimal "alternatives" are available, that there is unnecessary duplication of experiments, that experimentation is performed for frivolous reasons and that pets are used for research purposes.

Despite the existing oversight process, these and other concerns will continue to surface and fuel a continuing debate. A variety of legislative proposals designed to address perceived problems is virtually always pending at the federal and state level. Lawmakers have historically been cautious about restricting research but, as described in the preceding section, a rather elaborate system of animal research controls has been put in place over the years. Given the inevitable pressure to do so, legislators will consider additional measures to address public concerns and, no doubt, some further restrictions will be enacted.

The issues most likely to produce future action are discussed briefly below.

A. Use of Alternatives to Laboratory Animals

It is not unusual to read or hear that animals are no longer needed for research, safety testing, or education programs. The oft-quoted belief is that scientists can do most, if not all, of their work using computer models or cell and tissue cultures.

It is true that new technologies have the potential to reduce science's reliance on whole animal research models. In fact, the "Three Rs," replacement of laboratory animals, reduction in the numbers used, and refinement of existing procedures and techniques in order to minimize any animal pain or distress, are governing principles under which research is now planned and conducted [11]. While considerable progress has been made in reducing the number of animals used and refining procedures, the total replacement of animals is far from achieved. According to the Congressional Office of Technology Assessment Report Brief on Alternatives to Animal Use in Research, Testing and Educaiton, "For most areas of scientific experimentation, totally replacing animal use with non-animal methods, especially in the short term, is not likely" [12].

Nevertheless, legislation banning specific animal tests requiring the use of nonanimal methods or both, particularly for acute toxicity testing, has been introduced in the U.S. Congress and in eight states. Consumer and household product testing has most often been targeted; however, some proposals would apply to all acute toxicity testing. Thus far, legislation mandating nonanimal tests has not won passage, although some proposals have found considerable support. Attempts to legislate "alternatives" can be expected to continue.

B. Avoiding Unintended Duplication

Another recurring theme in the controversy surrounding the use of animals in research is that much work merely duplicates that which has been done before and represents, therefore, an unnecessary use of laboratory animals. The slow, often tedious process of science is not well-understood by the general public, hence there is ample opportunity for the appearance of duplication in the view of a layperson. Some confusion arises due to the public's unfamiliarity with the necessity to replicate research results in order to demonstrate their validity. Beyond this, a source of misunderstanding is the fact that more than one research team may be tackling the same unsolved problem or unanswered question, but using somewhat different approaches.

The possibility of unintended duplication certainly exists primarily because of the tremendous amount of information available from research laboratories around the world. However, in the United States, the peer review system for obtaining grants for biomedical research as well as industry's desire to contain research costs act as a screening mechanism to eliminate those research projects of less immediate value, or which might be unproductively repetitive.

Despite the research system's inherent mechanism for pursuing the most promising new avenue of inquiry and the fierce competition for research funding, legislation designed to further check duplication in research has been proposed in the last three U.S. Congresses. This has served to sensitize the scientific community to share data and information more extensively on a voluntary basis, rather than invite additional regulation.

A corollary issue to unnecessarily duplicative research is the belief by some that animals are used in frivolous research. In these cases, lack of knowledge regarding science and research is also a problem. As an illustration, animal activists often attack alcohol or other substance abuse projects saying we already know these substances are "bad" for us. They believe using animals to further study causes and treatments is unjustified since victims of addiction-related disease bring that problem upon themselves.

C. Species-Related Issues

The use of higher species for research is of particular concern to both scientists and the public. Perhaps greatest attention is paid to dogs and cats because the public identifies them as pet animals. Laboratory primates also receive much attention, especially the chimpanzee.

Restrictions on the research use of impounded dogs and cats is the most common type of legislation proposed at the state level. In addition, numerous county and local governments are considering the question of making unclaimed pound animals available to research institutions. Bills prohibiting the use of federal research funds to obtain pound animals have been introduced in both houses of the U.S. Congress. Public opinion polls and experience with local referenda indicate the majority of Americans support the use of unclaimed pound animals for research. Regardless of this fact, animal activists maintain that "pet" animals, even if they are no longer wanted and adoptive homes cannot be found for them, should not be used in research. Many groups campaign vigorously against allowing research facilities to acquire pound animals and often imply that there is a major problem with lost or stolen pets being used in research.

Unfortunately, according to the most recent American Humane Association estimates, at least 13 million dogs and cats are killed in this country's pounds every year. Currently research facilities accept approximately 200,000 of these animals for projects in which they can be appropriately used, including cardiovascular disease, cancer, diabetes, arthritis, asthma and other lung disorders, orthopedics, birth defects, hearing and sight loss studies. Given these statistics, the use of pets or adoptable animals is not only unnecessary, but the chance they will be used, even inadvertently, is very remote. Existing provisions of the federal Animal Welfare Act require research facilities and animal dealers to separately identify and keep appropriate records on all animals so that federal inspectors can insure that pet owners are protected. Current legal safeguards notwithstanding, because we are a pet-loving society, the use of dogs and cats in research will remain an emotional issue. The strong attachment the public has to companion animals will continue to be used by those who oppose the use of any animal in research.

Likewise, the use of nonhuman primates as experimental subjects is an issue of special concern. Chimpanzees, in particular, have received increasing attention in the media. An example of the growing interest in primates and the effect on research is the fact that the 1985 amendments to the federal Animal Welfare Act include a provision requiring a "research environment adequate to promote the psychological well-being of primates." Regulations describing how research facilities must comply with this new law were not yet published as of December 1989.

D. Public Accountability

General public interest in and familiarity with the research process should be encouraged, particularly in view of the questions being raised about animal research in the media. However, those individuals who completely oppose the use of laboratory animals are not really interested in factual information, but rather in ways to challenge and disrupt research.

Frequently, activist organizations claim that research and testing is carried out in secret behind closed doors. As a result, a variety of approaches to gain access to research facilities and information are employed by these groups.

The federal Freedom of Information Act (FOIA) [12] is used to obtain records pertaining to the regulation of research facilities, such as inspection reports maintained by the U.S. Department of Agriculture, NIH grant information and Food and Drug Administration files. Materials so obtained can be given to the media or presented in other public forums. All too often information is distorted and misused. FOIA provides nine exemptions from mandatory disclosure, including protection for proprietary information; however, the agency always has the discretion to release the information unless it is specifically exempted from disclosure by another statute, such as the Trade Secrets Act. It is, therefore, important to identify information submitted to federal agencies that should be kept confidential, as well as to monitor FOIA requests made concerning your facility.

Most states also have enacted freedom of information statutes which provide access to records held by state and local agencies. Many statutes closely parallel the federal FOIA, while others offer greater protection for personal or certain other categories of records. Thus, familiarity and monitoring of state FOIA's are equally essential.

Access to academic research facilities, and meetings at state institutions is being sought by activists either under state open meetings or sunshine laws. Because such statutes do not usually offer the desired entry, legislative proposals creating state level registration, regulation, and inspection for all animal research facilities have been increasingly common.

Shareholder resolutions are being proposed to gain entry into commercial research facilities and to require disclosure of industry information concerning animal research. In 1987-1988, a half dozen such resolutions surfaced in major corporations, some having been repeatedly introduced. To date, all have been defeated.

IV. CONCLUSION

There will be no complete conclusion to the animal testing question, at least not in the foreseeable future. Animal rights activists are not satisfied with the existing laws and the process for protecting laboratory animals from unnecessary pain or suffering. As quoted in the May 23, 1987, issue of *Newsweek*, one movement leader states, "The animal-rights philosophy is abolitionist rather than reformist. It's not better cages we work for, but empty cages. We want every animal out." Obviously, as long as scientists must rely upon some live animal models, the debate will continue.

REFERENCES

1. Singer, P. 1977. *Animal Liberation: A New Ethic for Our Treatment of Animals*. New York: Avon Books.
2. United States Code (USC) as amended by P.L. 99-198 effective December 23, 1986. Title 7; Section 2131-2156.
3. Code of Federal Regulations (CFR). 1984. Title 21; Part 58.1, U.S. Department of Health and Human Services, Food and Drug Administration, Nonclinical Laboratory Studies, Good Laboratory Practice Regulations; and 1983. Title 40; Parts 792.1 and 160.1, Environmental Protection Agency, Toxic Substances Control and Pesticide Programs Good Laboratory Practice Standards.
4. U.S. Department of Health and Human Services, Public Health Service, Office of Protection from Research Risks. Revised September, 1986. *Public Health Service Policy on Humane Care and Use of Laboratory Animals*. Bethesda, MD: National Institutes of Health.
5. U.S. Department of Health and Human Services, Public Health Service, National Institutes of Health. 1985. *Guide for the Care and Use of Laboratory Animals*. Bethesda, MD: NIH Pub. 85-23.
6. U.S. Interagency Research Animal Committee. 1985. *Principles for the Utilization and Care of Vertebrate Animals Used in Testing, Research and Training*. Bethesda, MD: NIH Pub. 85-23, p. 81.
7. Foundation for Biomedical Research (FBR). 1987. *The Biomedical Investigators Handbook for Researchers Using Animal Models*. Washington, DC: FBR.
8. Code of Federal Regulations (CFR). 1989. Title 9; Subchapter A, Animal Welfare, Parts 1, 2, and 3. 54 Federal Register 168:36112-63. Washington, DC: Office of the Federal Register.
9. National Association for Biomedical Research (NABR). 1987. *State Laws Concerning the Use of Animals in Research*, 2nd Ed. Washington, DC: NABR.
10. American Association for Accreditation of Laboratory Animal Care, 9650 Rockville Pike, Room 2504, Bethesda, MD, 20814.
11. Russell, W. M. S. and Burch, R. L. 1959. *Principles of Humane Experimental Technique*. Springfield, IL: Charles C. Thomas.
12. U.S. Congress, Office of Technology Assessment. *Alternatives to Animal Use in Research, Testing and Education*. OTA-BA 273. Washington, DC: U.S. Government Printing Office.
13. United States Code (USC). Title 5, Section 552. Freedom of Information Act (FOIA).

36

New Challenges in Medical Product Sterilization

ROBERT F. MORRISSEY *Johnson & Johnson, New Brunswick, New Jersey*

G. BRIGGS PHILLIPS *Petra, Inc., St. Michaels, Maryland*

I. INTRODUCTION

When, before the turn of this century, Fred Kilmer designed and commissioned a novel industrial sterilization process he did so reacting to the challenges of that era [1]. Later, the research and development work of Charles Artandi introduced accelerated electrons [2] and gamma radiation [3] as effective and practical sterilization methods for medical products. New challenges have arisen over the decades because of the continuous sophistication of invasive and noninvasive medical procedures and the development of new health care products for these modalities. Moreover, the introduction and rapid expansion of disposables shifted the application of sterilization methods away from hospitals toward industry, where high-speed manufacturing, packaging, consistent quality, and high assurance of sterility became business realities.

The challenges in the closing years of this century in medical product sterilization technology are perhaps even more demanding for us than for our predecessors.

As illustrated in many chapters of this "survival" handbook, every aspect of doing business in our industry is rapidly changing, requiring, in our opinion, cultural changes in all facets of product design, manufacture, regulatory compliance, research and development (R&D), marketing, and, indeed, in sterilization technology. In approaching the identification of the sterilization challenges of tomorrow, it is necessary to discuss the projected environment of the medical product manufacturing industry.

II. THE ENVIRONMENT OF THE FUTURE

The projected climate in the years ahead will evolve out of the following assumptions and needs:

An increased level of regulatory involvement
A broader public awareness of the environment
Restrictions in the supply of chlorofluorocarbons
A diversity in chemical sterilization methods
Increased overseas production by multinational companies
Restrictions concerning cobalt-60
Reuse of sterile single-use devices
Continued emphasis on cost containment
Normalization of sterility requirements

A. An Increased Level of Regulatory Involvement

In 1976 the Medical Device Amendments to the U.S. Food, Drug, and Cosmetic Act were enacted. During the past 12 years there has been a slow but steady increase in good manufacturing practices (GMP) regulations covering both devices and drugs throughout Europe, South America, and Asia. These new GMP regulations frequently focus on sterile product manufacturing, that is, sterilization processing.

The regulations in the United States and Europe have been expanded to include more specific details as a result of the issuance by the regulatory agencies and by private concensus groups of guidelines, which, by practice, become de facto standards. These "standards" are then used to measure the degree of compliance with GMPs. As an example, the Food and Drug administration (FDA) has recently issued guidelines on process validation that tighten the control that the agency exercises over industry's product sterilization procedures [4]. Likewise, organizations such as the Association for the Advancement of Medical Instrumentation (AAMI) have issued a variety of recommended guidelines covering ethylene oxide (EtO) and gamma sterilization that have rapidly become the standard of practice by industry and the gauge of compliance by FDA [5,6]. Some of the AAMI's suggested guidelines and practices have been adopted in other countries and translated into several languages.

Ethylene oxide (EtO) and cobalt-60 (^{60}Co), the two processes that helped make possible low-cost disposable sterile medical devices, continue to be the focus of increased regulatory attention. Concerns over EtO personnel exposure and by-product residues have given way to environmental issues such as manufacturing facility emissions, the depletion of atmospheric ozone by EtO diluents (chlorofluorocarbons) (see Section III), and, for ^{60}Co, the transportation and safe disposal of the isotope (see Section VI).

The U.S. Environmental Protection Agency (EPA) has a track record of continuing to tighten regulations concerning airborne emissions of hydrocarbons, and has recently promulgated specific maximum emission standards for EtO under the National Emissions Standards for Hazardous Air Pollutants (NESHAPS). EtO emissions will be regulated for both suppliers and users. It appears that there will not be an exemption for small steiliizers such as those found in hospitals and laboratories. The EPA is currently reviewing data on existing EtO emissions control equipment to evaluate its technical and economic feasibility. Indications are that the control standard will be based on a percent efficiency level for the control equipment and not on the actual pounds of EtO emitted. Final action should take place by 1989. It is anticipated that this regulatory trend will continue in the future with more invovlement by both state and municipal EPA organization.

Even more aggressive than EPA, the Occupational Safety and Health Administration (OSHA) continues to be challenged by consumer lobbyists on one hand and industry on the other. Emission control levels were reduced to an 8-hr 1 part per million (PPM) time-weighted average (TWA) in 1984 [7]. This tightened requirement (the former level was 50 ppm) had a major impact on the health care industry and resulted in a much tighter control over worker exposure during and after product sterilization. More recently, OSHA has issued a 5 ppm 15-min excursion limit (EL) as a result of constant pressure from the lobbying groups [8]. Since these groups maintain that even the 1 ppm TWA and the 5 ppm EL are too high, there is the possibility that in future years more restrictions will be forthcoming.

For large companies that have expended significant resources on engineering controls, personnel training, continuous monitoring equipment, medical examinations, and administrative procedures, the impact of present and future EPA and OSHA regulations will probably be minimal. For companies that have not made such investments, for the small medical device company who cannot afford the investment, and for the hospital sector, the challenge of meeting these requirements will result in increased costs, which inevitably will be passed on to the consumer. Such increased costs would result from (a) one-time and ongoing costs for upgrading EtO equipment and control programs or (b) costs associated with selecting and validating substitute sterilization methods. It is not inconceivable that some companies, faced with these choices, will elect to sell their products in a nonsterile state, which would then pass the responsibility for sterilization on to the health care provider.

Since 1978 we have seen little activity from the FDA in the area of sterilant residue by-products [9]. Professional organizations such as AAMI have devoted considerable time and resources in developing consensus guidelines for ethylene oxide residue analysis. The Health Industry Manufacturers Association (HIMA) has recently funded the Environ study to analyze the risk associated with the use of EtO sterilized products [10]. The Environ report, based on the maximum residue limits published by FDA, shows that the risk associated with such products is well within the accepted norm.

A hightened awareness of EtO residue by-products has resulted from California Proposition 65 (the "Safe Drinking Water and Toxic Enforcement Act of 1986"), which requires the labeling of "products" that contain substances found on the state list of carcinogenic and reproductive toxicants. This regulation went into effect on February 27, 1988, and was immediately challenged by both industry and environmentalists. According to the regulation, if a patient is "exposed" to more than 20 μg EtO per day, the medical product must contain a warning label. This regulation has far-reaching consequences since it may set a precedent for other states. The potential for confusion on the part of the public appears to be great since numerous medical products used to support and sustain our overall health and, in many instances, life itself may be associated with adverse publicity as a result of this legislation.

It is anticipated that regulations will expand not only from the federal and state level, but also from local municipalities. As an example, Hillsborough County in Florida is taking the lead away from both state and federal EPA organizations in its tightened restrictions on chlorofluorocarbon emissions to the atmosphere.

B. A Broader Public Awareness of the Environment

We expect the trend for greater public awareness of environmental issues to continue. This is in part due to the enactment of worker and community right-to-know legislation. This public awareness will manifest itself by requiring even tighter controls on the transportation, handling, and storage of sterilizing agents such as EtO and ^{60}Co.

All of us are consumers and want to know that the air we breathe, the water we drink, and the products that we consume are safe. Unfortunately, little attention has been devoted to educating the public on the concept of risk assessment analysis, and how such analyses can be used to characterize risk-benefit situations.

Risk assessment is concerned with uncertainty and probabilities, and is therefore difficult to grasp. Unfortunately, the level of public concern is generally not closely aligned to the true risk level of hazards, and the dominant perception for most Americans is that they face more risk today than in the past, and that future risks will be even greater [11]. This view is not held by professionals involved in risk assessment analysis [12].

The potential hazards of agents used to sterilize health care products must be weighed against the substantial benefits derived from these products.

In a related area, the public's reaction, particularly at the state level, against the use of radiation to preserve/sanitize/sterilize various food products spills over to the use of radiation to sterilize medical products. This is indeed unfortunate because, whatever the consumer advocates' arguments may be (that irradiated foods are not safe), they have no relevance at all to the use of ionizing radiation for the sterilization of medical products. As we pointed out earlier, the use of radiation for product sterilization has proven to be one of industry's safest and most reliable form of sterilization [13].

C. Restrictions in the Supply of Chlorofluorocarbons

The significance of this issue to our industry is that much of the EtO used for industrial sterilization (and for hospital sterilization) is diluted with chlorofluorocarbons (CFCs) to achieve a nonexplosive mixture. The CFC does not react with the products being sterilized and, upon completion of the cycle, is released to the atmosphere. In time it rises high into the stratosphere, where it reacts with the upper-strata ozone layer. The ozone layer exerts a protective effect against harmful cosmic radiations [14].

Concern over depletion of the ozone layer has resulted in international cooperation in attempting to control the production and consumption of CFCs. The signing of the Montreal protocol on September 17, 1987, by 31 countries, marked what we predict will be the demise of CFCs in the medical device industry [15].

CFCs are used in several industries. The major users, which make up approximately 70%, consist of refrigeration/air conditioning and foam blowing. Approximately 18% of CFC use is associated with solvent cleaning, while sterilization makes up 5.4%.

As mentioned, CDCs and also halons (fluorocarbons containing bromine in place of chlorine) migrate to the upper atmosphere where they react with ozone, thus reducing the protective effectiveness of the ozone layer. The reduction in ozone permits an increase in the amount of ultraviolet (uv) radiation reaching the surface of the earth. The effects of this increase uv radiation are varied, with predictions ranging from increases in skin cancer and cataracts, to reduced crop yields and increased degradation of polymers exposed to sunlight.

In order to become effective, the Montreal protocol must be ratified by 11 nations. To date the United States and Mexico have signed the accord. When ratified, the protocol was expected to take effect on January 1, 1989. It calls for freezing the level of CFC consumption to 1986 values by July 1989. It further calls for a 20% reduction in CFC consumption based on 1986 levels by mid-1993, and a 50% reduction by mid-1998. Halon consumption will be frozen at 1986 levels in 1992. Approximately 70% of all CFCs are consumed by developed countries, with the United States consuming approximately 30%. Developing nations and the USSR/Eastern Block comprise 28%, with China and India approximately 2%.

Consistent with the Montreal protocol, the U.S. EPA on December 1, 1987, issued proposed domestic regulations to implement the protocol in the United States. The proposed regulation calls for an allocated quota system for companies producing or importing (or both) CFCs and halons based on their production and consumption levels in 1986. EPA points out that the cost of CFCs should increase over time and that this will foster the development of alternative chemicals and processes. What EPA does not point out is that the inevitable result of this will be increased costs passed on to the customer. Not only will existing sterilization equipment and facilities have to be modified, but process development and validation will be needed to demonstrate that the new gas mixture is compatible with existing products and packages. On August 12, 1988, EPA made the proposed regulation official by issuing a final rule limiting the production and consumption of CFCs. On the same day, EPA published an advance notice of proposed rulemaking requesting comments on the appropriateness of agency controls to reduce industry-specific reliance on CFCs. Included on the list of 10 industries targeted for regulatory action are "medical sterilization" and "hospital sterilization."

Chemical substitutes consist of a variety of EtO/inert gas mixtures or the use of a new non-ozone-depleting fluorocarbon such as FC 134a. The DuPont Company has indicated that substitute fluorocarbons will take 5-7 years to develop and that the probability of success is not guaranteed. Some mixtures of EtO and carbon dioxide result in extremely high-pressure cycles with a resulting low gas concentration. Most exisitng EtO sterilizaers are not equipped to withstand such high pressures. The use of "pure" EtO requires that facilities be made explosion-proof. Such retrofitting of existing facilities would result in a major capital outlay. Some return on investment would be anticipated, however, due to the reduced cost of pure EtO versus the standare 12/88 EtO-CFC mixture.

In reviewing new sterilization technologies, it appears that only one has *potential* application to replace EtO at least in part. This technology is gaseous chlorine dioxide sterilization. Developmental work has been conducted on this modality over the last several years with favorable results. However, a considerably amount of work is necessary to apply the technology to a commercial situation.

Any alternative must take into consideration the type of sterilization cycle employed and the technical complexity of the alternative procedure.

The most recent data on the so-called "ozone hole" above the South Pole have been startling. The National Aeronautics and Space Administration (NASA) Ozone Trends Panel reported in March 1988 that extreme losses of ozone took place in October 1987. In some cases the loss was 95% of the normal ozone level in the lower atmosphere. Record losses were also detected outside of the hole. The depleted area was also shown to have a longer-lived effect with a lower temperature resulting in a year-round depletion. Several theories as to the cause of this depletion were discarded as erroneous, and CFCs were identified as the causative agent [16].

This recent NASA information convinced one major CFC producer to make a major announcement. On March 25, 1988 the *Wall Street Journal* reported that the DuPont Company, producer of 25% of the world's CFCs, would phase out CFC production over several years. CFC sales totaled approximately $600 million for DuPont. DuPont believes that other producers may also phase out production [17].

D. A Diversity in Chemical Sterilization Methods

The unfolding events described above are creating and will continue to create candidate new sterilization methods for the future. Some are likely to meet the high-volume needs of industry, while others are aimed at the requirements of hospitals, and still others may have only specialized applications.

Although the principles of product sterilization by radiation produced by isotopes such as ^{60}Co and by accelerated electrons have been known for many years, newer versions of electron accelerators are being proposed. Isotopic ^{60}Co will likely remain the most frequently used radiation sterilization method, with more than 140 irradiators in use worldwide. However, newer versions of accelerators are expected to be used for some industrial sterilization applications. Cleland described the various types of accelerators that have been developed [18]. Of these, two types are currently being used for sterilization: alternating current (AC) accelerators with ranges up to 20 kW of power and 5-12 MeV of energy, and direct current (DC) machines having ranges of 30-100 kW with 0.5-5 MeV. High voltages are used to generate and accelerate electrons, which are then "sprayed" on the product to be sterilized. The higher the power of the machine in kilowatts, the greater the number of electrons produced per unit of time and therefore the greater the product throughput. As the energy level or megavolts increases, the speed of the electrons is increased and they penetrate more deeply into the material to be sterilized [19]. Because of penetration limitations, electron beam sterilization works best with products of homogenous density. If the densities within products passing in front of the electrons vary widely, problems may be encountered due to overdosing of some parts of the product while attempting to achieve a minimum sterilization dose throughout.

In spite of these limitations, we expect that electron beam-type machines will see increased utilization in the coming years. This is primarily because advances and new innovations in this technology are appearing rapidly. Cleland and Pageau [20], for example, have recently suggested the use of an X-ray machine wherein accelerated electrons bombard a tungsten target to produce the sterilizing rays. Such X-rays are equivalent to gamma photons and provide a high degree of product penetration.

New gaseous sterilization methods as possible substitutes for EtO are in various stages of development. These gaseous methods employ oxidizing agents, as opposed to EtO, which is an alkylating agent. Currently there is interest in the use of chlorine dioxide (ClO_2) gas, which appears to be an effective antimicrobial agent. In former years the use of ClO_2 as a gaseous sterilant had been considered impractical primarily because the gas could not be safely shipped and stored. Recently, however, there have been developments that allow for *in situ* generation of the gas. This enhances the future potential for ClO_2 [21]. ClO_2 gas will probably not prove to an across-the-board substitute for EtO, but it is likely, we believe, to find selective applications for industrial and hospital sterilization.

Another oxidizing gas being considered for sterilization is ozone (O_3) [22, 23]. The emphasis has been on small chambers for hospital use. Improved methods for the generation of O_3 from air or oxygen have apparently made these developments possible. The use of this agent for large-scale commercial sterilization of packaged products does not appear promising at this time.

A new sterilization process based on gas plasma technology was patented by Surgikos, Inc., a Johnson & Johnson Company, in 1987 [24]. The process utilizes radiofrequency energy and hydrogen peroxide vapor to create a low-temperature hydrogen peroxide gas plasma and achieve relatively rapid sterilization. The sterilization cycle consists of the following steps: Articles to be sterilized are palced in the chamber, the door is closed, and a vacuum is drawn within the chamber. A solution of hydrogen peroxide is injected into the chamber and vaporized. The vapor diffuses throughout the chamber and surrounds the items to be sterilized. At the end of the diffusion phase, radiofrequency energy is applied to the chamber. The radio waves break apart the hydrogen peroxide vapor into reactive species (free radicals), which form a gas plasma. The gas plasma is maintained for sufficient time to interact with and kill microorganisms. During this gas plasma phase, the hydrogen peroxide is continuously depleted as the active (free radical) components recombine into oxygen and water. At the completion of the plasma phase, the source of the radio waves is turned off, the vacuum is released, and the chamber is returned to atmospheric pressure by introducing filtered air. Since the process temperature does not exceed 40°C, the technology is particularly well suited to materials that cannot be steam sterilized.

Vapor-phase hydrogen peroxide (H_2O_2), also an oxidizing agent, may find applications as a surface sterilant, but it lacks the ability to penetrate between mated surfaces as does EtO [25]. No equipment has been developed to generate large quantities of the vapor needed for the decontamination of surfaces in, for example, clean rooms.

For hospitals, small sterilizers based on the use of oxidizing agents in liquids are being investigated. Among the agents that have been suggested are O_3 and peracetic acid. Such applications appear to be quite limited.

Undoubtedly, the challenges we face in sterilization will result in other novel ways and means for product sterilization. As an example, in completely automated production facilities, sterilization machines with continuous throughput and that are "in-line" with production may offer advantages. Low-energy electron accelerators have been proposed for such applications for many years [26]. Also, it is possible that combination methods could be developed in which two or more sterilants are used simultaneously or in tandum to sterilize product. One example would be ^{60}Co sterilization used in combination with elevated temperatures [27].

E. Increased Overseas Production by Multinational Companies

While the adage that faster and better transportation shrinks the world's borders may be true, other factors are at play that will increase the production of sterile medical devices overseas. Not the least of these is the gradual penetration of modern health care techniques into developing countries and the demand for better health care in many developed nations. Also, it should be remembered that the major cause of human morbidity and mortality in other regions of the world are not necessarily the same as those on the North American continent. Thus, the needs may vary as to the types of sterile medical devices used for health care applications.

Several challenges are apparent for companies who produce or plan to produce sterile products overseas. We have already pointed out that foreign laws, regulations, and standards for sterilization processing are on the increase, and in some countries these may turn out to be more rigid than in the United States. Even with the lack of such regulations, most multinational firms will elect to operate sterilization facilities in a manner at least equivalent to U.S. standards. Substandard sterilization in terms of sterility assurance levels, process validation, and cycle development and dose-setting procedures will not be tolerated for a product characteristic as critical as sterility.

Given the above, those involved in operating remote sterilization facilities will face the challenges of supply acquisition, proper sterilizer maintenance, adequately trained personnel, etc. Wherever possible, we believe that radiation will be the preferred sterilization method in new overseas locations. Any new EtO installations will probably be completely automated to significantly reduce the possibility of human exposure. Of course, if and when developing sterilization methods, such as chlorine dioxide gas, are perfected, they would have advantages in these new facilities.

Finally, we believe that manufacturers planning overseas sterilization facilities would be wise to consider all mechanisms to control and reduce hazardous environmental emissions. The technology is now available to prevent worker exposure and to drastically reduce environmental emissions.

F. Restrictions Concerning Cobalt-60

In spite of its wide use and advantages for future use, some factors are operating that, to an unknown extent, could restrict the expanded use of ^{60}Co for radiation sterilization.

First, in some countries, states, or localities there may be resistance to the presence of a facility containing this radioactive isotope. Such resistance may be partly due to ignorance and the tendency to equate ^{60}Co sterilization plants with atomic power reactors. In any case, it has been widely recognized by our industry that an initial hurdle to overcome is that of securing required permits and licenses for the construction of ^{60}Co plants. The extent to which this factor will affect electron accelerator plants is not known, but logically, any contemplated restrictions should be less severe. With ^{60}Co, another reason for objecting to the presence of a plant in a particular locale is recognition of the need to transport the isotope back and forth over the highway systems.

A second restrictive factor may be the availability of the isotope. Demand for ^{60}Co will almost certainly increase as irradiation for foodstuffs is approved, and as the food industry begins to reap the advantages of prolonged shelf-life for perishable foods. In effect, if demand exceeds supply, our industry may be competing for the available cobalt isotope. Even if we compete successfully, the basic cost of the isotope is very likely to increase.

The third potential restrictive factor relates to the disposition of spent ^{60}Co. Disposal sites for spent isotopes are, to say the least, very limited. The filling up of present sites, regulations against their continued use, and the lack of the opening of new sties are potentials that could make disposition more difficult.

In sum, in planning for new ^{60}Co sterilization facilities, manufacturers should consider these potential restrictions as part of their overall master plan.

G. Reuse of Sterile Single-Use Devices

When health care providers or others decide to reuse medical devices designed and manufactured for single use, many complex issues are raised. The most important issue, however, is whether reuse is in the best interest of the patient and whether the remanufacture and resterilization of a single-use item results in a device whose quality meets the quality of the original manufactured device. To adequately prepare a used product for resterilization requires the elimination of all bioburden contamination, a task that is virtually impossible. We believe that there will be an increasing realization among reusers and increasing pressure from patient groups to ensure that reuse practices do not create a double quality standard that, of course, is not in the best interest of the patient. Also, the question of patient consent is of concern.

The quality issue as well as other issues will present challenges to manufacturers and users in the years to come. A legal issue, for example, is whether reusers should be required to register with the FDA and conform to GMPs and other regulations that govern the activities of the original manufacturer. Another issue, of particular interest to our industry, concerns liability implications resulting from hospital resterilization and reuse of single-use items. Will these practices ultimately affect the OEM (original equipment manufacturer)?

A surprising observation about reuse is that, except in several areas such as the reuse of dialyzers, there is often little or no properly documented evidence that cost savings occur. Cost savings, of course, is usually the motivating force for reuse.

The FDA has raised another issue by requesting the manufacturers of hemodialyzers to delete the single-use designation on labeling and to supply instructions for reprocessing. It remains to be seen whether this move will ultimately result in requests for similar labeling changes on other single-use sterilie devices.

HIMA has adequately documented its opposition to reuse and has provided some evidence of patient harm resulting from reuse [28]. It is difficult at this point to believe that any future increases in the reuse of single-use medical items will contribute to the common objective shared by us all of quality health care.

H. Continued Emphasis on Cost Containment

In order to maintain a competitive edge over foreign suppliers of health care products, each manufacturing process will undoubtedly come under scrutiny. In fact, our overall approach to manufacturing, sales, R&D, and marketing will be affected by the increased concern for quality and productivity. Although initially considered fads, such programs as Crosby's Quality Improvement Process and books such as *Quality Is Free* have fostered approaches to cost cutting and waste reduction by "defining the requirements" and adopting a policy of prevention [29].

This philosophy will challenge the cost effectiveness of sterilization processing and push for improved process control and optimization. EtO cycles have typically been established using an extreme overkill approach: that is, the inactivation of resistant biological indicators during a specified time. More attention will focus on reducing and thus optimizing the exposure time using more precise methodology such as those developed by AAMI. The concern over residues will add to the need for optimization and will result in a balance between exposure time and EtO gas concentration. Such an approach will result in process conditions that provide the necessary safety as well as the required economics.

A logical question is: What does the process of sterilization contribute to the total cost of manufacturing a sterile medical product? In general, we believe that in most companies the total budgeted cost of the sterilization process equates to 3-6% of the manufacturing costs. With such a small percentage, it is natural that cost reduction efforts will generally focus on larger segments of the manufacturing cost. However, the avoidance of a major capital appropriation, such as the purchase of a new sterilizer, has frequently been an incentive to scrutinize the process and to push to reduce cycle times. With this in mind, process optimization must be approached with caution. It should remain an objective only after careful consideration of the next question: Are efforts to optimize impeded or destroyed by unforeseen costs associated with improper sterilization, such as costs for scrapped product, recalls, returned goods, reworked goods, etc.? If efforts are made to improperly reduce total sterilization costs, the stakes are very high. Clearly, a technically sound program based on validated data is a prerequisite.

We must conclude from this discussion that very careful consideration is cost containment/cost reduction msut be given to programs to reduce operating costs associated with or resulting from the sterilization process.

I. Normalization of Sterility Requirements

Jorkasky [30] recently identified several sterility assurance-related issues in which significantly different requirements exist from country to country: sterility-testing methodology, cobalt-60 dosage, acceptability of EtO as a sterilant, EtO residual levels, expiration dating, and labeling.

While some degree of normalization may result from agreements between the United States, the United Kingdom, and Canada regarding reciprocal GMP inspections, much more will be required in the future concerning sterility assurance requirements. The European Economic Community (EEC) is attempting to provide a logical route for the normalization of sterility requirements among its 12 member nations. The European Confederation of Medical Suppliers Association (EUCOMED) is taking the initiative by preparing proposals for consideration by the EEC. Normalization across continents, however, will undoubtedly occur more slowly. Certainly organizations such as AAMI and HIMA should expand their efforts in this direction. The European Center for Standardization (CEN) is actively preparing standards that in themselves will not become legal requirements, but if a product is in compliance with a particular technical standard, then it will be deemed to be in conformity with the specific EEC Directive. The format of EEC Directives is to define certain essential requirements to be met by the products in question. For medical products, the requirements relate to safety and quality. Any medical device that meets the essential safety requirements will be permitted to be sold in every EEC market, and no EEC country may refuse its sale on grounds of safety or quality. A manufacturer who demonstrates compliance with the requirements set forth in a CEN standard is entitled to place an "approved" designation called the "CE Mark" on the product. This signifies conformance to the EEC Directive. Such conformity will become a prerequisite for survival as the European Community becomes a single market in 1992 and previously protected domestic markets are eliminated.

III. CONCLUSIONS

The challenges that lay ahead for medical product sterilization are numerous and complex, and therefore must be viewed from both a short-term and long-term perspective. The near-term effects are generally negative, and range from increased costs for medical products to the shutting down of substandard operations that fail to comply with regulations concerning such matters as worker exposure and environmental pollution. The long-term perspective is much more positive, with the many obstacles eventually being transformed into opportunities for the future. Tightened regulations will continue to foster innovative solutions that will advance the science of sterilization.

Over the next several years we expect to see the emergence of specialty chemical sterilants with specific product niches, a new generation of machine sources of radiation, and a continued shift away from EtO processes. The small device company with limited resources will rely more heavily on contract sterilization. For those contract operations with a strong technical staff, business hsould get a boost. Those possessing the know-how, who modify their in-house faiclities, equipment, and processes, will tend to move more toward automation with the associated benefits of decreased operating costs obtained from reduced human intervention. Overall, the technology of sterilization will be enhanced, causing a higher degree of confidence for this critical step in the delivery of quality health care products to all of us.

REFERENCES

1. F. B. Kilmer, Modern Surgical Dressings, *Am. J. Pharm.*, *69*:24-39 (1897).
2. C. Artandi and W. Van Winkle, Electron-Beam Sterilization of Surgical Sutures, *Nucleonics*, *17*:86-90 (1959).
3. C. Artandi, Successful and Promising Applications of Radiation Processing-Sterilization, *Radiat. Phys. Chem.*, *9*:183-191 (1977).
4. Guideline on General Principles of Validation, FDA, May 1987.
5. Association for the Advancement of Medical Instrumentation, *Guideline for Industrial Ethylene Oxide Sterilization of Medical Devices*, AAMI Recommended Practice, OPEO-87, AAMI, Arlington, VA, 1987.
6. Association for the Advancement of Medical Instrumentation, *Process Control Guidelines for Gamma Radiation Sterilization of Medical Devices*, AAMI Recommended Practice, RS-3/84, AAMI, Arlington, VA, 1984.

7. Occupational Safety and Health Administration, *Occupational Exposure to Ethylene Oxide: Final Standard, Code of Federal Regulations*, Title 29, 1910.19, U.S. Government Printing Office, Washington, DC, 1984.
8. Occupational Safety and Health Administration, *Occupational Exposure to Ethylene Oxide: Final Standard, Code of Federal Regulations*, Title 29, 1910.19, U.S. Government Printing Office, Washington, DC, 1987.
9. Food and Drug Administration, HEW/FDA EO, CH, and EG Proposed Maximum Residue Limits and Maximum Levels of Exposure, *Fed. Reg.*, *43*(122), June 23 (1978).
10. Environ Corporation, *Ethylene Oxide Residues on Sterilized Medical Devices*, Washington, DC, 1988.
11. P. Slovic, Perception of Risk, *Science*, *236*:280-825 (1987).
12. L. B. Lave, Health and Safety Risk Analyses: Information for Better Decision, *Science*, *236*:291-295 (1987).
13. J. A. Beck and R. F. Morrissey, An Overview of Radiation Sterilization Technology, in *Sterilization of Medical Products*, Vol. IV, ed. E. R. L. Gaughran, R. F. Morrissey, and Wang You-sen, Polyscience, Montreal, 1986, pp. 115-131.
14. United States Environmental Protection Agency, *CFCs and Stratospheric Ozone*, Office of Public Affairs, Washington, DC, 1987.
15. B. Rosewicz, U.S. Notifies Treaty to Curb Chemicals Depleting Ozone Layer in Atmosphere, *Wall Street Journal*, March 15, 1988.
16. R. A. Kerr, Stratospheric Ozone is Decreasing, *Science*, *239*:1489-1491 (1988).
17. M. L. Carnevale, DuPont Plans to Phase Out CFC Output, *Wall Street Journal*, March 25, 1988.
18. M. R. Cleland, International Developments in Electron Accelerators for Industry and Medicine, presented at the Symposium on Electron Accelerators in Mexico, Salazer, Mexico, Sept. 26, 1986.
19. W. J. Mayer, Application of Electron Beam Equipment for the Sterilization of Medical Devices, *Radiat. Phys. Chem.*, *15*:99-106 (1980).
20. M. R. Cleland and G. M. Pageau, Comparisons of X-Ray and Gamma-Ray Sources for Industrial Irradiation Processes, Ninth Conference on the Application of Accelerators in Research and Industry, Denton, TX, Nov. 10-12, 1986.
21. A. Rosenblatt, et al., Use of Chlorine Dioxide Gas As A Chemosterilizing Agent, U.S. Pat. 4,504,422, Scopas Technology Corporation, 1985.
22. E. L. Karlson, Sterilizer, U.S. Pat. 4,517,159, May 14, 1985.
23. B. Gurley, Ozone: Pharmaceutical sterilant of the future?, *J. Parent. Sci. Technol.*, *39*:256-261 (1985).
24. P. T. Jacobs and S. M. Lin, Hydrogen Peroxide Plasma Sterilization System, U.S. Patent 4,643,876, 1987.
25. R. K. O'Leary, *Vapor Phase Hydrogen Peroxide Sterilization*, American Sterilizer Company, Erie, PA. 1984.
26. S. V. Nablo, Progress Toward Practical Electron Beam Sterilization, in *Sterilization of Medical Products*, Vol. II, ed. E. R. L. Gaughran and R. F. Morrissey, Multiscience, Montreal, 1981, pp. 210-222.
27. M. A. Tumanyan, Perspective on the Use of Radio-sterilization in Medicine, in *Sterilizaiton of Medical Products*, Vol. IV, ed. E. R. L. Gaughran, and R. F. Morrissey, Polyscience, Montreal, 1986, pp. 108-114.
28. The Reuse of Single-Use Medical Devices: Issues and Impacts, HIMA Document 6-1, HIMA, Washington, DC, March 1984.

29. P. B. Crosby, *Quality Is Free*, McGraw-Hill, New York, 1979.
30. J. F. Jorkasky, Medical Product Sterilization: Changes and Challenges, *Med. Device Diag. Ind.*, 9:7, 32-37 (1987).

37

Legislation Needed to Improve the Medical Devices Law

ROBERT S. ADLER *University of North Carolina, Chapel Hill, North Carolina*

I. INTRODUCTION

A. Background

As the Food and Drug Administration (FDA) moves into its second decade of enforcing the 1976 Medical Device Amendments [1] to the Food, Drug and Cosmetic Act [2], it seems appropriate to reflect on the effectiveness of the Amendments as implemented by the agency. The Amendments, drafted at the end of the so-called "consumer decade," a period when numerous pro-consumer laws were enacted [3], sought through innovative means to provide the FDA with a comprehensive, yet flexible, legislative mechanism for regulating medical devices. Unfortunately, experience with the Amendments has shown that innovation, by itself, cannot guarantee a successful regulatory outcome. In fact, some of the most innovative provisions of the Amendments have proven to be the least workable.

The medical device industry supplies the American health care system with a staggering array of products "from bedpans to brainscans" [4]. In all, roughly 1700 different types of medical devices are produced in over 7000 establishments, which turn out over 41,000 separate products [4].

The Medical Device Amendments were passed in 1976 to fill what was widely viewed as a great regulatory void. Unlike with drugs, the FDA prior to 1976 could not review medical devices—no matter how risky or invasive—for safety and effectiveness before they were marketed unless the agency could convince a court to treat the devices as drugs [4]. Instead, the agency could only wait until sufficient evidence of injury or illness accumulated to enable it to claim that a device was "adulterated" or "misbranded" and seek to seize the device or enjoin its distribution [4,5].

B. The Medical Device Amendments

Congress realized that the tremendous variety of medical devices presented a different regulatory challenge from that presented by drugs. Drugs generally are absorbed into human or animal bodies in limited ways: they are ingested, injected, or inhaled. Because of the great potential for adverse consequences from unsafe or ineffective drugs, the law requires virtually all new drugs to undergo premarket approval [6]. Devices present a broader spectrum of risks. Some devices, such as bedpans, present minor and obvious risks, and should not require premarket approval. Other devices, such as pacemakers, are implanted into consumers' bodies and present an enormous potential risk. They should undergo an extensive examination before being marketed. Accordingly, Congress concluded that some way to deal with the enormous variety of devices was necessary.

In order to provide the FDA sufficient flexibility to regulate appropriately, Congress established a three-tiered scheme. Under this scheme, the agency must place all medical devices into one of three classes:

Class I devices may be sold without premarket approval and need not conform to performance standards [7,8]. They must meet specified general controls, such as good manufacturing practices (GMPs) [9]. Although the agency has promulgated regulations implementing its GMP authority [10], and claims that these regulations have produced product quality improvements on a "grand scale" [11], in fact, they provide little meaningful authority beyond that available to the agency before passage of the medical device amendments. Among the devices classified in Class I are tongue depressors, elastic bandages, ice bags, and bed pans.

Class II devices require performance standards because general controls alone under Class I are insufficient to provide a reasonable assurance of safety and effectiveness [12]. In providing for the promulgation of performance standards, Congress copied (and added to) an elaborate set of procedures known as the "offeror" process from a sister agency, the Consumer Product Safety Commission [13]. Once a performance standard has been promulgated for a type of medical device, no device of that type that fails to comply with the requirements of the performance standard may be marketed. Examples of Class II devices are syringes, bone plates, hearing aids, resuscitators, and electrocardiograph electrodes.

Class III devices must undergo premarket approval (PMA) [14]. This means that no device in Class III may be marketed until its manufacturer has submitted extensive data, including clinical data, that provide a reasonable assurance that the device is safe and effective. Under the law, each manufacturer of a Class III device must secure a PMAA in order to sell the device. In effect, the premarket approval process grants individual licenses to manufacturers of individual devices [8]. (As noted in ref. 8, data from one premarket approval application cannot be used to support another PMA without the express consent of the manufacturer.) Among the types of devices in Class III are pacemaker, IUDs, artificial hearts, and artificial joints.

Although Congress could have "grandfathered" (i.e., exempted from regulation) all devices on the market when it passed the Amendments in 1976, it did not do so [11, at 337]. Instead, it directed the FDA to develop an inventory of devices being sold, to place those devices into one of the three classes mentioned above, and eventually to regulate the devices [15].

Permitting devices on the market to remain there until the FDA classified and regulated them conceivably could have produced a tremendous competitive advantage for devices produced *before* the Medical Device Amendments. This would have occurred if the post-Amendment devices could not be sold until the FDA completed the lengthy regulatory proceedings called for in the

Amendments [8,16]. To address this problem, Congress established a provision referred to at various times as "substantial equivalence," "the 510(k) mechanism," and "premarket notification" [17,18]. Under section 510(k), manufacturers who wish to distribute and sell a device never marketed before the Medical Device Amendments must notify the FDA ninety days in advance of doing so, indicate to which of the three classes the device belongs, and certify that it complies with all applicable regulatory requirements. Section 513(f)(1) [19] allows a company to freely market a device if it is "substantially equivalent" to a preenactment device. Most devices that reach the market do so through 510(k) procedures rather than by certifying compliance with a performance standard or following the premarket approval requirements [12, at 338-339].

Although the FDA has no statutory authority to conduct drug recalls [20], Congress established a detailed set of procedures for device repair, replacement, or refund in the Amendments. Section 518(b)(1)(A)(ii) [21] provides that the agency may seek these remedies of the agency can demonstrate, inter alia, that a product presents an unreasonable risk of substantial harm to the public health and "there are reasonable grounds to believe that the device was not properly designed and manufactured with reference to the state of the art as it existed at the time of its design and manufacture [22].

A final section that bears mention is section 519 [23]. This section provides the FDA with the authority to require manufacturers, importers, and distributors to maintain records and make reports to the agency. Under this authority, the FDA has issued a medical device reporting rule, known as MDR, that requires manufacturers and importers to notify the FDA whenever they obtain information that reasonably suggests that one of their marketed devices (1) may have caused or contributed to a death or serious injury or (2) has malfunctioned and would likely cause death or serious injury if the malfunction were to recur [24].

II. THE FDA's IMPLEMENTATION OF THE MEDICAL DEVICE AMENDMENTS

To say the least, the FDA has not implemented the Medical Device Amendments as Congress contemplated. Congress envisioned that within a relatively short period—certainly less than a decade—the agency would have classified all medical devices, moved most, if not all, Class III devices on the market prior to 1976 through the premarket approval process, and written a substantial number or performance standards. As noted in the medical device hearings held by the Subcommittee on Health and the Environment in 1987, the FDA has not come close to meeting these goals [11]. As of the date of these hearings, the agency had not completed a single PMA proceeding for a pre-1976 Class III device nor had it promulgated a single medical device performance standard [11 at 340-347].

Although some of the FDA's troubles, especially in its early years of implementation, may have been managerial, the agency's implementation difficulties clearly stem from more than management problems. As noted by the Subcommittee on Oversight and Investigations in 1983:

> Six years have passed since the device amendments were enacted. For over 2 years—mid-1980 to mid-1982—[the] FDA's Bureau of Medical De-

> vices had no permanent Director to command the attention of the agency's management. The questionable quality of much of the Bureau's work led to a system where the agency's chief counsel in advance had to review and approve virtually everything the Bureau produced for public dissemination. The number of substantive steps to implement the device amendments accomplished by the Bureau were extremely few [2 at 2].

The FDA simply does not have the resources to meet all of the statutory demans of the Amendments. Moreover, resources aside, some provisions of the statute present virtually insurmountable implementation problems.

The difficulties faced by the agency present difficulties for Congress as well. Until the law is changed, Congress cannot simply say "don't enforce it." On this point, Rep. John Dingell (D-Mich.), Chairman of the House Energy and Commerce Committee, stated during the medical device hearings:

> [o]ur oversight responsibilities require the full and faithful execution of laws. If it is determined that there are deficiencies in existing laws, our responsibility then is to amend those laws. There exists no option which provides for an agency to determine the adequacy of existing laws and then to vary the implementation of those laws according to the whim of the agency. [11 at 335]

It is logical, therefore, to look to Congress to provide some relief and guidance for the agency. Congress should focus on the following issues.

A. "Substantial Equivalence" Determinations: The Great Loophole

As previously noted, most devices—even Class II and Class III—reach the market through "substantial equivalence" determinations. They are able to do so because the FDA must permit post-1976 devices to be sold without regulation until the agency regulates the pre-1976 devices to which they claim substantial equivalence. In 1986, for example, 4338 devices reached the market through premarket notifications while 72 devices did so through the PMA process [11]. Included within the premarket notifications were 2652 Class II and 281 Class III devices [11]. Until the pre-1976 predicate devices to which these Class II and Class III devices claim substantial equivalence are regulated, these devices will also not be regulated.

The attraction of substantial equivalence to manufacturers is clear. A 510(k) notification requires little information, rarely elicits a negative response from the FDA, and gets processed very quickly [16, at 515-519]. On the other hand, Congress clearly never intended that the substantial equivalence procedures, which were added "almost as an after-thought" [16, at 510], would become the primary mechanism by which devices reach the market [4, 8,16].

Unfortunately, substantial equivalence determinations provide little protection to the public. These determinations simply compare a post-1976 device to a pre-1976 device to ascertain whether the later device is no more dangerous and no less effective than the earlier device. If the earlier device poses a severe risk or is ineffective, then the later device may also be risky or ineffective [4 at 35]. Notwithstanding the risk, the FDA must permit the later device to be sold.

Compounding this problem is the fact that the FDA at times has taken a very expansive view of substantial equivalence. In cases where the agency is convinced the risks associated with a product are minor and the benefits substantial, it has determined that technologies not even developed before 1976 were substantially equivalent to pre-1976 devices. For example, the agency determined that new bioengineered *in vitro* diagnostic tests for chlamydia and the herpes virus were equivalent to earlier diagnostic tests for the same diseases even though the biotechnology used in the later tests did not exist in 1976 [8 at 360]. This expansive approach to substantial equivalence has led to a certain degree of confusion about where the FDA views the limits of 510(k) to be.

The FDA has clarified some of the confusion about its intentions by publishing a memorandum of guidance regarding its interpretation of substantial equivalence [25,26]. The memorandum makes it clear that the agency will continue to interpret its authority in a very broad manner. For example, it claims the right to demand clinical data from manufacturers to document assertions that post-1976 devices are as safe and effective as pre-1976 devices [25,26]. In similar fashion, the FDA rejects the notion that different materials, design, or energy sources between a post-1976 device and a pre-1976 predicate device preclude a substantial equivalence determination. The FDA reaches this conclusion notwithstanding House Report language stating that "differences between 'new' and marketed devices in materials, design, or energy source . . . would have a bearing on the adequacy of information as to a new device's safety and effectiveness, and such devices should be automatically classified into Class III" [27]. Whether the FDA's interpretation has stretched the law to its absolute outer boundaries or whether the agency has crossed the line into illegality has yet to be determined by a court.

One additional question about the 510(k) process is whether the law permits what some call "510(k) creep" and others refer to as "piggybacking" [28]. That is, can a post-1976 device be compared to an earlier post-1976 device that was found to be substantially equivalent to a pre-1976 device? According to the FDA, the answer is no [11 at 385, 27 at 130]. But, critics contend that, notwithstanding its denials, the agency does in fact permit "piggybacking" [11 at 360].

Given existing doubts about the underlying legality of the FDA's current approach to substantial equivalence determinations, it is clear that some congressional clarification of the Amendments is essential both for manufacturers and consumers. This clarification is essential for another reason. As the years pass, manufacturers are going to find that many new devices simply will have no predicate pre-1976 device to which they can be compared. Equally compelling, consumers will face unnecessary risks from devices that may have reached the market using twenty- or thirty-year old technology.

At one point, the General Accounting Office (GAO), a congressional watchdog office, recommended that Congress consider revising the law to bar the use of substantial equivalence determinations for post-1976 Class III devices [29]. The FDA rejected this recommendation in part because of the administrative burdens that would be created by such a change [16 at 524 (quoting a letter form Margaret Heckler, Sec. of Health and Human Servs., to Charles Bowsher, Comptroller General, Feb. 21, 9184)]. The FDA simply does not have the resources to process the 300-plus Class III devices that seek to reach the market each year.

Another, less drastic, change would be to retain the current scope of substantial equivalence determinations, but modify the law to expand protections for consumers while permitting manufacturers the use of 510(k) procedures in the future [30]. Under this approach, manufacturers of post-1976 devices not identical to pre-1976 predicate devices would have to submit information demonstrating that the post-1976 device was as safe and effective as comparable devices currently being sold in interstate commerce.

The advantage of such an approach is clear. It would not burden the FDA with premarket approval applications, yet it would enable the agency to ensure that devices currently marketed would be safe and effective according to modern technology. Although consumers would not have the full protection of FDA premarket approval for Class III devices, it would be a major step beyond the existing law, which requires only that new devices be equivalent in safety and effectiveness to pre-1976 devices.

B. Reporting Adverse Experiences: Improved But Still Inadequate

In response to severe congressional criticism regarding its failure to require the submission of reports with respect to manufacturers' adverse experiences with medical devices [4], the FDA, in 1984, issued its MDR rule [31]. As previously noted, this regulation requires manufacturers and importers to notify the FDA whenever they obtain information indicating that one of their products may present a serious risk of injury [32].

From the outset, MDR produced a significant change in the way manufacturers reported adverse device experiences to the FDA. In 1984, under a voluntary reporting system known as the Device Experience Network (DEN), the FDA received 2104 adverse experience reports; in 1985, the first year of operation for MDR, the FDA received 17,979 reports [11 at 347, 33]. The size of the reporting increase suggests that a number of manufacturers had been extremely cautious under the DEN system about sharing adverse safety information with the FDA.

Notwithstanding the increased flow of adverse experience reports from manufacturers under MDR, a serious deficiency remains in the system. As documented by the GAO in a recent study, adverse incidents involving medical devices in hospitals rarely get reported to the FDA [34]. According to the GAO, less than 1% of the incidents involving medical devices in hospitals traced in its study ever made their way to the FDA's files [34 at 54-55; also 11 at 367 (statement of Eleanor Chelimsky, Dir., Program Evaluation and Methodology Div., GAO)]. Even more disturbing, the GAO found that the more serious an incident with a medical device, the less likely hospitals were to report the incident [11 at 379].

The GAO's report is distressing for several additional reasons. First, many of the riskiest uses of medical devices occur in hospitals. Second, many of the newest and therefore least known devices are used in hospitals. Third, MDR will not have a significant impact on hospitals because hospitals are not covered by existing law with respect to adverse experience reports [11]. Fourth, and perhaps most important, the fact that most new devices—including Class III devices—currently reach the market through substantial equivalence determinations with only the barest scrutiny for safety and effectiveness means that the FDA must have extremely effective information feedback systems to ascertain when safety problems arise with consumers. Until a suitable reporting mechanism for hospitals is in place, the FDA will be operating blind with respect to this critical area.

The simplest solution to this problem is to extend MDR to hospitals (and nursing homes and other settings where medical devices are regularly used). Hospitals may have legitimate concerns about being unfairly dragged into the liability problems of device manufacturers when hospitals clearly have their own liability concerns to attend to. Crafting a degree of confidentiality into legislation for reporting hospitals should reduce or eliminate any potential for mischief in this direction. The essential point remains. Hospitals must fully participate in the identification and elimination of dangerous medical devices.

C. Premarket Approval: A Cumbersome Process

Under the Medical Device Amendments, gaining premarket approval is often a lengthy and difficult process [16 at 379]. Section 515(c) sets forth a detailed set of requirements that a manufacturer must comply with in order to secure FDA approval [30]. Included within these requirements are the need for submitting a "full" report of all information known to, or which should reasonably be known to, the applicant concerning investigations which have been made to show whether or not such a device is safe and effective [31], a "full" statement of the components, ingredients, and properties of the device [32], a "full" description of the methods used in, and the facilities and controls used for, the manufacture, processing, and, when relevant, packing and installation of the device [33], and numerous other requirements.

These requirements have been imposed to ensure that Class III devices have been determined to be safe and effective before they are sold [34]. This is obviously an essential public need and one that must be undertaken whenever a device presents a substantial risk to the public. Inevitably, the process consumes an enormous amount of agency resources and time. Although the Amendments require the agency to process PMAAs within 180 days [35], in fact, the agency usually takes about a year—and lately has slipped beyond a year [36]. In contrast, the amount of time for the agency to process a 510(k) substantial equivalence notification is generally a fifth of that required for PMAAs [36 at 3]. Little wonder, then, that given a choice, most manufacturers will opt for reaching the market through section 510(k) rather than through section 515.

One would imagine that any approach that would increase the FDA's flexibility in the PMA process while retaining its obligation to ensure the safety and efficacy of medical devices, would command an instant following among device manufacturers. But that is not always the case. Once a manufacturer has gone through the PMA process, it gains what has been termed a "regulatory patent" [11 at 381-382]. That is, until other manufacturers of similar products go through the PMA process, the manufacturer with a PMAA for its device is the only one authorized to sell it in the United States. In some cases, a regulatory patent can provide more protection against competition than a patent issued by the Patent Office.

Given the existence of the regulatory patent phenomenon, streamlining the PMA process may provoke substantial opposition from manufacturers. For example, many small manufacturers of rigid gas-permeable contact lenses claim that the large manufacturers of these lenses have opposed removing them from Class III simply to maintain the large manufacturers' competitive position [4 at 50-53].

Specific reforms of the PMA process must be undertaken cautiously in order to retain the FDA's ability to protect the public. Accordingly, it would be a mistake to cut back on the agency's authority to require information where the agency feels such information is necessary. On the other hand, the agency should be given flexibility to modify the PMA process in those instances where full compliance with section 515 would require a manufacturer to submit unnecessary materials. Because it is impossible to forecast situations where this would occur, the best approach to reforming section 515 is simply to grant the FDA the authority to waive the requirements of this section whenever the agency determines that a waiver is warranted.

D. Preamendment Class III Devices: A Need for Reevaluation

Congress directed the FDA to classify devices on the market before passage of the Medical Device Amendments and required premarket approval for those placed in Class III [37]. As a result of its review, the FDA has classified or proposed for classification nearly 150 generic types of pre-1976 devices into Class III [36 at 19]. Of these, the agency has identified thirteen generic types of devices for the establishment of regulations calling for PMAAs [36 at 19]. Of the thirteen generic types of devices, the FDA has proposed regulations calling for PMAs for four types of devices and finalized a regulation for one device: implanted cerebellar stimulators [36 at 19; On December 30, 1987, FDA issued another final regulation under 515(b) for the contraceptive tubal occlusion device and introducer]. The fact that the FDA has issued a final regulation under section 515(b) does not mean that those devices have gone through the PMA process. Rather, it means that they must begin the process.

The FDA's progress toward requiring PMAs for preamendment Class III devices has been painfully slow. Again, this does not necessarily mean that the agency has mismanaged the process. It means that the agency simply does not have the resources to act more quickly. But the implication of this slow progress is enormous. Not only do preamendment devices remain on the market without substantial scrutiny for safety and effectiveness, but all newly marketed Class III devices that can demonstrate substantial equivalence ot pre-1976 devices also can be sold without the necessity of a PMAA. To say the least, this strikes the FDA's critics as unacceptable [8 at 362].

One suggested solution for the preamendment Class III backlog is to permit the FDA to exempt those preamendment Class III devices with good safety records from the requirement for PMAAs [8 at 364]. (One might conclude that by setting priorities for the calling of PMAAs, that has been the FDA's *de facto* policy during the past 12 years anyway.) But this seems to contradict the basic concept of Class III. If a device presents such a small risk that it could be exempted from the requirement for a PMAA, it likely does not belong in Class III.

Another approach would be to reclassify those preamendment Class III devices that do not present serious risks of injury into Class II or Class I. Presumably, if done before any manufacturer has actually gone through the PMA process and secured a regulatory patent, reclassification would be a relatively simple and noncontroversial solution. Unfortunately, the FDA's view of reclassification—at least until recently—has seemed to present a problem. The FDA apparently was of the view that the evidence necessary to justify a reclassification request did not differ significantly from that for obtaining premarket approval [8 at 362]. Given the FDA's attitude, it would obviously make no sense for a manufacturer to seek reclassification since the manufacturer would have to develop the same documentation for reclassification as for a PMAA, but would not get the benefit of a regulatory patent.

Although there are indications that the FDA has modified its view regarding the amount of evidence needed for reclassification [8 at 361 (citing a criticisms task force convened by the FDA to determine whether the agency should adopt different criteria with respect to reclassification)]; the agency maintains that its reclassification reviews continue to consume "significant" amounts of staff time [36 at 18]. As discussed later, Congress may have to clarify the FDA's responsibilities with respect to reclassification. Similarly, Congress should provide a special simplified reclassification scheme for the agency to permit it to reclassify preamendment Class III devices to lower classes. Once the agency has completed its reclassification, Congress should then insist that the agency move expeditiously to process PMAAs for those preamendment devices that remain in Class III.

E. A Lack of Performance on Performance Standards

One of the most innovative actions taken by Congress in the Medical Device Amendments was the inclusion of an elaborate set of procedures under Section 514 for the setting of performance standards. Included among the procedures was the "offeror" process, a novel approach by which outside groups would prepare draft standards for the FDA and submit them to the agency for promulgation [13,38]. Regrettably, Section 514 demonstrates that novelty is no substitute for clarity, simplicity, and workability.

At the time this section was drafted, Congress was still enamored with the idea that it could incorporate participatory democracy into administrative rule-making. Unfortunately, as the Consumer Product Safety Commission, the only other federal agency encumbered with the offeror process, discovered, the process makes sense only in theory [13 at 323-326]. In actual practice, the process turned out to be expensive, time-consuming, and ineffectual [39].

As noted in the report by the Subcommittee on Oversight and Investigations, promulgation of performance standards under Section 514 is a cumbersome, highly detailed process involving as many as five separate *Federal Register* notices spread out over a lengthy period of time [4 at 12]. But, given that the FDA has placed roughly 1100 generic device types in Class II, the agency cannot afford a complex process that operates at such a leisurely pace. The result of this statutory complexity is predictable. The FDA has not promulgated a single performance standard in 12 years and seems unlikely to promulgate one in the near future. Unfortunately for the public, this means that preamendment and postamendment Class II devices are regulated as though they were Class I devices. Again, this is clearly not what Congress had in mind when it passed the Medical Device Amendments.

Given the FDA's lack of resources, it will never promulgate significant numbers of performance standards. Nor is it likely that Congress will ever provide the 50,000-plus staff years that it might take to promulgate standards for all of the devices currently classified in Class II [8 at 362]. Accordingly, some other approach must be taken to deal with the performance standard backlog.

One suggestion for reforming the process is to abolish Class II altogether [8 at 362; 40]. The problem with this approach is that it tends to lump a number of very low risk Class I devices with many higher risk Class II devices. Even if the FDA finds itself unable to develop performance standards for the higher risk devices, manufacturers and trade associations often have the resources to do so. Classification of a device in Class II sends a message to those who use the device that someone should develop a safety standard for it. A consumer injured by a Class II device should be able to ask its manufacturer whether there was an applicable performance standard that the device met.

A preferable approach to abolishing Class II is to develop simplified procedures for the promulgation of standards and to clarify that the FDA's obligation with respect to promulgating performance standards is discretionary, not mandatory [41]. This approach would retain the ability of the agency to identify higher risk products, to stimulate voluntary standards activity with respect to the products, and to take mandatory regulatory action where necessary.

F. Reclassification: The FDA's Mixed Signals

In addition to the enormous delays by the FDA in its original classification of medical devices [8 at 361], the agency has found itself in a self-imposed bind with respect to reclassifying devices. In 1982, the FDA tentatively concluded that it should reclassify certain contact lenses from Class III to Class I [42]. When challenged by manufacturers who had obtained premarket approval and who would lose their regulatory patents if these lenses were reclassified, the FDA reversed itself and adopted an extremely conservative position with respect to reclassification [43]. Essentially, the FDA's position was that the manufacturer seeking reclassification had to demonstrate that its device was safe and effective, a burden similar to that required for premarket approval [8 at 361]. Although expressing skepticism about the FDA's position, the D.C. Court of Appeals, in *Contact Lens Manufacturing Association v. FDA*, deferred to the agency's judgment and upheld its refusal to reclassify the disputed contact lenses [44,45]. Noting that the contact lens manufacturers seeking reclassification might have received inequitable treatment, the court suggested that Congress was the proper party to clarify the law [46].

The aftermath of the contact lens dispute has been that manufacturers file very few petitions for reclassification [47]. For example, in 1985, the FDA received 93 PMA applications, over 5000 510(k) notifications, but only 15 reclassification requests [47].

Recently, the FDA has indicated that it might change its position with respect to the degree of evidence needed to obtain reclassification [8 at 361; 48]. Under the FDA's new approach, a manufacturer seeking reclassification would no longer have to establish that a device was safe and effective before it could be reclassified form Class III. Rather, the manufacturer would have to establish only that the proposed new class would provide adequate regulatory controls [48].

Although the FDA's new position seems much more reasonable than its previous one, members of the public undoubtedly will continue to find the agency's approach to reclassification unclear. Not only is the statute complex and confusing,[1] but the agency has not yet demonstrated that it will maintain its more liberal interpretation in the face of a challenge by manufacturers intent on retaining regulatory patents.

The best approach to clarifying the FDA's obligations and authority with respect to reclassification is congressional action. Congress should provide a reasonable scheme that requires manufacturers only to demonstrate that their devices could be adequately regulated under the new classification they seek. Such a change would mean, for example, that a manufacturer seeking to reclassify a device from Class III to Class II would not have to demonstrate that there is sufficient information to *establish* a performance standard for its device; the manufacturer would have to show only that its device *could* be adequately regulated for safety and effectiveness under a performance standard.

G. Transitional Devices: Too Many Resources Devoted to Too Few Risks

Before passage of the medical Device Amendments, the FDA occasionally made a deliberate decision to regulate as new drugs products that resembled devices as much as, if not more than, they resembled drugs [49], such as contact lenses. In some cases, manufacturers resisted the agency's efforts to subject their products to premarket approval, forcing the agency to litigate to establish its authority [50].

When the Medical Device Amendments were passed, the definition of the term "medical device" was broad enough to include a number of devices previously regulated as drugs [51].[2] In virtually all cases, Congress chose to designate these products as devices and regulate them under the Medical Device Amendments. Section 520(e) of the Medical Device Amendments directs the FDA to place into Class III devices previously regarded as new drugs [52,53]. These products have come to be known as "transitional" devices. Under the statute, they are regarded as new devices and, accordingly, cannot claim substantial equivalence to pre-1976 devices. This means that they cannot be sold unless they have obtained premarket approval from the FDA [54].

---

[1] As noted by the Criticisms Task Force, opportunities for manufacturers to petition the FDA to reclassify devices are provided in five sections in the Medical Device Amendments, with three separate procedures specified [45 at 290-291; 48].

[2] Under the Medical Device Amendments, a medical device is an instrument, apparatus, implement, machine, contrivance, implant, in vitro reagent, or other or similar or related article . . . which is . . .

> (2) intended for use in the diagnosis of disease or other conditions, or in the cure, mitigation, treatment, or prevention of disease, in man or animals, or
>
> (3) intended to affect the structure of any function of the body of man or other animals, and

which does not achieve any of its principal intended purposes through chemical action within or on the body of man or other animals and which is not dependent upon being metabolized for the achievement of any of its principal purposes.

Transitional devices account for roughly sixty percent of all PMAAs filed with the FDA [48 at 10] and consume an enormous amount of agency resources. In many cases, they do not present serious risks, especially when compared to other Class III devices. Many FDA staff members believe that a number of transitional devices, for example contact lenses, could be adequately regulated under a Class II or Class I designation [48 at 10]. Unfortunately, the agency does not have the ability to place these devices on a "back burner" because they cannot be sold until they have obtained premarket approval.

The simplest solution to deal with the less risky transitional devices would be to reclassify them. But, as previously noted, the FDA has yet to achieve clarity, efficiency, or expeditiousness in the reclassification process [8,41-48]. In addition to its normal problems with reclassification procedures, the FDA faces a particularly difficult challenge with transitional devices because a number of manufacturers have otained premarket approval for their devices and fiercely oppose reclassification and the accompanying loss of regulatory patents. Given these obstacles, it seems clear that the only way to reform the process is for Congress to mandate a simplified reclassification process for transitional devices.

In this regard, the most appealing approach would be for Congress to mandate that all transitional devices would be automatically reclassified to Class I (or Class II as appropriate) after a review period by the FDA *unless* the agency affirmatively determined that certain devices should remain in Class III. Under such an approach, appeal rights would be limited to the filing of petitions for reclassification under section 513(e), where the burden of proof would be on the petitioner.

H. Repair, Replacement, or Refund Authority: Hogtied and Ignored

Although the FDA has explicit authority under the Medical Device Amendments to compel repair, replacement, or refund [20-22], it is so circumscribed by necessary findings and procedures that it may well be no more effective than if the agency had no specific authority. For example, without such authority for drugs, the FDA is often successful in inducing voluntary recalls through the use or threat of seizures, injunctions, criminal penalties, and adverse publicity [20 at 458].

One difference between drugs and devices that seems to warrant greater authority for devices is the fact that, unlike drugs, many devices are durable items that are used and reused. Thus, recalls from consumers, as opposed to distributors and retail outlets, may well be more frequent for devices than for drugs. But manufacturers may well resist recalls that extend to consumers because they are more expensive and produce more adverse publicity—thus, the need for greater authority.

Unfortunately, the FDA's greater authority over devices than drugs is more apparent than real. Under Section 518 the FDA may not simply seek the remedies of repair, replacement, or refund once it has determined that a product is unreasonably dangerous, that is, it presents "an unreasonable risk of substantial harm to the public" [25 at 470; 55]. The Amendments require the agency to determine (a) that notification to the public would not eliminate the hazard [56], (b) that the unreasonable risk was not caused by seller or user error [57], and (c) that there are reasonable grounds to believe that the device was not properly designed and manufactured with reference to the "state of the art" as it existed at the time of its design and manufacture [25 at 473; 58]. Only after these determinations can the agency seek a remedy under Section 518.

These requirements are much more stringent than those contained in other health and safety statutes, such as the Consumer Product Safety Act [59] or the National Traffic and Motor Vehicle Safety Act [60]. No other federal health and safety agency must make a determination with respect to "state of the art" before seeking statutory remedies.

Nothing in the legislative history of the Medical Device Amendments defines the term "state of the art." The only explanation of Congress's intentions lies in the House Committee Report, which states that Section 518 action is inappropriate in those instances in which "a device presented a reasonable risk according to the state of the art at the time of its manufacture which becomes unreasonable due to a change in technology" [61].

Although Congress's intentions may have been noble in writing this provision, the actual effect of this language is to tie the FDA in knots in product defect situations. One immediate stumbling block is the difficulty the agency faces in demonstrating what the state of the art was at the time a product was designed and manufactured. Since years may have passed between these events and the accumulation of evidence that a device presents an unreasonable risk of injury, the FDA may find it extremely difficult to prove that the manufacturer could have produced a safer product under the then existing state of the art.

But problems of proof are not the major obstacle to recalls in the Medical Devide Amendments. The essential flaw is that the Amendments exonerate a manufacturer from all repair, replacement, and refund obligations because it met the state of the art when its device was designed and manufactured. That is bad policy. State of the art is a concept that is useful in the context of determining liability in lawsuits brought by injured consumers against device manufacturers [62]. It is misapplied in the context of protecting the public health and safety.

An example will explain this point. Assume that all experts in the field would conclude that a manufacturer of pacemakers met the state of the art when it designed and manufactured its product. If the device was implanted in thousands of consumers and, notwithstanding its compliance with the state of the art, then was found to have a faulty design resulting in the deaths of 50% of its users, Section 518(b) would prevent the FDA from requiring the manufacturer to take any remedial action other than notifying consumers that their lives were in great danger.

No other health and safety agency under similar circumstances would be barred from requiring repair, replacement, or refund. While it is debatable whether the pacemaker manufacturer should have to pay tort damages to consumers for injuries that it could not have foreseen, it is unconscionable for a statute to relieve a manufacturer from all responsibility for saving lives of those at risk from its product once the manufacturer knows that the product presents an unreasonable risk.

In providing "state of the art" as a defense to actions under Section 518(b), the drafters of the Amendments improperly grafted a product liability concept onto a product safety statute. Product liability addresses legal obligations *after* an injury has occurred; product safety focuses on the avoidance of *future* injuries. State of the art may be an appropriate defense to one who has been sued for innocently causing a past injury, but it should not insulate one who has placed innocent lives at risk from taking adequate steps to protect those individuals from future injury.

The FDA has rarely considered using Section 518(b)'s authority.[3] This likely reflects, at least in part, the agency's reluctance to use such an obstacle-laden procedure. But the real impact of the poor drafting of Section 518(b) cannot be measured simply by counting the number of formal actions undertaken by the agency. The more immediate, but less measurable, effect of such a section arises when agency staff seek to negotiate voluntary recalls with manufacturers. Given that the agency faces enormous burdens in bringing formal action, recalcitrant manufacturers know that they have substantial bargaining power against the agency. This may well result in less effective recalls.

Removing the "state of the art" language from the statute presents a formidable challenge to Congress. Whether or not the language should have been placed in the statute, manufacturers now realize that it provides them a substantial advantage in dealing with the agency. Nevertheless, in order to protect the public, Congress should make every effort to remove it.

I. Excessive Procedures

In order to understand a regulatory statute properly, one must know more than what authority it provides to an agency. One must also understand how easily the agency may invoke its authority. If the agency must hold endless hearings, consult numerous advisory panels, make an infinite number of determinations, and publish reams of *Federal Register* notices, then the agency's authority is illusory. As previously noted, the FDA's authority to process PMAAs, promulgate performance standards, reclassify devices, and require repair, replacement or refund of defective devices all suffer from excessive and unnecessary procedural requirements.

In some cases, procedural requirements are essential. One would deem it unfair and probably unconstitutional to permit a agency generally to publish a regulation in the *Federal Register* without first permitting those affected—manufacturers and consumers—to comment on the provisions of the proposed regulation. On the other hand, the Medical Device Amendments contain many procedures that go far beyond what fairness and due process require. Why they are there seems obvious. In order to neutralize the opposition of manufacturers to expanded regulatory powers for the FDA, the Amendments' drafters agreed to add extra procedures.

Unfortunately, the additional procedures have had the effect of emasculating much of the FDA's regulatory authority. Congress should engage in a thorough review of the Amendments and remove excessive procedural requirements. One good model for such a review is H.R. 5516, introduced by Representatives Henry A. Waxman (D-Cal.) and John Dingell in the 99th Congress, which essentially makes discretionary all mandatory advisory panels and informal hearings currently contained in the Medical Device Amendments [30].

---

[3] As of 1983, the agency has turned to these provisions only once [4 at 12; 22 at 469, 470, 476].

J. Inadequate FDA Resources: Should User Fees Be Required?

Virtually all observers [8,11,28,29], including medical device manufacturers [63], share the view that the FDA's resources are inadequate to meet its obligations under the Medical Device Amendments. Manufacturers sometimes are as upset as consumers by the FDA's poverty because scarce agency resources mean that premarket approval of new Class III devices must proceed slowly, reducing the manufacturers' profits and lessening their ability to compete in world markets [63 at 16]. Although the Reagan administration sought to provide modest increases in funding for medical device approval [11 at 387], the increase will not be sufficient to permit the agency to improve the regulatory process dramatically [11 at 387].

One highly controversial proposal to augment the FDA's resources would be to charge manufacturers user fees for agency services, such as processing premarket approval applications. Although user fees for medical devices would be a new agency function, in fact, the FDA has charged user fees for many years for certifying antibiotics for human and animal uses [64]. The FDA, prodded by the Office of Management and Budget, has sought to impose user fees for the premarket approval review of new drug and antibiotic applications, and for certain supplemental applications [65]. In its proposed fiscal year 1987 budget, the Department of Health and Human Services indicated its intention to raise $34 million in user fees from PMA applications for drugs and devices. But in all instances, the agency has been blocked by Congress from imposing such fees [66]. (One should not, however, conclude that advocates of user fees have given up [67].)

Although most of the opposition to the specific HHS and FDA proposals for user fees has come from manufacturers, they have actually had mixed feelings regarding user fees. The Pharmaceutical Manufacturers Association has expressed qualified support for user fees as long as the fees are used to supplement the FDA's budget and are specifically targeted for improving the premarket approval process for drugs [68]. Device manufacturers, although sympathetic to the Pharmaceutical Manufacturers Association's position, have generally expressed greater reservations regarding the concept [69].

It seems desirable to expand the FDA's resources by charging user fees so long as two conditions are met: (1) user fees should *supplement*, not substitute for, current appropriations. Otherwise, no improvement in the FDA's resources would be achieved and the political opposition to such an approach would doom it from the start. And (2) user fees should not be so large as to discourage manufacturers, especially small ones,[4] from marketing useful devices.

Whether user fees, if adopted, should be assessed only for PMA applications and supplements is debatable. Given the infrequent resort to PMA applications by manufacturers, one could easily conclude that charging user fees only for PMAAs would hardly be worth the bother. On the other hand, given the thousands of 510(k) notifications filed every year, it would seem to make sense to charge user fees for these notifications, assuming that such charges were reasonable.

---

[4] According to the President of the Health Industry Manufacturers Association, about one-half of HIMA's members are companies with gross annual medical device sales of $5 million or less [69 at 4].

III. CONCLUSION

As the years have passed since the passage of the Medical Device Amendments, numerous groups, including the Office of Technology Assessment, the General Accounting Office, congressional committees, independent observers, and the FDA itself, have examined the Amendments and concluded that legislative changes are necessary in order to make the law work effectively. Notwithstanding this unanimity, modifying the law will take extreme dedication and perseverance on the part of all interested parties.

One of the obvious difficulties in modifying the law is that of competing interests. Manufacturers who have regulatory patents as a result of having gone through premarket approval may have little enthusiasm for streamlining the reclassification process. Others may take comfort in the fact that the agency's repair, replacement, or refund authority is tightly circumscribed.

On the other hand, most manufacturers probably feel uncomfortable about the long-term viability of 510(k) notification procedures and the lack of premarket approval resources for the agency. The key to developing new legislation is to promote measured and creative approaches that improve the process overall. If this is done properly, no group will like all of the provisions of a reform bill, but each will see sufficient benefit to warrant its support.

In this regard, one must be frank in noting that recent developments have not been encouraging. After 3 years of negotiation, industry, consumer groups and members of Congress have not yet been able to break a legislative impasse.[5] This has triggered some hardening of positions and finger pointing, leading to predictions that more extreme measures may be taken and some parties will be excluded from negotiations during the legislative process [70].

One can only express dismay at the intransigence of some interested parties regarding new medical device legislation. Although in the short run they have managed to block needed reforms, in the long run, it is unlikely that they will be able to prevent such powerful members of Congress as John Dingell and Henry Waxman from succeeding. New legislation is needed and will be enacted.

REFERENCES

1. Pub. L. No. 94-295, 90 Stat. 539 (1976). For a legislative history of the amendments see D. O'Keefe & R. Spiegel, *An Analytical Legislative History of the Medical Device Amendments of 1976* (1976).
2. Pub. L. No. 75-717, 52 Stat. 1040 (1938), as amended 21 U.S.C. 301-392 (1982).
3. Of 47 federal consumer protection laws enacted between 1891 and 1972, fewer than half, or 21 statutes, were enacted in the first 75 years; the remaining 26 were enacted during the period from the mid-1960s to the mid-1970s. This led some to call the latter period the "consumer decade." See Schwartz, The Consumer Product Safety Commission: A Flawed Product of the Consumer Decade, 51 *Geo. Wash. L. Rev.* 32, 34 (1982).

---

[5] Legislation to amend the medical device amendments passed the House of Representatives in the 100th Congress, but was killed in the Senate.

4. Staff of House Subcomm. on Oversight and Investigations of the House Comm. on Energy and Commerce, 98th Cong., 1st Sess., Report on Medical Device Regulation: The FDA's Neglected Child 1 (Comm. Pring 98-F 1983).
5. Despite the "after-the-fact" regulation, the FDA did succeed in removing some grossly hazardous devices from the market. See Munsey and Samuel, Medical Device Regulation, In Transition, in *Seventy-fifth Anniversary Commemorative Volume of Food and Drug Law* 350, 351 (1984).
6. 21 U.S.C. 355; 21 C.F.R. 314 (1988).
7. 21 U.S.C. 360c(a)(1)(A). Other general controls include regulations requiring registration, premarket notification, record-keeping, labeling, and reporting of adverse experiences.
8. Kessler, Pape and Sundwall, The Federal Regulation of Medical Devices, 317 *New Eng. J. Med.* 357, 358 (1987).
9. 21 U.S.C. 3670c(a)(1)(A)(i).
10. 21 U.S.C. 820.
11. Hearings on Medical Devices and Drug Issues Before the Subcomm. on Health and the Env't of the House Comm. on Energy and Commerce, 100th Cong., 1st Sess. 386-387 (1987) (statement of James Benson, Deputy Dir., Center for Devices and Radiological Health, Food and Drug Admin.).
12. 21 U.S.C. 360c(a)(1)(B). See also Boguslaki, Classification and Performance Standards under the 1976 Medical Device Amendments, 40 *Food Drug Cosm. L. J.* 421 (1985).
13. In 1981, Congress repealed the requirement for the Consumer Product Safety Commission to use the "offeror" process. Consumer Product Safety Amendments of 1981, Pub. L. No. 97-35, sec. 1202(b), 95 Stat. 703, 703-704 (1981). See generally Schwartz, Performance Standards under the Medical Device Amendments: A Flawed Process in Need of Reform, 39 *Food Drug Cosm. L. J.* 318 (1984) (describing the CPSC's experience with setting product safety standards).
14. 21 U.S.C. 360c(a)(1)(C).
15. See generally 21 U.S.C. 360c.
16. Kahan, Premarket Approval versus Premarket Notification: Different Routes to the Same Market, 39 *Food Drug Cosm. L. J.* 510, 514-515 (1984).
17. 21 U.S.C. 360(k).
18. For a thorough discussion of the 510(k) provision see Kaplan, Through the Maze of 510(k)s, 39 *Food Drug Cosm. L. J.* 160 (1984).
19. 21 U.S.C. 360c(f)(1).
20. See Schwartz and Adler, Product Recalls: A Remedy in Need of Repair, 34 *Case W. Res.* 401, 446 (1983-84).
21. 21 U.S.C. 360h(b)(1).
22. 21 U.S.C. 360h(b)(1)(A)(ii). See generally Mannen and Basile, The "Repair, Repalce, or Refund" Provision of the Medical Device Amendments of 1976, 40 *Food Drug Cosm. L. J.* 464 (1985).
23. 21 U.S.C. 360i.
24. 21 C.F.R. 803.1. See also 49 *Fed. Reg.* 36,326 (1984).
25. Center for Devices and Radiological Health, Food and Drug Admin., Premarket Notification Review Program, 510(k) Memorandum #86-3 (June 30, 1986).
26. Benson, Eccleston, and Barnett, FDA Regulation of Medical Devices: A Decade of Change, 43 *Food Drug Cosm. L. J.* 495, 512 (1988).

27. H. R. Rep. No. 94-853, 94th Cong., 2d Sess. 36-37 (1976). This report language provides the only congressional guidance with respect to substantial equivalence determiantions.
28. Office of Technology Assessment, Federal Policies and the Medical Device Industry (1984).
29. General Accounting Office, Federal Regulation of Medical Devices—Problems Still to Be Overcome, HRD-83-53, at 62-63 (Sept. 30, 1983).
30. Such an approach was suggested in H. R. 5516, 99th Cong., 2d Sess. (1986).
31. 49 *Fed. Reg.* 36,326 (1984).
32. 21 C.F.R. 803.1(a).
33. For a compilation of some interesting statistics on MDR reporting after the regulation's first years, see Basile, Medical Device Reporting: The Good, the Bad, and the Ugly, 42 *Food Drug Cosm. L. J.* 83 (1987).
34. General Accounting Office, Medical Devices: Early Warning of Problems is Hampered by Severe Underreporting (1986).
30. 21 U.S.C. 360e(c)(1).
31. 21 U.S.C. 360e(c)(1)(A).
32. 21 U.S.C. 360e(c)(1)(B).
33. 21 U.S.C. 360e(c)(1)(C).
34. 21 U.S.C. 360e(d)(2).
35. 21 U.S.C. 360e(d)(1)(A).
36. Office of Device Evlauation, Center for Devices and Radiological Health, Food and Drug Admin., Annual Report—Fiscal Year 1986 (Nov. 10, 1986). In fiscal year 1986, the agency took 395 days to review and approve PMAs. In fact, most PMAs took even longer because the FDA does not include in its calculation of time delays caused by deficient PMA applications. By FY 1988, FDA had taken measures to shorten the review time. The average time in this year was 261 days [4].
37. 21 U.S.C. 360e(b)(1).
38. See generally 21 U.S.C. 360d.
39. Hearings Before the Subcomm. on Consumer Proteciton and Finance of the Comm. on Interstate and Foreign Commerce, 95th Cong., 2d Sess. 59-64 (1978) (testimony of R. David Pittle, CPSC Comm'r).
40. Kahn, The Ten Year Record, 8 *Med. Device and Diagnostic Indus.* 60, 63 (1986).
41. As the law reads now, it seems to mandate the promulgation of performance standards; see 21 U.S.C. 360c(a)(1)(B).
42. 47 *Fed. Reg.* 53,404 (1982).
43. 48 *Fed. Reg.* 56,778 (1983).
44. *Contact Lens Mfrs. Ass'n v. Food and Drug Admin.*, 766 F.2d 592 (D.C. Cir. 1985).
45. Kahn, Medical Device Reclassification: The Evolution of FDA Policy, 42 *Food Drug Cosm. L. J.* 288, 292-293.
46. 766 F.2d at 603.
47. Remarks by Jonathan S. Kahn, "Medical Device Reclassification—An Uncertain Route to the Market," 29th Annual Educational Conference, The Food and Drug Law Institute (Dec. 10-11, 1985).
48. Food and Drug Admin., Executive Summary of the Criticisms Task Forces' Reports (1985).
49. 42 *Fed. Reg.* 63,472, 63,473 (1977).

50. *U.S. v. An Article of Drug . . . Bacto-Unidisk,* 394 U.S. 784 (1969), *rehearing denied,* 395 U.S. 954 (cardboard disc impregnated with various antibiotics used to determine a patient's antibiotic sensitivity); *AMP, Inc. v. Gardner,* 839 F.2d 825 (2d Cir. 1968), *cert. denied,* 393 U.S. 825 (suture product used to stitch blood vessels together during surgery).
51. 21 U.S.C. 321(h).
52. 21 U.S.C. 360j(l)(1).
53. 42 *Fed. Reg.* 63,472 (1976).
54. 32 *Fed. Reg.* at 63,474-475.
55. 21 U.S.C. 360h(b)(1)(A)(i).
56. 21 U.S.C. 360h(b)(1)(A)(iv).
57. 21 U.S.C. 360h(b)(1)(A)(iii).
58. 21 U.S.C. 360h(b)(1)(A)(ii).
59. Pub. L. No. 92-573, 86 Stat. 1207 (codified in scattered sections of 5 U.S.C. and 15 U.S.C. (1982)).
60. Pub. L. No. 98-620, 98 Stat. 3358 (1968) (codified at 15 U.S.C. 1415 (1982)).
61. House Comm. on Interstate and Foreign Commerce, Medical Device Amendments of 1976, H. R. Rep. No. 853, 94th Cong., 2d Sess. 23 (1976).
62, Murphy, Santagata, and Grad, *The Law of Product Liability Problems and Policies* (1982).
63. Health Indus. Mfrs. Ass'n, Report and Recommendations of the HIMA Device and Diagnostic Product Approval Task Force (Oct. 1985).
64. 21 C.F.R. 431.53.
65. 50 *Fed. Reg.* 31,726 (1985).
66. See FDA User Fee Proposal is Targeted by Devices Firms, *Generic Line,* Feb. 13, 1987, at 6.
67. User Fee Idea Raises Its Ugly Head—Again, *Generic Line,* Nov. 13, 1988, at 7.
68. Letter from Gerald Mossinghoff, Pres., Pharmaceutical Mfrs. Ass'n, to Frank E. Young, Comm'r of Food and Drugs, Food and Drug Admin. (Mar. 6, 1986).
69. Comments by Frank Samuel, Pres., Health Indus. Mfrs. Ass'n, to the Food and Drug Admin., Docket No. 84N-0101 (Nov. 4, 1985).
70. HIMA May Be Excluded From Device Bill Negotiations, *Med. Dev. Diag. Ind. Rep.* (December 5, 1988) at 10. (According to Patrick McLain, Counsel to the House Energy and Commerce Oversight Subcommittee, unless HIMA can keep members such as Pfizer, Lilly, and Bristol-Myers in support of agreements that HIMA reaches, it's "a virtual waste of time" to negotiate with HIMA.)

38

1989 Waxman-Dingell Medical Devices Bill

C. STEPHEN LAWRENCE *Hogan & Hartson, Washington, D.C.*

I. INTRODUCTION

Representatives Waxman and Dingell have introduced a bill, H.R. 3095, entitled The Safe Medical Devices Act of 1989. The bill's introduction was coordinated with a press conference staged by Ralph Nader's Public Citizen group announcing the release of a report that stated, in predictable fashion, that medical device companies are not reporting deaths, injuries, and malfunctions under Food and Drug Administration's (FDA's) medical device reporting (MDR) regulations and that the agency is not disciplining companies that are found to be noncomplying. In introducing the bill Mr. Waxman echoed those sentiments, criticizing FDA's enforcement efforts and stating "this situation has gone unresolved for much too long, and caused harm, and even death, to many people."

In keeping with this tone, H.R. 3095 is much harsher in its proposed treatment of the medical device industry than previous "medical device improvements" bills, such as H.R. 4640, which passed the House, but not the Senate, in 1988. While the administration is opposed to the new bill, the vigor of Mr. Waxman's offense, the current crisis of confidence at FDA, and the fact that a medical device bill nearly passed in 1988, all cause concern that the Safe Medical Devices Act of 1989 may become law to the detriment of the medical device industry and without significant benefit to the public health.

This chapter contains a summary of H.R. 3095 and a discussion of the potential impact of several of the bill's key provisions. It should be kept in mind that the bill is not well drafted, and the interpretation of several provisions is open.

II. SUMMARY OF H. R. 3095's PROVISIONS

H. R. 3095 contains the following provisions:

User Reporting—User facilities, including hospitals, clinics, and nursing homes, but not doctors' offices, would be required to report deaths, serious injuries, and certain malfunctions involving medical devices to both FDA and the device manufacturer.

Medical Device Reporting (MDR)—Distributors would be required to report under MDR and send copies of the reports to device manufacturers. Manufacturers, distributors, and importers would have to certify annually that they have reported all events to FDA as required.

Premarket Notification [510(k)]—FDA's current practices regarding determination of substantial equivalence and submission of safety and effectiveness data with 510(k)s would be codified into the statute. Summaries of safety and effectiveness similar to those currently required for PMAs would be required for 510(k)s.

After five years, market clearance could not be had by claiming substantial equivalence to a preamendments class III device. Before that time, a company would have to research and submit adverse data involving a class III preamendments device to which it was claiming substantial equivalence.

Preamendments and Transitional Class III Devices—The bill would require FDA to institute proceedings to revise classifications. As part of the review process, a company would have to report to FDA any adverse data about a device. A preamendments device left in class III by a classification revision regulation would need a premarket approval application (PMAA). Any preamendments class III device not already subject to a PMA requirements and not subject to a revision regulation after five years would require a PMAA.

Class II—Class II would be changed from "performance standards" to "special controls." In addition to performance standards, "special controls" would include guidelines, recommendations, and/or other "appropriate actions" necessary to provide reasonable assurance of safety and effectiveness. These probably could be adopted without formal procedural safeguards.

State-of-the-Art Exemption—This exemption would be stricken form the "repair, replacement, or refund" provision of section 518 of the Federal Food, Drug, and Cosmetic Act (the Act).

Clinical Investigations—"Well-controlled investigations" could be used to support a 510(k) notice or a PMAA regardless of the sponsorship of the investigation. While this appears to be a "paper PMA"-type provision, it is unclear whether it might also allow FDA to use a company's proprietary data in reviewing a competitor's device.

Traceability of Class III Devices to User Level—Manufacturers, distributors, and retailers of class III devices would be required to keep records of distribution down to the user level.

Reporting of Removals and Field Repairs—Companies would be required to report removals and field repairs (i.e., recalls and most field corrections) that are not already reportable under MDR.

Remanufacturers—Certain previously unregulated companies performing a variety of functions falling under the ambit of "remanufacturing" would be required to register and submit 510(k)s as device manufacturers.

Loss of Procedural Safeguards—A number of miscellaneous provisions would remove procedural safeguards now enjoyed by device companies. For instance, FDA could ban a device without a panel review or an opportunity for an informal hearing.

III. SIGNIFICANCE OF H. R. 3095 TO MEDICAL DEVICE MANUFACTURERS

Clearly, if the provisions described above were to become law, they would have serious impact on medical device manufacturers. In addition, the new requirements would strain FDA's resources and could ultimately delay device market clearance. In nearly every case, the important provisions of H.R. 3095 are more onerous than the provisions of previous medical device bills, such as H.R. 4640.

A. Medical Device Reporting

Manufacturers currently must report instances of death, serious injury, or malfunction that could cause death or serious injury. H.R. 3095 would require FDA to include distributors of medical devices in the MDR regulation within one year. The distributors would have to report events both to FDA and to device manufacturers. In addition, each manufacturer, importer, and distributor would have to certify annually to FDA that it has made all the MDR reports required of it. There was no such certification requirement in H. R. 4640. The implications of a false certification, whether or not intentional, are serious both from a civil and a criminal liability perspective.

Medical device users currently need not report deaths, serious injuries, or malfunctions. However, under the Waxman-Dingell bill, each hospital, nursing home, and diagnostic, treatment, or surgical clinic would have to report such events. Doctors' offices would, however, be exempt. Users would be required to report the same kinds of events manufacturers currently must report under MDR, except user reporting would also encompass information that reasonably suggests a device "will" malfunction. Thus, as contrasted to the current MDR regulation, it would appear that a user report would be required merely on the suspicion that a malfunction will occur.

The Waxman bill includes whistle-blower protection to encourage user facilities and their employees to report events and provides a special civil penalty of $10,000 per day, up to a maximum of $500,000, for a user's failure to report. The civil penalty would not become effective immediately. Instead, H.R. 3095 would require FDA to perform a study of compliance with the new user reporting requirements, to be completed four years after the bill's passage. If FDA's study shows there is "substantial compliance" with the reporting requirements, the special penalty for user reporting would be placed in abeyance and the penalties presently provided for by the Act would apply to any violation. Under the current provisions of the Act, fines are available only if FDA brings criminal action. The General Accounting Office (GAO) would be required to perform a concurrent study of reporting compliance and report to both the Waxman and Dingell Committees.

While H. R. 4640 also contained user reporting requirements, the previous proposals were overall less burdensome. For instance, under H.R. 4640, only device-related deaths would have been immediately reportable directly to FDA, outpatient clinics would have been excluded, and the identities of both the device and its manufacturer would have been kept confidential.

B. Premarket Notification

H.R. 3095 would for the first time statutorily define "substantial equivalence." The proposed definition would harmonize with FDA reviewers' current internal instructions for determining substantial equivalence. The bill also would bestow statutory authority for FDA's current practice of requiring safety and effectiveness data for some 510(k)s.

Currently, a manufacturer can obtain market clearance by claiming substantial equivalence to a marketed class III preamendments device as long as FDA has not promulgated a regulation calling for a PMAA for the device. H.R. 3095 proposes to eliminate this current practice and to speed the process of calling for PMAAs for class III preamendments devices. The proposed changes are drastic and, because of the tortuous language of section 513(f) of the Act, could have consequences that might not be fully known or understood until played out in practice.

During the initial five years following passage of H.R. 3095, FDA would be required to institute proceedings to review and, if appropriate, revise the classification of all preamendments class III devices that are not already subject to a regulation requiring a PMAA. As part of the review, a device's manufacturer would be required to submit all adverse safety or effectiveness data known to the manufacturer. If a device were not reviewed within five years, a PMAA would be required automatically. During the five-year review period, unreviewed preamendments class III devices could continue to be used as predicate devices for new 510(k) filings, but any manufacturer claiming substantial equivalence to such a device would be required to research all published and unpublished data available to the company and describe to FDA any adverse safety or effectiveness information concerning both the predicate device and the new device. After the five-year phase-in period, 510(k) market clearance could not be based on a claim of substantial equivalence to a preamendments class III device. Instead, a new device would need a PMAA unless it could be shown to be equivalent to a preamendments or postamendments class I or II device. H.R. 4640 would have allowed manufacturers to claim substantial equivalence to preamendments devices in the same way as at present.

C. Traceability of Class III Devices to User Level

The proposed legislation would require every manufacturer to record the consignee of each device sold and the date sold. In addition, intermediate and retail sellers would have to keep records by which class III devices could be tracked to the user level. Currently, manufacturers may keep the required distribution records to satisfy good manufacturing practices (GMP) requirements, but there is no general obligation for distributors and retailers to maintain records to the user level.

D. Class II Special Controls

Class II controls would be enlarged to include guidelines, recommendations, and other "appropriate actions" necessary to provide reasonable assurance of safety and effectiveness. Currently, FDA's practice increasingly is to issue guidelines when in the past the agency would have promulgated a notice-and-comment rule. The guidelines often are developed by at most a handful of individuals within the agency and sometimes without significant input by manufacturers, who often are in the best position to judge a rule's practicality. As a consequence, FDA guidelines often are found to be unworkable when actually applied to real-world medical device manufacturers. The Waxman proposal would appear to place greater emphasis on guidelines, "recommendations," and "appropriate actions," and less reliance on performance standards, which have been almost impossible for FDA to develop. (The bill does, however, contain expedited procedures for performance standards.) It is unclear how the agency would apply a guideline or recommendation as a control, and this concept certainly needs to be clarified. Regulation by guideline and recommendation could deny device companies procedural safeguards that attend the development of performance standards.

E. State-of-the-Art Exemption

Currently, medical device manufacturers are not subject to the "repair, replace, or refund" provisions of section 518 of the Act unless "there are reasonable grounds to believe that the device was not properly designed and manufactured with reference to the state of the art as it existed at the time of its design and manufacture." Thus, if the company was not at fault in its design or manufacture of the device, the draconian remedies envisioned by section 518 will not be invoked. The Waxman bill would strike the clause containing the state-of-the-art exemption from section 518. Thus, the repair, replace, or refund remedies could be applied even when the company could not have avoided a defect by design or manufacturing controls. By contrast, H.R. 4640 contained a compromise measure which would have allowed FDA to seek a court order if the company would not take voluntary action involving at state-of-the-art product.

F. Removals and Field Repair

The new bill would require companies to report to FDA any "removal or field repair" of a device taken to eliminate a risk to health or remedy a violation of the Act. Recalls, many field corrections, and, potentially, certain nonrecall situations would become reportable. Companies would have to keep records of nonreportable removals and field repairs. Routine servicing would be excluded from the reporting and record keeping requirements, and companies would not have to repeat reports required under MDR. Currently, manufacturers are requested, but not required, to report recalls to FDA. Other current regulations already may compel companies to keep records of, investigate, or report certain situations that might ultimately result in a removal or field repair of a device.

IV. CONCLUSION

In sum, the new medical device bill is substantially harsher than either current medical device regulation or previous bills. Moreoever, because of the present atmosphere at FDA and on Capitol Hill, this legislation may stand a better chance of success than previous attempts to change the Act. Whether or not this particular bill passes, it is important to understand the issues raised and implications for manufacturers and the FDA.

39

Software Regulation

RICHARD C. FRIES *Ohmeda Anesthesia Systems, Madison, Wisconsin*

GEORGE T. WILLINGMYRE *American National Standards Institute, Washington, D.C.*

DEE SIMONS *Health Industry Manufacturers Association, Washington, D.C.*

ROBERT T. SCHWARTZ *Industrial Designers Society of America, Great Falls, Virginia*

I. INTRODUCTION

Advances in computer technology over the past several decades have significantly improved the delivery of health care. Medical devices previously composed entirely of hardware elements are now produced with fewer discrete components, allowing them to be more compact. Today's software-controlled devices perform more varied and complex functions and have a higher reliability than their predecessors.

Information systems have greatly enhanced the reliability and efficiency of health care delivery; they have replaced manual order-entry and record-keeping systems. Self-diagnostics, preoperative self-checkout, and watchdog timers have become an integral part of medical devices, rendering them safer and more powerful for the user and the patient. Software is more versatile than hardware components; changes in functionality are quicker and easier than for hardware.

II. SOFTWARE AS AN EMERGING TECHNOLOGY

Software is a set of instructions (computer program) that, when executed, allows the desired function and performance to be realized. The term software includes data structures that enable the program to manipulate information and documentation that describes the operation and the use of the program [1].

Software differs from hardware in several important respects. Software is a logical rather than a physical element. Software is developed rather than manufactured in the classical sense. Software does not wear out as hardware does. Software costs are concentrated in engineering.

Software quality is a primary concern of software developers. Poor quality software is costly, damages the company's reputation, reduces its market share, and puts the manufacturer at risk for liability. Quality attributes that are important characteristics of any software product include usefulness—that is, satisfying user needs; reliability—that is, performing a designated function under stated conditions for a specified period of time; preciseness—that is, clearly written and easily understood; efficiency—that is, doing what it has been specified to do; and cost effectiveness.

Jorgens [2] has listed four categories of software and host devices:

1. Software resident in a specific medical device not intended to be altered by the user and required for the device to function. The software performs functions previously accomplished by other components and is actually a modernization of old technology.
2. Software temporarily installed in a specific medical device that has the capacity to alter the function or performance of the device. This category includes software in read-only-memory (ROM), programmable read-only-memory (PROM), tape, disk, and similar media.
3. Software designed for use on a single general purpose computer, that is, a computer not specifically dedicated to one particular medical device. This includes software that runs on specified computers such as an ECG arrhythmia detection program.
4. Software designed for use on several general-purpose computers. Medical programs that can be used on a variety of computers, such as those for patient record keeping, exemplify this type of software.

As the complexity of medical equipment increases and health-care delivery becomes more dependent on computer technology, it is paramount that manufacturers ensure the safety, effectiveness, and reliability of medical software.

III. THE CASE FOR SOFTWARE QUALITY ASSURANCE AND RELIABILITY

Software designed to control a medical function must be safe, effective and reliable for its intended use. These criteria may at times conflict with business decisions. For example, because of marketing conditions, there may be pressure to ship a software-controlled device within a certain time frame. This may put constraints on the software validation activity. In another situation, there may be tradeoff between the timeliness of coding without using structured design techniques versus the cost in time, money, and reputation if software errors must be detected and eliminated in the field. In all cases, safety, effectiveness and reliability must not be compromised. One method of assuring this is the establishment of a software quality assurance program.

Software quality assurance is a planned and systematic pattern of activities performed to assure that the procedures, tools, and techniques used during software development and modification are adequate to provide the desired level of confidence in the final product. The purpose of such a program is to ensure that the software is of such quality that it does not reduce the reliability of the device.

Assurance that a device works reliably has been classically provided by a test of the product at the end of its development process. Due to the nature of software, however, no test appears sufficiently comprehensive to adequately test all aspects of the program, especially when the program is one million lines of code in length or larger. Software quality assurance has taken the form of directing and documenting the development process itself and includes a system of checks and balances. Activities that comprise such a program include developing a software specification that details what the software is expected to do; implementing a structured approach to the design, including choice of model, language, and design reviews; detailing the interaction of the various software modules and the interaction of the hardware and software; developing validation protocols and reporting test results; and implementing reliability activity.

The goal of the reliability activity is to assure that the software does not reduce device reliability. There are four reasons to assure the software is reliable: safety, cost, the marketplace, and the Food and Drug Administration (FDA).

The primary criterion for any medical software package is the safety of the patient and the user. The software, whether during normal functioning or in a failure mode, must never cause harm to the patient or user. Software must be designed to be safe, tested to assure this during normal use and anticipated misuse, and must be fail safe. Liability cases, where safety was not a priority, are becoming more frequent and costly.

Software must be designed in such a manner as to be cost effective. It was estimated [3] that in 1955 approximately 18% of the total development cost of a software product was attributable to software, while 82% was attributable to the hardware (Fig. 1). Due to the many technological advances in hardware, this trend has reversed itself. By 1985 82% of the total development cost was attributable to software.

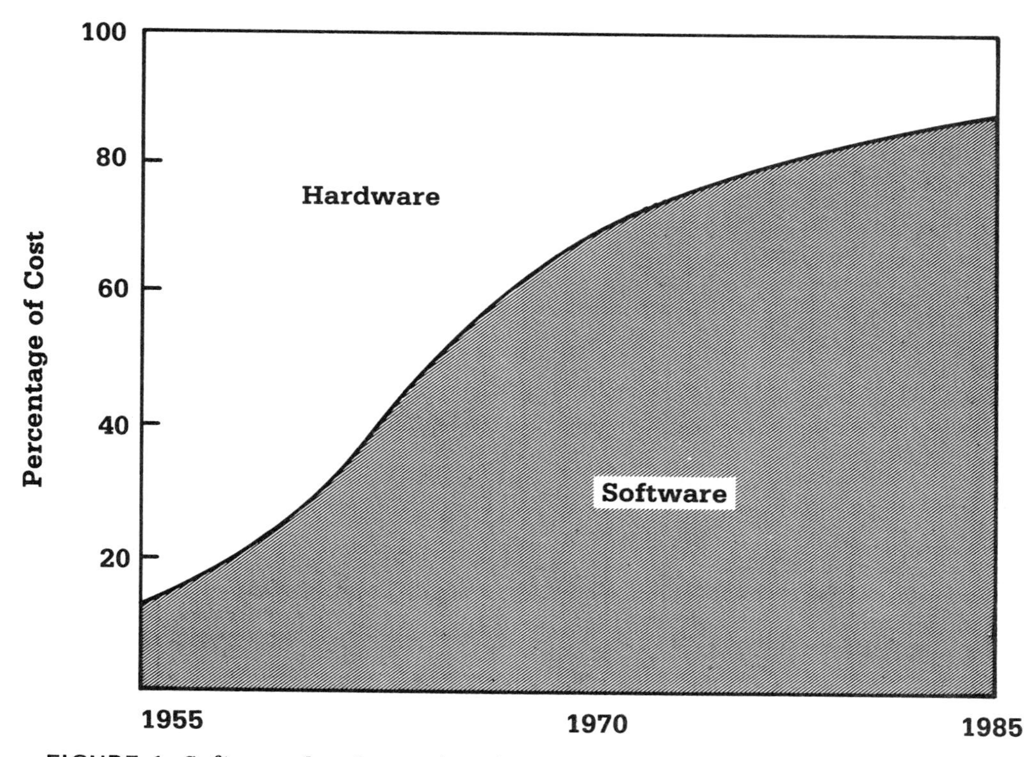

FIGURE 1 Software development cost.

Poor design can be costly in terms of detecting and correcting errors. It has been estimated [3] that the cost to detect and correct a software error in normal use is approximately 100 times greater than if the error had been detected and corrected in the design phase (Fig. 2). Current literature advocates the avoidance of errors through the use of techniques such as structured design rather than detection and correction after coding.

Software errors are costly in the marketplace. They are not only an inconvenience to the customer, but serve to damage the reputation of the manufacturer as well. As reputation suffers, so does market share.

Lastly, the FDA has heightened its regulatory scrutiny with regard to software products, particularly since 1985.

IV. THE FDA AND MEDICAL SOFTWARE

The FDA has been interested in the safety and reliability of software in medical devices for nearly a decade. Agency concerns fall into three major categories: medical device software, software used in manufacturing, and software information systems used for clinical decision making. The issues involved with each category differ.

In the case of device software, the FDA is responsible for assuring that the device utilizing the software is safe and effective. It only takes a few alleged serious injuries or deaths to sensitize the FDA to a particular product

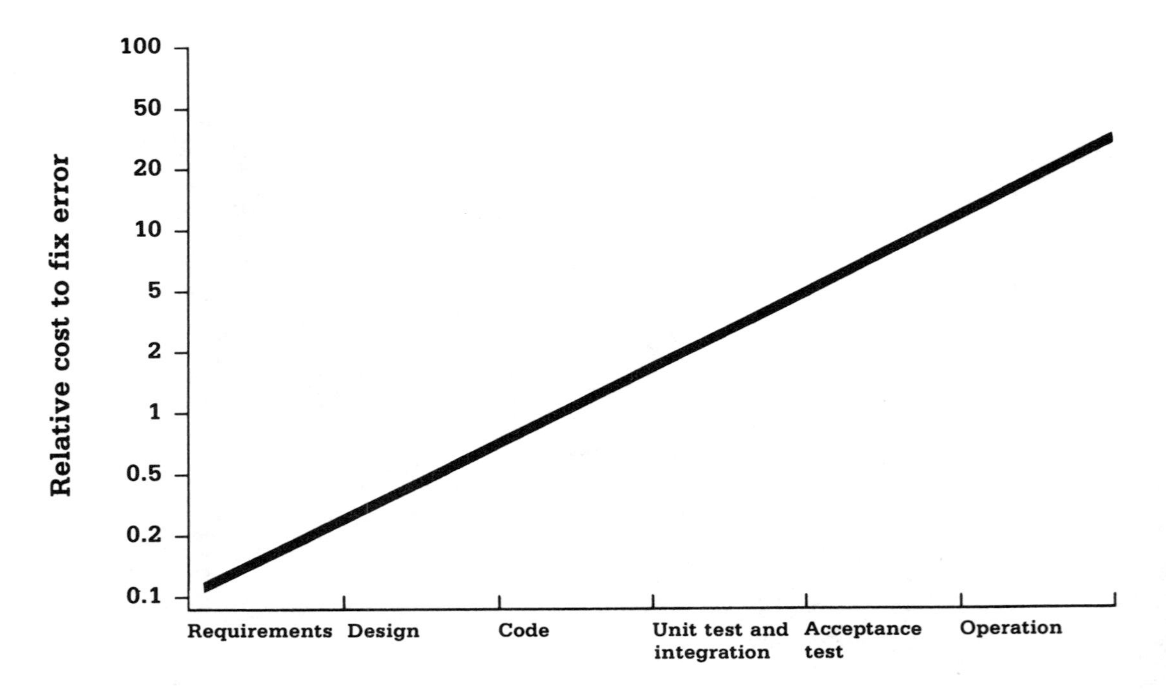

FIGURE 2 Relative cost of software correction.

or generic component that deserves attention. The FDA's review of medical device reporting (MDR) incidents and analysis of product recalls has convinced the agency that software is a factor contributing to practical problems within devices. In a much publicized 1985 incident, software in a radiation treatment therapy device purportedly caused a lethal overdose. Other problems noted in the literature have been calculation errors in programs for diagnostic use and misprogramming of the operating instruction for lasers, infusion devices, and electrosurgery devices.

In 1986 the FDA reviewed all of the reasons for recalls of medical devices from 1983 to 1985. The agency discovered that of the 41 product recalls related to software, 37 were caused by software errors designed into the program prior to production. From these data they concluded that an "increased" emphasis on the qualification of design and software validation might have prevented these recalls. The agency also observed that four of the 41 recalls were due to errors in revisions of the original software. From this the FDA concluded that "compliance with GMP change control procedures . . . should have prevented these recalls."

While these facts may be reason for increased regulatory attention, the FDA, unfortunately, has not placed the data into proper perspective with regard to the number of problems that might have occurred with medical devices. More importantly, the agency has not compared the positive effects of computer and software technology to the problems that were associated with precomputer technology.

When software is used during manufacturing, the FDA is more concerned with whether or not the software controlling an automatic drill press, sterilization chamber, or automatic tester is performing as expected. In this instance, the FDA's perceptions are rooted in experiences with the good manufacturing practice (GMP) inpsections of pharmaceutical manufacturers, where computers are heavily depended upon for control of manufacturing processes. There are far fewer incidents (as noted above) of device or manufacturing problems traceable to flaws in manufacturing software. Nevertheless, GMP inspections over the last several years have focused intensively on validation of software programs used in industry for control of manufacturing operations.

With regard to stand-alone software used to aid clinical decision making, the FDA is concerned with hypothetical problems rather than extensive records of adverse incidents. While most commercially available health care information systems replace manual systems that had a far higher potential for errors, the FDA believes that regulations should apply to the kinds of systems that may influence clinical treatment or diagnoses. The FDA has observed academic work on "expert systems" used by medical professionals and is concerned that such systems may be commercialized without sufficient controls.

The parallel concerns cited above have combined to trigger increasing regulatory attention to software issues and to the September 1987 FDA publication of a draft policy for the regulation of computer products [4]. The FDA has subsequently released a guidance document for field inspectors conducting GMP inspections [5] and a guide for product approval reviewers. The latter includes questions on software that should be asked duirng 510(k), premarket approval, and investigational device submissions [6].

The significant contributions of software in improving the overall reliability of health care through product innovations and data management must not be frustrated through misplaced regulatory preoccupation with 100% error-free programs. The FDA must balance the positive aspects of software with the negative effects of undue regulation such as delayed product introductions, stifling of small business, and increased cost.

V. THE BASIS FOR REGULATORY CLASSIFICATION

The Medical Device Amendments of 1976 to the Federal Food, Drug and Cosmetic Act (hereafter referred to as the Act) gave the FDA the responsibility and authority to assure that medical devices are safe, effective and reliable. The basis for regulatory requirements for software derive from this 1976 law.

Medical devices are divided into three classes with regulatory requirements specific to the individual calsses.

Class I devices are subject to only the general controls in the Act and to a comprehensive set of regulatory authorities applicable to all classes of devices. These include premarket notification, registration and listing, prohibitions against adulteration and misbranding, and rules for good manufacturing practices.

Class II devices are those that will eventually be subject to the requirements of a performance standard in addition to the requirements of Class I.

Class III devices are those for which premarket approval is or will be required. This class is represented by devices that are life-sustaining or life-supporting.

Software classification, as listed in the FDA draft policy, follows a similar pattern to devices. Medical software is divided into three classes with regulatory requirements specific to each.

Class I software is subject to the Act's general controls relating to such matters as misbranding, registration of manufacturers, record keeping, and good manufacturing practices. An example of Class I software would be a program that calculates the composition of infant formula.

Class II software is that for which general controls are insufficient to provide reasonable assurance of safety and effectiveness and for which performance standards can provide assurance. This is exemplified by a computer program designed to produce radiation therapy treatment plans.

Class III software is that for which insufficient information exists to assure that general controls and performance standards will provide reasonable assurance of safety and effectiveness. Generally, these devices are represented to be life-sustaining or life-supporting and may be intended for a use that is of substantial importance in preventing impairment to health. They may be implanted in the body or present a potential unreasonable risk of illness or injury. A program that measures glucose levels and calculates and dispenses insulin based upon those calculations without physician intervention would be a Class III device.

VI. REQUIREMENTS FOR REGULATION

The FDA's Draft Software Policy states "to the extent that computer products used in medicine are intended to affect the diagnosis and treatment of patients and are medical devices, the Food and Drug Administration must provide reasonable assurance that these products are safe and effective."

Computer products are subject to regulation as medical devices when they meet the following definition [4]:

> an instrument, apparatus, implement, machine contrivance, implant, in vitro reagent, or other similar or related article, including any component, part or accessory, which is . . . intended for use in the diagnosis of disease or other conditions, or in the cure, mitigation, treatment or prevention of disease in man or other animals . . . intended to affect the structure or any function of the body of man or other animals.

When a computer product is a component, part, or accessory of a product recognized as a medical device in its own right, the computer component is regulated according to the requirements for the parent device unless the component of the device is separately classified. This would include any computer product that is intended to have a direct interface with a medical device or one whose primary function is to provide input data intended to control the functioning of a medical device.

Computer products that are medical devices, and not components, parts, or accessories of other articles that are themselves medical devices, are subject to one of three degrees of regulatory control depending on their characteristics. These products are regulated with the least degree of control necessary to provide reasonable assurance of safety and effectiveness.

Computer products with uses excluding competent human intervention will not be exempt from the premarket notification requirement and manufacturers will be required to notify the FDA prior to marketing.

Computer products that are substantially equivalent to a device previously classified will be regulated to the same degree as the equivalent device.

Those devices that are not substantially equivalent to a preamendment device or that are substantially equivalent to a Class III device are regulated as Class III devices. The safety and effectiveness of the new Class III device must be demonstrated by the manufacturer before marketing, usually through a premarket approval application (PMAA). If the manufacturer believes a PMAA is unnecessary, he/she may submit a petition to reclassify the product to a lower class.

VII. EXEMPTIONS FROM REGULATION

The statutory definition of a medical device, if read literally, has a very broad scope. It could encompass virtually every item in a hospital including calculators, medical textbooks, and a variety of record-keeping forms. The FDA has historically not regulated all articles used in a health-care setting even if they are, in the broadest sense, used to diagnose or treat disease. Similarly, the FDA is not required to regulate every computer product that is used in a health-care setting, although some could be technically defined as a medical device.

The FDA's proposed policy on the regulation of medical software recognizes that the agency's regulations and authorities do not and should not apply to computer products used only for traditional library functions, general accounting, communication functions, or educational purposes. Computer products that perform these functions will not be regulated as medical devices.

All other software used in health care will be considered a medical device subject to at least some degree of regulatory oversight. Some software will be exempt from the most rigorous regulations. The FDA has proposed that manufacturers of the following categories of medical computer products should be exempt from the requirements for registering their establishments and listing their products with the FDA, for reporting adverse effects under the medical device reporting regulation, and for premarket notification.

1. General purpose articles [21 CFR 807.65(c)]: A general purpose article is a product that is not labeled or promoted for medical uses but that by virtue of its application in health care, meets the definition of a medical device. These devices either pose little or no risk, or are appropriately the sole responsibility of the health-care professionals who have used them in medical applications.
2. Computer products manufactured by licensed practitioners for uses in their practice [21 CFR 807.65(d)]: This exemption applies to "Licensed practitioners including physicians, dentists and optometrists who manufacture or otherwise alter devices solely for use in their practice."
3. Computer products used in teaching and nonclinical research [21 CFR 807.65(f)]: This exemption applies to "Persons who manufacture, prepare, propagate, compound or process devices solely for use in research, teaching or analysis . . . ."
4. Computer products including, for example, many software products known as "expert" or "knowledge-based" systems that are intended to involve competent intervention before any impact on human health occurs (e.g., where clinical judgment and experience can be used to check and interpret a system's output) will also be exempt from registration, listing, and premarket notification. This will be accomplished by the FDA through the routine exemption-granting procedure.

All four categories of software programs are, however, still subject to the misbranding and adulteration provisions of the Act and possibly subject to GMP regulations. The Health Industry Manufacturers Association (HIMA) has commented to the FDA on the agency's draft proposal that products that "provide the opportunity for competent human intervention" should not only be exempted from most FDA regulations, but should not be considered as medical devices at all. The imposition of regulation on these categories of medical software would have only one inevitable result: to slow down innovation and thus delay improvements in patient care. The only justification for this delay is that without it, greater harm will occur due to premature introduction of new products. This justification is worth attention sometimes. It is of little utility, however, when applied to software that only collects, stores, analyzes, and provides information more rapidly and more reliably than the human mind without substituting human judgment. The FDA's policy to exempt these programs from most regulation is a positive step. It is not total protection from unnecessary regulation that would result from an FDA decision that this kind of software is not a medical device, however.

VIII. GMPs AND INSPECTIONS AS A CONTINUING ISSUE

The current good manufacturing practice (GMP) regulation requires the establishment of an organized quality assurance program for all software-controlled test fixtures and tools used in manufacturing. Each tool or test fixture should have a specification that details what it is supposed to do, a test protocol that details how the function will be validated, and a test report that summarizes the results of the testing. In addition, each tool or test fixture is subject to revision level control procedures which assure the traceability of each revision.

The tools and test fixtures can be examined during a routine GMP inspection. The amount of detail asked for during the inspection depends upon the particular inspector and his or her philosophy of software quality assurance and reliability.

Current regulations do not address how to inspect medical device software. The FDA has inspectors in the field who have been trained in computer and software technology. The goal of the training is to expose the field investigators to computer terminology and basic operations and to provide more indepth exposure to computers used in manufacturing [7]. The training course is based on the software life cycle and includes requirements and specifications, design, coding and debugging, verification and validation, installation and checkout, and operation and maintenance. The course also incldues problem areas, sensors, interfaces, security, applicable GMPs, and software quality assurance. Training materials and guidelines reference the American National Standards Institute and the Institute for Electrical and Electronic Engineers software standards.

In 1987, the FDA produced a guide for inspectors that attempted to provide a checklist of questions to ask and material to inspect when examining medical software in or as a medical device. After comments from HIMA, the document was recalled and reissued several months later without reference to GMPs or inspections [8].

The GMP issue is further complicated because the FDA exercises little control over inspectors. Most are free to pursue whatever audit trail they believe will give them the confidence they seek during an inspection. Inspectors have brought regulatory action against firms who were not conducting software quality assurance activities the way the inspector thought was appropriate.

Two divergent philosophies exist regarding medical software and GMPs. On the one hand, the FDA perceives software as a component of a medical device, and as such treats it as it has hardware components. The amount of detail covered and requested will depend upon the education and experience of the investigator.

On the other hand, industry believes that software development is a research and development activity; thus, it cannot be inspected under current regulations. In addition, software documents, if made available to competitors via the Freedom of Information Act, would lead to a loss of competitive advantage. Manufacturers have declined to release such documents to an inspector because of this. Many believe that software GMP inspection should begin at the process of copying the program onto the appropriate media, as this is the starting point for manufacturing.

An FDA Center for Biologics Evaluation and Research (CBER) memorandum concerning GMPs for blood products that was issued to all registered blood establishments in September 1989 illustrates the regulatory complexities that may confront certain segments of the software industry in the future. The memorandum states that "when personnel depend on the computer system in making decisions regarding the suitability of donors and the release of blood or blood products for transfusion or further manufacture, the computer system is considered to have control functions." The CBER memo requires that the user ensure the consistency of control function performance and maintain documentation of validation criteria and procedures. In addition, computerized blood establishments msut incorporate certain items into their standard operating procedures, to include "validation protocols to independently test and challenge the critical functions and procedures."

The CBER memo, while addressing a very narrow regulatory area, may set a precedent for computer technology applications in the future. In this case, CDRH will regulate the manufacture of blood bank software, while CBER will regulate software after installation. This type of arrangement shifts significant responsibility of software operations to the user. CBER has indicated to HIMA that the agency will hold the manufacturer partially responsible for user compliance with software-related GMPs.

Industry organizations such as HIMA and the National Electrical Manufacturers Association (NEMA) are working to resolve the differences in opinion between the FDA and industry and are assisting the agency in producing regulatory documents that are appropriate and acceptable.

IX. SOFTWARE PRODUCT SUBMITTALS

The two primary routes to market for software devices are the 510(k) route and the premarket approval path. The advantage of the 510(k) notice provision of the Amendments is that a new device that is substantially equivalent to another device in commercial distribution prior to May 28, 1976, can be accepted for distribution with the submission of much less information than would be required for a device using the premarket approval route. The average time for the FDA to review a 510(k) is approximately 20 hours.

The premarket approval provision of the Amendments that applies to devices that are new and not substantially equivalent involves a much more detailed review of the product to include data from clinical studies and detailed analysis by FDA reviewers. The average time to review a PMA submittal is approximately 1300 hours.

The FDA's expectations on software product approval submittals, as all software regulatory initiatives, are in a state of change. The FDA issued a draft guidance on computer-controlled devices for product approval reviewers in 1988 [6]. The guide establishes classes of software based on safety concerns if the software device were to fail. It also establishes three levels of concern ranging from minor to major. Each class is subject to certain documentation requirements and to certain questions related to the level of concern.

The Medical Device Industry Computer Software Committee (MDICSC), a combined HIMA/NEMA task force, responded to this draft in November 1989 by rewriting the draft to incorporate criteria important to industry. The MDICSC noted that while industry agrees with the concept of level of concern, the FDA draft "presents ambiguous definitions of the various levels." To define these levels, the MDICSC recommended that:

1. Software should not be a separate component to be reviewed on its own. To establish a level of concern, the device should be considered first, independent of the role of software. The device level of concern then serves as a cap for the level of concern for any software element.
2. Different software functions may have different levels of concern, with none having a greater level of concern than the overall device. Determining factors are the role of the software function in presenting or controlling a hazard and the degree to which functionality is implemented by hardware, software, or a combination thereof.

Of particular concern to industry are parts of the draft that focus on details of the software development process. Emphasizing that 510(k) premarket notification clearance is based on the safety and efficacy of a device, the MDICSC commented that the 510(k) review "is not intended to scrutinize or approve development processes . . . . Scrutiny of a particular element of a software development process is justified only if the analysis of potential hazards identifies that element as a means of validating a hazard control." Moreover, because software development processes evolve over the life cycle of a product, the MDICSC believes that a 510(k) review is only "a snapshot in time" and should not be used to determine if a device is safe or effective in most instances.

Industry emphasized to the FDA that the MDICSC rewrite of the reviewers' guidance was a draft document only. The committee will meet with the agency for further clarification and discussion of the document in early 1990.

Reviewers are currently using the draft guidance as policy. Questions have been asked of industry quoting page and paragraph number from the draft. Some of the questions that have been asked include:

What equations of the formulas are used to generate the results produced by the instrument?
If a result is outside the expected range, what does the instrument do?
Can the user change the software in an instrument, and if so, what change controls exist?
Provide a narrative that explains the methods and procedures used for validation, verification, and certification.
What self-testing features exist within the software?
Describe recommended user practices for monitoring computer performance.
When the user turns on the instrument, how does he know it is working properly?

X. INTERNATIONAL REGULATIONS

The United States is ahead of other countries in establishing guidelines for medical software development. There is, however, movement within several international organizations to develop regulations and guidelines for software-controlled devices.

The Department of Health and Social Services (DHSS) in the United Kingdom is a regulatory agency similar to the FDA. The DHSS and FDA currently have a reciprocity agreement wherein the FDA and DHSS will accept each other's inspection reports and documentation to satisfy their own requirements.

The DHSS has also published a guideline addressing the functional safety of software-controlled devices [9].

Technischer Uberwachungs-Verein in Germany has also developed a guide for safety in software-controlled devices [10]. The document covers techniques in software design, requirements for documentation, hardware and memory requirements, and CPU (central processing unit) tests. It is a much more detailed guide than the DHSS document.

The International Electrotechnical Commission (IEC) is the group that has developed guidelines that are currently used in Europe. IEC-601 specifically addresses medical equipment. There are currently working gorups within the IEC addressing medical software issues.

The Canadian Standards Association published four documents on software for new and previously developed software [11-14]. These guides, published in January 1989, are the first Canadian documents on software.

Software regulations are also being addressed as part of the formation of the European Community (EC) scheduled for 1992. One of the first documents published for comment in this regard is the ISO 9001/2 document on software [15]. Though not specifically addressing medical software, it contains requirements that medical software must meet.

XI. THE FUTURE OF SOFTWARE REGULATION

Unfortunately, the FDA has limited resources and only a small number of staff members who have sophisticated computer technology expertise. The political and legislative environments surrounding the agency also have their effects. These elements create uncertainty about the FDA's ability to keep pace with technical advances. It also calls into question whether the agency will have an adequate scientific foundation upon which to base future regulations in the software area. As such, standards groups, industry, and academia alike will have an even greater obligation in the future to provide the FDA with timely information on the direction software technology is taking.

A key factor is the rapid emergence of electronic linkages between devices and information systems. A myriad of standards are being developed for open architecture systems without traditional electronic communications barriers. Theoretically, all software-driven devices and information management systems will have the capability to communicate between themselves and each other. Most of the standards activities in this arena are progressing without consideration for their regulatory implications.

At the present time, the FDA does not have a consistent policy on how such linkages will be regulated. For example, and as described earlier, CBER currently considers linked devices and information systems used in blood banks as total systems, and not separate components that are in communication with each other. Such linkages will become more commonplace throughout health care, and dozens of devices and information systems from different manufacturers will be interconnected. It will become impractical to regulate these as single systems. The FDA will need to recognize that it is the interface between the components that is critical. The easiest way to accomplish this is to consistently include FDA officials in standards activities. This may help to lessen their concerns, keep them up-to-date, and provide them with a tangible sense of co-ownership about the standards that are produced.

The establishment of regulations and standards for medical software that will satisfy both the producer and the regulator has had a good beginning. However, the road to completion is long and filled with many hazards. It is to the credit of leaders within medical industry and government that the producers and regulators continue open discussion to resolve difficult issues. It is only through awareness of each other's problems and concerns that a meaningful solution can be achieved. As long as those discussions are based on logic, professionalism, and practicality, both sides will be satisfied with the results. The real winner will be the patient who will benefit from the enhanced capabilities which software can provide, while being assured that the software device will function safely, effectively, and reliably.

REFERENCES

1. R. S. Pressman, *Software Engineering*, McGraw-Hill, New York, 1987.
2. J. Jorgens III, Computer Hardware and Software as Medical Devices, *Medical Device and Diagnostic Industry* May (1983), p. 62.
3. B. W. Boehm, Software Engineering, *IEEE Transactions on Computer C-25* 2 (1976), p. 1226.
4. Food and Drug Administration, *FDA Policy for the Regulation of Computer Products*, Washington, D.C., 1987.
5. Food and Drug Administration, *Medical Device GMP Guidance for FDA Investigators*, Washington, D.C., 1987.
6. Food and Drug Administration, *Reviewer Guidance for Computer-Controlled Medical Devices (Draft)*, Washington, D.C., 1988.
7. R. M. Garwood, FDA's Viewpoint on Inspection of Computer Systems, *HIMA Conference on the Regulation of Medical Software Proceedings*. Rockville, Md., 1987.
8. M. Schrage, FDA Sees Regulation for Medical Software, *St. Louis Globe Document*, August 28, 1984.
9. Department of Health and Social Services, *Functional Safety of Programmable Electronic Systems: Generic Aspects*, London, 1987.
10. TUV Study Group on Computer Safety, *Microcomputers in Safety Technique: An Aid to Orientation for Developers and Manufacturers*, 1987.
11. Canadian Standards Association, *Quality Assurance Program for the Development of Software Used in Critical Applications*, Toronto, Ontario, 1989.
12. Canadian Standards Association, *Quality Assurance Program for the Development of Software Used in Noncritical Applications*, Toronto, Ontario, 1989.
13. Canadian Standards Association, *Quality Assurance Program for Previously Developed Software Used in Critical Applications*, Toronto, Ontario, 1989.
14. Canadian Standards Association, *Quality Assurance Program for Previously Developed Software Used in Noncritical Applications*, Toronto, Ontario, 1989.
15. International Standards Organization, *Quality Systems—Guidelines for Software Quality Assurance (Draft)*, Geneva, Switzerland, 1989.

40

Maintaining Your Competitive Edge at Home and Abroad

MARCEL P. FAGES *Association for Manufacturing Excellence, Wyoming, Ohio*

I. INTRODUCTION

Many U.S. manufacturing industries today are struggling for their very survival. The main reason for this precarious position is the tremendous strides other countries have made. They have forged ahead of the United States in producing attractive goods at high quality and attractive prices, thus resulting in a tremendous market in the United States and for other world users of these goods. The market share of these foreign-made goods is increasing at an astounding rate. A vivid example is automobiles. Japan's U.S. market share of automobiles has risen from zero to 28% in the last few years. After World War II the United States was the major producer of automobiles, while Japan basically did not produce automobiles until 1950. Today Japan outproduces the United States in total automobile production, 12.3 million to 11.6 million, respectively.

When you consider all the varied products the United States produced in the past, it is disheartening how few of these products we manufacture in the United States today. These products include, for example, radios, watches, binoculars, textiles, bicycles, trucks, steel, televisions, cameras, clocks, shoes, motorcycles, ships, tape recorders, telescopes, and timers.

It may seem to some that the United States is abdicating its leadership as an industrial giant. As we rely more and more on the manufacturing capabilities of other countries, our position as a world leader continues to diminish. This is true not only in manufacturing but in economic, social, and possibly military areas, as we lose critical high-tech industries. As we lose manufacturing jobs and replace them with service jobs the very fiber of our society is weakened.

II. FACTORS CONTRIBUTING TO LOSS IN U.S. LEADERSHIP

A basic contribution to America's loss of leadership in manufacturing, during the past decade, has been the inability of four major institutions to integrate a common national strategy. These institutions are government, business, labor, and education. If there is one thing that is overwhelmingly obvious when a U.S. businessman visits Japan, it is the ability of the government, labor, and business to work together. Their major efforts are directed toward designing a strategy for leading industries, with the objective of making them world leaders.

The Japanese educational system, prior to the university level, is one of the best in the world, and visitors to Japan are continually impressed by this system. Japan continues to lead the world in teaching science and mathematics to preuniversity students. Eighth graders in Japan attend 243 days of school per year, versus 180 days in the United States.

It becomes imperative that the United States build an effective coalition between labor, business, and government. This coalition should be directed toward developing a national strategy for industrial leadership in the world today. Each of these institutions will have to make compromises in order for the United States again to become an industrial leader in the world. Executives will have to settle for less in order to send the right message to labor. Labor must also compromise and not take itself out of the world labor market, and government must be more of a partner and not an adversary of business. Although this will not be easy, it must be done in order for the United States to compete with its major competitors who already effectively use this approach.

By looking at some recent statistics, we may better appreciate the tremendous challenges facing us in our struggle to regain the leadership of the industrial world.

These six points are worth considering [1]:

1. The United States is now borrowing $100 billion a year from abroad.
2. We are paying $360 million per day in interest on the national debt.
3. The 1987 trade deficit was $170 billion dollars.
4. The financial security of the nation is currently being threatened.
5. We are in the most alarming economic crisis since the depression.
6. We are a debtor nation for the first time since 1918.

Some government officials, in cooperation with six other industrial countries, suggested reducing the value of the dollar. But recently, although the dollar has decreased more than 60% compared to the yen, the trade gap has not yet been significantly reduced. Other parties have cried for protectionism, resulting in the 1988 trade law. It is highly questionable how much this will effect our ability to compete in the industrial world. Today's major industries must be competitive in the world markets; it is no longer safe to "be at home" or "stay at home." We must now compete with the world on the basis of our manufacturing technology as well as our design technology.

It should be made clear that protectionism has four very dangerous adverse effects on our society. First, it takes away the incentive from industrial leaders to be competitive in the market place; second, it subsidizes high labor rates; third, it creates retaliation from other countries. The best example of retaliation was when the United States established quotas on steel, causing the affected countries to stop buying our agricultural goods, thereby causing U.S. farmers to suffer.

Our position as the unchallenged leader of the industrialized nations is no longer secure, due to old plants, too little automation, high labor rates, poor quality, international competition, lower worker productivity, and excess inventories.

III. STRATEGIES FOR MEETING THE CHALLENGE

During the last 15 years, we have been bombarded with information about the success of the Japanese management system. Japan's success can no longer be questioned, since it is now a major industrial competitor throughout the world. We are well aware of all the products where Japan is unsurpassed with respect to quality and price. Japan's amazing progress during the past few years is well accepted by all its major competitors. The pressing question now in the industrial world is, how have the Japanese accomplished this phenomenal success? Businessmen, government leaders, and educators have flocked to Japan during the past few years to search for the secret of Japan's success.

A major contributor to Japan's success can be placed under an umbrella called "Just in Time" and "Total Quality Control." This total program can be viewed as strategic and tactical, with three major areas being strategic:

A. Management attitude, philosophy, and commitment
B. Quality
C. Employee involvement

A. Management

Top management has the prime responsibility for forging manufacturing into a competitive weapon by formulating a manufacturing strategy based on the total philosophy and strategy of the company. No longer is manufacturing a silent and unequal partner to the other functions of the business. This is a fact that will be difficult for many top American executives to accept because, many times, these top executives are financing or marketing oriented with limited operational experience. This is in contrast to their Japanese peers, who are engineering and manufacturing oriented.

In order for our industries to become more competitive in the world, it will require a complete change in the attitudes and philosophies of company chief executive officers, as well as the employee on the floor. This change requires a major commitment from top management. One of the most difficult obstacles for management will be to become long-term oriented. Short-term results have been the driving force behind American industries. This emphasis will have to change, as too much short-term planning has adversely affected the strategy of our industries.

In order to make significant changes in business cultures, management will have to become more involved than ever before. Attempting to delegate the major changes required will not produce the results needed. Management will need to spend more time with customers and employees to exhibit its commitment to change.

Some U.S. corporations have successfully competed in the world market during the past few years as a result of top management's commitment and involvement in change. Many of these top managers have been personally involved with hourly employee groups in reviewing problem-solving activities. This is done by a very structured and disciplined process, where these employees feel management is truly committed to improvement.

B. Quality

"Some U.S. industries will never be able to compete with the Japanese in quality." "Some U.S. industries will be able to compete quality wise within 5 to 7 years if then."

Both of the above statements are from two world authorities on quality, who have worked closely with the Japanese in establishing their quality strategy in the 1950s. Both are very strong statements. Progress in improving quality has come very slowly with some industries in the United States during the past few years.

The major reason for these observations on quality by these leading authorities is the lack of full commitment and dedication by top management. These changes would require a complete change in attitude, philosophy, and culture of an organization, required at all levels of management. During the past 5 years, there have probably been as many failures as successes in companies attempting to change the culture for improved quality. Top management has enthusiastically started new programs, but lacked the discipline for involvement and follow-up. This lack of attention to detail has often turned off the work force to its commitment to improvements. As soon as management's short-term results led to adverse effects, it lost interest in its quality commitments.

Approximately 85% of quality problems are the responsibility of management and not the employees on the shop floor. Forty percent of quality problems are caused by poor product design, 30% by the manufacturing process, and approximately 20% by suppliers' defective goods. All of these corrections require leadership by management, and yet management still looks to the work force to correct quality problems. Therefore, quality improvement programs have to be directed to the process, product, sourcing, and then training of personnel. Greater emphasis has to be given to designing quality into the process, and less in trying to inspect quality into the product.

A quality strategy involves creating change, communication, and use of new tools. Creating change involves changing the attitudes of all employees to eliminate a certain mind-set of doing things the same way as in the past. This is not a short-term activity and is considered a strategic approach, requiring greater teamwork than has ever existed before. When it comes to communicating the quality strategy, top management must play an active role. Only through management commitment and visibility will the work force also make its full commitment. Related to communication is the importance of education and training required to institute an effective quality program.

When it comes to new tools, greater use of such things as statistical process control, Pareto charts, fish-bone diagrams, and other similar applications is important. These tools are some of the basics used by the Japanese to improve their quality to the high level that exists in many of their products today.

C. Employee Involvement

This subject is probably the greatest challenge American industries have today. For many years, we have looked on the worker as a doer and not a thinker. In order to fully utilize our greatest asset, the work force, it must be included in the process of problem solving. One of our major competitive countries, Japan, has utilized this so well that the degree that employees are involved in decision making and problem solving is absolutely unbelievable. As example of this is a major world competitor, Toyota, in Japan, which receives 38 suggestions per employee per year at a 90% acceptance rate. Another company claims to have received 100 suggestions per employee per year. This compares with the U.S. industry average of one suggestion for every six employees per year.

Is it any wonder that these companies are world leaders in their product lines? They completely overwhelm their U.S. competitors. To change the utilization of our work force will require a complete work culture change and a long-term commitment. It will not be easy to change our view of the work force after 50 some years of neglect of the employee's input. To involve the work force into a partnership of change will challenge the most astute of U.S. management. Many of our first- and second-line managers may not be able to make the transition. It will require relinquishing power that they have exerted for so many years. They will now have to look on employees as partners in improving operations to become a world-class manufacturer. In spite of these difficulties, some U.S. companies have moved successfully in this direction during the past few years. The U.S. companies that have been successful in more fully utilizing their employees have either regained world leadership or have truly become world-class manufacturers.

IV. IMPROVING COMPETITIVENESS

The following four items can be considered tactical to improve competitiveness: supplier programs, preventative maintenance, reducing set-up times, and reducing inventories.

A. Supplier Programs

It is always advisable before attempting to improve suppliers' performance to do a good job in-house to be a respectable supplier to our customers. This way, we can share with our suppliers the kind of improvements we have instituted. One prime emphasis is to have small numbers of suppliers, or, better yet, single sourcing wherever possible. American companies who have been successful in pursuing fewer suppliers have had tremendous success in improving quality, shortening delivery time, and reducing costs. Another area to be pursued is that of obtaining long-term contracts with renegotiable clauses to take advantage of cost reductions. Stable long-term relationships have proven to be a very effective tool, and emphasis should be applied to quality certification by all suppliers. Short lead times to rolling forecasts should be mandatory. Daily and weekly deliveries instead of monthly ones should be pursued continuously. A program for on-site audits should be applied wherever possible. Local sourcing wherever possible should be pursued whenever securing new vendors, and quality expectations should always be 100% with no compromising. Parts should be delivered on time with no inspection necessary. A vendor training program should be established whenever possible to assure that suppliers are adhering to required standards. Last, but not least, there should exist a close working relationship between vendor and user.

Each of the above programs will take time to institute, but they are absolutely necessary in order to be a world-class manufacturer. Some of our major international competitors have implemented each of these programs successfully and reaped the benefits of outstanding operational results.

B. Preventive Maintenance

In order to become more competitive, it is important to have the highest possible use of all manufacturing equipment—in other words, minimum downtime of equipment. This must be approached through a greatly improved preventive maintenance program. The most effective way this has been done is by shifting responsibility for maintaining equipment to the operators using the equipment. They know best how the equipment is operating. Given proper training, education, and the feel of ownership of their equipment, they are better able to keep their equipment operating. Where possible, aircraft-pilot-style equipment checks should be installed. When it is practical for operators to do their own maintenance, preventive maintenance should be done on off shifts. It is important for operators to keep their own equipment clean and lubricated and to perform other miantenance when possible. Where effective preventive maintenance programs have been instituted, they have resulted in high product quality, long machine life, and avoidance of stoppages.

C. Reducing Setup Times

A very important element in being a manufacturing leader in the world today is the ability to respond quickly to the marketplace. This means minimum lead time, that is, the time required between when a customer desires a product and when it is delivered to him. The companies that can produce on demand will lead in their industries. One of the major elements contributing to long lead times is "setup time," the time required to change from one product to another in the manufacturing cycle. Some of our foreign competitors have reduced setup times from double-digit hours to single minutes.

A vivid example of this is the Indianapolis race track where pit crews can change a tire in seconds, while a passenger-car flat tire may take 20-40 min to change. The difference is that the pit crews have the proper tools at the right place when needed and the proper techniques to change the tire. Compare this to the person changing a passenger car's flat tire with minimum knowledge of how to make the change. This is what happens many times when operators are required to change dies and tools on manufacturing equipment.

It is possible to reduce setup time by as much as 95%. This requires being organized for the change, by eliminating all waste in the process and reducing the labor required to a minimum. This also requires workplace organization, having everything ready and in its place when the change is necessary. The Japanese program is called "single minute exchange of dies." This approach was the first major project Toyota used to introduce "Just in Time" to the automobile industry 25 years ago, and we all know how successful that has been.

Setup time reduction has received minimum attention during the past many years in American industry, and yet this is the most important element of reducing lead times.

D. Reducing Inventories

A modern Japanese proverb states, "Inventories are the root of all evil." I am sure most world management would agree.

The major need for inventories is to cover problems; therefore, inventories should be reduced to surface problems. This is a major strategy used by the Japanese. This has resulted in the highest inventory turns in the world. The Japanese operate with 20-30 inventory turns per year, compared to similar U.S. industries of 6-8 turns. This results in very low inventory costs, which give Japanese companies tremendous price advantage in the world market. U.S. management has to be encouraged to solve problems rather than cover them up with inventories. Some of the major causes of high inventories that need to be addressed through problem-solving programs are:

1. Quality problems
2. Excessive setup times
3. Improper scheduling
4. Unreliable suppliers
5. Machine breakdowns
6. Inflexible workforce
7. Bad housekeeping

Each of these causes have to be specifically addressed by management incorporating "employee involvement" activities. It takes significant time to train and educate to this new approach to problem solving. It also takes top management commitment to fully utilize all employees solving these problems to lower inventories.

V. CONCLUSION

In summary, American manufacturers will have to address each of the seven areas discussed above. These are:

1. Management—commitment, philosophy, attitudes
2. Quality
3. Employee involvement
4. Supplier programs
5. Preventive maintenance
6. Reducing setup times
7. Reducing inventories

In order to meet this challenge, it will take dedication, stick-to-it attitudes, patience, and a certain mentality.

There have been a number of U.S. industries who have attempted changes but failed because they did not have the above characteristics necessary to succeed. However, a number of U.S. companies have made great strides in becoming world-class manufacturers by addressing each of the suggestions made in this discussion.

REFERENCE

1. *U.S. News & World Report*, February 1988.

41

Accessing New Technologies: The Federal Connection

JEROME S. BORTMAN *Naval Air Development Center, Warminster, Pennsylvania*

NORMAN F. ESTRIN *Health Industry Manufacturers Association, Washington, D.C.*

I. INTRODUCTION

Companies looking for new technology, processes, services, and cooperative research and development (R&D) partners to stay competitive will often overlook the largest, nonprofit, tax-supported R&D resource in the United States—federal laboratories. However, with recently passed laws and a Presidential Executive Order, access to over 700 federal laboratories, employing one-sixth of the country's scientists and engineers with a $20 billion annual R&D budget, represents an opportunity that few companies should ignore. A *Wall Street Journal* article [1] headlined "Promising Partnership" reported that "the trickle of new technologies leaving the federal laboratories could turn into a steady and profitable stream."

The new public laws (PL96-480 and PL99-502) and Executive Order 12591 make technology transfer an explicit responsibility of all federal labs, reduces bureaucratic obstacles, and provides incentives to federal researchers to have their work commercialized. Federal laboratories are now authorized to enter into mission-consistent, cooperative research and development agreements (CRDA) with universities and the private sector, and negotiate licensing agreements for inventions made at the laboratories. This significant shift in government policy means that the "feds" are moving beyond the role of regulator or contracting agency to a new role as innovator and partner with industry and universities. The reason for this change is the recognition that public-private cooperation in the development and deployment of technology is necessary for the United States to remain competitive in the world marketplace. The Council on Competitiveness [2], an organization of leading executives from industry, organized labor, and higher education, has recommended that "U.S. companies must become as much hunters and gatherers of technology as they are generators" to remain competitive. This chapter describes the variety of technology transfer opportunities available and suggests some contacts and information resources to assist you.

II. OPPORTUNITIES

The saying "the opportunities are unlimited" certainly applies to the field of domestic technology transfer. Technology and expertise are available from federal laboratories in virtually every area of science and engineering, including:

Biotechnology
Medicine
Advanced materials
Microelectronics
Computers
Superconductivity
Communication
Manufacturing technology

The latest legislation, the Federal Technology Transfer Act of 1986 (PL99-502), encourages cooperative research and development agreements in which industry and laboratory researchers work together on projects of mutual interest. Other technology transfer mechanisms that can be tailored to individual situations include:

Exclusive or nonexclusive licensing
Technical assistance from laboratory personnel with unique expertise
Use of laboratory facilities where unique capabilities are present that are not commercially available
Personnel exchange programs in areas of mutual interest
Workshops, seminars, and conferences
Information dissemination

The publication "Putting Technology to Work" [3] describes specific case histories of federal technology transfer, including the way in which the transfer occurred and the participating laboratories. The examples include:

The spinoff of chromosome karyotyping systems, chromosome analysis systems, and systems for cytogenetic imaging based on image processing technology from NASA's Jet Propulsion Laboratory.
Technical assistance by the Los Alamos National Laboratory in solving a product development problem of a biomedical company working in laser angioplasty procedures.
Licensing of a sickle-cell anemia test kit by the Lawrence Livermore National Laboratory to a biotechnology firm.
Joint research and information dissemination of *in vitro* techniques and computer models to provide alternatives to using animals in toxicology testing (U.S. Army Chemical Research, Development and Engineering Center).
Technical volunteers from the Naval Underwater Systems Center and the Naval Air Development Center have participated in the retrofit of computer/display systems to aid severely handicapped individuals.

III. THE PEOPLE CONNECTION

Finding your way through the federal lab system to locate the right resource requires persistence and patience. The Federal Laboratory Consortium (FLC) for Technology Transfer, a network of technology transfer representatives from most federal laboratories, was formed to help facilitate technology transfer from the laboratories and to help industry connect with the appropriate resources [4]. A good way to approach the FLC is through the FLC Regional Coordinators (Table 1). The FLC Clearinghouse is a brokerage service that assists in the matching of user technical requests with appropriate federal research capabilities and can also be reached directly (although it primarily services requests from individual laboratories or the Regional Coordinators).

TABLE 1 Federal Laboratory Consortium for Technology Transfer

Regional Coordinators	
Mid-Atlantic Region: Mr. Nick Montanarelli DOD/SDIO/T/TNO Office of the Secretary of Defense Washington, DC 20301-7100 (202) 693-1562	Northeast Region: Dr. William Marcuse Brookhaven National Laboratory Building 475-B Upton, NY 11973 (516) 282-2103
Southeast Region: Mr. Robert M. Barlow Code HAOO Stennis Space Center, MS 39529 (601) 688-1929	Midwest Region: Ms. F. Sinclair Ingalls Wright R&D Center WRDC/XO Wright-Patterson AFB, OH 45433-6523 (513) 255-2788
Midcontinent Region: Mr. Robert Stromberg Sandia National Laboratories Organization 4031 Albuquerque, NM 87185-5800 (505) 844-5535	Far West Region: Mr. Charles Miller Lawrence Livermore National Lab. P.O. Box 808 L-795 Livermore, CA 94550 (415) 422-6416
FLC Washington, D.C. Rep: Mr. Lee W. Rivers 1550 "M" Street, NW, 11th Floor Washington, DC 20005 (202) 331-4220	
FLC Clearinghouse: Mr. Allan A. Sjoholm DelaBarre & Associates, Inc. 1007 5th Avenue, Suite 810 San Diego, CA 92101 (619) 544-9033	
FLC Administrator: Mr. Del M. DelaBarre P.O. Box 545 Sequim, WA 98382-0545 (206) 683-1005/1828	

PL96-480 required that each federal R&D laboratory establish an Office of Research and Technology Applications (ORTA), and PL99-502 elevated the ORTA function to the laboratory management level to emphasize its importance. Any of the FLC contacts or agency headquarters can provide listings and/or individual contacts at federal labs (ORTAs). Of course, if you know researchers at a particular lab, you should contact them directly. Table 2 provides a list of some ORTAs and other offices that should be of interest to the health industry when looking for specific technology transfer or technical assistance contacts.

IV. INFORMATION RESOURCES

A. P. Garvin stated in the book *How to Win with Information or Lose without It* [6] that information makes the difference between a decision and a guess, between success and failure . . . knowledge is power. Information is particularly critical to companies struggling to survive in the marketplace.

The National Technical Information Services (NTIS) is the central clearinghouse for U.S. scientific and technical information. All U.S. agencies use NTIS to disseminate "approved for public" reports, journals, and computer tapes available from R&D and engineering activities. The NTIS publishes free guides to all of their information resources [7]. Some of the key publications are:

Catalog of Government Inventions for Licensing
Federal Technology Catalogs: Guides to New and Practical Technologies
Directory of Federal Laboratory and Technology Resources
Directory of Federal and State Business Assistance
Government Inventions for Licensing Abstract Newsletters
Directory of Computer Software

Table 3 provides a list of some of the key information clearinghouses/resources, but there are hundreds of other special libraries. There are some very good privately developed handbooks available in most bookstores that provide a guide to U.S. government experts and resources [8].

If you have a personal computer and modem, there are many online databases available that permit you to access directly major government databases, including NTIS, Federal Applied Technology Database, Federal Research in Progress, MEDLINE, NIOSHTIC, DOE Energy, and others. The free NTIS guide [7] describes these databases and how to access them. The online databases are often the quickest and cheapest way to locate specific government expertise, resources or patents. Private information brokers are available to perform online searches for you [9] if you do not have the equipment or experience.

V. CONCLUSION

New legislation has opened up opportunities for domestic businesses to access and/or work with federal resources. The importance of these resources to companies, particularly small and medium-size companies, is potentially enormous. Time will tell if U.S. industry takes full advantage of this opportunity.

TABLE 2 Research and Technology Transfer Contacts

Office of Intramural Affairs
Dr. Philip Chen
National Institute of Health
Bldg. 1, Room 140
Bethesda, MD 20892
(301) 496-3561

National Institute for Occupational Safety and Health
Mr. T. F. Schoenborn
4676 Columbia Parkway, DPSE-R2
Cincinnati, OH 45226
(513) 841-4305

Naval Air Development Center
Mr. J. S. Bortman, Code 024
Warminster, PA 18974-5000
(215) 441-2033

Army Laboratory Command
Mr. C. E. Lanham
2800 Powder Mill Road
Adelphi, MD 20783-1145
(202) 394-4120

Army Biomedical R&D Laboratory
Mr. Lee Merrell
Building 568, Fort Detrick
Attn: 5GRD-UBZ-RA
Frederick, MD 21701-5010
(301) 663-2024

Office of Naval Technology
Dr. David L. Woods (ONT263)
800 N. Quincy St., BCT-3
Arlington, VA 22217-5000
(202) 696-5991

Naval Research Laboratory
Dr. George Abraham
4555 Overbrook Ave., S.W.
Code 10032
Washington, DC 20375-5000
(202) 767-3744

Wright R&D Center
Ms. F. Sinclair Ingalls
WRDC/XO
Wright-Patterson AFB, OH 45433-6523
(513) 255-2788

Air Force Human Resources Lab
Douglas Blair
Code AFHRL/PRT
Brooks AFB, TX 78235-5601
(512) 536-3426

Center for the Utilization of Federal Technology
Mr. Edward Lehmann
5285 Port Toyal Rd., #300B-F
Springfield, VA 22161
(703) 487-4838

NASA Headquarters
Mr. Len Ault
600 Independence Ave., S.W.
Room 100, Code CU
Technology Utilization Div.
Washington, DC 20546
(202) 453-8377

National Institutes of Health
Office of Technology Transfer
Reid Adler, Director
Room B1 C36, Bldg. 31
Bethesda, MD 20892
(301) 496-0750

Food and Drug Administration/HHS
Mr. Edward Mueller
Div. of Mechanics & Material Science
12200 Wilkens Avenue
Wilkens Science Center
Rockville, MD 20852
(202) 443-7003

Strategic Defense Initiatives Organization–Technology Applications
Mr. Nick Montanarelli SDIO/T/TA
Office of Secretary of Defense
Washington, DC 20301-7100
(202) 693-1562

TABLE 2 (continued)

National Center for Toxicological Research Mr. Arthur Norris Country Road #3 (HFT-2) Jefferson, AR 72079-9502 (501) 541-4516	Lawrence Livermore National Laboratory Mr. Charles Miller P.O. Box 808, L-795 Livermore, CA 94550 (415) 423-6416
Jet Propulsion Lab/NASA Mr. Gordon Chapman 4800 Oak Grove Drive MS 180-801 Pasadena, CA 91109 (818) 354-8300	Rehab R&D Center (153) Dr. A. H. Sacks VA Medical Center 3801 Miranda Avenue Palo Alto, CA 94304 (415) 858-3991

TABLE 3 Government Information Resources

National Technical Information Service
5285 Port Royal Road
Springfield, VA 22161
(703) 487-4650 (Technical Reports)
(703) 487-4630 (Subscriptions)
(703) 487-4763 (Computer Products)
(703) 487-4780 (General Assistance)

NASA Scientific and Technical Information Facility
Technology Utilization Office
P.O. Box 8757
Baltimore, MD 21240-0757
(301) 859-5300, Ext. 242, 243

Office of Scientific and Technical Information
(Dept. of Energy)
P.O. Box 62
Oak Ridge, TN 37831
(615) 576-1303

Defense Department Technical Information Center
Technical Library
Cameron Station, Bldg. #5
Alexandria, VA 22314
(703) 274-6833

FLC Clearinghouse
1007 5th Avenue, Suite 810
San Diego, CA 92101
(619) 544-9033

It is well known, for example, that Japanese and Soviet R&D organizations routinely order all scientific papers published by the National Technical Information Service (NTIS).

The "hunters and gatherers of technology" will certainly gain a competitive edge in the worldwide marketplace. The Federal Laboratory, emphasized in this chapter, is one bountiful source of R&D. Other sources, such as university, corporate, and other research laboratories, as well as the services of state and local governments, deserve investigation. However, expressions of sporadic intense interest by the private sector will not do. United States manufacturers must include accessing of new technologies from all sources as an integral part of their strategic planning processes.

The goal is not merely technology transfer but rather commercial success of the product. For small businesses especially, the way to reach this goal is not by merely screening volumes of data, but by establishing personal relationships with researchers working in the company's general area of interest. For this relationship to be most effective, the researcher must be involved in developing the corporation's market research strategy while learning enough about the technology to guide the research to optimum future directions. This confluence of knowledge, ideas, trust, and shared vision can keep the medical device and diagnostic industry as the world leader in technology innovation.

REFERENCES

1. *Wall Street Journal*, Nov. 10, 1986.
2. Council on Competitiveness, "Picking Up the Pace: The Commercial Challenge to American Innovation," 1988.
3. Federal Laboratory Consortium for Technology Transfer, "Putting Technology to Work," FLC Administrator's Office, Fresno, CA, March 1988.
4. G. F. Linsteadt and L. L. Doig III, Maintaining the High-Tech Edge: Government-Industry Technology Transfer and the Federal Laboratory System, *Journal of Technology Transfer, 12*(1) (1987).
5. J. P. Bagur and A. S. Guissinger, Technology Transfer Legislation: An Overview, *Journal of Technology Transfer, 12*(1) (1987).
6. A. P. Garvin and H. Bermont, *How to Win with Inforation or Lose Without It*, Bermont Books, 1980.
7. NTIS Products and Services Catalog, PR-827, National Technical Information Services/U.S. Dept. of Commerce, Springfield, VA, 1988.
8. M. Lesko, *Information U.S.A.*, Viking Penguin, New York, 1986.
9. Directory of the National Federation of Abstracting and Information Services (NFAIS), Philadelphia.
10. Using Market Research to Convert Federal Technology Into Marketable Products, *Journal of Technology Transfer, 13*(1), 27 (1988).

42

Responding to the Challenge of Education of Nurses and Allied Health Personnel on Safe Use of Medical Devices

JUNE C. ABBEY *Vanderbilt University School of Nursing, Nashville, Tennessee*

MARVIN D. SHEPHERD *University of California Medical Center, San Francisco, California*

I. INTRODUCTION

The early influx of medical devices into care was relatively slow and consisted primarily of means for maintaining life, such as dialysis, respiratory support, suctioning, and transfusions. Although, in all probability, the thrust for development of the new advances was the outcome of discoveries made during World War II, one of the major contributions to both medical devices and computers was the invention in 1948 of the transistor, which permitted the miniaturization of electronic circuitry and the high-speed electronic computer.

Where technical advances were made, increasingly daring medical innovations occurred, which required more sophisticated instruments and precise monitoring. Regional medical programs, instituted in the mid 1960s, promoted updating of both physician and nurses in cardiac care on the national scene. In a brief period of time, probably less than 5 years, clinical specialties in nursing developed when hospitals redefined themselves into acute, intermediate, and diagnostic care settings. The devices found in each depends on the level of acuity of care, the purpose of stay, and the patient's condition and diagnosis.

As more and more medical devices became part of health settings, a concern about safety and iatrogenic illness became increasingly apparent [1]. The fact that nurses were a prime user of these tools in the clinical care arenas, and hence could contribute to the problem, was not readily acknowledged. Nursing, however, that ever-present health-care profession, contains 1,486,000 licensed practicing nurses, of which 60 percent work in hospitals caring for acutely ill patients [2]. In fact, the reason for hospitalization is nursing care over the full 24 hours of a day. Therefore, it should come as no surprise that nurses are primary users of the preponderance of medical care devices available today. The ubiquitous nurse acts at the patient care interface between the patient and the physician, and the patient and the bioengineer, always present, always watchful, using the information for care, for reporting, and for instituting change. These activities require that attention be paid to four safety issues: (1) safety of the patient, (2) safety of the information, (3) safety of the personnel, and (4) safety of the device. In other words, the concerns of nursing include not only that the instrument be safely used on the patient, but also that the information yielded by the device be accurate and thus safe to use for care. This requires that the devices (1) be stable in the environment without a tendency to fail or break, (2) work reliably and accurately over long periods of time without interruption or decrement of function, and (3) be free from hazards to the user, such as sharp edges or current leakage. The device needs an informed user for protection of the instrument from abuse and misuse, one who can determine and describe the elements of malfunction so that replacement and repairs can be made. The question then becomes, "Can nurses address these safety issues?"

II. THE PROBLEM

Health care, perhaps more than any other industry, has embraced advances in technology. The changes have been quick, dramatic, and continuous. At first the devices increased in numbers and kinds. Then came streamlining, and design change as other technological advances were incorporated. The knowledge provided by discovery and innovation supported further exploration and extension of medical research, which, in turn, necessitated greater effort in, and development of, technology. The instruments, were instantly more complex when computer linkages, telemetry, multichannelled monitoring, and miniaturization became an accepted part of health-care delivery.

A. Nursing Education

Nursing education, as an entity, was not prepared for these changes. Currently, there are three different types of nursing educational programs: (1) 3-year hospital schools, commonly called diploma schools, that are maintained and run by hospitals; (2) the 2-year junior, or community, college programs, which grant an associate of arts degree; and (3) the university schools, or 4-year programs, which award a baccalaureate degree. Students who complete these programs all take the same state licensure examinations. Graduates from the 2- to 3-year programs work primarily in hospitals, whereas baccalaureate-degreed nurses can also work in public health or teaching agencies. All schools teach about intensive care, which includes monitoring and sophisticated treatment devices. The content taught varies widely in breadth and depth, and in the amount of actual clinical time spent using the clinical devices. Few schools offer courses on the principles of equipment use. In the main, use of medical devices is taught in hosptial-sponsored inservice or continuing education programs [3].

One of the solutions to the technology problems by nursing education has been to focus on interpretation of the information obtained from the devices, rather than on the four safety concerns, or the principles on which the devices are based. In other words, emphasis has been on the indicated nursing care. Although this action is logical, and even commendable, it negates important aspects of care. The same pattern, a focus on clinical interpretation exclusively, holds in general in clinical specialist programs at the graduate level. A 1980 survey revealed only four programs that offered courses in bioinstrumentation at the graduate level. Little interest was expressed by any of the other 254 collegiate programs that responded to the questionnaire. A survey in 1986 [4] showed minimal change. Seventy-three percent (73%) or 299 out of 408 accredited baccalaureate programs responded. Of these, five offered an elective course specifically in clinical instrumentation. Forty-nine percent (146) reported a course that included an instrumentation component. Descriptive course materials (N=106) were then examined by three raters for inclusion of theoretical principles underlying the equipment. Course content emphasis remained on information gathering, for examples, patient assessment, critical EKG interpretation, how to's or procedure, and instrument-specific setups for hemodynamic monitoring.

Of the 38,000 nurses with master's or doctoral degrees, 8000 are certified clinical specialists. The Abbey et al. survey [5] of accredited graduate programs resulted in a 66% response (50/77). Nine out of 15 schools that stated that an instrumentation course was taught also sent course materials. Review by three raters showed that only two course descriptions mentioned underlying principles basic to clinical devices and instruments. Four course descriptions purportedly containing components relative to instrumentation did not mention devices, equipment, monitors, or instruments. All course material did, however, emphasize the nursing care, patient assessment, and the dependence of interpretation and evaluation of patient progress on the obtained information. The latter finding led to an exploratory, small survey designed to discover if practicing nurses thought a formal course that focused on medical devices, their characteristics, how they worked, the purpose of use, operational procedures, and safety concerns in hospitals, clinics, and home use was needed in addition to the existing inservice programs [6]. The questionnaire was sent to 15 urban and suburban agencies for distribution to three nurses each. The sites included three large teaching hospitals of greater than 400 bed, four urban hospitals (200-400 beds), five suburban hospitals (100-200 beds), one urban rehabilitation hospital (200 beds), one urban speciality hospital (100 beds), and one Visiting Nurses' Association. Each site had a representative response within the 35 out of 45 returns (77.7%). Twenty-six of the respondents worked in acute care or step-down units, and the remaining nine were nurses involved in home care delivery or supervision. The nursing educational backgrounds included eight diploma graduates, 11 baccalaureate degrees, and 14 master's degrees. There were no associate degree nurses in this sample. Formal education in use of the devices was limited. Agency inservice programs taught by biomedical technicians or manufacturer's detail people were cited as the primary sources of information. The respondents named 77 different pieces of equipment (medical devices) used in patient care. Of the 35 returns, 34 felt a clinical instrumentation course was needed. Of probably more significance was that 29 requested and received agency agreement for paid time off to attend class, and 14 of 15 agencies agreed to reimburse tuition fees. These results corroborate Breu and Dracup's finding [7] that critical care nurses include equipment management and care as part of their responsibility.

The interest of the home care nurses and the support of their agencies with time and money to learn about devices reflected the mounting changes in home care due to the tightening of hospital budgets, increasing average inpatient severity of illness, earlier discharge to home care or nursing homes, and development of treatment protocols that permit and facilitate the use of complex technology in the home or nursing home [8]. The Secretary's Commission on Nursing [9] reports an increase of 52% of Medicare-supported home visits from 1980 to 1987. The skilled nursing visits also increased by 47% during the same time interval. Swanson [9] reports that the growth of use of complex technologies has caused a greater need for professional nurse care in ambulatory settings. The same trend is occurring in nursing homes where patients who are not ill enough to stay in hospitals, and not well enough to go home, receive nursing care.

B. Home Care

Technology advances and changes in reimbursement have created new markets, which have expanded the volume of business and the variety of services in home care. The size of the market for home care equipment is worthy of note. Annual sales are estimated to be between $1.0 and $2.6 billion [10]. Tehan et al. [11], when addressing the problems of managing a program for servicing home health care equipment needs, approach the issue from marketing aspects and the variety of services required. These authors unequivocally state that "The education component is the most important function [of the program] because inadequately trained staff, patients or caregivers, adversely affect productivity, liability exposure and reimbursement." Documentation for reimbursement requires that standards defining patient need be clearly established. The criteria established by the Health Care Financing Association (HCFA) include that the equipment (1) is able to withstand repeated use, (2) is appropriate for home use, (3) is useful only in the event of illness or injury, (4) is usually and primarily used only to meet a medical purpose, and (5) is capable of being used repetitively without breakdown or failure [12]. Based on previous utilization experience, these definitions readily determined equipment that qualified for reimbursement prior to the massive influx of medical devices into home care; however, changes in hospital patient discharge patterns altered intensity of care, treatment, and monitoring in home care settings. Reimbursement standards lagged behind introduction of the equipment into the home. This increased financial risk and emphasized the need for knowledgeable user assessment of the reliability of design and manufacturing, and for servicing of the devices for home care.

Use of medical devices in home health care includes a considerable number of interarticulating groups of people for implementation of safe care. Primary among these are the manufacturers, supply buyers, personnel from regulating agencies, bioengineers, biomedical equipment technicians, physicians, nurses, physical therapists, respiratory therapists, nutritionists, infusion therapists, and, some would say, the liability insurers [13]. All of these concerned and involved people depend on the knowledge and skills of each other to use the devices in a safe, dependable way with the sick or infirm patient. It is here at the juncture of machine and patient that the acute hospital delivery system differs most from home care. Whereas in the hospital the equipment is used mainly by nursing and other trained personnel on very ill patients, in the home, the patient and family members govern device use. Whereas in the hospital, resource personnel and equipment are available for rapid replacement, in the home, where malfunction of high technology equipment can also be life-threatening when used improperly, backup services and resource personnel are usually not in close proximity. In other words, the hospital is a controlled environment with a variety of experienced, easily accessed, knowledgeable professionals. The home lacks constant access to widely experienced and informed professionals.

A careful reading of the literature reveals the complexity and interdependence of various group relationships in using medical devices in home health care. The common denominator is, of course, the patient, and the process is teaching the patient and family to accept and use the equipment safely. The content reflects input from each of the groups, as a consensus selects the essentials for a protocol that will provide sufficient knowledge about the equipment and its maintenance to secure maximum benefit to the patient without failure due to breakdown or disuse. Periodic followup for assessment of the patient, and evaluation of the efficacy of equipment performance, when documented, affords a record of whether or not the developed protocol addresses the concerns of each of the involved groups and the patient.

Table 1 depicts some of the equipment considerations that determine costing available services, appropriate staffing mix, and backup units—in other words, the view of the equipment from the manufacturers' and buyers-suppliers' viewpoints. Pragmatic and succinct, the columns clearly demonstrate the need for (1) professional, licensed personnel, (2) patient teaching by knowledgeable professionals, (3) attention to maintenance of the device, and (4) its category and estimate range of patient risk.

III. SAFE USE OF MEDICAL DEVICES

As all health-care settings become replete with medical devices that involve diagnosis, therapy, monitoring, and recording, the roles of the involved caregiving personnel change. New responsibilities replace or are added to old. Patient responses and the expectations of society modify. Educational requirements are reevaluated and often altered. The language of each discipline changes to incorporate the novel from a different perspective, bias, knowledge base, interest, and purpose. As Dyro [14] aptly states, "Solutions to the problems caused directly or indirectly by the introduction of technological advances in health care are to be found in technology catching up to itself." This might well be extended to the health-care disciplines during catchup as well, where the technology is outdistancing the caregivers.

Currently the use of devices is taught to health-care professionals primarily through inservice provided by (1) the manufacturer's representatives, (2) employers for use of recently purchased equipment, or (3) continuing education. The common difficulty occurring in each situation is lack of a comparable knowledge base among the learners. Experience, educational preparation, and goals are disparate even within specific classes. Yet the need to know how to use the devices correctly is acknowledged by all involved groups.

Each of the contributors to the development of device utilization, from manufacturers to the patient, requires a basic amount of knowledge for understanding. For instance, the manufacturer, by definition, controls the design of the equipment from components, circuitry, accessories, and packaging. This includes the acceptable degrees of sensitivity, precision, and reliability of the instrument. The manufacturer determines the quality of construction, the estimated amount of maintenance, the required support personnel for continued function, and initial cost of the equipment. The manufac-

TABLE 1 Home Health Care Equipment Utilization Concerns[a]

Category	Use	Example	Reimbursement	Risk
1. Durable med equipment (DME)	Mobility Comfort self care	Walker Wheel chair Commode	Yes	Low
2. High-technology equipment	Infusion therapy Variable term of use	Volumetric pumps Dialysis Insulin pumps	Yes	Moderate-high
3. Repiratory care products	Support respiratory function Long-term use	Apena monitors Respirators Volumetric Intermittent positive pressure nebulizers Liquid and gaseous oxygen concentrators	Yes	Moderate-high
4. Rehabilitation equipment	Strengthening exercise Short-term use	Traction Continuous passive motion Variable resistance	Variable	Moderate
5. Pain management devices	Control of transient pain Short- to long-term use	Transcutaneous electrical nerve stimulation (TENS) Massage Cooling pads Endothecal Crystalline M.S.	Variable	Moderate-high
6. Self diagnostic aids (last 5-7 years)	Measurement Pregnancy Blood glucose Urine constituents Colon cancer Variable term of use	Pregnancy testing Glucose monitors Alb Glucose Blood in stools Blood pressure	Not usual	Low

[a]Constructed from ref. 11.

Patient teaching	Professional followup	Instrument maintenance	Insurance and liability	Failure points
Tech professional optional	Minimal short term	Minimal	Minimal	Wear falls
Professional required RN and Tech	Moderate to high Licensed personnel as long as in use	Moderate to high	High	Failure of misuse Lack of understanding purpose, use and function by patient or professional
Tech professional required RT, RN	Moderate to high	Moderate to high	Moderate to high	Failure Misuse Abuse Disuse
Tech professional required RN, PT	Moderate	Minimal-moderate	Moderate (depends on problem and stage of recovery)	Patient instruction and understanding
Tech professional required RN, PT	Variable Low to high Depends on purpose and duration Licensed personnel	Minimal-high	Minimal-high	Failure of patient instructor and understanding Professional and staff knowledge
Usually minimal Moderate Overtime	Usually minimal professional monitoring M.D., RN Nutritionist	Usually minimal	Minimal	Lack of instruction and patient understanding Calibration loss Expiration dates/storage

turer also tests the prototypes in a controlled setting, be it laboratory or clinical. Informed and carefully trained personnel study the device according to carefully stated protocols that ensure as much replicability of setting, equipment, clinical problem, and utilization as is possible. The manufacturer develops the labeling, training manuals, and schematics for maintenance; sets standards for care of the device; and finally, conducts training sessions for the technicians and professional users.

At this point the manufacturer is the sole source for delivering information regarding the four safety concerns for use in clinical care: (1) safe for use with patients, (2) safe, or accurate, information, (3) safe for personnel, and (4) the methods for use that are safe for the device itself. The directions must be comprehensible to the users and they must be understandable by technicians, nurses, and patients in the absence of teachers. The sequence for introduction of a device is manufacturer, representative or supplier, technician or maintainer, user or operator, such as nurse, physician, physical therapist or respiratory therapist, family member, and patient. Each of the participants in the linkage is from a different discipline with greater or lesser preparation and experience with devices in a clinical situation. The perspective, language, and communication patterns differ between groups, yet each is required to teach other members of the linkage about the device, so that each can become participant and instructor in the chain. The information is therefore modified by the instructor-participant according to his knowledge base and his interpretation of requirements, understandings, and goals of the next learner-participant.

Through this instructor-learner-instructor process, natural levels of informational complexity occur that develop from the need of the manufacturer to give comprehensible stand-alone instructions for equipment use and the need of the various user groups for access to the information. These levels have not been defined or utilized by the manufacturers and are therefore not available for incorporation into any formal educational program.

Reports of studies of iatrogenic illness related to medical devices began to be published in greater numbers early in the 1980s, and interest in identifying the causes continues [14-16]. Abramson et al. [17] report a 5-year study that showed that (1) misuse of the medical device accounted for 30% of the incidents; (2) an incident clustering pattern was associated with inexperienced ICU physicians and nurses; and (3) other human factor problems included communication errors and nurse understaffing. Steel et al. [1] reinforced Abramson's findings by showing that 35% of the adverse occurrences in a specified hospital were related to medical devices. One source of the problem cited most often is the lack of formal education for nurses [14,18-21].

The multifaceted problem ensuring the safe use of medical devices thus includes these primary factors:

1. The rapid entrance of technology into health care has greatly modified the roles, job expectations, and settings in which health-care givers work.
2. New members from other disciplines accompanied the technology into the health-care settings.
3. Formal educational institutions have not kept up with the changes and therefore teaching does not include essentials for adequate preparation for (a) communication between disciplines and (b) optimum, safe use of the devices in health-care settings.

4. Current methods of institutional training are piecemeal without standardized organization of new knowledge and without recognition of different user defined levels of content.
5. The influx of technology will continue into all areas of health care for the foreseeable future.
6. The influx is driven by a generalized expectation of society.
7. Nursing occupies a critical, pervasive interface position in the use of medical devices in health care in all settings.

IV. DEVELOPMENT OF THE MODEL

In late 1984, the FDA Center for Devices and Radiological Health began to investigate operators' roles in the problem of safe use of medical devices. A review of the literature revealed an awareness of a lack in academic preparation of the principal users of the devices [21]. Student nurses quickly recognized the need and at their annual national convention passed a resolution to "support the inclusion of basic principles of biomedical instrumentation and technology as part of the undergraduate curriculum in nursing" [22]. The following year, in March 1986, the FDA held an invitational meeting, "Nursing and Medical Devices." Fifty representatives from nursing, medicine, home care, bioengineering, the medical device industry, related organizations, and government met to examine their respective roles in the safe and effective use of medical devices. The meeting objectives were to: "1) develop an understanding of device-related injuries and deaths; 2) identify factors that interfere with the safe and effective use of devices; and 3) develop strategies to address those factors." The meeting resulted in a number of recommendations for further study and future meetings. Planning ans steering committee memberships were established. The latter committees met during 1986 and 1987 to plan for a subsequent conference that would address salient topics about problems inherent in the implementation of technology and medical devices into health-care settings and nursing [23]. The decision was made to include the development of an educational model as a takeoff point for the conference. The Abbey-Shepherd Educational Model is based on characteristics identified by the steering committee. It is designed as an unambiguous guide for development of course content, and relates primarily to organizing content not currently taught in nursing (see Fig. 1). The content framework, however, addresses an orderly progression through information required by all personnel involved in the use of medical devices. No attempt has been made to develop behavioral objectives or performance evaluations. These are well within the province of current educational practice of the concerned disciplines and should reflect that discipline's integration of knowledge and skills as they relate to the device and health care. The recommended characteristics of the model were [24]:

1. Characteristics of each device
2. Operating principles
3. Common user errors
4. Adverse patient reactions
5. Device failures and frequency
6. Safety concerns

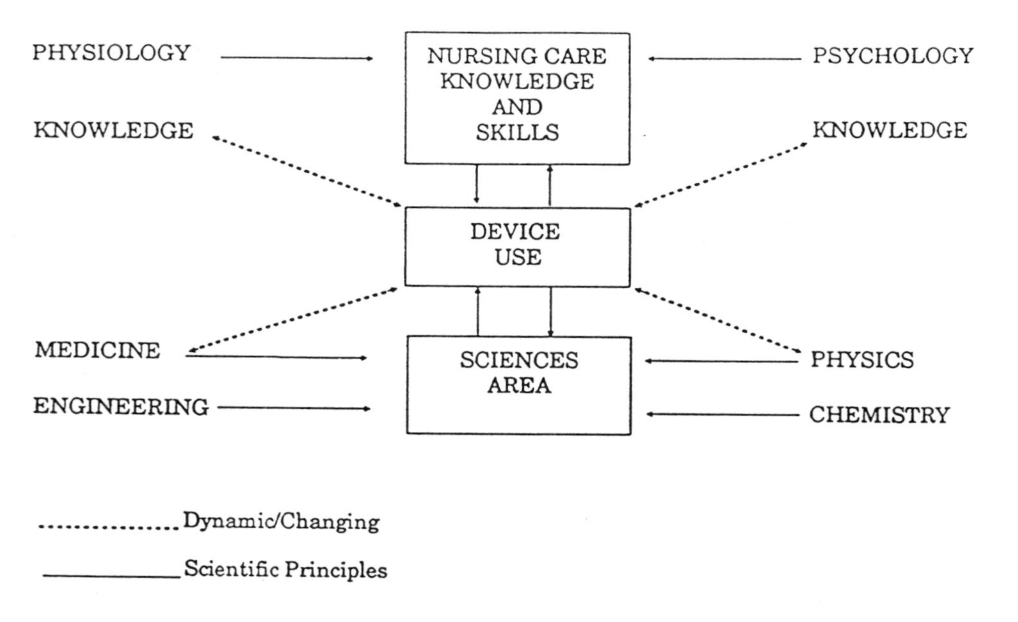

Fig. 1 Nursing knowledge and skills for device use [24].

7. All factors related to device function and safe use
8. Care of the instrument

The model should:

1. Be additive, realizing that individual course content will vary according to the knowledge base of the user and purpose of the instruction
2. Be applicable to all medical devices
3. Provide structure for generation of specific information and content levels
4. Allow for constant continued updating
5. Permit incorporation into ongoing curricula relating to patient care
6. Be based on scientific principles

The model should address:

1. Cost-effectiveness
2. Risk

With the exception of measurements of risk and cost-effectiveness, the Abbey-Shepherd model meets the above recommendations [24]. It is intended to be usefully relevant to any technology and at several levels of user competence . . . from the fundamentals of science and technology to a simple sequence of instructions. A step-by-step use of the model was published in the Plant Technology & Safety Management Series of the Joint Commission on Accreditation of Health Care Organizations [25].

Although the Device Educational Model was generally endorsed by the FDA conference participants for use by nursing educators and by the FDA in its labeling practices, it will be useful to manufacturers as well, since it:

1. Applies to any device
2. Can be used even in the early planning phases of a new device as a checklist for "Operator" information
3. Allows some predictability as to the specific information that will be needed and/or requested by the device users
4. Will assist nursing trainers employed by manufacturers to communicate with nursing educators using the model
5. Does not need any new information not already required by the FDA
6. Provides a standardized simplified format deemed appropriate by *users* and the FDA
7. Ensures all information is covered for each device and knowledge level of user
8. Increases user understanding, which should minimize product-liability claims
9. Reduces numbers of medical device reports filed with the FDA

V. THE DEVICE EDUCATION MODEL: AN OVERVIEW

The Abbey-Shepherd model approaches course content with a focus on four levels of complexity:

1. Operational procedures
2. Fundamentals of procedures
3. Science and technology
4. Systems principles

Briefly stated, the levels progress through the purpose and operational procedures for a specific device, followed by the fundamentals on which the procedures are based. The next level is that of the basic sciences, oftentimes physics, which undergird the principles of design and specifications, such as accuracy, precision, noise, and sensitivity. The fourth level includes systems operational principles that incorporate the interaction of the device, patient, nurse, facility, and environment.

The first and primary requisite in the ordering of the device education model is a clear statement of the medical purpose of the device. (More recently we have also been including a "nursing purpose" as well. The nursing purpose may be different from the medical purpose.) A device can be of single or multiple purposes. Most monitoring instruments not only monitor, but also record physiological activity. The purpose statement defines the use of the device in such a way that user expectations and demands are congruent with what the device can do. In other words, the user selects the instrument best suited to obtain the information or to deliver the treatment required. The purpose statement is then followed by a general description of how the device accomplishes that purpose. The medical purpose statement, along with a descriptive statement of how the device accomplishes this purpose, provide the basis for utilization of the four levels of complexity in the development of course content.

The operational procedure, the first level, relates to operational steps necessary to use the device. Minimal content is included about how the equipment works. This section, however, also contains sufficient understandings about routine safety and hazard concerns to protect the patient, the device, and the nurse. Upon completion of this content the learner is able to correctly use the device.

The second level develops the fundamentals on which the operational procedures are based. The content is presented in greater detail relative to the overall procedures involved in use of the equipment. The emphasis is on understanding relationships between the parts of the device, the device and methods of application, the method of application and the patient, and the patient and the information obtained. At this level, block diagrams are used to show the fundamental workings of the device. Emphasis is placed on measurement techniques, including accuracy, precision, reliability, calibration, and baselines. Passive and active devices are introduced, and the development of relationships of different energies to device function and the effects of these energies on tissue is initiated.

The principles of related science and technology comprise the third level of content, which builds on the operational fundamentals to include the basic science principles that permit generalizations. The content incorporates principles of physics related to pressures and volumes, transduction, potentials, levers, heat generation and transference, and mechanical energy. It also includes engineering aspects such as electrical components, transducers, and circuits. Attention is paid to design characteristics. The physics principles taught are directly applicable to many aspects of patient care. Table 2 lists a brief sample of clinical topics in which physics principles are applicable.

The fourth level is, in reality, a system interaction of components that can be incorporated according to the level being taught. There are five interacting components of the system. These are the device, the nurse, the patient, the facility, and the environment (see Fig. 2).

TABLE 2 Clinical Topics Incorporating Physics Principles

Clinical topics	Scientific and technical topics
Fractures: types of accident procedures in emergency dept., in wards; traction	Elasticity; stress and strain; ultimate strength; vectors
Rehabilitation after fracture	Center for gravity; moments; force diagrams
Pressure sores: causes, avoidance, treatment	Fluid pressure; Pascal principles; capillary pressure
Drainage; injuries and disease causing accumulation of fluid in pleural cavity	Positive and negative pressures in respiration
Respiratory distress	Measurement of flow, pressure, residual volumes, elasticity, compliance
Oxygen therapy	Bernoulli; Venturi
Electrosurgical burns	Current density, grounding connections; heating effects
Piped gases	Rotameter; Archimedes principles

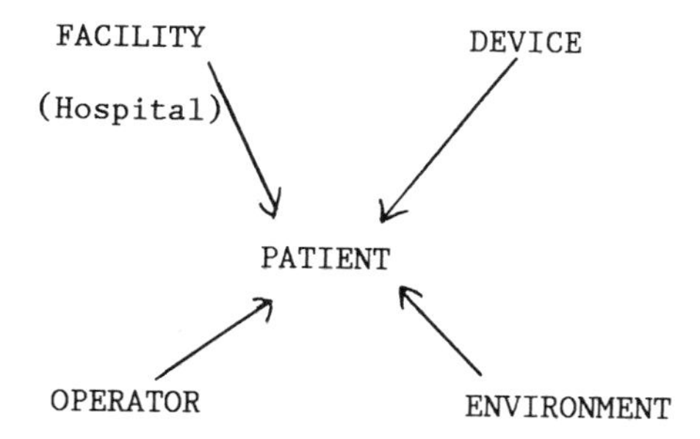

Fig. 2 Components of the clinical environment.

The question is, what should be taught to assure the safe and proper functioning and interaction of the device in the clinical setting? The device itself presents three areas of vulnerability to failure: design, deterioration, and maintainer. Of these, the maintainer is one source of failure that is often neglected in course content. The nurse can create failure through errors in application, incorrect interpretation of data, abuse of the equipment, and inadequate patient teaching. Patients contribute to instrument failure both passively and actively. Examples of these are dry skin, anxiety with muscle tremors, and/or actively picking at electrodes. The facility contributes to failure and errors through design faults, deterioration of the systems, and maintainer errors. Power fluctuations are an example of facility failure. The environment can contribute to failure through smoke, heat, and dust, as well as electromagnetic interference from power lines or radiation.

The model is intended for nursing educational opportunities in undergraduate, graduate, inservicing, and continuing education. The undergraduate is expected to complete levels 1-4. The graduate student will master the content included in the four levels but in greater depth and breadth. Inservice trainers will select whatever levels are indicated depending upon learner competence. For instance, level 1 would be sufficient if an updated version of a monitor already in use were purchased. An operational procedure update is all that is necessary if the learner is experienced. Continuing education presents a similar situation wherein the content level taught should include baseline information relative to all of the elements. If this content is known, then more advanced material is indicated.

The greatest numbers of devices are used in acute-care settings; however, as previously mentioned, with earlier hospital discharges more and more instruments are becoming a part of home care, where the patient, family members, and often the visiting nurse determine use, abuse, and function of the instrument. Whereas various types of medication pumps, breathing devices, pacemakers, and dialysis readily come to mind, more ubiquitous equipment, such as glucose monitors and sphygmomanometers, are becoming increasingly common home health-care monitors and treatment determinants. This example reinforces the need for the nurse to understand fundamentals of the operational procedures in sufficient depth to teach the patient how to use and care for the instrument. Examples of the use of the model are included at the end of this chapter.

VI. THE DEVICE EDUCATION MODEL: THE SPECIFICS

The device education model (DEM) rests on a standard engineering method—systems safety engineering. A brief review of this method will aid in the understanding of the DEM.

A. Fundamentals of a Systems Model

Systems analysis is a method that identifies the components of a system and explains the interaction of the components as it relates to the outcome(s) of the system. This type of analysis stresses the fact that a nursing device *is only one component of the system* and that for successful delivery of the device purpose, all components of the system must be functioning normally. The identified components must include any factors that may affect the function or safety of the nursing device while it is in operation. All complaints by nursing device users that have been received by the FDA, ECRI, The United States (USA) Pharmacopeia, and the manufacturers can be analyzed by this method. Common failures that are associated with specific components of the system can be identified and categorized by component failure. Corrective actions can be introduced, or operator 3education/training can be provided to minimize the occurrences.

It should be recognized that is a nursing device were of perfect design, it would never fail unexpectedly, would not deteriorate, would not require maintenance, would not be affected by its environment or the facility of use, and could never be misused or abused. In our imperfect world, nurses have some (if not all) of these deficiencies to contend with.

The use of a system approach has a number of advantages:

It provides a logical and complete model of the system so that when an analysis is complete the model user is certain that all critical factors of the system have been ocnsidered.
The system becomes a simple "check list" for the user.
It can be used for the purpose of evaluation of devices prior to purchase.
It can be used to determine what changes to a system's components are required if a new technology is introduced into the hospital.
It can be used to analyze nursing device failures to assure a reasonable correction.

The device education model is a modification of a device systems model. While the device systems model was designed to identify component failures and suggest reasonable corrective actions, the device education model uses the systems model to identify the basic and practical knowledge that a nurse requires to use a nursing device with confidence and competence. It looks first at the nursing device as a part of a system and it focuses second on the device in the context of its purpsoe, operational procedures, and fundamentals of operation. By analyzing each device by this method, the knowledge needed to safely operate it is identified. From this knowledge, nursing lesson plans can be written.

A device systems model for medical devices was described in a 1983 monograph [26]. An abbreviated, modified version is presented in the following text.

B. Components of the Device Systems Model

The functions of a nursing device and the hazards associated with it are related to one or more of the basic components of the clinical environment (see Fig. 2) and the interaction between each component and all other components. The component interaction noted in Fig. 2 focuses on the patient as the receiver of the benefits of the system.

A systems model emphasizes that if only these five components were available, the successful use of any device would relate to the interaction of the five components. The *facility* relates to the design of the immediate location of the patient, that is, exam room, OR, catheterization room, ICU bed area, and other. The *device* may be any nursing device. The *patient* may be any type patient, whether in good health or bad. The *operator* may be a nurse, technologist, or doctor, dependent upon circumstances (but of prime interest to us is the nurse).

In the real and more complex clinical environment, we may have more than one device and more than one operator (see Fig. 3). Then our concerns are increased by the number of components. We must consider the interaction between the devices themselves and the levels of knowledge and training of each operator.

The next step is identification of the subcomponents that also affect the functioning or safety of a device, as shown in Fig. 4.

C. Explanation of the Subcomponents

Each component will have two or three subcomponents. The subcomponents are factors that directly affect the functioning or safety of that component of the system.

1. Operator

For a medical device to function and function safely, the nurse operator must (1) know how the device is intended to function, (2) know the facility factors that may affect its function or safe use, and (3) be knowledgeable of any patient particulars or sensitivities that might either qualify or contraindicate its use. The lack of education or experience in any one of these three areas could result in the loss of function of the device and an increased level of risk to the patient or operator. The subcomponents of the *operator* component are (1) misuse and (2) abuse.

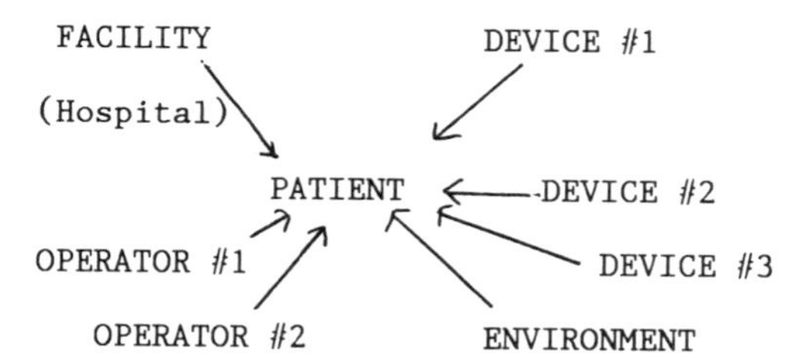

Fig. 3 A multicomponent clinical environment.

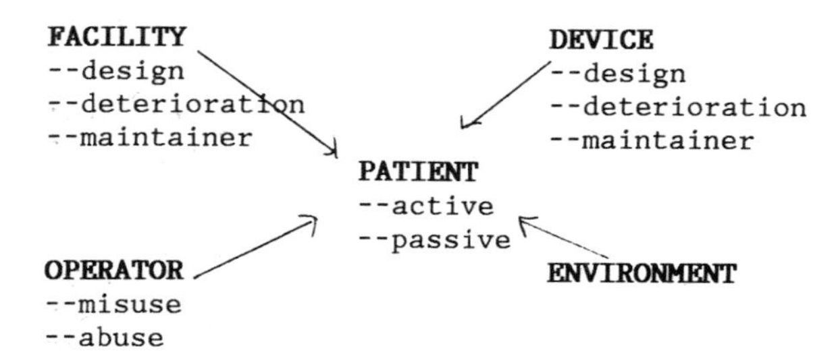

Fig. 4 Subcomponents affecting device function and safety.

1. Misuse. For the safe and proper use of any one device, the nurse is expected by the manufacturer to perform certain functions. Failure to know or to perform these functions constitute misuse of the device. An example of misuse might include the improper loading of EKG paper into an EKG machine, measuring blood pressures with bubbles in the system, or trying to measure a Lead I EKG when all three leads are placed on the patient's legs.
2. Abuse. During normal use a medical device could be dropped on the floor, have fluids spilled on it, be rammed by other devices, or be generally stressed in a manner not intended by its designers. The results of this stress may be ignored or go undetected. Use of the device following this stress can place the patient or operator at greater risk, or the device performance may suffer. It should be noted that the excessive stress placed on a device or its component parts is not necessarily intentional but occurs during normal operation. This type of stress damage can be repaired by a maintainer. Prevention, however, depends on the appropriate training and motivation of the device operator.

2. Patient

Two subcomponents are associated with the patient component: (1) active and (2) passive.

1. Active. The active patient may shut off an alarm, remove an IV, or throw EKG telemetry in the toilet. He or she may "twiddle" with the leads of an implanted pacemaker or incorrectly connect the accessories of a home dialysis machine. That is, a specific action on the part of the patient may directly affect the proper functioning of a device intended to assist in health care.
2. Passive. The passive patient is one whose conditions or sensitivities may contribute to the improper functioning of a device if their conditions/sensitivities go unrecognized. Patients with implanted pacemakers, pain controllers, and the like may have their functions affected by diathermy machines or radio transmitters. The CCU patient may have externalized catheters that are available for contact with electrified conductors. Patients with poor peripheral blood circulation are more likely to receive a burn from a hyperthermia unit. Each of these sensitivities or conditions must be recognized by the nurse and actions taken to prevent untoward incidents.

3. *Device*

From the systems viewpoint, the device has three subcomponents that can prevent it from functioning properly or cause it to be unsafe. The subcomponents are (1) design, (2) deterioration, and (3) maintainer.

1. Design. This "design" is not related to human factors of the operation but to the design of the components, circuitry, accessories, and the way all the parts are mechanically put together. By definition, the manufacturer has complete control over the design of a device and the nurse accepts the device and accessories as received on the unit.
2. Deterioration. Once the device is placed into operation, it begins to deteriorate, knobs loosen, electrodes break, dust gets into the machine, and parts wear. A biomedical equipment technician (BMET) or clinical engineer (CE) must monitor this deterioration to assure the device does not deteriorate below a safe level or below a level that prevents the machine from performing its expected function. Nurses have an operational-level maintenance responsibility that can affect safety and performance. For example, if reusable defibrillator paddles or EKG electrodes are not kept clean, device malperformance may occur.
3. Maintainer. It is generally assumed that the BMET or CE is educated and trained to properly maintain the devices. If they are inexperienced, undertrained for a particular device, or not properly motivated, they may not perform a proper maintenance function. The result may be that they have to return to repair the device a short time later, the performance of the device is not that expected by the nurse, or, in fact, the device is unsafe to use. This subcomponent is analogous to the operator "misuse" subcomponent in that only proper education and training can prevent this subcomponent from becoming a significant factor.

4. *Facility*

Like the device component, the facility has three subcomponents that can prevent it from functioning properly or being unsafe. The subcomponents are (1) design, (2) deterioration, and (3) maintainer.

1. Design. This term "design" describes the components used in the construction of the hospital, the type of circuits and systems in which the components are placed, and the mechanical placement of them. Operating rooms as well as some special care areas have isolated power systems, while general care areas do not. Some receptacles are high-quality, hospital grade, while others are of a lower quality. Hospital designs are determined largely by the architects, contractors, and appropriate codes and standards.
2. Deterioration. With use, the components of the hospital begin to deteriorate. Receptacles lose their pin contact tensions or become mechanically loose in the wall; gas fittings begin to leak due to worn gaskets. These components must be replaced or tightened before their function is affected or before they become unsafe. On these occasions, an individual called a hospital engineer, or sometimes a building services maintainer, arrives to provide repair.
3. Maintainer. Analogous to the person maintaining the device, the hospital maintainer must be properly educated and trained to be effective in his

maintenance activity. Otherwise he may replace a red "emergency" receptacle with a white receptacle, fail to replace a receptacle that has low pin contact tensions, or use oils in oxygen-line fittings.

5. *Environmental*

Environmental factors include heat, light, humidity, sound, vibration, air, EMI (electromagnetic interference), and the like. Any one of these environmental factors may occasionally cause a component of the system to malfunction or become unsafe. These factors could also be included under the facility component, since they are frequently controlled by facility adjustments. However, we find it useful, for analytical purposes, to consider them separately.

VII. USING THE MODEL: LEVELS 1, 2, AND 3

When a device first arrives at the hospital and is readied for use, the impression given is that if the operator follows the manufacturer's operational procedures, the device will achieve the desired results. There is an assumption by the manufacturer that the device will be used in the same context in which the manufacturer has tested it. Unfortunately, this is not always true. Newly invested devices that can interfere with the device in question may not have been included in the original test circumstances, and certainly environmental interferences in one hospital can significantly differ from those in a second.

With the device systems model presented, we understand level 4 of the DEM. We can now expand that model to include levels 1-3. To do this we must focus on the device itself—its medical purpose, its operational procedures, the fundamental reasons for the procedures, and the science and technology on which the fundamentals are based.

A. General

The educational model is device-specific. That is, each device may vary depending on features selected during purchase or modifications made later. In addition, devices with identical names (e.g., infusion pumps) but different manufacturers may have fundamental design differences. For this reason the educational model must address each device of each manufacturer by model and even by serial number.

B. Medical Purpose

A clear statement of the medical purpose or purposes of the device must be made. Some devices have a single purpose, such as an EKG machine. Its purpose is to diagnose abnormal electrical activity of the heart. Some have multiple purposes, such as a defibrillator/synchronizer/monitor. Note that these purposes are medical and/or nursing purposes and not device purposes. The device purpose of an EKG machine is to sense and display the electrical activity of the heart.

C. Device Purpose

A brief, general description should be provided of how each purpose of the device is accomplished. For the single-channel EKG machine, it is by detecting surface electropotentials at several points on the body, amplifying them, differentiating heart potentials from other body potentials, and displaying the hear potentials on a paper recorder.

Both medical purpose and device purpose for a few devices are given in Table 3.

D. Operational Procedures

Step procedures are required that assure the device accomplishes its purpose(s). These steps must be documented for each purpose. Normally these steps are identified by the manufacturer. Some of the steps are general and may not have to be repeated for each purpose of the device. That is, general steps make the device ready for operation, such as turning the power switch "on," calibrating the system, and placement of the "hot" pen in the center of the paper.

Procedures then lead the operator through a series of operational steps to acquire the desired knowledge. This is the minimum level of knowledge that is required to obtain the desired results from the device.

E. Procedure Fundamentals

For each of the operational steps, there is a fundamental reason for the action taken. The educational level required to understand these fundamentals is

TABLE 3 Purposes of Devices

Device	Device purpose	Medical purpose
EKG	Detect the heart's electrical activity and display its analogue on paper	Diagnose heart disease
Defibrillator	To provide a high-voltage DC, msec electrical pulse	To defibrillate the heart
Glucose monitor	To measure light intensity changes for a particular constituent of blood	To analyze a blood sample for glucose concentrations
Thermometer (mercury)	To sense and display temperature	To diagnose diseased body conditions
Manometer (mercury)	To measure pressure	To diagnose cardiovascular disease
Pacemaker	To provide a mV, msec, DC pulse	To assure proper pacing of the heart
IV pump	To provide a constant flow of fluids to the body	To provide nourishment or drugs to the body

such that they can be understood without a strong technical background. It is based more on the effect the action has on the results than on a highly technical explanation that includes formulas, physics, or circuitry. This is level 2. The next lower level provides the very basics of the technology involved.

F. Science and Technology

The fundamental for each operational step is rooted in the areas of science and technology: physics, math, chemistry, biology, physiology, engineering, etc. To fully understand the operational fundamentals, we must teach physics and engineering. These basics must be taught if a thorough understanding of the reasons behind the actions are desired.

A flow diagram for the DEM is given in Fig. 5. Examples of nursing devices analyzed by using the device education model are included in the Appendix.

The major steps to follow in using the DEM are as follows:

1. Obtain the device and a copy of the operating manual for the device to be analyzed.

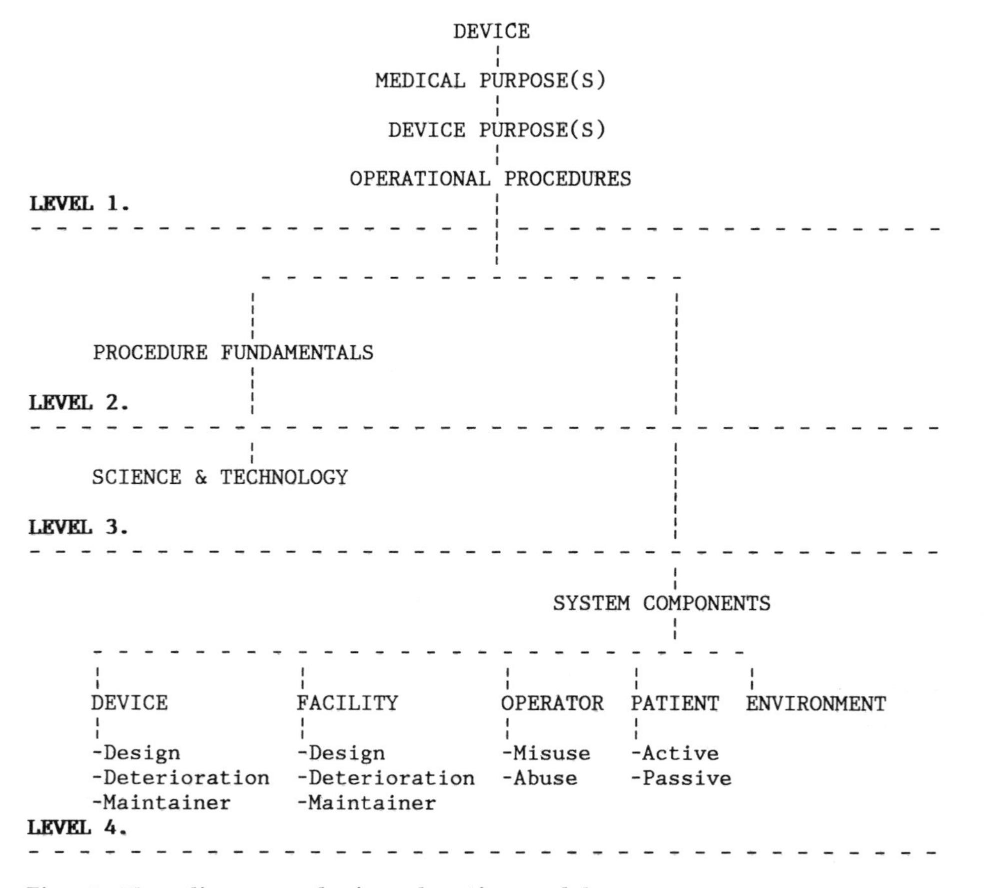

Fig. 5 Flow diagram: device education model.

2. Observe the actual use of the device (if possible) and talk to your colleagues about their own experiences.
3. Check with ECRI, FDA, nursing colleagues, and other device sources for any failures associated with this device.
4. Answer the questions posed by the flow diagram of Fig. 5:

   State the medical/nursing purpose.
   State the device purpose(s).
   List the operational steps for each purpose.
   Identify the fundamentals for each operational step.
   Identify the science and technology principles on which the fundamental is based.
   For each purpose of the device list the known and predicted failures; associate each failure with a component and subcomponent.

5. Write a training lesson plan based on the gathered material.

Two approaches may be used to develop lesson plans while using this model. The first selects a device or multiple purpose device and develops content sequentially based on levels 1-4. The second method selects a group of devices that represent the overall required purposes for study and content. The latter method assumes a certain knowledge base, understanding, and experience on the part of the student. The model permits either approach.

VIII. CONCLUSIONS

The problem of sharing information on incroproation and utilization of ever-changing informational and therapeutic devices into health care is great, and education of the user is essential for solution. Technology will continue to make possible the vision of science in the very real world of the ill, inform, and the care-givers, who must communicate with the inventors, the manufacturers, the marketers, and the maintainers. The manufacturer as the developer of the device controls the purpose, quality, design, and components. The manufacturer then assures the user, the FDA, and the public that the equipment is safe to use in clinical care if used according to the information and accompanying directions included in the manuals and labeling. The process assumes that (1) these materials are understood in such detail by the learner-user, who can then become the user-teacher to other staff or patients; (2) the level of content matches the level of user knowledge, experience, and purpose; and (3) the level of information required is the same for all users at all times.

A mismatch in communication occurs when the different disciplines interface to use the device, one of many others, in a given clinical setting, to care for a patient. The purposes of the device become singular in application when used on an individual, and the focus of the user changes from device use to patient care. Any misunderstanding or lack of knowledge becomes readily apparent.

This standard simplified model ensures that all of the information is covered for each device in a format deemed appropriate by the *users* and the FDA. No new information is necessary. The model by design is pragmatic and practical for ultimate use in clinical settings by people with a need to know about a device at a given level of complexity. The content structure is based on information supplied by the manufacturer. If the manufacturers gear up *now* to provide the device information to their customers in the model format, the users will benefit from previous use of the format during their formal professional education. The manuals and directions used by the maintainers and users would thereby be congruent with user education and regulatory requirements. Examples of model use are included in the Appendix. Nursing is a prime user and teacher of users of medical devices and must be an informed participant in the creation of an educational system that provides for a logical, cohesive development of information and content. The Abbey-Shepherd device education model provides such a construct.

REFERENCES

1. K. Steel and P. M. Gertman, Iatrogenic Illness on a General Medical Service at a University Hospital, *N. Engl. J. Med. 304*(11):638 (1981).
2. American Nurses' Association (ANA), *Facts About Nursing 1986-1987*, Table 4.1, American Nurses Association, Kansas City, MO, 1987.
3. E. Hay, Equipment Inservice: A Nurse's Perspective, Proceedings of the 23rd Annual Meeting & Exposition of Association for the Advancement of Medical Instrumentation, Washington, D.C., May 17, 1988.
4. J. C. Abbey, J. C. DePalma, and C. Rome, Survey of Bioinstrumentation Instruction in NLN Accredited Baccalaureate Programs (unpublished), 1986.
5. J. C. Abbey, J. C. DePalma, and C. Rome, Survey of Bioinstrumentation Instruction in NLN Accredited Master's Programs (unpublished), 1986.
6. J. C. Abbey, J. C. DePalma, and C. Rome, Exploratory Opinionnaire of the Value of a Formal Course on Understanding Bioinstrumentation (unpublished), 1986.
7. C. Breu and K. Dracup, Survey of Clinical Care Nursing Practice, Part III: Responsibilities of Intensive Care Unit Staff, *Heart and Lung 11*(2): 157 (1982).
8. U.S. Department of Health & Human Services (USDHHS), *Secretary's Commission on Nursing Final Report Vol. I.* Washington, D.C., p. 7, December 1988.
9. U.S. Department of Health & Human Services (USDHHS), *Secretary's Commission on Nursing Final Report Vol. I.* Washington, D.C., p. 8, December 1988.
10. *Home Health Line XI*:30 (February 21, 1986).
11. J. Tehran, R. Thornburg, and D. Grundfast, Managing a Home Care Equipment Program, in *Plant Technology & Safety Management Series: Medical Equipment and Quality Care*, D. Bushelle and O. R. Keil, eds., pp. 25-31, Joint Commission on Accreditation of Hospitals, Chicago, 1987.
12. *Medicare Home Health Agency Manual* (HIM-11), Section 206.3(b), US Department of Health and Human Services, Washington, D.C., February 1985.
13. DME Dealers Face Grim Liability Choice, *Home Health Line XI*:104 (March 31, 1986).
14. J. F. Dyro, *The Imapct of Technology in Biomedical Engineering & Nursing*, Association of Medical Instrumentation, Arlington, VA, 1983.
15. S. K. Agarwall, Infusion Pump Artifacts: The Potential Danger of Spurious Dysrhythmia, *Heart & Lung 9*(6):1063 (1980).

16. L. H. Amundson, Iatrogenesis: A Review, *Continuing Education June*: 411 (1985).
17. N. N. Abramson, K. S. Walk, A. N. A. Grenuik, D. Robinson, and J. V. Snyder, Adverse Occurrences in Intensive Care Units, *JAMA 24*(14): 1582 (1980).
18. H. R. Harton, Iatrogenic Complications: The Nurse Device User, *Proceedings AAMI 17th Annual Meeting*, Arlington, VA, Association for the Advancement of Medical Instrumentation, p. 47, 1982.
19. J. C. Abbey and M. D. Shepherd, Nursing and Technology: Moving into the 21st Century, *Dean's Notes 10*(3):1-2 (January 1989).
20. J. M. A. Lenihan and J. C. Abbey, Symbiosis: Nursing & the Bioengineer, *Nursing Clinics of North America 13*(4):589-595 (December 1978).
21. R. Smith and G. C. Brdlik, Medical Devices: Where Should Learning Begin, *Dean's Notes 6*(3):1-2 (January 1985).
22. National Student Nurses Association, Biomedical Instrumentation & Technology Education for Student Nurses. (Proposition No. 1), *Proceedings of the 33rd Annual Convention*, April 17-21, 1985; Anthony J. Jannetti, Pitmau, N.J., 1985.
23. U.S. Department of Health & Human Services, Food & Drug Administrations (USDHHS-FDA), *Nursing & Technology: Moving into the 21st Century Conference Proceedings*, U.S. Department of Health & Human Services, Rockville, MD, HHS Publication FDA 89-4231, April 1, 1989.
24. J. C. Abbey and M. Shepherd, The Abbey-Shepherd Device Education Model, *Proceedings, Nursing Technology; Moving into the 21st Century*, May 16-18, 1988, Annapolis, MD (HHS Pub. FDA 89-4231), 1989.
25. M. D. Shepherd and J. C. Abbey, Training Device Operators: A Method of Identifying Course Content, in *Plant Technology & Safety Management Series: Educating Medical Device Users*, Joint Commission on Accreditation of Health Care Organization, Chicago, 1989.
26. M. D. Shepherd, *A Systems Approach to Hospital Medical Device Safety*, Monograph, Association for the Advancement of Medical Instrumentation, Arlington, VA, 1983.

APPENDIX FIVE EXAMPLES OF THE ABBEY-SHEPHERD DEVICE EDUCATION MODEL

A. Example 1

DEVICE NOMENCLATURE: Thermometer, mercury
MFG: MODEL:

A. A clear statement of the medical purpose(s) of the device:
   To diagnose diseased body conditions
B. Device purpose(s): A general statement of how the purpose(s) is accomplished:
   The metal reservoir of the mercury-in-glass thermometer is placed against body tissue; the mercury rises in the glass tube a distance proportional to temperature.
C. What are the operational procedures for each purpose of the device?
   1. Shake thermometer down.
   2. Place the reservoir in contact with the body surface.
   3. Wait 2 minutes for stabilization.
   4. Read gradations on glass and record temperature readings.

D. What are the fundamentals on which the operational procedures are based?

1. At room temperature, the mercury will be at some height in the tube. "Shaking the thermometer down" lowers the mercury to a temperature point lower than body temperature.
2. The reservoir is metal and allows the heat to conduct readily to the mercury; the mercury expands with heat and rises in the tube.
3. The mercury will continue to expand until the mercury temperature is the same as the tissue. This would be an equilibrium point. The time for temperature equilibrium for most mercury thermometers is about 2 minutes.
4. The glass column has been calibrated in temperature C or F.

E. What are the science and technology areas on which the fundamentals are based?
Heat; temperature expansion of metals.

F. As related to systems operational principles—the interaction of device, patient, nurse, facility, and environment—what should be taught about each factor to assure the safe and proper functioning of the device?

1. Device (design, deterioration, maintainer):
   a. Accuracy and precision
   b. Calibration drifts with age due to air leaks
2. Nurse (misuse and abuse):
   a. Operational procedures of mercury thermometers
   b. Cleaning between patients
   c. Rectal perforations
   d. Measurement location
   e. Subject to breakage
   f. Clean up of broken thermometer and toxic effects of mercury
3. Patient (active and passive):
   a. No drinking or smoking before measurement
   b. Breath through nose during oral temperatures
   c. Children vs. adults
   d. Circadian effects
4. Facility (bed area) (design, deterioration, maintainer):
5. Environment (heat, light, sound, EMI, etc.): Length of time for measurement may vary with room temperature.

B. Example 2

DEVICE NOMENCLATURE: Sphygmomanometer, mercury
MFR: W. A. Baum MODEL: #33

A. A clear statement of medical purpose(s) of the device:
To diagnose cardiovascular disease.

B. Device purpose(s): A general statement of how the purpose(s) is accomplished:
Arterial flow is cut off due to pressure in a pressure cuff; cuff pressure is slowly reduced; mercury pressure levels in the cuff are related to arterial pulse sounds.

C. What are the operational procedures for each purpose of the device?

1. Select proper cuff size for the patient
2. Mount with cuff bladder centered on right brachial artery.
3. Squeeze bulb to increase pressure in cuff; do not exceed about 30 mm Hg above systolic level.
4. Place stethoscope tip over artery just below cuff.
5. Slowly reduce cuff pressure at about 2 mm Hg/sec until first sounds are heard; record systolic reading.
6. Continue reducing pressure until last sound is heard; record diastolic pressure reading.

D. What are the fundamentals on which the operational procedures are based?

1. Three cuff sizes are generally available; the cuff bladder should cover at least 75% of the circumference of the arm or be replaced by a cuff with a larger bladder. Small cuffs make normotensives appear hypertensive. Large cuffs make hypertensives appear normal.
2. Proper mounting of the cuff relates to accurate pressure readings.
3. The limit on bladder pressure is adequate to shut off the brachial pulse but, for the short measurement time, not sufficient to be harmful.
4. The stethoscope bell needs to be at the optimal location to hear sounds.
5. Listening for sounds while the pressure drops too fast could result in a greater error in pressure reading. At 2 mm Hg/sec drop in pressure, the error is minimal. When the first sound is heard, the pressure within the artery is equal to the pressure in the cuff.
6. As the bladder pressure continues to drop the pulse sounds fade away. At the last sound, the bladder pressure is equal to diastolic pressure.

E. What are the science and technology areas on which the fundamentals are based?
Mechanics (hydrostatics/hydrodynamics); sound.

F. As related to systems operational principles—the interaction of device, patient, nurse, facility, and environment—what should be taught about each factor to assure the safe and proper functioning of the device?

1. Device (design, deterioration, maintainer):
   a. Leak free
   b. Valves allow controlled fall of pressure
   c. Ill fitting connections
   d. Different cuff sizes available
   e. Clean mercury
   f. Manometer "zeros" OK
   g. Accuracy and precision
   h. Size of cuffs
   i. Subject to mercury spills
2. Nurse (misuse and abuse):
   a. Operational procedures of device
   b. Relaxed nurse
   c. Arm at heart level
   d. Mercury "0" at heart level

e. Pressure drop about 2 mm Hg/sec
f. Read meniscus at eye level
g. Mercury cleanup and toxic effects

3. Patient (active and passive):

a. No recent smoking/eating
b. Anxious
c. Relaxes patient; about 6 hours after admittance

4. Facility (bed area) (design, deterioration, maintainer):
Bed available (semirecumbent position most relaxing to patient)
5. Environment (heat, light, sound, EMI, etc.):
A noisy environment disturbes the patient

C. Example 3

DEVICE NOMENCLATURE: Electrocardiograph
MFR: Hewlett-Packard MODEL: 1511A

A. A clear statement of the medical purpose(s) of the device:
To diagnose heart disease.
B. Device purpose(s): A general statement of how the purpose(s) is accomplished:
Detect surface potentials at several points on the body, amplify them, and display on paper recorder.
C. What are the operational procedures for each purpose of the device?

1. General device conditions: The following features are to assure the device is ready to perform its intended purpose(s):

a. Turn power switch to "on."
b. Center baseline with position control.
c. Pause at each dot on the selector switch, before advancing to the next lead.
d. Clean electrodes after EKG has been taken.
e. Calibrate electrocardiograph.
f. Adjust stylus heat.

2. EKG:

a. Prepare the patient's skin as follows: Expose the patient's forearms and calves and clean the skin with rubbing alcohol. Squeeze a spot of conductive gel about the size of a dime onto the inside of the forearm about midway between wrist and elbow and onto the inside of each calf midway between ankle and knee. Rub gel into the skin until a slight reddening occurs. Leave slight moist residues of gel, and fasten electrodes with rubber straps. Hook straps over the pins so electrodes make good contact, but not so tightly as to cause discomfort. Connect the patient cable to the electrodes and to the electrocardiograph. Follow the standard color codes on the cable leads when making electrode connections.
b. Set paper speed.
c. Set sensitivity to 1/4, 1/2, 1, or 2.
d. Set power switch to "Run."

e. Advance lead selector to Lead I, and record a length of EKG; repeat with Lead II and other leads. The V leads are taken after the suction bulb is placed in each of the standardized positions.
f. Code the different lead records with the Mark button.

D. What are the fundamentals on which the operational procedures are based?

1. General device conditions:

When the power switch is turned "on" electricity is made available to various components inside the electrocardiograph. Essentially this warms them and places them and their circuits in a state in which the are ready to perform their expected task.

After the power switch is turned to "on," the internal components of the electrocardiograph warm up. This warming of components may place the baseline of the paper recorder off center. The position control allows the positioning of the baseline back to the center of the paper.

When moving the selector switch from one lead to another, the switch is selecting from the different electrodes. Because electrode potentials may be slightly different, the change may attempt to provide an artifact on the baseline. This artifact can be minimized if a pause occurs between the leads.

Electrode gel left on the silver electrodes causes corrosion of the electrodes. This corrosion provides a resistive coating on the electrodes and would not allow good electrical contact with the patient's skin. Cleaning with warm water after each use and occasionally buffing the electrodes to a shiny appearance will assure better electrical contact.

If the selector switch is in Std position, when pressing the "Standard" button, a 1-mV calibration pulse is generated within the electrocardiograph. If the sensitivity control is in the 1 mV/10 div position, the pen will deflect upward 10 divisions on the paper recorder. When an EKG potential of 1 mV exists under the electrodes to be measured, it will give the same 10-division deflection on the paper recorder as did the 1-mV calibration pulse.

Adjusting the stylus heat to its proper temperature sets the darkness of the EKG recording.

2. EKG:

a. To obtain clear EKG recordings, the electrical connections between the patient and the machine must be as good as possible. The steps followed in the skin preparation are intended to assure good connection; rubbing alcohol will remove oils or creams that are not very good electrical conductors. Conductive gel is designed to allow the easy conduction of electricity across the electrode skin junction. The rubbing of the gel into the skin penetrates the outer, drier layer of skin. Dry skin is also a poor conductor of electricity. Placing silver electrodes over the prepared sites and fastening them with straps completes the electrical connections. Connecting the patient cable leads to the electrodes completes the electrical circuit that includes heart activity.

b. Paper speed is set at 25 mm/sec or 50 mm/sec depending on the waveform. If the waveform is rapid, the faster paper speed may be selected to provide greater resolution of the waveform.
c. The sensitivity switch controls the vertical amplitude of the EKG on the paper recorder. It has four positions—1/4, 1/2, 1, of 2 mV/10 div. The numbers indicate that 1/4 millivolt (mV) at the electrodes will provide a 10 division (div) vertical deflection on the recording paper; 1 mV/10 div has been found to be the most frequently used. On occasion unusually large or small EKG potentials on the body surface will require a change in the sensitivity switch.
d. When the power switch is set to "Run," electricity to the paper recorder motor may be applied through the lead selector switch. The electricity will only be applied when the lead selector is in Lead I, II, or other lead positions. The paper speed will be determined by the paper speed switch.
f. When the lead selector swtich is advanced to Lead I, electricity is applied to the paper recorder motor while, simultaneously, the electrical activity occurring under the left arm and right arm electrodes are amplified and sent to the recorder to deflect the pen in a vertical direction. After a section of that lead is recorded, the lead selector is placed in the Dot position, stopping the recorder motor and the electrical activity. This procedure is repeated by proceeding to the next lead position.
g. For each lead position, the standardized Mark codes must be used so the EKG reader will be sure which lead has been recorded.

E. What are the science and technology areas on which the fundamentals are based?

Electricity and magnetism: Included should be batteries and AC electricity; circuits; terms; conductivity of materials; components (i.e., capacitors, resistors, etc.); Ohm's law, measurement techniques, galvanometers, oscilloscopes, and motors.

F. As related to systems operational principles—the interaction of device, patient, nurse, facility, and environment—what should be taught about each factor to assure the safe and proper functioning of the device?

1. Device (design, deterioration, maintainer):

   a. Hot pen of recorder is burnt out.
   b. Calibrate battery dead.
   c. Dirty lead selector switch.

2. Nurse (misuse and abuse):

   a. Operational procedures for the purpose
   b. Operational maintenance (i.e., paper replacement, electrode cleaning, etc.)
   c. Patient cable lead breakage
   d. Abused cords and plugs
   e. Dirty electrodes
   f. Loose electrodes
   g. Electrodes in wrong location
   h. Electrode connection identifiers and differences between countries

3. Patient (active and passive):

   a. Hairy/greasy/dry skin
   b. Muscle tremor
   c. Tense patient
   d. Pacemaker or other implants
   e. Mechanical movement
   f. Conversation

4. Facility (bed area) (design, deterioration, maintainer):
   Facility designs that may affect operation include lack of electricity, poor grounds, and isolated power.
5. Environment (heat, light, sound, EMI, etc.):

   a. Electromagnetic interference (EMI), radiated
   b. Impluses from other devices (i.e., controllers, external pacemakers, etc.)
   c. General environmental factors (i.e., dust and heat)

D. Example 4

DEVICE NOMENCLATURE: DEFIB/SYNC/MONITOR
MFR: Physio-control MODEL: LifePak-7

A. A clear statement of the medical purposes(s) of the device:

   1. To convert the electrical activity of the heart from ventricular fibrillation to normal sinus rhythm
   2. To continuously monitor the heart for abnormal electrical activity
   3. To convert less serious abnormal heart rhythms to normal
   4. To sense and provide a permanent record of the electrical activity of the heart
   5. To sense and display patient heart rate

B. Device purpose(s): A general statement of how the purpose(s) is accomplished by the device:

   1. By providing an impulse current to the heart of such a magnitude as to depolarize all cells of the heart
   2. By detecting and displaying electrical activity of the heart on the screen of a cardiac monitor
   3. By synchronizing the current impulse from the defibrillator with one of the repetitious cycles of the heart
   4. By directing the electrical activity to a paper recorder
   5. By electronically counting the EKG and displaying the results in digital form

C. What are the operational procedures that accomplish each purpose of the device?

   1. General device conditions: The following reatures are to assure the device is ready to perform its intended purpose(s):

      a. Check each battery level indicator to determine if batteries are properly charged.
      b. Check the "mon/Defib Bat Charge" lamp to see if it is "on."

c. Follow normal defibrillator use procedure and discharge the energy into the test load.
d. Snap-on pedi-attachments (if appropriate).
e. Turn LP7 Power switch to "on."

2. Defibrillator:

a. Select energy setting to be used.
b. Prepare paddle electrodes by applying gel to paddle surface.
c. Charge defibrillator with either the "Charge" button on the paddle handles or on the defibrillator panel. The lamp on the side of the Apex paddle lights when the defibrillator is ready to be discharged.
d. Expose patient's chest and place paddles firmly against chest wall.
e. Discharge the energy into the body by pressing both "Discharge" buttons on the paddle handles.

3. Monitor:

a. Prepare the patient's skin and place the EKG disposable electrodes on the patient; connect the patient cable to the electrodes and to the "Pt Conn" on the monitor.
b. Change the "Select" switch to view Lead I, II, or III or Paddle. NOTE: If the "Select" switch is turned ot the paddle position, the EKG that exists on the body and between the paddles will be displayed on the monitor screen.
c. Press Calibrate button and adjust "EKG Size" control for acceptable amplitude of EKG on the monitor screen.
d. Press "Freeze" button if arrhythmic changes viewed on the monitor screen are to be studied more closely.
e. If EKG is available, QRS lamp (to left of digital Heart Rate) will flash once with each complex; an audible beep can be heard if the QRS Volume is turned up.

4. Synchronized cardioverting:

a. Press "SYNC" button on panel.
b. View electrical activity as described in MONITOR, above.
c. Look at monitor screen to assure that the defibrillator marker is visible at the appropriate location on the screen.
d. (Follow all other steps under DEFIBRILLATOR, above.)

5. Heart rate: Heart rate is automatically displayed in digital form if either the patient electrodes or the paddle electrodes are in contact with the patient and the power switch is "on."
6. Paper recorder: Turn "on" the recorder on/off switch.

D. What are the fundamentals on which the operational procedures are based?

1. General device condition:

The battery level indicators provide a rough indication of the charge on the batteries. If battery level indicators are in the "Green" region, the batteries are probably in good repair. If in the "Red" region, the batteries may just need charging or they may be deteriorating badly and need replacing.

The "Mon/Defib Bat Charge" lamp indicates that the power plug has been inserted into an active electrical outlet and the mon/defib batteries are charging.

The readiness state of the defibrillator may be checked with the resistance of the "dummy" patient built into the defibrillator paddle holder. Upon discharge the lamp on the paddle will flash, indicating that energy has passed through the defibrillator cable and paddles to reach the dummy patient.

If the patient is a pedi patient the size of the electrodes placed on the chest will be smaller. This is appropriate because a lesser pulse current intensity is required for children and because the adult size electrodes will not fit on the child's chest.

When the power switch is turned "on" electricity is made available to various components inside the defibrillator. Essentially this places the components and circuits in a state in which they are ready to perform their expected task.

2. Defibrillator:

   a. The energy selected is proportional to body size and the location of the electrodes. The larger the body, the more energy required. Transthoracic electrode positioning requires more energy than AP electrode positioning.
   b. It is most efficient if all the electrical energy provided by the defibrillator reaches the patient. For this to occur, the electrical contact at the chest-paddle interface must be good. The conductive gel provided with the defibrillator is designed to conduct electrical currents with minimal loss of energy at the chest-paddle interface.
   c. Charging the defibrillator means that the defibrillator's energy-storing capacitor is filled with electrical energy and ready for discharge.
   d. Placing the gelled electrodes on the chest completes the electrical circuit in which the defibrillator is the energy source and the patient is the load (energy consumer).
   e. Pressing the "Discharge" button allows the energy to be transferred from the defibrillator capacitor to the patient. Both buttons are required to be pressed simultaneously to discharge the defibrillator. This is intended to prevent a discharge if one buttone is accidentally pressed.

3. Monitor:

   a. Disposable electrodes are regelled; preparing the skin by cleaning and rubbing assures a good conductive contact when the electrode is in place. Connecting the patient cable completes the electrical circuit that includes heart activity following the patient cable into the input resistances of the monitor.
   b. The "leads" noted here are not the standard Lead I, II, or III normally referred to when taking diagnostic EKGs. The EKGs observed on the monitor screen are dependent on the placement of the EKG electrodes.

      On occasion, if there is enough time to place the EKG electrodes on the patient, a quick look at the EKG activity can be done via the paddle electrodes. If none is observed, defibrillation can begin immediately.

c. When pressing the "Calibrate" button, a 1-mV pulse is generated within the monitor. By adjusting the "EKG Size" control, the 1-mV calibrate pulse can be adjusted such that it deflects the monitor electron beam a number of centimeters vertically. When an EKG is measured at the end of the patient cable, 1 mV activity will give the same deflection.

d. Certain arrhythmic changes are of particular interest to attending physicians/nurses. When they are observed on the monitor screen, the screen can be frozen for further study of the incident; if of great interest, the frozen section can be directed to the printer for a hard copy.

e. The QRS audible and visual indicators have some value when a repetitious EKG waveform is available. While performing other tasks, the changes in sound may be an indicator of changing patient conditions.

4. Synchronized cardioverting:

   a. With "Sync" button pressed, the defibrillator will discharge on the first pulse after the defibrillator paddles "Discharge" buttons are next pressed. The "Sync" button must be pressed once for each cardioverting impulse.
   b. By viewing the electrical activity on the monitor screen, the marker will identify the part of the cardiac cycle where the discharge will occur. The intent is to assure that the impulse will not fall during the "T" wave (vulnerable period) and result in ventricular fibrillation.

5. Heart rate:
6. Paper recorder: Turning "on" the recorder on/off switch applies voltage to the paper recorder motor and allows the EKG to be printed on the paper. The motor moves at a constant speed of 25 mm/sec.

E. What are the science and technology areas on which the fundamentals are based?

Electricity and magnetism. Included should be batteries and AC electricity; high voltage vs. low voltage; circuits; terms; conductivity of materials; components (i.e., capacitors, resistors, etc.); Ohm's law, measurement techniques, galvanometers, oscilloscopes, and motors.

F. As related to systems operational principles—the interaction of device, patient, nurse, facility, and environment—what should be taught about each factor to assure the safe and proper functioning of the device?

1. Device (design, deterioration, maintainer):

   a. Paddle designs to protect user against shock
   b. Battery failures

2. Nurse (misuse and abuse):

   a. Operational procedures for each purpose
   b. Defib testing to assure readiness
   c. Adequate but not excessive gel quantity used on paddle
   d. Care of the batteries

   e. Proper placement of paddle electrodes on chest (not on top of EKG electrodes, NTG patches, internal pacer or other foreign bodies)
   f. Discharging defibrillator in air or with paddles pressed together is not advised

3. Patient (active and passive):
   a. Avoid hirsute areas
   b. Patient with an implant
4. Facility (area of defibrillation) (design, deterioration, maintainer):
   a. Facility designs that may effect operation (i.e., circuit breakers, electrical outlets, etc.)
   b. Conductive fluids surrounding patient
5. Environment (heat, light, sound, EMI, etc.):
   a. Patient on O_2 therapy (potential for fire)
   b. Difference between several manufacturers' defibrillators
   c. Back-up defibrillators available
   d. Burns under EKG electrodes due to defibrillation energy following EKG electrode/patient cable pathways

E. Example 5

DEVICE NOMENCLATURE: Glucose monitor
MFR: Lifescan, Inc. MODEL: 3000

A. A clear statement of medical purpose(s) of the device:
   1. To analyze a blood sample for glucose concentration
   2. To sotre more than 1 glucose reading in a memory
B. Device purpose(s): A general statement of how the purpose(s) is accomplished:
   1. A drop of blood reacts with chemicals on a test strip; a light reflectance meter measures the amount of glucose on the test stip and displays it on a digital meter.
   2. A microcomputer stores a series of glucose test results.
C. What are the operational procedures for each purpose of the device?
   1. General device conditions: The following features are to assure the device is ready to perform its intended purpose(s):
      a. Turn power siwtch "on"
      b. Calibration check
      c. Metric conversion switch
      d. Volume control switches
      e. Test Start Method switch
   2. Glucose monitoring:
      a. Press power/start switch.
      b. Obtain a large hanging drop of blood.

c. Press power/start switch again; during countdown, prepare to apply blood to test strip.
d. When window displays "60" apply blood; cover entire pad.
e. When window displays "23" prepare to blot.
f. When window displays "20" blot twice.
g. Place test strip into holder and close door any time before window displays "00."
h. After window displays "00," test result will be displayed.

3. Memory storage:

NOTE: All test results displayed in the window at the end of a full 60-sec test procedure are automatically stored in the monitor's memory bank. Only the last 29 test results will be stored.

a. To recall stored test results of your last test, press the memory recall button on the meter and read the result in the display window.
b. To recall each test result, prior to the one in the display window, press the memory recall button once.

D. What are the fundamentals on which the operational procedures are based?

1. General device conditions:

When the power switch is turned "on" electricity is made available to various components inside the glucose meter. Essentially this warms them and places them and their circuits in a state in which they are ready to perform their expected task.

The calibration of an instrument may vary depending on the failure of components, aging, dirty components, abuse, heat, humidity, and the like. For this reason, to assure accurate glucose readings, it is necessary to check the calibration on a regular basis. Although the calibration can not be adjusted on this particular glucose meter, it can be checked to assure measurement accuracy. A calibration checkstrip comes with the instrument, and the manufacturer recommends daily checking, before adjusting your medication based on its readings, and when your results do not seem to reflect the way you feel.

In the United States the units of measurement are mg/dl while in Canada the units are mmol/l. The metric conversion switch allows you to select either.

The glucose meter sound cues are primarily a reminder that a certain time interval has passed or an event has occurred. Some patients like them and some don't. Their use is optional. Volume control switches located on the bottom of the glucose meter set the sound level of the cues from Loud to Soft to Off.

To being the glucose test the patient can either use (a) the Countdown Start or (b) the Instant Start method. With the Countdown Start, the patient is given 4½ sec before they need to apply blood to the test pad. This 4½ sec may be needed during the intial use of the glucose meter or for persons whose physical condition require it. Once the procedure is understood and well practiced, the impatient patient may wish to use the Instant Start method. Since the Instant Start method does not have a 4½ sec delay, the patient must be ready to immediately apply blood to the test pad. Test Start Method switches are available on the bottom of the glucose meter that allow changing from one method to the other.

2. Glucose monitoring:

   a. The pressing of the power/start switch again begins the countdown period. During the 4½-sec countdown, the patient is preparing themselves to apply a drop of blood to the pad on the test strip.
   b. The penlet can be used to prick the finger and obtain a large, hanging drop of blood.
   c. When the glucose meter window displays "60," blood is applied to the entire pad. This begins the 40-sec time period required for chemicals on the test pad to interact with the blood. The interaction separates out the red blood cells and allows them to be removed during the blotting process.
   d. Because the 40-sec timing is critical, when the glucose meter window displays "23" a beep is heard and this warns the patient that they should prepare to blot the pad.
   e. When the glucose meter displays "20," a long audible beep is heard and the blotting process begins. This step removes the red blood cells from the test pad. Blotting must occur twice and for 2 sec per blot.
   f. The test strip is placed into the test strip holder and the door is closed any time before the window displays "00." The light reflectance section of the monitor measures the glucose level of the blood.
   g. After the window displays "00," the glucose test result will be displayed on the monitor.
   h. Record your test results in your log book for future reference by your doctor.

3. Memory storage:

   Memory storage is accomplished through use of a microcomputer. The microcomputer stores in its memory the previous 29 test results.

E. What are the science and technology areas on which the fundamentals are based?

   1. Electricity/magnetism, optics
   2. Electrical circuits, microcomputers, reflectance measurements

F. As related to systems operational principles—the interaction of device, patient-user, home, and environment—what should be taught about each factor to assure the safe and proper functioning of the device?

   1. Device (design, deterioration, maintainer):

      a. Batteries dead
      b. Meter out of calibration

   2. Patient/operator (misuse, abuse, active, passive):

      a. Operational procedures for purpose(s)
      b. Operational maintenance
      c. Batteries not properly inserted

d. Couldn't get enough blood to make test
e. Could get enough blood but not enough placed on strip
f. No calibrations made
g. Glucose meter used even when error message read
h. Using outdated disposables/controls
i. Using obviously contaminated disposables/controls
j. Warranty card not sent in to allow manufacturer to update device
k. Other manufacturer's test strip used
l. Abused instrument
m. Fluids in monitor
n. Dirty instrument

3. Facility (home) (design, deterioration, maintainer):
4. Environment (heat, light, sound, EMI, etc.):

a. Electromagnetic interference (EMI)
b. Temperatures above 95°F during instrument use
c. Controls effected by humidity, temperature, heat

part six

Historic Overview/Legal Framework of Occupational and Environmental Safety and Health

43

An Overview of Regulations Governing Occupational Exposure

SARA D. SCHOTLAND and JOHN M. BREDEHOFT *Cleary, Gottlieb, Steen & Hamilton, Washington, D.C.*

I. INTRODUCTION

The most significant regulation of worker health and safety in the United States is the responsibility of the federal Occupational Safety and Health Administration (OSHA). A relatively large number of states have federally approved occupational health and safety programs, and operate their state programs under the framework of the federal Occupational Safety and Health Act of 1970.

This chapter provides an overview of the regulation of occupational exposure to potentially harmful substances under the Occupational Safety and Health Act. General standards imposed on workplace exposure are examined, and the regulation of occupational exposure to ethylene oxide is discussed. This chapter confines its discussion of the ethylene oxide issue to the actual federal regulatory standards and requirements imposed on workplaces using ethylene oxide. Subsequent chapters will deal in some detail with the manner in which occupational exposure to ethylene oxide may be managed and reduced.

II. REGULATION OF OCCUPATIONAL EXPOSURE TO CHEMICALS UNDER THE OCCUPATIONAL SAFETY AND HEALTH ACT

This section briefly reviews the structure of federal statutory protection of worker health and safety. The general federal statutory duty to provide a safe and healthful working environment and the specific standard-setting process that attempts to set criteria for such a working environment are discussed. Finally, this section discusses the federal regulation of both short-term and shift-long occupational exposure to a specific chemical of interest to the medical device industry, ethylene oxide.

A. The Occupational Safety and Health Act of 1970

The Occupational Safety and Health Act of 1970 (the "OSH Act" or the "Act," codified at 29 U.S.C. 651 et seq.) is the basic statutory authority for federal regulation of occupational exposure to hazardous agents in the workplace. The goal of the Act is "to assure so far as possible every working man and woman in the Nation safe and healthful working conditions" [1].

Section 2 of the Act authorizes the Secretary of Labor (functionally, OSHA) to, among other things:

Set mandatory occupational safety and health standards applicable to businesses affecting interstate commerce [2].

Provide medical criteria that will assure insofar as practicable that no employee will suffer diminished health, functional capacity, or life expectency as a result of his work experience [3].

Provide for the development and promulgation of occupational safety and health standards [4].

The Act defines "employer" as "a person engaged in a business affecting commerce who has employees" [5].

Section 5 of the Act sets forth the general statutory duty to provide a safe working environment. Section 6 of the Act describes the process whereby more specific health and safety standards are established.

B. The General Duty Clause

Section 5(a)(1) of the Occupational Safety and Health Act contains what is known as the "General Duty Clause" of the Act. That section provides that each employer "shall furnish to each of his employees employment and a place of employment which are free from recognized hazards which are causing or are likely to cause death or serious physical harm to his employees."

1. *Use of the Clause in the Absence of a Specific Standard*

OSHA has contended that the General Duty Clause imposes a duty on the employer to provide a safe workplace even in the absence of specific standards and regulations promulgated by OSHA. The United States Court of Appeals for the Second Circuit examined the scope of the General Duty Clause in *Carlyle Compressor Co. v. Occupational Safety and Health Review Commission* [6]. In that case a company that manufactures air-conditioning compressors was cited by an OSHA compliance officer for failing to provide a guard on a grinding machine. The OSHA inspector's action led to the filing of a complaint by the Commission that alleged violations both of the OSHA standard for guarding machines and of the General Duty Clause. The Court agreed with the company that the specific OSHA regulatory standard for providing guards on machines did not apply to the type of machine, or to the type of hazard, identified by the OSHA inspector. Nonetheless, the Court held that even in the absence of a specific health or safety standard, due to the absence of a guard on the machine the "place of employment" was not "free from recognized hazards which are causing or are likely to cause death or serious physical harm to his employees," within the meaning of the General Duty

Clause. The Court identified four requirements in order to establish a violation of the General Duty Clause:

A hazard must exist.
The employer or the industry must recognize the hazard.
The hazard must be causing or likely to cause death or serious physical harm.
OSHA must be able to identify the specific steps the employer could take to alleviate the harm, including demonstrating the feasibility and likely utility of the measures [7].

The Court held that remedial measure suggested by OSHA "need not completely solve the problem as long as it reduces the danger," and opined that "[c]ourts are entitled to base conclusions upon common sense where the facts so warrant" [8].

2. *Use of the Clause Where a Specific Applicable Standard Also Exists*

The General Duty Clause traditionally has been used by OSHA only in the absence of a substance-specific standard. This is consistent with, and reflected by, OSHA regulations that provide that "[a]n employer who is in compliance with any standard in this part shall be deemed to be in compliance with the requirements of Section 5(a)(1) of the Act [the General Duty Clause], but only to the extent of the condition, practice, means, method, operation, or process covered by the standard" [9]. The United States Court of Appeals for the District of Columbia Circuit has, however, interpreted the scope of an employer's duty under the General Duty Clause in a situation where a specific OSHA standard already exists respecting the substance at issue. In *International Union v. General Dynamics Land Systems Division* [10], the Court held that an obligation of an employer to take action under the General Duty Clause beyond that required by a specific standard arises when "an employer *knows* a particular safety standard is inadequate to protect his workers . . ." (emphasis added).

OSHA itself has relied on the General Duty Clause to enforce a "standard" more stringent than the OSHA standard for cadmium. In a December 1987 "Instruction" issued to OSHA Regional Administrators and Area Directors, OSHA stated that it considered exposures to cadmium that are in excess of the American Conference of Governmental Industrial Hygienists (ACGIH) threshold limit value (TLV) to constitute a violation of the General Duty Clause. The ACGIH TLV was, however, substantially more strict than the then-existing OSHA standards for exposure to cadmium fumes and cadmium dust. The "Instruction" also set forth requirements for engineering controls, respiratory equipment, hazard communications, and other items. It appears that OSHA thus was using the General Duty Clause to apply a new substance-specific standard prior to formal revision of the standard through notice-and-comment rule-making procedures. This procedure appears to conflict with language in the legislative history of the Act, which suggests that the General Duty Clause was intended to protect employees "who are working under such *unique* circumstances that no standard has yet been enacted to cover this situation" [11; emphasis in original]. "The general duty clause . . . would simply enable the Secretary to insure the protection of employees who are working under special circumstances for which no standard has yet been adopted" [12].

The implications for employers of the General Duty Clause are several. First, an employer may run afoul of the Act even in the absence of a specific OSHA-promulgated regulatory standard for health or safety. Moreover, under OSHA's view of the Clause, compliance with an existing standard may not provide a "safe harbor" for a workplace which is—notwithstanding compliance—unsafe. There also are several limitations on the use of the General Duty Clause: the hazard must be known to or obvious to the employer or others in the industry, and it must be a relatively severe hazard, one that is causing or is likely to cause death or serious physical harm to employees.

C. Safety and Health Standards

Section 2 of the Act [2] authorizes the Secretary of Labor (i.e., OSHA) to set mandatory occupational safety and health standards applicable to businesses affecting interstate commerce. Section 5 of the Act [13] provides that each employer "shall comply with occupational safety and health standards promulgated under this Act." In practice, and despite the broad sweep of the General Duty Clause, the establishment and enforcement of "standards" has been OSHA's primary means of implementing its mandate from Congress to secure safe and healthful workplaces.

Section 6 of the Act [14] sets forth three different ways in which OSHA may set a standard. First, when the Act was passed the Secretary of Labor was authorized, without regard to standard constraints on federal rule making, to promulgate "national consensus standards" and "established Federal standards" as federal occupational safety and health standards. The Act required OSHA to promulgate those standards within 2 years after the effective date of the Act (essentially by the end of 1972). Those standards were in fact established, using extant standards from many sources, in 1971. The vast majority of those standards (some 400 of which concern occupational exposure to specific chemical substances) remained in effect until 1989. Second, OSHA may—using normal federal "notice and comment" rule-making proceedings and procedures as set forth in Section 6(b) of the Act—"promulgate, modify, or revoke" an occupational safety or health standard. This procedure has been used in establishing the currently applicable standards for exposure to substances such as vinyl chloride, arsenic, and asbestos, as well as in establishing the currently applicable standards for occupational exposure to ethylene oxide, which are discussed in greater detail later in this chapter. Third, under Section 6(c) of the Act, OSHA may establish an "emergency temporary standard" or ETS for exposure, where it is determined that "employees are exposed to grave danger from exposure to substances or agents determined to be toxic or physically harmful or from new hazards," and the emergency standard is "necessary to protect employees from such danger."

1. Standards under Section 6(b) of the Act

Section 6(b) authorizes OSHA to promulgate occupational safety and health standards, including standards that regulate permissible exposure to a specific substance. OSHA has also used the authority granted in Section 6(b) to promulgate "generic" standards, which apply not to users of or exposure to a particular substance, but rather apply across-the-board to all employers in a particular class or group. An example of a "generic" standard is OSHA's Hazard Communication Standard, discussed below. Substance-specific standards include the limits for exposure to ethylene oxide, also discussed below.

More generally, OSHA has acted under Section 6(b) to promulgate standards limiting occupational exposure to airborne concentrations of specific substances in two broad manners: a permissible exposure limit (PEL) for exposure as calculated for an entire working day (an 8-hr time-weighted average PEL), and short-term exposure as calculated as a short-term exposure limit (STEL) or most recently as an excursion limit (EL).

There are a number of specific limits on OSHA's ability to establish a standard. An important limitation is found in Section 3(8) of the Act, the definitional section. Since Section 6(b) authorizes OSHA to set an "occupational safety and health standard," the manner in which the meaning of that term is delimited is itself a limit on OSHA's authority. Section 3(8) provides that the term "occupational safety and health standard" means a "standard which requires conditions, or the adoption or use of one or more practices, means, methods, operations, or processes, *reasonably necessary or appropriate* to provide safe or healthful employment and places of employment" [15; emphasis added]. The U.S. Supreme Court, in a case arising from OSHA's regulation of occupational exposure to benzene (and thus often referred to simply as the *Benzene* case) interpreted this definitional section as a substantive limit on OSHA's rule making authority. Supreme Court Justice Stevens, writing for a plurality of the Court and announcing the judgment of the Court, wrote:

> [W]e think it is clear that § 3(8) does apply to all permanent standards promulgated under the Act and that it requires the Secretary [of Labor, i.e., OSHA], before issuing any standard, to determine that it is reasonably necessary and appropriate to remedy a *significant* risk of *material* health impairment. . . .
>
> [W]e think it is clear that the statute was not designed to require employers to provide absolutely risk-free workplaces whenever it is technologically feasible to do so, so long as the cost is not great enough to destroy an entire industry. Rather, both the language and structure of the Act, as well as its legislative history, indicate that it was intended to require the elimination, as far as feasible, of *significant* risks of harm. [16; emphasis added]

Justice Stevens held that OSHA, before promulgating any health or safety standard, must make a "threshold finding that a place of employment is unsafe—in the sense that significant risks are present and can be eliminated or lessened by a change in practices" [17]. The Court provided one concrete example to illustrate what it meant by a "significant" risk. A one-in-a-billion chance of dying from drinking chlorinated water, wrote Justice Stevens, "clearly could not be considered significant." On the other hand, a one-in-a-thousand chance of dying from inhaling vapors containing benzene "might well" be considered significant. Justice Stevens wrote, "Although the Agency has no duty to calculate the exact probability of harm, it does have an obligation to find that a significant risk is present before it can characterize a place of employment as 'unsafe'" [18].

Another limit on OSHA's power to promulgate permanent standards is the language in Section 6(b) of the Act itself respecting "promulgating standards dealing with toxic materials or harmful physical agents." That subsection provides that, in such circumstances:

> [OSHA] shall set the standard which msot adequately assures, *to the extent feasible, on the basis of the best available evidence*, that no employee will suffer *material* impairment of health or functional capacity even if such employee has regular exposure to the hazard dealt with by such standard for the period of his working life. [19; emphasis added]

In a case involving an appeal of OSHA's standard for occupational exposure to cotton dust, the U.S. Supreme Court held that this language required OSHA to undertake a "feasibility analysis," but did not require OSHA to calculate and weigh the costs and benefits associated with any particular standard [20] (the *Cottondust* case). The Court held that in passing the Act, "Congress viewed the costs of health and safety as a cost of doing business" [21].

2. *Ancillary Requirements*

OSHA promulgates not only standards providing for exposure limits, but also for certain ancillary requirements such as medical surveillance and labeling of assertedly hazardous substances in the workplace. The United States Court of Appeals for the District of Columbia Circuit examined OSHA's ability to impose "ancillary" requirements on employers in connection with setting permanent standards in *National Cottonseed Products Association v. Brock* [22]. The Court of Appeals upheld OSHA's promulgation of an occupational health standard for the cottonseed processing industry that required medical examination and testing of cottonseed processing workers; OSHA imposed the examination and testing requirements even though it did not impose a PEL, and despite its failure to find any significant risk of material health impairment. OSHA determined that the risk of harm to cottonseed workers would not be "material" even without a PEL, as long as medical surveillance was retained as a "back-stop." The Court found OSHA acted within the scope of its discretion.

Another example of standard-setting under Section 6(b) [23] illustrates the type of inquiry made by OSHA in certain circumstances. In July of 1986, OSHA issued revised standards governing occupational exposure to asbestos. OSHA lowered the 8-hr TWA PEL for asbestos, but did not adopt a short-term exposure limit (STEL). A number of groups appealed OSHA's decision to the United States Court of Appeals for the District of Columbia Circuit. In *Building and Construction Trades Dept., AFL-CIO v. Brock* [24], the Court upheld OSHA's findings respecting significant risk and the feasibility of the new PEL, but remanded to OSHA a number of issues, including whether a short-term limit should be imposed and whether a lower PEL should be established for certain industrial subgroups. Of particular interest is the Court's decision of OSHA's duty to issue the most stringent feasible standard. Industry groups challenged OSHA's decision, "based on policy considerations," not to impose as an ancillary requirements more stringent antismoking measures for workers exposed to asbestos. The Court recognized that synergism between smoking and asbestos meant that the risk posed by asbestos exposure is 10 times greater for smokers than for nonsmokers, but refused to require OSHA to impose more stringent requirements regardless of the magnitude of the risk to health. The Court states:

> In some abstract sense a regulation can always be tightened. It will be rare that any given extra twist will push the cost up to the levels necessary for economic infeasibility. . . . But it does not follow from this that

> the statute mandates every proposed tightening not shown to be [economically or technically] infeasible . . . just because it is theoretically capable of conferring a health benefit. OSHA's enforcement and employers' complianceeduties both depend on ordinary human beings. Multiplication of requirements tax their abilities. . . . At some point the confusion resulting from an extra burden is likely to outweigh any practical benefit [25].

Accordingly, the Court held, "a party challenging an OSHA standard must bear the burden of demonstrating that the variations it advocates will be feasible to implement and will provide more than a *de minimis* benefit for worker health."

In light of the Supreme Court's holding in the *Benzene* and *Cottondust* cases, this type of health benefit analysis—a limited cost/benefit evaluation by OSHA, weighing health costs and health benefits, as contrasted to a more general cost/benefit analysis weighing health benefits against financial costs—may be appropriate. It remains to be seen whether other Courts of Appeals will adopt a similar "*de minimis*" criterion for such ancillary requirements.

3. *Emergency Standards under Section 6(c) of the Act*

Emergency temporary standards promulgated under Section 6(c) of the Act [26] are effective only until superseded by a permanent standard (which must be promulgated within 6 months of the emergency temporary standard). Because an ETS is effective immediately upon publication in the *Federal Register*, without regard to "notice-and-comment" rulemaking procedures, the courts have recognized that such a standard should issue only where there is a grave danger that normal rulemaking proceedings are not possible and substantial harm may be expected in the absence of the standard. OSHA's 1981 decision not to issue an emergency temporary standard for ethylene oxide, for example, was upheld by the United States Court of Appeals for the District of Columbia Circuit in 1983 [27]. Similarly, another U.S. Court of Appeals overturned OSHA's decision to issue an ETS for asbestos (lowering the PEL) because the record did not contain sufficient evidence of risk or evidence indicating the necessity for the standard [28].

4. *National Consensus Standards under Section 6(a)*

Many "national consensus" and "existing federal" standards adopted as federal occupational health and safety standards by OSHA in 1971 pursuant to Section 6(a) of the Act [29] remained in effect through 1989, although the apparent intent of Congress in providing for their adoption without formal rulemaking was to provide some concrete standards pending OSHA's formal adoption of standards under Section 6(b). These standards were taken by OSHA from sources such as the federal standards for federal contractors under the Walsh-Healey Act, standards proposed by the ACGIH, and standards recommended by the American National Standards Institute.

A number of problems existed with these national consensus standards, and OSHA recently amended them *en masse*. Many of the standards were old even when OSHA adopted them in 1971; OSHA had not changed the standards adopted in 1971 to reflect changes made by the standard-setting bodies themselves over the past two decades or to reflect changes in industrial hygiene practice without instituting a formal rulemaking under Section 6(b).

While the vast majority of the national consensus standards adopted by OSHA were "safety" as opposed to "health" standards, some 400 occupational exposure standards were adopted for "toxic" substances. Those standards, which generally set PELs for various substances, are found in Subpart Z of Part 1900 of Title 29 of the Code of Federal Regulations. While occupational exposure to ethylene oxide is probably the most pressing issue for manufacturers and users of medical devices, these "Z-table" PELs do contain exposure limits for a variety of other substances that might be found in the medical device user or producer workplace. Such substances include acetone, ammonia, soluble compounds of barium, ethyl alcohol, methyl alcohol, and carbon tetrachloride.

In June of 1988, OSHA proposed extensive revisions to a number of the national consensus standards for "air contaminants," and proposed to set PELs for approximately 200 additional substances not currently regulated by OSHA [30]. These revisions became final in January, 1989 (although a judicial challenge is pending in the United States Court of Appeals for the Eleventh Circuit). OSHA reduced existing PELs for approximately 100 substances currently listed in the "Z-tables" and added or changed STELs for 70 substances. OSHA based this massive revision on a review of "health evidence" for each substance; OSHA concluded that "the new limits substantially reduce a risk of deleterious health effects among American workers, including cancers, central and peripheral neuropathies, lung disease, liver and kidney damange and other systemic effects." OSHA used the ACGIH TLVs as "starting points" in its review of these health effects, but did not adhere strictly to them. OSHA noted that, in addition to the wholesale revision of the Z-tables, "OSHA will continue its practice of rulemaking for individual substances when regulations of that type are necessary and appropriate."

5. *Generic Hazard Communication Standard*

As originally promulgated in 1983, OSHA's generic Hazard Communication Standard [31] (the Generic Standard) required manufacturers and importers of hazardous chemicals to assess the hazards of their products, and required certain employers to provide information to their employees in the form of labels, material safety data sheets, and training. In 1987, OSHA extended the reach of the Standard to all employers [32]. The Office of Management and Budget (OMB) disapproved that rule, and OSHA revised it to take into account some—but not all—of OMB's objections.

The Generic Standard generally covers any hazardous chemical known to be present in the workplace to which employees may be exposed under normal conditions of use or in a forseeable emergency. The Standard imposes a number of obligations.

MSDSs. Manufacturers or importers of a hazardous chemical must prepare a material safety data sheet (MSDS) listing all health hazards of the substance [33]. Manufacturers, importers, and distributors must provide an MSDS to customers with the first shipment of the chemical; all employers must make available to their employees an MSDS for each hazardous chemical used in the workplace. Employers may prepare an MSDS or rely on the MSDS provided by a manufacturer, importer, or distributor.

Labeling. The Standard requires containers of hazardous chemicals in the workplace and containers of such materials leaving the workplace to be labeled with "appropriate" warnings [34]. For certain substances (e.g., ethylene oxide, as discussed below), substance-specific labeling requirements overlay the Generic Standard requirements. Where mixtures of chemicals exist in a liquid or gaseous form, labels must be provided where more than 1% of the mixture is a hazardous chemical (the cutoff level is 0.1% in the case of carcinogens).

Employee Information and Training. The Generic Standard requires employers to provide information and training on the hazards of chemicals in the workplace to those employees who are actually or potentially exposed to hazardous chemicals [35].

Written Program. All employers must develop, implement, and maintain a written hazard communication program. The program must contain a list of all hazardous chemicals present in the workplace, procedures used for complying with the MSDS, labeling, and training requirements of the Standard, and methods used to inform employees of the hazards of nonroutine tasks [36].

Future Direction of the Generic Standard. Recent enforcement activities under the Standard and a proposed rule that would revise the Standard, 53 *Fed. Reg.* 29822 (August 8, 1988), suggest that OSHA may be moving away from the performance orientation of the Standard toward a more traditional specification approach. For example, although the Standard authorizes employers to use their professional judgment to make "some assessment of the weight of each hazard" in selecting appropriate hazard warning language for the label, OSHA has taken the position (and the proposed rule would require) that hazards may not be omitted from the label based on an employer's assessment of the potential risks associated with exposure. In essence, OSHA appears to be moving to the view that all hazards listed in the MSDS must be included in the label. OSHA also has declined to implement reasonable *de minimis* exemptions to the Standard, and has taken a rather stringent position regarding the definition of containers that must be labeled and regarding inapplicability of the 1%/0.1% mixture rule thresholds to solid products. Although these trends should be regarded as preliminary, and as of September 1989 OSHA had not yet promulgated final revisions to the Standard, employers should be aware that OSHA is taking an increasingly stringent view of the requirements of the generic Hazard Communication Standard.

D. Ethylene Oxide Exposure—PEL

Ethylene oxide is a colorless gas with an ether-like odor. Its chemical formula is C_2H_4O. Ethylene oxide is one of the two dozen chemicals of highest production value in the United States, and approximately 99% of the ethylene oxide produced in the United States is used in the manufacture of other chemical products (principally ethylene glycol). Less than 1% of the ethylene oxide produced in the United States is used as a fumigant or sterilant. However, it is this latter use of ethylene oxide that is of primary importance to employers in the medical devices industry: ethylene oxide is used to sterilize many medical devices, particularly those devices that are delicate and are heat or moisture sensitive. For such devices, ethylene oxide has no ready substitute.

OSHA has established a permissible exposure limit for exposure to ethylene oxide of 1 ppm. In addition, a short-term excursion limit (discussed in the next subsection of this chapter) of 5 ppm for 15 min has been promulgated. Methods of controlling exposure to ethylene oxide in the workplace will be discussed in another chapter. This chapter focuses on the development of OSHA's ethylene oxide standards, and details the regulatory requirements of most importance to the medical device industry. OSHA's regulation of occupational exposure to ethylene oxide is the single regulation of most pertinence to the industry, and at the same time provides an excellent "case study" overview of the process of regulation under the Occupational Safety and Health Act of 1970.

In 1971, only a few months after the effective date of the Act and substantially more than a year before the statutory deadline for such action, OSHA adopted a "national consensus standard" for occupational exposure to ethylene oxide of 50 parts ethylene oxide per 1 million parts of air (50 ppm), which had been derived from the 50-ppm threshold limit value (TLV) for exposure to ethylene oxide recommended in 1968 by ACGIH.

A public advocacy group unsuccessfully petitioned OSHA to issue an emergency temporary standard (ETS) for ethylene oxide in 1981; as mentioned above, OSHA's refusal to issue that ETS eventually was upheld by the courts.

In January of 1982, OSHA issued an Advanced Notice of Proposed Rulemaking announcing its intention to review the existing 50 ppm PEL standard. OSHA issued a proposed rule in April of 1983, which among other things proposed a reduction in the 8-hr TWA PEL from 50 ppm to 1 ppm. OSHA's final rule for occupational exposure to ethylene oxide was published in the *Federal Register* the following June [37].

OSHA's final PEL standard for ethylene oxide [38] incorporates an 8-hr time-weighted average 1 ppm PEL, and utilizes the concept of an "action level" of 0.5 ppm, below which most requirements of the standard do not apply. OSHA found that some residual risk to health existed even at the 1 ppm PEL, but that the 1 ppm level "will achieve a 98 percent reduction in cancer mortality risk." OSHA found that the 1 ppm level "reduced the risk to the extent feasible" [39]. OSHA explicitly stated that "the 1 ppm level is feasible for *most* operations in *most* workplaces" that used ethylene oxide [39, emphasis added].

The final PEL standard requires the use of engineering controls, changes in work practices, and the use of respirators where necessary (that is, in certain specified workplace situation where work practice changes or engineering controls are not practicable) to reach the 1 ppm PEL. The standard applies to any workplace where ethylene oxide may be found, unless the workplace meets the exemption criterion of the standard. At the time of its adoption, that exemption provided that the standard:

> does not apply to the processing, use, or handling of products containing [ethylene oxide] where objective data are reasonably relied upon that demonstrate that the product is not capable of releasing [ethylene oxide] in airborne concentrations at or above the [0.5 ppm] action level under the expected conditions of processing, use, or handling that will cause the greatest possible release. [40]

(In 1988, this exemption was amended, as discussed below, to incorporate the newly promulgated "excursion limit" of 5 ppm for 15 min.) In workplaces where employee exposure to ethylene oxide occurs only below the "action level," certain provisions of the standard—most notably monitoring and medical surveillance—do not apply.

In addition to revising the PEL downward from 50 ppm to 1 ppm as an 8-hr TWA, the final PEL standard implemented a number of ancillary requirements for workplaces covered by the standard.

1. Employee Monitoring

First, the standard requires employers to monitor employee exposure to ethylene oxide. Any employer covered by the standard is required to undertake initial monitoring of breathing-zone air samples that are representative of the 8-hr TWA exposure for each employee. If the initial monitoring reveals employee exposure for a particular employee or class of employees at or above the action level of 0.5 ppm (but below the 1 ppm PEL), the employer must monitor each such employee at least every 6 months. If the initial monitoring reveals employee exposure for a particular employee or class of employees at or above the PEL of 1 ppm, the employer must monitor each such employee at least every 3 months. The employer may cease periodic monitoring if that monitoring reveals that employee exposure (as indicated by at least two consecutive measurements taken at least 7 days apart) is below the action level. Even where such monitoring indicates exposure below the action level, the employer is required to initiate monitoring whenever there has been a change in the production process, control equipment, personnel, or work practices that may result in new or additional ethylene oxide exposure (or where the employer has reason to "suspect" that a change may result in new or additional exposures). Employers must notify employees of monitoring results in writing [41].

2. Regulated Areas

Second, the standard provides that the employer shall establish a "regulated area" wherever occupational exposure to ethylene oxide may exceed the 1 ppm PEL. Access to regulated areas must be restricted to "authorized persons" (defined as a person specifically authorized by the employer, whose duties require the person to enter the regulated area). Regulated areas must be identified "in any manner that minimizes the number of employees within the regulated area" [42].

3. Respirators

Third, the standard requires the employer to establish and implement a written compliance program to reduce employee exposure by means of engineering and work-practice controls (and through use of respirators where permitted). Plans must be updated at least yearly, and must be made available upon request to OSHA, affected employees, and the representatives of affected employees [43].

4. Medical Surveillance

Fourth, the standard establishes a system of medical surveillance for employees exposed above the 0.5 ppm action level for at least 30 days per year. Employers must "make available" medical examinations and consultations to covered employees, without cost to the employee, without loss of pay, and at a reasonable time and place. All medical examinations must be performed by or under the supervision of a licensed physician. The medical surveillance system does not provide that employees must take advantage of the available medical examinations, nor does it provide that a physician must perform the examination herself [44].

5. Hazard Communication

Fifth, the standard provides for communication of information regarding potential ethylene-oxide-related hazards to employees. Precautionary signs are required indicating regulated areas. Precautionary labels are required to be affixed to all containers of ethylene oxide (except tanks, reaction vessels, and pipes and piping systems) whose contents are "capable of causing employee exposure at or above the action level." The standard requires employers who are manufacturers or importers of ethylene oxide to develop material safety data sheets (MSDSs), as specified in OSHA's generic Hazard Communication Standard. Employers also are required to provide employee training, including informing employees of (1) the requirements of the standard, (2) those areas in the workplace where ethylene oxide is present, (3) the location and availability of the final rule promulgating the standard, and (4) the details of the required medical surveillance program. Employees must be trained to detect the release of ethylene oxide in the workplace, must be informed of the physical and health hazards of ethylene oxide, must be trained in the measures employees can take to protect themselves from ethylene-oxide-related hazards, and must be made familiar with the details of the hazard communication program developed by the employer [45].

The standard also imposes certain record-keeping requirements, and provides that the affected employees (or their designated representatives) may observe any monitoring of employee exposure to ethylene oxide [46].

E. Ethylene Oxide Exposure—Excursion Limit

In 1984 OSHA explicitly declined to implement a short-term exposure limit (STEL) as an adjunct to the 8-hr TWA PEL. As OSHA stated, it was:

> reserving decision [in 1984] on whether the standard should contain a STEL. OSHA takes this action largely in response to reservations expressed by the Office of Management and Budget . . . concerning:
>
> Quantification of the risk avoided by issuance of the STEL;
> The appropriateness of relying on [certain] studies [in the record] . . . as partial support for the issuance of a STEL;
> A decision by ACGIH not to recommend a STEL for [ethylene oxide]; and
> The economic and technical feasibility of a STEL without the use of respirators. [39]

OSHA announced in January 1985 that, in its opinion, the available health data did not necessitate establishment of a STEL to supplement the 8-hr TWA of 1 ppm [47].

A peition for review of OSHA's decision not to promulgate a STEL was filed in the United States Court of Appeals for the District of Columbia Circuit. On July 25, 1986, that Court upheld OSHA's establishment of a 1 ppm TWA PEL, and also upheld OSHA's finding that the evidence in the record did not establish the existence of a "dose-rate" relationship for the health effects of ethylene oxide. The Court did, however, reject OSHA's position that a STEL was therefore unnecessary; the Court stated that

> If in fact a STEL would further reduce a significant health risk [found by OSHA to still exist at the 1 ppm PEL level] and is feasible to implement, then the OSH Act compels the agency to adopt it (barring alternative avenues to the same result) [48].

The Court remanded the issue to OSHA to "either adopt a short-term limit or explain why empirical and expert evidence on exposure patterns makes a short-term limit irrelevant to controlling long-term exposures" [49].

1. The New Standard

Responding to the Court's order, OSHA proposed an "excursion limit," or EL, for ethylene oxide in January of 1988 [50]. OSHA stated that it was not using the term STEL because that term implied a short-term limit "dictated by specific toxicologic or hazard data" whereas the proposed EL was not based on such data [50]. OSHA proposed an excursion limit of 5 ppm ethylene oxide as measured over a 15-min sampling period, as an adjunct to the 1 ppm PEL. Following comments and a March 1988 hearing, OSHA finalized the rule in April 1988 [51].

The final excursion limit rule provides:

> The employer shall ensure that no employee is exposed to an airborne concentration of [ethylene oxide] in excess of 5 parts of [ethylene oxide] per million parts of air (5 ppm) as averaged over a sampling period of fifteen (15) minutes. [52]

The final rule makes clear that the product exemption applies where the product "may not *reasonably be forseen* to release EtO in excess of the excursion limit. . . ." As OSHA explained, "The intent of the product exemption (paragraph (a)(2)) and labeling requirement (paragraph (j)(1)) is not to encompass products which under extremely unusual circumstances may unexpectedly or unpredictably release EtO above the excursion limit. . . ."

The EL rule also triggers compliance with the product labeling requirements of the existing EO standard [53]; that requirement was, as noted above, amended to include the "reasonable forseeability" standard to prevent imposition of labeling requirements in circumstances where release was extremely unlikely. Labeling must comply with the requirements of OSHA's generic Hazard Communication Standard [54].

Thus, the OSHA standard on occupational exposure to ethylene oxide currently imposes the following requirements:

Exposure at or above "action level" of 0.5 ppm (8-hr TWA)	Precautionary labeling Periodic monitoring *No* regulated areas
Exposure at or above PEL of 1 ppm (8-hr TWA)	Precautionary labeling Periodic monitoring Regulated areas

Exposure at or above EL of 5 ppm (15-min)	Periodic monitoring Precautionary labeling Regulated areas

In addition, the type and frequency of monitoring itself changes as exposure changes:

Below action level Below excursion limit	No monitoring required
Below action level Above excursion limit	Short-term monitoring required four times per year
At or above action level At or below TWA PEL At or below excursion limit	8-hr TWA monitoring required twice yearly
At or above action level At or below TWA PEL Above excusion limit	Both 8-hr TWA and short-term monitoring required
Above TWA PEL At or below excursion limit	8-Hour TWA monitoring required four times yearly
Above PEL Above excursion limit	Both 8-hr TWA (4 times) and short-term monitoring required four time yearly

An appeal to the Court of Appeals for the District of Columbia Circuit has been filed, challenging the final rule. The appeal, *Public Citizen Health Research Group v. Pendergass* (No. 88-1283), still pending in September 1989, contends that a short-term limit of 3 ppm rather than 5 ppm would be reasonable and feasible.

2. *Respirators*

One of the most important issues in the EL rule-making process was the degree to which the final EL standard would permit compliance through the use of respirators, particularly in the loading and unloading of sterilizers. A number of industry commenters informed OSHA that, for many sterilizer operations, engineering and work-practice controls (traditionally favored by the agency over use of personal protective devices) would be technologically and economically infeasible for achieving compliance with the EL standard. (In 1984, OSHA did note that for certain operations, including removal of biological indicators from sterilized materials, loading and unloading of tanks cars of EtO, and vessel cleaning, use of respirators would be permitted to achieve compliance with the 8-hr TWA PEL of 1 ppm.)

OSHA declined to approve categorically the achievement of the excursion limit through the use of respirators for sterilizer loading and unloading. OSHA stated that while "[i]t is true for the EL, as it is with the TWA, that some operations may need to supplement engineering and work practice controls with respirators[,] . . . OSHA does not believe that the need to use respirators in unloading sterilizers under certain conditions justifies a wholesale finding that engineering and work practice controls for all unloading of sterilizers are, therefore, infeasible" [55].

On the other hand, OSHA did recognize that for certain sterilizer unloading operations, work practice controls suggested by OSHA's contractor (such as allowing additional offgassing time in the sterilizer for sterilized materials) would not be feasible. In such circumstances (evaluated on a case-by-case basis), respirators could be used. OSHA noted:

> OSHA recognizes that the work practices recommended by [OSHA's contractor] would not work in all sterilization installations under all operating conditions. In particular, OSHA acknowledges that there may be difficulty in one type of sterilization operation: where sterilization of medical products and other materials is to be performed in large, room-sized sterilizers into which employees must enter to remove those materials after sterilization, and where those facilities are operated in consecutive 8-hour shifts throughout the full 24-hour day, where it is not possible to provide additional offgassing time for the sterilized materials in the sterilizer without severely disrupting the production process and capacity. [54]

Although OSHA does not inform its enforcement staff in advance of the precise conditions under which work practice and engineering controls will be infeasible and respirators may be used to achieve compliance with the excursion limit, the preamble language makes clear that personal protection may be used for compliance under certain circumstances. Indeed, OSHA (in a response to a request by the Health Industry Manufacturers Association to clarify the discussion of respirator use in the EL preamble) has reaffirmed the ability of employers to use respirators where engineering controls are not feasible to achieve compliance with the standard. OSHA emphasizes that the determination of feasibility must be made on a case-by-case basis.

In June of 1989 OSHA initiated a rulemaking proceeding with respect to a proposed rule on "Health Standards, Methods of Compliance," 54 *Fed. Reg.* 23991 (June 5, 1989). OSHA proposed to "incorporate additional flexibility in its methods of compliance requirements," 54 *Fed. Reg.* 23991, by permitting the use of respiratory proteciton in lieu of engineering or work practice controls under five enumerated circumstances. While the current rule permits use of respirators whenever the employer can demonstrate that engineering controls and work practice controls are infeasible, the new rule would permit the use of respirators in the listed circumstances without requiring the employer to make that showing. The proposed rule, for example, includes exemptions for "shutdowns," "repair," and entry into unknown atmospheres (intended to cover "confined spaces or vessel entry and tank cleaning and vessel cleaning" 54 *Fed. Reg.* 23995).

F. Enforcement

OSHA is authorized by the Act, "in order to carry out the purposes of this Act," to enter "without delay and at reasonable times any factory, plant, establishment, construction site, or other area, workplace or environment where work is performed by an employee of an employer," for the purpose of inspecting those premises for compliance with the Act and the standards promulgated by OSHA pursuant to the Act [57]. If the OSHA inspector believes a violation to have occurred, the inspector may issue a citation to the employer, fixing a schedule for abatement of the violation [58]. OSHA may also propose an administrative penalty in connection with a violation [59]. Review of OSHA's enforcement actions may be had in the Occupational Safety and Health Review Commission (OSHRC).

The Act also specifically grants OSHA the authority to seek an injunction in federal court against any "conditions or practices in any place of employment where "a danger exists which could reasonably be expected to cause death or serious physical harm immediately or before the imminence of such danger can be eliminated through the enforcement procedures otherwise provided by this Act" [60].

Section 17 of the Act [61] provides for the assessment of civil and criminal penalties. Willful or repeated violations may be punished by civil penalties not to exceed $10,000 for each violation. An employer receiving a citation for a "serious violation" may be assessed a civil penalty of up to $1000 for each violation; an employer receiving a citation for a violation "not of a serious nature" may also be assessed a civil penalty of up to $1000 for each violation. An employer who, having been issued a citation directing abatement of a hazard, has not corrected the violation is subject to a civil penalty of not more than $1000 "for each day during which such failure or violation occurs."

In interpreting these provisions, the courts have held that the employer's knowledge that a condition was unsafe is sufficient to prove a "serious" violation [62].

A willful violation leading to the death of an employee is punishable by a fine of up to $10,000 or imprisonment for up to 6 months.

Employers receiving citations or the notification of a proposed penalty under Section 10 of the Act [61], alleging violations of the Act, should be aware of the procedural steps appropriate to challenging a citation. Within a fixed period after receiving a citation or a notification of proposed penalty, the employer may notify the Secretary of Labor of the employer's intention to challenge the citation. The matter is then referred to an OSHRC administrative law judge. A hearing is conducted, affording interested parties the opportunity to participate.

The administrative law judge issues a decision, which may affirm, vacate, or modify the original citation. At that point, the employer (or any other party) may petition for review of the judge's decision by the full Review Commission. Final orders of the Review Commission may be appealed to the United States Courts of Appeals, pursuant to Section 11(a) of the Act [61]. Section 11(a) provides, however, that argument in the Court of Appeals is limited to points properly raised during the administrative enforcement proceedings, and that any finding of fact made during the administrative process that is supported by substantial evidence on the record is deemed conclusive; proper preparation and presentation of an employer's case to the administrative law judge is therefore imperative.

Contesting a citation through the administrative process will stay any remedial order or penalty proposed by OSHA. That automatic stay is in effect throughout the administrative appeals process—that is, until OSHRC issues a final order. Filing for an appeal to the Court of Appeals, or even a successful appeal to the Court of Appeals, does not automatically stay imposition of penalty and remedial provisions (although OSHRC or, under Section 11(a), the Court of Appeals may be petitioned for, and may grant, a stay as a matter of discretion).

Settlement or compromise during the course of OSHA enforcement proceedings is always a possibility.

III. STATE AND LOCAL REQUIREMENTS

An analysis of state and local regulation governing occupational exposure is beyond the purview of this chapter. Employers should be aware, however, that many states and some localities may have regulatory schemes independent of the federal occupational health system. The Occupational Safety and Health Act explicitly provides that it does not "prevent any State agency or court from asserting jurisdiction under State law over any occupational safety or health issue with respect to which no standard is in effect under section 6" of the federal Act [63].

A prominent example of such allegedly independent regulation of employee health and safety is the imposition by state agencies of occupational "warning" and "labeling" requirements. Certain states impose independent labeling requirements on certain chemicals used in the workplace. A federal court ruled in *New Jersey State Chamber of Commerce v. Hughey* [64] that such state-imposed labeling requirements are not necessarily preempted by federal labeling requirements. California's 1988 imposition of occupational warning requirements for chemicals under Proposition 65 is the most prominent example of such state-imposed labeling schemes. The regulations implementing the Proposition 65 occupational "warning" requirements state that the requirements do not apply in the case of "[a]n exposure for which federal law governs warning in a manner that preempts state authority" [Section 25249.6(a)]. Where the only potentially significant exposure is expected to be in a workplace subject to OSHA's jurisdiction, the generic federal Hazard Communication Standard [65] may itself preempt Proposition 65. In this regard, the California regulations state that a "warning to the exposed employee about the chemical in question which complies with all information, training and labeling requirements of the federal Hazard Communication Standard" satisfies the Proposition 65 requirements for warnings for occupational exposures [Section 12601(c)(1)(B)]. However, it remains unclear how this provision would be applied by the state of California to chemicals present at concentrations below the mixture rule thresholds of the federal Standard and where no warning would be required under the federal Standard.

On another level, state agencies themselves play a substantial role in the federal statutory scheme in implementing and enforcing OSHA standards. The federal Act provides a mechanism for a state agency to assume "responsibility for development and enforcement" in that state "of occupational safety and health standards relating to any occupational safety or health issue with respect to which a Federal standard has been promulgated" [66]. State agencies may assume such responsibility by submitting a "state plan" to OSHA. OSHA may approve a state plan if the plan "will be at least as effective" as the federal program [67]. This language permits states to develop plans more stringent than the federal standard. Approximately half the states currently operate approved state plans, and accordingly are responsible for enforcing OSHA's occupational safety and health standards in their respective states. In addition, two states (Connecticut and New York) administer OSHA-approved plans for state and local government employees only.

REFERENCES

1. 29 U.S.C. 651(b)2(b).
2. 29 U.S.C. 651(b)(3).
3. 29 U.S.C. 651(b)(7).
4. 29 U.S.C. 651(b)(9).
5. 19 U.S.C. 652(5).
6. 683 F.2d 673 (2d Cir. 1982).
7. 683 F.2d 676.
8. 683 F.2d 677 and n.9.
9. 29 C.F.R. 1910.5(f).
10. 815 F.2d 1570 (D.C. Cir.), *cert. denied*, 108 S. Ct. 485 (1987).
11. Report of the House Educaiton and Labor Committee, H.R. Rep. No. 1291, 91st Cong., 2d Sess (1970) at 22.
12. Report of the Senate Labor and Public Welfare Committee, S. Rep. No. 1282, 91st Congr., 2d Sess. (1970) at 10.
13. 29 U.S.C. 654(a)(2).
14. 29 U.S.C. 655.
15. 29 U.S.C. 652(8).
16. *Industrial Union Department, AFL-CIO v. American Petroleum Institute*, 448 U.S. 607, 639, 641 (1980).
17. 448 U.S. at 642.
18. 488 U.S. at 655.
19. 29 U.S.C. 655(b)(5).
20. *American Textile Manufacturers Association v. Donovan*, 452 U.S. 490, 509 (1981).
21. 452 U.S. at 520.
22. 825 F.2d 482 (D.C. Cir. 1987), *cert. denied*, 108 S. Ct. 1573 (1988).
23. 29 U.S.C. 655 6(b).
24. 838 F.2d 1258 (D.C. Cir. 1988).
25. 838 F.2d at 1271.
26. 29 U.S.C. 655 6(c).
27. *Public Citizen Health Research Group v. Auchter*, 702 F.2d 1150 (D.C. Cir. 1983).
28. *Asbestos Information Association v. OSHA*, 727 F.2d 415 (5th Cir. 1984).
29. 29 U.S.C. 655 6(a).
30. 53 *Fed. Reg.* 20,960 (June 7, 1988).
31. 48 *Fed. Reg.* 53,279 (November 25, 1983).
32. 52 *Fed. Reg.* 31,851 (August 24, 1987).
33. 29 C.F.R. 1910.1200(g).
34. 29 C.F.R. 1910.1200(f).
35. 29 C.F.R. 1910.1200(h).
36. 29 C.F.R. 1910.1200(e).
37. 49 *Fed. Reg.* 25,733 (June 22, 1984).
38. 29 C.F.R. 1910.1047
39. 49 *Fed. Reg.* 25,775.
40. 29 C.F.R. 1910.1047(a)(2).
41. 29 C.F.R. 1910.1047(d).
42. 29 C.F.R. 1910.1047(e).
43. 29 C.F.R. 1910.1047(f)(2).
44. 29 C.F.R. 1910.1047(i).
45. 29 C.F.R. 1910.1047(j).

46. 29 C.F.R. 1910.1047(k).
47. 50 *Fed. Reg.* 64 (January 3, 1985).
48. *Public Citizen Health Research Group v. Tyson*, 796 F.2d 1479 1505 (D.C. Cir. 1986).
49. 796 F.2d at 1507.
50. 53 *Fed. Reg.* 1724 (January 21, 1988), n. 1.
51. 53 *Fed. Reg.* 11,414 (April 6, 1988).
52. 29 C.F.R. 1910.1047(c)(2).
53. 29 C.F.R. 1910.1047(j)(1)(ii).
54. 29 C.F.R. 1910.1200(f).
55. 53 *Fed. Reg.* 11,434 (April 6, 1988).
56. 53 *Fed. Reg.* 11,434-435.
57. 29 U.S.C. 657(a)(1).
58. 29 U.S.C. 658(a).
59. 29 U.S.C. 659(a).
60. 29 U.S.C. 662(a).
61. 29 U.S.C. 666.
62. *East Texas Motor Freight, Inc. v. Occupational Safety and Health Review Commission*, 671 F.2d 846 (2d Cir. 1982).
63. 29 U.S.C. 667(a).
64. No. 84-3255 (D.N.J. February 5, 1988).
65. 29 C.F.R. 1910.
66. 29 U.S.C. 667(b).
67. 29 U.S.C. 667(c).

44

Right-to-Know Regulations

ROBERT C. BRANDYS *Occupational and Environmental Health Consulting Services, Hinsdale, Illinois*

I. INTRODUCTION

The OSHA Hazard Communication Standard (Employee Right-to-Know Law) is the most significant piece of legislation on occupational safety and health since the passage of the OSHA Act. *It essentially requires review and re-establishment of your chemical safety program from employee exposure monitoring to emergency procedures.*

Right-to-know regulations evolved out of two regulatory issues from a state and federal level. There was a need first to communicate potential hazard information to employees about the chemical substances with which they worked, and second, to communicate to the local community (emergency services) the potential chemical hazard situation to which they might respond. This chapter discusses these two different "rights-to-know" areas separately because of their differing regulatory evolution and governmental enforcement bodies. In addressing the right-to-know area, it is important to note that if employees are exposed to a hazardous chemical in a facility, the community is also probably exposed to a lesser degree.

A. Employee Right-to-Know Regulations

The first regulatory impetus started with NIOSH developing a draft recommended labeling system for chemicals in 1976. This system had number codes and hazard categories. The Occupational Safety and Health Administration (OSHA) proposed to issue a mandatory label standard in 1980 based on this system, but it was never enacted into regulation. With the advent of an anti-regulatory philosophy from Washington, many states with OSHA approved state safety plans went ahead with this type of hazard regulation. However, instead of just a labeling standard, the state regulations went much further and required specific training of employees in the hazards of the chemicals with which they work.

In the period from 1980 to 1983 some 22 states issued worker right-to-know regulations. These state regulation were not uniform, and significantly different requirements were being mandated and enforced by each state. As a result, industry went to Washington and said to OSHA, "If we must have a regulation, we would rather deal with one federal standard than 52 different state standards."

OSHA quickly responded, and by December 1983 a right-to-know standard for the industrial sector (Standard Industrial Classification codes 20-39) was promulgated. This federal regulation preempted the state standards in most aspects. However, some individual state requirements, which are more stringent than the federal standard, still exist and are being enforced. Some of these state standards require MSDSs be included with each shipment and may also cover "physical agents" (noise, heat, etc.), radiation, and infectious hazards. These are not addressed in the federal regulation.

A number of lawsuits were filed regarding the federal standard, the most significant of which challenged the limitation of the new regulations to only the industrial sector. After all, employees in agriculture, custodial, maintenance, and construction work also used chemicals on a daily basis. In addition, a number of state standards at this time already covered all employees.

As a result of this action, in December 1987, OSHA issued an expanded hazard communication standard to cover all employees. To date, the expanded OSHA hazard communication standard and individual state requirements compose the regulatory basis of employee right-to-know regulations.

The major responsibilities and requirements of the federal OSHA regulation are discussed below.

B. Three Federal Right-to-Know Regulations

1. OSHA 200 Log

Employee "rights to information" or "right-to-know" in some respects actually started with the OSHA Act in 1970. This act allowed employees access to the OSHA "Log and Summary of Occupational Injuries and Illnesses" (OSHA 200 Form). This form lists all occupationally related illnesses and injuries that occur at each facility each year. Employers must maintain these lists (logs). Failure to do so has recently resulted in substantial fines. In some cases the fines exceeded $1,000,000.

2. Medical/Exposure Records

In 1978, OSHA issued another right to information standard called the "Employee Access to Medical and Exposure Records" standard. This standard gave employees the right to any health or exposure data (industrial hygiene monitoring) that the employer has, accumulates, or researches about an employee or their occupation exposures.

3. Chemical Hazards

Finally, with the promulgation of the hazard communication standard, employees were guaranteed a third right—a right to information on the potential hazards of the chemicals in their workplace and procedures to insure that they are not overexposed. In addition, employees may have additional rights depending on specific state or local requirements (e.g., notification of overexposure, access to personnel files, etc.)

Right-to-know training should cover all of these rights in one program. It saves considerable training time and covers training requirements that most employers simply overlook. Table 1 details some of the requirements of these federal rights.

II. EMPLOYER RIGHT-TO-KNOW COMPLIANCE STEPS

A. Fourteen Basic Steps

The following list details some of the major steps necessary to develop and implement a comprehensive right-to-know training program. Each of these steps is discussed in detail in the following sections. *Additional steps may be necessary due to special characteristics of the facility or additional state/local requirements.*

1. *Review all standards related to employee right-to-know and post a list of employee rights:*

   OSHA 1904.7—Access to the Log of Occupational Injuries and Illnesses Standard
   OSHA 1910.20—Access to Employee Medical and Exposure Records Standard
   OSHA 1910.1200—Hazard Communication Standard
   State Right-to-Know Laws or Regulations
   Local Right-to-Know Laws or Regulations

2. *Survey each operation for chemicals used, number of employees involved and frequency of use.* Note for each operation:

   Any previous industrial hygiene data or current industrial hygiene needs
   Any safety procedures/equpiment in use (e.g., SOPs, ventilation systems, eye washes/safety showers, personal protective equipment for both routine and emergency situations, etc.)
   Adequacy of labeling of hazardous material containers or piping systems
   The presence or absence of specific chemical spill clean-up procedures

3. *Develop a list of hazardous chemicals used routinely* and send notification to the state and the local fire department as required.

4. *Obtain material safety data sheets (MSDSs) for all chemicals used make sure they are dated.* (As of September 23, 1987, chemical manufacturers must supply MSDSs with initial shipments and updates as necessary. In some states inclusion with each shipment is mandatory.) Include information on chemical constituents that warrant industrial hygiene monitoring.

5. *Review existing MSDSs to identify and revise:*

   The most current worker exposure standards [e.g., OSHA permissible exposure limits (PELs) or exposure guidelines such as threshold limit values (TLVs) published by American Conference of Governmental Industrial Hygienists]
   Information relative to "safe use" procedures on each sheet

6. *Perform industrial hygiene monitoring* as necessary.

TABLE 1 Federal (OSHA) Employee Right-to-Know Laws

Employee right	Access provisions	Copy of records	Training requirement	Records retention
Log and summary of occupational injuries and illnesses	Upon request	Free copy	Annually by posting in February	5 years
Medical and exposure records	Written request (includes former employees and employee representatives)	Free copy	New employees and annually by posting or training	30 years + (including MSDS)
Chemical hazard communication (written program required)	Upon request	Free copy	Periodic[a]/update; new, transferred, temporarily assigned and contract employees	30 years

[a]Some state laws require annual retraining.

7. *Upgrade container and piping system labeling* as needed.
8. *Develop a written employee hazard communication training program*, emergency response training program, and respiratory protection program as required by state, local, and/or OSHA requirements.
9. *Develop a master file of MSDSs* for all chemical substances used in the facility and MSDS booklets for chemical substances used in specific work areas.
10. *Train all regular employees* and document the training for the various programs.
11. *Train all contract, new, and transferred employees* as appropriate, and document this training.
12. *Update master MSDS file, MSDS booklets, and employee training* as significant new information becomes available on chemical hazards.
13. *Retrain employees* periodically (annually by some state laws).
14. *Perform periodic industrial hygiene monitoring* as necessary.

One can see from the above list that right-to-know compliance is a time-consuming process that demands significant commitment of staff resources. In fact, most facilities have organized in-house committees representing a variety of disciplines to develop appropriate training materials.

The use of a qualified outside consulting firm and/or commercially available training films can assist in this process. A qualified consulting firm can handle the program from start to finish or just do certain parts. For example, they can:

Survey for hazardous chemicals
Computerize lists of chemicals and MSDSs
Review and upgrade MSDSs specific to your operations
Develop required written emergency response procedures
Develop required written respiratory protection policy
Make recommendations as to where your overall safety program may need improvement

B. Implementation and Management Techniques

Below is a detailed discussion of each step in the development and implementation of a comprehensive Right-to-Know Program.

1. *Review all standards related to employee right-to-know and post a list of employee rights*:

OSHA 1904.7—Access to the Log of Occupational Injuries and Illnesses Standard
OSHA 1910.20—Access to Employee Medical and Exposure Records Standard
OSHA 1910.1200—Hazard Communication Standard
State Right-to-Know Laws or Regulations
Local Right-to-Know Laws or Regulations

In some facilities, the information system for handling OSHA logs, employee exposure records, and medical records may be fragmented and hence not readily available to employees as the standards require. For example, OSHA logs may be in Personnel, medical records in the medical department, and exposure information in Engineering or Facilities. Since OSHA standards require the medical and exposure records be maintained for 30 years past employment, all of these records should be brought together and properly managed under one function. Otherwise, specific procedures must be developed, outlining each department's responsibilities for maintaining and archiving this information.

State laws provide additional responsibilities in some cases (e.g., notification of overexposure, access to personnel files, etc.). *If you are in a state that has its own regulations, it is essential that you obtain a copy of these regulations and review them in detail.* In most cases, employees have a right to a copy of these specific laws, and therefore they must be kept on the premises.

2. *Survey each operation for chemicals used, number of employees involved and frequency of use.* Note for each operation:

Any previous industrial hygiene data or current industrial hygiene needs
Any safety procedures/equipment in use (e.g., SOPs, ventilation systems, eye washes/safety showers, personal protective equipment for both routine and emergency situations, etc.)
Inadequately labeled hazardous material containers or piping systems
The need for a chemical spill clean-up team
The need for a fire response procedure

The regulations for a written hazard communication program require "(e)(1)(i) a list of the hazardous chemicals. . . ." Identifying all the hazardous chemicals in a facility is a time-conuming and frustrating task. Many facilities have tried the memo approach, such as "Please send me a list of all hazardous chemicals used by your department." Unfortunately, most personnel are not aware that the regulatory definition of hazardous chemical includes nearly all chemicals used by employees. Most personnel exclude consumer products, "ordinary" laboratory chemicals, and materials used infrequently. Some may actually reply, "We don't use any." One company listed that they only used five hazardous chemicals. Upon an actual physical inspection, over 100 different hazardous chemicals were found. Based on this typical experience, it is recommended that one do a physical survey of all work areas, if at all possible, to ensure that a comprehensive chemical list is developed.

In managing these chemical exposure risks, it is usually necessary to prioritize the allocation of resources available to assess these risks. In order to allocate these resources effectively, identifying the chemical exposures that have the highest potential for overexposure to the largest number of employees is necessary. Therefore, the number of employees potentially exposed to the chemical and the average length of time they use the chemical are critical data to be gathered during the survey. The combination of these two parameters can be used to rank which chemicals departments should be addressed first with respect to both training and industrial hygiene monitoring.

3. *Develop a list of hazardous chemicals used routinely and send notification to the state and the local fire department as required.*

Once the chemical survey has been completed, it is suggested that one enter the data into a personal computer data base. In this way, the list can be easily updated when new products are purchased. By including a field for the department, number of employees, potential risk of overexposure, and other information, one can develop individual departmental chemical lists and can rank order the departments that present the larger potential risks. These higher-ranking departments can be scheduled for more frequent and more in-depth training than others, as necessary.

4. *Obtain material safety data sheets (MSDSs) for all chemicals used and make sure they are dated.*

The OSHA Hazard Communication Standard states:

> "(g) Employers shall have a Material Safety Data Sheet for each hazardous chemical which they use."

As of September 23, 1987, chemical manufacturers, suppliers, and distributors were to supply Material Safety Data Sheets (MSDSs) with initial shipments of hazardous chemicals. They were also to provide updated MSDSs within 90 days of significant new information being available on the chemical or with the next shipment to all employers. In some states, inclusion with each shipment is mandatory. Unfortunately, not all manufacturers are keeping up with this regulation. In addition, many MSDSs arrive at different places within the facility because of different mailing practices by chemical suppliers. This leads to many MSDSs becoming lost, misfiled, and not properly maintained as required by the regulations.

In attempting to assemble all the MSDSs for your facility, you will probably have to request quite a few MSDSs from your suppliers. This is a time-consuming task and can be delegated once the list of chemicals has been compiled in step 3. If MSDSs are not dated by the manufacturer, they should be dated by the facility for future reference for keeping the MSDS file up to date.

If you have problems obtaining MSDSs from a chemical manufacturer or supplier, follow the procedures for documenting your requests as described in the OSHA standard. If after these steps the manufacturer still refuses to supply the MSDS, contact your local OSHA office. The office has specific procedures to follow to obtain MSDSs from reluctant manufacturers or suppliers of hazardous materials.

Are MSDSs required for FDA labeled items? One confusion that exists regarding MSDSs for drugs and medical devices is that the regulations appear to exempt "FDA labeled items." *This exemption only applies to changing or upgrading existing labels* on drug and device products with regards to label warnings. This exemption *does not* apply to the need for these manufacturers to develop MSDSs for their products. However, a proposal has been made to allow the use of the package insert to serve this purpose. No regulatory action on this proposal is expected on this issue until late 1989 or 1990.

5. *Review existing MSDSs to identify*:

All chemical constituents that warrant industrial hygiene monitoring

If the most current worker exposure standards are listed [e.g., OSHA permissible exposure limits (PELs) or exposure guidelines such as threshold limit values (TLVs) published by American Conference of Governmental Industrial Hygienists]

If information relative to "safe use" procedures is on each sheet

Approximately 60-70% of all Material Safety Data Sheets are incomplete in one or more pieces of important information, such as:

a. Not listing the most recent exposure standard or guideline (PEL, TLV, etc.)
b. Not listing the chemical substance as a carcinogen pursuant to the National Toxicology Program (NTP), the International Agency for Research on

Cancer (IARC) or OSHA standards.*

c. No reproductive hazard information listed (teratogen or mutagen)
d. No specificity on the type of safety glove (e.g., PVC, PVA, nitrile, butyl)
e. Incomplete information on disposal of the chemical substance
f. Not dated
g. Inappropriate trade secret claim
h. Blank spaces
i. Not listing hazardous components that are less than 1%, but by definition still render the mixture "hazardous"

It is suggested that a person/consultant familiar with the regulations and the hazards of chemicals review and upgrade the MSDSs. This should include updated information as needed and specific information as relates to your operations (e.g., where to dispose of wastes, what type of glove to use, etc.).

6. *Perform industrial hygiene monitoring as necessary.*

The Need for Employee Exposure Chemical Monitoring (Industrial Hygiene). The Hazard Communication Standard specifically states that employers must have some method to be able to identify and quantify to what extent employees are exposed to hazardous chemicals. Employee training should include:

> (2)(i) Methods and observations that may be used to detect the presence or release of a hazardous chemical in the work area (such as {industrial hygiene} monitoring conducted by the employer, visual appearance or odor of hazardous chemicals when being released, etc.)

Understandably, the first question employees ask after being informed of their rights about potential hazards of chemical exposure is, "How much am I exposed to?" Without employee chemical exposure monitoring data it is not possible to provide a quantitative answer to this question. Employee chemical exposure monitoring is called "industrial hygiene monitoring." In Europe, it is known as "occupational hygiene monitoring." *If your facility does not have an industrial hygiene monitoring program, one should be started in conjunction with the development of the right-to-know training program.*

When the facility makes the effort to obtain this chemical exposure information and makes the necessary improvements to reduce exposure levels, it shows to employees that their health and safety is of real concern to the employer. It can also help in future litigation of excess exposure claims. *Industrial hygiene data is also important because most worker compensation insurance policies now exclude coverage of an employee's illness caused by chemical exposure in excess of OSHA standards.* This means that if industrial hygiene monitoring is not conducted, and an employee claims an occupational disease at some point in the future, your liability may not be covered by your worker compensation insurance and your company could be totally liable for all costs.

7. *Upgrade container and piping system labeling as needed.* Adequate labeling of potentially hazardous materials has been a confusing and controversial subject since the enactment of the OSHA Hazard Communication Standard. It still is undecided. This is because the standard directly links the

---

*It is also suggested to include NIOSH and ACGIH carcinogens under this category.

adequacy of labels to the amount of training employees receive. If more information is on the label, generally less training time is required. The important considerations of this issue are discussed below.

Different Labeling Systems. An important aspect to consider in the development of a hazard communication training program is what labeling systems are used on hazardous chemicals. If many different types of hazard warning labels are used and if they are cryptic, then training will need to be more extensive. Examples of cryptic labeling systems are the NFPA and HMIS labels. In these systems, red background is fire hazard, blue background is health hazard, and yellow background is reactivity hazard. A 4 is extremely hazardous, a 3 is moderately hazardous, etc., down to a 0 which is relatively nonhazardous. Training employees on the *specific* hazards of chemicals using these systems can be difficult, if not impossible. If labels are more uniform and descriptive, then training can be streamlined.

Portable Containers. Another important part of the labeling requirements of hazardous chemicals relates to the containers into which they are dispensed. The standard states,

> "(7) The employer is not required to label portable containers into which hazardous chemicals are transferred . . . and which are intended for the immediate use of the employee who performs the transfer."

This has been interpreted to mean that if the container of hazardous material is unlabeled, the employee must have the container in his or her possession at all times. If, for example, the container is to be left unattended or used the next shift, it must be labeled. One area that is significantly impacted by this requirement is housekeeping chemical containers. Typically, plastic bottles and buckets of diluted hazardous materials are used from shift to shift. These containers must now be labeled.

Pipe System Labeling. The Hazard Communication Standard does not directly state that piping systems that contain hazardous materials have to be labeled. However, some state laws do require that piping systems be identified. State statutes should be thoroughly reviewed on this issue. The issue here is one of protecting the worker who may be trying to fix a leaking pipe or containing the leaking material. Without proper labeling, exposure to hazardous chemical could easily occur.

Consumer Products. The Hazard Communication Standard has a provision that exempts consumer products from additional labeling as follows:

> where the employer can demonstrate it is used in the workplace in the same manner as normal consumer use, and which use results in a duration and frequency of exposure which is not greater than exposures experienced by consumers;

However, in the case of commercial products, nearly all of these products are used more frequently in the workplace than by consumers in their homes. (An example would be the frequent use of toilet bowl cleaners by custodial personnel.) Therefore, you may need to add additional warning information to some consumer products.

Laboratory Chemicals. OSHA standards for labeling of laboratory chemicals only state that "(3)(i) Employers shall ensure that labels on incoming containers of hazardous chemicals are not removed or defaced." As labels become contaminated with chemicals they may need to be replaced. Individual laboratories should manage this responsibility along with other quality control practices.

8. *Develop a written employee hazard communication training program, emergency response training program, and respiratory protection program* as required by state, local, and/or OSHA requirements.

The regulations state the following requirements for a written hazard communication program.

> (e) *Written hazard communication program.* (1) Employers shall develop, implement, and maintain at the workplace, a written hazard communication program for their workplace which at least describes how the criteria specified . . . for . . . labels and other forms of warning, material safety data sheets and employee information and training will be met, and which includes the following:
>
> (e)(1)(i) a list of the hazardous chemicals. . . .
> (e)(1)(ii) The methods the employer will use to inform employees of the hazards of non-routine tasks. . . .

The written program should be available to employees at all times. It should also be realistic in what it claims to accomplish and to whom it delegates responsibilities. It is suggested that individual responsibilities be included in a particular person's job description to prove to regulatory bodies that the program is adequately staffed. This also establishes accountability on paper for their responsibilities in the overall program.

In addition, under (2) *Training*, it also states that employee hazard communication training must include:

> "(ii) The measures employees can take to protect themselves from these hazards including . . . emergency procedures, and personal protective equipment. . . .

These latter two items have specific OSHA standards that describe how they must be addressed. These standards are 1910.155 Fire Protection (and emergency response teams) and 1910.134 Respiratory Protection. They specifically require the development of written programs to be in compliance with these regulations. Therefore, written emergency procedures and a written respiratory protection program must either be part of or in the appendix of the written hazard communication program. The development of emergency response programs and respiratory protection programs is beyond the scope of this book. Refer to OSHA information sources for more guidance in these areas. However, the importance of these is discussed below.

The Need for a Chemical Spill Team and Special Personal Protective Equipment. If a leak of spill occurs, a specially trained team of employees may be necessary to properly handle the situation as required by OSHA regulations. These specially trained personnel would also have to be fitted and trained in the use of various personal protective equipment. This can include self-contained breathing apparatus, full-body protection suits, application of spill neutralization materials, clean-up of hazardous materials, proper disposal of hazardous wastes, etc.

Foreseeable Emergency Procedures. The Hazard Communication Standard requires employers to explain to employees the procedures to follow in case of a foreseeable emergency. The "easy way out" on this question is to simply evacuate employees from the affected areas. However, many questions must be answered to develop such plans, such as: What areas will be affected? Where do you evacuate people from the building? What effect does wind direction have on airborne chemicals? Are employees expected to fight small fires? Who calls for emergency services? Are employees expected to clean up small chemical spills?

These procedures should be developed and coordinated with local emergency authorities as necessary. A lack of proper emergency preparedness may lead to employee injury with needless negative publicity for a company. Also, some state and local governments have community right-to-know laws that require a written chemical safety contingency plan.

9. *Develop a master file of MSDSs for all chemical substances used in the facility and MSDS booklets for chemical substances used in specific work areas.* The hazard communication standard states:

> (8) The employer shall maintain copies of the required material safety data sheets for each hazardous chemical in the workplace, and shall insure that they are readily accessible during each work shift to employees when they are in their work area(s).

What is usually done to handle this requirement is to maintain one complete copy of all MSDSs in a master file. However, if these are kept in the safety managers office, or other person responsible for the MSDS program, MSDSs may not be "readily accessible" to all employees on all shifts. Therefore, it is recommended to have separate MSDS files or booklets in each department that contain duplicates of the MSDSs for the chemicals used in that department. These booklets can be kept in the supervisor's office or in the general work area. These additional files can then be used to ensure employee access to MSDSs on all shifts as required by the standard.

10. *Train all regular employees and document the training for the various programs.* Hazard communication is a training and management challenge. This is largely due to the vast range of departments that are impacted by the standard, the numerous chemicals used in each department, the complexity of the information to be conveyed, and the general level of education of the work force. Conducting comprehensive training of all affected employees that is specific to the types of exposures on their job is a monumental task. Also, many departments have their own training programs and political hierarchies that may not sympathize with efforts to provide a unified approach to chemical hazard training. Also, department training programs may have already addressed many of the issues required in the right-to-know and other hazards such as physical agents, radiation, and infection. Some training redundancy is inevitable.

The Hazard Communication Standard has specific training information that must be conveyed to employees as shown below:

(h) *Employee information and training.* Employers shall provide employees with information and training on hazardous chemicals in their work area at the time of their initial assignment, and whenever a new hazard is introduced into their work area.

(1) *Information.* Employees shall be informed of:
(i) The requirements of this section;
(ii) Any operations in their work area where hazardous chemicals are present; and,
(iii) The location and availability of the written hazard communication program, including the required list(s) of hazardous chemicals, and material safety data sheets required by this section.

(2) *Training.* Employee training shall include at least:
(i) Methods and observations that may be used to detect the presence or release of a hazardous chemical in the work area (such as monitoring conducted by the employer, continuous monitoring devices, visual appearance or odor of hazardous chemicals when being released, etc.);
(ii) The physical and health hazards of the chemicals in the work area;
(iii) The measures employees can take to protect themselves from these hazards, including specific procedures the employer has implemented to protect employees from exposure to hazardous chemicals, such as appropriate work practices, emergency procedures, and personal protective equipment to be used; and,
(iv) The details of the hazard communication program developed by the employer, including an explanation of the labeling system and the material safety data sheet, and how employees can obtain and use the appropriate hazard information.

C. Major Aspects of a Comprehensive Training Program

Since right-to-know training is a complex process, there are several important areas in which advanced planning and prior understanding is necessary. Further, in developing and conducting this training, a number of potential areas need to be addressed. These are discussed below.

Requirement for Understandability. Though not specifically mentioned in the expanded federal law, the initial federal law for manufacturers and many state laws have the requirement that the right-to-know training be "understandable." In addition, from an effectiveness viewpoint, if the training is not understandable then it was not worth the time and effort expended. Interestingly, some states mandate specific minimum times for employee training, such as 3-5 hours. Further, some states require that employees be given tests after their training to measure the effectiveness of the training. From a liability standpoint, being able to demonstrate some understandability by employees after the training session is desirable.

Making right-to-know training understandable for all employees is perhaps the most difficult part of the whole right-to-know program. The plant workers, housekeeping staff, orderlies, and maintenance personnel have varying levels of education and may even have English as a second language. College-level concepts of exposure thresholds, industrial hygiene monitoring, toxicology, epidemiology, and other multisyllabic jargon may not be easy for these individuals to grasp. For training to be fairly universally understandable by the average worker, it should be simplified, as much as possible, to a seventh or eighth grade level or lower.

It is recommended that the training staff commit a special effort in making right-to-know training for all employees as understandable and comprehensive as possible, especially for workers who do not have a high level of education or who do not have English as a native language.

Bilingual Training. Probably the most difficult part of the issue of understandable training is how to properly handle the employee who is not fluent in English. Many employers have been asking whether this means that training should be in an employee's native language. There is no clear-cut answer to this question since there are many variables to consider. However, we do know that toxicology, industrial hygiene, and toxic properties are difficult concepts for many English-speaking employees to understand. Compound this with English being a second language, and right-to-know training can be of little value to these employees. Therefore, by not providing the training in their native language, an employer may not be adequately fulfilling the training requirement.

An employer also may be more likely to become involved in litigation because of a lack of understanding that results in an employee injury or illness. It is, therefore, very important to consider whether your employees adequately understand the information being conveyed. Cases of employees claiming that Engligh training was not "understandable" have already occurred in other types of litigation.

It has been the author's experience that bilingual training may be warranted for some employees. Bilingual training is generally preferred to a strictly native-language training session. Many English terms such as industrial hygiene, time-weighted average, hazardous chemical, and even Material Safety Data sheet have little meaning when translated directly into a foreign language. Bilinuual training should present key concepts in English with an explanation of the concept in the native language. The employees should then be able to recognize the English word when it comes up in future conversations in their work environment. They should also be better able to remember what it means because it was explained to them in an understandable form. It has been the author's experience that employees who receive this type of bilingual training in both English and their native language indicate that they get a much better understanding of chemical hazards and the proper precautions for handling them safely.

Videotaping bilingual training sessions will save considerable training time and expense for new and transferred employees, especially if you had hired a consultant to perform this training. In case of future litigation it can show that an effort was made to make the training understandable.

Perspective on Relative Risk. During right-to-know training sessions, it is extremely important to communicate the relative risk of hazardous chemicals used on the job to other risks in everyday life. This can give employees the proper perspective on the health risks of the chemicals used in the workplace, and can help prevent them from overreacting to the information provided.

Chemicals are a fundamental part of life. For many chemicals, certain amounts are not harmful. They can be found in the human body or in foods. Other chemicals are found in commonly used consumer products. If these perspectives are not brought out in the training sessions, employees can develop an inappropriate fear of the chemicals with which they work. Comparing their exposure to chemicals on the job to the relative risks from driving a car, smoking, and tornados, for example, can also minimize fears of working with controversial chemicals like carcinogens. An experienced industrial hygienist can usually provide this perspective to employees.

Question and Answer Periods. One of the major keys to employees understanding their rights and gaining a proper perspective on the use of chemicals is their ability to ask questions freely without fear of reprisal. A fairly good measure of this is if the question and answer period takes up as much time as the actual right-to-know training. This employee question and answer period can also provide excellent feedback to management. In some cases, the new training brings out particular aspects of a job that may have gone unnoticed by management and that may present an excess exposure that warrants further evaluation. Management can use this opportunity to demonstrate its concern for employee health and safety by addressing or assigning responsibility to assess the potential problems brought up in these training sessions. Do not pass it up!

Qualified Training Staff. One employee complaint that may come up during the training session is, "We have raised these questions before, but have never been given a satisfactory answer." In some cases, this is because supervisors respond to industrial hygiene or safety questions with, "Don't worry about it, it's not a problem," or "I don't think it is unsafe." In all fairness to employees and supervisors, *most supervisors are not safety professionals* and realistically should not be relied on to make such judgments. They are usually not familiar enough with the concepts of chemistry, industrial hygiene, and toxicology to properly interpret employee exposures or MSDSs. The liability that supervisors impose on their employers in making such judgments should be of great concern to management.

A recommended option to deal with this potential problem during the training sessions would be to have the employee questions answered by an independent expert, such as an industrial hygienist experienced in your type of chemical exposures. Interestingly, employees usually open up and ask more questions of an outsider than of their own management staff. The reason for this additional openness is because employees fear that if they ask questions they may appear ignorant to management or be viewed as a troublemaker.

Consultants can also answer questions as to current industry practice and better ways to do things if necessary. An outside consultant can also be viewed as a more credible source of information and can objectively state if a problem exists or not, since this is a third party who does not want to assume any liability by saying an operation is safe when it isn't.

Adequate Time and Funding. The above discussion details much of the work required to adequately address the right-to-know requirements. However, it is typical to underestimate the time and resources necessary to comply with the law. The author's experience has shown that a good training program costs at least $50-125 per employee, depending on the level of compliance at the start of the program. This includes lost production time during the traning sessions, staff time to survey the operations, and the development of a written program. It does not include industrial hygiene survey costs, which can run between $1000 and $1500 per day. Employee training sessions should also be limited to 20-25 employees so that questions can be properly addressed. This may have further impact on the cost of training.

Developing and implementing a right-to-know program for 500 employees is at least 2 months of full-time work, assuming you are familiar with the regulations. Delegating this responsibility to someone who already has numerous responsibilities will probably make it impossible for them to find the time to adequately handle the right-to-know program. Using a consultant may increase these costs slightly, but you trade your management time for an experienced consultant's time.

Going Beyond Generic Training Programs. Generic slide or video presentations are available to help in the right-to-know training process. A word of caution is necessary about generic programs. Generic presentations are designed to cover only general rights and definitions. Detailed information on the various types of exposures in the facility and the specific hazards of the individual or group of chemicals used must still be communicated to employees. Therefore, you will still need to do some training program development.

Inclusion of Contractors in Training Efforts. The right-to-know standard requires that contractors who work in areas of the facility where hazardous chemicals are present be apprised of the potential hazards they may encounter. The expanded hazard communication standard refers to this situation as a "multiple employer site." *This situation requires an exchange of information between the employers, usually in the form of MSDSs.* In some cases it may be advisable to include certain contractor employees in the formal training sessions. This would be warranted for long-term contractor employees and those who may be working around hazardous materials that the contractor's employer would have a difficult time covering adequately in their own right-to-know training programs.

Appropriate Interpretation of the Standard. Though the Federal Hazard Communication Standard was designed as a performance standard, the government uses a number of specific inspection criteria that, in essence, have the force of regulation. OSHA utilizes the Hazard Communication Compliance Directive CPL 2.238, which is available to the general public through your local OSHA office. Obtaining a copy of this directive can be helpful in deciding how well your facility would do if its hazard communication program was inspected by OSHA.

It should also be noted that noncompliance with OSHA standards and compliance directive components can have serious ramifications. This can include time-consuming inspections of the entire facility, significant financial penalties, and citation appeal conferences, to name just a few should you be cited by OSHA.

D. Use of Outside Consultants or Can You Do It Yourself?

Right-to-know compliance is a time-consuming process that demands significant commitment of staff resources. Delegating this responsibility to someone who already has a full schedule will probably make it impossible for them to find the time to adequately handle the right-to-know program. You must insure that this person has 3-4 months of time to develop your program.

11. *Train all contract, new, and transferred employees as appropriate, and document this training.* The hazard communication standard states that employees shall be provided with information and training on hazardous chemicals in their work area *at the time of their initial assignment, and whenever a new hazard is introduced into their work area.* This can be a difficult aspect of the regulation to comply with. In practice, new employees would have to receive right-to-know training as part of their in tial orientation training. Also, any employee who may be rotated to a different job within a facility (which has different chemical exposures) or who has a chemical with new hazards introduced into a current working environment will also have to receive additional training.

For example, if an employee is transferred from a department where no carcinogens are used to a department where they are used, he or she will have to be trained on the specific hazards of carcinogens and procedures for handling them. Also, if a material currently in use in a number of departments is determined to be a carcinogen, notification and training of employees will be necessary. If carcinogens are already being used in the affected department, then employees should be notified of the new hazard class for this material. If the material is also being used in a department where carcinogens are not already used, then these employees will have to be trained on the specific hazards of carcinogens and procedures for handling them as described earlier. It is easy to see that these various training requirements can be a significant ongoing effort. One should, therefore, be aware of the requirement for timely and appropriate training when organizing their training strategies.

Videotaping Training Sessions. Some facilities find it beneficial to videotape various right-to-know training sessions. In this way, when new or transferred employees need specific training for a certain type of exposure or for a particular department, they can receive a part of the training on video. Employee questions and other aspects not covered in the video can then be addressed by a qualified in-house trainer. This method can ensure some uniformity in the training presentations and optimize the training efforts of the staff.

12. *Update master MSDS file, MSDS booklets, and employee training as significant new information becomes available.* The Hazard Communication Standard requires that significant new information on a chemical be communicated to an employee as soon as the information becomes available to the employer. Specifically, the standard states: "(h) Employers shall provide employees with information and training on hazardous chemicals . . . whenever a new hazard is introduced into their work area." In order to assure that this new information is provided to emploeyrs the regulations required that chemical manufacturers provide this information to employers within 3 months after they become aware of the information. This is quoted below:

> (5) . . . If the chemical manufacturer, . . . becomes newly aware of any significant information regarding the hazards of a chemical, . . . this information shall be added to the MSDS within three months.
> (6) and employers are provided (with an MSDS) . . . with the first shipment after the MSDS is updated.

One problem that sometimes occurs is that an employer receives a new MSDS but is unaware that new information has been added. Without directly comparing the old versus the new MSDS, you will not know if it has been updated or not. Further complicating this situation is the fact that some chemical suppliers utilize MSDS software that prints a new date on the MSDS with each order without actually changing any information making line for line comparison mandatory. In addition, some chemical manufacturers do not keep their MSDS up to date. Therefore, you cannot totally rely on all manufacturers to provide the necessary updated information.

In some respects, it is almost easier to keep up with current occupational safety and health research than to try and compare each MSDS with an old one and hope the manufacturers are prudently updating the MSDS. Fortunately, there are not that many chemicals studied each year and keeping up with the latest information is not that difficult. Another way to handle this potential problem is to contract with an outside consultant to keep you apprised of any new information regarding chemicals used at your facility. Either way, the issue of monitoring for new information can be very costly and time consuming.

13. *Retrain employees periodically (annually by some state laws)*. Retraining is required when ever a new chemical is introduced in an operation or when new information about a chemical becomes available. For new chemicals, this training should occur either prior to or when the chemical is introduced. With new information about an old chemical, the employer has 90 days to provide the new information to employees.

In regards to the need for retraining, as of this time, OSHA has not issued any guidance on this subject. But it is their feeling that annual retraining is most productive. In addition, it is required by some states. Therefore, it appears prudent to retrain employees annually.

14. *Perform periodic industrial hygiene monitoring* as necessary.

Recommended Industrial Hygiene Monitoring Frequency. One of the purposes of industrial hygiene monitoring is to document that employee exposure levels are blow established limits. If exposure levels are not below established limits, then this informaiton can be used to indicate what ventilation improvements or changes in work practices may be needed. However, climatic changes and variability in employee work practices can cause significant differences in employee exposure levels from day to day. As a result, more than one sample of employee exposure levels is typically taken. As an example, sampling employee exposure during the summer and then again during the winter could be suggested since ventilation system's operating conditions can be significantly affected during these periods.

Once a sufficient number of tests are completed and statistically analyzed, then a decision as to the need for periodic monitoring can be made using Table 2.

E. Summary

Employee "right to know" or "hazard communication" is probably the most pervasive safety and health regulation of the present and near future. It encompassed all aspects of your safety program and your operations. The regulations cover not only the workers who actually use the chemicals but also those employee who may enter the area where the chemicals are used on an occasional basis. Therefore, planning to comply with these regulations should be part of the overall philosophy of management. The sooner this

TABLE 2 Industrial Hygiene Monitoring Testing Frequency

Exposure level	Testing frequency
≤ 20% of the TLV	Every 2 years
> 20% but ≤ 50% TLV	Every year
> 50% and ≤ 100% TLV	Every 6 months
> TLV	Quarterly

training is integrated into the overall training program, the easier it will be to deal with and promote employee understanding.

Finally, the potential for legal liability of not following these regulations is always hanging in question. In particular, the lack of compliance right to know regulations have the potential for significantly increasing ones liability due to such legal theories as "willful neglect" and "willful intent to harm." Though, at this point, these theories have only been used to force settlement on many cases, they could cause significant financial burden to a company. *Doing it "right" in the case of "right-to-know" is not redundant!*

III. COMMUNITY RIGHT-TO-KNOW

A. Historical Background

"Community right-to-know" is the other half of the right-to-know regulations. Employee right-to-know was discussed in the first half of this chapter. Community right-to-know regulations are intended to inform the community about potential catastrophic and long-term chemical exposure risks that are coming from the vairous facilities operating in their area.

Community right-to-know regulations had their start many years ago with local governing bodies regulating the design and placement of underground storage tanks and pressure vessels. Even though most authorities paid little attention to the chemical contents of these vessels, they did check to see that safety systems were installed and operable.

The first set of regulations to actually look at the potential hazards of chemical discharges of hazardous materials were enacted in Philadelphia in the early 1970s. These regulations required that industry within the city provide the city regulating agency with annual discharge volumes to the air of known or suspect carcinogens and other hazardous chemicals as defined in the regulations. Ethylene oxide was one of these chemicals.

One of the main concerns expressed by industry about such community right-to-know laws is that they provide a list of chemicals that could be used by plaintiffs to sue a company based on a cause of action that exposure to these chemical(s) resulted in disease in a specific person or an increased incidence in a disease in the general population. That same fear was expressed about Philadelphia's regulations. However, to date, no such legal actions using this information have occurred. This lack of any legal actions may be due to the lack of a sufficient latency period for the disease to evolve in a population or that no significant health effects are occuring in the population.

The community right-to-know laws that have been enacted in the last few years on a state, local, and federal level were a direct result of the chemical incident in Bhopal, India, where over 2000 people died after accidental exposure to methylene diisocynate. Other international incidents, such as the Sandoz chemical spill in Europe, reiterated the need for such regulations.

The United States federal government's response to the Bhopal chemical incident was for the Environmental Protection Agency (EPA) to expand the existing hazardous waste regulations to include a right for communities to know if toxic chemicals are in their local areas. These regulations were specifically devel-oped as the Superfund Amendments and Reauthorization Act (SARA) Title III. "Superfund" is the act that was passed by Congress to clean up the hazardous waste sites in this country.

Consequently, on a federal level, the community right to know law is also known as SARA Title III regulations. Under SARA Title III, states were required to set up a state emergency response commission (SERC) as of March 1987. In addition, local committees were required to set up a local emergency planning commission (LEPC) to handle chemical emergencies as of September 1987. These agencies are required by federal law to develop plans to handle potential chemical emergencies that may result from an accidental release of the toxic chemical present in their local jurisdiction. Companies using hazardous chemicals were to provide lists of these chemicals to the federal EPA, state emergency response commissions (SERC), local emergency planning commission (LEPC), and local fire department.

The vairous state and local governmental bodies also enacted community right-to-know regulations. The major difference between the federal regulations and state and local regulations is the minimum quantities of vairous chemicals that msut be reported. For some chemicals it is as little as 1 pound, and for other it is more than 10,000 pounds. This reporting quantity is dependent on the toxicity and reactivity of the material. In addition, the federal regulations, in some cases, require the development of an emergency chemical contingency plan and reporting of total discharges of some chemicals to the environment.

Generally, state and local community right-to-know regulations additionally require that chemical users or manufacturers provide the local governmental body or fire department a detailed inventory all potentially hazardous chemicals in quantities as low as 5 gallons and a map showing the location of the chemicals within the facility. Since these local regulations are so numerous, they are beyond the scope of this book. However, it is best to obtain and review a copy of all the applicable regulations before starting a facility compliance program to eliminate performing redundant tasks.

In complying with the federal regulations there are two major steps that must be completed. The first step is completing the notification forms. The second is developing the emergency contingency plan. (Note: Additional reporting regulatory requirements may also exist from local and state agencies). A general discussion of the reporting requirements of the federal community right-to-know regulations and contingency planning follows.

B. Reporting Requirements of Federal Community Right-to-Know Regulations (SARA Title III)

Under SARA Title III, there are four reporting requirements that are governed by different chemical lists. These are:

1. *The list of extremely hazardous substances (EHS List)*
2. *All hazardous chemicals for which you have an MSDS*

3. *The list of chemicals regulated under CERCLA, RCRA, and other federal regulations (Tier I and Tier II forms)*
4. *The list of chemicals for which all environmental discharges must be reported* (Form R)

Each of these chemicals lists has special reporting requirement and forms that need to be completed on an annual basis. Also, under SARA Title III, if a facility falls under these regulations, they must have a system to provide 24-hr response capability to address questions from emergency response personnel in case of an emergency. This can either be an expert available at all times, or a phone service that can immediately contact one. This is one of the main pieces of information on each form and is checked by the PEA upon submission of the forms. A brief discussion of each list and notificaiton requirements follows:

1. *The list of extremely hazardous substances (EHS List)*. This is a list of approximately 402 chemicals. Its foundation was based on a list of 125 extremely hazardous chemicals identified in European regulations. EPA added to this list other chemicals that in its opinion would also present a serious hazard to the surrounding community if discharged even in small quantities. In most cases, these are highly toxic or highly reactive materials. The minimum reporting quantity at which you must report that you have a chemical in your facility quantities varies from a low 1 pound to a high of 500 pounds (approximately one 55-gallon drum). This reporting can take place by simply sending a letter to the appropriate regional EPA office and the state emergency response commission (SERC).
2. *All hazardous chemicals for which you have an MSDS.* State emergency response commissions (SERC) require a list of chemicals that you use in excess of 10,000 lbs per year for which you have been provided an MSDS by the manufacturer. Again, a simple list sent to the SERC is sufficient.
3. *The list of chemicals regulated under CERCLA, RCRA, and other federal regulations (Tier I and Tier II forms)*. This list contains some 782 chemicals that may result in potential harm to the environment or to the public if released in sufficient quantity concentration. The reporting requirements have a phase-in period, based on quantity, for chemical manufacturers, and apply to any facility using more than 10,000 lbs* of a chemical on these lists per year. The use and inventory of these chemicals must be reported on a Tier I or Tier II form and sent to the SERC and the LEPC. These forms required detailed health and physical property information on each chemical. Considerable time is required to compelte these forms, particularly if one uses large quantities of many different chemicals.
4. *The list of chemicals for which all environmental discharges must be reported (Form R)*. This final list of chemicals is based on approximately 329 chemicals originally called the "State of Maryland List," since it was copied from Maryland's community right-to-know regulation. Again, the reporting requirements are based on those chemicals that are used in excess of 10,000 lbs per year. This list contains suspected and known carcinogens and other potentially toxic chemicals that when released to the air may pose a long-term (chronic) health risk.

---

*This drops to all quantity (0 lbs) in 1990. However, it is hoped the EPA will increase this to 500 lbs.

This last reporting requirement is the most complex of any of the above reporting requirements. The reporting form for this list is known as "Form R" or the "Release Form." This form requires extremely detailed and complete information on the facility and the chemicals used, including:

1. Exact latitude and longitude of the facility
2. Annual amounts of each chemical discharged to atmosphere
3. Annual amounts of each chemical discharged to hazardous waste sites
4. Annual amounts of each chemical discharged to sewer systems
5. Annual amounts of each chemical discharged as fugitive emissions
6. Annual amounts of each chemical used
7. Annual amounts of each chemical recycled

A form R must be completed for each chemical. Each of these forms is six pages long. The EPA requires that the information be based on actual measurements or engineering estimates. Because of the degree of exactness required, a mass balance on total chemical operations is usually the easiest way of completing the form; otherwise a risk of inconsistent numbers is possible.

C. Emergency Contingency Plan Development and Implementation

A secondary requirement of the SARA Title III is for employers who use or process sufficient quantities of toxic chemicals to develop a management and response plan to handle potential accidental chemical releases.

Development of a chemical contingency emergency response plan is the most important part of community right-to-know regulations. A facility's chemical contingency emergency response plan is the main tool for preventing a catastrophic accident.

Though few specific regulatory requirements exist on this subject of chemical spill contingency plan development, an outline for such a plan is found in EPA hazardous waste regulations (40 CFR 265). This outline recommends:

a. An internal communications or alarm system
b. A device, such as telephone or two-way radio, capable of summoning emergency assistance from local emergency response teams (i.e., police, fire, hospital, etc.)
c. The names, addresses, home telephone numbers, and business telephone numbers of individuals qualified to act as emergency coordinator
d. The location and capability of all existing emergency equipment
e. The emergency telephone numbers of local emergency response groups
f. An evacuation plan for the facility personnel
g. A description of arrangements agreed to by local emergency response groups, and
h. A description of the actions that facility personnel must take to respond to fires, explosions, or unplanned releases of hazardous materials

Once the emergency/contingency plan has been developed and implemented, copies should be disseminated to all appropriate emergency response groups (i.e., fire, police, hospitals, etc.). These groups should be familiar with the facility layout and the storage location of all hazardous materials.

One thing that this outline does not cover is the need to designate a person responsible for and trained to provide information to the local press. The importance of this additional requirement should not be underestimated. This person actually can maintain or destroy your company's image, depending on the way they handle the press.

1. *Cooperation with the Local Emergency Planning Committee*

SARA Title III does not directly impose any regulatory requirements on a facility to participate in actual local emergency planning committee activities, at this time. However, it is possible that you may be asked to participate in this planning process, seeing that your expertise may be the best available in the area. Facility risk managers and corporate legal counsel should be consulted to minimize your company's liability in these activities.

2. *Information That Employers Should Provide to Emergency Care Facilities*

As part of this plan, employers are required to designate a medical facility as the primary care facility for employees or other persons who are injured during a toxic chemical release.

The designation of a hospital as an emergency care facility by a company imposes numerous requirements on the hospital. It is encumbent upon the employer to assure that the hospital has the necessary facilities and knowledge to handle each specific chemical exposure health emergency that could forseeably occur.

The information that should be supplied to the hospital by the company is:

1. A list of the chemicals used at the facilities
2. A *complete* MSDS on each chemical
3. A medical history of each employee
4. Access to trained personnel who can estimate the amount of exposure

The first two pieces of information are preferred to be on a computer disk in a database software compatible with that available in the emergency room. In this way, when an injured employee(s) or person in the community is brought to the emergency room, the chemical and its MSDS can be quickly and easily retrieved.

One serious obstacle that hinders the hospital in its efforts to treat chemical exposure cases is the lack of chemical hazard information being passed on to the treating physician. One of the concepts of the Hazard Communication Standard is to insure that sufficient information is available to properly treat persons exposed to hazardous chemicals. The Material Safety Data Sheet (MSDS) was to be the communication medium for information on what hazardous chemical ingredients are found in the material, what are the acute and chronic health effects from overexposure, and first aid for overexposed employees. However, due to a number of reasons, one cannot rely on the MSDS for all this information.

Most MSDSs do not state what the long-term consequences may be from the chemical exposure and whether prophylactic treatment can help minimize these effects. And for many chemicals, particularly mixtures of numerous hazardous chemicals, we just do not know what is the best teatment for overexposured patients. Under Section 6 of the MSDS "Emergency and First Aid Procedures" many MSDSs contain statements such as "Wash with water for 15 minutes. Consult a physician." This puts a tremendous burden on the treating physician to know what to do in each case. And unfortunately, many physicians are not trained to handle chemical injuries.

Emergency rooms that handle chemical emergencies should have access to at least one source of chemical toxicity information to initially treat the overexposed patient. This can help alleviate many of the problems of insufficient MSDS information.

The medical history is another very important piece of information. For examle, if the injured employee has liver, kidney, hematopoietic, or other disease, he or she could be affected far more adversely than a healthy individual. In addition, if the employee has a particular chemical sensitivity, treatment may be far different than just a single overexposure.

Information on the estimated amount of exposure can also help the emergency-room physician to decide the best course of treatment in some cases. For example, for an ingested organic material the physician can weigh the risks of giving the patient charcoal to decrease the rate of absorption versus gastric lavage therapy.

D. Summary

Community right-to-know is a steadily evolving area of regulatory reporting and planning responsibilities. Each year the requirements get more specific and apply to more substances. Keeping up to date with the regulations, maintaining emergency response equipment, periodic personnel training, meetings with local emergency response personnel, annual reporting requirements, and inventory management are just a few of the items that require a commitment of sufficient staff and funds. It is important to properly manage this responsibility to minimize future liability in case of an accidental spill or leak.

One company, which did not have an adequate chemical contingency response plan, had a large leak of a hazardous material that resulted in multiple employee injuries and evacuation of their facility and local area. When one of the company executives was questioned about the incident by the local TV news he made the real-life mistake of saying, "I didn't know we had a tank that big!" *Don't let this happen to you.*

SELECTED READING IN EMERGENCY PAREPAREDNESS

The following list of selected readings is intended for general guidance and is not intended to be comprehensive.

Emergency Planning and Management

1. Emergency Program Manager, Federal Emergency Management Agency, SS-1, November 1983.
2. 1984 Emergency Response Guidebook, DOT P 5800.3, U.S. Department of Transportation, Research and Special Programs Administration, Materials Transportation Bureau.
3. Emergency Management USA, Federal Emergency Management Agency, HS-2, May 1986.
4. Hazardous Material Emergency Planning Guide, NRT-1, National Response Team of the National Oil and Hazardous Substances Contingency Plan, U.S. Coast Guard, March 1987.

5. Planning Guide and Checklist for Hazardous Material Contingency Plans, Federal Emergency Management Agency, FEMA-10, July 1981.
6. Chemical Emergency Preparedness Program, U.S. EPA, November 1986.

Chemical First Aid Information Sources

1. Poison Index (database), 660 Bannock Street, Denver, CO 80204-4506, 1-800-525-9083.
2. R. H. Dreisbach, *Handbook of Poisoning: Prevention, Diagnosis & Treatment*, Lange Medical Publications, Los Altos, CA, 1983.
3. Gosselin, Hodge, Smith, and Gleason, *Clinical Toxicology of Commercial Products*, 4th ed., Williams & Wilkins, Baltimore.

45

Safe Handling of Biologically Contaminated Medical Devices

JOHN H. KEENE *Biohaztec Associates, Inc., Midlothian, Virginia*

AMIRAM DANIEL *Health Industry Manufacturers Association, Washington, D. C.*

I. INTRODUCTION

Medical devices that have been used and are routinely serviced in the field or returned to the manufacturer for evaluation must be considered contaminated with potentially infectious biological materials. A Health Industry Manufacturers Association (HIMA) task force has recommended that, in general, contaminated devices be decontaminated at the health-care facility location prior to return to the manufacturer. Occasionally, it may not be possible or desirable (for legal reasons) to decontaminate such devices. As an example, one cannot destructively decontaminate a device that must be evaluated for malfunction or may be part of a liability law suit. Since these devices may not be decontaminated prior to return or service, it is important that personnel assigned to receive, evaluate, and repair medical devices be informed of their potential hazards and be appropriately trained in their handling in order to minimize the risk of injury and infection.

There are numerous government and state regulation that govern the use of medical devices and disposal of medical waste as well as hazardous risk to employees associated with servicing medical devices or handling medical waste, their rights to know such risks, etc. HIMA has published an extensive document entitled "Safe Handling of Biologically Contaminated Medical Devices" [1]. The document contains the following chapters:

I. HIMA's recommendations on service and repair of potentially contaminated medical devices
II. General recommendation and policy statements on the subject from variety of organizations (government, professional, and private) dealing with safety of health-care workers
III. General precautions procedures addressing health-care workers in general (OSHA) and laboratory safety manuals in particular (CDC, NIH)

IV. Specific precautions procedures—from hand washing to biological safety cabinets
V. Numerous postal authorities and Department of Transportation Regulations dealing with shipment of etiological agents
VI. Federal waste disposal regulation
VII. Some applicable international regulation

In the following pages we present a synopsis of those aspects of manufacturers' activities that involve handling potentially infectious devices and suggest some precautionary measures.

II. MINIMIZING EMPLOYEE RISK

The possible risk to employees who must handle potentially contaminated medical devices is that of infection caused by etiologic agents (i.e., bacteria, viruses, fungi, and parasites) that may be present on or in the device as a result of use. (Note: Devices that have not been used pose no infectious risk to employees.) Recent concern has focused on the possible transmission of AIDS and hepatitis, but it must be recognized that any blood-borne pathogen may be present on items contaminated with blood. Nor is the risk confined only to those devices that may be contaminated with blood, since devices that may be contaminated with other patient fluids, tissues, or excreta may also be contaminated with various human pathogens. Numerous articles and recommendations have been published that outline not only the methods of transmission of many of these agents, but also the appropriate means for handling or eliminating the agents to prevent possible transmission.

Each manufacturer has the responsibility to ensure that employees are informed of the hazards associated with their work and to provide the appropriate safety equipment and training to minimize the risk of infection. The initial step in fulfilling this responsibility is to develop a program for handling the contaminated devices. This program should consist of:

1. Instruction to users (health-care facilities) on methods for decontaminating and/or packaging contaminated devices prior to their return to the manufacturing facility
2. Written policies regarding the shipment, reception, evaluation, and ultimate disposal of contaminated devices
3. Establishment of a specific area for receiving and handling potentially contaminated devices.
4. Identification of appropriate personnel and provision of facility safety equipment to minimize possible personnel exposure.
5. Documentation of ongoing safety training for all personnel involved in the handling of contaminated devices.

III. GENERAL RECOMMENDED PROCEDURES

1. Whenever possible, potentially infectious medical devices should be decontaminated prior to evaluation or service.
2. Any work with known or suspected contaminated devices should be confined to designated areas.

3. Always wear protective gloves and lab coats (remember gloves do not protect against needle-sticks or cuts form sharp objects) and other appropriate safety equipment as needed. Hands should always be washed at the end of the procedure.
4. Procedures that may result in aerosolization of potentially infectious materials should be performed in a biosafety cabinet. If this is not possible, some method for elimination of aerosols or for minimizing potential aerosol formation should be used and a surgical mask should be worn by personnel performing the procedures.
5. All contaminated materials should be disposed of in an appropriate container and decontaminated prior to disposal. Once appropriately decontaminated, this waste material is no longer infectious and may be disposed of as normal solid waste. Some localities have specific regulations regarding the disposal of medical waste, and these should be followed.
6. When work with contaminated devices is completed, the work area should be decontaminated with appropriate disinfectants: all personal protective clothing should be placed in a designated area for disposal or laundering, and personnel should thoroughly wash their hands.
7. Personnel should be instructed to report all accidents no matter how insignificant they may seem. Appropriate medical assistance should be made available.
8. Other prophylactic measures such as hepatitis B vaccine have been recommended by the CDC for prevention of hepatitis B infection in health-care workers whose jobs include exposure to human blood and other body fluids. Although the risk of infection for field service representatives and returned goods specialists is lower when compared to hospital nurses or laboratory technicians, it is appropriate to offer hepatitis B vaccination to these people and encourage them to take it.
9. The appropriate staff should be educated as to the nature of the medical devices with which they work, the associated risks and the protective measures necessary. This education effort should be undertaken continuously and should be documented.
10. Impress on the industrial workers the potential for hazards in each area and ensure that they work and behave appropriately.

IV. DISINFECTION AND STERILIZATION

The terms "disinfection" and "sterilization" are not synonymous and often are used inappropriately. *Sterilization* is defined as a process that eliminates or destroys all forms of microbial life. Even the most resistant bacterial or fungal spores are killed by these processes. The processes of sterilization usually involve steam heat (autoclaving), incineration, radiation, or the use of certain gases (ethylene oxide) or chemicals (formaldehyde). *Disinfection* is generally defined as a process that eliminates all pathogenic microorganisms from an inanimate object with the exception of the bacterial or fungal spore. Disinfectants are usually chemicals that inhibit the physiological activity of the organisms in such a way that they cause the death or inactivation of the organisms. It should be noted that some chemical disinfectants, used in high enough concentration and for an extended period of time, may be considered sterilants. The term *decontamination* is often used to indicate that objects have been treated with disinfectants so that they are safe for routine handling.

Methods of sterilization may cause destruction of contaminated devices, which may not be appropriate when one requires the device to remain intact for failure evaluation associated with potential litigation or for use following servicing. In many instances, decontamination of contaminated devices with disinfectants may be the best way to minimize the risk of exposure for the workers who must work with such devices.

The appropriate method of disinfection or sterilization and cleaning should be determined by the manufacturer for each device. Once determined, this information should be made available to those who are required to perform the decontamination. During the design phase of medical devices, manufacturers should consider ease of decontamination as well as potential risk to the user and personnel who must service these devices. Manufacturers should attempt to provide designs that will minimize these risks and allow for easy decontamination.

V. CONCLUSION

There has always existed a risk of exposure to infectious agents as a result of handling contaminated devices. However, the emergence of the human immunodeficiency virus (HIV) and with it the heightened concern regarding possible transmission of all other blood-borne pathogens has put new emphasis on the safe handling of these devices by consumers and device manufacturer personnel. The risk of exposure exists, but with an understanding of the hazards, proper training, and appropriate techniques and procedures, this risk can be minimized. It is important to recognize that although some contaminated devices may not withstand sterilization procedures, they can be handled, evaluated, and repaired safely without harm to the industrial lab personnel.

REFERENCE

1. A. Daniel, L. Olsen, and N. Estrin, eds., "Safe Handling of Biologically Contaminated Medical Devices," Health Industry Manufacturers Association, Washington, D.C., publication 88-2, March 1988.

part seven

Current and Emerging Issues in Environmental Safety and Health

46

Government Initiatives in the Environmental and Occupational Safety and Health Arena

JAMES F. JORKASKY *Health Industry Manufacturers Association, Washington, D.C.*

I. INTRODUCTION

The previous chapters on regulations governing occupational exposure and worker right-to-know have provided a comprehensive discussion on some specific worker safety issues. This chapter will provide a much broader prospective on both environmental and occupational safety and health issues, which have become increasingly important to the medical products industry.

The chapter will begin with a discussion of some points that one should keep in mind when considering these issues. It will then present a brief review of how environmental and occupational legislation and regulation grew in importance in the 1970s and 1980s. This will set the stage for a comprehensive discussion of what will likely be the environmental and occupational changes and challenges in the 1990s.

One caveat—this chapter should not be used as a substitute for the careful reading of the legislation or regulation that is discussed. Specific questions or concerns should be addressed to appropriate counsel.

A. Points to Consider

It should be said at the outset that there are some generalizations or "points to consider" that can be made regarding environmental and occupational issues. Many of these also apply to other regulatory arenas, such as regulation of foods, drugs, cosmetics, and devices. An understanding of these points is important, since they have been instrumental in how legislation and regulation have developed in the past and will develop in the future. They are set forth below:

This chapter reflects the views of the author and does not necessarily represent those of HIMA or its member companies.

There is little doubt that everyone wants safe work places and a clean environment. How we achieve that goal while providing products and services is the pivotal question.

Environmental and occupational regulation must balance several concerns—sound science, public safety, technological and economic feasibility, business (especially small business) impacts, aesthetics, political considerations, etc. These concerns are often in competition, with little chance of complete resolution.

Environmental and occupational regulation is often fraught with controversy as legislators, regulators, scientists, industry, labor, and advocates struggle with ever-expanding scientific knowledge, technological advancements, changing legal concepts, and growing public awareness.

Environmental occupational regulation is no longer a domestic issue, but an international one. This is pointed out dramatically by the international concern over the ozone depletion/greenhouse effect and the attendant efforts to reduce reliance on ozone-depleting chlorofluorocarbons (CFCs).

Congress plays a large role in determining this country's environmental and occupational agenda as it translates public policy into government action. Congress can enact legislation for which EPA and OSHA must develop regulatory programs. Congress conducts oversight over the Environmental Protection Agency (EPA), the Occupational Safety and Health Administration (OSHA), and the National Institute for Occupational Safety and Health (NIOSH), and reviews agency policies and procedures (implementation of regulations, enforcement). Congress also has oversight over the agencies' budgets. Due to this high degree of involvement, environmental and occupational issues are by their very nature highly political and subject to extensive visibility and a great deal of controversy.

States and communities are not waiting for the federal government to take action—they are determining their own environmental destiny. For example, by the time OSHA promulgated its Hazard Communication standard, about half of the states already had right-to-know laws in place. State and local statutes and regulations make it increasingly difficult for manufacturers to comply with the myriad of different and possibly conflicting regulatory requirements.

Corporate America must now conduct its business in a fishbowl, not a vacuum. A good example is the worker and community right-to-know requirements that require companies to inform the government and the public about the hazards of chemicals they use. These requirements help form public perceptions of a company, its relations with its workers and the surrounding community, and its social responsibility.

A larger portion of each dollar used to conduct business will be going to environmental and occupational safety and health compliance. This represents not only the cost of filling out the plethora of forms and reports, but costs associated with new plants and equipment, waste minimization, development of substitute and alternative chemicals and processes, accounting for all wastes created during the manufacturing process, and worker/community education.

A company can *never* be *too* prepared for an environmental or occupational emergency. Therefore, it must not be complacent, but rather proactive in constantly evaluating the potential hazards associated with its processes and procedures.

A company must always be prepared to deal with tough questions from legislators, regulators, advocates, and the media regarding its environmental and occupational policies. Part of this preparation is a thorough understanding by all levels of management of a company's practices and policies. A company that is prepared can respond to questions as to why it does what it does, pointing out, for example, the lack of alternative processes, disadvantages of alternative processes, benefits of the products derived from certain manufacturing operations, etc.

The public awareness and concern over environmental and occupational issues has never been greater. Henceforth, it is likely that the resources a company devotes to these issues will increase dramatically. Public awareness is likely to remain high, especially as the media continues to highlight irreversible damage done to the earth from past practices.

The past 20 years have seen the emergence of a legitimate new professional career in the work place, the "environmental and occupational specialist." Although these individuals come from many different backgrounds (industrial hygiene, engineering, safety, academics, law, regulatory affairs), they understand the "points to consider" raised above and use these to guide them in their profession. Please keep these points in mind as you read the remainder of this chapter.

II. THE PAST TWO DECADES

To understand what environmental and occupational issues will be vital to the medical products industry in the 1990s, it is useful to review what has occurred in the past two decades with respect to regulation in general, as well as to regulation that has occurred in the medical products industry.

For some time, the medical products industry focused its regulatory efforts solely on compliance with Food and Drug Administration (FDA) reguirements. With the advent of environmental and occupational regulation in the early 1970s, the regulatory burden grew.

Medical manufacturers were at first faced with general OSHA work-place inspections and EPA reporting requirements. Much of the environmental focus was on the sterilants used by industry, which are regulated by EPA, such as ethylene oxide (EtO). As OSHA and EPA expanded their regulatory programs to include worker and community right-to-know, substance-specific exposure standards, and hazardous waste management, industry's focus broadened beyond the sterilant issues into a full-fledged regulatory program. Concurrently, as industry's use of literally thousands of chemicals and processes grew, industry's regulatory responsibilities grew. For example, many medical manufacturers became "chemical producers" through their manufacture of raw materials, solutions, and diagnostics.

A. Development of Environmental Regulation

The 1970s began with the creation of the Environmental Protection Agency (EPA) from 15 components of five executive departments and independent agencies. EPA was charged with the protection of the air, water, and land. The agency draws its authority from a variety of federal statutes authorizing it to develop regulatory programs to control polluting substances. These statutes also authorize various civil and criminal penalties. Table 1 presents a brief summary of six of the major environmental statutes enacted by Congress in the 1970s. Table 2 presents a brief summary of major regulatory activity under the statutes that affects the medical products industry.

If the 1970s were a time for establishing environmental statutes and attendant regulatory requirements, the 1980s were a time for revision of those statutes. Congress became aware of the effectiveness and, more often, the failure of the regulatory programs EPA established to meet the intent of the statutes. Environmentalists charged that the programs were ineffective, that compliance efforts were minimal, and that it was pretty much "business as usual" in industry, citing the necessity for revised and reinvigorated statutes. Science also grew in sophistication—in its ability to measure for pollutants, to assess the damage to the environment, and to make predictions about the future. The media discovered that environmental issues also made good stories, especially environmental problems of global proportions.

The Clean Air and Clean Water Acts were revised in the late 1970s. In 1984, Congress reauthorized the Resource, Conservation and Recovery Act (RCRA), making major changes. In its reauthorization, Congress acknowledged problems with the statute and EPA's hazardous waste management regulations, promulgated in 1980. Congress changed the statute to expand the definition of hazardous waste and to require more small-quantity generators to meet regulatory requirements. Congress also took two bold steps—establishing an explicit national policy in favor of alternatives to land disposal, and placing "hammer" clauses in the statute, which set up a time frame in which EPA had to develop a regulatory program and to report back to Congress on its implementation. RCRA's 1984 reauthorization affected the medical products industry in that a greater number of smaller companies had to meet EPA's expanded small-quantity generator requirements.

Congress, however, did not require EPA to take specific action on infectious waste, one of the six categories of hazardous waste. However, in late 1988, Congress enacted the Medical Waste Tracking Act of 1988 in response to the concern about beach wash-ups of medical waste that occurred earlier that summer. The act authorized EPA to establish, within 9 months, a demonstration program requiring medical waste management and tracking, as well as reports to Congress over a 2-year period on the success of the program. The demonstration program went into effect in June 1989 in New York, New Jersey, and Connecticut, states hardest hit by beach wash-ups. Although the act and attendant regulations are designed primarily to address hospital and health-care facilitiy waste, they do include potentially infectious waste from production facilities. The medical industry has been affected in three ways:

1. Industry's products, especially sterile medical disposables, become medical waste after use in health-care facilities.
2. The industry is a major provider of home health-care products, which may become medical waste after use.
3. Industry facilities that produce microbiological waste from laboratories or in-house medical functions may be required to dispose of these wastes as medical wastes.

In 1986, Congress reauthorized Superfund, adding Title III, the Emergency Planning and Community Right-to-Know Act, under the Superfund

TABLE 1 Overview of Key Environmental Statutes

The Federal Insecticide, Fungicide, and Rodenticide Act (FIFRA), 7 U.S.C. 136 et seq., requires EPA to register all pesticides marketed in this country and to classify them into general or restricted use categories. FIFRA's goal is to protect humans and the environment from the potential hazard of these substances. EPA evaluates data submitted by pesticide manufacturers concerning the risks associated with the use of the pesticides. Based on its conclusions, EPA may refuse to register a new pesticide or may cancel or suspend the registration of a product already on the market. EPA's concern is that the product, when used as directed, will be effective against pests and not cause unreasonable adverse effects to humans, animals, or the environment.

The Resource Conservation and Recovery Act (RCRA), 42 U.S.C. 6901 et seq., requires the EPA to institute a national program to control "hazardous waste." A federal-state permit program has been established to regulate hazardous waste from its point of generation through treatment, storage, and ultimate disposal. Regulatory requirements include specific monitoring, recordkeeping, and reporting for generators, transporters, and disposers of hazardous waste materials. RCRA also includes an underground storage tank program and the new medical waste tracking demonstration program.

The Clean Air Act (CAA), 42 U.S.C. 7401 et seq., authorizes EPA to establish national standards for air quality for certain criteria of pollutants. It also authorizes the agency to identify and set more stringent standards for "hazardous" pollutants. In addition, the agency has established separate performance standards for major new sources of pollution.

The Clean Water Act (CWA), 33 U.S.C. 1251 et seq., previously known as the Federal Water Pollution Control Act, requires compliance with technology-based effluent limitations and prohibits the discharge of pollutants from a point source into navigable waters without a permit. It also requires the pretreatment of wastes discharged into publicly owned treatment works. In addition, the act places a strong emphasis on the control of toxic pollutants and oil and hazardous substances spills.

The Comprehensive Environmental Response, Compensation, and Liability Act of 1980 (Superfund or CERCLA), 42 U.S.C. 9601 et seq., gives EPA emergency response and cleanup authority for releases of hazardous substances into the environment and provides for the cleanup of inactive waste sites. Atypical of most environmental statutes, which are implemented through specific regulatory programs requiring limits or controls for pollutants, this law directly imposes notice and liability requirements on industry for the cleanup of hazardous waste dumpsites. Liability is strict, joint and several. Guidelines and procedures for the cleanup of hazardous waste sites are set out in a document called the National Contingency Plan. Reauthorized in 1986 under the Superfund Amendments and Reauthorization Act (SARA) to include Title III, the Emergency Planning and Community Right-to-Know Act.

The Toxic Substances Control Act (TSCA), 15 U.S.C. 2601 et seq., mandates that EPA protect the public health and the environment from unreasonable chemical risks. It authorizes broad regulatory authority for chemicals produced, used, or imported into the United States. It gives the agency authority to gather information on chemicals, to identify harmful substances, and to control substances where the risks outweigh the benefits to society and the economy.

TABLE 2 EPA's Major Regulatory Programs Affecting the Medical Products Industry

FIFRA

Registration of "pesticides." A pesticide is defined (in part) as "any substance or mixture of substances intended for preventing, destroying, repelling or mitigating a pest." The definition of a "pest" includes viruses, bacteria, and microorganisms. Industry sterilants, disinfectants and decontaminants are regulated by EPA, licensed as pesticides (40 C.F.R. Parts 152-158).

Reporting requirements if new evidence found of adverse effects (44 *Fed. Reg.* 40716, July 12, 1979).

RCRA

Hazardous Waste Management Regulations (40 C.F.R. Parts 260-272).

Standards for the Tracking and Management of Medical Waste (40 C.F.R. Parts 22 and 259).

CAA

Section 112, standards for "hazardous air pollutants" (40 C.F.R. Part 61).

Protection of stratospheric ozone (40 C.F.R. Part 82)—includes CFC production limits.

CWA

Effluent limitations regulations (40 C.F.R. Parts 401-471).

Permit regulations (40 C.F.R. Parts 122-125).

CERCLA/SARA

"Strict liability" requirements—if a company's waste is found in a Superfund clean-up site, that company is responsible for the cost of cleaning up that waste, even if the company disposed of the waste in an appropriate manner.

Title III, Emergency Planning and Community Right-to-Know Act—establishes EPA regulatory authority to require reporting of inventories and emissions of hazardous chemicals (40 C.F.R. Parts 300, 355, 370, and 371).

TSCA

Notification requirements for manufacture or import of new chemicals or new uses of chemicals (40 C.F.R. Parts 720, 721, and 723).

Reporting requirements if new evidence found of adverse effects (43 *Fed. Reg.* 11110, March 16, 1978).

Chemical testing requirements (40 C.F.R. Parts 790-799).

Recordkeeping and reporting requirements for chemical manufacturers, importers, exporters, and processors (40 C.F.R. Parts 704, 707, 710, 712, 716, and 717).

Amendments and Reauthorization Act (SARA). Congress took this action to address emergency situations, such as the chemical incidents in Bhopal, India, and Institute, West Virginia, which had occurred in late 1984 and 1985.

In the SARA amendments, Congress again enunciated a specific time frame for the development of state and community emergency planning commissions, requirements for the provision of Material Safety Data Sheets (MSDSs) to local fire departments, and hazardous chemical inventory and emissions reporting requirements. These requirements have become more stringent with time—many of the threshold quantities for reporting have decreased, and companies must indicate methodologies used to assess reportable chemical emissions. The medical products industry is subject to all of these regulatory requirements.

In 1987, Congress reauthorized the Clean Water Act (CWA), adding new programs to deal with toxic pollutants and stiffer penalties for violations of the act.

In 1988, Congress reauthorized the Federal Insecticide, Fungicide and Rodenticide Act (FIFRA), mandating that all pesticides be reregistered over the next 10 years. This includes sterilants, such as ethylene oxide, chlorine dioxide, hydrogen peroxide, formaldehyde, and glutaraldehyde. EPA will receive efficacy and toxicity testing data from pesticide manufacturers to determine if it will continue to register the pesticide or ban or severely limit its continued use.

As the decade draws to a close, Congress is revising the Clean Air Act (CAA), which is likely to change significantly, with fortified provisions regarding automobile and public utility exhausts, the regulation of toxic air pollutants, and acid rain. These provisions will likely affect the medical products industry in the following ways:

Increased energy costs
Increased transportation costs
Extensive requirements for emissions control
New requirements for accidental/emergency releases

Actions already taken under the Clean Air Act regarding ozone-depleting CFCs have resulted in a call for the elimination of all CFCs by the year 2000. EPA has already put into place regulations that would limit CFC production and encourage the development and use of substitutes and alternatives. EPA is also considering legislation that would establish research programs into the causes and control of indoor air pollution.

There has also been impetus for further revisions to the Toxic Substances Control Act (TSCA), to place a greater burden of responsibility on manufacturers to prove the safety of their chemicals, and to RCRA, limiting landfilling, developing municipal waste incineration emissions limits, and developing federal medical waste regulations well beyond the three-state medical waste tracking system currently in place.

B. Development of Occupational Regulation

In 1970, the Occupational Safety and Health (OSH) Act created the Occupational Safety and Health Administration (OSHA) within the Department of Labor. The OSH Act gives OSHA the mandate to assure, as far as possible, safe and healthful working conditions for employees. OSHA's jurisdiction applies to all industry employers and employees in the United States. In carrying out its duties, OSHA is responsible for promulgating legally enforceable standards for the physical working environment. The agency is authorized to conduct workplace inspections to enforce its standards. With very few exceptions, inspections are conducted without advance notice. The OSH Act provides for various penalties depending on the nature of the violation, including civil monetary penalties and imprisonment.

Table 3 lists OSHA's responsibilities. Table 4 lists some of the major regulations OSHA has issued pursuant to the OSH Act.

In addition to specific standards and guidelines, the OSH Act also contains the general duty clause, which states in part that "employers shall furnish to each employee employment and a place of employment which are free from recognized hazards that are causing or likely to cause death or serious physical harm to employees." In the absence of a specific standard, OSHA has often cited the general duty clause as justification for enforcement actions.

OSHA has adopted many industry standards and guidelines since 1970. Sections of the OSH Act require that all health and safety standards must be "reasonably necessary or appropriate to provide safe or healthful employment and places of employment." Other sections require that a standard dealing with toxic materials or harmful physical agents must be "feasible." OSHA and the courts have construed the term "feasible" to mean economically and technologically feasible. Thus, in developing a standard, OSHA must first find a significant risk of harm to workers and then set a feasible standard to reduce or eliminate that risk. A determination of feasibility must balance worker protection, technology, and economics.

In 1971, OSHA adopted as federal requirements (legally enforceable standards) a number of previously existing federal and national consensus standards and guidelines, including the American Conference of Governmental Industrial Hygienists (ACGIH) threshold limit values (TLVs) and the American National Standards Institute (ANSI) standards for exposures to toxic substances and harmful agents. The ACGIH TLVs were voluntary standards developed and updated by ACGIH. Henceforth, ACGIH's voluntary TLVs became OSHA's permissible exposure limits (PELs) for a large number of chemical substances. These limits were set forth in OSHA's "Z-Table," so called

TABLE 3 OSHA's Responsibilities Pursuant to the OSH Act

The OSH Act gives OSHA the mandate to assure, as far as possible, safe and healthful working conditions for employees. In so doing, OSHA shall:

Promulgate legally enforceable standards for the physical working environment. These include permissible exposure limits (PELs) for occupational exposure to a variety of circumstances, including chemicals (vapors, liquids and solids), noise, and radiation.

Issue regulations regarding work-place practices (e.g., medical surveillance, exposure monitoring).

Conduct work-place inspections to enforce compliance.

Provide training and education concerning the dangers posed by toxic substances in the work place.

TABLE 4 Major OSHA Regulations

Hazard Communication (29 C.F.R. Part 1910)—requirements for worker right-to-know, informing workers of the hazards of the chemicals with which they work.

Occupational Exposure to Ethylene Oxide (29 C.F.R. Part 1910)—1 ppm 8-hour time-weighted average; 5 ppm, 15-minute excursion limit, and ancillary requirements.

Hazardous Waste Operations and Emergency Response (29 C.F.R. Part 1910)—occupational standard for workers in hazardous waste operations and requirements for all workers in chemical emergencies.

Permissible Exposure Limit (PEL) Update (29 C.F.R. 1910.1000)—updates over 100 PELs and sets PELs and short-term exposure limits for 200 other substances. It is mostly based on 1988 ACGIH threshold limit values (TLVs) or NIOSH recommended exposure levels.

Access to Employee Exposure and Medical Records (29 C.F.R. Part 1904).

In progress:

- Bloodborne pathogens standard—control of potential exposures to bloodborne pathogens such as hepatitis B virus and the human immunodeficiency virus.
- Respiratory protection standard—fit-testing and maintenance of respirators.
- Methods of compliance—situations in which respirators can be used as primary means of exposure control.
- Methylene chloride exposure standard—a revised PEL for this chemical, used extensively as a stripping agent, solvent, and gluing agent for plastics.
- Occupational exposure to hazardous chemicals in laboratories—a standard dedicated solely to controlling potential exposures in the laboratory.

---

because the PELs are codified in Subpart Z of OSHA's regulations at 29 C.F.R. Part 1910.1000.

Through 1988, OSHA had revised its PELs for workplace exposure to chemicals by way of separate rulemakings for each chemical, resulting in detailed rules for only about 25 chemicals (including asbestos, benzene, cotton dust, ethylene oxide, formaldehyde, and lead). The rulemakings for these chemicals took upward of 5-10 years, most often due to litigation by the affected industries, labor, or public safety advocates. OSHA had repeatedly come under fire by a variety of critics for moving slowly in the regulation of potential work-place hazards. In response to this criticism, OSHA, in early 1989, promulgated a final rule on air contaminants that adopted en masse many of ACGIH's 1988 TLVs as PELs. In so doing, OSHA revised its Z-Tables by reducing the 8-hour time-weighted average PELs for about 100 of the listed air contaminants, set PELs for about 205 substances not previously regulated, and established short-term and ceiling concentrations for various other substances.

In the roughly 20 years since OSHA was established, it has done more than simply set standards for chemical exposures. Table 4 reveals that OSHA has also promulgated far-reaching standards, such as its Hazard Communication standard, which established the worker right-to-know requirements; standards for workers at hazardous waste operations; and regulations on access to employee exposure and medical records.

As the end of the decade approaches, OSHA is involved in a myriad of activities. It is revising its respirator standards to acknowledge new fit-testing technologies. It is also reevaluating its traditional hierarchy for exposure control, acknowledging that in situations where engineering controls and workplace practices are not feasible to control potential exposure, respirators can be a primary means of exposure control. OSHA is developing guidelines for medical surveillance and employee exposure monitoring. And, significant for the health-care and medical product industries, OSHA is finalizing a rule that would establish a program for control of potential exposures to blood-borne pathogens.

As indicated above, OSHA has come under fire for moving slowly in developing exposure standards. OSHA's compliance policies, especially the quantity and frequency of its inspections and its significant reductions in penalties sought for violations, have also been under close congressional scrutiny. OSHA has also been taken to task for not keeping up with the times by developing standards that reflect the changes that have occurred in the work place over the last 20 years—potential hazards associated with robotics, biological agents, biotechnology, computers (video display terminals), musculoskeletal problems from repetitive motions, and stress. As a consequence, there is a growing movement, spurred on by the unions and prolabor legislators, to significantly revise the Occupational Safety and Health Act as it approaches its twentieth anniversary.

C. Research into Occupational Hazards

In addition to establishing OSHA as a work-place enforcement agency in the Department of Labor, the OSH Act also established the National Institute for Occupational Safety and Health (NIOSH), which is part of the Centers for Disease Control (CDC) within the U.S. Public Health Service in the Department of Health and Human Services. NIOSH is the CDC research arm that studies chemicals and recommends maximum exposure levels for substances in the workplace. NIOSH conducts research and makes technical recommendations regarding chemical exposure, while regulation writing and enforcement are left to such agencies as OSHA, EPA, and the Consumer Product Safety Commission (CPSC).

One of the Institute's most important responsibilities under the OSH Act is to transmit recommended standards to OSHA. The NIOSH recommendations are intended to serve as the basis, along with other information, for assisting in developing new standards and in reviewing existing standards. The NIOSH recommendations have often been found in publications called Criteria Documents. The documents include an environmental limit for workplace exposure, as well as recommendations on the use of labels and other forms of warning, type and frequency of medical examinations to be provided by the employer, air sampling and analytical methods, procedures for technical control of hazards, and suitable personal protective equipment.

NIOSH also publishes Current Intelligence Bulletins on specific hazards as data become available. These bulletins describe recent research on specific chemical substances and frequently provide useful hazard information in advance of regulatory action. NIOSH also makes hazard evaluations of the workplace. A hazard evaluation is often the result of a specific employee complaint. Based on the findings of a hazard evaluation, NIOSH can make available to OSHA information on a company's compliance with a specific OSHA standard.

Throughout the 1980s, NIOSH has been conducting a study in the medical products industry to determine if there is a correlation between occupational exposure to ethylene oxide (EtO) and the incidence of leukemia or other cancers. Called a retrospective mortality study, the investigation is based upon information from employee work records from the early 1950s to the early 1980s at 14 industrial sterilization facilities. From the employee information, NIOSH established an "exposed population." It then sent the names of the exposed individuals to the Social Security Administration to determine vital status (i.e., if the individual is dead or alive). For those that have died, NIOSH has obtained death certificates from states to determine the cause(s) of death, specifically leukemia or other cancers. NIOSH then compares the number of deaths from cancer in the exposed population to the number that would be expected in a similar population in the general U.S. population.

At the time of the writing of this chapter, the NIOSH study had not issued—it is expected in mid 1990. The study is important since it is the largest and most comprehensive investigation into EtO of its kind. If the study finds a correlation between EtO exposure and cancer, the medical products industry's continued use of EtO, its predominant sterilant, could be jeopardized. The U.S. health system could also be jeopardized by any restriction in the continued use of ethylene oxide, since it is the predominant industry sterilant and there currently is not sufficient capacity in alternative methods of sterilization to produce the sterile products needed.

Throughout the 1980s, NIOSH has been most active in supporting legislation that would establish federal requirements for informing workers of the hazards associated with toxic chemicals. Although OSHA's Hazard Communication or worker right-to-know standard requires material safety data sheets (MSDSs) to be updated to reflect new evidence of the dangers of toxic chemicals, labor, advocates, and many state and federal legislators are calling for more information. The legislation introduced to date, the High Risk Occupational Notificaiton and Prevention Act, would require the federal government to notify workers at risk of illness from occupational exposures to hazardous materials or chemicals. Workers would be told of the nature of their risk, the diseases or conditions associated with it, and their option to seek medical monitoring to detect any disease symptoms.

Although, as of late 1989, Congress has not enacted legislative requirements for high-risk disease notification, NIOSH has implemented internal requirements to notify workers in any of its epidemiology studies of adverse findings and risks associated with use of the chemical being studied. For example, if the NIOSH study of industrial EtO workers finds a correlation between EtO exposure and cancer, NIOSH intends to inform all study participants (including current and past workers and relatives of deceased employees) of the study results.

D. The Reagan Years

No survey of environmental and occupational regulation in the United States throughout the 1970s and 1980s, no matter how brief, could be complete without a discussion of the Reagan administration's policies on occupational safety and health and the environment. This section is not intended to be a critique of that administration's activities, but a brief review of some of the most controversial issues raised during that era. It is through understanding what has (or in many cases, has not) occurred in the 1980s that one can predict what actions the Bush administration, Congress, and the states will take in the 1990s.

One of the first actions the Reagan administration took in its early days in 1981 was to put in abeyance the plethora of proposed and final regulations issued in the last days of the Carter administration. This included, for example, rules on laboratory safety, access to employee exposure and medical records, and production limits on ozone-depleting CFCs. The president also organized a Task Force on Regulatory Reform, headed by then Vice President George Bush. The Task Force requested from industry examples of how government regulation was unduly burdensome—many companies complained about overzealous OSHA inspectors, burdensome EPA reporting requirements, etc.

Another major action taken by the Reagan administration was to expand the role the Office of Management and Budget (OMB) played in the regulatory process. The OMB reviewed proposed regulations to determine what economic impact they would have on business, especially "major regulations" with economic costs of $400 million or more. The Reagan administration also espoused the "New Federalism" in which states would take an increasing role in government, including environmental and occupational regulation.

Finally, Reagan administration appointees were seen as more responsive to industry and less responsive to labor and advocates than their predecessors in the Carter administration.

What affect did the Reagan administration have on these issues? One could argue this question ad infinitum—and, quite frankly, many continue to do so. Unfortunately, there are so many ways to assess whether the Reagan administration helped or hurt the environment and workplace—number of regulations published, occupational death/injury rates, number of enforcement actions and penalties, etc.—that it is difficult to develop a specific response. It is more instructive instead to make some general observations:

Regardless of the criticism, there were a number of major rules promulgated during the Reagan years—worker and community right-to-know, hazardous waste management, CFC production limits, PEL revisions, and substance-specific standards.

Although the Reagan administration attempted to streamline the regulatory process, its "New Federalism" added to industry's compliance concerns. For example, in the absence of federal rules, many states moved forward in developing legislation and regulation. Examples of this include state initiatives on worker and community right-to-know, medical waste, and chemical emissions. In most instances, companies have had to comply with a variety of regulatory requirements (often conflicting).

The OMB's role in the 1980s regulatory process was controversial. Labor and advocates often litigated against OSHA and EPA over standards they felt were inadequate, claiming that the OMB exerted undue influence over the regulatory process. A good example in the medical products arena is OSHA's EtO standard. Several advocate groups and unions filed suit against OSHA for not including a short-term exposure limit in its 1984 revision of the EtO standard downward from 50 ppm to 1 ppm. OSHA finally set a 5 ppm, 15-minute excursion limit in 1988.

There were significant budget cuts in major OSHA and EPA programs, which limited the regulatory and research activities of those agencies. The budget cut for NIOSH was so deep that the agency was almost abolished. By the end of the Reagan administration, most budget items had been restored.

Global concern for the environment required the Reagan administration to focus on such issues as acid rain, ozone depletion, and global warming.

The medical products industry became much more highly regulated during the Reagan administration. Medical manufacturers had to meet requirements for worker and community right-to-know, hazardous and medical waste, revised EtO worker standards, and CFC production limits.

The Bush administration has some tough issues to face and a very critical audience to satisfy, since public awareness of environmental and occupational issues is at an all-time high. President Bush campaigned as the "environmental candidate," and the Democratic leadership in the Congress is prolabor and pro-environment. In the era of "no new taxes," the president also faces some tough tradeoffs between deficit reduction and increasing domestic and international efforts for environmental protection and occupational safety.

What does the future hold for industry in general and the medical products industry in particular?

III. ENVIRONMENTAL AND OCCUPATIONAL ISSUES OF THE 1990s

The facts and figures are somewhat staggering:

EPA estimates that it has already listed about 68,000 chemicals under TSCA and is adding 7000 more per year. The National Research Council estimated that the degree of toxicity of 80% of these substances is unknown [1].

As many as 70,000 deaths a year may be caused by exposure to toxic substances in the work place, and 350,000 new cases of chemically induced work-place illness may occur annually [1].

These two brief examples emphasize how increasingly complex regulation of the environment and workplace is becoming. This complexity will mean additional burdens for the already taxed regulatory agencies. It will also mean additional responsibilities for companies to perform testing, inform workers, establish programs for workplace and environmental safety, and investigate alternative chemicals and processes.

What specific regulatory programs can manufacturers expect, especially those in the medical products arena? The medical products industry has become increasingly proactive throughout the 1980s. This is also expected to grow through the 1990s. For example, companies have voluntarily reduced exposure limits, redesigned products for sterilization by alternatives to EtO, mandated that substitutes for certain chemicals be found or developed and implemented, and investigated ways to redesign products for easier disposal and recycling. The medical products industry has also worked with governmental agencies to develop accurate information upon which to base regulatory actions.

A. Future Environmental Regulation

Since EPA already has such a complex statutory framework in place, much of what happens in the 1990s will focus around changes to this framework and attendant regulatory actions. There will likely be major rewrites of the Clean Air Act and TSCA, as well as further revisions to RCRA (which has already been reauthorized once). Concurrently, as science advances, EPA will likely develop new policies regarding toxicity testing and risk assessment. Below appears a listing of the major statutes and what we might expect to see, with an emphasis on the effect on the medical products industry.

1. *FIFRA*

In the next ten years, EPA will reregister all pesticides—this process may bring into question the adequacy of chemical testing and subsequent interpretation of the data. EPA will employ its "special review" process to analyze those products posing the greatest potential risk and will likely cancel the registrations of a number of major products or severely restrict their use. Sterilants, disinfectants, and decontaminants will be included in the reregistration process. If the NIOSH study of industrial EtO workers shows a correlation between EtO use and cancer, EPA could consider this finding the basis for limitations to continued EtO use.

2. *RCRA*

There already is impetus in Congress for further revisions, especially limits on landfilling, further regulation of municipal waste incineration and ash dumping, and expansion of medical waste regulation. Companies will find it increasingly difficult and costly to dispose of waste as landfills are closed and states erect barriers to the transport and disposal of hazardous waste. Congress is likely to spur EPA into taking faster action on waste minimization initiatives. EPA will also have to resolve the conflict between requirements to disinfect or decontaminate and incinerate medical waste and simultaneous agency efforts that may ban or restrict key sterilants and limit hospital incineration.

3. *CAA*

Congress will likely have reauthorized the Clean Air Act in early 1990. It will probably revise Section 112, dealing with regulation of hazardous air pollutants, with respect to timing (expediting the process) as well as the basis on which regulation is based (i.e., economic and technological considerations as opposed to risk alone). A number of chemicals used in the medical products industry—EtO, methylene chloride, chloroform, carbon tetrachloride—will likely be regulated in the 1990s under this scheme.

EPA will likely take whatever actions are necessary to expedite the elimination of ozone-depleting CFCs. Armed with the Montreal Protocol, an international agreement for CFC reductions, and regulations that specify a phaseout of all CFCs within the next 10 years, EPA will actively promote the development and use of substitutes and CFC reclamation systems. EPA may also take action to expedite the use of CFC substitutes and alternative sterilization methods—substitute CFCs for sterilization uses are expected to be available in 1992.

4. *CWA*

EPA will likely continue to tighten up on what can be discharged into rivers, lakes, and municipal water systems. These limitations will probably be based on the toxicity of the pollutants as well as the control technology.

5. *CERCLA/SARA*

EPA will continue to grapple with the dilemma of too many waste sites to clean up and too little time and resources to carry out the job. As more sites are designated for cleanup, more companies will be responsible for the costs of the cleanup, per the statute's strict liability requirement. As SARA's community right-to-know requirements come into maturity, they will require a complete mass balance—accounting downstream for all resources that entered a facility. Companies will continue to face increased scrutiny from the local community regarding their inventories and emissions of hazardous chemicals. Reporting this information will no longer be sufficient—companies must take all appropriate actions to reduce dangers associated with inventories of chemicals and reduce emissions.

6. *TSCA*

TSCA will likely be reauthorized, putting greater onus on producers to show that a chemical is safe.

B. Future Occupational Regulation

NIOSH was already looking well into the 1990s when it convened its National Symposium on the Prevention of Work Related Diseases and Injuries, held in 1985 [2]. At this convocation, several hundred public health officials identified 10 major causes of occupational illness and outlined a strategy for their prevention. These ten major causes include:

1. Occupational lung diseases
2. Musculoskeletal injuries
3. Occupational cancers
4. Occupational traumatic injuries
5. Occupational cardiovascular disease
6. Reproductive disorders
7. Neurotoxic disorders
8. Noise-induced hearing loss
9. Dermatological conditions
10. Psychological disorders

The Bureau of National Affairs (BNA), a Washington-based publishing company that produces several newsletters in the environmental and occupational arena, has also prepared its "Seven Critical Issues for the 1990s" analysis, based upon discussions with regulators, legal counsel, union representatives, advocates, and industry [1]. The BNA has concluded that seven major issues in the 1990s will be:

1. Criminal prosecution, especially efforts by states to prosecute employers for work-place deaths
2. High-risk notification, expanding the worker and community right-to-know requirements already in place
3. Construction industry, especially preventing tragedies such as the 1987 L'Ambience Plaza collapse in Bridgeport, Connecticut
4. New chemical exposure standards, especially efforts by OSHA to expedite rulemakings and to keep standards current with technology and feasibility
5. Recordkeeping, especially the ever-expanding burden on industry to report, inform, and communicate
6. Regulating blood-borne diseases, especially OSHA efforts to develop a "Biohazards" standard
7. The right-to-act, which would require employers to expeditiously correct occupational exposure problems, and would give employees the right to refuse to do what they consider dangerous work assignments without retribution from employers

Another issue for the 1990s not listed above is substance abuse and how it affects workplace safety. This is a particularly sensitive topic, since it has technical, legal, moral, and ethical considerations.

The medical products industry will be greatly affected by many future OSHA actions. The industry will need to keep current with all exposure standards, especially if OSHA develops a system for updating its PELs. OSHA's general policies on exposure monitoring, medical surveillance, and respirator use and testing will affect how companies manage their workplaces to reduce and eliminate potential exposures. This may require the development and use of alternative chemicals and processes, as well as increased automation. OSHA's imminent blood-borne pathogens rule will require companies to manage potential exposures to biohazards, and to develop policies for providing vaccinations and testing for viral exposure. Testing and vaccination are sensitive issues, since they deal with legal and ethical considerations.

IV. CONCLUSION

This chapter has provided basic information on the regulation of environmental and occupational safety and health. This is an area of great importance for industry, and is likely to become a major issue in the 1990s as the public becomes increasingly concerned about the workplace and the environment.

Each company should make itself keenly aware of federal, state, and local requirements and impending initiatives and take appropriate actions. Companies should also marshal the appropriate resources, in terms of staff and plant and equipment. In this era of right-to-know, the costs of being unprepared are simply just too high.

ACKNOWLEDGMENTS

The author would like to express his thanks to Stanley H. Abramson and Kathryn L. Rhyne of the legal firm of King & Spalding for their editorial assistance.

REFERENCES

1. Occupational Safety and Health, *Seven Critical Issues for the 1990s*, Bureau of National Affairs, Washington, D.C., 1989. (The estimates of deaths from occupational exposure were made by Philip J. Landrigan, director of environmental and occupational medicine at Mount Sinai School of Medicine in New York City on May 18-19, 1989, at a BNA-sponsored conference.)
2. *Proposed Strategies for the Prevention of Leading Work-Related Diseases and Injuries*, prepared for NIOSH by the Association of Schools of Public Health, Washington, D.C., 1986.

part eight

Historic Overview/Legal Framework of Coverage, Payment, and Marketing

47

Medicare's Policy Perspective on Coverage and Payment for Medical Technology

KATHLEEN A. BUTO *Health Care Financing Administration, Baltimore, Maryland*

I. BACKGROUND

In recent years, manufacturers of medical technology have discovered what hospitals and physicians have known for a long time—that Medicare coverage and payment must be considered if you plan to succeed in the health-care business. Medicare accounts for 26% of spending for hospital services and 20% of spending for physicians' services. The use of medical technology is an integral part of the medical services that generate those levels of spending. The awareness of manufacturers and other representatives of the medical technology industry that they need to understand Medicare coverage and payment rules has been heightened as federal financing policy has moved in the direction of prudent purchasing and "bundling" payment for services, for example, through the prospective payment system for inpatient care and through capitated, prepaid health plans such as health maintenance organizations (HMOs). Where in the past providers of care had a clear incentive under the Medicare cost reimbursement system to acquire medical technology and pass costs along to the federal government, under new payment rules medical technology may have to compete with other costs of doing business, such as cost of provider staff and routine operating costs.

As a result of this new environment, the Medicare program has been approached with greater urgency on the part of technology representatives to obtain a national Medicare coverage decision as quickly as possible. Increasingly, manufacturers approach the Health Care Financing Administration (HCFA), which administers the Medicare program, before they have received approval for marketing from the Food and Drug Administration (FDA) in the hope that Medicare coverage can coincide with, or follow closely on, FDA approval. From HCFA they seek both assurance of Medicare coverage and decisions about specific payment levels. Where previously the medical technology industry was content to let dissemination take its own course following FDA approval, it increasingly sees that the Medicare program is a lucrative market that can "make or break" new products.

In view of the pressure brought by industry for speedy coverage of new medical technology, what should be the posture of the Medicare program vis-à-vis assuring the availability of new technology? The principal consideration has to be the program's underlying responsibility to assure beneficiaries access to quality care. This responsibility involves making certain that new and important medical technology becomes available as quickly as possible under Medicare. At the same time, the technology must satisfy basic safety and effectiveness criteria because of the certainty that Medicare coverage will encourage the use and diffusion of a new technology and thereby directly impact, for better or worse, the quality of care to beneficiaries.

II. MEDICARE'S COVERAGE PROCESS

As originally structured, the Medicare program was modeled after traditional health insurance programs and was set up fundamentally to protect the elderly against the costs of catastrophic acute care costs. With certain statutory exceptions, such as outpatient self-administered drugs and routine screening services, Medicare coverage has been available broadly for a wide range of health services. As with traditional insurance, benefits have been extended as medical technologies have reached general acceptability in the medical community, though breakthrough technologies have been accorded special and often expedited consideration. Because Medicare was designed to be a decentralized program that would take into account local practice patterns, decisions about medical acceptability have been made predominantly by the private contractors who pay claims for the program, such as Blue Cross and Blue Shield plans or the Travellers Insurance Company. A small number—about 10-20 per year—of coverage issues involving technology will be decided on a national basis. These are usually technologies that are likely to have a large impact on the program, are significant new technologies, or seem to be the source of considerable confusion among the contractors who pay claims. This overall approach, which leaves much discretion to local contractors, has meant that the Medicare program has often moved more slowly and less uniformly than the industry might have liked to cover new medical technology. But as a result, the vast majority of coverage decisions have stood the test of time. This fact is important because, in reality, it is very difficult to withdraw or limit coverage of a technology or procedure once broad coverage has been extended, even though new information may become available.

Despite the importance of assuring that Medicare benefits keep pace with new developments in medical science, the process has not been open to the public or well understood. A number of studies have been undertaken, including those by the Health Industry Manufacturers Association, the Administrative Conference of the United States, the Secretary of Health and Human Services Council on Health Care Technology Assessment, and several consulting groups. One common misconception highlighted in some of the studies is that HCFA's coverage analysis duplicates FDA's review of safety and effectiveness. Another is that HCFA's staff-level Physicians Panel makes coverage decisions rather than acting as a screening device for issues that HCFA might potentially refer to the Public Health Service for review of the medical/scientific merits. Critics of the process complain that there is an unreasonably long lag in providing Medicare coverage and have urged steps be taken to speed it up. Some have challenged the notion that Medicare needs to wait until a technology is widely accepted and would like coverage to be provided at an earlier stage in the state of knowledge about a technology, either through using the Medicare Trust Fund to fund clinical trials or through providing coverage on an interim basis and shifting the burden to Medicare to show why coverage should be withdrawn. Numerous recommendations have been made to make the process more open.

In one sense, these concerns overstate the place of national coverage decision making in ensuring that new technologies and procedures are paid under Medicare. As previously stated, Medicare's contractors make the vast majority of coverage decisions affecting technologies. That means that, with respect to new FDA-approved technologies, many contractors will begin to cover them immediately. Thus, in a sense, there already exists a means for "interim coverage" pending a national coverage decision. What is lacking in the contractor-level decision making is any broad assurance that a technology will be covered to the extent that the manufacturer believes is justified, as well as a decision for inpatient technologies on which diagnosis-related group (DRG) to assign to them. Nevertheless, with few exceptions involving those technologies for which there is an existing noncoverage instruction, the review time involved in a national coverage decision should not preclude dissemination of the technology. Obviously, an affirmative decision, reached promptly, does enhance the marketability of any technology.

The issue of whether Medicare has a further responsibility to finance clinical trials for investigational therapies or technologies presents a clear dilemma for the program. The availability of funding from the Medicare Trust Fund would certainly be well received by researchers and would probably stimulate the development of technologies that would improve medical care to the elderly. But problems with the growing Medicare budget and the fundamental purpose of the Medicare program have to be weighed in considering any new expansion. Given the potential size of the market for technologies suited to Medicare beneficiaries, the technology industry should have ample reason to finance clinical trials. In addition, the debate that surrounded the Medicare Catastrophic Coverage Act of 1988 illustrates that any proposal to use the Trust Funds to finance clinical research must compete with consideration of expansions in Medicare expenditures in other areas. It is also not at all clear that Medicare should pay for services that are not yet regarded as "reasonable and necessary" medical care.

As distinct from device technologies, medical procedures often diffuse with very limited clinical information and so present a different problem. HCFA, along with provider groups and the research community, is considering what should be done to promote more effective data collection and research in the area of measuring effectiveness in medical care.

We in HCFA are receptive to taking a fresh look at Medicare's coverage process. We continue to believe that the process should remain one that principally is decentralized—in other words, that within the broad rules of Medicare coverage, Medicare contractors should continue to gauge medical necessity in accordance with local practice patterns and the individual circumstances of beneficiaries. But HCFA is reassessing the need to be more proactive in the area of national coverage decision making and is examining whether changes should be made in the overall process for covering medical technologies. On January 30, 1989, HCFA issued in the *Federal Register*, proposed regulations that describe in detail, for the first time, the coverage process, including criteria used nationally and by contractors in making decisions. The proposed regulations reflect several objectives: (1) articulate what HCFA believes is the proper balance between making medical technology available under Medicare and not fostering inappropriate use of a technology; (2) make the coverage process clear, including the criteria used and the difference between contractor and national issues of coverage; (3) draw distinctions among different types of technologies and procedures (e.g., devices approved by FDA based on clinical trials or accepted as equivalent to existing devices and new or existing medical procedures) and specify the different nature of Medicare review; and (4) propose ways of allowing more public participation in the coverage process. This rule-making process provided an opportunity for interested parties to comment on the need for any changes to streamline the national coverage process and to recommend changes in the definition of medical necessity or the criteria for judging safety, effectiveness, appropriateness, or medical acceptability of new technologies.

III. MEDICARE PAYMENT FOR MEDICAL TECHNOLOGIES

Despite the interest in the coverage process, coverage of a new technology is only the threshold decision for Medicare. Once a technology is covered, an equally important question is how payment will be made. Here a new set of issues arises. Payment for medical technology is illustrative of the patchwork nature of Medicare payment policy in general. Under Part A of Medicare, which pays for inpatient care, the prospective payment system for hospitals accommodates payment for new technology as part of the overall increase in payments determined by Congress each year. In addition, when a new technology is associated with a new therapeutic service, such as shockwave lithotripsy, the new procedure is assigned to a DRG, which determines the payment rate associated with individual inpatient admissions. DRG payment rates are based on a methodology of relative weights that reflect the relative resource costs associated with each of more than 400 diagnostic categories. In accordance with available cost and charge information, a new procedure is assigned to the DRG that is clinically the most similar to the new procedure and that will result in a fair level of payment. These weights are updated each year, in an effort to keep the payment structure as up-to-date as possible in reflecting such factors as new technology or increasingly more severely ill patients within diagnostic categories. In addition, under the Prospective Payment System (PPS), hospitals may receive capital payments associated with the acquisition of medical equipment on a reasonable cost basis, subject to reductions recently enacted by Congress.

The adequacy of the PPS mechanisms for accommodating new, expensive technology has been called into question from time to time. Most recently, the hospital industry argued that failure to make a special payment accommodation for the high cost of the new heart attack drug TPA (tissue plasminogen activator) put hospitals at extreme financial risk and could ultimately lead to lower-quality medical care for Medicare patients. HCFA has resisted providing a special payment for TPA, largely because it determined that the overall payment update enacted by Congress would be adequate to accommodate both cost-increasing and cost-saving new technology. In addition, the cost and savings impact associated with the drug's use is not clear, and the extent to which Medicare patients would be likely candidates is also unknown, given the numerous warnings of contraindicated coexisting conditions. This recent instance is illustrative of a basic conflict within PPS that can pit the need to preserve the prospectivity of the system against the concern about its impact on the use and diffusion of individual new technologies.

In contrast, a technology may be subject to very different rules under Part B of the Medicare program, which provides for payment of physicians' services and other outpatient services. Under Part B, payment for every physician procedure is based on the physician's individual charges and the prevailing charges of other similar physicians. When a new procedure is used or a procedure involves a new technology, HCFA's contractors might construct a payment amount from a composite of factors. This "gap filling" is done until enough charges associated with the procedure are available to establish a reasonable charge. If a technology is associated with a procedure that is a diagnostic service, the physician may be able to bill a technical component to cover some of the costs of the equipment itself. If the technology is associated with an ambulatory surgery center (ASC) service, its payment may be considered to be included either in the physician's charge or in one of the fixed payment rates under which the procedure is paid. If the technology is associated with an outpatient hospital service, it would be paid under completely different rules, either on a reasonable cost basis or, if it is also an ASC surgical procedure, possibly on the basis of a blend of the ASC payment rate and reasonable cost. This is so unless the procedure is associated with an inpatient admission to the same hospital within 24 hours, in which case it would be considered paid under the DRG for the admission.

There are also several other very specific payment rules that affect actual payment rates, including, for example, the use of wage indices associated with payments to providers; differing rules regarding technology used as part of ambulance services, depending on the ownership of the ambulance; and limitations on physicians' markups for diagnostic services, depending on whether the equipment and testing is owned and done by the physician or a supplier or laboratory. Finally, items of durable medical equipment (DME) are paid under a fee schedule.

IV. INCENTIVES FOR APPROPRIATE UTILIZATION

The existence of these different sources of payment and different payment amounts creates a variety of incentives. Under PPS, hospitals have a clear incentive to adopt cost-saving technology. On the other hand, hospitals will be more selective in acquiring technology that adds to their costs, although quality of care considerations and status in the medical community will continue to influence hospital purchasing decisions. The constraints placed on inpatient hospitals by PPS, the ability to avoid certificate of need limitations, and reductions in production costs have led to a proliferation of technologies in freestanding settings, which historically would have been available principally in hospitals, for example, magnetic resonance imaging (MRI). Recently, hospitals have been pursuing joint ventures with Part B groups that will allow them to annex new medical technology capacity without having directly to absorb the costs under PPS. While hospitals have an incentive under PPS to keep cost effectiveness in mind when using technology, there is no such financial incentive under Part B. In fact, the incentives are all in the other direction. For example, the ability of physicians to profit from clinical and physiological laboratory services is a substantial reason for the use of these services and for the growth of Part B expenditures.

As perhaps the single largest source of payment for medical technology, the Medicare program has developed over the years numerous rules and systems for identifying, covering, reviewing appropriateness, and paying for medical technology. HCFA relies heavily on health-care providers, physicians, and its contractors, including peer review organizations, to help ensure the appropriateness of medical technology used and the services rendered. Because of the size of the system and the number of individual medical services rendered within it, the overall ability of the federal government to monitor quality of care and guard against inappropriate use of medical technology is quite limited.

This is particularly true under Part B, where lack of effective constraints on utilization of services and technology has contributed to the explosive growth in expenditures. These in turn have been accompanied by efforts to limit Part B spending, including a physician fee freeze. But such efforts have been short-term actions designed to slow the rate of spending and have not fundamentally changed the structure or incentives inherent in the Medicare payment system.

Instead, HCFA believes, to the extent possible, that decision making about appropriate care and use of technology should be decentralized as long as the right incentives for appropriate care are in place. The Medicare program would like increasingly to pay for services on a prepaid, capitated basis, allowing individual providers to make decisions about the right and most efficient mix of services for any given individual. At the same time, it is important to keep in mind possible pitfalls when it comes to assuring that new and important technology is available under these plans. For example, continued availability of technology presumes that payment to providers will be set at a level that is adequate to allow them to achieve cost savings and to acquire cost-increasing new technology. In addition, prepaid plans may also be concerned that they may experience adverse selection by beneficiaries needing an expensive new technology if the plans are significantly ahead of their competitors in adopting technologies.

We recognize that, even though we would like Medicare beneficiaries to select prepaid health plans, they will, and should continue to, have a choice and that there will continue to be a fee-for-service delivery system as part of the Medicare program. Thus, we are trying to improve our ability to identify and reduce inappropriate utilization or medically unnecessary care. These efforts include targeted review by peer review organizations of certain medical procedures for appropriateness; "rebundling" back to outpatient departments the costs associated with diagnostic services provided in conjunction with hospital outpatient services to help guard against duplication and unnecessary services; enforcing prohibitions against physician markups of diagnostic tests provided by other suppliers; and supporting more "effectiveness" research that will evaluate medical practice patterns and perhaps lead to medical practice guidelines that can be used not just in the Medicare program but by the medical community in general. HCFA continues to believe that PPS, though it has limitations and only applies to inpatient services, properly allows hospitals and physicians to make decisions about the medical technologies and procedures to use in individual cases and provides incentives to deliver inpatient services more efficiently. HCFA is also considering sponsoring demonstrations of Medicare Preferred Practice Organizations (PPOs), which would be made up of physicians with prudent practice patterns and provide incentives for beneficiaries to elect to receive physicians' services through a PPO. HCFA would likely select organizations that would have their own utilization review systems and leave more of the service-by-service oversight to the PPO. The hope would be that the Medicare program would be able to select good PPOs and that there would be a reduction in unnecessary services and a more prudent use of medical technology as a result.

V. ISSUES FOR THE FUTURE

Given the overall Medicare coverage and payment environment, what are the prospects for the medical technology industry? The answer is uncertain. Unless the nature of the Medicare program fundamentally changes, and it does not appear it will, the coverage of new medical technology will remain an important responsibility of the program. To ensure that beneficiaries do have access to breakthrough and new life-saving technologies, the Medicare program will continue to make decisions to assure national coverage. HCFA is currently reexamining how to focus that process in a way that will speed up decisions where substantial benefits to the Medicare population are involved, while continuing to allow our contractors to make judgments about technologies that promise to provide more marginal improvements. We in HCFA will continue to promote Medicare beneficiaries' selection of capitated risk plans, which are in a better position to make decisions about which technologies to adopt, when, and what tradeoffs to make with other quality enhancements, acquisitions, or expenditures. It is important that during this time of rapid change in the Medicare program, proponents of medical technology be well acquainted with the basic system of coverage and payment at both the contractor and national levels and that they participate in helping shape the changes in the rules that affect them. As the Medicare program seeks increasingly to pay only for appropriate care, it will not be enough for the industry to learn the rules in order simply to seek ways of maximizing payment or assuring a large market for products. The medical technology industry needs to plan for working in the Medicare environment of the future—one driven by providers concerned about cost-effective delivery of services and the ability to compete in a system that increasingly will be characterized by managed care. Recently, the industry has focused more on developing concrete proposals for modifying Medicare coverage and payment policy as it relates to medical technology. This active and constructive role is important, because the stakes are very high, not only for the industry, but for Medicare beneficiaries and for the future of high-quality medical care in this country.

48

The Impact of Cost Control in Restructuring of the Health-Care Industry

JAMES A. RICE *Health One Corporation, Minneapolis, Minnesota*

PAMELA M. GARSIDE *The Newhouse Group, New York, New York*

I. INTRODUCTION: A CONCEPTUAL FRAMEWORK TO EXAMINE COST CONTAINMENT

The U.S. health-care industry experienced a profound restructuring in the late 1970s and early 1980s. After a period of significant growth in the system since the introduction of government financing programs in the 1960s, unprecedented escalations in system costs have occurred, resulting in significant actions directed at controlling those costs. These actions have been taken by private and public payors, by insurance companies and individuals, and have resulted in systemic and structural changes to the health sector overall.

In examining cost-control mechanisms in the U.S. health care system, we must recognize some important underlying features. The first is that the U.S. health services sector does not function as a system with defined goals or well-balanced, interactive component parts. It is rather a quasi-system, a partly regulated marketplace depending on subparts, some of which are public, some private, part government financed, part privately financed with both nonprofit and for-profit delivery systems. While it has thus been impossible to impose across-the-board cost-control measures—initiatives have been taken by each of the various payor groups as discussed below.

Regardless of political party in power, or geographic region, or era of inflation, society has sought to gain the maximum return in terms of quality of care, access to care, and level of services provided for the funds invested, by either government agency, insurance carrier, employer, or individual. During the past 30 years, the U.S. experience in maximizing its return on investment (ROI) within the health sector has been checkered. In the health services sector of the U.S. economy, this return on investment is maximized by increasing health status while lowering societal expenditures for health services. As illustrated in Fig. 1, lowering expenditures and raising health status are a function of many interrelated variables.

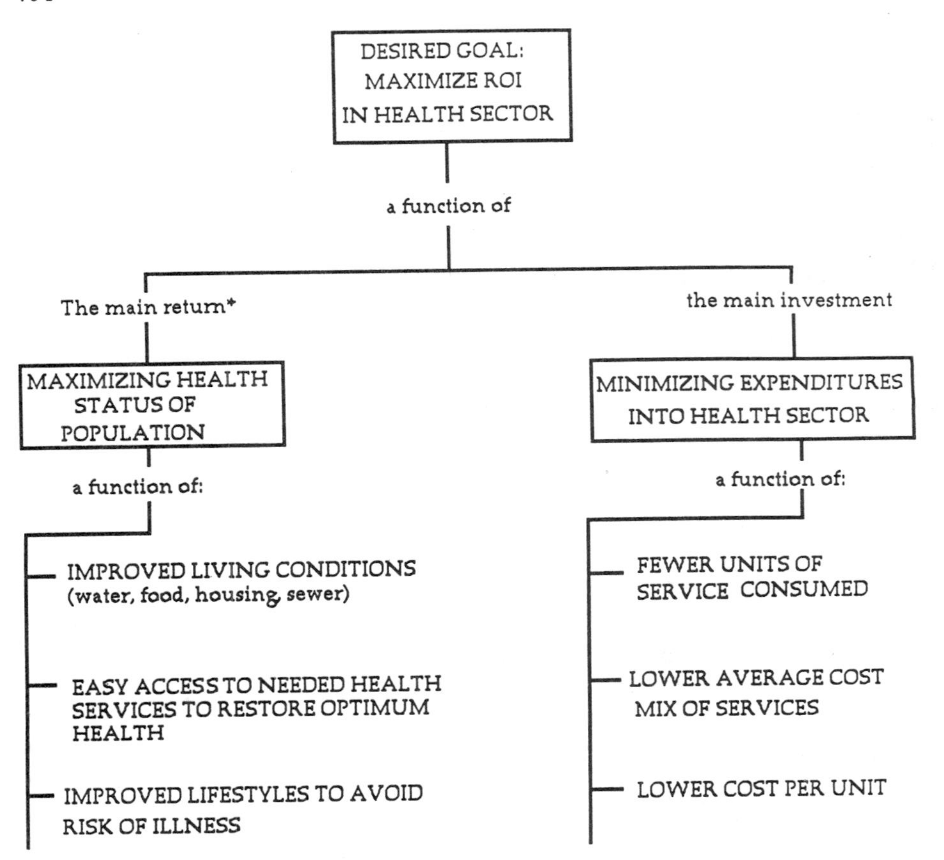

FIGURE 1 Graphic overview of societal challenge to maximize return on investment from expenditures in U.S. health-care sector.

Increasing health status requires (a) improved quality of living conditions, (b) easy access to quality health services to repair damages brought on by injury, illness, or poor life styles, and (c) proper personal behavior to live a lifestyle that minimizes exposure to illness, that is, avoids disease or accidents, or builds resistance through exercise and diet.

Lowering health service expenditures requires (a) consuming fewer units of health services, (b) consuming a mix of services that has a lower average unit cost, and (c) lowering the cost of each unit of service consumed. Efforts at constraining or lowering health-care expenditures in the United States have addressed all three of these arenas, but most have mistakenly been targeted at lowering unit costs. Furthermore, until relatively recently, the majority of these attempts at controlling unit costs have concentrated upon controlling the behavior and cost structure of *providers* (notably hospitals), rather than the consumers of care. Incentives for providers to change either their underlying *cost structures* or their behavior to seek the most efficient way to produce the units of service have traditionally relied on negative regulatory sanctions, for example, through withholding approval to add a service or facility or piece of equipment or withholding approval to raise prices charged per unit. Meaningful attempts at cost containment in the future will need to address all facets and variables in the equation.

This chapter examines the causes of the cost spiral in the health care industry, and both past and current attempts to control expenditures, and the relative failure of past attempts to lower expenditures.

A profile of the restructuring health care industry is presented, and conclusions are drawn about the implications of cost controls and industry restructuring on the use of medical technologies and products. These implications yield both threats and opportunities for manufacturers and suppliers.

II. THE ESCALATION OF COSTS IN THE HEALTH-CARE INDUSTRY

A. The Facts

Expenditures on health care in the United States increased over 800% between 1960 and 1986, while the consumer price index increased 110.6% during the same period. In 1987, U.S. health care spending reached $512 billion. In 1986 it was $465 billion, 10.9% of the GNP, or $1870 per capita. Expenditures are projected to be $1.5 trillion by the year 2000. Federal government spending for personal health care has risen at an average annual rate of nearly 13% since 1980, somewhat more slowly than the 1970, but faster than the growth rate of other nondefense federal spending.

These levels of spending are widely considered to be unacceptably high, and rising too quickly. A summary profile of who pays for what and where the money goes on a relative basis is required. In 1986, 59% of all health-care expenditures were privately paid, either by insurance (31%) or directly by individuals (25%). While the delivery of medical care is overwhelmingly private, government directly financies more than 40% of this care and indirectly subsidizes the rest through various tax incentives. In 1986 Medicare financed 17% of health care expenditures and Medicaid approximately 10%. More than half of Medicaid spending is federally financed. In the same year, 39% of expenditures were for hospital care and 20% for physician services. Hospitals rely on private insurance as their most important single source of revenue, representing around 35%. Approximately 70% of Medicare spending is for hospital provided care (Fig. 2).

B. The Causes of the Cost Spiral

The reasons behind the significant cost spiral are multiple. General inflationary forces affect the health care industry as they do others. General labor costs have increased dramatically over the last two decades, contributing significantly to health-care inflation. If the major environmental forces affecting the health industry during this time period are examined, we see that they have each been inflationary—demographic shifts resulting in increased demand, rapid technological development, and the medical malpractice crisis. Both population growth and, more importantly, the aging of the population contribute to increase demand on the health-system infrastructure. The over-65 age group, which accounts for more than one quarter of the

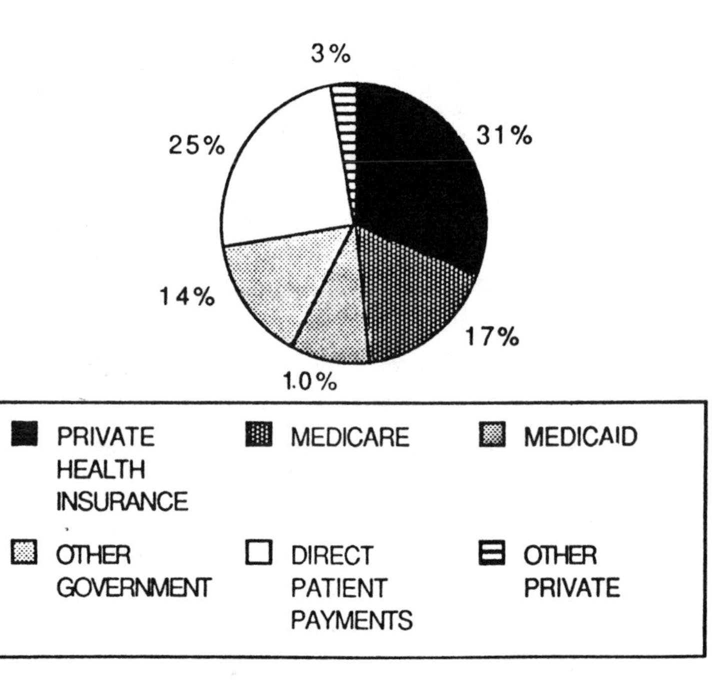

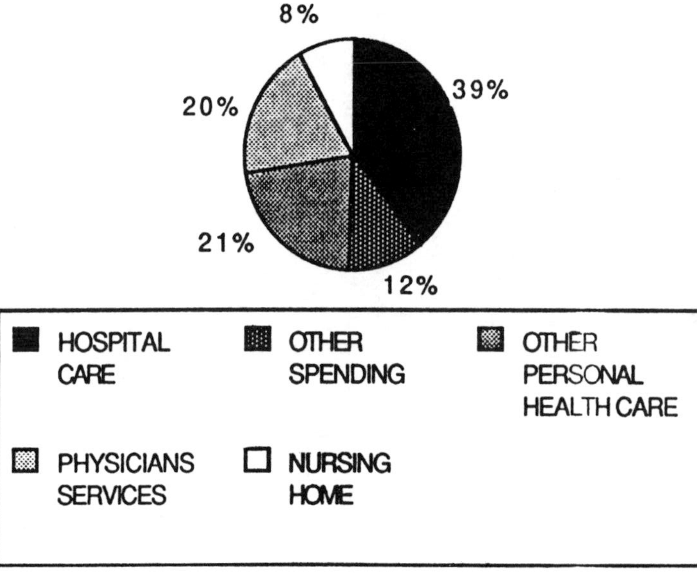

FIGURE 2 National health care expenditures, 1986, by source of funds and type of expenditure: total expenditure $465 billion.

overall population growth from 1980 to 1990, accounts for three and one half times as much medical care expenditures as the under sixty-fives. Approximately 10% of the elderly population accounts for 70% of Medicare expenditures for health care for the over-65 age groups. Technological advances contribute to rising health-care costs due to necessary capital outlays and associated labor, training, and per procedure costs. In addition, the malpractice crisis has created a move to defensive medicine, with attendant overprescription and oversupply of services to patients. In a 1984 study the American Medical Association estimated that defensive medicine and associated administrative costs contribute $15.1 billion annually to the cost of health care.

It is interesting to observe that market forces peculiar to the health care industry also contribute to cost escalation. As the majority, albeit a decreasing segment, of health care is still financed retrospectively by third-party payers based on provider costs or charges, incentives are not in place for providers to be conservative and prudent in the supply of services. Both the patient as ultimate consumer and the physician as provider are insulated from the economic consequences of their decisions, as both are reimbursed by the traditional system.

III. INITIATIVES TO CONTROL HEALTH CARE EXPENDITURES

A. Historical Initiatives

The rapid and continued growth of health care expenditures has spurred action to contain costs among all payers of the health care bill—government, employers, insurance carriers, and individuals. Initiatives began primarily with government agencies. Because hospitals are the principal recipient of health-care expenditures, most cost-control initiatives have focused on controlling expenditures for hospital care.

Table 1 provides an overview of the historical attempts to lower expenditures in the health-care sector. The earliest attempts by government to control health-care costs sought to prevent "unnecessary" capital investments and the provision of "unnecessary" services, ranging from the regional medical program of 1965 through the comprehensive health planning act of 1966 to the passing of the HMO act of 1972. PL293-641, enacted in 1974 by the federal government, focused its regulatory limitations on the supply side of the equation by limiting facility and equipment expansions rather than improving labor productivity. As these efforts disappointed expectations, more comprehensive forms of regulation were proposed and adopted in many states.

B. Contemporary Initiatives

Table 2 profiles "contemporary" cost-control initiatives by various types of payer. These payers cannot tolerate the level of increases seen in the past, and have adopted serious control measures. Initiatives are various and far-reaching, and have affected every component of the health-care delivery system and the way in which those components fit together.

The most far-reaching cost containment initiative has been the prospecttive payment system (PPS) based on diagnostically related groups (DRGs) introduced for the Medicare program by the federal government in 1983. The impact of this legislation has been the most dramatic. Medicare has, in addi-

TABLE 1 Overview of Past Attempts at Lowering Expenditures in the Health Sector

Historical initiative		
Date	Events	Apparent focal point
	Public health measures	↓ Units required by avoiding illness
1965	Regional medical program (RMP)	↓ Mix by movement or patient to appropriate level of care
1966	Comprehensive health planning (CHP)	↓ Unit cost by avoiding capital costs
1966	Professional standard review organization (PSRO)	↓ Units used by reducing duration of care
	Environmental Protection Acts	
1971-1974	Price controls	↓ Unit costs
	State rate review agencies	↓ Unit costs
1972	HMO Act	↓ Units and mix by avoiding illness and seeking less intensive settings for care
1974	Health Planning Act	↓ Capital as way to ↓ unit costs
1977-1985	Medicare PPS	↓ Unit costs
	Medicare DRG	↓ Unit costs by less incentive for resource consumption by providers
	OSHA	↓ Units by avoiding injuries and illness
	Private or second opinion preadmit certification concurrent review retrospective review	↓ Units and mix by reducing duration and setting for care
	Private insurance copayments deductibles	↓ Units and mix by incentives for patients to either avoid care or seek less costly care
	Tax benefits for health promotion	↓ Units
	Taxing health hazards smoking alcohol	↓ Units

TABLE 2 Contemporary Cost-Control Initiatives

Buyers		
Federal government	Utilization review/professional review organization Prehospital certification Promotion of ambulatory care	DRGS for Medicare Prospective payment system Freeze on DRG rates Freeze on physician rates Limitations on access to capital
State governments	Utilization review Prehospital certification Promotion of ambulatory care	Contracting with discounts PPS/DRGs for Medicaid Discounting Rate setting for all payors
Insurance companies	Utilization review "Managed care," "Triple-option" offerings involving HMOs, PPOs, traditional insurance Second-opinion programs	Managed care—HOMs/PPOs PPS/DRGs for reimbursement Discounting rates to providers
Employers	Cost charing with employees Promotion of managed care programs Utilization review Second-opinion programs Wellness progrmas	Preadmission testing Promotion of ambulatory care, care in home, extended-care facilities over hospital care Self-insurance In-house medical programs Discounted purchasing of health care direct from providers

tion, frozen physician fees, and the prospective determination of physician payments is the likely next step.

Medicaid cost-control actions have varied extensively between the states. Several states have changed hospital payment methods and limit hospital coverage using DRGs, contractual "bidding out" to the most cost-effective providers has been adopted, led by California, and many have altered their program structures under waivers from Health Care Financing Administration (HCFA) to pursue cost containment strategies such as case management.

Nongovernmental payors have moved in different ways toward methods of payment that offer cost-control incentives for either provider or consumer. Private employers have aggressively pursued cost-control strategies, the most common being cost sharing, utilization review programs, preadmission testing, encouragement for ambulatory surgical care, treatment in extended-care facilities, and second opinions. These programs affect company health-care expenditures either by increasing the deductible or copayment cost to the employee, or by directing them to use lower-cost health-care alternatives. These incentives generally do not address improved lifestyles or wellness measures to avoid the need for health services.

Other employer strategies involve self-funding for health insurance to control plan administration and health-care costs more directly. Employers with 3-5 years experience with self insurance are now experimenting with the provision of wellness and "lifestyle" programs for company employees, and the discounted purchase of health-care services directly from providers.

Insurance companies, positioned between the payors of health care and the providers, have made significant changes in reaction to the demands for cost containment. They can no longer pass on provider cost inflation to the employers through annual premium hikes. They have responded by developing varied new products in the health insurance field, most falling under the rubric of "managed care systems."

Traditional insurers, including Blue Cross/Blue Shield plans at the state level, have developed HMOs (health maintenance organizations) and PPOs (preferred provider organizations) and offer them in combination with more traditional indemnity insurance options. The features and growth of these alternative health insurance and delivery combinations will be discussed later in this chapter. Insurers have adopted extensive utilization review programs and disincentives for the use of costly care alternatives, and have formed national networks in order that "alternative" insurance offerings can be made throughout the country in a national plan format.

C. The Apparent Failure of Cost-Control Measures

Despite the rhetoric about "competitive restructuring," most of the changes that have occurred during the 1980s have been the result of regulation rather than competition. Although the proliferation of HMOs, PPOs, fixed-price contracting, and increased insurance copayments is encouraging competition among providers, the most common cost-containment strategies entail utilization controls, a form of private regulation. While competition was obviously increased during the 1980s, regulation remains a major containment tool for both the government and private sector. Much of the responsibility for the lack of clear success of past attempts to lower expenditures is due to the mistaken focus on cost per unit of service, rather than on the restriction of number of units consumed.

A low inflation rate, moderate oil prices, lower hospital inpatient admissions, competitive forces, and a freeze on Medicare hospital payments and physician fees have, in fact, had very little effect in reducing health care spending. Despite this failure, the cost-control initiatives in both private and public sector have had wide-ranging effects on the structure of the health industry as a whole.

IV. THE RESTRUCTURING U.S. HEALTH CARE INDUSTRY

The "next generation" of the health-care industry began to evolve in the late 1980s as a result of the significant environmental change of the previous 7-10 years. Much of this restructuring can be attributed to be a direct result of cost containment efforts outlined in the first part of this chapter. Features of this new generation are illustrated in Fig. 3 and are described further in the remainder of this chapter.

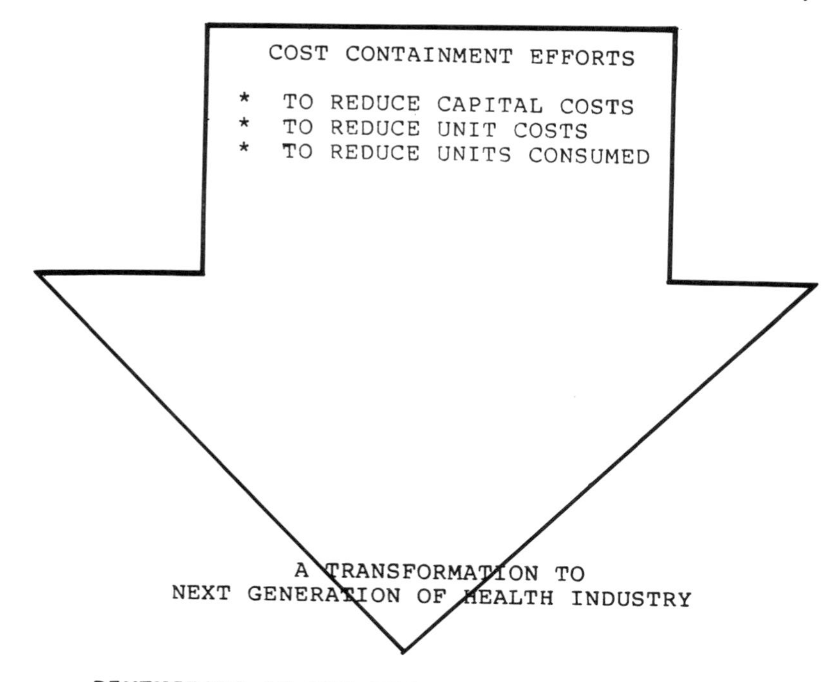

DIMENSIONS OF THE NEW HEALTH SERVICES SECTOR:

- HOSPITAL CONSOLIDATION AND ALLIANCES
- RESTRUCTURING OF U.S. HEALTH INSURANCE INDUSTRY
- DEVELOPMENT OF MANAGED CARE SYSTEMS
- HOSPITAL AS FINANCIAL RISK
- COMMERCIALIZATION OF HEALTHCARE AND "NICHE" MARKETING

FIGURE 3 Cost containment driving transformation of U.S. health-services sector.

A. Hospital Consolidation and Alliances

The hospital industry continues to consolidate, with hospitals joining hospital systems. Hospital consolidations and alliances are now organizing and behaving as integrated "systems" of health-care providers, encompassing significant diversified activities. In 1986, over 40% of hospitals and beds in the United States were affiliated with multihospital systems. These systems have formed due to the strong imperatives of economies of scale, such as with purchasing programs, improved access to capital, and the ability to move resources through the system components to be able to offer the most cost-effective care alternatives. Diversified activities include psychiatric servcies, alternative delivery systems, HMOs, and long-term care. As in the managed care system environment, despite the move to increasingly large groups of hospitals and providers, the provider systems that retain strong local control through locally developed managed networks have been the successful ones able to offer integrated, lower-cost, quality care. Hospitals and other diversified providers in systems and alliances are more easily able to offer structured packages of care to the new cost-conscious buyers in the marketplace.

B. The Changing U.S. Health Insurance Industry

The private health insurance industry is undergoing dramatic change as the health-care provider industry restructures and its major customer, business, demands new products and cost accountability. Large employers have discovered the limitations of traditional insurance companies in terms of data availability and the ability to achieve economics in the new marketplace of competing health-care plans and multiple options.

Health maintenance organizations (HMOs) have experienced extremely rapid growth rates in terms of both enrollees and numbers of plans. At the end of 1986 there were 626 HMOs in the United States, up a phenomenal 234% from 1981, with an enrollment of 25.7 million members. By 1990 there are projected to be 700-750 HMOs covering 40 million enrollees. The large HMOs are becoming even larger, with extensive vertical integration occurring. However, despite these phenomenal growth rates, many face strong competitive pressures and operational deficits. Costs of health-care delivery have continued to rise, affecting HMO premiums, which cannot be kept artificially low. Preferred provider organizations (PPOs) continue to grow in number. These are groups of providers who offer programs of care directly to employers where cost containment is achieved through the number of health-care providers in the group at discounted rates and utilization reviews. Hospitals and physician groups were formerly the major sponsors of PPOs and are now being taken over in numbers by PPOs sponsored by indemnity insurance carriers.

The substantial shift in the bearing of insurance benefit risk away from indemnity insurance companies, to hospitals, physicians, and subscribers, has made insurers' ability to generate revenue and profit more heavily dependent on the performance of specific business functions. These include claims processing, benefit design and coordination, and actuarial services. In addition, faced with intense competition from HMOs and PPOs and the resulting threat of loss of business, most major insurance companies are developing new competitive strategies, including intensified cost-containment efforts, experienced-rated plans, the investigation of fraudulent claims, revised hospital discounts, and the movement toward managed care.

C. The Development of Managed Care Systems

Managed care can be defined as the linkage of integrated insurance products with select-provider networks who are at risk, in order to more effectively control the use and price of health care for large-size groups. A managed care system incorporates care provision with health promotion, some form of managed payment (discounting or capitation), utilization review, and quality accountability. The blurring of the financing and the delivery of health care into the evolving managed care system has occurred because of two major forces: the success of HMOs and utilization-controlled PPOs, resulting in employer pressures on insurance companies to provide similar capabilities, and the providers' response as they watch significant portions of their patient revenues disappear.

A managed care system takes the strengths of traditional insurers and HMOs and PPOs and combines them to overcome the disadvantages of each. A managed care company, then, encompasses sales and marketing, financing, operations, provider selection, care management, and, in some cases, care delivery itself.

Many managed care companies have diversified in order to offer an integrated package of managed care products, including an HMO, a PPO, and traditional fee-for-service insurance that employs utilization management. These packages are known as triple-option plans. Employers favor triple-option plans because the plans allow them to offer a broad range of health-plan options to their employees while reducing the concern that HMOs are skimming off low-cost employees.

As managed care systems have become a major product for the employers market, insurers, providers, and third-party administrators are joining together to structure these systems on a national and regional basis. The boundaries between insurer and provider are blurring as hospital and physician groups and insurers move toward vertical integration. The "managed care system" is the linchpin of the restructured insurance industry. Examples of major alliances to develop such systems are the Voluntary Hospital of America organization with AETNA insurance to offer "Partners Health Plan," and Hospital Corporation of America with Equitable to develop the Equicor Insurance Company and programs. These alliances allow the sponsors to offer package products to buyers in the marketplace. Insurers need stronger control over health-care costs and vehicles through which they can offer packages and tailored cutbacks to employers.

D. The Hospital at Financial Risk

Pressures, principally resulting from cost-containment efforts, forcing the delivery of health-care services from an inpatient to an outpatient setting continue to undermine the financial performance of hospitals. A deemphasis of traditional acute care delivery caused by pressures from payors to keep inpatient care to a minimum and continued loss of revenues to alternative care-delivery systems threaten profitability and cash flow. From 1981 through 1986, inpatient days fell by 50 million, or 18%. Further drains on cash flow are found in hospital attempts at diversificaiton and new program start-up operations; in the increasing number of uninsured or underinsured patients; and in increased bad debt as insurers shift more responsibility for payment onto patients. Hospitals' credit ratings are following a declining trend, negatively affecting their potential to raise capital at reasonable cost. Financial difficulties faced by hospitals will have a ripple effect on other providers, such as nursing homes and home health care companies, as patients are discharged more rapidly and in a sick state due to reimbursement constraints. All providers are therefore expected to dramatically change their approach to the purchase of medical technologies, equipment, and devices.

E. The Commercialization of Health Care and "Niche" Marketing

As the environment has changed and cost containment has become such a strong imperative, multiple opportunities have developed in the changing marketplace at the same time as hospitals as the cornerstone institution have faced economic risks. The industry is becoming more cost- and consumer-driven and market-sensitive. New entrants to the marketplace are capitalizing as those new "niches" in the system and on the profit potential of the industry as a whole. These new entrants encompass everything from for-profit hospital chains to "instant" walk-in medical clinics, freestanding centers for a variety of types of care including surgery, rehabilitation centers, home health care, birthing centers, durable medical equipment, and clinical laboratories. Corporations can often capitalize on economics of scale when concentrating on a very specialized marketplace niche, offering low-cost but at the same time profitable services. These developments suggest new client targets for manufacturers and suppliers serving the health-care industry.

V. IMPLICATIONS OF RECENT COST-CONTAINMENT INITIATIVES FOR THE MEDICAL-DEVICE INDUSTRY

As illustrated in Fig. 4, the impact of the preceding trends and developments is a new health-care marketplace characterized by:

A changing customer base—significantly changed reimbursement incentives for hospitals, the continued formation of buying groups, and the growth of alternative site care.

New criteria and decision makers for purchasing products and services—the "location" and methods of the purchasing decision are changing as cost effectiveness drives the processs.

New product requirements for the restructured health-care delivery system—the move to standardization, sophisticated technology to improve accuracy and cost-effectiveness of diagnosis and treatment, and encouragement of out-of-the hospital uses.

The changing customer base will dictate a new "buyer profile" and changed incentives to buy more prudently. Key implications are:

Hospitals will not only scrutinize supply costs by individual DRG, but will be increasingly aware of the total cost of acquisition, storage, and distribution of supplies in the organization.

The prospective payment system contains built-in disincentives for the adoption of new technological advances. There will be careful consideration of the best application of new products or new technologies either in the hospital or the laternate site setting, particularly for capital acquisitions after 1990.

Hospitals, while remaining a principal customer, will lose relative priority as new physician clinics, HMOs, and specialty provider companies enter the market. The "new" health corporation represents a conglomeration of the various components of the industry.

The new hospital customer has a significantly changed set of needs: cost effectiveness and efficiency will drive decisions. Generic and standardized products and labor-saving technologies will be the most valuable

FIGURE 4 Inpact of trends in the health-care provider behavior.

and marketable. Inventory asset management and consulting services are of great importance to the manager.

Figure 5 emphasizes the changing characteristics of decision making now required in the transformed health sector. Product selection will be more formal, more sophisticated, and more centralized. Medical suppliers will therefore need to consider the following issues:

Decisions of product selection committees will be principally cost-based, and only new products with medical benefits and demonstrated cost efficiency will find acceptance and escape severe price pressure. Developers of these products will have to invest—and therefore will have to risk—more time and money than in the past.

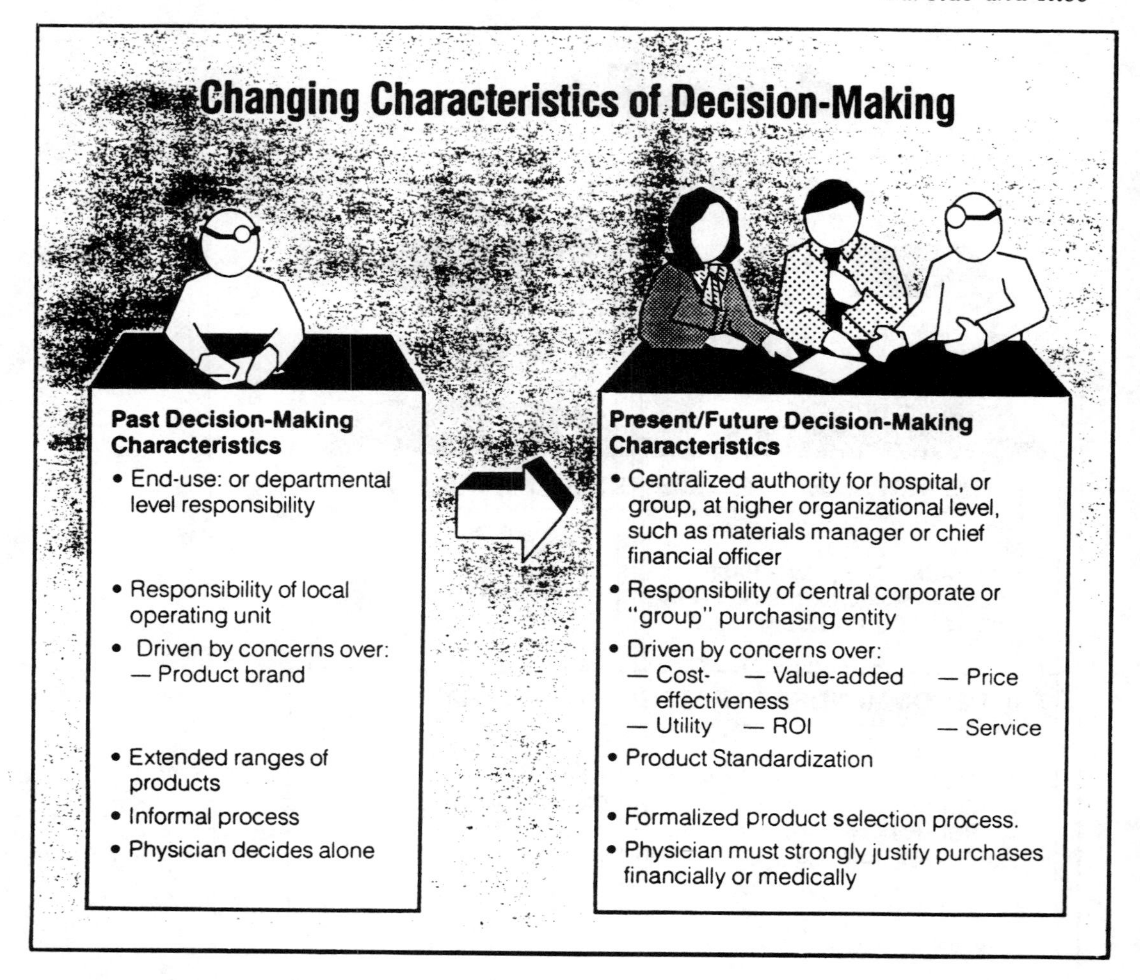

FIGURE 5 Changing characteristics of decision-making.

With the clinical staff pressured to care for patients with fewer personnel and greater productivity, materials managers are dominating the process of product standardization, product evaluation, and competitive bidding.

Suppliers will be increasingly negotiating with and selling to different types of representatives—the "specialized" group purchasing officer in the group setting, the materials manager in the hospital setting, and the delegated administrator or business manager for physician groups. It is to these types of individuals that sales must be made.

Suppliers and their sales forces must be able to operate in an environment where customer interaction is required at multiple levels—for example, product selection committees will make decisions at one level of the organization while customer service is handled at another, lower level.

Chief financial officers of health-care providers and provider gorups will particularly look for "hard" savings in terms of direct price savings and inventory reduction potential over "soft" savings in terms of promised labor-saving potential.

Potential denial of payment for services through government financial review programs has implications for suppliers in specific program areas selected for review. Denials of payment can decrease utilization in this area and will negatively impact manufacturers and suppliers of cardiac care equipment and supplies through reduction in sales.

New product requirements will also need to be considered by medical manufacturers as they refine their manufacturing and sales strategies. In the past a multiplicity of product lines and sources was acceptable to the market. Now and in the future, standardization of products and sources has become a requirement of certain market segments. There are, in addition, needs for products to be packaged for easy administraiton and use; for example, bar coding, or products bearing explanations of cost effectiveness and recommended use by DRG. In general, these are products that have the characteristics of contributing to reduced hospital labor costs or overall costs, or that aid in the promotion of outpatient procedures or high quality patient care. Impacts on the medical supply and manufacturing sector can be summarized as follows:

In addition to the "generic" and standardized product requirements addressing the cost-containment needs of the industry, there is at the same time a need for innovative products which have a greater value than current products. The drive for price pressures is also counterbalanced with a need for high-volume products.

Unit demand for certain products will decrease, affecting market share and pricing arrangements.

With purchases based on brand loyalty declining, marketing and sales strategies to physicians, purchasing officers, and materials managers will change to address the newly segmented market. This market includes both segments with reduced choice or "standardized" product needs, and segments where the "value" and quality of certain products retain high priority.

Companies with greater financial resources are the ones likely to be able to afford the long-term research and development needed to produce new cost-efficient products. They will also be the ones that can afford the cost of automation to become the low-cost producers.

The supplier's customer and the "end users" of the health industry will be increasingly willing to cooperate with the manufacturers in product design and modification.

Product packaging to respond to requirements for explanations and bar coding will require analysis and investment.

Clearly a difficult and challenging future awaits suppliers to the health care industry. Change within the industry is more pervasive than it has ever been during the past 20 years. This change has been not only extensive in scope, but extremely rapid. Increasing competition exists within the health-care industry and among firms attempting to serve the industry. There are expected to be constraints on growth in revenue and cash flow because of new price sensitivity to unit costs and to changing consumption patterns. There could be increasing pressure for higher capital investments and operating costs in the short term, and a corresponding reduction in profit margins and stock values. These challenges will require new organizational and management structures and systems. All this indicates that some existing firms attempting to serve the health care industry will fail or be consolidated into larger, more diversified supplier organizations.

The future, however, does not hold only potential difficult challenges and frustrations. The changing marketplace and market dynamics also offer positive opportunities.

As a result of an increasing sophistication in the end users, as well as the changes in the customer base to larger buying groups and more specialized market segment-oriented providers, a variety of new markets will become open to suppliers. Aging services, self care and noninstitutional care, and new diagnostic and therapeutic procedures of a noninvasive nature all represent new market segments and business opportunities for firms attempting to serve the health-care industry.

By understanding the forces that are changing the way the health care system operates, suppliers will be better able to create new strategies to transform a challenging period into a profitable and successful one.

VI. SUMMARY

This chapter observes that significant changes within the structure and function of the U.S. health-services sector have been brought about by the presence of numerous cost-containment measures that could be characterized as regulatory in nature and focused on reducing unit costs. More recent initiatives have attempted to embrace free-market competition to provide incentives for improved cost effectiveness. These efforts, however, have in general failed to remove the call for continuing cost containment. The health-care industry has transformed into a curious mix of large health-care corporations and a plethora of small niche providers. All sectors buy medical products with a greater attention to the products' role in cost effectiveness. The medical technology, manufacturing, and supply industries must continue to innovate to respond to the changed imperatives of the health-care market.

REFERENCES

1. J. A. Meyer, ed., *Incentives vs. Controls in Health Policy*, American Enterprise Institute, Studies in Health Policy, Washington, D.C., 1985.
2. D. Abernathy and D. Pearson, *Regulating Hospital Costs: The Development of Public Policy*, AUPHA Press, Washington, D.C., 1979.
3. W. B. Schwartz, The Inevitable Failure of Current Cost Containment Strategies, *Journal of the American Medical Association, 257*(2) (1987).
4. Environmental Assessment publications, Health One Corporation, Minneapolis, MN, 1984-1989.

49

The Structure of the Medicare Program: Legal Considerations of HCFA Regulations

GORDON B. SCHATZ *Reed, Smith, Shaw & McClay, Washington, D.C.*

I. INTRODUCTION

The Medicare program pays and sets standards for health care delivered to Americans who are 65 years of age and older, who are disabled, or who have end-stage renal disease. Medicare is the largest single payer for health-care services in the United States. For many manufacturers of medical technology, Medicare coverage and payment policies have become critical in developing and marketing new products. This chapter will examine the structure of the Medicare program and the most significant legal considerations of Health Care Financing Administration (HCFA) regulations for manufacturers of medical devices, diagnostic products, and health-care information systems.

II. MEDICARE, MEDICAID, AND OTHER HEALTH INSURANCE PROGRAMS

As noted above, Medicare is a federal health insurance program for senior citizens and the disabled. By contrast, Medicaid is a health insurance program paying for care to indigent persons. Medicaid is funded through federal and state monies, but is administered by the 50 sates. In 1987, Medicare paid $21 billion for health benefits to 32.4 million Medicare beneficiaries [1].

During the same time period, Medicaid paid $49 billion for health care services to 24 million Medicaid enrollees. Since coverage and payment may vary from Medicare to Medicaid, it is important for manufacturers to distinguish which health insurance program is paying for a particular product. In addition to these programs, there are a number of other governmental health programs, such as the Veterans Administration, Civilian Health and Medical Plans of the Uniformed Services (CHAMPUS), and private health insurance—Blue Cross/Blue Shield, Prudential, Aetna, and health maintenance organizations—whose policies and payment amounts will vary and that need to be understood (see Fig. 1).

If the patient population for a particular technology is elderly, then Medicare may be the most appropriate insurance program for the manufacturer to understand. Medicare decisions are, on occasion, followed by private insurers. Since Medicare can be a leading policymaker on technology payment, familiarity with Medicare's organizational structure and decisions can be an important first step in solving reimbursement problems.

A. Organizational Framework of the Medicare Program

The Health Care Financing Administration (HCFA) is an agency of the U.S. Department of Health and Human Services. HCFA administers the Medicare program from its central offices in Washington, D.C., and Baltimore, Md (see Fig. 2). There are also 10 regional offices of HCFA throughout the country (see Table 1). On a local level, HCFA contracts with private health insurance companies, such as Blue Cross/Blue Shield, Aetna, or Prudential, to process and pay Medicare claims. A Medicare claim is generated by a hospital, clinical laboratory, nursing home, physician, or durable medical equipment supplier when services or products are delivered to a Medicare beneficiary.

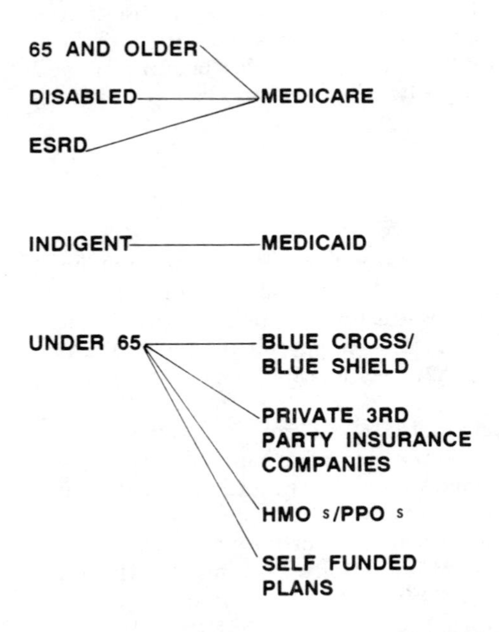

FIGURE 1 Medicare, Medicaid, and other health insurance programs.

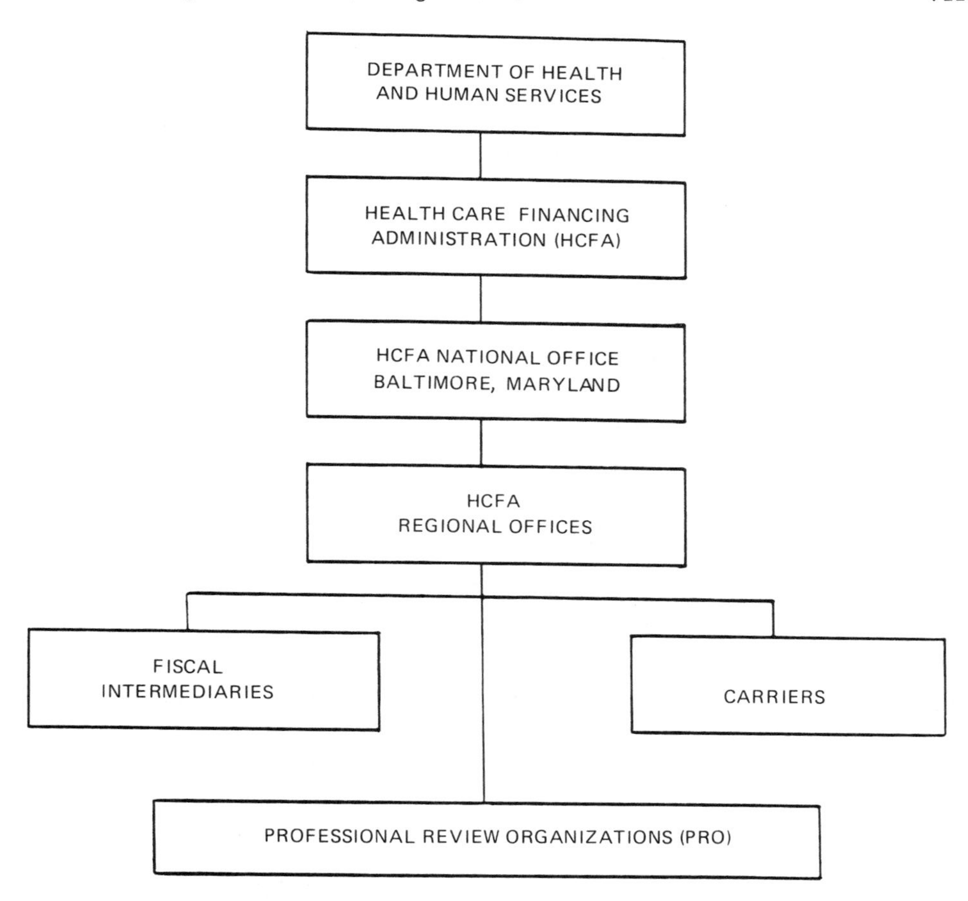

FIGURE 2 Framework of Medicare program.

There are two different types of Medicare contractors. The first type is termed a fiscal intermediary (FI). The fiscal intermediary processes and pays claims for inpatient hospital services. The second type of contractor—a carrier—processes and pays claims for physician and outpatient services, such as clinical laboratory tests. There are 37 Medicare carriers and 54 fiscal intermediaries. In addition to fiscal intermediaries and carriers, HCFA contracts with professional review organizations (PROs), which review the appropriateness of physician care in the hospital.

Recognizing this hierarchial structure is important since many policy decisions are made at a national level, but implemented locally. For example,

TABLE 1 Department of Health and Human Services HCFA Regional Offices

Region	States	Main office
I	Connecticut, New Hampshire, Massachusetts, Maine, Rhode Island, Vermont	Boston, Massachusetts
II	New York, New Jersey, Puerto Rico, Virgin Islands	New York, New York
III	Delaware, Maryland, Pennsylvania, Virginia, West Virginia, District of Columbia	Philadelphia, Pennsylvania
IV	Alabama, Florida, Georgia, Kentucky, Mississippi, North Carolina, South Carolina, Tennessee	Atlanta, Georgia
V	Illinois, Indiana, Michigan, Minnesota, Ohio, Wisconsin	Chicago, Illinois
VI	Arkansas, Louisiana, New Mexico, Oklahoma, Texas	Dallas, Texas
VII	Iowa, Kansas, Missouri, Nebraska	Kansas City, Missouri
VIII	Colorado, Montana, North Dakota, South Dakota, Utah, Wyoming	Denver, Colorado
IX	Arizona, California, Hawaii, Nevada, Guam, Trust Territory of Pacific Islands, American Samoa	San Francisco, California
X	Alaska, Idaho, Oregon, Washington	Seattle, Washington

HCFA's national office may determine that a technology, such as magnetic resonance imaging, is a covered Medicare service. It is the FI and carriers who will pay hospital and doctor bills for these imaging services. In addition to implementing national policies, coverage and payment decisions can be at a local level through Medicare contractors or the regional offices.

B. Agencies within HCFA

Within HCFA there three key agencies: the Bureau of Policy Development (BPD), the Bureau of Program Operations (BPO), and the Health Standards and Quality Bureau (HSQB) (see Fig. 3). HSQB sets the standards that institutional providers, such as hospitals, nursing homes, and independent clinical laboratories, must meet in order to participate in the Medicare program. HSQB also oversees the PRO program for physician review.

BPO has primary responsibility for establishing contracts with fiscal intermediaries and carriers and managing the operations of the contractors.

BPD is the HCFA agency that sets national policy dealing with who is eligible for Medicare benefits, what services will be covered and under what

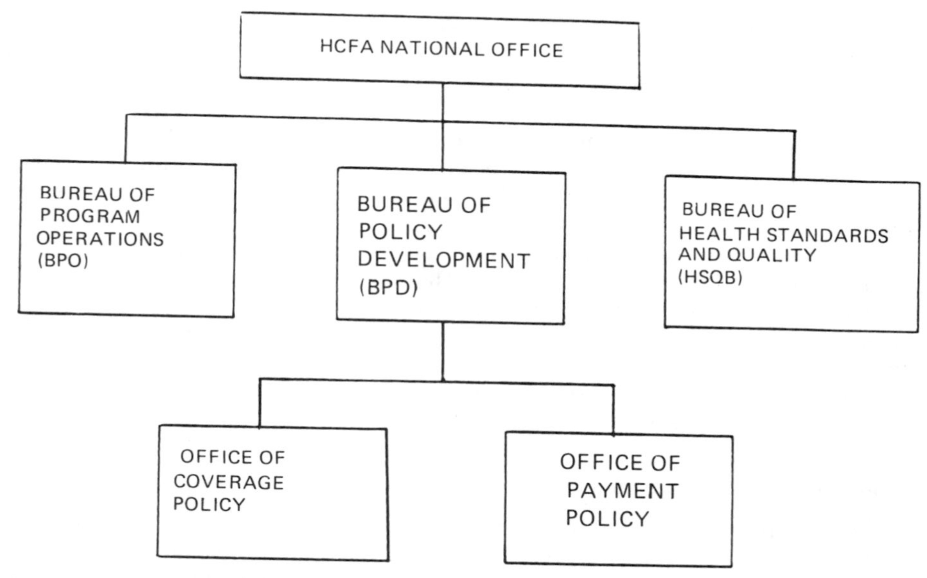

FIGURE 3 Agencies within HCFA.

circumstances, and how much Medicare will pay for those services. Within BPD, there are several offices. For device and diagnostic product manufacturers, the Office of Coverage Policy and the Office of Payment Policy are especially vital. The Office of Coverage Policy issues national Medicare coverage decisions on medical technology [2].

C. Role of Congress and HCFA

An increasingly significant player in the Medicare program is the U.S. Congress. In 1965, Congress passed the Medicare Amendments to the Social Security Act, establishing the Medicare program [3]. Congress fundamentally restructured Medicare payment for inpatient hospital services with the creation in 1983 of the prospective payment system and diagnosis-related groups (DRGs). Congress has played an active role during the 1980s, legislating substantial and detailed changes in the Medicare program. For example, in the Omnibus Budget Reconciliation Act of 1987 [4], Congress provided for specific reductions in payment for 12 overpriced physician procedures. The procedures included:

Bronchoscopy
Cataract surgery
Dilation and curretage
Carpal tunnel repair
Coronary artery bypass surgery
Knee arthroscopy

Knee arthroplasty
Suprapubic prostatectomy
Upper gastrointestinal endoscopy
Pacemaker implant surgery
Transurethal resection of the prostate

Further, Congress reduced the Medicare payment by 8.3% for 28 outpatient clinical laboratory tests involving automated multichannel laboratory equipment.

The development of Medicare policy can be seen, in many cases, as a progression of actions starting with Congress, which enacts statutes. Next, HCFA proposes, then finalizes regulations implementing the statute through publication in the *Federal Register*. Further policy details are developed when HCFA issues fiscal intermediary or carrier instructions. The instructions guide contractors in how to implement the statutory and regulatory requirements at a local level.

For a manufacturer, development of Medicare policy should be recognized at all three levels, as shown in Table 2.

D. Congressional Committees with Jurisdiction over Medicare

Although the full Congress, both the Senate and House of Representatives, votes to approve federal legislation, there are several congressional committees with jurisdiction over Medicare whose roles are especially important in the development of Medicare laws.

In the House, the Health Subcommittee of the Energy and Commerce Committee has jurisdiction over Part B of Medicare. The Health Subcommittee of the House Ways and Means Committee has responsibility for both Parts A and B.

On the Senate side, the Finance Committee has lead jurisdiction for Medicare legislation.

E. The Prospective Payment Assessment Commission and the Physician Payment Review Commission

When Congress created the prospective payment system (PPS), it also established an advisory commission, the Prospective Payment Assessment Commission (ProPAC), which recommends changes in PPS DRGs and the update factor. ProPAC has made recommendations on a number of technology-specific DRGs, including pacemakers, implantable cardiac defibrillators, cochlear implants, and magnetic resonance imaging. ProPAC meetings are open to the

TABLE 2 Inputs for Medicare Policy Development

Agency	What the agency does
Congress	Statutes: for example, P.L. 100-203, Social Security Act, section 1861, 42 U.S.C., section 1395.
HCFA	Regulations: for example, *Federal Register*, 42 CFR, section 205.
HCFA	Carrier Manual Instructions: HCFA Publication 6, section 5203.
Contractors	Notices to hospitals, providers, and suppliers.

public, and manufacturers with concerns about new technologies and the need for new DRGs or changes in DRGs to accommodate the new technologies may direct an inquiry with ProPAC staff [300 7th Street, S.W., Washington, D.C. 20024; tel. (202)453-3186]. Since HCFA is the final authority on changes in the prospective payment system (except when Congress steps in), manufacturers should be prepared to work closely with HCFA's Office of Payment Policy, as well as ProPAC.

A parallel advisory committee, dealing with physician payment, is the Physician Payment Review Commission (PPRC). The Commission also makes recommendations to HCFA on changes in payment for physician services, often with technology-specific concerns. [PPRC is located at 2120 L Street, N.W., Washington, D.C. 20037; Tel. (202)653-7200.]

III. CODING

New technologies are often first recognized in the Medicare program through the assignment of an ICD-9-CM, CPT, or HCPCS code. The International Classification of Diseases, ninth revision, Clinical Modification (ICD-9-CM) procedure codes describe inpatient hospital procedures and diagnoses. When a new technology is being considered for coverage and payment, or when a technology expands its indications for use, there may be a need for assignment of new ICD-9-CM codes.

The ICD-9-CM Coordination and Maintenance Committee, an HCFA agency, is responsible for reviewing new or changing procedures to recommend revisions in the existing procedure codes. For example, when cochlear implants and implantable defibrillators were approved as new Medicare-covered services, new ICD-9-CM codes were assigned to the procedures for these products.

Manufacturers can present information to the committee and attend its meetings to discuss the need for a new code [5]. The Committee will make recommendations to HCFA, which publishes a notice of proposed changes to the DRG classification system in the *Federal Register*. HCFA has final authority in deciding on new ICD-9-CM codes.

There is a second system of codes for physician services, clinical laboratory tests, and durable medical equipment. The American Medical Association (AMA) has developed a comprehensive list of physician services and clinical laboratory tests known as the Current Procedural Terminology (CPT) codes. An AMA CPT Editorial Panel meets quarterly in Chicago, considers new codes, and makes revisions on a yearly basis [6].

Additionally, HCFA has established a coding system for durable medical equipment. These codes, the HCFA Common Procedure Coding System (HCPCS), identify specific products such as wheel chairs, oxygen concentrators, canes, and items used in the home environment.

Coding is significant since it links products and procedures with Medicare payment. When a hospital submits a bill for a procedure performed on a Medicare beneficiary to the fiscal intermediary, the hospital identifies the procedure by putting an ICD-9-CM code on the claim form. The fiscal intermediary translates the code into a DRG, which in turn establishes the payment amount to the hospital for the procedure.

Similarly, when a physician or supplier bills the Medicare carrier for a procedure or product, a CPT or HCPCS code is used on the claim form. The carrier will identify the product and the payment amount based on the code.

For manufacturers, establishing a new, accurate code or communicating the proper code to the provider can be vital in facilitating recognition of the product and appropriate reimbursement.

IV. HCFA AND OHTA: COVERAGE FOR NEW TECHNOLOGY

The above discussion of Congress, HCFA, and ProPAC applies mainly to changes in Medicare payment. With respect to coverage for new technologies, different agencies may be involved. Medicare carriers, in processing claims, will make a number of local or regional decisions on coverage for new or changing medical technology. In some cases, the carriers will not make a coverage decision if the technology presents a question of national significance, or the technology involves unique medical features or high costs. Carriers look to the Medicare coverage issues manual for guidance [7]. This HCFA publication reports national medicare coverage policy, but does not have determinations on all technologies. When there is no national policy and a carrier decides not to make a local decision, the carrier may refer a coverage question to the Regional Office or to HCFA's Office of Coverage Policy (OCP).

The OCP may make a decision based on its own review and analysis. When there are substantial questions dealing with the technology's safety, effectiveness, or efficiency, OCP may refer the matter to the HCFA Physicians Panel, which in turn can make a decision or seek a formal technology assessment from the Office of Health Technology Assessment, in the Public Health Service. OHTA is a branch of the Public Health Service that makes recommendations on coverage of new technology for Medicare and other federal health programs such as the Veterans Administration or the Civilian Health and Medical Program of the Uniformed Services (CHAMPUS). This process is described in greater detail in Chapter 53.

V. STANDARDS, QUALITY CONTROLS, AND UTILIZATION REVIEW

In addition to Medicare regulations on payment for technology and coverage of technology, Medicare standards, quality controls, and utilization review can have a significant impact on the use of technology.

A. Institutional Standards

Medicare requires institutions that provide care to Medicare beneficiaries to meet specified conditions of participation. Hospitals, skilled care facilities, end-stage renal dialysis facilities, ambulatory surgical centers, clinical laboratories, and other providers, such as health maintenance organizations, must meet standards for personnel, medical recordkeeping, quality controls, equipment maintenance, and other operational features. These are significant for technology manufacturers whose products may help institutions comply with Medicare standards.

HCFA's Health Standards and Quality Bureau (HSQB) is the agency primarily responsible for establishing institutional provider standards. HSQB also has responsibility for survey and certification of facilities. The certification process for certain hospitals has been delegated to the Joint Commission on the Accreditation of Health Care Organizations.

B. Utilization Review

HCFA contracts with professional review organizations to examine hospital admissions and lengths of stay for medical appropriateness. Certain procedures receive a high level of scrutiny from the PROs, such as pacemaker implants and cataract surgery.

Utilization review, which affects most directly the physician, will still have an indirect impact on manufacturers when screens place restrictions on the frequency, appropriateness, and location of certain procedures involving medical technology.

VI. ENFORCEMENT OF THE MEDICARE LAWS

When products or services are billed using the wrong code, or excessive charges are billed, or if there are improper rebates or kickbacks to providers, there may be a violation of the Medicare/Medicaid fraud and abuse laws. The Department of Health and Human Services, Office of Inspector General is charged with investigating and imposing sanctions on wrongdoing that impairs the integrity of the Medicare program.

VII. CONCLUSION

The Medicare program is a complex network of congressionally made statutes, HCFA-issued regulations, and carrier instructions. The legal structure is made that much more diverse by the numerous agencies, committees, and advisory groups that play important roles in deciding what technologies to cover, how much to pay, which codes to use, and what standards to follow in providing care to Medicare beneficiaries. For manufacturers who seek efficient introduction of new medical technology, a working understanding of the Medicare program is essential.

REFERENCES

1. S. W. Etsch, K. R. Levit, and D. R. Waldo, National Health Expenditures, *Health Care Financing Review*, *10*(2):112 (1988).
2. See especially August 21, 1989, *Federal Register*, listing the 207 national coverage decision. The Office of Payment Policy sets payment amounts for procedures performed in the hospital and in other setting such as ambulatory surgical centers, laboratories, and physician offices.
3. 42 U.S.C. 1395, et seq.
4. P.L. 100-203, December 22, 1987, Social Security Act, section 1842(B)(10), 42 U.S.C. 1395 U (b)(10).
5. ICD-9-CM Coordination and Maintenance Committee, Office of Coverage Policy, HCFA, 6325 Security Boulevard, Baltimore, Md. 21207.
6. *Physicians Current Procedural Terminology*, Fourth Edition, American Medical Association, Chicago, 1989.
7. HCFA publication 6, available from Government Printing Office (GPO), or see *Federal Register*, August 21, 1989.

50

Basics of Payment for Hospital and Physician Services

GARY S. SCHNEIDER* Pharmaceutical Manufacturers Association, Washington, D.C.

I. INTRODUCTION

Medicare payment methods have been changing rapidly in recent years, as policymakers attempt to encourage providers to better manage the use of resources and to hold down medical costs. Payment changes have altered incentives for providers of care and restructured the marketplace for manufacturers of medical devices and diagnostic products.

In the past, Medicare payment methods encouraged providers to expand their use of medical products. Today the incentives have changed, forcing providers to reevaluate their response to medical product use.

Manufacturers should seek the latest information on Medicare payment methods to evaluate market response to their products and to develop policies for research, development, marketing, and sales. Keeping up with the substantial changes is difficult because Medicare payment approaches are rapidly assuming new forms.

This chapter provides a fundamental understanding of Medicare hospital and physician payment policies as of April 1989. Since the policies are subject to continual change, readers are encouraged to monitor these changes and update the information in this chapter accordingly.

II. BENEFIT CATEGORIES AND PAYMENT OVERVIEW

Medicare benefits are covered under the program's two complementary but distinct parts: Part A, Hospital Insurance, and Part B, Supplementary Medical Insurance. Part A covers inpatient hospital care, skilled nursing facility

*Deceased.

services, home health care services, and hospice services. Part B covers physician services, outpatient laboratory and radiology products and services, ambulatory surgical services, and other outpatient products and services. The Health Care Financing Administration (HCFA) administers the Medicare program nationally. Intermediaries and carriers are insurance companies that contract with HCFA to administer the program on a local level.

Historically, hospital payment under Part A was based on reasonable costs for necessary patient care as determined after the delivery of services. The first shift in this approach occurred in 1982 when the Tax Equity and Fiscal Responsibility Act (TEFRA) extended existing per diem Medicare payment caps on operating costs. TEFRA also established a hospital-specific maximum limit on inpatient operating costs per case. The prospective approach was solidified in 1983 when the Medicare payment system for acute-care hospitals was firmly based on a fixed price per admission, established for separate diagnosis-related groups (DRGs) of patients.

The Medicare payment method for physician services and other noninstitutional suppliers of products and services is based on reasonable charges determined according to actual billed charges and fee screens (i.e., a limit used to determine payment for a particular product or service). Payment is based on the lowest of three figures: the actual charge, the customary (median) charge of an individual physician or supplier, or the area's prevailing charge. The prevailing charge is the 75th percentile of the distribution of customary charges of all area physicians or suppliers.

III. PAYMENT FOR INPATIENT HOSPITAL OPERATING COSTS: THE MEDICARE PROSPECTIVE PAYMENT SYSTEM

The Medicare prospective payment system (PPS) pays hospitals for nonphysician inpatient care, based on a prospective payment amount per DRG, additional outlier payments, an adjustment for hospitals serving a disproportionate share of low-income patients, and an indirect medical education adjustment. PPS applies to short-term, acute-care hospitals. Certain hospitals are exempt. Several types of hospitals and hospital units are excluded from PPS and continue to be reimbursed on a retrospective cost basis with limits.

The prospective payment amount is the core of PPS. It is a per case rate based on a federal standardized amount, which is multiplied by the appropriate DRG relative cost weight. The federal standardized amounts, the system's base prices, were determined by computing 1981 Medicare allowable operating costs per discharge for inpatient hospital services for hospitals under PPS. The amounts were standardized and updated for inflation through 1984. The amounts were then updated for inflation, changes in productivity and technology, and other factors through fiscal year 1989.

The federal standardized amounts are separated into labor-related and non-labor-related components. The labor-related portion is adjusted by an area wage index, which measures the level of hospital wages in the area in which the hospital is located relative to the national average of hospital wages. Separate standardized amounts are calculated for large urban areas, other urban areas, and rural areas. Large urban areas are defined as urban areas with a population of more than one million.

The standardized amounts for large urban, other urban, and rural areas are subject to "regional floors." In those regions where the regional standardized amounts exceed the national standardized amounts, the federal standardized amount is based on a blended amount of 85% of the national standardized amount and 15% of the regional standardized amount.

Table 1 presents the Medicare PPS national federal standardized amounts for discharges occurring on or after October 1988 through September 1989.

A hospital's prospective payment rate is determined by assigning each discharge to a DRG and multiplying the standardized amount by the cost weight for the DRG. Patients are assigned to one of 477 DRGs based on their principal diagnosis (the diagnosis chiefly responsible for causing the patient's admission), secondary diagnosis, procedures performed, age, sex, and discharge status. Each DRG is assigned a weighting factor that reflects the cost of hospital care for discharges classified in a DRG relative to the average cost for all DRGs. For example, the weight for DRG 106, Coronary Bypass with Cardiac Catheterization, is 5.5493. The PPS payment for discharges in a hospital located in a large urban area is $17,841.94 (5.5493 × $3215.17).

In addition to the prospective payment rate, Medicare pays hospitals for "outliers," or cases that have either unusually long lengths of stay or unusually high costs in relation to other cases in the DRG. Cases are deemed outliers when they exceed specific length of stay and cost threshold points for the DRG. Outlier payments may not be less than 5% or more than 6% of total payments for PPS hospitals per year.

A case becomes a length of stay outlier when the number of days in a patient's stay exceeds the average length of stay for discharges in that DRG by the lesser of 24 days or 3.0 standard deviations. The additional payment is a per diem amount equal to 60% of the average per diem federal rate for the DRG. This amount is multiplied by the number of additional patient days over the threshold to determine the total outlier payment adjustment.

A case that does not qualify as a length of stay outlier may qualify as a cost outlier if the charges of covered services (adjusted to cost by applying a hospital-specific cost/charge ratio) exceed a threshold that is the greater of 2.0 times the federal rate for the DRG or $28,000. The additional payment is equal to 75% of the difference between the hospital's adjusted charges for the discharge and the outlier threshold.

Medicare provides up to a 25% adjustment to the federal standardized amount for hospitals that service a disproportionate share of low-income patients.

Medicare also provides for an adjustment to allow for the higher indirect costs incurred by teaching hospitals, such as added tests and procedures

TABLE 1 Medicare National Federal Standardized Amounts for Discharges Occurring on or after September 30, 1988 and before October 1, 1989

	Labor	Nonlabor	Total
Large urban (i.e. >1 million)	$2374.22	$840.95	$3215.17
Urban	2351.10	832.75	3183.85
Rural	2219.89	614.82	2834.71

ordered by residents and increased staff time required for teaching. Hospitals with an approved graduate medical education program receive a 7.7% increase in federal payment rates for each 0.1% increase in the hospital's ratio of full-time equivalent interns and residents to its bed size.

A. Costs Excluded from PPS Rates: Capital Costs

Capital-related costs are excluded from prospective payment and are reimbursed separately on a reasonable cost basis. All hospitals must treat capital-related costs under PPS in a manner consistent with the way identical or similar costs were treated prior to PPS.

Medicare payments for capital are reduced by 15% for fiscal year 1989. Medicare thus pays 85% of the reasonable costs for capital equipment. Capital-related costs are to be added to PPS as of October 1, 1991.

Capital-related costs reimbursable by Medicare are limited to:

Net depreciation expenses
Taxes on land or depreciable assets used for patient care
Leases and rentals for the use of depreciable assets
The costs of betterments and improvements
The costs of minor equipment that are capitalized
Insurance expense on depreciable assets
Interest expense
The capital costs of related organizations

B. Depreciation on Assets

Medicare will provide an "appropriate allowance" for depreciation on buildings and equipment. Depreciation costs must be based on the historical cost of the asset and be prorated over the estimated useful life of the asset with the straight-line depreciation method or, in certain circumstances, the accelerated depreciation method. The estimated useful life of a depreciable asset is its normal operating or service life. Providers must use the useful life guidelines published by HCFA or, in their absence, the relevant edition of the American Hospital Association (AHA) useful life guidelines.

HCFA's *Provider Reimbursement Manual* identifies the types of equipment that can be depreciated, including the following:

1. Building equipment, which includes attachments to buildings such as wiring, electrical fixtures, plumbing, elevators, heating systems, air conditioning systems, etc. The general characteristics of building equipment are that it is affixed to the building and not subject to transfer and it has a fairly long useful life.
2. Major movable equipment, which includes such items as beds, wheelchairs, X-ray machines, etc. The general characteristics of this equipment are a relatively fixed location in the building; capable of being moved as distinguished from building equipment; a unit cost sufficient to justify ledger controls; sufficient size and identity to make control feasible by means of identification tags; and a minimum life of approximately 3 years.
3. Minor equipment, which includes such items as waste baskets, bed pans, etc. The general characteristics of this equipment are no fixed location and subject to use by various departments of the provider's facility, com-

paratively small in size and unit cost; subject to inventory control; fairly large quantity in use; and a useful life of approximately 3 years or less.

C. Sales, Leaseback, and Lease-Purchase Agreements

When a provider enters into a sale and leaseback agreement for plant or equipment with a nonrelated purchaser, the incurred rental is an allowable cost and can be depreciated. Some lease agreements for equipment are essentially equivalent to purchasing the equipment. In such cases, the lease is a "virtual purchase" and the rental charge is an allowable cost and can be depreciated only to the extent that it does not exceed the amount that the provider would have included in allowable costs if it had legal title to the asset.

D. Interest Expense

Medicare pays for necessary and proper interest on both current and capital indebtedness. To be considered "necessary," interest expense must be incurred on a loan made to satisfy a patient-care-related financial need and be reduced by the provider's investment income.

E. Costs Excluded from PPS Rates: Direct Medical Education

Direct costs of approved medical education activities are reimbursed separately from PPS rates. Allowable costs are those approved for intern, resident, and nursing school and paramedical education programs.

IV. MEDICARE PART B BENEFITS AND PAYMENT METHODS

A. Overview

Part B benefits supplement those provided under Part A. Part B covers services and products in a variety of settings. For example, Part B covers outpatient diagnostic laboratory services in a hospital laboratory, physician's office, or independent clinical laboratory.

Several methods are used to pay for products and services under Part B. These methods include free screens, fee schedules, inherent reasonableness criteria, reasonable costs subject to limits, and prospective rates. Payment for most products and services is based on reasonable charges. HCFA and carriers determine reasonable charges by considering the following factors and paying the lowest figure:

The actual charge.

The customary charge for similar services generally made by the individual physician or other person furnishing the services.

The prevailing charge in the locality for similar services.

The carrier's usual amount of reimbursement for comparable services to its own policyholders under comparable circumstances.

In the case of medical services, supplies, and equipment (including equipment servicing) that, in the judgment of the Secretary of the Department of Health and Human Services, do not generally vary significantly in quality from one supplier to another, the charges determined to be reasonable may not exceed the lowest charge levels at which these services, supplies, and equipment are widely and consistently available in a locality, except to the extent and under the circumstances specified by the Secretary.

The inflation-indexed charge limit for nonphysician services, supplies, and equipment.

The charge limit for physician outpatient services.

Other factors that may be found necessary and appropriate with respect to a specific item or service to judge whether the charge is inherently reasonable.

The factors used to determine reasonable charges are discussed below.

B. Customary Charge

The customary charge is the charge most frequently made by a physician or supplier for a given medical service or product provided to his or her patients in general. The customary charge is based on the actual charges both for Medicare patients and for other patients on which the carrier has data. The customary charge is the lowest actual charge that is sufficient to cover the median of charges arrayed in ascending order.

C. Prevailing Charge

The prevailing charge is derived from customary charges weighted by the frequency that a physician or other person has rendered the service or provided the product. The prevailing charge is established at the 75th percentile of the customary charges made for similar services or products in the same locality. (For physician services, the prevailing charge level cannot exceed that determined by an economic index.) Carriers can develop and use different prevailing charge screens that consider differences in physician specialties.

An example of the calculation of the prevailing charge is presented in the Medicare *Carriers Manual* and displayed in Table 2. Each customary charge is arrayed in ascending order and weighted by how often the physician or other person rendered the service. The lowest customary charge that is high enough to include the customary charges of the physicians or other persons who rendered 75 percent of the cumulative services is then

TABLE 2 Example of Calculation of Prevailing Charge

Customary charge	Number of services rendered by physicians with customary charges as indicated	Cumulative services
$50	1402	1402
$60	1115	2517
$70	1680	4197
$80	803	5000

determined as the prevailing charge for the service (subject to the economic index limitation).

In this example, 75% of the total 5000 services equals 3750 services. The prevailing charge is therefore $70. A total of 2517 services were rendered by physicians with $50 and $60 customary charges, and an additional 1680 services were rendered by the physicians with $70 customary charges. The 3750 service was thus rendered by a physician with a $70 customary charge.

D. Payment for Physician Services

Medicare currently employs a fee-for-service system to pay for physician services. As shown above, the system is based largely on historical fees originated by doctors themselves. For established procedures, payments are based upon the "customary, prevailing, and reasonable charge" (CPR) for localities in the country. For new procedures, payments initially are based upon fees that physicians submit to local Medicare carriers. Annual increases in Medicare reimbursement levels are limited by an economic (inflation) index. However, physicians have the right to "balance bill" patients if Medicare reimbursement is less than their established fee. Through 1990, there is some limitation on total fees that can actually be charged, called the maximum allowable actual charge (MAAC).

The flexibility built into Medicare's physician payment system allows for reimbursement levels that vary widely by geographic area and for higher payments for specialists compared to general practitioners for identical services.

As noted above, the Medicare reimbursement formula for a billed service pays the lowest of four possible amounts:

1. The physician's submitted charge—the individual physician's actual billed amount.
2. The customary charge—the individual physician's median charge for the service during the previous year.
3. The "unadjusted" prevailing charge for that service in the locality—the 75th percentile of the distribution of customary charges for all physicians in the area during the previous year.
4. The "adjusted" prevailing charge—the prevailing charge for the service as of June 1973 inflated by the current Medicare economic index (MEI). The MEI is calculated by the Health Care Financing Administration and applied as an annual cap on prevailing charges. Since 1976, the prevailing charge has been capped at the corresponding adjusted prevailing charge—the product of that year's MEI and the prevailing charge for the procedure in 1973, the MEI's base year.

The result of the CPR payment methodology is that reimbursement levels reflect three major factors:

1. Historical fee patterns: Once a claim is submitted to Medicare (either by the beneficiary or the physician) the billed charge for a particular service becomes part of that physician's charge history (or "profile"), and is used to determine his or her customary charge. The CPR methodology gives physicians an incentive to increase fees annually, since the next year's reimbursement levels are, in part, determined by the prior year's values.

2. Medicare fee constraints: Initially, physician reimbursement levels were limited only by the doctor's own charge patterns or those of his peers. This being inflationary, however, Congress took steps to curb the growth in payments. The most important action was the development of the MEI, which ties increases in Medicare prevailing charges to practice costs and wage rates throughout the economy.
3. Carrier claims processing policies: The carriers that pay claims are allowed to define geographic, specialty, and other subcategories that can affect payment levels.

E. Fee Schedules

Medicare payments to radiologists are based on a relative value scale (RVS). There is a specific payment level for each radiology procedure. The payment is the product of a weight, which indicates the relative value of each procedure, and a conversion factor, which translates the weight into a fee. The procedure-specific weight is based on the time to do the procedure, its complexity, and radiologist practice costs.

F. Assignment and Balance Billing

Two additional concepts are important to Medicare physician payment: "assignment" and "balance billing." Assignment works as follows: Physicians who accept assignment bill Medicare directly and accept Medicare's CPR calculation as payment in full (except for the deductible and coinsurance). Alternatively, for nonassigned claims the doctor bills his full fees to the patient (subject to some interim limits), who in turn is responsible for the difference between the doctor's bill and what Medicare allows. This difference is referred to as a balance bill. Physicians are permitted to decide on a claim-by-claim basis whether to accept "assignment."

During the early 1980s, Medicare policymakers were concerned that balance billing was placing beneficiaries at increased financial risk. In response, Congress took steps to raise assignment levels. Specifically, they froze physician fees and used the action to encourage physicians to accept Medicare assignment. Under the current program, physicians are asked to "participate" in Medicare. Participating physicians (known as PARs) accept assignment for all beneficiaries served in their practice. In exchange for becoming PARs, physicians are eligible for somewhat higher payments. These initiatives are credited with increasing the assignment rate on Medicare claims from 59% in 1984 to 73% in 1987.

51

Payment Planning for New Medical Diagnostics: An Action Plan for the 1990s

WAYNE I. ROE *Health Technology Associates, Washington, D.C.*

I. INTRODUCTION

The decade of the 1980s produced fundamental changes in the system for financing medical care in the United States, with concomitant impacts on medical device and diagnostic product markets. Historically, the major insurers of medical care in America—Medicare, Medicaid, Blue Cross/Blue Shield, and commercial carriers—were passive payers of insurance claims. When new medical products were developed and purchased by hospitals and physicians, payers reimbursed either provider costs or charges with little regard for the medical necessity of the associated services. In large part, this economic dynamic allowed technological development to drive health-care financing, and led to two decades of unparalleled success for device and diagnostic product manufacturers.

Along with the success of medical innovation came an explosion of health-care costs. Consistent year-to-year increases of hospital and physician expenditures exceeding 15% forced the U.S. Congress and private health insurers to take steps to control medical inflation. The weapons in this fight include diagnosis-related group (DRGs), relative value scales, preferred provider discounts, coverage restrictions, and utilization review programs designed to manage medical care. The growth of "managed health care" signals a major reversal in the market dynamic for new medical products. Rather than medical innovations driving health-care financing, payer policies are beginning to dictate the markets for new medical products. Consequently, payer policies have become regulatory barriers equal in impact to FDA approval. Increasingly, insurers determine the kinds of products that can be sold, who they are used for, the settings where they are applied, and the profit margins that can be earned. Confronting these new market forces requires companies to develop and implement strategic payment plans for new products.

II. PAYMENT POLICY—THE FIFTH "P" OF MARKETING

Health insurers finance 93% of all hospital purchases of devices and diagnostics and 67% of physicians' services in using them. This collective market power requires that product manufacturers consider payment policy as an essential element of marketing planning. Experience to date with managed care insurance policies demonstrates that payers impact each of the four components of traditional marketing—product, price, placement, and promotion.

A. Product

The incentives inherent in insurance payment policies are directing companies to develop products that improve the efficiency of diagnostic and therapeutic regimens, move patients from expensive institutional settings to outpatient and home care, and reduce the invasiveness of traditional interventions. For example, in the mid 1980s, several manufacturers developed continuous passive motion machines to prevent deep-vein thrombosis in patients after total knee replacement surgery. The technology demonstrated the capacity to reduce inpatient stays by several days, and insurers rewarded the industry with a home rental market of upward of $50 per day. In contrast, product markets for additive and expensive inpatients' disposables (e.g., coated Foley catheters, IV sets) have been significantly depressed.

B. Price

Insurance reimbursement levels have been radically reformed and consequently have impacted the prices that providers will pay for new medical devices and diagnostics. For example, Medicare's decision to prospectively pay for outpatient radiological scans has forced companies introducting new radiopharmaceuticals to price them at levels near those of older products despite their clinical advantages. Unless manufacturers factor reimbursement levels into their product pricing decisions, they risk establishing prices at levels that are unprofitable for hospital or physician users.

C. Placement

Manufacturers have discovered that payer policies for medical procedures and diagnostics differ significantly by the setting where they are used—hospital, physician office, ambulatory surgery center, etc. Thus, decisions about where to launch a new device must consider the associated reimbursement levels to assure the most profitable initial market. For example, pulse oximetry studies can be reimbursed under a fixed DRG operating payment system for inpatients, on a cost basis for hospital outpatients, and on a charge basis in physicians' offices. Deliberations on launch strategies for new generations of sensor technologies should be made in light of these site differentials.

D. Promotion

Physicians and hospitals are increasingly interested in knowing not only that new products have acceptable clinical performance, but also that they are reimbursable. Consequently, manufacturers must demonstrate to customers that their products are properly paid, and provide customer support geared to enhancing reimbursement. "Reimbursement promotion" has become an industry standard in the launch of expensive new drugs like nonionic contrast media, thrombolytic therapy, Epoetin Alfa for dialysis, and hormonal therapy for prostate cancer. As these programs demonstrate marketing success, they will expand to new device launches as well.

III. TOOLS FOR PRODUCT MANAGEMENT

In planning for third-party payment, marketing executives at medical product firms must understand the four major policy tools that insurers can use to influence technology markets. These are product/procedure coverage policy, diagnosis and procedure coding, reimbursement rate setting, and utilization review. Although related, each policy is unique and subject to a separate set of decision-making criteria. Furthermore, policies vary among payers, which further complicates a careful product payment analysis. The critical component of each policy are as follows.

A. Coverage

Coverage policy reflects decisions by payers to consider a particular product or service as medically appropriate for the plan's beneficiaries. Coverage determines the extent of a product's market, and can be limited to selected patient indications or settings. Payers have differing criteria for deciding on coverage, but typically these include FDA approval, proven safety and efficacy, cost effectiveness, and medical necessity. Insurers look carefully at FDA labeling and published peer-reviewed journal articles as guides to coverage, and manufacturers must link their regulatory and investigator strategies to a company plan to secure coverage acceptance.

B. Coding

Insurers use a number of complex systems of procedure codes and associated identifiers to process medical claims. Likewise, providers complete claims forms with the same codes to identify the types of patients they treat and services they provide. Since both coverage reviews and reimbursement rate setting are tied to codes, unless proper codes exist to identify a medical device or diagnostic, it may be improperly treated by the ultimate payer. The most important coding systems are:

1. The ICD-9-CM diagnosis codes for identifying patient clinical conditions on all claims
2. The ICD-9-CM procedure codes for identifying medical and surgical procedures on hospital inpatient claims
3. The CPT-4 procedure system for identifying all physician services regardless of setting and most clinical laboratory tests
4. The HCPCS coding system to identify a range of physician, laboratory, pharmaceutical, and durable medical equipment services uniquely covered by the Medicare Program

Each coding system has its own sponsoring organization, which follows a separate set of criteria to decide whether to establish a unique code for a new interact with the critical coding authorities at an early stage, since new codes take a year or more to become established.

C. Reimbursement

Insures establish specific formulas and associated dollar levels to pay for medical products and services covered by their plans. Reimbursement amounts differ widely by payer, provider, and site of service, and must be carefully evaluated for prospective new products. For example, for new inpatient hospital services Medicare reimburses on a prospective DRG basis, Medicaid on a per diem basis, Blue Cross on a discounted charge basis, and commercial carriers at full institutional charges. These amounts can differ by several hundred percent depending upon the product in question.

D. Utilization Review (UR)

Managed care insurance plans increasingly use sophisticated clinical guidelines and written protocols to direct or influence the course of care for patients. These utilization review systems may include prior surgical certification, concurrent plan of care review, postdischarge audits, or case management of specific chronic conditions like diabetes or cardiovascular disease. The UR programs have the potential either to limit the use of specific services or to promote the substitution of new and cost-effective items for old ones. Manufacturers should examine existing UR programs to determine whether they will pose a barrier or opportunity to the launch of a new item. UR firms are highly competitive and are willing to consider adding new items to their protocols when they fit the objectives of the sponsoring organization.

IV. A MODEL FOR PAYMENT PLANNING

Medical device and diagnostic product manufacturers must develop and execute a carefully drawn plan to assure favorable third-party payment for new products that are brought to markets. Over the years, the staff at Health Technology Associates (HTA) has developed a successful model for payment planning that has been validated on products ranging from cardiac implants to interventional catheters, to radionucleide diagnostics. The approach follows two phases—payer market planning and subsequent payment plan implementation.

A. Steps in Payer Market Planning

1. Market Segmentation

The first step in payment planning involves segmenting the insurance market for a new product. This involves answering the question, Who pays for the patients on whom this product will be used? By understanding the principal payers for a product, the manufacturer can target data collection, insurer education, and other launch strategies to the unique criteria of Medicare, Blue Cross, CHAMPUS, or other insurers. Companies can determine market shares using the National Hospital Discharge Survey and other public data bases containing demographic and claims information.

2. *Product-Specific Policy Evaluation*

Once a company has established the payer mix for a product, the next step is to evaluate the current coverage, coding, reimbursement, and utilization review policies for the key payers. Typically, one considers policies for current competitive diagnostics or therapies, as well as the policy criteria the company must address should a product-specific change be necessary to accommodate the new introduction.

3. *Product Plan Development*

Combining the results of the market segmentation analysis and policy evaluation, the manufacturer must develop a third-party payment action plan. This outlines specific action steps, data needs, contact points, and time frames for securing favorable payment. The goal of the plan should be to obtain adequate coverage, codes, and reimbursement levels as soon as possible after receiving FDA product approval. One important element of this plan is the approach to payers that the company must pursue. In certain circumstances, it is prudent to seek favorable payment from individual local insurers, while in other situations companies will have to deal directly with national Medicare, Blue Cross, or other policymakers. Actions that may be required include a national coverage decision, a product-specific procedure code, or establishment of a new DRG payment or other reimbursement adjustment.

B. Steps in Payment Plan Implementation

1. *Policy Development*

Once the plan is established, companies should act immediately to secure favorable policies from the key target payers. This typically requires preparing written materials about the technology for insurance policymakers, securing support from leading clinicians or medical specialty societies, assuring the publication of peer-reviewed clinical articles, and coordinating the inputs to payers with those to the FDA. Policy development is a strategic marketing exercise that requires careful preparation and persistence.

2. *Policy Promotion*

At product launch manufacturers should take steps to assure that the market is fully informed regarding the specific payment policies secured for the product. This involves, first and foremost, sales force education on the fundamentals of coverage, coding, reimbursement, and claims processing for the technology. Additionally, physicians, hospitals, and other customers should be provided with detailed materials on product payment and steps to assure adequate reimbursement. For high-margin and visible items like new implants or expensive diagnostics, companies should establish customer support hotlines to help answer questions and assist in resolving problem claims. Policy promotion is likely to emerge as one of the most desirable "value added" services in the 1990s.

3. Policy Assurance

Both payer policies and medical product markets are evolutionary, and subject to continuous change. Consequently, even the best-implemented payment plan demands continual refinement to assure maximal customer payment over time. Manufacturers should establish long-term monitoring programs to identify specific regions and payers posing problems in the market so that targeted remedies can be implemented. Often, this monitoring is provided via an insurer data base developed by the hotline. In situations where the product is launched for limited indications but is to be expanded to others, companies should conduct phase IV clinical studies designed to inform payers of the added value of the technology. As long as the device or diagnostic continues to have a viable medical market, payment policy must be an active marketing focus.

V. CONCLUSIONS

The era of managed insurance plans has ushered in a radically new set of market incentives for medical technology that require careful planning. Because it can require 2 years or more to secure favorable payment policies, insurer strategies need to be established early in the product research stage and continually refined through launch. In many situations, an early look at the policy environment for a new product will suggest design changes or entire new research alternatives. Coordination with associated regulatory activities is likewise critical, since payers rely on FDA approval as a key input to policymaking.

Proper input to insurers requires well-documented clinical and economic data on the product, as well as supportive professionals. Assembling these inputs is a strategic marketing responsibility, not a sales function. As medical markets are increasingly driven by payer concerns, industry success will be measured by an ability to deliver favorable policies. Customers will look to companies for the ultimate in value-added service—a guarantee of favorable coverage and reimbursement. Success in this endeavor will require policy sophistication, sound clinical and economic fundamentals, and a company approach that integrates payment planning with existing R&D, regulatory, marketing, and sales functions.

52

Medical Standard Setting in the Current Malpractice Environment: Problems and Possibilities[1]

ELEANOR D. KINNEY *Indiana University School of Law, Indianapolis, Indiana*

MARILYN M. WILDER *Whitted & Buoscio, Merrillville, Indiana*

I. INTRODUCTION

The 1980s have witnessed sharp and dramatic increases in the size and frequency of medical malpractice claims and in the cost of malpractice insurance for health-care providers [1-3]. Some evidence suggests that physicians have responded by practicing "defensive" medicine and adopting many inefficient practice patterns [4]. These practices have contributed to overutilization of medical services in the American health care system [5].

---

[1]This chapter represents the collective work of many members of The Center For Law and Health at Indiana University School of Law, Indianapolis. The idea for the chapter came from William Garrison of Blue Cross and Blue Shield of Indiana, Inc. Initial work on the chapter was done by research assistants Gayle Reindl and Marti Baker, under the direction of Barbara McCarthy Green, former Assistant Director of The Center For Law and Health. Julie Ann Randolph, a research assistant for The Center, provided invaluable contributions to this chapter. We are also very grateful to Nancy Cahill, J. D., Visiting Associate Professor of Law, in residence at The Center For Law and Health during the summer of 1988, for her excellent comments on numerous drafts. We also thank Kenneth M. Stroud, Professor of Law at Indiana University School of Law, Indianapolis, and Charles M. Clark, Jr., M.D., Professor of Medicine, Indiana University School of Medicine, for their assistance with the article. Finally, we thank Phyllis Bonds, the Center's administrator, and Wendy Fisk for their clerical assistance with the article. Research on this chapter was funded by a Project Development Grant from Indiana University, Indianapolis. The chapter was previously published in the *U.C. Davis Law Review, 22*:421-451 (1989).

The 1980s have also witnessed continued inflation in the cost of health-care services at rates exceeding inflation in other sectors of the economy [7]. In response, federal and state governments and private health insurers have put significant pressure on physicians to cut costs. Many payment reforms, which public and private health insurers have adopted, transfer the financial risk of providing costly services from the health insurer to the health care provider [8].

Concerned that health-care providers cut corners in delivering care, or "game" utilization of health care services to maximize payment, health insurers and others who pay for care have developed methods to ensure that health-care services are utilized appropriately and are of high quality. For example, Congress created peer review organizations (PROs), to monitor the utilizaiton and quality of inpatient hospital services for Medicare beneficiaries [9]. State Medicaid programs and private insurance plans have instituted similar utilization and quality review programs [10,11]. These developments have precipitated a greater interest among payers in ways to measure the "outcome" of medical treatment to be sure that payees are purchasing high-quality services in the wake of cost-containment efforts [12]. Further, these developments have spawn scholarly speculation as to how these payment reforms and accompanying standard setting activities will influence the standard of care in malpractice cases [13-15].

Traditionally physicians resisted attempts to quantify or even define what constitutes quality in the performance of specific procedures or treatment of specific diseases. However, this situation has changed markedly in recent years. The leadership of American medicine is actively attempting to define quality medical care with respect to specific diseases and procedures as well as to delineate when health-care services are appropriately provided. In 1987, the Council of Medical Specialty Societies convened a conference on Standards of Quality in Patient Care: The Importance and Risks of Standard Setting [16]. A consensus emerged from the conference that the medical profession and specialty societies need to set standards to define quality medical care. The leading professional organizations for physicians, the American Medical Association (AMA) and the American College of Physicians (ACP), have adopted formal policies endorsing the development of standards for clinical practice [17,18].

These developments have resulted in an increase in the number and type of standards regarding the appropriate utilizaiton and quality of health-care services. These standards may have an important impact on day-to-day clinical practice and on the medical malpractice climate because, as explained below, negligence law requires that a defendant physician's conduct be measured against a standard of care established by the medical profession before liability can be imposed.

The purpose of this chapter is to analyze how medical standards may be used in medical malpractice litigation. Part I describes the types of medical standards in effect today. Part II (p. 759) explains how the standard of care is now established in medical malpractice lawsuits and how medical standards may be used in establishing the applicable standard of care. This chapter concludes that medical standards will and should play a prominent role in setting the standard of care in malpractice cases.

II. TYPOLOGY OF MEDICAL STANDARDS

Basically, two major types of medical standards[2] have evolved: (1) clinical practice protocols and (2) utilization review protocols. These two categories can be distinguished further according to the purpose for which the standard is to be used and the author or sponsorship of the standard.

A. Clinical Practice Protocols

Clinical practice protocols are disease- or procedure-specific protocols that describe or recommend specific steps in the diagnosis or treatment of particular diseases or conditions or the performance of medical procedures. Most clinical practice protocols are written by physicians for physicians in the medical management of their patients and are intended to be recommendations only.

However, payers are increasingly using clinical practice protocols and other medical standards as a basis for making payment decisions. For example, in the early 1980s, the Blue Cross and Blue Shield Association was concerned that admission testing of hospital patients was excessive. In response, the American College of Physicians developed standards in its Medical Necessity Program describing appropriate tests to provide patients upon admission to the hospital [18]. These standards are now used by Blue Cross plans around the country to determine what tests administered to hsopital patients should be reimbursed [19]. Similarly, in recent years, the Medicare program has based policies and decisions on coverage of medical procedures and technologies on clinical practice protocols, technology assessments, and other standards developed by medical organizations [20-22].

An important factor in the development of clinical practice protocols and their role in malpractice litigation is the sophistication of health services research conducted in the last decade by physicians on the efficacy and cost effectiveness of specific medical treatments. These researchers have described sharp differences in clinical practices among different geographic areas [23-27]. John E. Wennberg, M.D., has conducted extensive research on geographic variations in medical practice [28-31] and has exhorted the medical profession to apply the same scientific analysis in determining the efficacy of clinical procedures as is now required for new drugs [32,33]. Robert Brook, M.D., and his researchers at the Rand Corporation have convened medical experts to assess the appropriate application of certain medical procedures. Brook and his team have then conducted studies to determine the extent of the inappropriate use of certain medical procedures in practice [34-40]. David Eddy, M.D., Ph.D., of Duke University, has employed quantitative methodologies to predict the likely medical outcomes of certain clinical procedures [13,41,42]. Further, several prestigious physician researchers have called for an expansion of research to develop clinical prediction rules to guide physicians in the treatment of disease [43]. Finally, the Robert Wood Johnson Foundation is now funding research that develops clinical practice protocols specifically designed to prevent malpractice claims [44].

---

[2] The authors are aware that the term "standards" is controversial in the medical community and may have numerous meanings. The term is used in a very general sense in this Chapter.

For the most part, clinical practice protocols are developed in three ways: (1) by individual physicians for use in a local context such as a hospital; (2) by commercial enterprises with physician input; and (3) by national medical specialty societies, voluntary health organizations, or government agencies. The authorship and sponsorship of clinical practice protocols have important implications for their roles in mdedical malpractice litigation. Thus, the subsequent discussion of clinical practice protocols will distinguish between professionally sponsored and individually sponsored protocols.

1. *Professionally Sponsored Clinical Practice Protocols*

Perhaps the most important clinical practice protocols from a malpractice perspective are those developed by physicians under the auspices of national medical specialty societies, voluntary health organizations, or the Public Health Service (PHS). As noted above the development of clinical practice protocols by these organizations has been a relatively recent phenomenon, recognizing from health services research that standards by which to measure and assess quality of medical care can be developed more rigorously than in the past.

The medical professional organizations have played a prominent role in this development of clinical practice protocols. The American Medical Association (AMA) has been quite active in developing recommendations regarding the proper provision of medical care in a variety of contexts. The AMA's Council on Scientific Affairs, using panels of experts, conducts studies and publishes reports that outline proper procedures for the treatment of many diseases and conditions and include informal standards and recommendations [45-77]. The AMA's Diagnostic And Therapeutic Technology Assessment (DATTA) program conducts assessments of medical technologies. Specifically, through a formal process of synthesizing expert opinion and literature reviews, DATTA develops and publicizes conclusions about the safety and effectiveness of medical technologies [48,49]. The AMA publishes the DATTA reports and the reports of its Council on Scientific Affairs in its journal, the *Journal of the American Medical Association* (JAMA).

The activities of medical specialty societies in developing standards are diverse, but almost all are engaged in some way. The ACP has been extremely active, indeed innovative, in the development of medical standards. Through its Clinical Efficacy Assessment Project, ACP has published a wide range of standards and guidelines for the use of medical technologies and the management of medical conditions [50,51]. The standards and guidelines developed in the Clinical Efficacy Assessment Project are disseminated through its Clinical Efficacy Reports and published in the *Annals of Internal Medicine* [52-56]. In addition, as noted above, ACP, working with the Blue Cross and Blue Shield Association, developed the Blue Cross Medical Necessity Program to assess costly, but commonly used, medical tests and procedures for determining coverage of services under Blue Cross health insurance programs [57]. Finally, ACP has recently published a manual specially addressed to primary care physicians on the use and interpretation of common diagnostic tests [58].

Other specialty societies in the field of internal medicine are also active in developing clinical practice protocols. The American Society of Internal Medicine has developed some guidelines for the treatment of disease in outpatient settings [41,42]. The American College of Cardiology, through its Cardiovascular Norms Committee, has developed guidelines for the use of medical technologies, such as cardiac pacemakers, in the treatment of heart disease [59]. Other medical specialty societies in the internal medicine field have few clinical practice protocols or other medical standards, but are engaged in some standard setting activity. For example, the American Academy of Neurology has a formal process for responding to inquiries from insurers and others about appropriate uses of medical procedures [60-62].

The American Academy of Pediatrics (AAP) has also been extensively engaged in the development of medical standards for many years. For example, the *Report of the Committee on Infectious Disease*, colloquially called the "Red Book," provides current information and consensus on the effective control of pediatric infectious disease [63]. Pediatricians use the AAP's Red Book extensively in the management of infectious disease. AAP now publishes guidelines and manuals for a wide variety of subjects and will soon publish guidelines for perinatal care and supervising the health of children [64].

The American College of Surgeons (ACS) has published a statement of principles, some of which address appropriate management of surgical patients [65]. In addition, ACS publishes manuals on the practice of surgery in specific settings [66-68] and other publications containing recommendations for good surgical practice in the treatment of cancer, trauma and other conditions [69]. The American Society of Anesthesiologists has also become especially prolific in publishing standards and recommendations for the organization of anesthesia practice [70] and for proper management of anesthesia in various clinical settings [71,72].

The American College of Obstetrics and Gynecology (ACOG) is in a state of transition with respect to the publication of standards and recommendations for the practice of obstetrics and gynecology. ACOG is presently revising its clinical practice protocols and other medical standards and recommendations [16]. ACOG also publishes technical bulletins, written by practitioners under ACOG's direciton, designed to provide practitioners with information on the latest proven medical procedures [73].

The American Academy of Family Physicians does not publish clinical practice protocols or other medical standards because the academy maintains that there is wide variation throughout the country in the practice of family medicine and medical standards are best developed at the local level [74]. However, the Academy convened a task force in August of 1988 to consider whether to develop clinical practice protocols that are national in scope. This task force will convene in early 1989 to identify several specific disease areas for which protocols would be useful and begin developing such protocols [74].

Voluntary health organizations have also gotten involved in the development of clinical practice protocols. The three major voluntary health organizations—the American Heart Association, the American Cancer Society, and the American Diabetes Association—are excellent examples of this effort. The American Heart Association has developed standards and recommendations for the management cardiac patients in a variety of settings, for example, recommending appropriate diets for the prevention of Coronary Artery Disease and appropriate exercise for patients with heart disease or at risk of heart disease [75-83]. The American Heart Association has also published an extensive guide, *The Exercise Standards Book*, for use in conducting exercise tests on heart patients [84]. The American Diabetes Association (ADA), in direct consultation with other medical specialty societies, has prepared guides with protocols for the diagnosis and treatment of both juvenile and adult diabetics [85,86]. The ADA's *Long-Range Plan* calls for the adoption and development of standards for the treatment of diabetes [87]. The American Cancer Society has also been active in setting medical standards. For example the American Cancer Society has adopted protocols for the use of Pap smears in diagnosing cervical cancer, the use of Hemoccult and other techniques in diagnosing colorectal cancer, and the use of mammography in diagnosing breast cancer [88-93].

2. *Government-Sponsored Clinical Practice Protocols and Other Medical Standards*

The federal government, in its capacities as chief financier of biomedical research and major health insurer, has played an important role in the development of medical standards. The Public Health Service (PHS) within the Department of Health and Human Services (DHHS) and the Health Care Financing Administration (HCFA), also within DHHS and which administers the Medicare and Medicaid programs, have been the most prominent government agencies involved in the development of medical standards. The Department of Defense health care programs have also been particularly active [57].

The National Institutes of Health (NIH), within PHS, have played a critical role in developing medical standards. Periodically, the Office of Medical Applications of Research (OMAR), with the NIH, convenes consensus development conferences to evaluate new and established medical technologies and associated treatment modalities [94-97]. In these conferences, the NIH gathers the leading experts in the field to develop conclusions about the appropriate uses and efficacy of the modality or technology used in the treatment of disease or injury [98,99]. Using the information from some of these consensus development conferences, the NIH has published a guide for the prevention of venous thrombosis and pulmonary embolism and the role of diet and exercise in non-insulin-dependent diabetes mellitus [100,101]. Other organizations within the NIH have also been actively involved in the development of standards and guidelines. For example, the National Diabetes Advisory Board has published a guide for the treatment of diabetes for primary-care physicians [102].

The PHS also engages in some technology assessment to advise federal health insurance programs on coverage of new technologies [103]. The PHS Office of Health Technology Assessment (OHTA), in the National Center for Health Services Research and Health Care Technology Assessment (NCHSR/HCTA), coordinates the statutorily mandated assistance of PHS in making recommendations on whether federal health insurance programs should pay for specific technologies [104]. OHTA often evaluates the uses of a technology using data, studies, and opinions of other federal agencies as well as the medical community and other interested parties [105-111]. In particular, OHTA considers actions of the Food and Drug Administration (FDA) in its review of new medical devices under the Medical Device Amendments of 1976 [112]. Under these amendments, the FDA must approve all medical devices that are marketed in the United States and, in so doing, determines whether a device is safe and effective for the purposes stated on the manufacturer's label [112,113].

Most recently, NCHSR/HCTA has initiated the National Program for the Assessment of Patient Outcomes [114]. The purpose of this research program is to evaluate patient outcomes to determine the appropriateness of these treatments and procedures, to create data bases and improve research methods used to evaluate patient outcomes, and to disseminate study results to modify clinical practice [114,115].

HCFA, in developing coverage policy for the Medicare program, has become increasingly involved in the development of medical practice standards. Since the inception of the Medicare program, HCFA and its predecessors have made approximately 200 national coverage policies regarding medical technologies and procedures and continue to make about 10-20 coverage policies each year [8,57]. Most of these national coverage policies have been developed in consultation with the PHS technology assessment activity described above.

In recent years, HCFA has demonstrated considerable and increasing interest in medical standard setting [116]. Just recently, HCFA announced that, in conjunction with PHS, it would become directly involved in the effort to develop standards for clinical practice within its "Health Care Effectiveness Initiative" [115, p. 1198]. One key activity of this initiative is the funding of clinical research to examine "the appropriateness and effectiveness of various procedures and interventions" [115, p. 1198]. HCFA has already funded some studies of procedures and interventions based on PRO data, conducted by the Rand Corporation [117-119]. One objective of these HCFA efforts and the NCHSR/HCTA initiative referred to above, according to HCFA's administrator, is to improve guidelines for clinical practice and thus provide more protection from malpractice liability for physicians complying with guidelines [13; 115, p. 1202].

Finally, the activities of the Institute of Medicine (IOM) in the development of medical standards are noteworthy. In 1984, Congress authorized federal support to the IOM to support creation of the Council on Health Technology [118]. The council's purpose is to promote the development and application of appropriate health-care technology assessments, to review existing health care technologies and identify those that were obsolete and inappropriately used, and to fund private-sector initiatives in this area [118]. In 1985, the IOM conducted a seminal study of technology assessment that renewed existing technology assessment activities and methodologies [119]. The IOM and its Council on Health Technology are now developing an initiative to coordinate technology assessment activities in the public and private sectors, and to assess medical technologies for clinical use [120,121].

3. *Individually Sponsored Clinical Practice Protocols*

Many clinical practice protocols are developed by individual physicians—often for use in a local context, such as the physician's hospital. These protocols are of varying degrees of sophistication, and most are not published or widely disseminated. Some individually sponsored protocols, however, have been developed by nationally known physicians for prestigious hospitals and are published in national medical journals. Examples of such protocols are the standard for minimal patient monitoring during anesthesia that the Department of Anesthesia at Harvard Medical School developed for use in its teaching hospitals [122] and the standards for arterial blood gas analysis developed at Stanford University School of Medicine [54].

One type of individually sponsored protocol that is proliferating in recent years is the commercially prepared consultation guide for treatment or diagnosis of specific disease. Such consultation guides are now available for a number of diseases including diagnosis and treatment of glaucoma [123] and diagnosis of particularly challenging cases in internal medicine [124-128].

Some of these services are intended to operate like an electronic textbook of medicine [129]. Subscription services are also available for selection of drugs to treat specific conditions [130,131]. While technically not clinical practice protocols, these consultation aids suggest normative approaches to making diagnosis or treatment decisions and have generated concern within the medical profession about their proper use from a legal and ethical perspective [132].

B. Utilization Review Protocols

Utilization review is an established quality assurance and cost containment strategy. Its goal is to assess whether the health-care services in a given case are medically necessary and appropriate. Basically, the utilization review process involves comparing the use of health-care services for a particular patient against some established norm for the utilization of similar services for comparable patients.

A wide variety of utilization review protocols is in use today. Like clinical practice protocols, many are developed by or for individual payers either commercially or, in limited cases, with HCFA funding. Predominantly, payers use the protocols to control the utilization of hospital inpatient and ancillary services. Utilization review protocols are chiefly used in two different contexts: (1) before or during the course of an admission to determine whether continued stay or a particular treatment is medically appropriate, and (2) after discharge, to determine for payment purposes whether the stay or services rendered were medically appropriate. The protocols have an impact on clinical practice because physicians often feel pressure to conform their treatment of patients to the utilization review protocol.

HCFA has served as a strong catalyst for the development and use of utilization review protocols. By law, peer review organizations (PROs) must develop standards for the utilization of inpatient hospital services in their areas and for quality of care services [133]. PRO regulations and program instructions require PROs to specify norms, criteria, and standards for review and prescribe the manner in which they must be established [134]. Also, other Medicare contractors, with HCFA's encouragement, and in the case of home health benefits by statute [135], often use "screens" to identify claims for more conventional services in which the items or services provided exceed preset norms as to durations, frequency or intensity of the utilization of such services [136,137]. These screens are used for a wide variety of Medicare benefits including physician, home health, and skilled nursing home services.

Recent years have witnessed the development and widespread dissemination of several utilization review protocols. One of the most widely used standards is the Appropriateness Evaluation Protocol (AEP), developed by Gertman and Restuccia at Boston University with HCFA funding [138,139]. The AEP measures whether a particular admission or day of stay is appropriate given the patient's treatment and use of services. The AEP is generally used by Blue Cross plans and other insurers, particularly those using cost-based reimbursement payment methodologies, for pre-admission or concurrent review of hospitalization as well as postdischarge review for payment purposes.

Several utilization review protocols also measure severity of illness and account for more variables regarding the patient's condition in order to identify more accurately the appropriateness of hospitalization and the use of services for particular patients. For example, the ISD Review System developed by J. Lamprey at InterQual for use by hospital utilization review committees not only measures whether services are medically necessary and whether services provided are rendered in an appropriate setting, but also whether quality of services meets profesionally recognized standards of care. The ISD Review System expressly takes into account the patient's diagnosis as well as the intensity of service needed. The AS-SCORE severity of illness classification, developed by Roveti, Horn, and Kreitzer [140-142], uses five factors—patient age, organ system involved with the patient's disease, disease state, complications, and response to therpay—to measure the severity of illness for purposes of utilization review.

III. MEDICAL STANDARDS AND THE LEGAL STANDARD OF CARE

The increasing number and use of clinical practice and utilization review protocols raise critical questions about their potential role in medical malpractice litigation, including whether and how they may be used to establish the standard of care for a defendant physician in a medical malpractice lawsuit. This aprt of the chapter explains how the standard of care is now established in medical malpractice lawsuits and analyzes how the various types of medical standards described above could be used in establishing the applicable standard of care.

A. Establishing the Legal Standard of Care in Medical Malpractice Litigation

Medical malpractice is the tort of negligence committed by physicians and other health care professionals. The *Restatement Second of Torts* defines negligence as "conduct which falls below the standard established by law for the protection of others against unreasonable risk of harm" [143]. In order to recover damages in negligence, three elements must be present: the plaintiff must be damaged, the defendant must be at fault, and the defendant's fault must be the legal cause of the plaintiff's damage [144]. To show fault, a plaintiff must prove that a defendant breached the legal duty not to expose the plaintiff to a reasonably foreseeable risk of injury [144].

1. The Standard of Care in Conventional Negligence

In conventional negligence actions, breach of a defendant's duty not to expose the plaintiff to a reasonably foreseeable risk of injury is demonstrated by establishing that the defendant did not meet the applicable standard of care. There are several ways to establish the standard of care in conventional negligence cases including statute, regulation, or prior judicial decision [143, Sec. 285].

The most prevalent standard of care by which the defendant's conduct is measured is that of the "reasonable person" under like or similar circumstances [143, Sec. 283]. The trier of fact, generally the jury, has the responsibility of determining compliance with the reasonable person standard. The jury decides whether the defendant should have recognized the potential risk of his conduct [143, Secs. 289, 290] and whether his judgment is assessing the potential risk was reasonable [143, Sec. 291]. In making this determination, the jury assesses whether a reasonable person would have determined that the utility of the conduct giving rise to the accident outweighed the magnitude of the risk posed by the conduct [143, Secs. 291-293] and whether the probability of the risk was significant enough to modify his conduct [145,146].

Industry customs and standards of industrial organizations have long been used as evidence to assist a jury and court to establish the standard of care [147-149]. The theory for this approach was aptly stated by Justice Oliver Wendell Holmes: "What usually is done may be evidence of what ought to be done" [150]. However, courts need not find that custom is conclusive evidence of the standard of care, for as Justice Holmes continued: "[W]hat ought to be done is fixed by a standard of reasonable prudence, whether it is usually complied with or not" [150]. Indeed, courts tend to view industry custom with circumspection and many courts have found that a custom may itself be found to be negligent [151-155; also see next subsection].

2. *The Standard of Care in Medical Malpractice Cases*

The standard of care, as well as its proof and application, in a medical malpractice case is markedly different than in the conventional negligence case. Negligence law leaves the definition of the standard of care and the determination of breach of the standard to members of the medical profession. As one commentator observed: "The open-ended task of planning reasonable medical care is not attempted in court, but is delegated to the collective managerial authority of the medical profession" [156].

The reason for this approach is that the practice of medicine involves a body of knowledge outside the scope of knowledge of the average lay person. Thus, only a physician, as an expert witness, may testify as to the applicable standard of care and give an opinion as to whether or not the defendant breached that standard. The plaintiff must have a medical expert witness testify at trial to establish a prima facie case of negligence against the physician defendant. The defense is then generally under considerable pressure as a practical matter to put on contrary expert testimony. The judge and jury have no role in evaluating the defendant physician's conduct directly but rather only evaluate the persuasiveness of the expert testimony in light of all other evidence [154, p. 364].

In general, the standard of care for a physician in a medical malpractice case is that degree of care exercised by physicians of good standing and of the same "school" as the defendant physician [149]. But most states impose specific limitations as to whom the defendant physician will be compared with based on either the defendant physician's geographic location or, more recently, the defendant physician's field of specialization [157].

Until the middle of this century, most states established the standard of care by comparing the defendant physician to other physicians in the same locality [149,158]. This "strict locality rule" sometimes proved unworkable because local physicians would often refuse to testify against their colleagues, making it impossible for plaintiffs to establish the local standard of care and its breach. Moreoever, a small group of practitioners in an isolated community could establish a standard of practice in that community that was lower than that required in other communities—an unattractive possibility given the development of transportation and communication that should enable physicians in remote communities to learn about and adopt mainstream medical advances.

Most states have now modified the strict locality rule. They hold physicians to the standard of physicians practicing in the same or a similar locality [157,159,160]. It has been suggested that, because of modern communication and transportation, all physicians should be held to national standards of care [160]. Although no state has moved wholly to such national standards for all physicians, almost all states have adopted a national standard of care for specialists [161].

The standard of care in malpractice cases is actually based much more on "industry" custom than it is in conventional negligence cases [154, p. 358; 162, 163]. In present times, this situation is perhaps inevitable since physician testimony is necessary to establish the standard of care. Many physicians rely on how they would have conducted themselves or how they believe other physicians in the applicable comparison group would have conducted themselves in the particular situation at issue. This is particularly true if there are no standards, recommendations, or guidelines published by medical specialty societies, physician groups, or an acknowledged medical text to guide the testifying physician. As a result, a defendant physician often is held to a standard of care that reflects the "habit" of the medical expert testifying [154, pp. 361-362]. Some commentators have decried this situation and have exhorted courts to move toward a standard of care based on what ought to be done, rather than relying on "customary" practice [163, p. 1213].

Not all courts, however, accept "customary" practice as an absolute determinant of the standard of care in medical malpractice cases, and in some instances courts have concluded that "customary" practice in itself is negligent [154, pp. 456-457; 164, 165]. In *Helling v. Carey* [166], the most famous example of such a judicial departure from the usual rule, the trial court ruled that compliance with customary practice within the medical profession is not conclusive evidence that a physician was not negligent. In *Helling*, defendant physicians did not test the young female plaintiff for glaucoma for 5 years while she was under their care and, when she was finally tested at age 32, some vision had been irreparably lsot. Both plaintiffs' and defendants' medical experts testified that physicians do not customarily test for glaucoma for patients under 40 years of age. Nevertheless, the Supreme Court of Washington concluded that the jury could find the defendants liable for negligence even though the physicians followed an established professional custom [167].

3. *Governing Evidentiary Principles*

A very important aspect in understanding the current and potential use of medical standards in malpractice litigation is the rules of evidence that govern the introduction of evidence of any standards at trial. As indicated above, only a medical expert witness can testify to the applicable standard of care, whether the defendant's conduct did or did not breach that standard, and, in many cases, whether the defendant's breach was the legal cause of the plaintiff's damage.

The applicable evidentiary principles depend on how the medical standard is to be used at trial. The first distinction is whether a witness will testify as to the standard of care and use the medical standard as a resource, or will testify that the standard itself should be admitted as documentary evidence of the standard of care. The second important distinction is whether the medical standard is to be used to establish the standard of care or to impeach the testimony of an opposing expert witness. In any event, two requirements must be met before the medical standard, or any other evidence, can be used as proof at trial: the evidence must be demonstrated to be both authentic and admissible under the principles of evidence law for the jurisdiciton [168,169].

If a document contains a standard that a party wishes to offer as substantive evidence, the normal procedure is to use a witness with knowledge of the standard and its origin to verify the authenticity of the document's origins and content and to explain how the document should be used [168, Secs. 321, 324.2; 169, Sec. 1694]. This witness, whom the court must qualify as an expert witness, will describe the standard, its development, and that the standard is accepted by the applicable professional group sponsoring the standard [168, Sec. 321, p. 899; 169, Sec. 1694].

The critical barrier to admissibility of evidence regarding industry customs is the hearsay rule. This rule bars all out-of-court statements offered as proof of the matter contained therein as evidence at trial [168, Sec. 321, p. 899; 169, Sec. 657]. Documents describing standards are technically "hearsay" evidence because they are effectively out-of-court statements offered for the truth of the matter that the standards are industry customs [168, Sec. 321, p. 899; 169, Sec. 657]. However, most courts generally admit documentary evidence of industry standards under an exception to the hearsay rule called the learned treatise exception [168, Sec. 321, p. 900; 169, Secs. 1690-1700]. This exception is defined in the Federal Rules of Evidence as follows:

> To the extent called to the attention of an expert witness upon cross-examination or relied upon by him in direct examination, statements contained in published treatises, periodicals or pamphlets on a subject of history, medicine, or other science or art, established as a reliable authority by the testimony or admission of the witness or by other expert testimony or by judicial notice, if admitted, the statements may be read into evidence but may not be received as exhibits. [170]

Most states, either judicially or legislatively, have adopted the learned treatises exception to the hearsay rule [168, Sec. 321; 169, Secs. 1690-1698]. But jurisdictions tend to permit evidentiary use of established treatises, including medical treatises, only when the major policy considerations for invocation of the hearsay rule can be met [168, Sec. 321; 169, Secs. 1690-1698; 171].

Some states have statutes that specifically permit, at the court's discretion, the admission of learned medical treatises in medical malpractice litigation without accompanying expert testimony [172]. States with such statutes include Massachusetts, Nevada, Kansas, and Rhode Island [173]. The express purpose of these statutes is to permit plaintiffs in medical malpractice cases to introduce a treatise into evidence liberally in order to establish the proper standard of care for a physician without the need for direct expert testimony to establish the treatise as a reliable authority [173-176].

The oldest statute is that of Massachusetts [177], and it appears that the statute, in keeping with legislative intent, has not operated to permit admission of treatises into evidence without accompanying medical testimony [178]. Specifically through the statutory provision according judicial control over the operation of the exception, Massachussetts courts have severely limited the ability of trial counsel to introduce treatises to establish the standard of care or other facts without accompanying testimony of a medical expert [178, pp. 24-26]. For example, in *Reddington v. Clayman* [179], the Massachusetts Supreme Court upheld the exclusion of medical treatises on grounds that plaintiff's counsel failed to prove that the authors of the treatise were experts on the subject despite the proffer of biographical data from the *Directory of Medical Specialists* and an English version of *Who's Who* [178, pp. 21-24].

Clearly, it is possible for medical standards to get into evidence in a malpractice suit to establish the standard of care in a malpractice lawsuit. Indeed, in the seminal case, *Darling v. Charleston Community Memorial Hospital* [180], recognizing malpractice liability for hospitals, the Illinois Supreme Court ruled that the hospital accreditation standards for the Joint Commission for the Accreditation of Hospitals (JCAH), were admissible as evidence of minimum standards for the hospital's conduct [180, p. 332, 211 N.E.2d at 257; 181]. Subsequent to *Darling*, the Supreme Court of Minnesota ruled JCAH standards were admissible evidence on the issue of accepted hospital practice [181, pp. 729-730; 182].

However, the learned treatise exception to the hearsay rule is viewed conservatively by most courts, including its statutory corollary aimed specifically at the use of learned treatises in malpractice litigation to prove the standard of care. This conservatism results in judicial reluctance to admit learned treatises when clear evidence of the statute and veracity of the treatise does not mitigate the basic policy of the hearsay rule—prohibiting out-of-court statements as evidence for the truth of the matter asserted when there is no opportunity for cross-examination. Thus, and most important, courts are reluctant to admit treatises as evidence without accompanying expert testimony as to the qualifications of the treatises' authors to serve as experts.

As a practical matter, whether the learned treatise exception and its corollary would permit admission of a medical standard would depend largely on the standard's authorship. Certainly the standard would have to be written by a physician. Further, the physician author would probably have to be in the medical specialty that dealt with the medical problem addressed. It would likely be necessary to have a medical expert testify as to the authenticity of the standard, the expert nature of the author's credentials, and the reputation of the author as expert. Also, it would be necessary in most cases to have a medical expert witness testify as to whether the standard would actually apply to the medical situation involved in the malpractice case and then as to whether it would establish the standard of care in that situation.

It is essential to appreciate that at this point it would be unlikely that most courts would accept medical standards as evidence of the standard of care without accompanying medical expert testimony. Thus, as a practical matter, the "mechanics" of putting on proof of the standard of care in a malpractice lawsuit are fundamentally unaffected by the use of medical standards to establish the standard of care.

B. Use of Medical Standards in Medical Malpractice Litigation

Nevertheless, medical standards can be used in medical malpractice trials as evidence of the standard of care. In the past, expert witnesses would testify as to the standard of care and breach of the standard. A medical standard brought to the attention of a court or jury may have a much greater impact because it was developed by physicians and, perhaps, endorsed by a large number of prestigious physicians.[3] Thus, with the development of clinical practice and utilization review protocols, new "experts" have emerged.

The influence of a given standard in a medical malpractice suit is dependent upon three factors: the standard's intended use, its sponsorship, and its character. The purpose of a standard may be critical in determining its role in malpractice litigation. For example clinical practice protocols prescribing the course to take in a clinical situation will be persuasive in a malpractice case involving the same clinical situation much in the way that industry standards are used in conventional tort litigation. But if the purpose of the standard is not to describe specifically what steps to take in the medical care of a particular disease, then it may be less useful in establishing what standard of care is required of a physician in terms of specific steps in the management of the plaintiff patient. However, a standard that prescribes certain requirements that should be present in a facility could be probative evidence of negligence on the part of the institution if a sufficient causal link can be demonstrated between the deficiency and the injury. For example, if the standard prescribes certain requirements as to the level of nursing staff that should be present when a general anesthetic is administered, a deficiency of nursing staffing could be probative evidence of negligence if the deficiency is shown to be causally related to the patient's injury from anesthesia.

It is less likely that utilization review protocols would be used effectively to establish the standard of care in malpractice litigation because they define standards for the use of resources rather than prescribe procedures for managing a particular disease or condition. However, this is not to say that they could never be used to establish the standard of care in a malpractice suit. For example, if a reputable utilization review protocol recommended inpatient hospital care for a plaintiff's condition and the plaintiff was not admitted to the hospital and suffered injury as a result, the utilization review protocol might be persuasive evidence that hospitalization was both customary and medically necessary.

In one important recent case, *Wickline v. State* [183-185], the plaintiff was discharged inappropriately but in accord with a utilizaiton review protocol of the California Medicaid program. The plaintiff, who developed complications following discharge that resulted in the amputation of her leg, sued the California Medicaid program for bad faith in denying her benefits. The California appellate court, reversing the trial court, refused to impose liability on the state. The court emphasized that the treating physician had an obligation to request the extended hospitalizaiton he thought necessary, even if he felt the Medicaid officials would not authorize it. The court stated:

> There is little doubt that Dr. Polonsky was intimidated by the Medi-Cal program but he was not paralyzed by [the reviewing physician's] response nor rendered powerless to act appropriately if other action was

---

[3]This situation may apply to a standard adopted by a medical specialty society, government, or a voluntary health organization, as discussed in Section II.A.1.

> required under the circumstances. If, in his medical judgment, it was in his patient's best interest that she remain in the acute care hospital setting for an additional four days beyond the extended time period originally authorized by Medi-Cal, Dr. Polonsky should have made some effort to keep Wickline there. . . . It was his medical judgment, however, that Wickline could be discharged when she was. . . . [186].

Thus, *Wickline* suggests that a doctor may not defend himself by claiming that his own professional judgment was swayed by fear of an adverse utilization review decision.

The persuasiveness and, in some cases, the use of both utilization review and clinical protocols at trial is largely dependent on their sponsorship. Protocols developed by national medical specialty societies and other groups of prestigious physicians may be highly influential because of their origins and imprimatur of approval. It is intuitively obvious that a jury of lay people would be impressed by standards sponsored or authored by organizations with which they were familiar, such as the American Cancer Society, or by organizations of prestigious physicians like the American College of Physicians.

Individually sponsored, institution-specific standards are generally intended to be minimal standards of quality care for *that* institution [54, p. 390; 122, p. 1017] and, because they are institution-specific, may not be generally accepted within the medical profession as a whole. Thus, it would be easier to vitiate the persuasiveness of these standards as one physician's opinion of quality and not as a generally accepted standard within the profession. In some cases, however, because of the prestige of the institution for which they are developed, the prestige of the authors, or the fact of their publication in national professional journals, these standards may be persuasive evidence of the standard of care in a medical malpractice suit outside their intended area of influence. Still, such standards might not serve as conclusive evidence of the appropriate standard because they are institution-specific and their authors have presumably made no attempt to identify conventional practice in the applicable geographic area.

On the other hand, individually sponsored standards may play a crucial role in a suit against a physician in the institution for which the standard was created, particularly if the institution can impose some kind of sanction for failure to comply with the standard. For example, an anesthetist at a hospital affiliated with Harvard Medical School is required by hospital policy to follow the standards developed for the anesthesiology department by Harvard Medical School physicians. In fact, courts have found a defendant negligent for failure to follow his own hospital's policy [187,188].

Commercial consultation guides, if developed by physicians, may have the effect of other clinical practice protocols. However, if they are not established by a nationally recognized association of physicians or an acknowledged expert in the field, they should be more readily challenged as hearsay evidence. In any event, they have to be introduced by expert testimony at trial from a physician witness who would essentially endorse the commercial consultation guide.

The final critical factors influencing the persuasiveness of a medical standard as evidence of the standard of care in a medical malpractice suit are the character of the standard and whether or not compliance with the standard is somehow enforceable. Most medical standards are not enforceable but are, in effect, only guidelines or recommendations. On the other hand, the designation of a standard by a national medical specialty society for a particular clinical practice is persuasive as to what "ought" to be the practice, if it is not in fact the accepted or customary practice in the nation or the community.

The medical profession's increasingly active role in the development of medical standards and particularly clinical practice protocols and the use of these medical standards to establish the standard of care in malpractice cases are highly desirable. The medical profession, the health-care industry, and patients are well served by this development. As discussed earlier, too much deference is accorded to "custom" evidence in medical malpractice cases [154,163-165,189]. Without stated medical standards, there is no assurance that the standard of care being applied in a malpractice case reflects the preferred or proven mode of practice with respect to the medical situation presented in the case. Indeed, as pointed out above, in the worst but not unusual case, the defendant physician may be held to a standard of care that may only reflect the "habit" of the medical expert testifying rather than an established standard of the profession.

If the leaders in the medical profession established standards for the diagnosis and treatment of various diseases as well as for the performance of specific procedures and medical experts relied on those standards in testifying at trial, defendant physicians would have a greater chance of being held to an appropriate standard of care. Further, over time physicians would become better informed as to what these standards are and would have greater incentive to conform their clinical practice to these standards. In addition, payors would have better information about what constitutes quality health care in specific situations and could tailor their utilization review and payment policies to promote desirable practices.

IV. CONCLUSION

Utilization review and clinical practice protocols as well as other medical standards are now an established factor in the practice of medicine. Their role in medical malpractice litigation is one of increasing influence. Their use and influence depend on the sponsorship or authorship of the standard and the degree to which it applies to the given treatment situation at issue in the lawsuit. Given the constraints imposed by the rules of evidence in most states and the long history of using expert witnesses to establish medical "custom" and, thus, the standard of care in medical malpractice cases, the procedures by which medical standards are introduced into evidence and used in medical malpractice cases will not change substantially from current practice.

Nevertheless, because of the availability of medical standards and the increasingly enthusiastic embracing of these standards by medical professional organizations, medical standards will certainly play an increasingly important role in establishing the standard of care in medical malpractice cases. This is a positive development. Medical standards, especially those clinical practice protocols authored or sponsored by medical specialty societies and organized medicine, are more likely to reflect standards for high-quality care.

POSTSCRIPT AND COMMENT ON NATIONAL PRACTICE STANDARDS

NANCY E. CAHILL *McDermott, Will & Emery*

When the editor first sketched the contents of this book, he decided that a piece on medical technology assessment (beyond that done by the FDA) and its implications for device manufacturers would be useful. I had just started to write that chapter in the summer of 1988, when Eleanor Kinney, a colleague for a year of guest teaching at the Indiana University Law School, asked me to review an early draft of the chapter above. After reading the paper, I shifted gears and called the editor to tell him that he would be getting a somewhat different, but important, chapter.

When the editor read the Kinney/Wilder chapter he loved Section I—The Typology of Medical Standards. It is a marvelous catalogue of the standard-setting activities developing in the academic, medical, and payor communities. But what would device industry people get out of reading Section II—Medical Standards and the Legal Standard of Care? My answer was immediate: first, a better understanding of the changing environment in which their product users must operate, and second, an appreciation of the impact that clinical practice protocols (developed by medicine or payers) may have on product planning and markets.

After all, the accuracy or inaccuracy of the central thesis of this piece is very important to device manufacturers, particularly those committed to innovation: Will medicine's practice protocols "more likely reflect standards for high quality care" *and* should they be imposed on physicians and other providers by the courts even if a particular standard (in the words of Kinney and Wilder) "is not in fact the accepted or the customary practice in the nation or the community"? I have several concerns with the authors' policy position and comments about alternative scenarios that may be equally likely to occur.

First, how comfortable should we be with medicine setting new standards for patient care, especially given our constricting financial resources? Can we be confident that medicine's standard-setting programs will be competent and unbiased? Since third-party payors are a primary user of medicine's published technology assessments, there may be an unavoidable economic bias in some organizations' recommendations. The classic example is the long-stand-standing refusal of the American College of Obstetrics and Gynecology (ACOG) to move from a 1-year to a 3-year interval for recommended Pap testing despite compelling empirical data and countervailing professional opinion supporting the safety and cost effectiveness of this change in the standard of care.

My second concern is even more fundamental and should really shake device manufacturers. What if cost containment policies cause a "technology backlash" among both physicians and payers? Testimony offered by the Director of the Office of Management and Budget (OMB) during 1989 hearings on the federal budget for fiscal years 1990 and 1991 clearly indicated that government analysts believe that the 1980s cost spiral was technology driven. OMB's official estimate was that

Growth of the beneficiary population accounted for 1.9% of the yearly increase in federal spending for physician services.

Aging of the Medicare population may have accounted for another 1% of the increase.

The increasing volume of services per beneficiary accounted for the remaining 6% of the growth not accounted for by inflation, producing a total estimated 15% a year increase from 1980 to 1989.

One long-time health policy analyst has even suggested that highly visible rationing of new technologies (and expensive established technologies) will be needed to contain this cost spiral [1].

In the past, physicians and the allied health professions have unswervingly opposed attempts by government or the private-sector payers to write a "cookbook" for medicine or force physicians to write one for their cohort. Medicine has usually won these policy battles, and industry and innovation have been the beneficiaries of those victories. Now the balance of power in the health policy world is changing rapidly and dramatically. The payors, not providers, have the upper hand.

Physicians are suddenly eager to write clinical practice and utilization guidelines. In 1989 congressional negotiations, the American Medical Association (AMA) offered substantially more effort to develop recommendations about "appropriate" practice in order to try to head off the imposition of expenditure limits ("expenditure targets") for physician services under federal programs. Though budget policy is still in flux at this writing, it is highly likely that physicians will face mandates for *both* more efficiency via practice guidelines and expenditure ceilings in publicly funded programs.

Is it possible that medicine and the health-care professions will join the payors in technology bashing? If so, how will they do it? One possibility, already evident in some of medicine's statements opposing the adoption of new technologies,[1] would be a call for a rollback in some standards of practice. Perhaps Kinney and Wilder were wrong about the inevitably beneficial consequences for patients of a move to *national* standards of care. Perhaps medicine will join the payors in opposing innovation in care that may be quality enhancing because the costs of any incremental gains are just too high.

Given their overriding medical ethic (i.e., to help without harming), physicians are unlikely to move wholesale to such rollbacks. On the other hand, the possibility of less than enthusiastic medical acceptance has to be considered whenever a new technology is being developed that may put physicians in a cost-benefit squeeze. To avoid being caught in the middle of such controversies, some of medicine's standard-setting bodies are moving, as the payors are demanding, to a requirement that proponents of a new technology demonstrate both positive "health outcomes" and the cost-effectiveness of their offering before the "experimental" label will be removed and third party payment encouraged.

---

[1]ACOG opposed FDA approval of a home uterine activity monitoring device in 1989, despite study data showing a significantly lower incidence of preterm births with home monitoring, because of cost concerns: "if 30% of pregnant patients had such a program the cost would be in the range of $5.6 billion per year in the U.S." Similarly, the American College of Radiology (ACR) has attempted since their introduction in 1985 to ration the use of the (six- to tenfold safer) nonionic contrast agents because of many payors' refusals to cover radiologists' greatly increased drug costs.

These shifts in medicine's policies will have dramatic consequences for innovators in the device industry. For example, under long-standing FDA policies, premarket approval of new technologies may be obtained by a showing that the technology provides clinical benefits considered significant by the professional community. Under payors' "outcomes" rhetoric, however, many FDA end points are merely "intermediate outcomes," benefits important to clinicians such as lowered blood pressure or increased blood flow that may or may not have any impact on final outcomes important to patients (and payors) such as avoiding a stroke or staying alive. In addition, sponsors of a new technology subject to FDA review usually choose the end points (label claims) they wish to make and have to prove in clinical trials.

Under "outcomes" policies being promoted by medicine and payors, manufacturers and other innovators face both a loss of control over the end points to be proven (ie.., the benefits deemed important) and significant new burdens of proof before they can hope to begin recouping their investment in product research and development. Indeed, the payors' policy push toward "outcomes" analysis is so dominant that it increasingly is entering the FDA's processes as well. Genentech's experience in seeking approval of tissue-type plasminogen activator (t-PA) illustrates what can happen.

Did the firm have to "prove merely that t-PA opened clogged arteries or also that its use reduced illness and death?" [3]. In a move that shocked industry and the FDA leadership (and cost Genentech many months and dollars), FDA senior staff and the advisory panel opted for the latter standard. DA staffers "found 'a paucity of conclusive data that either mortality or morbidity is significantly altered by the reestablishment of blood flow' [and] expressed doubt whether the FDA should clear [the drug] simply on the 'suspicion' that it prolongs life" [3, p. 16]. Anecdotal reports from both the drug and device industry suggest that FDA reviewers increasingly may be moving toward this type of "outcomes" analysis, despite the objections of product sponsors and the concerns of the agency leadership.

Let's summarize the concerns expressed so far with the Kinney/Wilder thesis, that increased standard setting activity by medical organizations and national standards of care are positive developments:

1. Some of medicine's technology assessment and standard setting programs may not be sufficiently fair, open, and unbiased, given all of the forces at play in the policy world.
2. Medicine's practice standards may increasingly reflect a "technology backlash," particularly in a world where product expenditures may compete directly with physicians' income under an expenditure target *and* physicians are given almost complete control over how the fund will be spent.
3. Medicine may respond to concerns about aggregate costs and cost increases by attempting to roll back (or hold back) a standard of care in some instances, even where there may be demonstrable patient benefits.
4. Medicine and the payors may combine to force proponents of new technologies to prove positive "health outcomes" and the cost-effectiveness of their offerings before a technology will be eligible for third party payment.

This list should be sufficiently daunting to most manufacturers without my adding one last concern: In those instances where medicine's clinical protocols do attempt to advance the state of the art, will the payors concur in their coverage and payment decisions? The coverage and payment arena is truly where physicians' malpractice concerns and manufacturers' market concerns mesh most closely. Experience to date suggests that we have reason to be concerned.

Despite radily apparent and appreciable patient benefits, many payors are stonewalling coverage of highly attractive new technologies, favoring cost containment over patient preferences (e.g., laser angioplasty for peripheral vascular disease to replace bypass graft surgery or amputation). As one bemused physician has noted, "for carriers, inefficiency (bungling, confusion and delay) is profitable" [4]. This is why it is important for manufacturers to work closely with the relevant medical specialty groups to make sure that payors are held accountable to patients and policymakers for their noncoverage policies and payment limitations.

In just 3 years, many payors' threshold criteria for accepting a new technology have risen dramatically, moving from the first to the third question presented here and substantially increasing the time and capital needed to obtain a favorable coverage decision:

1. *Can* this technology be used safely and effectively (the FDA standard)?
2. *Is this technology being used* safely and effectively?
3. *Does the refereed literature "prove conclusively"* that this technology is safe and cost-effective?

What happens to physicians, patients, and product vendors when the professional community and the payor community do not agree on what constitutes "medically necessary" care? For example, on an individual level, what if the American Diabetes Association says that a brittle juvenile diabetic needs a blood glucose meter with data analysis capabilities in order to assure optimum patient management, but Medicaid or private reimbursement for durable medical equipment is inadequate to pay for that device?

What is such a child's treating physician to do? Does he or she disregard the patient's economic circumstances and insist that the family make up the insuracne shortfall? Of is it better for physician and patient to adapt and make do with what the system will provide? If so, at what risk? And imagine the pressure on manufacturers of such devices to bring prices down to meet what the lowest-level payor is willing to pay.

If we extend this example and analysis to the community or regional level, we begin to understand why medical malpractice law has historically avoided national standards of practice in favor of some sort of "locality" standard. We never have had and may never have uniform health-care resources throughout this country or even within one community. Therefore, a "resources-based caveat" to national standards of practice may be the dominant theme of health policy in the 1990s.

Let me end on a point of agreement. I do agree completely with Professor Kinney and Ms. Wilder that we will all need to monitor developments in the courts, as well as the legislatures, to determine precisely which way standards of care will be permitted to move. Device manufacturers in particular will need to stay abreast of these developments in order to adjust their R&D and market plans accordingly.

REFERENCES

1. U.S. General Accounting Office (GAO), Medical Malpractice: No Agreement on the Problems or Solutions (1986).
2. GAO, Medical Malpractice Characteristics of Claims Closed in 1984 (1987).
3. Department of Health and Human Services (DHHS), Report of the Task Force on Medical Liability and Malpractice 3-6 (1987).
4. Harris, Defensive Medicine: It Costs, But Does it Work?, *J.A.M.A.* *257*:2801 (1987).
5. Zuckerman, Medical Malpractice: Claims, Legal Costs, and the Practice of Defensive Medicine, *Health Affairs*, Fall:128 (1984).
6. American Medical Association (AMA), *Professional Liability in the 1980s* *3* (1985).
7. Arnett, Freeland, McKusick, and Waldo, National Health Expenditures, 1986-2000, *Health Care Financing Review*, Summer:1 (1987).
8. Payment reforms in recent years are of three types. See Kinney, Making Hard Choices Under the Medicare Prospective Payment System: One Administrative Model for Allocating Medical Resources Under a Government Health Insurance Program, *Indiana Law Review, 19*:1151 nn.1, 2, & 4 (1986). The first is rate regulation by a public authority or private insurer which, directed chiefly at institutional providers, regulates the amount paid for a unit of services, i.e., the price per case as under the Medicare prospective payment system, or even the entire amount the program will annually pay an institution under revenue caps or budget review strategies (p. 1151 n.1). In many but not all of these rate regulation schemes, the provider is put at risk for services provided patients in excess of preset norms upon which payment is based. The second type of payment reform is the preferred provider organization (PPO), which is an arrangement between selected providers and at least one group purchaser whereby the services of the providers are purchased for a specified group of individuals at a negotiated rate (p. 1151 n.2). The third type of payment reforms are those characterized by prepaid health plans, in which the consumer or someone on her behalf pays a fixed amount to the provider, and in return, the provider furnishes any volume of covered health care services regardless of cost (p. 1151 n.4). A Health Maintenance Organization (HMO) is an example of a prepaid health plan. A prepaid health plan is distinguished from conventional health insurance in that the provider rather than the health insurance company is at risk for the cost of services to beneficiaries over and above the premiums (p. 1151 n.2).
9. Peer Review Improvement Act of 1982, Pub. L. No. 97-248, 141-150, 96 Stat. 381, 385 (1982) (enacted as part of the Tax Equity and Fiscal Responsibility Act of 1982, Pub. L. No. 97-248, 141-150, 96 Stat. 324 [amending 42 U.S.C. 1320c-1 to 1320c-12 (1982 & Supp. III 1985); 42 C.F.R subpts. 461-478 (1988)]).
10. 42 U.S.C. 1396a (1982); 42 C.F.R. Subpart 456 (1988).
11. Jost, The Necessary and Proper Role of Regulation to Assure the Quality of Health Care, *Houston Law Review, 25*:525, 565-568 (1988).
12. Ellwood, Shattuck Lecture—Outcomes Management: A Technology of Patient Experience, *New England Journal of Medicine, 318*:1549 (1988).
13. Eddy, Costs: Impact on the Standard of Care, in *Medical Malpractice* (D. Yaggy and P. Hodgson, eds.), 1986.

14. Morreim, Cost Containment and the Standard of Medical Care, *California Law Review*, *75*:1719 (1987).
15. Note, Rethinking Medical Malpractice Law in Light of Medicare Cost-Cutting, *Harvard Law Review*, *98*:1004 (1985).
16. Council of Medical Specialty Societies, Standards of Quality in Patient Care: The Importance and Risks of Standard Setting, Sep. 25-26, 1987.
17. Department of Public Policy, American College of Physicians, Quality Assurance and Utilization Review (1984).
18. AMA, Diagnostic and Therapeutic Technology Assessment, DATTA: The Facts (1987).
18a. Technology Management Department, Health Benefits Management Div., Blue Cross & Blue Shield Association, Medical Necessity Program Diagnostic Testing Guidelines (1987).
19. Freudenheim, New Guidelines on Giving Tests in Medical Care, *The New York Times*, April 2, 1987, p. A12, col. 1.
20. Medicare Program, Health Care Financing Administration, Procedures for Medical Services Coverage Decisions; Request for Comments, *Fed. Reg.* *52*:15,560 (1987) (not codified in C.F.R.).
21. Kinney, National Coverage Policy under the Medicare Program: Problems and Proposals for Change, *St. Louis University Law Journal*, *32*: 869 (1988).
22. Ruby, Banta, and Burns, Medicare Coverage, Medicare Costs, and Medical Technology, *Journal of Health Politics, Policy and Law*, *10*:141 (1985). On January 30, 1989, HCFA issued a proposed rule outlining criteria and procedures for making medical services coverage decisions that relate to health care technology. [*Fed. Reg.*, *54*:4302 (1989).]
23. Brook, Lohr, Chassin, Kosecoff, Fink, and Solomon, Geographic Variations in the Use of Services: Do They Have Any Clinical Significance, *Health Affairs*, Summer:63 (1984).
24. Caper, Variations in Medical Practice: Implications for Health Policy, *Health Affairs*, Summer:110, 113-118 (1984).
25. Eddy, Variations in Physician Practice: The Role of Uncertainty, *Health Affairs*, Summer:74 (1984).
26. Wennberg, Dealing with Medical Practice Variations: A Proposal for Action, *Health Affairs*, Summer:6 (1984).
27. Schwartz, The Role of Professional Medical Societies in Reducing Practice Variations, *Health Affairs*, Summer:91 (1984).
28. Fowler, Wennberg, Timothy, Barry, Mulley, and Hanley, Symptom Status and Quality of Life Following Prostatectomy, *JAMA*, *259*:3018 (1988).
29. Wennberg, Freeman, and Culp, Are Hospital Services Rationed in New Haven or Over-Utilized in Boston?, *Lancet I*, *7*:1185 (1987).
30. Wennberg, Mulley, Hanley, Timothy, Fowler, Rous, Barry, McPherson, Greenberg, Soule, Bubolz, Fisher, and Malenka, An Assessment of Prostatectomy for Benign Urinary Tract Obstruction: Geographic Variations and the Evaluation of Medical Care Outcomes, *JAMA*, *259*:3027 (1988).
31. Wennberg, McPherson, and Caper, Will Payment Based on Diagnosis-Related Groups Control Hospital Costs?, *New England Journal of Medicine*, *311*:295 (1984).
32. Wennberg, Improving the Medical Decision-Making Process, *Health Affairs*, Spring:99 (1988).

33. Wennberg, Commentary: On Patient Need, Equity, Supplier-Induced Demand, and the Need to Assess the Outcome of Common Medical Practices, *Medical Care, 23*:512 (1985).
34. Brook and Lohr, Efficacy, Effectiveness, Variations, and Quality: Boundary-Crossing Research, *Medical Care, 23*:710 (1985).
35. Kosecoff, Brook, Fink, Kamberg, Roth, Goldberg, Linn, Clark, Newhouse, and Delbanco, Providing Primary General Medical Care in University Hospitals: Efficiency and Cost, *Annals of Internal Medicine, 107*:399 (1987).
36. Merrick, Fink, Park, Brook, Kosecoff, Chassin, and Solomon, Derivation of Clinical Indication for Carotid Endarterectomy by an Expert Panel, *American Journal of Public Health, 77*:187 (1987).
37. Park, Fink, Brook, Chassin, Kahn, Merrick, Kosecoff, and Solomon, Physician Ratings of Appropriate Indications for Six Medical and Surgical Procedures, *American Journal of Public Health, 76*:766 (1986).
38. Winslow, Solomon, Chassin, Kosecoff, Merrick, and Brook, The Appropriateness of Carotid Endarterectomy, *New England Journal of Medicine, 318*:721 (1988).
39. Chassin, Kosecoff, Winslow, Kahn, Merrick, Keesey, Fink, Solomon, and Brook, Does Inappropriate Use Explain Geographic Variations in the Use of Health Care Services?, *JAMA, 258*:2533 (1987).
40. Chassin, Brook, Park, Keesey, Fink, Kosecoff, Kahn, Merrick, and Solomon, Variations in the Use of Medical and Surgical Services by the Medicare Population, *New England Journal of Medicine, 314*:285 (1986).
41. Fink, Siu, Brook, Park, and Solomon, Assuring the Quality of Health Care for Older Persons: An Expert Panel's Priorities, *JAMA, 258*:1905 (1987).
41a. Eddy, Neugent, Eddy, Coller, Gilbertsen, Gottlieb, Rice, Sherlock, and Winawer, Screening for Colorectal Cancer in a High-Risk Population: Results of a Mathematical Model, *Gastroenterology, 92*:682 (1987).
42. Eddy, Hasselblad, McGivney, and Hendee, The Value of Mammography Screening in Women under Age 50 Years, *JAMA, 259*:1512 (1988).
43. Wasson, Sox, Neff, and Goldman, Clinical Prediction Rules: Applications and Methodological Standards, *New England Journal of Medicine, 313*:793 (1985).
44. Tomich, AFIP Unit to Study Liability Issues, Develop Protocol for "Prevention," *U.S. Medicine*, December:3, 17 (1987).
45. Council on Scientific Affairs, AMA, Dementia (1985).
46. Council on Scientific Affairs, AMA, Toxic Shock (1985).
47. Council on Scientific Affairs, AMA, Statement on Liver Transplantation (1986).
48. AMA, DATTA: An AMA Program in Medical Technology Assessment (1987).
49. Jones, The American Medical Association's Diagnostic and Therapeutic Technology Assessment Program, *JAMA, 250*:387 (1983).
50. Clinical Efficacy Assessment Program, American College of Physicians, Procedure Manual (1986).
51. Clinical Efficacy Assessment Project, American College of Physicians, Recommendations 1981-86 (1987).
52. Centor, Meier, and Dalton, Throat Cultures and Rapid Tests for Diagnosis of Group A Streptococcal Pharyngitis, *Annals of Internal Medicine, 105*:892 (1986).

53. Goldberger and O'Konski, Utility of the Routine Electrocardiogram Before Surgery and on General Hospital Admission, *Annals of Internal Medicine, 105*:552 (1986).
54. Raffin, Indications for Arterial Blood Gas Analysis, *Annals of Internal Medicine, 105*:390 (1986).
55. Sox, Probability Theory in the Use of Diagnostic Tests, *Annals of Internal Medicine, 104*:60 (1986).
56. Tape and Mushlin, The Utility of Routine Chest Radiographs, *Annals of Internal Medicine, 104*:663 (1986).
57. Lewin and Associates, Inc., A Forward Plan for Medicare Coverage and Technology Assessment B.23-B.25 (1987).
58. American College of Physicians, Common Diagnostic Tests: Use and Interpretation (1987).
59. Joint American College of Cardiology/American Heart Association Task Force on Assessment of Cardiovascular Procedures, Guidelines for Permanent Cardiac Pacemaker Implantation (1984).
60. Letter from H. D. Garretson, M.D., Ph.D., President of the American Association of Neurological Surgeons, to Mr. Francis Ohnsorg, Blue Cross & Blue Shield of Minnesota, Medical Policy Comm. (Oct. 26, 1987) (discussing reimbursement criteria for extracranial/intracranial bypass (EC-IC) procedures).
61. American College of Obstetricians and Gynecologists, Technical Bulletins (1988).
62. American College of Physicians, Position Statements (1986).
63. American Academy of Pediatrics, Report of the Committee on Infectious Disease (1988).
64. American Academy of Pediatrics, Publications and Services for Pediatric Health Care Professionals (1988).
65. American College of Surgeons, Statements on Principles (1985).
66. Pre- and Postoperative Care Communication, American College of Surgeons (ACS), Manual on Preoperative and Postoperative Care, 3rd ed. (1983).
67. Pre- and Postoperative Care Communications, ACS, Manual of Surgical Intensive Care (1977).
68. Subcommittee on Control of Surgical Infection, Pre- and Postoperative Care Communications, ACS, Manual on Control of Infection in Surgical Patients, 2d ed. (1984).
69. ACS, Publications and Services 1988 (1988).
70. American Society of Anesthesiologists, The Organization of an Anesthesia Department (1987).
71. American Society of Anesthesiologists, Ambulatory Surgical Facilities (1983).
72. American Society of Anesthesiologists, Guidelines for Critical Care in Anesthesiology (1986).
73. American College of Obstetrics and Gynecology, Technical Bulletins (1988).
74. Telephone interview with Karen Carter, Manager, Information Services Department, American Academy of Family Physicians (Jan. 8, 1989).
75. American Heart Association (AHA), Diet and Coronary Heart Disease (1978).
76. AHA, Diet Modification to Control Hyperlipedmia (1978).

77. AHA, Risk Factors and Coronary Disease: A Statement for Physicians (1980).
78. Committee on Electrocardiography and Cardiac Electrophysiology, Council on Clinical Cardiology, AHA, Task Force Recommendations for Standards of Instrumentation and Practice in the Use of Ambulatory Electrocardiology (1985).
79. Committee on Exercise, AHA, Exercise Testing and Training of Individuals with Heart Disease or at High Risk for its Development: A Handbook for Physicians (1975).
80. Committee on Rheumatic Fever and Infective Endocarditis, Council of Cardiovascular Disease in the Young, AHA, Prevention of Rheumatic Fever (1985).
81. Nutrition Committee, AHA, Dietary Guidelines for Healthy American Adults (1986).
82. Subcommittee on Exercise/Cardiac Rehabilitation, AHA, Statement on Exercise (1981).
83. Working Group to the Subcommittee on Smoking, AHA, Public Policy on Smoking and Health: Toward a Smoke-Free Generation by the Year 2000 (1988).
84. Subcommittee on Rehabilitation Target Activity Group, AHA, *The Exercise Standards Book* (1979).
85. American Diabetes Association (ADA), Guides; ADA, Physician's Guide to Insulin-Dependent (Type I) Diabetes: Diagnosis and Treatment (1988).
86. ADA, Physician's Guide to Non-Insulin-Dependent (Type I) Diabetes: Diagnosis and Treatment, 2d ed. (1988).
87. ADA, Long-Range Plan (1988).
88. American Cancer Society, Colorectal Polyps: Pathologic Diagnosis and Clinical Significance (1985).
89. American Cancer Society, Detecting Colon and Rectum Cancer (1984).
90. American Cancer Society, Dysplasia, Carcinoma in Situ, and Early Invasive Cervical Carcinoma (1984).
91. American Cancer Society, Mammography: Two Statements of the American Cancer Society (1984).
92. American Cancer Society, Mammography Guidelines 1983: Background Statement and Update of Cancer-Related Checkup Guidelines for Breast Cancer Detection in Asymptomatic Women Age 40-49 (1983).
93. American Cancer Society, Summary of Current Guidelines for the Cancer-Related Checkup: Recommendations (1988).
94. Office of Medical Applications of Research, DHHS, NIH, Guidelines for the Selection and Management of Consensus Development Conferences (1986).
95. Greenberg, Health Care Technology: A Small Office vs. a Big Problem, *New England Journal of Medicine, 302*:243 (1980).
96. Kosecoff, Kanouse, Rogers, McCloskey, Winslow, and Brook, Effects of the National Institutes of Health Consensus Development Program on Physician Practice, *JAMA, 258*:2708 (1987).
97. Perry and Kalberer, The NIH Consensus-Development Program and the Assessment of Health Care Technologies: The First Two Years, *New England Journal of Medicine, 303*:169 (1980).
98. Bernstein, National Institutes of Health Consensus Development Program, *Connecticut Medicine, 48*:513 (1984).

99. Mullan and Jacoby, The Town Meeting for Technology: The Maturation of Consensus Conferences, *JAMA*, *254*:1068 (1985).
100. Office of Medical Applications and Research, NIH, Prevention of Venous Thrombosis and Pulmonary Embolism (1986).
101. Office of Medical Applications and Research, NIH, Diet and Exercise in Noninsulin-Dependent Diabetes Mellitus (1986).
102. The Prevention and Treatment of Fine Complications of Diabetes: A Guide for Primary Care Practitioners, 1983.
103. Health Promotion and Disease Prevention Amendments of 1984, Pub. L. No. 98-551, 5, 98 Stat. 2815 [codified as amended at Public Health Service Act 305(e), 42 U.S.C. 242c(e) (1982 & Supp. IV 1986)].
104. 42 U.S.C. 242c(e)(1)(Supp. IV 1986).
105. National Advisory Council on Health Care Technology Assessment, Office of the Assistant Secretary for Health and Human Services, The Medicare Coverage Process (1988).
106. National Center for Health Services Research, Public Health Service, DHHS, Anglechilk Anti-Reflux Prosthesis (1986).
107. National Center for Health Services Research, Public Health Service, DHHS, Continuous Positive Airway Pressure for the Treatment of Obstetrical Sleep Apnea in Adults (1986).
108. National Center for Health Services Research, Public Health Service, DHHS, Full Automated Ambulatory Blood Pressure Monitoring of Hypertension (1986).
109. National Center for Health Services Research, Public Health Service, DHHS, Liver Transplantation (1983).
110. National Center for Health Services Research, Public Health Service, DHHS, Public Health Service Procedures for Evaluating Health Care Technologies for Purposes of Medicare Coverage (1983).
111. National Center for Health Services Research, Public Health Service, DHHS, Single Photon Absorptiometry for Measuring Bone Mineral Density (1986).
112. Pub. L. No. 94-295, 90 Stat. 539 [codified as amended at Federal Food, Drug, and Cosmetic Act 513-521, 21 U.S.C. 301, 360(c)-(k) (1982 & Supp. III 1985)].
113. Cahill, Technology Assessment: One View from American Medicine, *International Journal of Technology Assessment in Health Care, 3*: 693 (1985).
114. National Center for Health Services Research and Health Care Technology Assessment, Program Note/Patient Outcome Assessment Research Program Extramural Assessment Teams (Nov. 1988).
115. Roper, Winkenwerder, Hackbarth, and Krakauer, Effectiveness in Health Care: An Initiative to Evaluate and Improve Medical Practice, *New England Journal of Medicine, 319*:1197, 1200 (1988).
116. Roper and Hackbarth, HCFA's Agenda for Promoting High-Quality Care, *Health Affairs*, Spring:91 (1988).
117. Brook and Lohr, Monitoring Quality of Care in the Medicare Program, *JAMA*, *258*:3138 (1987).
118. Chassin, Kosecoff, Park, Winslow, Kahn, Merrick, Keesey, Fink, Solomon, and Brook, Does Inappropriate Use Explain Geographic Variations in the Use of Health Care Services, *JAMA*, *258*:2533 (1987).
119. Winslow, Solomon, Chassin, Kosecoff, Merrick, and Brook, The Appropriateness of Carotid Endarterectomy, *New England Journal of Medicine, 318*:721 (1988).

119a. Health Promotion and Disease Prevention Amendments of 1984, Pub. L. No. 98-551, 8, 98 Stat. 2815 [codified at Public Health Service Act 309(2), 42 U.S.C. 242n (1982 & Supp. 1985)].
119b. Institute of Medicine (IOM), Assessing Medical Technologies (1985).
120. Committee for Evaluating Medical Technologies in Clinical Use, IOM, Assessing Medical Technology 39 (1985).
121. Council on Health Care Technology, IOM, Interim Report (1987).
122. Eichorn, Cooper, Cullen, Ward, Maier, Philip, and Seeman, Standards for Patient Monitoring During Anesthesia at Harvard Medical School, *JAMA*, *256*:1017 (1986).
123. Weiss, Kulikowski, and Safir, Glaucoma Consultation by Computer, *Computer Biology and Medicine*, *8*:25 (1978).
124. First, Weimer, McLinden, and Miller, Localize: Computer-Assisted Localization of Peripheral Nervous System Lesions, *Computers and Biomedical Research*, *15*:525 (1982).
125. Masarie, Miller, and Myers, Internist-I Properties: Representing Common Sense and Good Medical Practice in a Computerized Medical Knowledge Base, *Computers and Biomedical Research*, *18*:458 (1985).
126. Miller, Internist-I/Caduceus: Problems Facing Expert Consultant Programs, *Methods in Information and Medicine*, *23*:9-14 (1984).
127. Miller, Poole, and Myers, Internist-I, An Experimental Computer-Based Diagnostic Consultant for General Internal Medicine, *New England Journal of Medicine*, *307*:468 (1982).
128. Van Melle, Mycin: A Knowledge-Based Consultation Program for Infectious Disease Diagnosis, *International Journal of Man-Machine Studies*, *10*:313 (1978).
129. First, Soffer, and Miller, QUICK (Quick Index to Caduceus Knowledge): Using the Internist-I/Caduceus Knowledge Base as an Electronic Textbook of Medicine, *Computers and Biomedical Research*, *18*:137 (1985).
130. Shortlife, Axline, Buchanan, Davis, and Cohen, A Computer-based Approach to the Promotion of Rational Clinical Use of Antimicrobials, *Clinical Pharmacy and Clinical Pharmacology*, 259 (1876).
131. Shortlife, Davis, Axline, Buchanan, Green, and Cohen, Computer-Based Consultations in Clinical Therapeutics: Explanation and Rule Acquisition Capabilities of the MYCIN System, *Computers and Biomedical Research*, *8*:303 (1975).
132. Miller, Schnaffner, and Meisel, Ethical and Legal Issues Related to the Use of Computer Programs in Clinical Medicine, *Annals of Internal Medicine*, *102*:529 (1985).
133. 42 U.S.C. 1153(c)(7), 1154(a)(6) (1982 & Supp. IV 1986).
134. 42 C.F.R. 466.100 (1988); see also 2 Medicare & Medicaid Guide (CCH) para. 12,870 (1987).
135. 42 U.S.C. 1395y(f).
136. GAO, Improving Medicare and Medicaid Systems to Control Payments for Unnecessary Physicians' Services (1983).
137. Issues Related to Medicare Contracting: Hearings Before the Subcommittee on Health of the House Committee on Ways and Means, 99th Cong., 2d Sess. (1986).
138. Gertman and Restuccia, The Appropriateness Evaluation, Protocol: A Technique for Assessing Unnecessary Days of Hospital Care, *Medical Care*, *19*:855, 868-869 (1981).

139. AEP Review Manager's Manual (1984). Another utilization review protocol which refines and combines the approaches of ISD and AEP is the Standardized Medreview Instrument (SMI) 5 developed by SysteMetrics, Inc. (1983).
140. Roveti, Horn, and Kreitzer, AS-SCORE: A Multi-Attribute Clinical Index of Illness Severity, 1984 QRB 472, 476.
141. Conklin, Lieberman, Barnes, and Louis, Disease Staging: Implications for Hospital Reimbursement and Management, *Health Care Financing Review*, November:13 (1984).
142. Louis and Gonnella, Disease Staging: Applications for Utilization Review and Quality Assurance, *Quality Assurance and Utilization Review*, February:13 (1986).
143. Restatement (Second) of Torts Sec. 282 (1977).
144. G. Christie, Cases and Materials on the Law of Torts 109 (1983).
145. United States v. Carroll Towing Co., 159 F.2d 169 (2d Cir. 1947).
146. Helling v. Carey, 83 Wash. 2d 514, 519 P.2d 981 (1974).
147. Shafer v. H.B. Thomas Co., 53 N.J. Super. 19, 146 A.2d 483, 485 (1958).
148. Pan Am. Petroleum Corp. v. Like, 381 P.2d 70, 76 (Wyo. 1963).
149. W. Prosser and J. Keeton, *Handbook of the Law of Torts*, pp. 193-196, 5th ed., (1985.
150. Texas & Pacific Ry. v. Behymer, 189 U.S. 468, 470 (1903).
151. The T. J. Hooper, 60 F.2d 737 (2d Cir. 1932), cert. denied sub nom.
152. Eastern Trans. Co. v. Northern Barge Corp., 287 U.S. 662 (1932).
153. Dempsey v. Addison Crane Co., 247 F. Supp. 584 (D.D.C. 1965).
154. Keeton, Medical Negligence—The Standard of Care, *Texas Tech. Law Review, 10*:351-354 (1979).
155. Morris, Custom and Negligence, *Columbia Law Review, 42*:1147, 1149, 1158 (1942).
156. Henderson, Expanding the Negligence Concept: Retreat from the Rule of Law, *Indiana Law Journal, 51*:467, 480 (1976).
157. Annotation, Modern Status of "Locality Rule" in Malpractice Actions Against Physician Who Is Not Specialist, 99 A.L.R.3d 1133 (1980).
158. Waltz, The Rise and Gradual Fall of the Locality Rule in Medical Malpractice Litigation, *De Paul Law Review, 18*:408 (1969).
159. Annotation, Malpractice Testimony: Competency of Physician or Surgeon from One Locality to Testify, in Malpractice Case, as to Standard of Care of Defendant Practicing in Another Locality, 37 A.L.R.3d 420 (1971).
160. Annotation, Standard of Care Owed to Patient by Medical Specialist as Determined by Local, "Like Community," State, National, or Other Standards, 18 A.L.R.4th 603 (1982).
160a. Shilkret v. Annapolis Emergency Hospital Association, 276 Md. 187, 192-199, 349 A.2d 245, 248-252 (1975).
161. Annotation, Standard of Care Owed to Patient by Medical Specialist as Determined by Local, "Like Community," State, National, or Other Standards, 18 A.L.R.4th 603, 607, 614-620 (1982 & Supp. 1988).
162. Keeton, Professional Malpractice, *Washburn Law Journal, 17*:445, 455 (1978).
163. King, In Search of a Standard of Care for the Medical Profession: The "Accepted Practice" Formula, *Vand. Law Review, 28*:1213, 1235-1236 (1975).

164. Toth v. Community Hosp., 22 N.Y.2d 255, 239 N.E.2d 368, 292 N.Y.S.2d 440 (1968).
165. Morgan v. Sheppard, 188 N.E.2d 808 (Ohio Ct. App. 1963).
166. 83 Wash. 2d 514, 519 P.2d 981 (1974).
167. The Washington legislature sought to mitigate the decision by statute although the Washington Supreme Court subsequently opined that the statute did not overrule Helling v. Carey. See Gates v. Jensen, 92 Wash. 2d 246, 595 P.2d 919, Rev'd 20 Wash. App. 81, 83, 579 P.2d 374, 376 (1979) (in which appellate court ruled that Wash. Rev. Code Ann. 4.24.290 overruled Helling v. Carey).
168. J. McCormick, *McCormick on Evidence*, Sec. 51 (Cleary 3d ed. 1984).
169. W. Wigmore, *Wigmore on Evidence*, Sec. 9 (Tillers rev. 1983).
170. Fed. R. Evid. 803(18); see also Imwinkelried, The Use of Learned Scientific Treatises Under Federal Rule of Evidence 803(18), *Trial*, Feb.:56 (1982).
171. Pierce, Admissibility of Expert Testimony in Hearsay Form, *American Journal of Trial Advocacy*, *5*:277 (1981).
172. J. Waltz and F. Inbau, *Medical Jurisprudence*, pp. 85-87 (1971).
173. Kansas Statutes Annotated 60-460(cc)(Supp. 1987).
174. Massachusetts Annotated Laws Ch. 233, Sec. 79 (Law Co-op. 1986).
175. Nevada Revised Statutes Sec. 52.040 (1985).
176. Rhode Island Public Laws. 1833 (1969).
177. Massachusetts' statute provides in pertinent part:
Statements of facts or opinions on a subject of science or art contained in a published treatise, periodical, book or pamphlet shall, in [the discretion of the court, and if] the court shall find that [they are] relevant and that the writer of such statements is recognized in his profession or calling as an expert on the subject, be admissible in actions of contract or tort for malpractice, error or mistake against physicians, surgeons, dentists, optometrists, hospitals and sanitaria, as evidence tending to prove said facts or as opinion evidence; provided, however, that the party intending to offer as evidence any such statements shall, not less than thirty days before the trial of the action, give the adverse party . . . notice of such intention, stating the name of the writer of the statements, the title of the treatise, periodical, book or pamphlet in which they are contained . . . [Massachusetts General Laws Annotated, Ch. 233, Sec. 79(C)(West 1986)].
178. Kehoe, Massachusetts Malpractice Evidentiary Statute—Success or Failure? *Boston University Law Review*, *44*:10 (1964).
179. 334 Mass. 244, 134 N.E.2d 920 (1956).
180. 33 Ill. 2d 326, 211 N.E.2d 253 (1965), aff'g, 50 Ill. App. 2d 253, 200 N.E.2d 149 (1964).
181. Schockemoehl, Admissibility of Written Standards as Evidence of the Standard of Care in Medical and Hospital Negligence Actions in Virginia, *University of Richmond Law Review*, *18*:725, 728-731 (1984).
182. Cornfeldt v. Tongen, 262 N.W.2d 684 (Minn. 1977).
183. 192 Cal. App. 3d 1630, 228 Cal. Rptr. 661 (1986).
184. Smith, Insurance Carrier Liability As A Result of Pre-Admission Screening and Hospital Stay Guidelines, *Ohio Northwest University Law Review*, *12*:189 (1985).
185. Note, Wickline v. State: The Emerging Liability of Third Party Health Care Payers, *San Diego Law Reivew*, *24*:1023 (1987).

186. Wickline, 192 Cal. App. 3d at 1645-46, 228 Cal. Rptr. at 671.
187. French v. Fischer, 50 Tenn. App. 587, 362 S.W.2d 926 (1962).
188. Kalmus v. Cedars of Lebanon Hosp., 132 Cal. App. 2d 243, 281 P.2d 872 (1955).
189. D. Yaggy and P. Hodgson, eds., *Medical Malpractice*, 1986.

REFERENCES TO POSTSCRIPT

1. B. Schwartz. The inevitable failure of cost-containment strategies, *JAMA*, *257*:220-224 (1987).
2. ON device panel moves closer to cost-benefit advice, *Dickinson's FDA*, March 15, 1989, p. 9.
3. B. R. Schlender and M. Waldholz, Genentech's missteps and FDA policy switch led to TPA setback, *Wall Street Journal*, June 16, 1987, p. 15.
4. G. W. Grumet, Health care rationing through inconvenience, *New England Journal of Medicine*, *321*:607-611 (1989).

part nine

Responding as Companies to Coverage, Payment, and Marketing

53

Understanding and Securing Medicare Coverage for New Medical Technology

GORDON B. SCHATZ *Reed Smith, Shaw & McClay, Washington, D.C.*

ROBERT E. WREN *Health Care Financing Administration, Baltimore, Maryland*

I. INTRODUCTION

Manufacturers of medical devices and diagnostic products may need, in addition to FDA clearances or approvals, a coverage determination by the Medicare program to assure payment for new technology. This chapter will explain when coverage decisions may be needed, and alternate approaches to securing a coverage decision.

II. WHAT IS A MEDICARE COVERAGE DECISION?

A Medicare coverage decision is a determination by the Health Care Financing Administration (HCFA) or by a Medicare contractor on whether a particular medical device, diagnostic product, procedure, drug, or service will be paid by the Medicare program. In their broadest contexts, coverage determinations may involve questions of medical necessity, length of stay, qualifications of the provider, site of service, or frequency of service. This chapter, however, will only address the issues most directly facing manufacturers of medical technologies who are developing new products, and not the issues of provider and beneficiary status. HCFA coverage decisions can be quite detailed, such as the one for cochlear implants which states:

> 65-14 COCHLEAR IMPLANTATION (Effective for services performed on and after October 1, 1986.)
>
> A cochlear implant device is an electronic instrument, part of which is implanted surgically to stimulate auditory nerve fibers, and part of which is worn or carried by the individual to capture and amplify sound. Cochlear implant devices are available in single channel and multi-channel models. The purpose of implanting the device is to provide an awareness

and identification of sounds and to facilitate communication for persons who are profoundly hearing impaired.

Medicare coverage is provided only for those patients who meet *all* of the following selection guidelines:

1. Diagnosis of total sensorineural deafness that cannot be mitigated by use of a hearing aid in patients whose auditory cranial nerves are stimulable.
2. Cognitive ability to use auditory clues and a willingness to undergo an extended program of rehabilitation.
3. Post-lingual deafness.
4. Adulthood (at least 18 years of age).
5. Freedom from middle ear infection, an accessible cochlear lumen that is structurally suited to implantation, and freedom from lesions in the auditory nerve and acoustic areas of the central nervous system.
6. No contraindications to surgery. [1]

HCFA decisions may also be more general. For example, the national coverage instruction for laser procedures states:

35-52 LASER PROCEDURES

Medicare recognizes the use of lasers for many medical indications. Procedures performed with lasers are sometimes used in place of more conventional techniques. In the absence of a specific noncoverage instruction, and where a laser has been approved for marketing by the Food and Drug Administration, contractor discretion may be used to determine whether a procedure performed with a laser is reasonable and necessary and, therefore, covered. [2]

While significant HCFA coverage decisions are compiled in the HCFA Coverage Issues Manual, many local decisions (described below) that are made by Medicare contractors are developed individually by carriers and fiscal intermediaries.

III. DOES A TECHNOLOGY NEED A MEDICARE COVERAGE DECISION?

A threshold question for the manufacturer is whether a new medical technology needs a national Medicare coverage decision. The Medicare program pays for health services rendered to individuals who are 65 years of age or older, who are disabled, or who have end-stage renal disease. Certain medical products are designed exclusively for pediatric or neonatal use. Infants, children, or young adults who will use technologies, such as neonatal monitors, are often not generally eligible for Medicare. Accordingly, a manufacturer of such products may not need a national coverage decision from Medicare.

By contrast, products like pacemakers, implantable defibrillators, cochlear implants, or magnetic resonance imagers, which would likely be used by senior citizens or disabled persons, who are eligible for Medicare, may benefit from a national Medicare coverage decision.

Apart from consideration of the target population, certain supply-type items that are integral parts of a larger medical procedure, like sutures, bandages, or X-ray film, and that do not offer a significant new patient benefit or do not create substantial new costs for the Medicare program may not need a coverage decision.

Manufacturers developing breakthrough technologies that would pose additional costs for Medicare, such as extracorporeal shock-wave lithotripsy, may be assisted by a national coverage decision. Similarly, if there have been inconsistent local coverage decisions (either by different carriers or by the same carrier) with no clear reason for the difference, such as varying regional medical practice, then a national coverage decision may help resolve local inconsistencies.

The approach that Medicare follows to decide which products to consider for a coverage decision reflects the diverse technologies and sources of input. Medicare makes coverage decisions not only for medical devices and diagnostic products but also for medical procedures. A few Medicare national coverage decisions, as an example of the types of technologies subject to national coverage decisions are:

Cochlear implant
Automatic implantable cardiac defibrillator
Magnetic resonance imaging
Hemoperfusion and hemofiltration
Heart transplants
Transurethral ureteroscopic lithotripsy
Gastric balloon for treatment of obesity
Serologic testing for AIDS
Continuous positive airway pressure
Portable hand-held X-ray
Lymphedema pumps
Laser procedures

This list demonstrates the wide variety of medical devices, diagnostic products, drugs, and procedures that are subject to a national coverage decision by Medicare.

IV. NATIONAL OR LOCAL COVERAGE DECISION

There are two quite different types of Medicare coverage decisions involving new medical technology. One is the decision made by HCFA's Office of Coverage Policy in Baltimore, Maryland. These decisions are generally incorporated into the Medicare Coverage Issues Manual and may involve extensive review and assessment at HCFA and the Public Health Service.

The second type of Medicare coverage determination will be made by Medicare contractors in the course of processing claims for payment. HCFA estimates that more than 400 million individual claims were processed by carriers and intermediaries in fiscal year 1987. When questions of coverage do arise, they relate primarily to whether the item or service was medically necessary for the individual and was furnished in an appropriate manner and setting, rather than the broader coverage issues, raised in national determinations.

When the contractors identify services for which there appear to be questions of possible national significance, but for which there are no applicable national coverage decisions, they refer the services to HCFA for review. In general, a service will be considered for a national coverage decision when one or more of the following factors is present:

1. The service is likely to be used in more than one region of the country.
2. The service is likely to represent a significant expense to the Medicare program.
3. The service has the potential for rapid diffusion and application.
4. There is substantial disagreement among experts regarding the safety, effectiveness, or ethical judgment involved in the use of a service.
5. The service represents a significant advance in medical science.
6. The service represents a new product, that is, a device, drug, or procedure for which there is no similar technology already covered under Medicare.
7. Coverage under Medicare is otherwise permitted.

V. LEGAL FRAMEWORK FOR COVERAGE DECISIONS

A. Medicare Statute

The Medicare statute provides coverage for broad categories of health benefits, including inpatient and outpatient hospital care, skilled nursing facility care, home health care, physician services, clinical laboratory and diagnostic tests, durable medical equipment, prosthetic devices, and artificial limbs [3]. The statute also specifically excludes certain categories of services, such as cosmetic surgery, personal comfort items, custodial care, routine physical checkups, eyeglasses, hearing aids, orthopedic shoes, supportive devices for the feet, and certain dental services [4].

The statute does not furnish an all-inclusive list of specific items, services, treatment procedures, or technologies covered by Medicare. However, in addition to the broad coverage categories and the specific exclusions from coverage, the statute sets forth a general principle as follows:

> Notwithstanding any other provision of this title, no payment may be made under part A or part B for any expenses incurred for items or services . . . which are not reasonable and necessary for the diagnosis or treatment of illness or injury or to improve the functioning of a malformed body member. [5]

It is this general principle, together with the broad categories of coverage and exclusions, that guides the Medicare program in making coverage decisions for new technology.

B. Medicare Regulations

There are extensive regulations that implement the coverage and exclusion sections of the Medicare statute [6]. These, however, are not product specific, but rather set forth the conditions under which a hospital benefit or an item of durable medical equipment will be paid. For example, durable medical equipment (DME) is defined as equipment that can withstand repeated use, is primarily and customarily used to serve a medical purpose, is not generally useful in the absence of illness or injury, and is appropriate for use in the home [7]. Some definitions of covered items may also be found in regulations dealing with payment for the items.

As a result of these specific restrictions, some products that may be reasonable and necessary, such as a *disposable*, home-use infusion pump, may not be covered as an item of durable medical equipment. The reason is that the item, as a disposable product, fails the particular test of durability.

A regulation further detailing the coverage process and criteria used for technology was proposed in the January 30, 1989 *Federal Register*.

C. Site-of-Service Considerations

The site of service may be significant in determining coverage. For example, the use of an air fluidized bed as therapy for decubitus ulcers is a covered service when provided in the hospital. It is covered under the broad category of hospital services. However, the use of air fluidized beds is not currently covered when used in the home because its use has not been demonstrated to be safe and effective in the home setting. Site-of-service considerations increase in importance with the movement of services form the hospital setting to the home setting and with the proliferation of alternate care providers, such as ambulatory surgical centers, hospices, and dialysis facilities, which may involve special coverage rules for particular procedures.

VI. IMPLEMENTATION OF THE LAW

In the course of adjudicating a Medicare claim, intermediaries and carriers are charged with the responsibility to assure that payments are made only for items or services that are covered under Medicare Part A or B. Additional guidelines for the contractors are contained in the Medicare Intermediary, Carrier, Utilization and Quality Control, Peer Review Organization, and Coverage Issues Manuals. These manuals are updated periodically through the instruction issuance process and comprise the primary source of information from HCFA to the contractors.

The administration of the Medicare program, including coverage decision making, is highly decentralized. This means that the majority of coverage decisions are made through claims processing by carriers, intermediaries, and professional review organizations (PROs). However, a small number of items and procedures—20-30 per year—are the subject of a centralized process of technology assessment and coverage decision-making.

VII. THE CLAIMS PROCESS

A. Description of Claims Process

When a hospital, physician, laboratory, or durable medical equipment supplier renders services or provides a product to a Medicare beneficiary, the provider will submit a claim form to the Medicare fiscal intermediary or carrier for payment. The claims review process is designed to pay for appropriate care and to identify services that may not be covered or about which there may be a question of medical necessity or reasonableness. Resolving these issues is usually a matter of deciding whether the service or item meets statutory, regulatory, and other guidelines for the specific patient under the specific circumstances.

Medicare contractors have discretion, working within the legal framework, to decide coverage questions raised in the claims review process. All carriers and intermediaries have nurses on their claims review staffs and all have physicians available to provide advice and to consult with specialists and peer groups in the locality to analyze and resolve coverage issues on the basis of sound medical judgement.

If a national coverage policy decision has been issued, describing a particular item or service, contractors are bound by it and are required to apply it throughout the claims process.

If no national coverage policy decision has been issued, a contractor must decide whether the service is appropriate, taking into account the frequency, duration, and the setting in which it was furnished. These decisions are usually made in consultation with the contractor's own medical staff and local medical specialty groups. Since each contractor makes determinations only for its own claims, coverage of a particular item or service may vary among contractors.

B. Role of Manufacturer

If the manufacturer anticipates that hospitals and doctors will be submitting claims for new technologies, then advance analysis and preparation can help improve contractors' understanding of the technology and processing of claims.

Before the claim is submitted, the manufacturer together with the provider should examine the ICD-9-CM (International Classification of Diseases, 9th Revision, Clinical Modification) procedures codes for inpatient hospital services and the CPT (current procedure terminology) or HCPCS (HCFA common procedure coding system) codes for physician services, laboratory tests, and DME [9]. This examination will identify whether there is an existing code that describes the new technology or associated procedure. If not, the manufacturer may want to meet with the carrier to present information about the new technology, indicate that claims may be forthcoming, and discuss what codes would be proper for providers to use. Carrier medical staff may wish to examine scientific data about the product's safety and effectiveness as well as cost-related information to make determinations as to coverage and payment.

Critical information for contractors, as well as for national decisions, is the FDA product approval status of a technology. If a device has not yet been approved by FDA, the Medicare will not pay for it. The instructions to carriers are quite explicit in stating:

> *Devices not approved by FDA—*
>
> Medical devices which have not been approved for marketing by the Food and Drug Administration (FDA) are considered investigational by Medicare and are not reasonable and necessary for the diagnosis or treatment of illness or injury, or to improve the functioning of a malformed body member. Program payments, therefore, may not be made for medical procedures or services performed using devices which have not been approved for marketing by the FDA. [10]

This policy does not mean that FDA approval will assure Medicare coverage. FDA criteria do not necessarily match Medicare requirements. The manufacturer that has received approval to market a product under the 510(k) process may still have to present documentation to HCFA that a product is safe and effective for Medicare beneficiaries.

VIII. THE NATIONAL COVERAGE PROCESS

A. Referral and Identification

Coverage issues regarding medical items, services, or procedures come to HCFA's attention in a variety of ways. Some coverage issues are raised by individual Medicare beneficiaries, physicians, equipment manufacturers, physicians, professional associations, or governmental entities. Other coverage issues are raised during HCFA's ongoing review of current medical practice. Still others are raised in the course of hearings on disputed claims. However, the program's contractors are the most frequent source of referrals. Most of the individual claims processed by Medicare contractors do not raise serious questions about coverage. Where questions of coverage do arise, they relate primarily to whether the item or service was medically necessary for the individual and was furnished in an appropriate manner and setting, rather than to broader coverage issues. Even when a claim for a new, or otherwise questionable, service is received, the contractor is authorized to make "reasonable and necessary" decisions with respect to the item or service, in the absence of preexisting national policy.

In their review of claims for payment, Medicare carriers, intermediaries, and PROs are bound by all HCFA administrative issuances of general applicability [11], including all national coverage determinations that are published in HCFA rulings or manual instructions.

Medicare carriers and intermediaries also are bound by medical necessity determinations of local PROs under section 1154(c) of the Social Security Act, with respect to the application or use of a covered item or service in a particular case and with respect to items or services for which HCFA has not issued a national coverage determination. If the contractor cannot resolve the question satisfactorily, or believes a national coverage determination may be necessary, the issue is then referred to HCFA Central Office through an HCFA regional office.

Medicare coverage of drugs and biologicals is treated differently. In accordance with section 1861(t) of the Social Security Act, current Medicare manual instructions provide for coverage of drugs and biologicals that have been approved for marketing by FDA, except when a particular use has been expressly disapproved by the FDA or designated as not covered in a national HCFA instruction. Medicare contractors still must determine whether or not a treatment is reasonable and necessary for the individual patient, and whether other coverage requirements are met.

Most of the "reasonable and necessary" coverage issues referred to HCFA can be decided based on the statute, regulations, and policy precedents. Some issues arise that cannot be resolved without seeking additional professional medical expertise. Such issues, including questions of the safety and effectiveness of an item or service or its common acceptance by the medical profession, cannot easily be resolved. These questions may concern either new or unusual items or services, or items and services that are believed to be outmoded and no longer reasonable and necessary.

B. HCFA Analysis

1. Overview

HCFA has used various methods for seeking medical and scientific advice in determining whether an item or service is safe and effective or commonly accepted. These have included, at one time or another, the use of the Public Health Service (PHS), advisory councils, ad hoc groups of representative physicians, and various forms of consultation and liaison with national medical associations. At present, a formal process of review and assessment involves HCFA Central Office physicians, PHS representatives, and the national medical community. That review process is described below.

2. Initial Consideration

When medical advice is needed in order to evaluate a coverage question, the issue is referred within HCFA to the Office of Coverage Policy (OCP) within the Bureau of Policy Development (BPD). Staff in OCP/BPD conduct a medical literature search, determine the status of any FDA action, and prepare a backgorund paper on the item or service.

The background paper stresses the information obtained from the medical literature search and administrative aspects of the issue. Current Medicare coverage guidelines are discussed, as well as any information obtained regarding published determinations by other groups involved in technology assessment.

3. HCFA Physicians Panel

The next step in the process involves presenting the background paper to the HCFA Physicians Panel for review. This panel, an internal organization composed of physicians and other health professionals in HCFA's Central Office and counterparts from PHS, meets approximately once every 6-8 weeks in a closed session. These meetings are nonpublic staff meetings that are necessary for a full discussion and exploration of the medical aspects of HCFA's options in reaching a coverage decision. Under the January 30, 1989 proposed coverage rule, presentations by interested parties can be made prior to the closed meetings.

Upon weighing the facts and discussing the evidence, the panel may recommend to BPD that the item or service should be (1) referred to PHS on either an informal, inquiry basis or with a request for a full assessment as to safety and effectiveness, or (2) covered or not at the discretion of the individual Medicare contractor pending receipt of more information, or (3) subject to a special study or action. (For example, HCFA commissioned a special study on heart transplants, although this is a rare instance). In the past, the panel has also made recommendations on whether to cover or not cover a particular item or service.

C. PHS Assessment Process

On the recommendation of the Physicians Panel, HCFA may refer an item to the Office of Health Technology Assessment (OHTA), in the Public Health Service (PHS). In the case of an informal request, known as an inquiry,

OHTA conducts an up-to-date, intensive review of the medical literature, discusses the issues with other government components and medical groups, and responds to the specific questions raised by HCFA [12].

The handling of a request for a full assessment, by contrast, varies with the significance of the issue. In the more usual case, a full assessment of safety and effectiveness involves the publication by OHTA of a *Federal Register* notice announcing that an assessment of the issue is being undertaken and soliciting comments from interested parties [13]. As part of this process, OHTA seeks information and advice from other governmental agencies, for example, the National Institutes of Health (NIH), the FDA, and the Veterans Administration. OHTA also consults directly with concerned medical specialty groups and professional organizations to determine whether a consensus exists within the medical community on the safety and effectiveness of the item or service. Information may also be obtained from commercial and industrial groups and specific manufacturers. An exhaustive search of the published medical and scientific literature is conducted and the findings of all relevant studies and reports are carefully analyzed.

OHTA staff then synthesize the available information in order to develop the PHS recommendation. This involves four steps: (1) All pertinent information, including expert opinions, is summarized. (2) The various sources of information are weighted according to their comparative validity and significance. (3) To the extent possible, conclusions are formulated about the safety and effectiveness of the item or service. (4) OHTA's policy recommendations to HCFA regarding the appropriateness of Medicare coverage of the item or service are developed. OHTA may recommend that the item or service not be covered by Medicare, that it be covered with certain restrictions, that it be covered without restriction, or that it be assessed at a later date.

OHTA assessments and recommendations are not available to the public until after HCFA has made and issued its final decision. OHTA publishes a Health Technology Assessment Report for items that have received a full assessment, and has issued abstracts or summaries of the assessments and recommendations for items assessed in 1984 and 1985 [14].

D. Medicare Coverage Decisions and Notification

After HCFA receives an OHTA recommendation concerning the coverage of an item or service, HCFA considers the OHTA recommendation and makes a decision. HCFA ordinarily notifies Medicare carriers, intermediaries, and PROS of all national coverage determinations through publication of a HCFA ruling [15] or an instruction in the Coverage Issues Manual.

E. Reevaluation of Coverage Decisions

In addition to assessing new items and services, HCFA might reevaluate an item or service that is already excluded or covered under the Medicare program. This might occur, for instances, if in a previous assessment, PHS suggests a reevaluation of an item or service at a later date. Knoweldge of research being conducted and the eventual publication of the results may prompt reevaluation of an item or service.

The decision to reevaluate might also occur if an item or service is considered obsolete. Instances such as these come to HCFA's attention in a variety of ways—for example, other third-party payors, medical literature articles, or they might be self-generated within HCFA's Central Office. In all of these instances, the same coverage assessment process is followed as is described above for evaluation of new items and services, except that if the decision is to withdraw coverage of an item or service, HCFA ordinarily would publish a notice in the *Federal Register* announcing that intent.

At any time after publication of a Medicare coverage instruction, interested parties may submit evidence demonstrating that a reassessment of that coverage determination is warranted. A reassessment would only be done if there is acceptable information or evidence available that was not available at the time the most recent assessment was performed.

While administrative review is available, judicial review is limited. National coverage determinations, dealing with whether or not a particular type or class of items or services is covered, cannot be reviewed by administrative law judges or held unlawful for noncompliance with the Administrative Procedures Act. If the administrative record is incomplete, the matter must be remanded to the Secretary [16].

F. Coordination with Payment Decision

Although the payment decision by HCFA is a separate decision from coverage, manufacturers should be working with HCFA's Office of Payment Policy and perhaps the prospective Payment Assessment Commission (ProPAC) at the same time review is under way on the coverage decision. Contact with ProPAC is valuable for those items to be used in the hospital and paid under a diagnosis-related group as an inpatient service. Such coordination can link agency reviews and facilitate a prompt decision.

G. Federal Register Proposed Coverage Rule—January 30, 1989

The most significant development in Medicare coverage policy has been the issuance by HCFA of a proposed rule establishing the criteria and procedures for making medical services coverage decisions that relate to health care technology (January 30, 1989). The proposed rule, by and large is consistent with the process described above. It differs in its degree of detail and in the authority it will carry if and when finalized.

The proposed rule sets forth four criteria to be used by Medicare in making a national coverage decision: (1) The service or technology is safe and effective. (2) The service or technology is not experimental or investigational. (3) The service or technology is cost effective. (4) The service or technology is appropriate.

The key criterion is safety and effectiveness, which can be proven by the technology being generally accepted in the medical community as safe and effective or proven by authoritative evidence. Breakthrough technologies may qualify under less stringent standards of safety and effectiveness. FDA approval under a premarket approval or 510(k) with clinical data may satisfy the requirement for safety and effectiveness.

The most controversial criterion is cost effectiveness. A technology is considered cost effective if it is: (a) Very expensive to the program, but provides significant medical benefits not otherwise available; (b) less costly and at least as effective as an alternative covered intervention; (c) more effective and more costly than a covered alternative, but the added benefit is significant enough to justify the added cost; or (d) less effective and less costly than an existing alternative, but a viable alternative for some patients.

An item will be considered experimental if it is furnished for research purposes in accordance with predetermined rules or has not been approved by FDA. A service or item will be considered appropriate if it is furnished in a setting commensurate with the patient's medical needs and by qualified personnel.

The proposed rule is significant in establishing for the first time the specific criteria, in making concrete the procedures, in requiring carriers to follow the same criteria as the national Office of Coverage Policy, but generally in providing manufacturers and the medical community at large, a better road-map to be followed to obtain a coverage decision, either at a national or local level.

IX. RECOMMENDATIONS FOR MANUFACTURERS

The following are recommendations for manufacturers in approaching questions of Medicare coverage.

1. Plan clinical trials to develop data needed for OHTA/HCFA as well as FDA.
2. Review Medicare Coverage Issues Manual to identify whether there is any national policy.
3. Examine Medicare statute and regulations to see if product is excluded from coverage or otherwise limited as to when it will be covered.
4. Review ICD-9-CM and HCPCS procedure or product codes to locate comparable code.
5. Collect scientific data documenting product being reasonable and necessary for Medicare beneficiaries. Safety and effectiveness data as well as cost information can be developed during clinical trials in anticipation of coverage and payment questions.
6. Introduce relevant medical specialty groups to new technology.
7. Identify FDA product approval status and data for presentation to HCFA.
8. Consider whether a national or local decision is needed.
   a. Consult with area Medicare contractors.
   b. Consult with HCFA's Office of Coverage Policy.
9. Work with contractors or Office of Coverage Policy to develop information for coverage analysis and decision.
10. Communicate decision to providers along with related information dealing with coding.
11. Monitor claims processing to follow implementation of decision.

X. CONCLUSION

To understand the Medicare coverage process, a manufacturer must identify the patients who will benefit from a technology, the sites of service, and the regulatory framework for assessing and approving devices, diagnostic products, and procedures. If a coverage decision is needed, careful planning, data acquisition, and presentation are vital for either a local or national determination. Manufacturers can speed acceptance of new technologies by working closely with researchers, professional groups, Medicare contractors, or the national Office of Coverage Policy to assure that Medicare beneficiaries have prompt access to care that is reasonable and necessary.

REFERENCES

1. HCFA Coverage Issues Manual, HCFA Pub. 6, Section 65-14. Also available in *Commerce Clearing House* (CCH) *Medicare and Medicaid Guide*, Vol. IV.
2. HCFA Coverage Issues Manual, HCFA Pub. 6, Section 35-52.
3. Social Security Act, as amended, Section 1861, 42 U.S.C. Section 1395x, definitions of services and Social Security Act, Section 1812, 42 U.S.C., Section 1395d, scope of benefits.
4. Social Security Act, Section 1862(a)(6)(12), 42 U.S.C. Section 1395y(a)(6)-(12).
5. Social Security Act, Section 1862(a)(1)(A), 42 U.S.C. Section 1395y.
6. 42 CFR Sections 405.231 et seq. and 405.310.
7. 42 CFR Section 405.514.
8. 54 *Federal Register* 4302, January 30, 1980. Also 52 *Federal Register* 82, April 29, 1987, p. 15560. See also the related case, *Jameson v. Bowen*, no. CV-F-83-547 REC (E.D. Calif. February 19, 1987), order approving settlement.
9. ICD-9-CM codes are available generally in a three-volume set from the Commission on Professional and Hospital Activities, Ann Arbor, Michigan, the American Hospital Association, Chicago, Illinois, or the U.S. National Center for Health Statistics. Volume 3 contains the procedure codes. CPT codes are published by the American Medical Association. HCPCS codes are issued by HCFA.
10. HCFA Medicare Carriers Manual, Part 3—Claim Process. HCFA Pub. 14-3, Section 2303.1.
11. Public Health Service, Procedures for Evaluating Health Care Technologies for Purposes of Medicare Coverage, Office of Health Technology Assessment, National Center for Health Services Research April 1983.
12. HCFA intermediary letters nos. 77-4 and 77-5 (1976 Transfer Binder), Medicare and Medicard Guide CCH para. 28,152 (1976). This sets forth the principle that items should be generally accepted by the professional medical community as effective and proven treatments.
13. 42 U.S.C. Section 242c(e) authorizes the National Center for Health Services Research and Health Care Technology Assessment, the agency within the Public Health Service, and the parent of OHTA, to develop recommendations for HCFA.
14. Abstracts of Office of Health Technology Assessment Report, 1985. National Center for Health Services Research and Health Care Technology Assessment.
15. 51 *Federal Register* 201, October 17, 1986, p. 37164 Medicare program criteria for Medicare coverage of heart transplants, proposed notice, and 52 *Federal Register* 65, April 6, 1987, p. 10935, final coverage notice.
16. Social Security Act, Section 1869(b)(3).

54

Avoiding Fraud and Abuse

SANFORD V. TEPLITZKY and S. CRAIG HOLDEN *Ober, Kaler, Grimes & Shriver, Baltimore, Maryland*

I. INTRODUCTION

As members of the medical device industry struggle to stay competitive, the pressure to aggressively maximize revenues from all sources, including Medicare and Medicaid, as well as to aggressively seek out new customers continues to mount. As these pressures increase, so do the dangers of running afoul of the ever-growing number of Medicare and Medicaid fraud and abuse statutes. Conduct that is both lawful and common in other areas of the business world may constitute criminal activity when it in any way involves the Medicare or Medicaid programs. The purpose of this chapter is to familiarize the reader with the general areas of prohibited conduct and to highlight certain potential fraud and abuse pitfalls, so that when they do arise, expert advice can be sought to avoid violations before they occur. This chapter deals with the federal fraud and abuse prohibitions that govern all transactions relating to the provision of medical services and supplies to Medicare and Medicaid beneficiaries. In addition, there has been a growing trend in many states to adopt similar provisions pertaining to the entire health care field in that state. Thus, the reader should also be aware of the need to familiarize oneself with the laws of the state in which one is operating.

II. BACKGROUND

The original Medicare Act simply incorporated the misdemeanor provisions then applicable to the Social Security program that prohibited obtaining or attempting to obtain money by fraudulent means, and made them applicable to those providing care or services to Medicare beneficiaries [1]. In addition, since Medicare is a federal program, general criminal statutes relating fraud against the government were also applicable [2]. In 1972, Congress amended the Medicare statute to create criminal provisions specifically applicable to Medicare fraud [3]. These amendments marked the first time that the payment of kickbacks in the context of the Medicare and Medicaid programs was made unlawful.

In the mid 1970s, public hearings showed the existence of widespread fraud and abuse in both the Medicare and Medicaid programs. In addition, these hearings revealed the widespread practice of kickback payments in the health-care area, often thinly disguised as rental or other payments. In response to these abuses, Congress passed the Medicare and Medicaid Anti-Fraud and Abuse Amendments of 1977 [4]. These amendments expanded the scope of the antikickback provisions, mandated the disclosure of individuals having ownership or control of health-care entities, and upgraded criminal offenses aginst the Medicare and Medicaid programs from misdemeanors to felonies.

In 1981, Congress amended the Medicare statute to permit the imposition of stiff civil fines against persons who either intentionally or negligently filed false Medicare or Medicaid claims. In 1987, Congress passed the Medicare and Medicaid Patient and Program Protection Act, which, in addition to consolidating the various fraud and abuse authorities contained throughout the Medicare and Medicaid statutes, added a laundry list of conduct, including violations of the antikickback provisions, which can form the basis for being barred from any involvement in the Medicare or Medicaid programs [5]. Finally, in recent years, each year's budget reconciliation act has added miscellaneous civil fines and exclusion authorities to enforce various reimbursement rules under Medicare [6].

Perhaps the most significant development in the field of Medicare and Medicaid fraud and abuse was the 1976 passage of legislation creating the Office of the Inspector General of the Department of Health and Human Services (HHS) (then the Department of Health, Education, and Welfare) [7]. Prior to the creation of this office, the investigation of violations of the Medicare and Medicaid antifraud and abuse statutes were primarily vested in the Federal Bureau of Investigation (FBI), which has general investigative authority over all federal crimes, and the United States Postal Service, which has investigative authority over all crimes involving fraudulent schemes that utilize the U.S. mails. Actual prosecutions were conducted by the U.S. Department of Justice, typically through its local U.S. Attorneys.

The creation of the Office of the Inspector General changed this. The OIG has a number of statutory responsibilities, including maintaining security programs for the department, conducting and supervising audits of department programs, providing leadership and coordination and recommending policies and corrective actions to promote economy and efficiency and prevent and detect fraud and abuse in HHS programs. However, the primary impact of the creation of this office for the health-care industry has been the fact that the OIG is responsible for conducting both criminal and civil investigations of all violations of the Medicare and Medicaid fraud and abuse provisions. In addition, the Secretary of Health and Human Services has delegated all of his authority to impose civil fines and to bar persons from future involvement in the programs to the Inspector General [8]. This enables the OIG to both investigate and, with regard to civil sanctions, prosecute those who violate the fraud and abuse statutes. (The authority to initiate criminal prosecutions remains with the U.S. Department of Justice. In addition, the FBI, the Postal Service, and other federal agencies may continue to investigate fraud and abuse allegations.)

The results of the creation of an investigative body devoted exclusively to the HHS programs have been impressive. In the OIG's most recent semi-annual report to Congress on its accomplishments, it reported that for fiscal year 1987 alone it obtained 118 criminal convictions against individuals or entities who violated the fraud and abuse prohibitions [9]. In addition, the OIG took 440 administrative actions against health care providers. Of these, 79 were actions imposing civil fines, while 361 were actions barring health-care providers from future participation in Medicare or Medicaid. The total amount of the civil fines imposed by the OIG for 1987 alone totalled 11.7 million dollars.

In addition to the OIG, Congress has created financial incentives for states to set up Medicaid fraud control units consisting of both investigators and prosecutors dedicated exclusively to the detection and prosecution of fraud in state Medicaid programs. Although the quality of these investigative units varies from state to state, as a whole they have been extremely aggressive in their pursuit of Medicaid fraud. The most recent report by the OIG (which is responsible for overseeing their activities) indicates that the 38 units in existence in fiscal year 1987 reported 453 convictions of health care providers during that year alone [9]. In addition, these units obtained approximately 8.5 million dollars in fines and restitution.

The impetus for the initiation of an investigation by one of these agencies can come from a variety of sources. Many cases begin as a result of Medicare intermediary or carrier audits or computerized billing screens that identify aberrant billing practices. However, many are initiated as a result of specific complaints received from a variety of sources. Among these sources are patients who note discrepancies between the explanation of Medicare benefits form which they receive whenever benefits are paid on their behalf, and the services they actually received. Another fertile source of investigative tips is present or former employees of the health-care provider. It is not at all unusual for disgruntled employees, particularly those whose employment has just been terminated, to bring any marginally questionable activity of their employer to the attention of the OIG. Finally, perhaps the most fertile source of investigative leads for the OIG, particularly in the area of alleged antikickback violations, is competitors. It is not at all unusual for a competitor who sees referrals that once went to his or her business begin to go to another business as the result of financial payments made to the referral sources, to protest to the Inspector General.

The ever-growing arsenal of antifraud and antiabuse authorities available to the government, combined with the increasingly aggressive enforcement stance of the OIG and the large number of potential sources for tips initiating investigations, makes it crucial to be aware of the scope of the antifraud and antiabuse statutes and to structure operations accordingly to avoid being the next target of an OIG investigation.

The fraud and abuse statutes can be divided into three categories: those dealing with the payment of unlawful kickbacks or referral fees, those dealing with the filing of false or otherwise improper reimbursement claims, and the relatively new authorities that impose substantial civil penalties and exclusions from the Medicare and Medicaid programs for the violation of certain reimbursement rules. Each of these areas will be discussed in detail below.

III. THE ANTIKICKBACK STATUTE

Perhaps the most troublesome area of the fraud and abuse laws for those who are trying to stay competitive in the rapidly changing medical device market is the antikickback provisions [10]. While these provisions are commonly referred to as the "antikickback" provisions, they proscribe far more than what traditionally would be considered a kickback. Simply stated, they ban the payment of anything of value if it is even partially in exchange for the referral of business. They are so exceedingly broad that they prohibit transactions that are commonplace in virtually all other areas of the business world. For example, many major airlines offer their frequent flyers additional mileage bonuses if they can convince others to join that airline's frequent flyer club. Similarly, many book clubs offer free books to individuals who refer new members. Such arrangements are common and do not raise an eyebrow in these business sectors. However, were a manufacturer or retailer of medical devices to engage in such a promotion for any items that might eventually be paid for by either Medicare or Medicaid, he would be guilty of a felony and could also be barred from making any future sales to Medicare or Medicaid beneficiaries.

The key to avoiding these disastrous consequences is a basic understanding of what conduct is prohibited by the antikickback provisions. The starting place for such an understanding is the language of the statute itself. The statute provides as follows:

> Whoever offers or pays any remuneration (including any kickback, bribe, or rebate) directly or indirectly, overtly or covertly, in cash or in kind to any person to induce such person—
>
> (A) to refer an individual to a person for the furnishing or arranging for the furnishing of any item or service for which payment may be made in whole or in part under this Medicare or Medicaid, or
>
> (B) to purchase, lease, order or arrange for or recommend the purchasing, leasing, or ordering any good, facility, service of item for which payment may be made in whole or in part under [Medicare or Medicaid],
>
> shall be guilty of a felony and upon conviction thereof, shall be fined not more than $25,000 or imprisoned for not more than 5 years, or both. [10]

It is readily apparent from reading this language that it is extraordinarily broad. Indeed, it is so broad that it could be read to prohibit *any* payment or other transfer of things of value from persons doing Medicare or Medicaid business to any individual with the capacity to refer business to them.

There is some indication that, despite the breadth of the language chosen, Congress did not intend to prohibit all such payments. During the debate on the 1977 antifraud and antiabuse amendments to the Social Security Act that created the statute in its current form, Congressman Dan Rostenkowski, one of the sponsors of the legislation, made the following comments:

> In broadening these criminal provisions, the Committee sought to make it clear that kickbacks are wrong no matter how a transaction might be constructed to obscure the true purpose of a payment. In exempting

> reimbursement by employers to employees from employment in a provision of covered services the Committee intended to exempt only those payments that represented payment for legitimate employment. For example, if a distributor of equipment or supplies pays a retailer on a commission basis for the use of his store to sell a product, such payment would represent a legitimate payment to a legitimate agent employed in a traditional manner to sell a product. That is a simple extension of the conventional chain of sale. On the other hand, the payment by a laboratory of the salary of an employee of a physician or of a clinic as an inducement to the physician or the clinic to refer business to the laboratory uses the employment relationship simply as a guise to pay a kickback. We are in a complex area where right and wrong are often clouded with shades of gray. In such situations, the Committee stresses the need to recognize that the substance, rather than simply the form, of the transaction should be controlling. [11]

The agencies charged with enforcing the statute have similarly indicated that not *all* payments to referral sources violate the statute. The Health Care Financing Administration (HCFA) has stated that "the intent of these provisions is not to penalize individuals or entities participating in legitimate business transactions. Rather, it is to penalize those who engage in unethical and illegal financial arrangements (such as kickbacks, rebates, or bribes) that unnecessarily increase the cost of federal health care programs" [12]. Similarly, an attorney representing the Office of the Inspector General has stated that payments to referral sources for things other than the referral of patients do not violate the statute provided they are not a subterfuge for the payment of referral fees [13].

As can be seen from the above, there is general agreement from all sides that the antikickback provisions do not prohibit *all* payments to individuals who might also be referral sources. Unfortunately, there is no such consensus on the difficult question of which payments are prohibited and which are not.

A. Statutory Exceptions

Congress has enacted several express exceptions to the statute, in recognition of the fact that the antikickback provisions are so broad that they may be read to prohibit some conduct that is, in fact, beneficial to the Medicare or Medicaid programs and their beneficiaries. These exceptions permit three specific types of conduct that would otherwise be prohibited.

1. Discounts or Other Reductions in Price

The first exception is for discounts or other reductions in price obtained by a provider or other entity *if* the discount is properly disclosed and appropriately reflected in the cost claimed or charges made by them [14]. The classic example of this practice is a bulk discount given a large purchaser of supplies, such as a hospital. Such discounts would otherwise be a violation of the antikickback provisions, since they constitute remuneration (i.e., anything of value) in exchange for increasing the volume of such purchases. The existence of an exception for these types of discounts reflects the general statutory goal of reducing costs in the provision of health care, thereby reducing the costs to the Medicare and Medicaid programs. In the case of a provider of services that is reimbursed on a cost basis under Medicare (such as a skilled nursing home or a home health agency), the discounts in question must be reflected on the cost report filed by the provider, thereby passing along the discount to the Medicare program.

However, the statutory exception also indicates that such discounts must be disclosed and reflected in Medicare claims filed for services that are reimbursed on a *charge* basis. The statute is silent on just how one is to disclose and reflect these discounts when reimbursement is charge based. Currently, there exists no mechanism either for such disclosure or for the adjustment of charges to reflect such discounts. Notwithstanding the lack of such a disclosure mechanism, the Inspector General's Office has indicated that non-cost-based entities may avail themselves of this exception. However, it should be noted that the Inspector General's Office believes that any such reductions in price should be passed along to patients and ultimately to Medicare in the form of reduced charges. As a practical matter, the Inspector General lacks the authority to require such reduced charges. However, excessive markups on goods that are the subject of such discounts could eventually result in the lowering of allowed charges for given items, and allegations of abusive charging practices.

Finally, the Office of the Inspector General has recently given some indication that it does not believe that discounts given directly to patients are protected by this discount exception. Under such an analysis, any rebate program whereby patients who purchase a medical device could send in a rebate coupon with proof of the purchase and receive a check for an amount partially offsetting the purchase price would be unlawful. Such an interpretation is an unduly technical one, which fails to consider the fact that it is a physician who orders such medical devices, not the patient. This requirement for a physician's order effectively eliminates the potential of the rebate resulting in overutilization, the chief evil that the statute seeks to prevent. Be that as it may, any medical device supplier that engages in such a rebate program bears some risk of being the subject of an OIG investigation.

2. Group Purchasing Organizations

A second exception to the antikickback provisions also involves the concept of volume discounts. It is common in the health-care industry for health-care providers to pool their buying power through the use of group purchasing organizations. Typically, such organizations are funded in one of two ways. Under the first, the participating providers pay membership fees to the group purchasing agent, which in turn arranges for each member provider to receive volume discounts at a negotiated rate directly from suppliers of goods or services. When the group purchasing organization is funded in this manner, it clearly comes within the discount exception discussed above.

Under an alternative financing arrangement, the participating providers pay no dues to the group purchasing agent. Rather, the agent is compensated by receiving a rebate from the supplier of the goods or services, typically based on a percentage of the total sales to the member providers. Since such payments to the group purchasing agent are quite clearly in exchange for referring the business of its member providers, these arrangements fall within the broad scope of the antikickback provisions. After a long period of uncertainty and debate among the health-care community and federal enforcement agencies, in 1986, Congress created an express exception to the statute that made such arrangements lawful [15]. Interestingly, the creation of this exception was made retroactively, immunizing all past arrangements using this financing method from any enforcement action. The policy consideration underlying this change was the same as that for the discount exception: the use of volume discounts is common in the business world generally, and those who provide services or supplies under Medicare and Medicaid should be encouraged to provide services as economically as possible.

In order for this exception to apply there must be a written contract specifying the amount to be paid to the purchasing agent, which must be set at either a fixed amount or fixed percentage. In addition, the purchasing agent must disclose all amounts received from vendors to the participating providers [15].

3. *The Employer-Employee Exception*

The statute also expressly exempts any payment made to employees under a "bona fide" employer-employee relationship [16]. Thus, for example, a durable medical equipment supplier could pay a sales employee commissions based on the amount of business he or she generates for the company. Unfortunately, the applicability of this exception is not always clear. The exception is quite obviously applicable in cases where the employee receiving such bonuses is an employee in the strict sense; that is, subject to the total control of their employer, who sets working hours, pay salary and benefits, makes appropriate tax withholdings from compensation, and provides the employee with an IRS Form W-2 at the end of the year. In numerous public statements, the Office of Inspector General has taken the position that this is the *only* type of relationship that is protected under the employer/employee exception.

However, such an unduly narrow reading of this exception exalts form over substance. Why, for example, should it not also be applicable to sales representatives who are not employees in the strict sense, but rather are independent contractors who may represent a number of different suppliers? No sound policy reason for such an interpretation comes readily to mind. In fact, there is language in the legislative history underlying this statutory exception that supports a far broader interpretation. Congressman Dan Rostenkowski, one of the sponsors of the 1977 amendments to the antikickback provisions that were then being debated, provided the following example of the type of arrangement that he believed fell within the employer/employee exception:

> For example, if a distributor of equipment or supplies pays a retailer on a commission basis for the use of his store to sell a product, such payment would represent a legitimate payment to a legitimate agent employed in a traditional manner to sell a product. That is a simple extension of the conventional chain of sale. [17]

The example chosen by Congressman Rostenkowski quite clearly rejects the overly rigid interpretation adopted by the Inspector General and instead relies on an analysis of the substance rather than the form of a given relationship. Under Congressman Rostenkowski's formulation, a supplier of durable medical equipment would be free to employ commissioned sales representatives on an independent contractor basis, and to pay a commission based on sales, without fear of running afoul of the antikickback provisions. However, it is vital to remember that the chief enforcer of the antikickback provisions is the Office of the Inspector General, and it quite clearly has adopted the restrictive view that the employee exception applies only where the individual is an employee in the technical sense and receives a W-2 form at the end of the year.

The issue of whether independent contractor sales agents or retail outlets receiving sales commission come within the employer/employee exception raises the larger question of what constitutes an unlawful "referral" under the statute. Certainly the retail outlet that leases floor space to a wholesale supplier of durable medical equipment for a display and is paid based on a percentage of the revenues generated by that display does not "refer" patients within the normal usage of that term. However, it is important to remember that the statute pertains not just to the referring of patients, but also to payments made to induce individuals "to purchase, lease, order, or arrange for or recommend purchasing, leasing, or ordering" any item or service that might be paid under Medicare or Medicaid. Given the extreme breadth of this language, one could argue that if a supplier of durable medical equipment hired an advertising agency to promote its products, it would violate the antikickback provisions if the agency were paid on a volume-sensitive basis. Of course, such an interpretation would be absurd and would bear no relationship to the underlying goals of the statute.

The OIG has long taken the position that the types of "referrals" that are within the ambit of the antikickback provisions are those that take place within the context of what can be called a "traditional health care relationship" between the referrors and the referred. Just what constitutes such a "traditional health care relationship" is subject to debate. On the one end of the spectrum, an advertising agency clearly does not have such a relationship with the intended audience for the advertisement. On the other end, the physician-patient relationship quite clearly falls within this category. Probably the safest way to view the situation is that anyone engaged in the business of providing health care, in whatever form, possesses such a relationship. This would include physicians, hospitals, nursing homes, home health agencies, and durable medical equipment suppliers, as well as nurses, therapists, and other health professonals and employees who work for such entitites. It is these individuals and entities that have a relationship with individual patients such that their referral or recommendation of a given provider of services could be expected to exercise significant influence over the patient's choice.

4. Regulatory Exceptions

In 1987, Congress again amended the antikickback statute, this time to grant the Department of Health and Human Services the authority to issue formal regulations that would create "safe harbors" of conduct that, although within the broad prohibitions of the statute, would not violate it [18]. In the legislative history underlying this amendment, Cognress recognized the need for guidance in the health-care industry as to what is and what is not prohibited by the statute. The legislative history notes, with considerable understatement, that "It is the understanding of the Committee that the breadth of this statutory language has created uncertainty as to which commercial arrangement are legitimate, and which are proscribed" [19].

Shortly after the passage of this legislation, the Office of the Inspector General issued a formal notice requesting comments from members of the health care industry regarding what, in their opinion, should be included in these regulatory "safe harbors" [20]. Over 140 sets of comments were received. Proposed regulations were published on January 23, 1989 [21]. As of this writing, the OIG has yet to issue final regulations. Until final regulations are issued, the only exceptions on which members of the health care industry can rely are those set forth in the statute itself. Even afterward, only those in a "safe harbor" are without worry. Those not fitting into a safe harbor will still be left to interpret the statute as it pertains to their agreement. The following section discusses how this should be done.

B. Avoiding Kickback Violations—General Considerations

As already discussed, the extreme breadth of the antikickback provisions draws into question any arrangement whereby anything of value is given by an individual or entity to a person who has the ability to refer Medicare or Medicaid business to it. When such extremely broad statutory language is involved, lawyers usually refer to court decisions that have been issued in cases arising under the statute. Unfortunately, the case law in this area adopts such a broad interpretation of the statute as to provide virtually no guidance in discriminating between those arrangements that are prohibited and those that are not.

The leading case in this area is *United States v. Greber* [22]. The *Greber* case involved a cardiologist who also owned a small company that provided mobile diagnostic tests, such as Holter monitoring, to various internists and other physicians. Dr. Greber would pay 40% of what the company received as Medicare reimbursement back to the referring physician, up to a maximum amount of $65.00. Dr. Greber characterized these payments as "interpretation fees" that were paid to the physicians for interpreting the results of the Holter monitor tests. Leaving aside the inherent illogic of paying a physician to provide the company with a test interpretation that the company would then merely turn around and hand back to the physician, there were a number of factual flaws in Dr. Greber's assertion. First, the referring physicians were paid regardless of whether they actually interpreted the tests. Second, the amount they received was far in excess of the fair market value of performing such a test interpretation. Third, in a separate civil proceeding, Dr. Greber testified, under oath, that the true purpose of the "interpretation fees" was to provide the physicians with an economic incentive to refer patients to him. Given these facts, it is hardly surprising that Dr. Greber was convicted of violating the antikickback statute.

On appeal, the court, which upheld his conviction, took the opportunity to make an extremely broad and perhaps gratuitous statement on the scope of the antikickback prohibition. The court stated that, in its opinion, if payments made to a referral source are "intended to induce the [referral source] to refer patients, the statute was violated, *even if the payments were also intended to compensate for professional services*" [23; emphasis added]. In recent cases, two other federal courts of Appeal have concurred in this interpretation of the statute [24].

Given the extremely broad language of the statute as well as of the case law interpreting the statute, meaningful guidance for members of the industry trying to structure their transactions in conformance with law must be found elsewhere. The Office of the Inspector General has indicated that its review of specific transactions is intended to determine whether they result in the perceived evils the statute was seeking to prevent. The first of these potential evils is the overutilization at Medicare- or Medicaid-covered services. The theory is that if an individual is in a position to receive a financial reward every time he or she refers a patient for the provision of a particular service, a strong incentive is created to do so, even where that service is not necessarily warranted. Such overutilization needlessly increases the cost of operating the Medicare and Medicaid programs and exposes the beneficiary to unnecessary risks. Similarly, the second such evil is the additional cost to Medicare or Medicaid that would result if the cost of paying such financial incentives for referrals is eventually rolled into the total cost for providing health-care services, as it most likely would be. In this era of tightening budget constraints, such concerns have been given a very high priority. Finally, there is a legitimate concern that such referral fees serve to impinge on a patient's right to seek health-care services from whomever he or she pleases. When reviewing any proposed transaction to determine whether or not it violates the antikickback provisions, a good rule of thumb is that the more likely an arrangement is to cause one of these "evils," the more likely it is to be found in violation of the statute.

With these general considerations in mind, we will now turn to several specific types of transactions.

1. Fee-for-Service Payments to Referral Sources

The antikickback provisions prohibit those payments to referral sources that are intended in whole or in part to compensate for, or induce, the referral of patients. The statute clearly does not prohibit a medical device supplier from making *any* payments to an actual or potential referral source, specifically when those payments are in exchange for services rendered to the supplier by that referral source. Thus, for example, a supplier of inhalation therapy equipment could pay a respiratory therapist, who also makes referrals to the supplier, to perform such tasks as setting up equipment in a patient's home and training the patient in the proper use of the equipment. Such a payment would not violate the antikickback provisions, provided that it was not intended as an inducement to, or conditioned upon, future referrals of patients.

While the distinction between payments made to referral sources to induce referrals, as opposed to those made to compensate for services they render, may be easy to make in theory, it is a most difficult one to make in practice. Although the federal enforcement agencies have never stated that all such payments on a fee for service basis to referral sources violate the antikickback provisions, it is clear that such arrangements are viewed very skeptically. Perhaps the prime example of this is the case of Dr. Greber discussed above. The fact that the payments to referring physicians in that case were styled as "interpretation fees" did not prevent Dr. Greber's conviction because the government was able to establish that the true intent behind the payments was to induce referrals.

The key to avoiding possible violations of the antikickback statute in situations where a referral source provides services on a fee-for-service basis is to structure the transaction in such a way that it is not subject to the inference that its true purpose is to induce referrals. A number of considerations must be kept in mind if this is to be successfully accomplished. First, it is essential that the payments be only for legitimate services that are either medically or administratively necessary. Mere "make work" duties that serve no valid purpose, or that would not be performed but for the desire to create a vehicle to channel money to a referral source, do not meet this test. For example, paying a referral source to set up medical equipment in a patient's home and to train the patient in its proper use clearly constitutes a legitimate service that would have to be performed regardless of whether or not payments were being made to referral sources. Conversely, if the referral source receives the payments for more ephemeral duties such as acting as a "liaison" between the patient and the supplier, the arrangement is much more likely to be treated as a sham used to cover the payment of referral fees. An attorney for the Inspector General has specifically warned that payments to referral sources for unnecessary or nonexistent services invite the scrutiny of the Inspector General:

> [The] creation of new or unique services is also suspect. For example, a DME [durable medical equipment] supplier comes to you and offers to pay you for something like "family counseling" or "patient communication." These types of make work duties could be construed as payment for referrals. [13, p. 15]

Second, even if the services for which the referral source is being paid are necessary, the payments can still form the basis for a violation of the antikickback statute if the health care provider paying for the services is not the true beneficiary of those services. Stated differently, the services must constitute something other than normal "doctoring."

A good example of this potential problem is the practice of durable medical equipment (DME) suppliers paying physicians to complete certificate of medical necessity forms. Under Medicare rules, in order for a DME supplier to be reimbursed for equpiment provided to a medicare beneficiary, the patient's attending physican must prescribe the equipment for that patient by completing a certificate of medical necessity form. Many DME companies engaged in the practice of paying physicians to complete these forms for their patients, some to the extent of sending a completed certificate of medical necessity form to the physician together with a check and a request that he or she sign the form and return it.

In correspondence responding to an inquiry concerning this type of practice, the Inspector General himself expressed the opinion that such payments are unlawful:

> In our view when a physician bills Medicare for his services to his patients including examination, diagnosis, and ordering or prescribing treatment, any costs associated with his paperwork (such as medical documentation, prescription, and certifying medical necessity for DME) are covered by his Medicare reimbursement for physician services. If the physician's patient genuinely needs the item of DME, when the physician certifies that need, he is performing a service for his patient—not the DME supplier. . . . Where it can be shown that payments were knowingly and willfully solicited by a physician as a prerequisite for patient referrals, both the physician and any DME company that knowingly and willfully makes such payments are violating the law. [25]

The third consideration is whether paying the referral source to perform a particular service makes any medical or business sense other than as a means to induce referrals. A classic example of an arrangement that runs afoul of this consideration is the one that resulted in the conviction of Dr. Greber. In the *Greber* case, physicians would order certain cardiac diagnostic tests, such as Holter monitors, from Dr. Greber's company. Dr. Greber would pay these physicians an "interpretation fee" for each such test they ordered. Payment of such an "interpretation fee" to the ordering physician, who would also be the ultimate recipient of the interpretation report, makes absolutely no medical or business sense. In the normal course of practice, a cardiac diagnostic company might refer the test results to a cardiologist for an interpretation and report, and then forward that report togehter with the test results to the referring physician. Alternatively, the results of the test would be returned to the referring physician without an interpretation report, and the referring physician would interpret the results for himself. In Dr. Greber's situation, the referring physician was being paid to write a report to himself. The lack of any logical reason for such an arrangement provided strong support for the inference that these "interpretation fees" were in fact unlawful referral fees.

The final, and perhaps most important, consideration in fee-for-service arrangements is whether the amount being paid approximates the fair market value of the services rendered. Any time the payment to a referral source exceeds fair market value, the payment is subject to the inference that the difference between the amount of the payment and fair market value constitutes an unlawful referral fee.

For a time it appeared that the government would take the position that fee-for-service arrangements could violate the antikickback law even where the sole payments made were for medically necessary services and were at fair market value. In September, 1984, the Health Care Financing Administration (HCFA) issued Part B Intermediary Letter no. 84-9. This transmittal was issued in response to questions raised concerning arrangements where DME suppliers paid respiratory therapists, employed in hospitals or other institutional settings, to set up the necessary equipment in the patient's home, instruct the patient in the proper use of the equipment, and/or perform monthly equipment maintenance. The Intermediary Letter took an extremely expansive reading of the statute, and stated:

> Others have argued that payments under such arrangements are only made for services rendered and are not made in return for referrals. The statute, however, proscribes payment or receipt of any remuneration that is intended to induce a referral. The opportunity to generate a fee is itself a form of remuneration. The offer or receipt of such fee opportunities is illegal if intended to induce a patient referral. Thus, a supplier who induces patient referrals by offering therapists fee-generating opportunities is offering illegal remuneration, *even if the therapist is paid no more than his or her usual fee*. [emphasis supplied]

Not surprisingly, the industry responded very strongly and very negatively to this unduly broad interpretation of the statute. After several months of industry protest, HCFA issued Program Memo B-85-2, which superceded Intermediary Letter 84-9. This program memorandum deleted the language quoted above, noting that:

> We have received a number of inquiries regarding the meaning of the sentences, and have concluded that they unduly prejudged the legality of certain referral arrangements which cannot be determined without the

consideration of the relevant factors and practice patterns described above.

Among the factors to be considered, as set forth in the program memorandum, are whether the respiratory therapist provides services only to those patients which he or she refers, whether therapists are used to perform these services in the absence of a referral from a therapist, whether unusual geographic or medical considerations support using therapists in this manner, and how such equipment is customarily installed and maintained by other suppliers in the area.

2. *Joint Ventures between Durable Medical Equipment Suppliers and Physicians or Other Referral Sources*

An increasing practice in the durable medical equipment industry is the creation of new and innovative arrangements, such as joint ventures, between DME companies and other health-care providers. These arrangements present many competitive advantages. They may permit a medical device supplier to expand its service area and provide a vehicle for raising needed capital, while at the same time creating investment opportunities for others. They may also form the basis for ensuring a future revenue stream. However, since at least some of the joint venturers will typically also be in a position to refer patients to the venture, extreme care must be taken to ensure that payment of joint revenues to these individuals cannot be construed as the payment of unlawful referral fees.

The legal ramifications involved in each of the multitude of possible structures of health care joint ventures is a topic far beyond the scope of this chapter. Rather, the following discussion is intended to provide the reader with general guidelines to aid in the recognition of potential legal problem areas so that competent advice can be sought to assist in the structuring of the venture when appropriate.

At the outset, it is important to note what the statute does *not* prohibit. Both HCFA and the Office of the Inspector General have consistently interpreted the antikickback statute not to prohibit, per se, a health-care provider, such as a physician, from having an ownership interest in a health-care entity or joint venture to which he or she may make referrals. However, the reader should be aware that there is a trend toward the prohibition of such ownership interests.

In the absence of an express ban on physician or other referral source ownership in health-care joint ventures, the primary concern in structuring such ventures must be whether they run afoul of the existing antikickback provisions. While the ultimate resolution of these issues may be quite complex, the analysis to be used in reaching such a conclusion consists of two basic inquiries. The first is whether the joint venture vehicle is a bona fide business venture or simply a sham transaction constructed to camouflage the payment of unlawful referral fees. The second is whether the monies received by the investors constitute legitimate returns on investment or are compensation for the referral of patients.

Turning to the first inquiry, the bona fides of the venture, a number of factors are to be considered. The first is whether there is a legitimate business purpose underlying the joint venture. If a venture makes absolutely no business sense save for a desire to funnel money to referral sources, there is a high probability that it could be found in violation of the antikickback statute.

Even if the venture has a legitimate business purpose on its face, it may still be found to be an unlawful sham if it is not operated as a legitimate business. A number of factors must be considered in this regard. First, it is essential that the venture be adequately capitalized to carry out the purpose for which it was formed. Second, the investors in the venture, typically actual or potential referral sources, must have paid fair value for the percentage of ownership they are given, and the full amount of that investment must be at risk in the venture. Care must also be taken in choosing the individuals who will participate in the joint venture. If participation is limited to only those who are in a position to refer business to the venture, or if participants are required to sell back their interest in the event that they lose the ability to refer to the venture, that is, by retirement or relocation, there is some risk that the venture will be found to violate the antikickback provisions. Similarly, it is imperative that there be no requirement, either express or implied, that participants must refer all or even a part of their health-care business to the venture.

The next key concern is the amount and manner in which payments are made during the term of the venture. Simply stated, it is essential that such payments either be limited to a legitimate return on the participant's investment, or constitute the pay-ment of fees for services actually rendered. An attorney for the Inspector General has specifically addressed this issue:

> Any arrangement whereby individuals receive remuneration from an entity to which they refer Medicare beneficiaries for services will raise serious questions if the remuneration (1) is not a dividend based exclusively on a legitimate (i.e., purchased) ownership interest in the entity or (2) is not for actual services rendered to the entity. Even where one of the above conditions apply, the arrangement may be questioned if (1) the apparent ownership interest appears to be merely a facade to shield a so-called "limited partner" who is the source of referrals from potential liability under the kickback provision (i.e., the dividends actually paid do not bear a direct relationship to the individual's capital investment share) or (2) as was the case in *Greber* the payment, although justified on the basis of services rendered by the individuals, appears intended primarily to induce referrals and bear some relationship to the number of referrals. [13, p. 15]

In addition, the Inspector General has long indicated that where the return on investment given to referal sources in health-care joint ventures is excessive, when compared to rates of return for comparable investments, the arrangement will be subjected to close scrutiny.

Finally, in an effort to ensure the protection of the patient's freedom of choice, it is advisable to require that joint-venture participants who refer patients to the venture disclose the existence of their ownership interest to the patient and inform them that they are free to seek the needed services from this or any other provider of their choice. Any arrangement that abrogates this freedom by channeling patients toward one particular health care provider is subject to scrutiny .

In April, 1989, the Inspector General issued a "fraud alert" to the entire health care industry addressing the issue of health care joint ventures generally, as well as laboratory and DME joint ventures specifically. The fraud alert identified a number of what the OIG views as "questionable features," the existence of which would render a joint venture "suspect." These "questionable features" were divided into three categories: (1) the manner in which investors are selected and retained; (2) the business structure of the joint venture; and, (3) the method of financing and profit distribution.

Under the investors category, the OIG indicated that it views a venture as suspect if: (1) investors are chosen because they are in a position to make referrals; (2) physicians expected to make a large number of referrals are offered a larger investment share; (3) investors are encouraged both to make referrals and to divest their ownership if they fail to refer; (4) the venture distributes information regarding the referral patterns of investors to other investors; and, (5) investors must divest their ownership if they lose the ability to refer, e.g., by ceasing practice in the service area.

The fraud alert goes on to state that a DME joint venture is suspect if the business structure is such that it can be characterized as a "shell." The OIG's example of such a "shell" DME venture is one where one of the joint ventures is an existing DME supplier that is joint venturing with referral sources, the venture itself owns very little of the DME or other capital equipment, and the existing DME supplier is responsible for all day-to-day operations of the venture.

Finally, the fraud alert notes that the financing and profit distribution methodology of the joint venture may come under scrutiny if the amount of capital invested by referral sources is either nominal or disproportionately small in comparison with anticipated returns. Similarly, the OIG views it as suspect if returns are "extraordinary," i.e., over 50 to 100% annually. Finally, the OIG indicates that mechanisms whereby investors may "borrow" the amount of their investment from the entity itself, and pay back the loan through deductions from profit distributions, render the venture suspect.

3. *Waivers of Medicare Co-Insurance and Deductible Amounts*

An increasingly common technique used by DME suppliers to attract business is the routine waiver of Medicare deductibles and co-insurance amounts for Medicare beneficiaries who do not have supplemental insurance to cover the payment of these amounts. Medicare, like most other health insurance plans, requires the beneficiary to pay an annual deductible amount before benefits will be paid, as well as a "co-insurance" amount that is generally 20% of the Medicare allowed charge [26]. Routine waivers of any obligation to pay these amounts are often marketed to beneficiaries under the label of "no out of pocket expenses" or "no cost to you."

In recent years, this practice has been the subject of some controversy. Both Inspector General Kusserow and the Chief of the Criminal Division of the Department of Justice have publicly stated that routine waiver of co-insurance programs constitute at least a technical violation of the antikickback statute [27,28]. The theory underlying this interpretation is that the Medicare beneficiary is being offered something of value, the forgiveness of a debt, in exchange for utilizing the services of that specific health-care provider. While this interpretation does have support in the language of the statute, the issue remains an open question. The only federal court to be presented with this issue to date suggested, without deciding, that the fact that the statute does not *require* health-care providers to collect Medicare Part A co-insurance and deductible amounts presents a "rather convincing" argument that this type of program is not within the intended scope of the antikickback prohibition [29].

Despite this potential ambiguity, the federal enforcers have made increasingly strong statements condemning the practice under both Part A and Part B. In recent correspondence between the Secretary of Health and Human Services and Senator Edward M. Kennedy, Secretary Bowen stated his view that: "As a general matter, charging patients a co-insurance amount discourages overutilization of care and the provision of medically unnecessary items or services." He went on to state that because of this "[you] can be sure that the Inspector General will use his kickback exclusion authority to enforce the charging of co-insurance amounts in settings . . . where those charges are an important deterrent to overutilization, or where lack of enforcement would lead to unfair competition or improper cost to the program" [30].

Co-insurance waiver programs instituted by DME suppliers present a second and potentially more serious problem. Effective January 1, 1989, Medicare allowable payments for DME are determined as the lower of the supplier's actual charge or a state wide fee schedule amount based on the actual allowable during a historical data period. The Inspector General has publicly stated that: "Where a Part B supplier routinely waives collection of the co-insurance amount, he may be misrepresenting his 'actual' charges" [27, p. 34]. Under this theory, if a supplier billed $100 for a piece of equipment and then routinely waived the $20 co-insurance amount, the submission of a bill to Medicare for $100 would constitute a misrepresentation of the supplier's actual charge. This is so because the supplier never expected to collect more than the $80. The filing of such a claim, according to the Inspector General, constitutes a false claim, which can be the subject of a variety of very severe criminal and/or civil penalties. (The nature and type of these penalties are discussed in Section IV.)

Because of the foregoing, DME suppliers are well advised not to engage in any program involving the *routine* waiver of Medicare co-insurance and deductible amounts. However, it is important to note that co-insurance and deductible amounts can be waived on a case-by-case basis where the patient is unable to pay or where the costs of collection are unwarranted given the amount involved and the likelihood of successful collection.

4. Consumer Rebates

A relatively common retail marketing practice is to offer consumers a cash rebate when certain products are purchased. This practice is only recently being utilized in the health care industry. While this practice may constitute an effective marketing tool, it does present risks to the DME supplier who engages in it if the equipment will ultimately be paid for by Medicare or Medicaid.

Like the routine waiver of deductible and co-insurance amounts discussed above, one can construct an argument that this practice constitutes a technical violation of the antikickback statute. Rebates are expressly included in the definition of remuneration under the statute. Further, they are clearly offered in an attempt to induce the recipient to purchase a particular item of medical equipment. Moreover, the Office of the Inspector General has recently given some indication that, in their opinion, such programs do violate the antikickback statute.

In addition to the potential antikickback violation, such programs create the potential for allegations of false claims. As in the case of waived co-insurance, if the amount of the rebate is not reflected in the charge for the item that is submitted to Medicare of payment, it arguably constitutes a false claim for Medicare reimbursement. In view of this, DME suppliers should be extremely cautious in implementing any sort of rebate program.

IV. FALSE CLAIMS

The primary focus of the Medicare fraud and abuse enforcement effort has involved the filing of false or fraudulent claims for Medicare or Medicaid reimbursement. Such claims may take a variety of forms in the durable medical equipment business. For example, they may include situations where claims are filed for equipment that was not provided, claims where the medical necessity for the equipment was falsified, claims where the actual charge for the equipment is falsified, as in the case where co-insurance amounts are routinely waived, and "double billing" situations where more than one claim is filed for the same equipment. The filing of such claims can subject the DME supplier to a broad range of very severe criminal and civil penalties. The various criminal and civil penalty statutes that apply to these situations are discussed in detail below.

A. Criminal Penalties

It is a felony under the Medicare program to knowingly and willfully make or cause to be made any false statement or representation of any material fact in any claim for Medicare or Medicaid reimbursement. It is also a felony for an individual with knowledge of facts that might affect whether such a claim will be paid or the amount of such payment, to intentionally conceal or fail to disclose that fact. Violations of these provisions are punishable by imprisonment of up to 5 years and/or a fine of up to $25,000 [31].

In addition to the criminal statute specifically targeted toward Medicare and Medicaid false claims, there are a number of general federal criminal statutes that frequently are used by prosecutors in health-care cases. The first of these is the mail-fraud statute, which generally prohibits any scheme to defraud that in any way uses the U.S. mails, such as where a claim form or reimbursement check is sent through the mail [32]. Additionally, the criminal False Claims Act prohibits any false claim against the federal government [33]. Finally, there is a general federal statute prohibiting false statement to a government agency with regard to any matter under its jurisdiction [34]. In instances where more than one person is involved in the filing of false claims, prosecutors may also utilize the criminal statutes prohibiting conspiracies to defraud the United States as well as those that prohibit aiding and abetting the commission of a federal offense [35].

There has been a recent trend toward the use of federal antiracketeering statutes against health-care providers. These statutes prohibit the use of any monies derived directly or indirectly from a pattern of racketeering activity [36]. "Racketeering activity" is defined to encompass acts in violation of the federal mail fraud statutes, which, as noted above, includes the filing of false Medicare or Medicaid claims [37]. The penalties for violations of the antiracketeering laws are extremely severe. Violations are punishable by a fine of not more than $25,000 and/or imprisonment of not more than 20 years [38]. In addition, under the forfeiture provisions of the statute a convicted person may be forced to forfeit any property that was obtained through the use of funds derived from the racketeering activity [38].

In addition to these federal criminal statutes, every state has laws that prohibit fraud generally and that address the crime of Medicaid fraud specifically. These statutes increasingly are being used by state prosecutors in cases of health-care fraud.

Prosecution under these criminal statutes carries with it substantial procedural protections. First, in the federal system, no felony charges may be brought against a health-care provider until a grand jury has voted to return an indictment. In addition, an indicted individual has the right to a jury trial and cannot be convicted unless the jury finds him guilty "beyond a reasonable doubt." As will be discussed below, such procedural protections do *not* exist with regard to many of the civil penalties that can be imposed on health-care providers who file false claims.

B. Civil Penalties

In addition to the criminal penalties described above, there exists a panoply of severe civil penalties that can result from such conduct. The two most frequently used civil penalties are civil monetary penalties and exclusions from program participation. "Exclusion" means that no Medicare or Medicaid payment can be made to anyone (including the patient) for services or supplies rendered, ordered, or supervised by the excluded individual or entity [39]. The duration of such exclusions has ranged from 1 year to 25 years. For a DME supplier that relies in any significant part on providing supplies and equipment to Medicare and Medicaid patients, the imposition of such an exclusion is tantamount to being forced to close the business.

The impact of the imposition of civil monetary penalties can be even more devastating. Under the statute, a penalty of up to $2000 for each false line item on a claim for Medicare or Medicaid reimbursement may be imposed. In addition, an "assessment in lieu of damages" of up to double the amount *charged* (without limitation as to the amount of the charge that was improper) for each such line item may also be imposed [40]. As can readily be seen, a very small number of violations can rapidly generate a potential civil penalty far in excess of the amount of reimbursement that might have resulted from the filing of the claims in question. For example, in a case where the diagnosis code supporting medical necessity on a Medicare claim for five pieces of medical equipment was falsified, and the total charge for the equipment was $500, a penalty of $10,000 together with an assessment of $1000 could be imposed for that one claim. This amount could be imposed even if Medicare had denied payment on the claim. In one case, a court upheld the imposition of civil money penalties and assessments totalling over 1.8 million dollars in a case which the total amount of Medicare overpayment was less than $25,000 [41].

Further, while criminal penalties may only be imposed on those who "knowingly and willfully" file false Medicare or Medicaid claims, these civil penalties may be imposed upon those who file false claims through mere negligence, without any intent to defraud the Medicare or Medicaid programs. The statutory language setting forth liability under the statute permits the imposition of penalties and assessments on any individual or entity that either submits, or causes to be submitted, a claim for Medicare or Medicaid reimbursement that contains line items that they "know or *should know*" were either "not provided as claimed" or are "false, fictitous, or fraudulent" [42]. The use of the "know or should know" standard encompasses the full spectrum of conduct from the knowing and intentional submission of false claims, to the merely negligent submission of an incorrect claim in instances where the person had no reason to believe that the claim was incorrect, but could have determined that it was upon "reasonable" inquiry. The potential impact of this broad standard of intent is exascerbated by the fact that, in the recent Medicare Catastrophic Coverage Act of 1988 [43], the statute was amended to make a principal fully liable for any acts of his or her agent in violation of the statute [44]. The net effect of this amendment is to make a DME supplier, and possibly its principals, fully liable for severe civil money penalties and assessments based on errors by a billing clerk of which they had no knowledge.

The regulations that implement the civil money penalties law contain a listing of aggravating and mitigating factors that are to be used in determining how close the actual penalty and assessment imposed should be to the statutory ceiling of $2000 per line item plus double the amount claimed [45]. In addition to the statutory maximum amount of penalties and assessments, the regulations indicate that the *minimum* amount that will be imposed is twice the amount of the Medicare or Medicaid payment resulting from the filing of the claim at issue [46]. The factors considered in this analysis include the nature of the claims and the circumstances under which they were submitted, the degree of culpability of the individual (i.e., whether the false claim was filed intentionally or negligently), whether there were any prior related offenses, the financial condition of the person, as well as a general category of any "other matters as justice may require" [47].

A recent civil monetary penalty action brought against a DME supplier is illustrative of how these factors are applied by the Office of the Inspector General in the context of an actual case [48]. The allegations in this case were that a small DME supplier had submitted 50 claims containing 76 improper line items for the rental of certain equipment such as alternating pressure pads and pumps, quad canes, trapeze bars, wheelchairs, hospital beds and mattresses, bedrails, and commode chairs. The Inspector General alleged that the particular rental items at issue either had never been provided to the patients or had been returned to the company prior to the rental period for which payment was being sought. The total amount claimed for these rental items was $4158. However, apparently as a result of denials of payment by the Medicare carrier, the total overpayment resulting from these allegedly false claims was only $46.40. Despite this extremely small amount of financial harm to the Medicare program, the Inspector General sought penalties of $1000 for each line item as well as double the amount claimed for that line item, for a total proposed penalty of $76,000 and a total proposed assessment of $8317.

A number of factors were cited by the OIG to support this large amount. Among them were the fact that the conduct occurred over a relatively lengthy period of time (2 years) and that two separate billing schemes were involved. However, the Inspector General was most concerned by the marketing strategy utilized by this particular DME supplier, which, the Inspector General alleged, resulted in the provision of a large volume of medically unnecessary services.

As described by the Inspector General, this marketing strategy consisted of the "direct solicitation of beneficiaries, the representation to beneficiaries that these durable medical equipment services were provided without charge to beneficiaries, the provision of completed medical necessity forms to physicians for their signature after the services had been initiated, the failure to provide beneficiaries the option of purchasing (as opposing to renting) this equipment . . ., and the provision of multiple pieces of equipment without regard to medical necessity" [48].

Despite the potentially devastating impact of the imposition of such civil penalties, an accused individual does not have the same rights available to a defendant in a criminal matter. For example, since these cases are tried before an administrative law judge rather than in federal court, there is no right to a jury trial. In addition, guilt need only be proven by a "preponderance of the evidence" rather than the far more restrictive "beyond a reasonable doubt" standard applicable in criminal proceedings [49]. Finally, the strict rules of evidence applicable in federal court proceedings are not applicable in administrative proceedings such as these [50]. As a result, it is possible for the imposition of these penalties to be based exclusively on hearsay evidence.

Two other civil sanctions are applicable to the filing of false Medicare and Medicaid claims. The Program Fraud Civil Remedies Act is modeled on the civil monetary penalty statute discussed above and is applicable to any type of government claim, including Medicare and Medicaid claims [51]. It permits the imposition of civil monetary penalties of up to $5000 per claim, as well as damage assessments of twice the amount of the false portion of the claim [51]. As above, liability can be imposed for negligently filed false claims, and actions under it are brought administratively before an administrative law judge [51]. The Federal False Claims Act permits the imposition of penalties between $5000 and $10,000, plus treble the damages sustained by the government, for any false or fraudulent claim for payment presented to the United States, including Medicare and Medicaid claims [52]. Liability under this statute is limited to the knowing and willful presentation of false claims. Actions under it must be brought in federal court, where the accused is entitled to a jury trial and the proceedings are subject to the strict federal rules of evidence [53]. Actions under the False Claims Act are typically brought by the government through the Department of Justice. However, the False Claims Act has a unique provision that permits private citizens to bring actions on behalf of the government [54]. If such actions are successful, the person bringing the action is entitled to a "bounty" consisting of a percentage of the penalties and damage amounts imposed as well as attorney's fees [55]. Under this provision, known as the *qui tam* provision, actions could be brought against a DME supplier by a competitor or by a disgruntled former employee.

V. CIVIL SANCTIONS FOR VIOLATIONS OF MEDICARE PROGRAM RULES

There has been an ever-increasing trend in Congress to pass legislation that permits the imposition of civil penalties and/or exclusion for violations of certain discreet Medicare reimbursement or other rules. While most of these new penalty provisions have been directed specifically toward physicians, there are several that can have a direct effect on manufacturers and suppliers of durable medical equipment. These provisions are discussed below.

A. Violations of Medicare Assignment or Participation Agreements

Companies supplying durable medical equipment to Medicare beneficiaries may opt to accept "assignment" from the Medicare program on a claim-by-claim basis. By accepting assignment, the supplier agrees that it will not bill the patient more than 20% of the total Medicare-allowed charge for the equipment, the "co-insurance" amount, together with any applicable deductibles [56]. (Note that the *routine* waivers of this amount can also result in sanctions; see Section III.B.3.) In exchange for agreeing to limit charges to the patient, the supplier receives payment for 80% of the Medicare-allowed charge directly from the Medicare carrier. Suppliers may also elect to sign "participation agreements" [57]. By signing such an agreement, the supplier commits to accepting assignment on *all* of its claims for Medicare reimbursement.

The knowing and willful violation of an assignment or a participation agreement, if done repeatedly, is a misdemeanor that is punishable by up to 6 months of imprisonment and/or $2000 in fines [58]. In addition, violations of these agreements, even if not knowing and willful, are subject to a civil money penalty of up to $2000 per violation as well as exclusion from the Medicare and Medicaid programs [59].

Further, the reader should be aware that virtually all state Medicaid programs prohibit the billing of Medicaid beneficiaries for any item or service that is covered by the state Medicaid program. Persons who charge beneficiaries in violation of these statutes are subject to the criminal conviction and the same civil penalties applicable to participation and assignment agreement violations [60]. In addition, there may be other penalties under state law.

B. Violations of New Limits on Patient Charges for Rental of Medical Equipment

The Omnibus Budget and Reconciliation Act of 1987 enacted the so-called "Six-Point Plan" for Medicare reimbursement of durable medical equipment [61]. A full discussion of this new method of Medicare reimbursement is contained in Chapter 53. For the purposes of this chapter, the reader should be aware that the violation of certain requirements of this plan can result in the imposition of civil money penalties and/or exclusion. Specifically, under this new reimbursement scheme, where a supplier furnishes durable medical equipment on a rental basis, its charges are subject to a 15-month rental cap, which is based on the purchase price of the equipment in question. If the supplier charges the Medicare beneficiary for the rental of such equipment after the expiration of the 15-month cap (except for authorized servicing charges), it is subject to civil monetary penalties of up to $2000 per violation as well as an exclusion from Medicare and Medicaid participation for a period of up to 5 years [63].

C. Misleading Use of the Social Security or Medicare Names or Symbols

An increasing phenomenon in the health-care industry is the use of advertising that seeks to place the health-care provider in a favorable light by implying an official connection with the Medicare or Social Security programs. In the Medicare Catastrophic Coverage Act of 1988, Congress authorized the imposition of civil money penalties for those who engage in this practice [63]. Under this legislation, any person who, through advertisements or other communications, uses the terms Social Security or Medicare or other related terms in such a way as to create the false impression that the items or services they provide are "approved, endorsed, or authorized" by the Department of Health and Human Services is subject to a civil money penalty of up to $5000 per violation. Where the violation consists of a radio or television broadcast, the civil money penalty may be up to $25,000. These penalties are subject to a cap of $100,000 where they pertain to multiple substantially identical communications occurring in the same year [64].

D. Miscellaneous Other Bases for Exclusion from Medicare and Medicaid Participation

The Medicare statute sets forth a long list of conduct that can form the basis for the exclusion of a supplier of durable medical equipment from participation in the Medicare or Medicaid programs. These bases are divided into two categories; those that result in a mandatory exclusion of not less than 5 years, and those that, in the discretion of the Inspector General, may result in an exclusion.

Exclusion for a minimum of 5 years is required where any individual or entity has been convicted of a criminal offense relating to the provision of items or services under the Medicare or Medicaid programs, or has been convicted of a criminal offense relating to patient neglect or abuse [65]. The following circumstances can lead to a permissive exclusion:

1. Convictions relating to fraud, theft, embezzlement, breach of fiduciary duty, or other financial misconduct relating to the delivery of health care in general or to participation in any government program.
2. Convictions for obstruction of justice.
3. Convictions relating to controlled substance violations.
4. Revocation or suspension of a state license to provide health care.
5. Exclusion from any other federal health-care program.
6. Charging Medicare or Medicaid substantially in excess of "usual charges."
7. Providing items or services to patients that either are substantially in excess of their needs or that fail to meet professionally recognized standards of health care.
8. Entities in which an individual, who either has an ownership or control interest or is an officer, director, agent or managing employee, has been excluded from Medicare or Medicaid.
9. Failures to make certain required disclosures of information.
10. Failures to grant "immediate access" to a facility or to records or documents, upon "reasonable request" by the Secretary of Health and Human Services, the Inspector General, the Medicaid state agency, or the state Medicaid fraud control unit [66].

As with the other exclusions discussed above, the effect of these actions is to bar any Medicare payment for any services rendered by the excluded individual or, in the case of physicians, for services rendered on their order or under their supervision [67].

VI. CONCLUSION

The growing array of Medicare and Medicaid antifraud and abuse provisions has created a mine field for the durable medical equipment supplier who is struggling to stay competitive in today's market. Only by keeping abreast of new developments in the law and being aware of those types of transactions that can result in problems can this mine field successfully be negotiated.

REFERENCES

1. Pub. L. 89-97, 1872.
2. See, e.g., 18 U.S.C. 287.
3. Pub. L. 92-603, 242.
4. Pub. L. 95-142, 4.
5. Pub. L. 100-93, 2.
6. See, e.g., Pub. L. 99-509, Pub. L. 100-203.
7. 42 U.S.C. 3521.
8. See 50 *Fed. Reg.* 45,438 (October 31, 1985).
9. Richard P. Kusserow, Inspector General, DHHS, *Semiannual Report to the Congress*, April 1, 1987 to September 30, 1987.
10. 42 U.S.C. 1320a-7b(b).
11. 97 *Cong. Rec.* H9818 (daily ed., September 22, 1977).
12. Letter from Don Nicholson, Director, Office of Program Integrity, to Honorable Leon E. Panetta, dated October 30, 1978.
13. D. McCarty Thornton, Esq., Medicare Fraud and Abuse, "Let's Be Careful Out There!", *American Association of Respiratory Therapists Times*, August 1985.
14. 42 U.S.C. 1320a-7b(b)(3)(A).
15. 42 U.S.C. 1320a-7b(b)(3)(C).
16. 42 U.S.C. 1320a-7b(b)(3)(B).
17. 97 *Cong. Rec.* H9818 (daily ed., September 22, 1977).
18. 42 U.S.C. 1320a-7b(b)(3)(D).
19. S. Rep. No. 109, 100th Cong., 1st Sess. 27.
20. 52 *Fed. Reg.* 38,794 (October 19, 1987).
21. 54 *Fed. Reg.* 8033 (Jan. 23, 1989).
22. 760 F.2d 68 (3d. Cir. 1985).
23. 760 F.2d 72 (3d. Cir. 1985).
24. *United States v. Kats*, 871 F.2d 105 (9th Cir., 1989); *United States v. Bay State Ambulance and Hospital Rental Service*, 874 F.2d 20 (1st Cir. 1989).
25. Letter from Richard P. Kusserow, Inspector General, to Arunaba Das, M.D., dated October 27, 1987.
26. 42 U.S.C. 1842(b)(3)(B)(ii).
27. Letter from Richard P. Kusserow, Inspector General, to Stephen S. Trott, Assistant Attorney General, Criminal Division, Department of Justice, dated April 17, 1985.
28. Letter from Stephen S. Trott to Richard P. Kusserow, dated October 30, 1985.
29. *West Allis Memorial Hospital, Inc. v. Bowen*, Civ. Action 87-C-0053 (E.D. WIS. May 28, 1987), *aff'd* 852 F.2d 251, (7th Cir., 1988).

30. Letter from Otis R. Bowen, M.D., Secretary, U.S. Department of Health and Human Services to Senator Edward M. Kennedy, dated November 23, 1987.
31. 42 U.S.C. 1320a-7b(a).
32. 18 U.S.C. 1341.
33. 18 U.S.C. 287.
34. 18 U.S.C. 1001.
35. 18 U.S.C. 2, 371.
36. 18 U.S.C. 1962.
37. 18 U.S.C. 1961.
38. 18 U.S.C. 1963.
39. 42 U.S.C. 704(b)(6), 1395y(e), 1395(j)(2), 1396b(i)(2), 1397d(a)(9).
40. 42 U.S.C. 1320a-7a(a).
41. *Mayers v. DHHS*, 806 F.2d 995 (11th Cir. 1986) cert. denied, 108 S.Ct. 82 (1987).
42. 42 U.S.C. 1320a-7a(a)(1)(A) and (B).
43. Pub. L. 100-360.
44. 42 U.S.C. 1320a-7a(*1*).
45. 42 C.F.R. 1003.106.
46. 42 C.F.R. 1003.106(c)(3).
47. 42 C.F.R. 1003.106(a).
48. Letters from Eileen T. Boyd, Deputy Assistant Inspector General, to Medical Rental Unlimited, Inc., dated March 13, 1987, and August 13, 1987.
49. 42 C.F.R. 1003.114(a).
50. 42 C.F.R. 1003.118.
51. 31 U.S.C. 3801-3812.
52. 31 U.S.C. 3729.
53. 31 U.S.C. 3730.
54. 31 U.S.C. 3730(b).
55. 31 U.S.C. 3730(d).
56. 42 U.S.C. 1395u(b)(3)(B)(ii).
57. 42 U.S.C. 1395u(h)(1).
58. 42 U.S.C. 1320a-7b(e).
59. 42 U.S.C. 1320a-7a(a)(2)(A).
60. 42 U.S.C. 1320a-7a(a)(2)(B), 1320a-7b(d).
61. Pub. L. 100-203, 4062.
62. 42 U.S.C. 1395m(a)(11)(A).
63. Pub. L. 100-360, 428.
64. 42 U.S.C. 1320b-10.
65. 42 U.S.C. 1320a-7(a).
66. 42 U.S.C. 1320a-7(b).
67. 42 U.S.C. 1320a-7(b) n. 47.

55

Complying with Antitrust/ FTC Restrictions

ARNOLD C. CELNICKER *The Ohio State University, Columbus, Ohio*

I. INTRODUCTION

This chapter deals primarily with antitrust issues affecting distribution in the medical device industry. Antitrust law is invariably complex. It is often necessary to conduct an economic analysis of competitive conditions in specific markets to determine the legality of certain practices. It would be foolhardy for businesspeople to rely on this chapter to determine the legality of questionable practices. This chapter is not a substitute for antitrust counsel.

This chapter will alert the reader to the most common antitrust issues involved in the distribution of medical equipment and supplies. It will also provide some of the basic rules used by the law in resolving those issues. This should enable the reader to recognize potential problems and, sometimes with the help of counsel, prevent a potential problem from maturing into an actual problem. In addition, this chapter should help the businessperson communicate effectively with counsel regarding antitrust issues.[1]

Most businesspeople are aware that during the 1980s the antitrust laws have not been vigorously enforced by the federal government, with the exception of situations involving horizontal price fixing. In addition, a number of Supreme Court decisions in the past decade plus the changing complexion of the judiciary have been making antitrust judgments in favor of private plaintiffs fairly uncommon. These facts may lead businesses to take more legal risks than they might have found prudent a decade ago. In considering the optimal amount of antitrust exposure, however, there are other facts that should not be ignored.

[1]The importance of business ethics, which is interrelated with many antitrust legal issues, is not meant to be minimized, but is simply beyond the scope of this chapter.

First, contrary to the general trend, the health-care industry has been under increasing antitrust scrutiny during the past decade. Supreme Court cases starting in the mid 1970s have made it clear that health-care professionals and the industry enjoy no special status under the antitrust laws. This, coupled with continuing concern about the growing costs of health care and the role that competition might play in reducing those costs, has led to unprecedented antitrust activity in the health-care industry. The federal government has made antitrust enforcement in the health-care industry a top priority, and private cases abound.

Second, states have increased their antitrust enforcement activities. Many state attorneys general feel that the generally decreasing federal enforcement requires increased scrutiny by the states. Although the state antitrust laws vary from state to state, generally they closely reflect the federal laws. This chapter will discuss only federal law. However, it must be remembered that each state also has antitrust laws, and many states are increasing their antitrust efforts.

Third, while fewer private antitrust suits succeed now than did in the 1970s, antitrust suits continue to be highly disruptive and costly endeavors. Therefore, even with an increased probability of victory for the defendant, they are not to be courted.

Information regarding the risks and enforcement efforts in specific areas will be included in subsequent sections. Generally, members of the medical device industry would be well advised not to jettison antitrust compliance efforts prematurely.

II. OVERVIEW OF THE ANTITRUST LAWS

As America became an industrial nation during the late 1800s, Congress feared that economic and political power was becoming concentrated in the hands of a relatively few "robber barons." The response was the Sherman Act of 1890, and the Clayton Act and Federal Trade Commission Act of 1914.

A. The Sherman Act

The heart of the Sherman Act is section 1, which says: "Every contract, combination, . . . or conspiracy, in restraint of trade or commerce among the several states, or with foreign nations, is declared to be illegal." In essence, the law prohibits agreements that unreasonably restrain trade.

Note that section 1 of the Sherman Act does not come into play when one company is acting unilaterally. An agreement (contract, combination, or conspiracy) between two or more separate companies is always required before the law can possibly be violated. For those purposes, an "agreement" between a corporation and its wholly owned subsidiary is not an agreement between two separate companies and therefore can never be illegal under section 1.

If there is an agreement, it is illegal only if it *unreasonably* restrains trade. Determining which agreements unreasonably restrain trade and which are reasonable restraints is the crux of most section 1 analysis.

Over the years, the courts have decided that certain types of agreements are always unreasonable restraints. Examples of these are horizontal price-fixing agreements, horizontal allocations of territories or customers, and vertical price-fixing agreements. These are discussed in Sections III and IV.A, later in this chapter. Agreements of these types are labeled per se violations, meaning that the agreement by itself is illegal without any analysis of its actual effect on competition or the market. In other words, such agreements are presumed to always unreasonably restrain trade.

Other types of agreements may reasonably or unreasonably restrain trade depending on the facts of each particular case. For example, vertical allocations of customers or territories, and exclusive dealing arrangements fall in this category.[2] These are discussed in Sections IV.B and IV.C. These agreements may be reasonable or unreasonable, and therefore legal or illegal, depending on whether an analysis of the agreement's purposes and effects indicates that, on net, it enhances competition or lessens competition. These types of cases are labeled "rule of reason" cases because an analysis to determine if the agreement reasonably or unreasonably restrains competition determines whether it violates section 1 of the Sherman Act.

The other major prohibition contained in the Sherman Act is section 2, which makes monopolization illegal. Legitimate monopolization cases are relatively rare. Therefore, section 2 will not be examined in this chapter. Firms fortunate enough to have a dominant position in a market might wish to explore the contours of monopolization with counsel.

B. The Clayton Act and The Robinson-Patman Act

In 1914, Congress passed the Clayton Act to supplement the Sherman Act. Section 3 prohibits exclusive dealing and tying arrangements.[3] Exclusive dealing occurs when a manufacturer and its customer agree that the customer will not carry the manufacturer's competitors' lines. The legality of exclusive dealing is considered in Section IV.C.

Tying arrangements are agreements to sell or lease one product only on condition that the customer purchase another product from the same seller or lessor. For example, if a manufacturer of X-ray equipment requires its customers to obtain plates from it, that would be a tying arrangement. The legality of such agreements will be discussed in Section IV.D.

Section 2 of the Clayton Act is commonly known as the Robinson-Patman Act. It prohibits price discriminations under certain circumstances. Price discriminations occurs when a seller charges different prices to different customers. A detailed discussion is presented in Section V.

The Clayton Act also includes section 7, which prohibits mergers that may substantially lessen competition. Merger law will not be considered in this chapter.

C. The Federal Trade Commission Act

The Federal Trade Commission Act established the Federal Trade Commission (FTC) in 1914. The FTC is an independent federal agency headed by five

---

[2] Exclusive dealing cases are technically brought under either Clayton Act section 3 or Sherman Act section 1, depending on whether goods or services are involved. For our purposes, the distinction is unimportant.

[3] Exclusive dealing and tying arrangements involving services, as opposed to goods, are covered by section 1 of the Sherman Act.

commissioners. The commissioners are appointed by the President, subject to Senate approval.

The FTC has authority to enforce the Clayton Act and Federal Trade Commission Act, but not the Sherman Act. However, section 5 of the Federal Trade Commission Act makes "unfair methods of competition" unlawful. Because the courts have said that any violation of the Sherman Act also violates section 5 of the Federal Trade Commission Act, the FTC can challenge any practices that violate the Sherman Act by bringing a case under section 5 of the Federal Trade Commission Act.

D. Enforcement of the Antitrust Laws

At the federal level, the antitrust laws are enforced by the Federal Trade Commission and the Antitrust Division of the Department of Justice. Each agency has an active unit dealing with the health-care industry. The agencies cooperate with each other in deciding which will handle a particular case.

Sherman Act violations can be brought either criminally or civilly. Only the Department of Justice can bring criminal cases. Criminal cases are generally limited to horizontal price fixing. The penalty for corporations is a fine up to $1 million, and for individuals a fine up to $100,000, or 3 years in prison, or both.

Noncriminal violations, whether brought by the Department of Justice or the FTC, usually result in an order prohibiting future violations. Most cases are settled without a trial.

The antitrust laws are also enforced through private suits. In fact, about 90% of antitrust suits are private actions. A person or corporation that proves it was injured because of an antitrust violation can recover three times the amount of damage suffered plus attorney's fees and costs. This provides a significant incentive for private suits.

It is important to note that private suits and government suits are not mutually exclusive. A government suit will typically be followed by a private suit based on the same activity.

In limited circumstances, state governments also have authority to sue on behalf of the state's citizens who have been injured by a violation of the federal antitrust laws. Such suits are relatively uncommon. At the state level, it is more common for the state to sue for violation of the state's antitrust laws, which are often similar to the federal law.

Enforcement issues and trends relating to specific types of violations will be discussed in the subsequent sections dealing with those violations.

III. HORIZONTAL RESTRAINTS

This chapter deals primarily with antitrust issue arising out of the relationship between the manufacturer and its distributors or customers, called vertical or distributional restraints. Although the majority of antitrust suits involve vertical restrains, the most serious cases, especially from a federal enforcement perspective, involve horizontal restraints. This section will briefly consider horizontal restraints.

A. Horizontal Price Fixing

Horizontal restrains involve agreements among competitors at the same level of the distribution chain. The cadillac of antitrust violations, horizontal price fixing, is an agreement among competitors to increase, decrease, peg, stabilize, or otherwise affect the price of a product. Examples of agreements that would likely be considered price-fixing agreements include agreements among competing manufacturers not to extend more than 30 day's credit to distributors, or not to advertise except in trade journals, or not to produce generic or private-label products. Such agreements, while not direct agreements on price, would probably be found to have as their central purpose and effect the stabilizing or increasing of prices. While an individual manufacturer is free to adopt any of the foregoing policies, an agreement to the same effect among competing manufacturers is likely illegal.

Horizontal price-fixing agreements are per se illegal. Therefore, once it is shown that there was such an agreement, guilt is established. The fact that the defendants might want to show that the prices set were reasonable prices, or that the agreement was necessary to prevent a deterioration in quality, or that the agreement was terminated, is of no interest to the law. The law is violated the moment an agreement occurs. It is presumed that all such agreements unreasonably restrain trade and are, therefore, illegal.

Horizontal price fixing normally occurs when competing sellers enter into an agreement to increase or stabilize their selling price. However, in situations where purchasers act concertedly to affect the price of what they buy, they may also be guilty of horizontal price fixing. For example, if automobile manufacturers were to agree that they would not pay more than $X per ton for steel, they would be guilty of horizontal price fixing. Therefore, it is reasonable to ask why hospital buying groups are not considered to be engaged in per se illegal horizontal price fixing.

Cooperatives escape such condemnation when the members are integrating or combining to perform some legitimate function (such as purchasing), the pricing agreement arises out of that integration, and there is still ample competition in the overall market because the cooperative lacks substantial market power. On the other hand, if a buying group is not performing a legitimate function, or if it has substantial market power, it may be challenged for horizontal price fixing.

B. Horizontal Allocations of Territories and Customers

Competitors cannot agree among themselves as to the geographic areas each will serve, or the classes of customers to which each will sell. Thus, an agreement between two manufacturers of wheelchairs that one will serve the home market and the other will sell to hospitals would be an illegal horizontal allocation of customers. Similarly, an agreement that one will concentrate on sales in states east of the Mississippi River and the other will concentrate on western states is an illegal horizontal allocation of territories. Like horizontal price fixing, horizontal allocations of customers or territories are per se illegal.

C. Horizontal Group Boycotts

The term "group boycott" is used in antitrust analysis to describe a number of similar practices. The traditional horizontal group boycott occurs when a group of competitors is able to deny another competitor access to needed suppliers or customers. For example, an association of retail lumber dealers circulated a list of lumber wholesalers who also sold at retail to the public. The hope was that the retail dealers would boycott those wholesales, thus influencing those wholesalers to stop competing with the retail dealers. In 1914, the Supreme Court held this to be an illegal group boycott [1].

More recently, a group of dentists in Indiana agreed among themselves to withhold X-rays from their patients' insurance companies for use in benefit determinations. The purpose was found to be stifling the insurers' cost-containment programs. This too was held to be an illegal group boycott [2].

Regarding buying cooperatives, the circumstances under which expulsion of a member may be a group boycott was recently addressed by the Supreme Court in *Northwest Wholesale Stationers, Inc. v. Pacific Stationery & Printing Co.*, 472 U.S. 284 (1985). Northwest was an office supply wholesale buying cooperative, owned by retail office supply dealers. Pacific had been a member, but was expelled. Northwest claimed the expulsion was because Pacific failed to comply with a rule requiring members to notify the cooperative of any change in ownership. Pacific claimed it was because Pacific sold at both wholesale and retail, thus competing with the cooperative at the wholesale level while also competing with other members at the retail level. Although the case was sent back to the district court to resolve disputed factual issues, the Supreme Court did state that:

> Unless the cooperative possesses market power or exclusive access to an element essential to effective competition, the conclusion that expulsion is virtually always likely to have an anticompetitive effect is not warranted.

The Court recognized that buying cooperatives can achieve economies of scale in the purchasing and warehousing of goods, enabling smaller competitors to operate efficiently and to effectively compete with larger firms. These desirable effects can be achieved only if the cooperative is allowed to enforce reasonable rules necessary for its organization and operation. On the other hand, if the cooperative is large enough to possess significant market power, or is acting for anticompetitive purposes, excluded members may be able to bring a successful group boycott case against the cooperative.

D. Trade Associations

The opportunity to inadvertently enter into horizontal agreements is probably greatest at trade association meetings. Discussions of legitimate concern to the industry as a whole can stray into discussions that might be horizontal restraints of trade. Therefore, it is always advisable for the businessperson to have discussed with his or her counsel ahead of time what to do if that situation arises, to stay alert to the possibility, and to insist on the presence of trade association counsel to help prevent problems. (In the event of problematic discussions, I would advise leaving the meeting and immediately telephoning counsel to get further advice.)

One of the legitimate functions of trade associations is to try to influence government decisions. Such activity, whether aimed at the legislature, the executive, or the judiciary, may have anticompetitive potential. For example, certificate of need laws and licensing laws may make entry by new competitors more difficult, advertising restrictions may reduce price competition, and generic substitution laws may extend the market power obtained from a patent. However, the First Amendment to the Constitution guarantees the right "to petition the Government for a redress of grievances." Therefore, industry members may agree, through a trade association or otherwise, to try to influence the government without fear of antitrust violation. This is known as the Noerr-Pennington doctrine.

There are two important limitations on the Noerr-Pennington doctrine. First, it only covers attempts to influence the government. If a trade association or group of competitors attempts to influence standard-setting bodies, whose standards are, in turn, routinely adopted by the government, they are not protected from antitrust scrutiny. Second, it only protects legitimate, good faith efforts to influence the government, not shams. For example, if a trade association filed an opposition to all CON applications with a state agency, regardless of the merits, in the hope of increasing the cost of the process and thereby discouraging entry, the trade association activity might be viewed as a sham and subject to antitrust attack.

E. Enforcement Issues Involving Horizontal Restraints

Horizontal restraints have been vigorously pursued by the federal government during the past decade, and strict enforcement is likely to continue. Many of these cases are brought criminally by the Department of Justice. In addition, there has been more emphasis and success in obtaining jail sentences for the convicted felons.

IV. VERTICAL RESTRAINTS

A. Vertical Pricing Fixing and the Termination of Distributors

Unlike horizontal price-fixing agreements, which are between competitors, vertical price-fixing agreements, also called resale price maintenance, involve sellers and resellers, such as a manufacturer and its distributor. For example, if a hospital buying group negotiates a special discount with a manufacturer, but the manufacturer sells to independent distributors who then resell to the hospitals, any agreement between the manufacturer and the distributors regarding the price the distributors will charge the hospitals will raise a vertical price-fixing issue.

Like horizontal price-fixing agreements, vertical price-fixing agreements are per se illegal. In 1984, the Supreme Court reaffirmed their per se illegality in *Monsanto Co. v. Spray-Rite Service Corp.*, 465 U.S. 752 (1984). Spray-Rite was a distributor of Monsanto's herbicides. Spray-Rite was a discounter, and some of its competitors had complained to Monsanto regarding its discounting. Monsanto subsequently terminated Spray-Rite, and Spray-Rite sued claiming that it had been terminated as part of a vertical price-fixing agreement between Monsanto and its full-price distributors.

In addition to reaffirming the per se illegality of vertical price-fixing agreements, the Court examined the type of evidence necessary to prove that an agreement existed between the manufacturer and its full-price distributors. Specifically, the Court said that evidence of complaints by full-price distributors about a discounter, followed by termination of the discounter, could not be enough evidence to establish the existence of a resale price maintenance scheme. (As of this writing, Congress is considering legislation that would overturn this aspect of *Monsanto.*) Something more is necessary. For example, Monsanto had spoken to other price-cutting distributors telling them to maintain the suggested resale prices or they would not receive adequate supplies from some of Monsanto's herbicides. In addition, when Spray-Rite was terminated, it was told by a Monsanto official of the complaints about its discounting, and was not told that it had failed to adequately promote Monsanto's herbicide, which is why Monsanto alleged it terminated Spray-Rite at the trial. Therefore, the Court concluded that there was enough evidence, in addition to the complaints and termination, to prove a vertical price-fixing agreement and that Spray-Rite was terminated because it refused to go along with the agreement. Spray-Rite received over $10 million in damages.

Although a manufacturer cannot agree with its distributors as to the resale prices of its products, the manufacturer does have the right to choose who it will, or will not, deal with, as long as it makes that decision independently. Thus, a manufacturer can unilaterally refuse to deal with discounters (or with distributors that charge excessive prices). Moreover, the manufacturer can announce its policies in advance and refuse to deal with those who fail to comply with those policies.[4]

Assume a manufacturer announces the following policy to all of its distributors: "Any distributor that resells at a price other than list price will be terminated." At this point, the manufacturer has not violated the law; it has simply announced the terms on which it will deal. Assume that a distributor, not wishing to be terminated, decides to sell at list price. Is there a vertical price-fixing agreement? The answer is "no." The manufacturer and the distributor have each made unilateral decisions, there is no agreement. Assume that another distributor decides to sell at less than list, that full-price distributors complain to the manufacturer, and that the manufacturer terminates the discounter. This is legal under the *Monsanto* case. Some additional evidence of a price-fixing agreement would still be necessary. For example, if the manufacturer were to have its representative meet with the discounter to "explain the situation," and the discounter then "saw the light" and raised its prices to list, the jury would have enough evidence to find a vertical price-fixing agreement.

At this point, the reader may be confused. That reflects the fact that the law in this area is confusing. Therefore, some antitrust lawyers simply advise the manufacturer against trying to control the resale price of its product. The probability of stepping over the narrow line between unilateral be-

---

[4] This is called the *Colgate* doctrine [3]. Colgate had specified resale prices and refused to deal with any retailer who failed to maintain those prices. The federal government alleged vertical price fixing. The Supreme Court, finding no violation, held:

> In the absence of any purpose to create or maintain a monopoly, the [Sherman Act] does not restrict the long-recognized right of trader or manufacturer engaged in an entirely private business, freely to exercise his own independent discretion as to parties with whom he will deal; and, of course, he may announce in advance the circumstances under which he will refuse to sell.

havior and an agreement is considered too great. Alternatively, it is certainly possible for a manufacturer to effectively control the resale price of its product without violating the law. However, any such effort should be carefully reviewed by counsel.

B. Vertical Allocations of Territories and Customers

We have just seen that agreements between a manufacturer and its distributor regarding the resale price the distributor will charge are per se illegal. We now examine agreements between a manufacturer and its distributors regarding such matters as where the distributor will sell and to whom it will sell. These types of agreements, which do not directly affect the resale price, are called vertical allocations of territories and vertical allocations of customers.

In 1967, the Supreme Court had said that these types of restrains were per se illegal [4]. However, there was a lot of criticism of this approach, and in 1977 the Court reconsidered the question and overruled what it had said 10 years earlier.

Continental T.V. v. GTE Sylvania [433 U.S. 36 (1977)] involved a location clause in the agreement between Sylvania and its franchised T.V. dealers. The location clause required the dealer to sell only from locations agreed to by Sylvania. Continental, a dealer, sold from a location that was not approved, was terminated, and sued, claiming the location clause was a per se violation of the Sherman Act.

The Court explained that vertical restraints, such as territorial and customer restraints, have an impact on the marketplace that is complex. They certainly reduce competition among the manufacturer's dealers. This reduced intrabrand competition may be necessary, however, to implement a marketing program that will allow more effective competition between the manufacturer's product and the products of competing manufacturers. Thus, enhanced interbrand competition may be an outgrowth of the vertical restraints. The Court concluded that vertical territorial and customer restraints would be judged by a rule of reason analysis in which the reduced intrabrand competition would be balanced against any increased interbrand competition resulting from the restraint. The restraint would be illegal only if it failed to stimulate interbrand competition sufficiently to offset any reduced intrabrand competition.

Vertical allocations of customers or territories may stimulate interbrand competition by inducing competent and aggressive distributors to take on a new line, or enter a new market. The restraints may encourage the distributor to promote the product and develop the market, to provide good service, and to be concerned with the product's reputation. Absent restrictions, discounters might "free ride" off the market development, promotion, service, and product reputation developed by the distributor. In such situations, vertical restraints are almost invariably found to be legal.

The manufacturers with the largest legal risk from imposing vertical restraints are those with a very large market share, long-established products, and no reasonable fear that free riders will undermine their distribution systems. Such manufacturers should consult counsel regarding any attempt to control where, or to whom, its distributors resell.

C. Exclusive Dealing

Exclusive dealing refers to an agreement between a manufacturer and a distributor that requires the distributor to deal exclusively with that manufacturer regarding certain products. For example, if a rubber glove manufacturer and its distributor were to agree that the distributor would not carry rubber gloves produced by any other manufacturer, the agreement would be an exclusive dealing agreement.

Exclusive dealing agreements are usually legal. An antitrust problem only arises if the exclusive dealing contract forecloses other manufacturers from marketing their products. This will occur only if a major portion of the distribution network is foreclosed for a long period of time by exclusive dealing contracts. Thus, if a major rubber glove manufacturer were to have exclusive dealing agreements with the only four significant distributors in a given market, and those agreements were 2 years in length, they would probably be illegal. On the other hand, if the agreement was with only one of the distributors, or if it could be terminated by either party without cause on 30 days notice, it would not be illegal.

In 1982, the Federal Trade Commission applied this analysis to the distribution of hearing aids by Beltone Electronics Corp. [5]. Beltone entered into exclusive dealing contracts with its dealers. The contracts were not illegal because they covered only 7-8% of the nation's hearing aid dealers accounting for 16% of hearing aid sales, and they were of reasonable duration. Therefore, competitors were able to obtain adequate distribution.

D. Tying Arrangements

A tying arrangement is typically an agreement between a seller and a buyer whereby the seller agrees to sell one product (called the tying product) only on the condition that the buyer agrees to purchase a different product (called the tied product) from the seller. Tying arrangements can also involve the lease of one product on condition that the lessee buy certain products, such as supplies used with a leased machine, from the lessor.

A classic tying arrangement was IBM's requirement that lessees of its patented card tabulating machine use only IBM punch cards in the machine. IBM defended its practice by arguing that if punch cards of inferior quality were used, machine damage or malfunction could result. This would be undesirable and injure IBM's reputation. Regardless, IBM's tying arrangement was found illegal in 1936 [6].

For there to be a tying arrangement, there must, of course, be two separate products. The IBM tabulating machine (the tying product) and the punch cards (the tied product), for example, were two separate products. On the other hand, when Chrysler sells a chassis and an engine tied together, the car is considered one product, not two. But what if Chrysler would sell a car only if the purchaser also bought a car radio from Chrysler? Whether that is one product or two is an issue in a tying case recently filed against Chrysler.

In 1984, the Supreme Court decided a case arising out of a hospital's denial of staff privileges to an anesthesiologist because of an exclusive contract between the hospital and another group of anesthesiologists [7]. One of the charges by the anesthesiologist was that the contract was an illegal tying arrangement. The typing product was allegedly the hospital's operating rooms, and the tied product was anesthesia services from the group chosen by the hospital. The Court ruled that these were two separate products.[5]

---

[5]The Court reasoned that:

> the anesthesiological component of the package offered by the hospital

However, the anesthesiologist lost because a tying arrangement is not illegal unless the seller has considerable market power in the tying product. Here, the hospital had only 30% of the market for hospital services. Patients were not forced to buy the anesthesia services sold by the hospital: they could go to other hospitals where they could choose a different anesthesiologist.

By comparison, in the IBM case mentioned earlier, IBM was the only manufacturer of tabulating machines of that type. Thus, it possessed the market power necessary to force its customers to buy punch cards from it.

Rather than trying to force the purchaser to buy a tied product by conditioning sale of the tying product on purchase of the tied product, the seller might set its pricing policies to simply encourage (though not force) the same result. This form of tying, sometimes called leveraging, is illustrated by the case of *White & White, Inc. v. American Hospital Supply Corp.*, 723 F.2d 495 (6th Cir. 1983).

American Hospital Supply Corp. (AHSC) offered Voluntary Hospitals of America (VHA) member hospitals a year-end rebate if the aggregate volume of all merchandise sold by AHSC to VHA hospitals exceeded $41 million. The rebate would increase from 1% to 6%, as the aggregate volume increased to certain levels above $41 million. Competing distributors sued AHSC because these pricing practices led to hospitals buying all of their supplies from AHSC. The court ruled that AHSC was not involved in illegal leveraging because it lacked market power in any product. Hospitals were free to go to other distributors for any and all of their needs.

In situations where there are two separate products, and the seller has considerable market power in the tying product, attempts to force the purchaser to buy a tied product will normally be illegal. (Even this situation will not be illegal if the impact on competition in the market for the tied product is trivial. However, if there is an impact, it will rarely be trivial.) However, in situations where, for example, a diagnostic machine will not function properly without the use of supplies meeting certain specifications, and it is impractical to provide those specifications to others, then tying the machine to the supplies may be justified. However, if the supplies could be produced by others with access to the specifications, most courts would condemn the use of the tying arrangement.

E. Enforcement Issues Involving Vertical Restraints

During the past decade there has been considerable controversy regarding the treatment of vertical restraints under the antitrust laws. Previously, most such restraints were condemned. Now, with the exception of vertical price fixing, they are usually found to be legal. Moreover, since 1980 the federal government has not brought any cases challenging vertical restraints. It is, however, an area of the law that is still in flux, making it particularly

---

could be provided separately and could be selected either by the individual patient or by one of the patient's doctors if the hospital did not insist on including anesthesiological services in the package it offers to its customers. As a matter of actual practice, anesthesiological services are billed separately from the hospital services petitioners provide. . . . Customers differentiate between anesthesiological services and the other hospital services provided by petitioners. [8]

important that businesspeople get up-to-date advice when dealing with issues in this area.

V. PRICE DISCRIMINATION

The Robinson-Patman Act of 1936 amended section 2 of the Clayton Act. It is the only federal law dealing with price discrimination. A price discrimination is a price difference, where price is the net price after taking into account all discounts, rebates, allowances, and deductions. The Robinson-Patman Act delineates the situations in which it is illegal for a seller to charge different prices to different customers.

The law is aimed at two basic situations. The first, called primary line price discrimination or geographic price discrimination, occurs when a seller charges extremely low prices in one geographic area in order to drive competitors in that market out of business, while charging relatively higher prices in other markets (typically where competition is soft). If the seller succeeds in eliminating its competitors by its low pricing, it them presumably will raise prices. Cases alleging primary line price discrimination are relatively rare, and are generally brought in conjunction with a monopolization charge.

The second basic situation that the Robinson-Patman Act is concerned with is called secondary line price discrimination. This situation typically arises when a seller charges some resellers (called the favored customers) lower prices than it charges competitors of the favored customers. The disfavored resellers, who pay more to the seller, are at a disadvantage in their competition with the favored customers. This is not only unfair to the disfavored customers, but may result in their exiting the market (or not growing as much as they would otherwise) despite the fact that they may be as efficient as the favored customers. (In a distribution system with several levels of resale, injury may occur at the tertiary level and beyond. For simplicity, all of these levels will be referred to as the secondary level.)

It is important to recognize that the law does not make all price discriminations illegal. They are illegal only when certain criteria are all satisfied. We will now discuss those criteria.

A. Jurisdictional Requirements

The first set of criteria are sometimes called jurisdictional requirements. The Robinson-Patman Act only comes into play when (1) one seller (2) makes at least two contemporaneous sales (3) of commodities (4) of like grade and quality (5) in commerce.

The first requirement is that there be one seller. Assume that a manufacturer sells directly to certain end users, for example, large hospital chains, while reaching smaller hospitals through an independent distributor. The manufacturer may charge the large chain the same as, or even less than, it charges the distributor. This will lead to the small hospital invariably paying more to the distributor than the large hospital chain pays the manufacturer for the same products. If the small hospital were to sue the manufacturer, the manufacturer would successfully argue that the "one seller" requirement was not satisfied. The disfavored customer (the small hospital) is not buying from the same seller as the large chain.[6]

If, in the same situation, the manufacturer set up a wholly owned subsidiary to distribute to the smaller accounts, but sold to select accounts directly, then most courts would say that the one-seller requirement was satisfied. The manufacturer and its wholly owned subsidiary would be viewed as one entity.

The next jurisdictional requirement is that the seller must make at least two contemporaneous sales. If a syringe manufacturer sells to four distributors in a market and favors one of the four, the sales are "contemporaneous" if the sales are made close enough in time so that the distributors are competing on the resale.

In addition, the law only applies to "sales." It does not apply to consignments, leases, offers to sell, refusals to sell, or bids. Thus, if a syringe manufacturer offers to take on a new distributor but indicates it will charge the new distributor more than the established distributors, there can be no Robinson-Patman violation if the new distributor opts not to deal with the manufacturer on those terms.

Another situation that may arise regarding the requirement of two contemporaneous sales is when a manufacturer has both independent distributors and wholly owned distributors. The manufacturer can favor its wholly owned distributor because the transaction between the manufacturer and its wholly owned distributor is not a "sale" under the Robinson-Patman Act. It is viewed as an internal transfer. (Some cases have found a "sale" between a parent and its subsidiary when the transaction was at arm's length. This approach would probably no longer be followed.)

The next requirement is that the products involved in a Robinson-Patman case must be commodities. Services are not covered by the law. Commodities are generally tangible goods. When a manufacturer is selling a commodity coupled with services, the law will apply if the essence of the transaction is the sale of the commodity.

The fourth jurisdictional requirement is that the commodities must be of like grade and quality. The law is not limited to sales of identical products. If it were, a manufacturer could easily circumvent the law by making a superficial change in a product and then selling that "different" product to a favored buyer at a lower price. A meaningless change in the color of a product, or a change in the name put on a product from a "premium label" to a "private label," are examples of such superficial changes. The law applies despite these changes because the law applies whenever the products are of like grade and quality. If there are no bona fide physical differences affecting marketability, then the products will be of like grade and quality.

The final jurisdictional requirement is that at least one of the sales must be in interstate commerce. This means that a sale, either to the favored or disfavored buyer, must actually cross a state line.

If all of the jurisdictional requirements are satisfied—one seller makes at least two contemporaneous sales of commodities of like grade and quality

---

[6]Some cases have said that if the manufacturer exercises sufficient control over the distributor, then the disfavored customer is an "indirect purchaser" from the manufacturer and the one seller requirement may be satisfied. This situation may be uncommon, and some courts would not follow this approach.

in commerce—the next issue is whether the price discrimination may substantially lessen competition. This is known as the competitive injury requirement.

B. Competitive Injury Requirement

The test to determine whether a price discrimination may substantially lessen competition differs for primary line and secondary line cases. As discussed above, primary line cases typically involve a manufacturer charging different prices in different markets in order to eliminate competing manufacturers from certain markets by charging extremely low prices in those markets. Of course, charging lower prices in certain markets may simply be a manifestation of competition in those markets—not something the law wishes to stifle. Therefore, the law only becomes concerned when the low price is below the manufacturer's cost. In other words, only if a manufacturer sells below cost is it possible that competition, or equally efficient competitors, might be injured.

Before the late 1970s, the definition of "cost" for purposes of determining if sales were below cost was total or fully allocated cost. During the 1980s, most courts have been using average variable cost (excluding fixed costs). This change in the interpretation of the law has made it very difficult to prove below-cost sales.

Most price discrimination cases are secondary line cases. They typically involve favoring one distributor over another, resulting in possible injury at the distributor level (as opposed to the manufacturing level in primary line cases). Normally, if the favored and disfavored distributors do not compete with each other, there will be no competitive injury. Thus, if they sell in different markets or to different classes of customers, there usually will be no injury, and therefore no violation.

Whether price discriminations between competing distributors are likely to cause competitive injury will depend mainly on the size and duration of the price discrimination. A price discrimination of 1% is less likely to cause injury than one of 10%. Similarly, a price discrimination lasting 1 week is less likely to cause injury than one lasting 1 year. There are no "magic numbers"; each case is judged based on the market situation involved.

Although most secondary line cases involve favoring one distributor over a competing distributor, the law also applies to price discriminations given to users, as opposed to resellers, of a product. For example, if a manufacturer of blood analyzers sells the same model to a large hospital or a hospital in a buying group at a lower price than to a competing small hospital, the small hospital may claim that its ability to compete with the favored hospital has been injured by the price discrimination. Although cases based on this theory are rare, the wording of the statute covers such a situation. One reason such cases are rare is that the disfavored hospital still must prove that it was injured by the price discrimination. If the product is not being resold, but is being used as input into another product (for example, in-hospital health care), it would normally be difficult to prove competitive injury because of the price discrimination.

Another issue relating to competitive injury and hospital buying groups arises when the group negotiates a favorable discount with the manufacturer, but the manufacturer then sells to independent distributors who have been chosen by the group to resell to group members. Can the favorable discount given to the select distributors injure other distributors who wish to sell to member hospitals?

It can be argued that once the group has selected certain distributors to service its member hospitals, competition between the distributors as to who would be selected is over, and assuming the discount was available to whoever the group selected, there is no competitive injury. This argument loses some of its force, however, if the member hospital is contractually free to buy from others besides the select distributors. In such situations, there is ongoing competition between the distributors for sales to group members, and limiting the discount to the select dealers may impact that competition. The resolution of this issue is uncertain.

If it is established that one seller made at least two contemporaneous sales of commodities of like grade and quality in commerce at different prices, and that the effect of the price discrimination may be to substantially lessen competition, then a prima facie case of a violation of the Robinson-Patman Act is established. This means that the manufacturer must establish a defense for its price discrimination or it will be found to have acted illegally. There are three important defenses.[7] We will now examine each of them.

C. Cost-Justification Defense

If the seller's cost of manufacture, sale, or delivery in selling to various customers differs, the price charged to the customers may also differ. The most common example is the quantity discount. If a distributor orders by the truckload, that may result in a lower per-unit cost for the manufacturer in filling the order. For example, transportation cost, selling cost, and billing cost may be less per unit for the large order. The cost savings may be passed on to the customers. Of course, a cost savings of 5% would only justify an extra discount of up to 5%, not 10%.

Volume discounts, based on total purchases during a certain period such as a year, can rarely be justified on a cost-savings basis. Most cost savings accrue based on order size. Thus, a year-end 10% rebate for those who purchase over $1,000,000 of product during the year would normally have to be justified based on one of the two defenses discussed below, assuming it can be justified.

D. Meeting-Competition Defense

Section 2(b) of the statute provides that the seller has a defense if the lower price "was made in good faith to meet an equally low price of a competitor." The Supreme Court has explained the meeting-competition defense as follows:

> The meeting-competition defense requires the seller at least to show the existence of facts that would lead a reasonable and prudent person to believe that the seller's lower price would meet the equally low price of a competitor; it also requires the seller to demonstrate that its lower price was a good faith response to a competitor's lower price. [9]

---

[7]There is a fourth defense, the changing conditions defense, which is not important and therefore will not be discussed.

Assume that a syringe manufacturer has three distributors in a market. A competing syringe manufacturer offers to sell a similar product to one of the three distributors for 10% less in the hope of luring away that distributor. When that distributor informs its manufacturer of the situation and asks for an additional 10% discount, the manufacturer may agree. The result would be a 10% price discrimination in favor of the solicited distributor. If one of the manufacturer's other distributors sues the manufacturer, the manufacturer would raise the meeting-competition defense.

For the defense to be valid, the manufacturer would have to show that it was acting in good faith as a reasonable and prudent businessperson. Therefore, the manufacturer may ask the distributor for proof of the competitive offer, such as a price list or quotation sheet. In addition, the manufacturer's experience with the distributor, its knowledge of the market and its competitor's practices, and the specificity of the information provided by the distributor will be considered in determining whether the manufacturer acted in good faith.

The manufacturer is allowed to meet, but not beat, the competitor's offer. Here, too, the test is good faith. If it turns out that the better offer was only an additional 5% discount, but the manufacturer had a reasonable, good faith basis to believe it was 10%, it could justify the 10% discount even though it actually beat the competition.

Once the competitive offer that the manufacturer is responding to no longer exists, the defense no longer exists. The manufacturer will be able to justify a lower price to select distributors only as long as it reasonably believes in good faith that it is responding to a competitive situation, although that may be many years in certain circumstances.

The meeting-competition defense can be used to obtain new business as well as to retain old business. Assume manufacturer A is selling a certain product to its distributors for $25 per unit. Competing manufacturer B is selling the same product to its distributors for $22 per unit. Manufacturer A can offer manufacturer B's distributor a $22 price to try to get the distributor to switch manufacturers. If the distributor switches, it would be receiving a $3 price discrimination compared to manufacturer A's other distributors. That price discrimination would be justified under the meeting-competition defense even though it resulted from manufacturer A obtaining new business, as opposed to retaining old business.

In establishing a meeting-competition defense, one invariably is involved with checking on a competitor's pricing. However, it is not advisable to communicate with a competing manufacturer regarding its prices. If such communications are challenged as part of a horizontal price-fixing agreement under the Sherman Act, the manufacturer will not be able to defend the price-fixing charge by saying it was verifying its competitor's offer to a distributor to see if a meeting-competition defense would be available under the Robinson-Patman Act.

E. Availability Defense

The availability defense is not found in the Robinson-Patman Act itself, but has been created by the courts. An otherwise illegal price discrimination may be defended by showing that the better price was practically available (not just theoretically available) to the disfavored distributor, but that distributor simply chose not to meet the requirements for the better price.

Assume that a syringe manufacturer offers a 10% year-end rebate to any of its distributors in a certain market that make at least 50% of their syringe purchases from that manufacturer. All distributors are informed of the plan. A distributor that only purchases 30% of its needs from the manufacturer will pay more than a competing distributor that purchases 60% of its needs from the manufacturer. That price discrimination will not be illegal because it was practically available to all distributors. If the disfavored distributor was injured because it paid more, the injury resulted from its own choice not to meet the requirement for the additional 10% discount.

Assume instead that the syringe manufacturer offers a year-end 10% rebate to any distributor who purchases one million dollars worth of syringes from it during the year. Even though this offer is made to all distributors and is in theory available to all, it would probably not be available as a practical matter to smaller distributors. Therefore, there would be no availability defense to any resulting price discrimination.

In addition to the three defenses discussed above (cost justification, meeting competition, and availability), there are two other escapes for the manufacturer who engages in what would otherwise be illegal price discrimination. Sales to the federal government and to nonprofit institutions are exempt from the law. Although these exemptions are unimportant for many industries, they are important for segments of the medical equipment and supplies industry.

F. Federal Government Exemption

The Robinson-Patman Act itself does not contain an exemption for sales made to the federal government. However, the courts have said that Congress did not intend to cover sales made to the federal government. Therefore, a manufacturer may charge less (or more) to a federal government agency (for example, the Veterans Administration or the Department of Defense) than it charges to others without raising any problems under the Robinson-Patman Act.

Until 1983, it was widely believed that there was a similar exemption for sales made to state or local governments, such as sales to state or county hospitals. In that year, however, the Supreme Court said that there was no such general exemption [10]. (The Court left open the possibility of an exemption in limited circumstances.) The case was brought on behalf of pharmacies in Alabama who paid more to Abbott Labs and others for drugs than the University of Alabama Hospital paid to the same suppliers. The pharmacies claimed that the hospital's clinic resold the drugs in competition with local pharmacies. Abbott Labs claimed that sales to the University Hospital were exempt because it was a state agency. The Supreme Court held that:

> the sale of pharmaceutical products to state and local government hospitals for resale in competition with private pharmacies is not exempt from the prescriptions of the Robinson-Patman Act. [11]

There is a second exemption to the act that covers products sold to a nonprofit hospital for that hospital's own use. We will address the scope of that exemption next. Note, however, that the sales made to the University of Alabama Hospital were not covered by that exemption because the drugs were resold in competition with private pharmacies, and not used by the hospital itself.

G. Nonprofit Institutions Exemption

The Nonprofit Institutions Act of 1938 amended the Robinson-Patman Act by exempting products sold to nonprofit hospitals when the products were for the hospital's own use.[8] The scope of the "own use" concept was addressed by the Supreme Court in *Abbott Laboratories v. Portland Retail Druggists Association*, 425 U.S. 1 (1976).

The suit was brought on behalf of private pharmacies against various drug manufacturers. The manufacturers charged lower prices to nonprofit hospitals than to the pharmacies. The private pharmacies alleged this price discrimination injured their ability to compete with the hospitals' pharmacies. The manufacturers claimed that the sales to the nonprofit hospitals were exempt from the Robinson-Patman Act by the nonprofit institutions exemption. The exemption, however, is limited to sales to the hospitals "for their own use." The Court determined that:

> "their own use" is what reasonably may be regarded as use *by the hospital* in the sense that such use is a part of and promotes the hospital's intended institutional operation in the care of persons who are its patients. [13]

Therefore, the exemption included supplies utilized for on-premises treatment, or sent home with former patients, and supplies issued to a hospital's employees, students, or staff for their own use. On the other hand, the exemption did not include supplies sold to walk-in customers or former patients, and supplies used by staff members' private patients or dependents.

Faced with the argument that segregating drugs based on who bought them or the use to which they were put created an unworkable standard for hospitals and their suppliers, the Court responded that if the hospital chose to sell in ways which were at times outside the exemption, the hospital could:

> establish a recordkeeping procedure that segregates the nonexempt use from the exempt use. This would be supplemented by the hospital's submission to its supplier of an appropriate accounting followed by the price adjustment that is indicated. This, to be sure, is cumbersome, but it obviously is the price the Congress has exacted for the benefits bestowed by the controlling legislation, and it should be no more cumbersome than the accounting demands that are made on commercial enterprises of all kinds in our complex society of today. [14]

Regarding possible liability of suppliers who relied on the hospital's accounting, the Court stated:

> The supplier, on the other hand, properly may expect to be protected from antitrust liability for reasonable and noncollusive reliance upon its

[8] "Nothing in [the Robinson-Patman Act] shall apply to purchases of their supplies for their own use by schools, colleges, universities, public libraries, churches, hospitals, and charitable institutions not operated for profit" [12].

> hospital customer's certification as to its dispensation of the products it purchases from the supplier. But it is not unreasonable to expect the supplier to assume the burden of obtaining the certification when it seeks to enjoy, with the institutional purchaser, the benefits provided by section 13c. [14]

Thus, the manufacturer and the nonprofit hospital are expected, in effect, to work together to ensure that a discriminatory price is not given on products that are not for the nonprofit hospital's own use.[9]

H. The Diversion Issue

As discussed above, if the manufacturer gives a lower price to nonprofit hospitals, the manufacturer does have a legitimate need to know if the hospital is using the product "for its own use" or is reselling the product. The manufacturer can certainly obtain this information and adjust its pricing accordingly. Alternatively, the manufacturer can adopt and implement a policy to terminate nonprofit hospitals who resell any of its product, or to charge the higher price on all purchases by such hospitals.

The nonprofit exemption to the Robinson-Patman Act, however, is related to only one part of the diversion issue. The manufacturer may also be providing a lower price to proprietary hospitals. If the proprietary hospital resells in competition with other distributors, who pay more to the same manufacturer, that would raise a price discrimination issue. To prevent this problem, the manufacturer may have a policy of limiting the use made of the product by the hospital. Although this is analogous to the "for its own use" limitation regarding sales to nonprofit hospitals, its basis is totally independent of the nonprofit exemption.

When dealing with either nonprofit hospitals or proprietary hospitals, the manufacturer can unilaterally formulate its own pricing policies, which may limit favorable pricing to goods that are not resold by the hospital in competition with other distributors, or the offending hospital may be charged the higher price on all purchases, or even terminated. However, if the manufacturer enters into an agreement with the hospital restricting to whom the hospital can resell, that agreement will raise Sherman Act questions.

At a minimum, the agreement would probably be a vertical allocation of customers, which is discussed in Section IV.B of this chapter. As a practical matter, these types of agreements are usually held to be legal. It is also possible, though less likely, that the restriction on the hospital's resales may be challenged as part of a resale price maintenance scheme. This would depend, in part, on whether the primary purpose and effect of the restriction was to prevent price competition with distributors, thereby allowing distributors to maintain a certain margin. (See Section IV.A for additional discussion.) A third possibility is that restricting the hospital's resales may

---

[9]The nonprofit exemption is not limited to nonprofit hospitals, but applies to any nonprofit, charitable institution. For example, in *De Modena v. Kaiser Foundation Health Plan, Inc.*, 743 F.2d 1388 (9th Cir. 1984), the court applied the exemption to drugs sold at discriminatorily low prices to a nonprofit HMO, which resold the drugs to its members. Of course, drugs sold to the HMO at lower prices than pharmacies and then resold to nonmembers would not be covered by the exemption.

be viewed as an attempt to prevent arbitrage, which would undermine any price discrimination. In other words, the provision may be viewed as preventing the hospital from reselling at a small profit to distributors at a price lower than the manufacturer's price to the distributors.

I am not saying that an agreement between the hospital and the manufacturer restricting the use made of the product by the hospital is necessarily illegal. Moreover, the restriction may be economically desirable, resulting in lower prices than would otherwise exist. However, it is the type of agreement that can raise a number of complex issues and should be entered into only after careful consideration of the legal, and of course business, ramifications.

I. Other Sections of the Robinson-Patman Act

Although we have covered the major price discrimination issues relevant to the distribution of medical equipment and supplies, the Robinson-Patman Act has four other sections worthy of note.

The favored recipient of an illegal discrimination may be sued under section 2(f). It must be established that the favored buyer either knew, or should have known, that it was the recipient of an illegal price discrimination. This means that the favored buyer not only must know that it is the recipient of a lower price than its competitors, but also must know that all the jurisdicitonal requirements were met, that the discrimination may injure competition, and that none of the defenses are available. Although most suits are brought against only the manufacturer, suits can be brought against both the manufacturer and the favored customer, or against only the favored customer.[10]

Section 2(c) of the Robinson-Patman Act prohibits the payment of brokerage commissions, or discounts in lieu of brokerage, except in situations where actual brokerage services are rendered. This section has also been interpreted to prohibit commercial bribery. Commercial bribery involves secret "under the table" payments aimed at bribing a purchasing agent or other representative to provide business to the briber.

Finally, sections 2(d) and 2(e) involve promotional services and allowances. These sections are most important in many consumer goods industries where promotional programs are more common than in the medical device industry. The Federal Trade Commission has provided guidelines for complying with sections 2(d) and 2(e). The FTC summarizes the law's requirements as follows:

> Simply stated, what the law requires, in essence, is that those who grant promotional and advertising allowances treat their customers fairly and without discrimination, and not use such allowances to disguise discriminatory price discounts. [15]

---

[10]For example, in *Jefferson County Pharmaceutical Association v. Abbott Laboratories*, 460 U.S. 150 (1983), discussed earlier, the pharmacies sued Abbott Labs and other manufacturers under section 2(a) and the University of Alabama Hospital under seciton 2(f).

In essence, any promotional programs, whether they involve money or the provision of services such as display racks, signs, sales material, or training for the distributor's sales personnel, must be made available to all competing resellers on a proportionally equal basis. Moreover, all resellers must have notice of the program.

J. Enforcement Issues Involving Price Discrimination

The Robinson-Patman Act is enforced at the federal level only by the Federal Trade Commission.[11] It brings its cases as administrative complaints, which are heard by an FTC administrative law judge, with possible appeals to the Commission itself, then the United States Court of Appeals, and finally the Supreme Court if it chooses to hear the case.

The last Robinson-Patman Act complaint that resulted in a trial was issued in 1980. There have also been a few additional cases brought by the FTC and settled without litigation during 1980s. There is no indication that the FTC will change its current practice of minimal enforcement of the law. Of course, a change in administration may result in a change in enforcement approach.

The law can also be enforced by a private suit. Unlike an FTC suit, which results only in an order prohibiting future violations, a private suit carries the possibility of treble damages plus attorney's fees and costs. Although the number of private price discrimination suits is still significant, a number of cases during the past decade have made it more difficult for plaintiffs to collect damages. This, in turn, has reduced both the number of cases brought and the number in which damages are paid. Although most cases will be won by defendants, even in those cases the disruption to business and the costs will often be significant.

VI. CONCLUSION

Distributional arrangements in the medical device industry are complex and rapidly changing. The same can be said of antitrust law. Therefore, applying antitrust law to distributional arrangements in the medical device industry is a challenging task. Regardless, members of the industry must function within that environment. Hopefully, this chapter will help industry members understand the relevant legal issues, make better-informed decisions aimed at minimizing unnecessary legal risk, and recognize when consultation with counsel is warranted.

REFERENCES

1. *Eastern States Retail Lumber Dealers Association v. United States*, 193 U.S. 38 (1914).

[11] There is a criminal seciton to the law, which can be enforced by the Department of Justice. However, the Department of Justice chooses not to enforce the law.

2. *Federal Trade Commission v. Indiana Federation of Dentists*, 476 U.S. 447 (1986).
3. *United States v. Colgate & Co.*, 250 U.S. 300, 307 (1919).
4. *United States v. Arnold, Schwinn & Co.*, 388 U.S. 365 (1967).
5. *Beltone Electronics Corp.*, 100 F.T.C. 68 (1982).
6. *International Business Machines Corp. v. United States*, 298 U.S. 131 (1936).
7. *Jefferson Parish Hospital District No. 2 v. Hyde*, 466 U.S. 2 (1984).
8. 466 U.S. 22-23.
9. *Falls City Industries, Inc. v. Vanco Beverage, Inc.*, 460 U.S. 428, 451 (1983).
10. *Jefferson County Pharmaceutical Association v. Abbott Laboratories*, 460 U.S. 150 (1983).
11. 460 U.S. 171.
12. 15 U.S.C. 13c.
13. 425 U.S. 14.
14. 425 U.S. 20.
15. Federal Trade Commission, Guides For Advertising Allowances and Other Merchandising Payments and Services (1972) at 1.

part ten

Current and Emerging Issues in Coverage, Payment, and Marketing

56

Megamarketing: An Expanded Approach for the 1990s

PAUL M. CAMPBELL *The Lash Group, Washington, D.C.*

I. INTRODUCTION

The medical technology industry was quite successful in expanding the market for medical products as the health-care industry expanded over the last several decades. The United States medical supply industry was a major producer of medical products and the American health-care system was a major consumer. Once a product had received approval from the Food and Drug Administration (FDA), marketing staffs used the traditional marketing tools of product, price, placement and promotion. Until the 1980s, the FDA was the only real gatekeeper or market deterrent for most medical products. However, the rapid change in the economics of health care in the 1980s resulted in several new gatekeepers ranging from hospital material managers attempting to influence product selection to HMO medical directors taking an active role in managing care. These trends have forced medical technology companies to take a broader view of the marketing function by adding power and public relations to the traditional four Ps of marketing. Power and public relations are critical tools to assist the company to support the traditional customer and end-user. This expanded view of marketing was labeled Megamarketing by Phil Kotler in a *Harvard Business Review* article with that title. Kotler defines Megamarketing as the "strategically coordinated application of economic, psychological, political, and public relations skills to gain the cooperation of a number of parties in order to enter and/or operate in a given market." (See Figure 1.)

The Kotler Megamarketing approach is quite relevant as medical technology companies respond to the new health-care trends. The new financial incentives of insurers, government programs, and managed care plans continue to have a significant impact on the companies traditional customers and end-users of products. Like the FDA, these new players are not customers; nonetheless they need to be influenced in order to create a favorable market for the end-user. An example illustrates the point.

Phase 1: Clinical Marketing

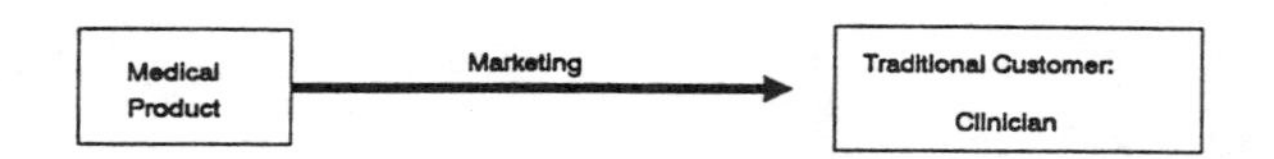

Phase 2: Budget Neutral Marketing

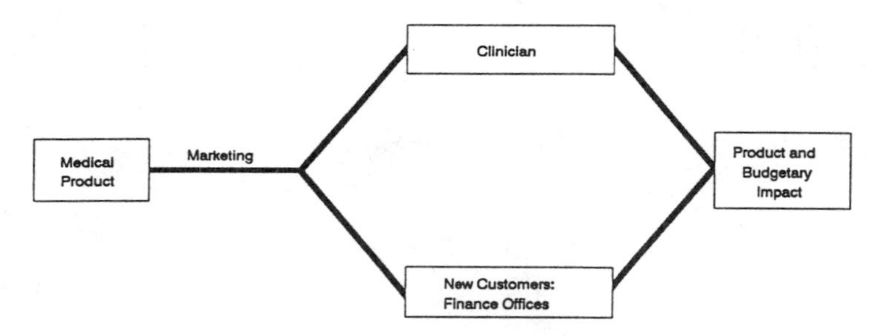

Phase 3: Financial Marketing

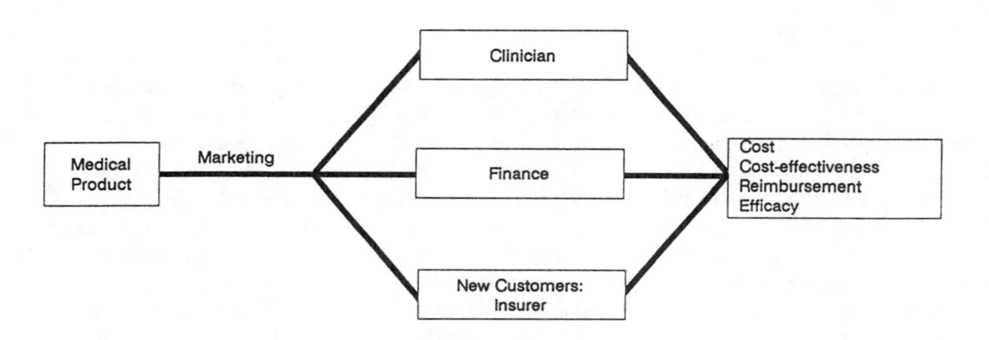

Phase 4: Megamarketing

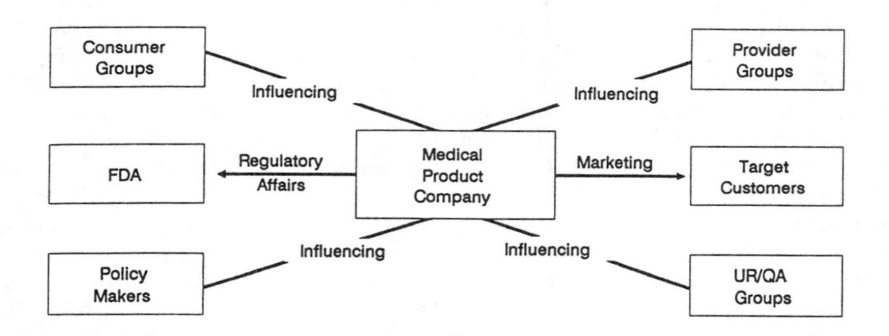

FIGURE 1 Four approaches to marketing medical devices.

In the 1970s, the computed tomography (CT) scanner was introduced to the American health-care system. With relatively few economic controls, the technology was quickly adopted by leading hospitals and the scanner diffused to most medium to large-sized hospitals within a three- to five-year period. However, when the magnetic resonance imaging (MRI) scanner was introduced in the same time period as the tumultuous health-care reforms, end-users were nervious about whether insurers would cover the new technology and, if so, whether the price would be adequate. Although the technology had received FDA approval, it was subsequently considered for assessment by the Congressional Office of Technology Assessment, the Federal Office of Health Technology Assessment, the Congressional Prospective Payment Assessment Commission, and a multitude of private assessment organizations. Each group was assessing the technology for different issues of effectiveness, cost-effectiveness and reasonable payment. As a result, MRI product managers needed to take a broader view of the market in order to help service their traditional customers in hospitals and physician offices.

MRI, along with other advances such as the cochlear implant and automated implantable defibrillator which were released from the R&D pipeline soon after the major health-care reforms started, shows the importance of megamarketing in the new health care environment. The traditional marketing tools and strategies continue to be appropriate for the customers and end-users, but the tools of power and influence are critical in developing an approach to work with the new gatekeepers. Kotler describes power as a push strategy that is used to win the support of the gatekeepers or market influencers. In the case of MRI, cochlear implants, and defibrillator, the Health Care Financing Administration was a key power broker in opening the market for these technologies. A favorable HCFA decision was particularly helpful since many of the private insurers followed their policy decisions. Public relations, on the other hand, is a pull strategy which recognizes that gatekeepers are influenced by the opinions of others and that, over the long term, public opinion can influence the marketplace and the ability of a company to maintain its place in the marketplace. In the case of cochlear implants, the companies worked closely with patient groups and health-care providers to be supportive of their efforts with the obtaining coverage. This proved to be invaluable during the early efforts with third-party payers and federal regulators.

II. MEGAMARKETING'S ADDED VALUE

Figure 2 outlines various gatekeepers who influence the health-care environment for traditional target customer. Megamarketing is an approach to use power and public relations strategies to influence the environment, making it more favorable for the target customer. Companies have traditionally committed a significant number of resources to staff a regulatory affairs function to ensure compliance with FDA regulations and a marketing and sales staff to support the organization. Recently, however, some companies have taken a broader view and created internal working groups or teams to market a new product which will face several gatekeepers. Lead by a senior director, marketing managers have joined forces with the regulatory affairs and public affairs departments in an attempt to shape the market for a new product. The teams have taken a longer term perspective than that used in traditional marketing and have accepted the challenge of actually changing the environment

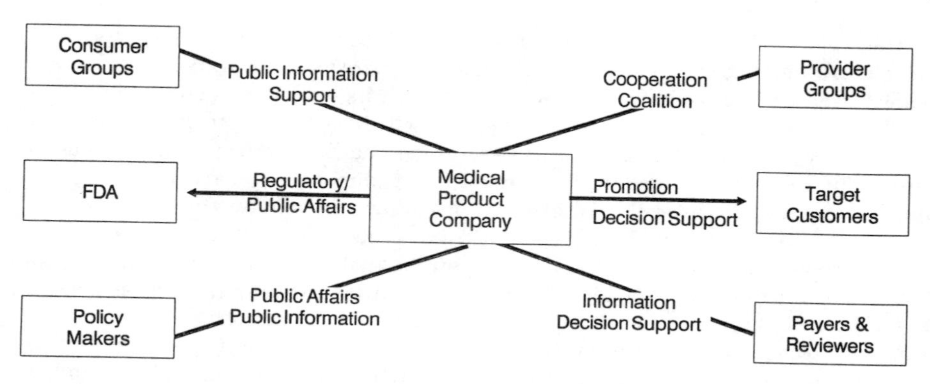

FIGURE 2 Marketing tools for gatekeepers and target customers.

instead of accepting it as a factor around which a marketing strategy is developed.

FDA has been the original gatekeeper since the Medical Device Amendments in 1976. Device companies rely on regulatory and public affairs strategies to meet their needs related to product approval and compliance with FDA regulations.

Third-party payers, including the new managed-care plans, are increasing their influence over the delivery of health care. Payers have become more aggressive in issues related to coverage and have attempted to limit the use of some technologies. Some research has shown that HMOs are less likely to cover a new technology during its early stage of diffusion. Prudential, a major national insurer, made national news when they began a centers of excellence approach for some high-risk procedures such as heart transplant. Prudential is relying on information about the quality of services offered by leading institutions and will only cover a procedure it it is performed at a center of excellence. Payers are also attempting to reduce expenditures by setting rates in advance or securing discounts from hospitals and physicians in exchange for assuring patient referrals.

Utilization-review organizations include both the federal peer-review organizations mandated to review Medicare admissions as well as the private sector firms. These organizations have been referred to as "fourth party" payers and are responsible for a range of activities including pre-admission certification, retrospective review, bill audits, and quality of care screens. The overall objective is to limit the use of unnecessary services or those provided at an inappropriate level.

Health-care policy makers are critical forces since they are attempting to control costs, assure quality, and increase access. In an era of persistent budget deficits, legislators are exerting more pressure on the health care system to find ways to be more efficient.

Consumers and consumerism are emerging as important trends in health care. Consumers are responding to information and are increasingly making informed decisions.

Professional/consumer organizations are important sources for building consensus and developing coalitions needed in a megamarketing strategy.

III. MEGAMARKETING STRATEGIES

Each of the market influencers are likely to respond to a different marketing strategy. The traditional customer will continue to respond to promotional strategies and products which meet clinical needs. However, they will want to know whether the product is covered and how it fits into the current reimbursement scheme for hospitals and physicians. This will require a manufacturer to have ready access to information about trends and policies of the key market influencers. In addition, for some products, it will be necessary to develop a megamarketing strategy to influence the decisions. As mentioned before, megamarketing relies on power and public relations. Power can be achieved using the following three strategies: information/education, coalition building, and cooperation.

Information and education strategies are becoming more important since decisions are becoming more information driven. Payers want to know whether a product or service is effective and how much will it cost. Consumers want to know whether it will make a difference. The use of information and education strategies will become more widely used as the effectiveness initiative begins to bear fruit.

Coalition building is a critical element to the adoption of a new policy that could create a market for new services. Coalition building is done most efficiently by using professional and consumer organizations and the power that they hold.

Cooperation strategies are an essential strategy to convert those who are neutral to a position to become an ally.

IV. TRENDS AFFECTING THE GROWTH OF MEGAMARKETING

The use of megamarketing concepts will be influenced by several trends in the health care environment.

The general health-care environment is marked by rising costs associated with cost inflation, population growth, and technology advances. During the last several years, there have been renewed attention at reforming the health-care financing environment. Medicare introduced prospective payment for hospital care, and private insurers and employers turned their attention on new mechanisms to control utilization and costs. Health-maintenance organications and preferred-provider organizations emerged as important players in the health-care field. In the next decade, Medicare is scheduled to reform the way in which it pays physicians. This is likely to have a major impact on the delivery system as physicians attempt to respond to major shifts among specialties. The effect could be as profound as the start of the diagnosis related group (DRG) system for inpatient hospital care.

V. POLICY TRENDS

Policy makers have spent the last decade implementing new approaches to control costs and reduce utilization. Recently, there has been increasing attention paid to the use of consumer choice to accomplish the same goal.

When Congress approved the Medicare Prospective Payment System (PPS), few realized how much power would be given to the congress and the Medicare program. Now through a series of complex formulae, program expenditures could be controlled on an annual basis. Each case had an average price which would be adjusted on an annual basis. The average price was scaled by the DRG weight to produce the actual hospital price. The hospital industry began to rely on its political power to lobby for adequate updates in the average price and use aggressive public relations strategies to convince consumers of the long-term problems of inadequate funding. The industry relied on a coalition of support from other health care providers and technology manufacturers to push for adequate hospital prices. Medicare PPS had another side effect which was to increase the use of nonhospital services when patients were discharged early. This lead to higher out-of-pocket costs and with it increased cost sensitivity on the part of the patient.

Starting in 1992, the Medicare program will begin a five year transition to a new payment system based on relative values driven in part by the resources needed to provide the service. Congress has mandated the use of voluntary expenditure targets which have the potential to control Medicare spending for physician services. As currently written, the new law would limit the amount that a physician could charge a Medicare beneficiary beyond the Medicare limit. With physician payment reform underway, policy makers will have accomplished most of the cost control techniques developed in the last two decades. As a result, they are likely to turn to new approaches involving consumer choice based on information.

The federally sponsored Effectiveness Initiative was created to fund research on procedures which are performed frequently despite questions about their effectiveness. It has developed into a full-fledged government agency called Agency for Health Care Policy and Research. The agency will fund research into the effectiveness of procedures and will target projects with new and creative ways to get the message to physicians and consumers giving them information about the effectiveness of procedures such as coronary bypass surgery and cataract surgery.

Related to the general trend of generating information to assist in decision-making, policy makers are attempting to require all providers to report medical-device problems to the FDA in an attempt to improve the quality and effectiveness of health care related to medical devices.

VI. PAYMENT TRENDS

In the 1980s, the federal government implemented new payment policies based on the concept that it should exert itself as a prudent purchaser in the health care market place. The best example of this approach is found in Medicare's taking the lead with the widescale introduction of a per-case payment system. Based on this approach, the Medicare program would control payments to the hospital industry. Congress and its advisory commission and the Medicare program have become important "gatekeepers" of medical technology used in the hospital setting. Several Medicaid programs and a limited number of Blue Cross plans have adopted a per-case payment approach. Most rely on the federal rules, however, some State Medicaid agencies are also important "market influencers." Most commercial plans and some managed care plans have relied on discounted charges as the major means of reimbursing hospitals. Negotiated per diem rates have also been used by some managed-care plans and several Blue Cross plans. Because the latter payment approach of discounted charges does not include the same incentives as a per-case approach, these payers also rely heavily on utilization review to control length of stay and utilization of ancillary services. As a result, these "fourth party" payers will be important targets of influence for some companies.

Physician payments continue to be based on charges which they set and are subsequently limited by Medicare and private insurers. However, the new payment system known as Resource-Based Relative Value Scale (RBRVS) would put the government in more control of expenditures and decisions about the relative monetary value of certain services. Congress and its advisory commission on physician payment and the Medicare program will become important gatekeeprs of physician services and technologies used by physicians.

VII. UTILIZATION TRENDS

Recent data show the recent shift from inpatient services to outpatient services. Outpatient and sub-acute care is becoming an important part of Medicare program expenditures and outpatient revenue is an increasing percentage of a hospital's total revenue. These shifts in utilization have resulted from research that has shown that some services are unnecessary while others could be provided at a lower cost setting. The patterns of care were changed by restructuring incentives—giving patients and providers an incentive not to use some services or to shifting the site of other services to the least costly setting. Several strategies have been used including preadmission certification by a UR company, higher deductibles and copayments for some services, or risk sharing arrangements of managed care plans such as HMOs.

The utilization trends highlight what companies have seen in recent years. There are new customers. Alternate site providers are becoming increasingly important in addition to the hospital. In addition, the traditional customer is influenced by new players which msut become targets of influence in a megamarketing approach.

VIII. IMPLEMENTING A MEGA APPROACH

Megamarketing is most relevant to products that are new, highly visible, and costly. These are the technologies likely to be watched by government and private payers. However, it can also be used for new products or services that can save money for the payers but will cause new tensions among some existing providers. For example, home antibiotic-infusion therapy can save money over the long term but could be viewed as a competitive threat by some hospitals. The approach could also be used for a product with lower-than-projected sales, particularly if the lackluster performance can be attributed to one or more "gatekeeper."

Kotler describes a "grand strategy" using the tools of information, coalition, and cooperation along with public relations to address the three key groups on any issue: the opponents who have to be neutralized if not convinced, the allies who need to be organized into a coalition, and the neutral groups who need to be turned into allies. He also describes two approaches to develop a tactical plan which can either be linear or multilinear. Traditionally, companies have used a linear plan where they waited for FDA approval before working with payers and managed care companies. Now, in order to gain market acceptance more expediently, companies are using a multilinear sequencing approach where several gatekeepers are contacted at key intervals.

IX. ROLE OF TRADE ASSOCIATIONS

Trade associations can play an important role in megamarketing, particularly when several companies are willing to work together to open a market broadly and where a givernment agency plays an important role. The Health Industry Manufacturers Association (HIMA) was able to serve a group of its members who were interested in creating a new Medicare benefit for home antibiotic-infusion therapy. The Association served as a central point for conducting the necessary research on the cost-effectiveness of the service and in drafting legislation that would be favorable to the participating companies. The association was able to be effective advocates for creating a new market on behalf of member companies. This approach required the support of the senior members of the association's executive staff and the board of directors.

X. CONCLUSION

Megamarketing is becoming increasingly relevant for medical technology companies as more and more of the traditional target customers are influenced in their purchasing by other players. Companies need to understand the growing set of health-care gatekeepers. The policy changes of the last decade will continue to exert new pressure, and physician payment for Medicare is the next target. The overall trend is to shift the burden of proof as to what is effective from the payer to the provider and manufacturer. More and more information is likely to be available in the future to show what procedures are effective, what providers provide high-quality services, and what devices tend to have a good track record of safety and effectiveness. In this environment, the four Ps may not always be enough. Companies will need to take a broader view, involve more senior management time to developing multi-departmental workgroups to open a market or sustain its influence.

part eleven

Historical Overview/Legal Framework of International Environment

57

Trends in International Regulations and Their Growing Impact

JOHN E. OLSSON *General Electric Company, Milwaukee, Wisconsin*

I. INTRODUCTION

Warranted or not, countries around the world are increasing their regulation of medical devices. And the rate of increase accelerates with time. One would hope that the new regulations were motivated by a concern for public health and safety. However, this is not always the case. Regulation shows up in many forms, affecting product design, product manufacturing, and product distribution. Regulations are being passed to control health care cost, protect local industry, protect national security, and raise tax revenue, as well as to protect public health.

In some areas, the concept of appropriate technology is entering the regulatory discussion. This concept is being advanced largely by the World Health Organization, principally for third-world countries. The concept is that underdeveloped nations should not acquire technologies, medical devices, or procedures that their infrastructure cannot support and effectively utilize. However, support and trial implementation are occurring in developed areas such as Scandinavia.

Obviously, a manufacturer wishing to market a device in a regulated environment must be aware of and follow the correct procedures in complying with the requirements. Several routes are available to understand the requirements, including direct involvement in their development, monitoring the development process, and utilization of experts or reference services. An important point to remember is that the product has to satisfy the needs of the market in addition to complying with all of the regulatory requirements if it is to be successful in the market.

II. TRENDS IN MEDICAL DEVICE REGULATION

Although regulation is increasing everywhere in the world, a close look at Europe will illustrate how fast. In 1977 there were 13 regulations for medical devices within 16 of the European countries. In 1987 there were 28 regulations and 19 more in the discussion phase: a 115% increase in 10 years with the prospect of an additional 145% increase in the foreseeable future. Admittedly, the formation of the common market in Europe, planned for 1992, had and will continue to have a profound affect on its regulatory environment. Since 1987, when the new approach was adopted requiring only a simple majority to adopt a new directive, there has been significant pressure within the member states to adopt regulations in expectation that such regulations would have to be accommodated in the 1992 harmonized scheme. After 1992, the 12 nations of the European Community are required by treaty to enforce the directives adopted by the Community.

At the present time, the European Community (EC) has four and possibly five medical device directives under consideration that would affect the ability to market a medical device. The Treaty of Rome requires the 12 signatories to pass legislation in agreement with the directives adopted by the Community. This will result in the cancellation of the current differing regulations. For the record, it is worth examining the variety of regulatory schemes in place in 1989.

France has a homologation scheme that affects the marketing of many types of electromedical devices. The affected devices must conform to French National Standards, complete a type test by a French national laboratory, undergo a clinical trial at two designated French hospitals, and be accepted by a French government review board. Operating controls and manuals must be in the French language.

Germany has a similar set of requirements known as the Medizinigeraet-everordnung (MedGV) law. Class 1 devices, some 26 types of electromedical devices, have to conform to German national standards, be type tested by one of the designated German laboratories, and be certificated by one of the German states. Operating controls and manuals must be in the German language. Powered implants must meet the above and also have a separate registration card. Manufacturers of sterile devices must comply with good manufacturing practices (GMP). At the present time, the German government will accept a certificate of inspection from the U.S. Food and Drug Administration (FDA) as evidence of conformance. X-Ray tube housings must be type tested by the German government Physikalisch-Technische Bundesanstalt (PTB) laboratory for radiation emissions.

Spain requires type testing and yearly GMP inspections for patient monitoring devices and video monitors. These devices must conform to Spanish national standards.

The United Kingdom has a manufacturer registration scheme. To qualify for inclusion on the bidder's list for national hospitals, a manufacturer must complete an evaluation form stating whether the device conforms to UK standards, international standards, or the equivalent. In addition, the manufacturer must pass a GMP inspection. A bilateral Memorandum of Understanding has been enacted between the FDA and the United Kingdom for the mutual recognition of inspection results. However the U.K. Department of Health is unwilling to accept FDA certificates for manufacturers of sterile devices.

The passage of the medical device directives will cause a single regulatory environment to exist throughout Europe. At the present time it is unclear whether that environment will be the sum or some average of the above requirements. The sum would consist of conformance to essential requirements (listed in an attachment to the directive) or European standards, type test by an acceptable laboratory (notified body), GMPS, electromagnetic compatibility (EMC) for emissions and susceptibility, local language for operating controls and manuals, clinical trials, and approval to market by one of the member states (competent authority). This would be comparable to the present regulatory scheme in Japan (except for EMC) but more stringent than the present U.S. scheme.

Although a scheme that is more stringent than that in any single European country is not exactly desirable, it nonetheless is an improvement over today's environment. To market some products in Germany, France, the United Kingdom, and Spain today the manufacturer or importer has to follow the above procedures separately for each of the four countries. Admittedly, he is not required to do anything for the other eight countires of the European Community. However, given the trend it would not be very long before the eight remaining countries felt compelled to institute some form of marketing regulation in order to avoid their homeland from becoming a "cesspool for devices" as declared by a health ministry official at the 1987 European Confederation of Medical Suppliers Associations (EUCOMED) conference. The common market would have the advantage of requiring the device to meet the requirements in only one of the member states to be legally salable in all of the member states.

Other directives will have similar impact. While imposing more severe requirements than some member states do currently, the requirement in all states will at least be the same. For example, the directive on customs would result in a single set of customs requirements, declaration forms, and tariffs in place of the present 12 different procedures. In some cases, directives are proposed that have no precedent in Europe for medical devices. Such a directive is the product safety directive, which requires manufacturers and distributors to inform competent bodies of recalls or retrofit programs initiated within the member state.

Another key element in the forthcoming European common market program is the role of standards. The European Committee for Standardization (Comité Européen de Normalisation, CEN) and the European Committee for Electrotechnical Standardization (Comité Européen de Normalisation Electrotechnique, CENELEC) have been assigned the task of developing the European Norms that will be required for the implementation of the common market. The standards will be considered voluntary in that a manufacturer can elect to pursue other means to meet the essential requirements specified in a directive. However, if a manufacturer does conform to the standards and otherwise satisfies the necessary conditions to qualify for labeling a product with the CE mark and so labels its product, a member state must presume conformance to the essential requirements and not interrupt the marketing of the device unless it has reason to believe that public health and safety is in jeopardy. The European Norms will supplant the individual national standards.

Third-party laboratories will play an important role in the 1992 Common Market. If so designated by a member state they will become a "notified body." As such, they will be responsible for type test of new devices for conformance to European Norms or, if none are applicable, to the essential requirements listed in the directives. They will also be responsible for GMP inspections and production sampling. The permissible combinations of tests and inspections that a manufacturer may use in getting product to market are described in a directive on certification and testing (see Fig. 1). The medical device directives specify which paths in the general directive are acceptable for medical devices.

A. EC declaration of conformity

Manufacturer
Keeps technical documentation at the disposal of national authorities

B. EC type examination

Manufacturer submits to notified body
- technical documentation
- type

Notified body
- ascertains conformity with essential requirements
- carries out tests, if necessary
- issues EC type examination certificate

G. EC unit verification

Manufacturer
- submits technical documentation

H. EC declaration of conformity (full QA)

Manufacturer
- operates an approved quality system (QS) for design

Notified body
- carries out surveillance of the QS
- verifies conformity of the design (1)
- issues EC design examination certificate (1)

A.

Manufacturer
- declares conformity with essential requirements
- affixes the CE mark

Notified body
- tests on specific aspects of the product (1)
- product checks at random intervals (1)

C. ED declaration of conformity to type

Manufacturer
- declares conformity with approved type

Notified body
- tests on specific aspects of the product (1)
- product checks at random intervals (1)

D. EC declaration of conformity to type (product QA)

Manufacturer
- operates an approved quality system (QS) for production and testing
- declares conformity with approved type
- affixes the CE mark

Notified body
- approves the QS
- carries out surveillance of the QS

E. EC declaration of conformity (product QA)

Manufacturer
- operates an approved quality system (QS) for inpsection and testing
- declares conformity with approved type, or to essential requirements
- affixes the CE mark

Notified body
- approves the QS
- carries out surveillance of the QS

F. EC verification

Manufacturer
- ensures conformity with approved type, or to essential requirements

Notified body
- verifies conformity
- issues certificate of conformity
- affixes the CE mark

G.

Manufacturer
- submits product

Notified body
- verifies conformity with essential requirements
- issues certificate of conformity
- affixes the CE mark

H.

Manufacturer
- operates an approved QS for production and testing
- declares conformity
- affixes the CE mark

Notified body
- carries out surveillance of the QS

(1) Supplementary requirements which may be used in specific directives.

FIGURE 1 Conformity assessment procedures in community legislation.

The criteria for qualifying a third-party laboratory for designation as a notified body is contained in the 45000 series of European Norms issued by CEN. The actual process of qualifying laboratories is planned to be under control of CEN/CENELEC. A framework for coordination of activities involved in qualifying laboratories to be notified bodies exists within CEN/CENELEC. The framework consists of a European Organization for Testing and Certification (EOTC) and Sector Certification Committees. The organization and committees will establish the criteria to use in the laboratory certification process and oversee the process. The actual qualification process will be managed through the cooperation of the participating testing and certification bodies.

A major concern with the above certification and testing system is access by a manufacturer. If the notified bodies have insufficient experience in a particular device or have insufficient staff to provide prompt service, a manufacturer could experience significant and costly delays in being able to market his product in Europe. This concern is magnified for the non-European manufacturer because of the added cost and time involved in travel from and to Europe if a European-based notified body is employed for the product and production certification. If queuing by manufacturers becomes necessary due to insufficient trained personnel within the notified bodies, an added concern for the non-European manufacturer is whether priority will be higher for European manufacturers. Although such unfair treatment by European notified bodies could be a General Agreement on Tariffs and Trade (GATT) issue, the case would be difficult to establish and would probably take considerable time to resolve. During any interim period, the advantage would go to European manufacturers. Because of the time, schedule and preference concerns, considerable effort will go into qualifying United States-based laboratories to be notified bodies and establishing the mechanism whereby they can be so designated. Hopefully, this will be accomplished by the time the directives are planned to be in effect, early 1994.

Clinical trials are required in France as part of the homologation process for many electromedical devices. Other countries require clinical trials for implantable devices prior to marketing. Clinical trials for nonimplantable devices may become a requirement in the directives as a result of trying to obtain a consensus among the member states. Another possibility at this time is that a manufacturer may be required to notify a national health authority prior to starting clinical trials within the country.

Notification prior to sale of a new device is another idea that was originally advanced for the directive but has been shelved for the present. Manufacturer or importer registration is a companion idea that has not been raised so far. While not having much impact on the manufacturer or importer, it presents a clerical nightmare and little utility to the regulatory agency based on FDA experience.

Local language requirements are not specifically addressed in the directives. However, IEC 601-1, Medical Electrical Equipment, Part 1: General Requirements for Safety, does require operating controls and instructions to be in a language acceptable to the user. Within the 12 member states there are nine official languages. It is unclear which local languages may be required or even acceptable. However, it is reasonably certain that even if there is no official requirement for a language other than English, the marketplace will demand it.

Europe is not the only area where new regulations are being created. Canada is in the process of enacting a GMP regulation. Saudi Arabia is writing standards. Japan is considering some form of GMP reciprocity with the FDA. The European Free Trade Area (EFTA) and some Eastern Block countries are actively considering participation in the Common Market.

III. TYPE OF REGULATIONS

Most of the prior discussion dealt with regulations affecting the ability to market a new product. The market approval requirements include certification of conformance to standards, GMP requirements for design and/or manufacture, possible clinical trials, and notification. Other regulations are of equal or even more importance to a manufacturer depending on the stage of a product's maturity. After introduction, the concern is over reimbursement, added standards in the bid specificaitons, bidding restrictions, local content requirements, appropriate technology guidelines, and pricing policies. All of these are market constraints. Another group of regulations concerns distribution of a product. These include export controls (designed to keep high-technology hardware out of the hands of potential enemies) and related restrictive trade prohibition regulations, import quotas including foreign currency exchange regulations, customs procedures, tariffs, import lot sampling, and flag carrier restrictions. Postmarketing regulations include reporting of adverse incidents associated with the use of a product, recalls, retrofits, higher tariffs on spare parts, and maintenance restrictions.

Virtually every country imposes some form of regulatory control over the import of medical devices. Countries with government-owned hospitals often exercise regulatory control through their purchasing practices. More often than not the regulatory control is motivated by economics rather than concern for public health and safety.

IV. CONCLUSION: COPING WITH INCREASING REGULATION

A manufacturer has several options available for coping with a changing regulatory environment. These range from working to shape the new regulation to responding to it upon completion. Ignoring it is not considered a viable option.

Many manufacturers feel it is in their best interest to be very involved in the development process for a new regulation. They seek to be a part of the group doing the development. By being part of the process they hope to minimize the impact of the new requirement on the development of a product and its production, sale, or distribution. It is their belief that they can present knowledgeable inputs to the discussion that will persuade the group to balance public interests with the manufacturers' need to make a profit.

Another approach, often attributed to the Japanese, is to regularly attend all of the meetings where new regulations are being discussed, take copious notes, say very little, return home, and quietly modify the process, product or procedure to comply with the new requirements. Clearly this gives them an edge on the competition in preventing delays in product flow. In addition it minimizes the expense entailed in bringing products into compliance because it permits needed changes to be incorporated with the longest possible lead times.

At the other end of the spectrum is the manufacturer that proceeds until someone or something calls the new requirement to its attention. At this point, a crash effort is implemented to comply with the new requirement so that shipments can be resumed. While the maximum effort is necessary to preserve corporate profitability, it is very taxing on the people and disruptive of the product development and production process. Further, it could result in significant scrap cost if a change was necessary to a product in production. Depending on when the change became known to the manufacturer, regulatory authorities may demand recall or retrofit of product that has already been delivered.

A manufacturer can keep abreast of changing regulations through a variety of ways. Trade associations try to keep their members advised of changing environments. More United States-based associations are establishing international programs in response to their members increasing globalization of their marketing plans. These programs seek to keep their members informed of developments and also to identify avenues for member participation in the process. Identifying avenues for participation in international regulatory development is a new and often frustrating role for most U.S. trade associations.

Many organizations offer seminars on new regulatory initiatives. Seminars represent a good opportunity to keep informed. The speakers are generally knowledgeable about the new regulation and often can suggest what the possible impact may be. Discussions with the other participants often lead to thoughts on how industry should respond to the regulatory authorities or how manufacturers should cope with the new requirements. Trade associations, the Food, Drug Law Institute, Regulatory Affairs Professional Society, and others organize such seminars.

Manufacturers with affiliates in the region where the regulations are being developed have a unique opportunity to keep informed and to participate in the process. However, this capability requires development to assure that it is in place when needed. Affiliates have to be encouraged to get involved and communicate with the United States-based operation. Expectations for influencing the process have to be realistic. Above all, the messenger should be praised rather than shot for bringing the new requirement to everyone's attention.

There are many publications that report on the regulatory scene and its changes, so many so that one wonders how can the time be spared to review a selected few. Conversely, how can the time not be afforded to avoid even one surprise? A list of publications follows in the next section.

Finally, a manufacturer can do some things now to better position itself and its products for a global market.

1. Design for compliance with International Electrotechnical Commission (IEC) and International Standards Organization (ISO) standards. This will minimize expense and delay in the event that third-party certification becomes a requirement.

    a. IEC spacings and insulation requirements are more stringent than Underwriters Laboratories (UL) or Canadian Standards Association (CSA) requirements. Components and power supplies should have recognition marks from Verband Deutscher Elektrotechniker (VDE), British Standards Institute (BSI), or another laboratory with a comparable reputation for testing to IEC or European Norms.

b. Use of an IEC-approved isolation transformer between primary and secondary circuits will avoid the need for IEC-approved components in the secondary circuits.
c. Protective earth (ground) wires in power cords and internally must be green and yellow. Green and yellow wire must not be used for any other purpose. Functional earth wires should be some other color such as orange.
d. Functional and protective earths require separate studs.
e. Provide ground studs with IEC ground marking to connect power cord. Provide stud to connect neutral for three-phase power.
f. Use international symbols, or easily recognizable symbols if no international symbol exists, for operating controls.
g. Power cords should be harmonized if connected to the mains.
h. Use a circuit breaker for mains protection. Fuse requirements for IEC and UL conflict. The circuit breaker must protect and open all mains legs but not the protective earth.
i. All Low-voltage wiring must be separated from mains wiring.
j. IEC tests are conducted at 90% and 110% of rated mains voltage. UL tests are conducted at rated voltage.
k. IEC is more severe than UL on pinch points.
l. A UL-approved cathode ray tube (CRT) requires an implosion test for IEC compliance.
m. Single faults including software errors must not lead to a safety hazard. A single wire leading to many cooling fans would be cut during an evaluation.

2. Video display terminals (VDTs) operating above 25 kV require PTB approval in Germany.
3. Plan for operator manuals and labels on controls without symbols to be in local language. Include VDT screens and responses. Local language function keys and programmable, read-only memories (PROMs) used as look-up tables for prompts and responses may be a viable approach.
4. United Kingdom mains voltage is 240 ± 10% V. The rest of Europe is 230 ± 10% V. Japan is 100/200 ± 10% V. All are 50 Hz. Mains voltage label must state the nominal voltage for which the equipment is setup. In the United Kingdom this must be at least 240 V.
5. Design and test for compliance with International Special Committee on Radio Interference (Comité International Special des Perturbations Radioelectriques, CISPR) EMC emissions and susceptibility standards.
6. Obtain experience with GMPs for design.
7. Participate in relevant IEC activities.

V. USEFUL SOURCES OF INFORMATION

A. Trade Associations

Health Industry Manufacturers Association (HIMA)
1030 Fifteenth Street, N.W.
Washington, DC 20005
Represents the medical device and diagnostic product industry.

National Electrical Manufacturers Association (NEMA)
2101 L Street, N.W., Suite 300
Washington, DC 20037
Represents manufacturers of medical diagnostic imaging and therapy systems.

National Committee for Clinical Laboratory Standards (NCCLS)
711 East Lancaster Ave.
Villanova, PA 19085

Pharmaceutical Manufacturers Association (PMA)
1100 Fifteenth Street, N.W., Suite 900
Washington, DC 20005

B. Other Organizations

U. S. Department of Commerce
Bureau of Export Administration
Exporter Assistance Staff
Room 1099D
Washington, DC 20230

C. Periodicals

International Drug and Device Monitor
Monitor Publications
1545 New York Avenue, N.E.
Washington, DC 20002

Medical Devices, Diagnostics & Instrumentation Reports (The Gray Sheet)
F-D-C Reports, Inc.
5550 Friendship Boulevard, Suite One
Chevy Chase, MD 20815

Business America
U. S. Department of Commerce
Superintendent of Documents
U. S. Government Printing Office
Washington, DC 20402

International Trade Reporter—Current Reports
The Bureau of National Affairs
1231 25th Street, N.W.
Washington, DC 20037

International Trade Reporter—Export Shipping Manual
The Bureau of National Affairs
1231 25th Street, N.W.
Washington, DC 20037

58

U.S. and Foreign Requirements: A Legal Overview

JOHN F. STIGI and ARTHUR C. KOHLER *Center for Devices and Radiological Health, Food and Drug Administration, Rockville, Maryland*

I. OVERVIEW OF IMPORT/EXPORT DEVICE REGULATIONS

When the Medical Device Amendments became law, some predicted these stronger controls over devices by the Food and Drug Administration (FDA) would cripple industry growth. No such dire consequences have come to pass. Of last year's $36 billion international trade market in devices, the United States has over half the market, with over $3 billion in exports [1]. It must also be noted, however, that foreign firms are increasing imports of medical devices into the U.S. as they become more familiar with FDA requirements.

Behind the trade figures is the growing international awareness of the value of devices in promoting public health. FDA's role, as always, has been to establish reasonable but effective controls over devices to assure they are safe and effective. We shall discuss these controls as they apply to devices imported and exported into the United States.

FDA regulates products under the federal Food, Drug, and Cosmetic (FD&C) Act. Regulations, policies, and guidance are issued by the agency to enforce requirements of that act. The Medical Device Amendments of May 28, 1976, amended the FD&C Act to broaden its controls over devices. Part of this expanded authority was a legal process of device classification by which all devices were to be placed into Class I, II, or III. The class determines what levels of control apply to the given device.

Imported and exported devices commercially distributed in the United States must meet FDA requirements for their class. There is a separate set of conditions by which a device not cleared for U.S. commercial distribution may be exported. These conditions are explained later, and a flowchart of the process is given in Figure 1. The numbers of requests to FDA in 1988 to export devices not commercially sold in this country are shown in Table 1.

Much of FDA's authority is based on the "Prohibited Acts" in Section 301 of the FD&C Act. Two often cited prohibited acts are misbranding and mislabeling, which cover a wide variety of circumstances. For example, a device

TABLE 1 Requests to Export Devices Not Commercially Sold in the United States, January-December 1988

Requests	Number of requests	Percent
By device		
Intraocular	124	31
In vitro diagnostic	58	15
Contact lens	51	13
General cardiovascular	25	6
Catheters	23	6
Other	23	6
Pacemakers	22	5
Lasers	17	4
Prostheses	16	4
Sutures	11	3
Bone/spine	8	2
Infusion pumps	7	2
Collagen related	4	1
Lithotriptors	3	1
Contraceptives	2	1
Total number of requests	398	
By country		
Canada	16	
France	8	
Netherlands	7	
Japan	7	
Australia	5	
England	5	
Sweden	5	
Switzerland	4	
Belgium	3	
Germany	3	
Denmark	3	
Italy	3	
Hong Kong	3	
New Zealand	2	
West Germany	2	
Others	25	
Total requests	398	

is misbranded if it requires premarket notification approval (a control explained in Section IV) and is sold without meeting this requirement.

FDA regulates radiation-emitting electronic products under the Radiation Control for Health and Safety Act, as well as under the FD&C Act. These can vary from medical devices such as computed tomography scanners to microwave ovens and television sets. All device regulations, including those for radiation-emitting products, are first published in the *Federal Register* and codified annually under Title 21 of the Code of Federal Regulations (CFR), Parts 800-1299. Other related regulations, such as good laboratory practices, or for color additives, would be found in a different title of the CFR. References to "21 CFR" in this chapter are to Title 21 of the Code of Federal Regulations. "Section" refers to the section in the FD&C Act.

The Food and Drug Administration has several components for regulation of medical products: for devices and radiation-emitting products, the Center for Devices and Radiological Health (CDRH); for blood or blood-related products, the Center for Biologics Evaluation and Research (CBER); and for drugs, the Center for Drug Evaluation and Research (CDER). At times, a device may be regulated by CDRH and may also contain a drug that is regulated by and that needs approval from CDER. In addition, FDA has field offices that conduct good manufacturing practices inspections and have other regulatory functions over the range of products regulated by FDA, including devices.

We will explain import and export device requirements in two different sections below. However, many of the same FDA regulations apply to both categories. Those requirements that apply equally to imported and exported devices are covered in Section IV. Specific requirements for imports and exports are in the narrative that follows.

II. EXPORTING DEVICES

We will consider two basic types of exported devices: (1) those that are legally in commercial distribution in the United States and (2) those that are not.

It is simpler to first describe requirements for "unapproved" (a term typically coined by the industry) devices, although they make up a small part of the export volume.

A. Exporting "Unapproved" Devices

A device intended for export will not be deemed adulterated or misbranded under the Food, Drug, and Cosmetic Act, and may be exported without FDA permission, provided the device:

1. Accords with the specifications of the foreign purchaser.
2. Is not in conflict with the laws of the country to which it is intended for export.
3. Is labeled on the outside of the shipping package that it is intended for export.
4. Is not sold or offered for sale in domestic commerce.

A device that is subject to, but does not comply with an applicable requirement under Section 514 (performance standards) or Section 515 (premarket approval), or a device that is a banned device under Section 516, or a device that is the subject of an Investigational Device Exemption (IDE) under Section 520(g), *may be exported only* if, in addition to meeting requirements 1-4 above, FDA expressly authorizes the exportation of the device. In order to obtain such authorization, a person must submit a request to FDA and include the following information.

A description of the device intended for export.
The status of the device in the United States (investigational, banned, etc.).

A letter from the foreign liaison (listed in Appendix 1), which must be either in English or certified English translation, stating the device is (1) in accord with the specification of the foreign purchaser and (2) not in conflict with the laws of the country to which it is intended for export.

The requester should flag the request "Export Request," and send it, along with any questions concerning the export of medical devices, to Office of Compliance (HFZ-323), Center for Devices and Radiological Health, Food and Drug Administration, 1390 Piccard Drive, Rockville, Maryland 20850.

The export procedure flowchart for products not in domestic commerce is outlined in Fig. 1. The number and type of actual export requests for unapproved devices for 1988, by device type and by country destination, are in Table 1.

B. Exporting Legally Marketed Devices

FDA permission is not required to export a device that meets other FDA requirements for commercial distribution in the United States. Laws that govern device export are in Sections 801(d)(1) and (d)(2) of the FD&C Act. For a device to be legally marketed in the United States, applicable requirements in Section IV of this chapter must be met.

Manufacturers sometimes receive a request for a Certificate of Free Sale from a country to which they wish to export devices. Such certificates are sought as a means of assuring that devices comply with the requirements of U.S. law for distribution in domestic commerce. FDA does not have the resources to provide the kind of continuous inspection and supervision that would be required to give the assurances that the issuance of a Certificate of Free Sale would imply; therefore, FDA does not issue such certificates. However, FDA will provide a general statement of the status of a specific product

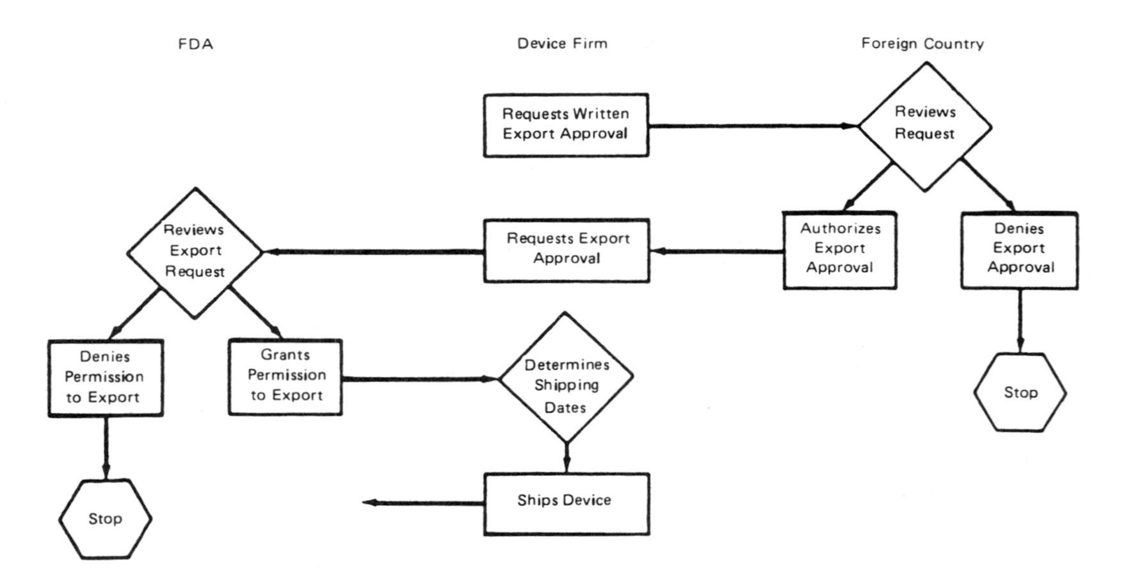

FIGURE 1 Export procedure flow chart for products not cleared for domestic commerce.

upon request. For medical devices, such a statement may be obtained from the Division of Compliance Operations, CDRH. These statements are called Certificates for Products for Export, and are issued with a cover letter. The certificates generally state only (1) the manufacturer's name and location (city and state), (2) that the device is in domestic distribution, (3) that FDA inspects (or has authority to inspect) the facility in which the device is manufactured, and (4) that no legal action is pending against the firm or device in question. Requests to FDA for Certificates of Products for Export in 1988 are listed in Table 2.

C. U.S. Agencies Regulating Exports

Regulation of exported medical devices is shared by three separate departments of the U.S. federal government. They are the Department of Commerce (DOC); Department of State (DOS); and Department of Health and Human Services (DHHS), which includes FDA.

The Department of Commerce regulates the export of medical devices through the Office of Export Licensing (OEL) under the Under Secretary for Export Administration. OEL issues licenses to firms who export medical devices that are highly sophisticated in design and operated by lasers, electron-

TABLE 2 Certificates for Products for Export Requests, January-December 1988

Country	Percent	Device	Percent
Taiwan	32	Cardiovascular	9
Germany	24	Needles and syringes	8
Spain	5	Catheters (various types)	7
Italy	4	*In vitro* diagnostics	7
Japan	3	Contact lens and solutions	7
Korea	2	Dental	6
Colombia	2	Intraocular lens	5
France	2	Anesthesia	5
Mexico	2	Orthopedic	5
England	2	Infusion pumps	3
Brazil	1	Implants (various types)	2
Belgium	1	Dialysis equipment	2
Norway	1	Lithotriptors	1
Turkey	1	Others	33
Others	18		
Total: 645		Total: 645	

ics, computers, and microprocessors. The Department of Commerce derives authority to monitor exported products in the Export Administration Act of 1979. Under this law, the OEL staff works in conjunction with the Departments of State, Defense, and Energy.

A copy of the Commodity Control List (which lists products controlled) may be found in Title 15 of the Code of Federal Regulations under Part 339.1 (15 CFR 399.1). The time frame involved in applying for a license is approximately 10 days if the product is being shipped to the free world. There is no cost for a license. For more information, contact Exporters Assistance Staff, Office of Export Licensing, Department of Commerce, Room 1099, 14th and Pennsylvania Avenue, N.W., Washington, D.C. 20230; telephone (202)377-4811.

Within the Department of Commerce is the International Trade Administration (ITA). ITA promotes and coordinates all export functions for all commodities throughout the industry in the United States. Medical device firms should contact ITA when researching to expand their market potential. However, contact your local U.S. Commerce District Office for assistance before calling the Department of Commerce in Washington, D.C.

ITA relies on three operating units to promote programs:

U.S. and Foreign Commercial Service (US&FCS).
Trade Development (TD).
International Economic Policy (IEC). This unit gives market-specific counseling to American business. A medical device firm should contact IEC to determine a given market potential for its device.

The Department of State regulates the export of medical devices through its Office of Trade, which works in conjunction with the Department of Commerce.

III. IMPORTING DEVICES

Requirements for importing a device can be considered in two parts: (1) the device must meet the same requirements as apply for any commercially sold in the United States, and (2) added requirements for importation.

Requirements for commercial distribution of any device in the United States are found in Section IV. Note that the registration/listing function described in that section is divided: the importer, as initial distributor, is the person who registers with FDA; the foreign firm normally lists the device.

A. FDA versus U.S. Customs Requirements

The major responsibility of the U.S. Customs Service is to administer the Tariff Act of 1930, as amended. Primary duties include assessment and collection of all duties, taxes, and fees on imported merchandise; administration and review of import entry forms; the enforcement of U.S. Customs and related laws; and administration of certain navigation laws and treaties.

Currently, there is a working agreement between FDA and U.S. Customs for the cooperative enforcement of Section 801 of the FD&C Act. Most cooperative action centers on violations to the FD&C Act.

1. Country of Origin Marketing

Among U.S. Customs requirements is that each imported article be legibly and conspicuously marked in English with the name of the country of origin. Exceptions to this rule are articles that are merely in transit through the United States; articles that are under a bond to insure their exportation; and articles that are otherwise specifically exempted from marking requirements. Certain articles may also require special marking.

The country of origin must be marked and must be legible on the product that reaches the "ultimate purchaser." It is not feasible to state who the "ultimate purchaser" will be in every circumstance. However, broadly stated, an "ultimate purchaser" may be defined as the last person in the United States to receive the article in the form in which it was imported. Generally, if an imported article will be used in manufacture, the manufacturer is the "ultimate purchaser." If an article is to be sold at retail in its imported form, the purchaser at retail is the "ultimate purchaser."

B. Steps in the Import Process

The steps by which an imported device is processed to be accepted into the United States, FDA and U.S. Customs roles, and forms involved are somewhat complex. Within the space limitations of this chapter, they are best presented in this outline form.

1. Importer or agent (customhouse broker) files entry documents such as:

   "Importers Entry Notice" (FDA 700 set) or "Land Port Entry Notice" (FDA 720 set)
   Copy of U.S. Customs form 7501, "Entry Summary"
   Copy of commercial invoice
   Bond to cover potential duties, taxes, and penalties.

2. FDA reviews the "Importers Entry Notice" (form FDA 701 and other documents) to determine if wharf examination, sample examination, or documentary sampling will take place.

3a. If no sample is collected, FDA issues "May Proceed Notice" (FDA 702) to U.S. Customs and broker. Shipment is released into domestic status for FDA purposes.

   "May Proceed Without FDA Examination" (FDA 702) releases the shipment from immediate FDA surveillance. FDA 702 is issued when conditions surrounding a shipment do not appear to warrant examining and/or sampling operations. This issuance moves the product to the same status as domestically produced goods. Regulatory action could still be taken if, at a later time, the product is found to violate FDA's domestic regulations.

3b. FDA decides to sample device due to nature of the device, FDA priorities, or past history of the device. In this case, FDA will issue a "Notice of Sampling" (FDA 712) to the broker or importer and the shipment must be held intact pending further notice. However, the shipment may be moved from the entry point to a local warehouse upon proper authorization.

   "Notice of Sampling" (FDA 712) states that a sample has been or will be collected. Whenever FDA takes a sample, FDA 712 is given to the importer of record, or to the customhouse broker who is acting as agent for the importer. There may be one or more notices for each shipment. Usually there is one for each product sampled; however, there may be only one notice issued and it may require that the entire shipment be held. This form:

Is also sent to U.S. Customs.
Directs that the entire shipment, unless otherwise noted, be held intact until released.
Requires the importer to hold the shipment in the local area.
Advises the importer that FDA will pay for samples if a copy of a "Release notice" (FDA 717) is presented to FDA, and the samples are found not to be in violation of any of the laws administered by FDA.

4. FDA examines and/or obtains samples and performs all necessary analyses. Methods of sample analysis are nationally or internationally recognized and published, or specifically given in an FDA regulation.

5a. FDA finds the sample to be in compliance with requirements and sends a "Release Notice" (FDA 717) to the broker or importer of record and U.S. Customs. As far as FDA is concerned, the shipment may be released into domestic status. Form FDA 717 allows the importer to release the bond and clear the portions of the shipment covered by that notice.
The samples items may be released via a form FDA 717 three ways:

A "Release Notice" (FDA 717) without comment is sent to the customhouse broker or importer of record with U.S. Customs advising them that the shipment need not be held any longer.
A "Release Notice" (FDA 717) with comment is sent to the customhouse broker allowing a specific lot to be released. This notifies the broker or importer of record and U.S. Customs that the lot is not in compliance, but violations are not sufficiently serious to refuse entry or require immediate correction. Future shipments however, *will* be refused entry unless violations are corrected.
A "Release Notice" (FDA 717) after detention is sent to the customhouse broker or importer of record and U.S. Customs, captioned "Originally Detained, Now Released."

5b. FDA determines the sample "Appears to be in Violation of the FD&C Act and/or Other Related Acts." In this case, a "Notice of Detention and Hearing" (FDA 718) is sent to the broker or importer of record specifying the nature of the violation. This notice gives the importer of record 10 working days to introduce testimony or file a written statement concerning why the imported product should be allowed entry into the United States, or to ask for an extension of the hearing date. Failure to respond promptly will cause products to be refused admission.

6. Although a response to a "Notice of Detention and Hearing," or a request for an extension of the hearing date, is not required, failure to do either will result in FDA's issuing a "Notice of Refusal of Admission" (FDA 772).
The customhouse broker or importer of record, or a designated representative, responds to a "Notice of Detention and Hearing." The response may be the introduction of testimony, either in person held during hearing with FDA or in writing, as to the admissibility of the shipment; or the response may be a request to recondition the violative product (see 7a and 7b).
The hearing held as a result of a "Notice of Detection and Hearing" is the importer's only opportunity to present, either in person or by a

written submission, a defense of the importation and/or evidence to show how the importation may be made eligible for entry. The hearing is an informal procedure conducted by an FDA compliance officer and is confined to matters pertaining to the violations alleged in the "Notice of Detention and Hearing." In the hearing, the importer is given opportunity for the presentation of relevant evidence (such as private laboratory analyses, proof of registration, filing of a required submission) and is given the chance to submit a proposal on form FDA 766 as discussed below.

After the importer's response during the hearing, the FDA district director or his designee (normally the compliance officer) will decide if the detained merchandise will be released, reconditioned, or refused entry.

7a. The importer of record presents evidence that indicates the merchandise is in compliance. Certified analytical results of samples, examined by a reliable laboratory, and data that are within the guidelines may be presented. This evidence will not necessarily establish that a product is acceptable, only that the private laboratory did not report finding any problem.

7b. The importer of record submits to FDA a written application form FDA 766, "Application for Authorization to Relabel o0 Perform Other Action," for authorization to recondition or to perform other action in order to attempt to bring the product into compliance. The importer must detail how he will bring the merchandise into compliance.

Relabeling or reconditioning is an action performed to bring into compliance all or part of the merchandise, which would otherwise have to be exported or destroyed. The request to relabel or recondition:

Must specify the time and place where reconditioning operations will be performed and an appropriate time for their completion.
Is authorized at FDA discretion.

These actions are a *privilege* and *not* a *right*. FDA must approve the reconditioning or relabeling plan before it can be implemented. The plan, which is discussed by FDA with the owner usually at the time of the hearing held as a result of the detention of the product, is approved in almost every instance when there are assurances that the operations can be effective and the violations are such that allow it.

8a. FDA reviews and accepts evidence submitted or samples to determine if the product is in compliance with the FD&C Act.

8b. FDA approves the importer's reconditioning application (FDA 766) if past experience or evidence shows that the proposed method will succeed.

8c. If FDA denies the original reconditioning proposal, the importer of record will be allowed to submit a second proposal. This second and final request will be considered if it contains meaningful changes in the reconditioning operation to assure success.

Denial of request to relabel or recondition is indicated on the original FDA 766, which is returned to the customhouse broker. The denial provides the reason for denial, and allows an additional period of time for response.

8d. FDA does not find the testimony adequate to clear the shipment. The importer may submit an application to recondition or FDA will issue a "Notice of Refusal of Administration."

9. Importer completes all reconditioning procedures and advises FDA that goods are ready for inspection or sample collection. When the applicant completes the reconditioning/relabeling, he or she returns the "Importer's Certificate," which indicates that the relabeling or reconditioning has been completed and all merchandise is ready for inspection or sample collection. FDA then may decide whether to examine the attempted relabeling or reconditioning.
10. FDA conducts follow-up inspection and may sample product to determine compliance with the terms of the reconditioning authorization.
11a. If FDA finds the sample to be in compliance, a "Release Notice" with the statement "Originally Detained and Now Released" is sent to the broker or importer of record.
11b. If FDA finds the sample to be out of compliance, an "Application for Authorization to Recondition or to Perform Other Action" may be submitted if not previously done, or FDA will issue a "Notice of Refusal of Admission."
12. If FDA issues a "Notice of Refusal of Admission" to the broker or importer of record, another "Notice of Refusal of Admission" advises the customhouse broker that the detained merchandise is not in compliance, and must be destroyed if not exported.

    A "Notice of Refusal of Admission" is issued:

    If FDA does not receive a response to a "Notice of Detention and Hearing" within the specified 10-day period and FDA has not granted an extension of time for responding.
    When efforts to relabel or recondition the detained shipment according to an approved application have failed.
    If the FDA hearing officer has ruled that the violations are valid and an application to relabel or recondition had not been submitted.

    The refusal of admission also applies to those devices from a detained shipment that could not be reconditioned or relabeled.
13a. The exportation or destruction of the merchandise listed on the "Notice of Refusal of Admission" is performed under the direction of U.S. Customs, who supplies FDA proof of exporting or destroying the shipment.
13b. The importer must export the product within 90 days of the issuance of the refusal. After the ninetieth day, U.S. Customs will only allow its destruction.

IV. FDA REQUIREMENTS FOR ALL DEVICES COMMERCIALLY DISTRIBUTED IN THE UNITED STATES

A. Device Classification

FDA is required under Section 513 of the FD&C Act to classify all devices intended for human use into one of three regulatory classes depending on the extent of control necessary to provide reasonable assurance of safety and effectiveness. This classification process is the cornerstone of FDA regulatory requirements. The three classes, briefly described, are:

Class I devices are those which present minimal potential for harm to the user and are often simpler in design and function than those in Class II or

Class III. Class I devices are subject only to General Controls. Examples are enema kits, elastic bandages, and pipetting and diluting systems for clinical use. General controls prohibit adulteration and misbranding and include establishment registration and device listing; FDA authority to ban certain devices; premarket notification submissions; record keeping and reporting by the firm; good manufacturing practices by the firm; and repair, replacement, or refund by the firm to the customer.

Class II devices are those that require more than general controls to assure safety and effectiveness and for which there is sufficient information for FDA to establish a performance standard. Devices placed into this class must meet general controls, and are required to meet an applicable mandatory standard if one is prescribed by FDA. Examples of Class II devices are powered wheelchairs, some pregnancy test kits, and condoms.

Class III devices must meet general controls, and currently or in the future will require an FDA-approved premarket approval application (PMA) for the firm to commercially distribute or market them in the U.S. Examples of Class III devices are heart valves, automated heparin analyzers, and infant radiant warmers.

Domestic firms, foreign manufacturers of medical devices, and initial distributors/importers must meet FDA requirements that vary depending on the class of the device. All devices must meet General Controls, described in detail later. The domestic firm and either the foreign manufacturer or its U.S. initial distributor submits premarket notification or a premarket approval for the device. Certain Class I devices, however, are exempt from two General Controls: premarket notification, and the good manufacturing practices regulation, except for general record provisions in 21 CFR 820.180 and complaint file requirements in 21 CFR 820.198.

Marketing clearance time frames are the same for imported and exported devices:

If Class I and exempt from premarket notification, the device can be imported/exported *at once*, if the exporter has properly registered and the device is listed, or if the importer is properly registered and the foreign firm properly listed.

If premarket notification is required, it will take *up to 90 days* for device approval, and perhaps longer if FDA requests additional information.

If premarket approval is required (for Class III devices), the time frame can be from months to years, depending on what information FDA needs about the device. Clinical studies, investigational device studies, or both may be required.

B. Transitional Devices

Under the expanded definition of devices in the 1976 Device Amendments, certain items previously regulated as drugs became devices. These are placed into Class III automatically under the transitional device provision of Section 520(1) of the FD&C Act, but are regulated in a way consistent with their prior status as drugs. In this regard, former drugs requiring new drug approvals became devices requiring premarket approvals. A specific list of transitional devices that require PMAs is available from CDRH.

C. General Controls

General Controls below apply to devices in all three classes.

1. *Establishment Registration*

Who Must Register. Each domestic firm, U.S. initial distributor (importer) is required to register its establishment with FDA in accord with 21 CFR 807.20(a). An establishment is a place of business under one management or at one general physical location at which a device is manufactured, assembled, or otherwise processed. If several initial distributors import the same medical device, each is required to register. Foreign manufacturers commercially distributing devices in the United States are *not* required to register, but they are encouraged by FDA to do so. Other persons who must register are those who develop initial specifications for, repackage, relabel, or sterilize devices, or who are contract manufacturers of devices. An importer, for example, would register with FDA as importer and as sterilizer, if it performs that latter activity.

How to Register. To register an establishment, the person fills out and submits to FDA form FDA 2891, "Initial Registration of Medical Device Establishment."

Those required to register should bear these points in mind.

Use actual street addresses. FDA will not accept post office (P.O.) box numbers as addresses on form FDA 2891.

The owner/operator is the corporation, subsidiary, affiliated company, partnership, or proprietor directly responsible for the activities of the registering establishment.

An official correspondent must be designated. This person is assigned by the owner/operator to be responsible for the annual registration of the establishment, contact with FDA for listing, and other official communications.

Establishments required to register must do so within 30 days of an activity requiring registration. This means initial distributors must register *before or within* 30 days of importing a medical device.

Make a copy of your completed FDA 2891 for your own records. *All* copies of form FDA 2891 are to be submitted to Device Registration and Listing Branch [DRLB] (HFZ-342), Center for Devices and Radiological Health, Food and Drug Administration, 1390 Piccard Drive, Rockville, MD, 20850. FDA will return a validated copy of your form, with your registration number.

Updating Registration Data. The owner/operator is responsible for keeping data current on the registration form. When changes occur in ownership, establishment name, official correspondent, or addresses, DRLB must be notified in writing within 30 days of such changes. In addition, registration is required each year. Each registrant receives annually a form 2891a, which is to be filled out and returned, even if there are no changes in information.

2. *Device Listing*

Who Must List. Domestic and foreign manufacturers must also list their devices by their generic category (an FDA classification name) with the agency. Initial distributors do not list themselves, but must send to FDA a letter listing their foreign suppliers and addresses. A device may be listed on behalf of a foreign firm, however, by its sole U.S. initial distributor, that is, the only initial distributor for the device in this country under certain conditions.

Foreign firms must list their devices with FDA before these can be imported. Domestic firms register first, then list devices after they have received their registration number along with listing instructions from FDA.

How To List. Listing of a device is done by completing form FDA 2892. The submitter should keep the bottom (yellow) copy of the form as proof of listing. The form has the actual listing number on its upper left area. FDA does not send back any validation.

Here are other factors to consider in listing:

Only one form FDA 2892 is required for each generic category (classification name) of device exported by a foreign manufacturer. Therefore, for example, for various sizes of syringes that fall within the same generic category or classification name, a foreign firm fills out only one FDA 2891.

Each foreign manufacturer or its *sole* initial distributor must list independently.

A properly completed form FDA 2892 must be submitted by the foreign manufacturer or its *sole* initial distributor before their medical device can be imported into the United States.

All but the bottom (yellow) copy of form FDA 2891 should be submitted to the following address: Device Registration and Listing Branch (HFZ-342), Center for Devices and Radiological Health, Food and Drug Administration, 1390 Piccard Drive, Rockville, Maryland 20850.

Updating Listing Data. The manufacturer or the *sole* initial distributor is responsible for keeping current the data on their FDA 2892. Unlike establishment registration, device listing data is updated by entering the new information on a "new" form FDA 2892. Device listing data must be updated when one or more of the following occurs:

A "new" device is marketed with a classification name that is not currently listed;

The intended use of a listed device changes in such a way that would result in its being more appropriately classified under a different generic category (classification name);

The marketing of all models or variations of the listed device is discontinued;

A discontinued device (not listed) is remarketed; or

Any information changes on form FDA 2892, other than changes in proprietary and common or usual name.

3. *Premarket Notification*

When Required. A premarket notification [510(k)] must be submitted to FDA 90 days prior to entering a device into commercial U.S. distribution. This submission enables FDA to determine if the device is substantially equivalent to another "predicate" medical device that was in commercial distribution before May 28, 1976, the date the Medical Device Amendments became law, or to a device reclassified from Class III to Class I or II. If found to be substantially equivalent, the device may be commercially marketed. If found not to be substantially equivalent, then the device requires an approved premarket approval (PMA) application or reclassification before it can be marketed. PMAs, which are not a general control, are discussed later.

A 510(k) submission is also necessary for a device currently in, or being reintroduced into, commercial distribution in the United States that is about to be significantly changed or modified, or for a major change or modification in the intended use.

There can be confusion as to when a premarket notification is required. A 510(k) submission is needed when any one of the following occurs:

A foreign manufacturer intends to export a medical device to the United States that the firm has never before shipped to the United States.
Either the foreign manufacturer or initial distributor changes the intended uses of devices that are legally being marketed in the United States.
Changes or modifications are made to a legally marketed device that could significantly affect its safety or effectiveness.

Premarket notifications are submitted by the device manufacturer. In the case of an imported device, either the initial distributor or the foreign firm may make this submission.

Who Submits. The following persons must or may submit a premarket notification [510(k)] submission:

Domestic manufacturers. Domestic firms *must* submit a 510(k).
Foreign firms and their initial distributors. *Either* the foreign firm or the initial distributor *must* submit the premarket notification submission, not both.
Others: In some cases, a 510(k) may be required from a U.S. specification developer having a device manufactured under contract abroad, and from repackagers and relabelers that may reprocess or significantly alter the labeling.

Contents of a Premarket Notification. A premarket notification should contain the following information:

Product name (including proprietary and common, usual or classification name).
Registration number, if any, assigned by FDA to a registered establishment.
Present class of the device, if any or if known, and the FDA classification panel that will consider the device.
If applicable, action taken to meet FDA performance standards.
Samples of proposed labels, labeling, and advertisements that describe the device, its intended use, and directions for use.
A statement of how the device is similar to and/or different from others commercially marketed. Include labeling, studies, or other material that supports the comparison.

Modified Devices Requiring Premarket Notification. If a device is changed or modified, a new premarket notification must be submitted if the changes significantly affect safety and effectiveness of the device.

4. *Good Manufacturing Practices (GMP)*

The good manufacturing practices (GMP) regulation promulgated under Section 520(f) of the FD&C Act requires that domestic and foreign manufacturers of medical devices commercially distributed in the United States have a quality assurance (QA) program. Adequate specifications and controls must be established and finished devices must meet these specifications. Thus, the GMP regulation helps assure that medical devices are safe and effective. FDA continually monitors data concerning problems with medical devices and inspects the operations and records of foreign and domestic manufacturers to determine compliance with the requirements of the GMP regulation.

To Whom GMP Applies. Domestic and foreign manufacturers that export medical devices to the United States for commercial distribution must comply with the requirements of the GMP regulation. Initial distributors who repackage, relabel, sterilize, change the condition or intended use, or develop specifications for the device are subject to applicable parts of the GMP regulation. Initial distributors that import a device to be commercially distributed in the United States and that do not alter the device or its labeling are not subject to the GMP regulation.

Who Is Exempt. The GMP regulation applies to the manufacture of finished medical devices intended to be commercially distributed unless there is an approved investigational device exemption in effect, or the device is exempted by a classification regulation. Certain classified Class I devices have been exempted with the exception of 820.180, general requirements concerning records, and 820.198, complaint files.

How to Comply. The GMP requirements found in 21 CFR 820 cover QA programs, personnel, buildings, equipment, components, production and process controls, packaging and labeling control, device distribution and installation, device evaluation, records, and complaints. In addition, the GMP regulation is two-tiered, with more stringent requirements for critical devices.

FDA's Foreign Firm Inspection Program. FDA inspects the operations and records of foreign as well as domestic manufacturers to determine GMP compliance. There is no difference between the foreign and the domestic inspection program, except that FDA must obtain permission of the foreign country and manufacturing facility before initiating an inspection.

5. Labeling Requirements

United States and foreign-made medical devices must be labeled in accord with the applicable labeling regulations for medical devices or *in vitro* diagnostic products (Sections 201 and 502, and see 21 CFR 800-1299), and the Fair Packaging and Labeling Act. Labeling must have adequate directions for use, and any warnings needed to assure the safe use of the device, unless it is specifically exempted from such requirements by regulation.

6. Banned Devices

If a device for human use presents a substantial deception or an unreasonable and substantial risk of illness or injury, FDA has the authority under Section 516 to ban it.

7. Notification, Repair, Replacement and Repair Provisions

This Section 518 of the FD&C Act has yet to be implemented.

8. *Adulteration and Misbranding*

Under Section 501, a device is considered adulterated if it does not comply with FDA regulations required for it, such as a performance standard or investigational device exemption, or if it is not manufactured in accordance with the good manufacturing practices.

Section 502 provides conditions under which a device would be deemed misbranded, such as for advertisements that do not comply with FDA requirements.

D. Mandatory Performance Standards (Class II)

Under Section 514 of the FD&C Act, FDA is authorized to develop and establish mandatory performance standards for Class II devices. No performance standard under Section 514 presently exists. Several devices, however, have been targeted as high priority for development of such an FDA standard. Standards for radiological devices do exist, but are not under this section. These can be found later in the discussion of radiation-emitting device requirements.

E. Premarket Approval (Class III)

Premarket approval (PMA) is the most stringent type of device marketing application required by FDA. A PMA is an application submitted to FDA by a manufacturer to request approval to market, or to continue marketing of a Class III medical device. Unlike premarket notification, PMA approval is to be based on a determination by FDA that the PMA contains sufficient valid scientific evidence that provides reasonable assurance that the device is safe and effective for its intended use or uses.

1. *Who Must Submit a PMA?*

The PMA applicant is the person who owns the rights, or otherwise has authorized access, to the data and other information to be submitted in support of FDA approval of the PMA. The PMA applicant is also the person who intends to market or continue to market the device.

2. *When Is a PMA Required?*

PMA requirements apply to all Class III devices marketed (commercially distributed) or to be marketed in the United States. However, not all Class III devices require an approved PMA before they can be marketed. Preamendment Class III devices and substantially equivalent postamendment devices may be marketed until FDA completes a special regulatory process to require PMAs for them. There are three types of Class III devices: transitional, preamendment, or *not* substantially equivalent postamendment.

Transitional Class III devices are devices regulated by FDA as new drugs before the May 28, 1976, enactment of the Medical Device Amendments to the FD&C Act

Preamendment Class III devices and postamendment devices determined to be substantially equivalent to them can continue to be marketed without submission and FDA approval of a PMA until such approval is required by regulation. These devices, other than transitional Class III devices, were in commercial distribution in the United States before the May 28, 1976, enactment of the Medical Device Amendments.

Not substantially equivalent postamendment Class III devices must have an approved PMA in effect before being marketed in the United States.

3. How to Submit a PMA

Information regarding the content requirements, formatting, procedures, and review process for PMAs can be found in the *Premarket Approval (PMA) Manual*, available upon request from CDRH.

F. Medical Device Reporting (MDR)

Manufacturers and initial distributors of medical devices must comply with the Medical Device Reporting (MDR) Rule, 21 CFR 803, if they are required to register (Part 607 and 807) with FDA.

Manufacturers and initial distributors of devices (referred to as an "importer" in the MDR rule) must report to FDA whenever the firm becomes aware of information that reasonably suggests that one of its marketed devices (1) may have caused or contributed to a death or serious injury or (2) has malfunctioned and that the device or any other device marketed by the domestic firm or initial distributor would be likely to cause or contribute to a death or serious injury if the malfunction were to reoccur.

Reports of death or serious injury must be telephoned to FDA at (301)427-7500 no later than 5 calendar days from the time the manufacturer or initial distributor receives the information. Additionally, telephone reports must be followed up by written report to FDA within 15 days from the time the manufacturer or initial distributor receives the information, for all reports of death and serious injury and all reports of malfunctions likely to cause or contribute to a death or serious injury.

All initial distributors of devices, as well as U.S. manufacturers, are required to establish and maintain a complaint file and to permit any authorized FDA employee at all reasonable time to have access to and to copy and verify the records contained in this file. FDA considers any expression of dissatisfaction, be it oral or written, regarding the identity, quality, durability, reliability, safety, effectiveness, or performance of a device, to be a complaint.

G. Investigational Device Exemption (IDE)

The Investigational Device Exemption (IDE) allows manufacturers to ship and use imported devices intended solely for investigational use on human subjects, without having to first meet some FDA statutory requirements. The IDE applies to all clinical studies that are undertaken on humans to gather safety and effectiveness data about a medical device.

The objective of clinical studies is to gather data needed to support premarket approval, premarket notification, or reclassification. An investigational device is one that has not been give 510(k) clearance or PMA approval for marketing in the United States but that is exempted from these requirements in order to collect safety and effectiveness data. Investigational use also includes clinical evaluation of certain modifications or new intended uses of legally marketed devices. FDA requires that all clinical evaluations of investigational devices, unless exempt, have an approved IDE before the study is initiated (see 21 CFR 812 and 813).

A person who imports or offers for importation an investigational device is required to act as the agent of the foreign exporter with respect to the investigation and either as the sponsor of the clinical investigation or to ensure that another person acts as agent and sponsor. A sponsor is responsible for submitting the IDE application to FDA. As long as the device is being studied, sponsors with an approved IDE need not comply with certain sections of the FD&C Act pertaining to misbranding, registration, listing, premarket notification, premarket approval, good manufacturing practices, color additive requirements, etc. Sponsors are *not* exempt from the adulteration provisions of the FD&C Act.

Investigations require Institutional Review Board (IRB) approval. An IRB means any broad, committee, or other group formally designated by an institution where the study is to be conducted to review, to approve the initiation of, and to conduct periodic review of, biomedical research involving human subjects.

The IDE regulation distinguishes between significant and nonsignificant risk devices. Procedures differ for each. For example, a significant risk device IDE requires FDA as well as IRB approval.

H. Radiation-Emitting Electronic Products

Under the Radiation Control for Health and Safety Act, manufacturers of radiation-emitting electronic products must submit initial reports, annual reports and models change reports to CDRH. Such reports are in addition to requirements which apply if these products are medical devices. CDRH has specific reporting guidelines for these products. As explained later, some radiation products also must meet performance standards.

Radiation-emitting products which require reports from manufacturers are listed in 21 CFR 1002.61. The list is as follows:

Ultrasonic products
Microwave heating equipment
High voltage vacuum switches
Rectifier tubes
Shunt regulator tubes
Cathode ray tubes intended to be operated at voltages greater than 5000 V but less than 15,000 V
Ultraviolet lamps and products containing such lamps intended for irradiation of any part of the human body by light of wavelength in air less than 320 nanometers to perform a diagnostic or therapeutic function
Television receivers that meet the Federal standard, provided the voltage of the cathode-ray tube cannot exceed 15,000 V
High voltage vacuum switches, rectifier tubes, shunt regulator tubes and cathode-ray tubes intended to be operated at voltages of 15,000 or greater
Products in addition to television receivers that are subject to Federal radiation standards
Diagnostic x-ray, cabinet x-ray, microwave ovens, laser products, sunlamp

products, high intensity mercury vapor discharge lamps, ultrasonic therapy products

Products intended to produce x-radiation including radiation therapy devices

Industrial dielectric heaters including radiofrequency (RF) sealers and electromagnetic (EM) induction heating equipment that operate in the frequency range from 2 to 500 megahertz

Microwave diathermy devices

CDRH has developed, and has copies of, reporting guidelines for each of the products requiring such report.

1. Radiation Control Standards

In addition to performance standards under the FD&C Act, CDRH may promulgate, under authority of the Radiation Control for Health and Safety Act, performance standards for radiation-emitting electronic products. So far, CDRH has developed performance standards for the following products:

Television Receivers—This standard became effective January 15, 1970, and has been recodified in 21 CFR 1020.10. It applies to television receivers designed to receive and display a television picture, and includes electronic viewfinders on TV cameras, TV projectors, and TV monitors used with X-ray and other medical systems.

Diagnostic X-Ray Equipment—This standard became effective August 1, 1974, and has been recodified in 21 CFR 1020.30-33, with subsequent amendments. It applies to complete diagnostic X-ray systems, as well as major components, including tube-housing assemblies, X-ray controls, high-voltage X-ray generators, fluoroscopic imaging assemblies, X-ray tables, cradles, film changers, cassette holders, and beam-limiting devices.

Cabinet X-Ray Systems—This standard became effective April 10, 1975, and has been recodified in 21 CFR 1020.40. In addition to baggage inspection systems, it applies to other X-ray machines in enclosed free-standing cabinets.

Laser Products—This standard became effective August 1, 1976, and has been recodified in 21 CFR 1040.10 and 1040.11, with subsequent amendments. It applies to all lasers and products containing lasers. Specific requirements for medical lasers are in 21 CFR 1040.11(a).

Sunlamp Products and Ultraviolet Lamps Intended for Use in Sunlamp Products—This standard became effective on May 7, 1980, and was amended September 8, 1986. It applies to all sunlamp products and ultraviolet lamps intended to induce suntanning.

Ultrasonic Therapy Products—This standard became effective February 17, 1979, and has been recodified in 21 CFR 1050.10. It applies to any device intended to generate and emit ultrasonic radiation for therapeutic purposes at frequencies above 16 kHz, or any generator or applicator designed or specifically designated for use in such a device.

Manufacturers and importers planning to import devices subject to the above performance standards should be aware that the standard applies to devices intended for commercial distribution as well as those imported for investigational use. *Note*: *An investigational device exemption does not exempt a device or a component from these radiation control standards.*

V. CONCLUSION

The requirements described may make it appear that importing or exporting devices involves quite intricate and burdensome legal steps. Actually, even the smallest business can easily be an importer or exporter. There is a wealth of aid in terms of printed materials and guidance that simplify importing and exporting obligations. Much of this assistance is available from the Center for Devices and Radiological Health, most notably its Division of Small Manufacturers Assistance.

REFERENCE

1. Medical and Dental Instruments and Supplies, *U.S. Industrial Outlook*, December 1987, p. 24, U.S. Department of Commerce, Washington, D.C.

APPENDIX 1 Foreign Liaison Listing, Center for Devices and Radiological Health

- A -

AFGHANISTAN
Afghanistan Embassy[1]
2341 Wyoming Avenue, N.W.
Washington, D.C. 20008

ALGERIA
Attache[1]
Embassy of Algeria
2118 Kalorama Road, N.W.
Washington, D.C. 20008

ARGENTINA
Secretaria de Estado de Salud Publica
Relaciones Sanitarias Internacionales
Defensa 120-4°-Of. 4027
1345 Buenos Aires, Argentina

AUSTRALIA
Assistant Secretary
Medical Devices and Dental Products Branch
Department of Community Services and Health
GPO Box 9848
Canberra ACT 2601
Australia

AUSTRIA
Bundesministerium fur Gesundheit und Umweltschutz
Stubenring 1
1010 Wien, Austria

- B -

BAHAMAS
Ministry of Health and National Insurance
P.O. Box N 3729
Nassau, N.P., Bahamas
Commonwealth of the Bahama Islands

BAHRAIN
Embassy of the State of Bahrain[1]
3502 International Drive, N.W.
Washington, D.C. 20008

BOLIVIA
Chief Medical Officer
Ministerio de Prevision Social y Salud Publica
Plaza Franz Tamayo
La Paz, Bolivia

BANGLADESH
Director
Drug Administration
Ministry of Health and Population Control (Health Division)
80 Motijheel Commercial Area
Dacca, Bangladesh

BARBADOS
First Secretary
Embassy of Barbados
2144 Wyoming Avenue, N.W.
Washington, D.C. 20008

BELGIUM
Economic Minister[2]
Embassy of Belgium
3330 Garfield Street, N.W.
Washington, D.C. 20008

BELGIUM
The Pharmaceutical Inspectorate[3]
Ministry of Public Health
Vesalius gebouw 3° Verdieping
Administratief Centrum
1010 Brussels, Belgium

BENIN
Directeur General de l'ONP
Office National de Pharmacie
BP 1255
Cotonou
Republique Populaire du Benin
Africa

BOTSWANA
Permanent Secretary
Ministry of Health
Gaborone, Botswana
Africa

BRAZIL
Ministerio da Saude
Secretaria de Vigilancia Sanitaria
Divisao Nacional de Vigilancia Sanitaria de Medicamentos—DIMED
Avenida Brazil, 4036, 6° andar sala 602
20930—Rio de Janeiro, RJ
Brazil

BULGARIA
Information Officer
Bulgarian Embassy
2100 16th Street, N.W.
Washington, D.C. 20009

BURMA
Chancelor
Embassy of Burma
2300 S Street, N.W.
Washington, D.C. 20008

BURUNDI
The Ambassador
Burundi Embassy
2717 Connecticut Avenue, N.W.
Washington, D.C. 20008

- C -

CAMEROON
The Ministry of Public Health
P.O. Yaounde
United Republic of Cameroon
Africa

CANADA
Director
Bureau of Medical Devices
Health and Welfare of Canada
Health Protection Branch
Environmental Health Directorate
Tunney's Pasture
Ottawa, Ontario
Canada K1A OL3

CHILE
Ministerio de Salud
Jefe de Coordinacion y Planes
Mac Iver 541 Piso 3
Santiago, Chile

CHINA
Embassy of the People's Republic of China[1]
2300 Connecticut Avenue, N.W.
Washington, D.C. 20008

COLOMBIA
Ministerio de Salud
Calle 16 N 7-39
Bogota, Colombia

CAPE VERDE
Ministry of Health and Social Affairs
Secretaria-Geral do Ministerio de
Saude e Assuntos Sociais
Praia, Cape Verde
Cape Verde Islands

CENTRAL AFRICAN EMPIRE
Directeur General de la
Sante Publique
au Ministere de la Sante et de la
Population
Banqui
Empire Centrafricain
Africa

CHAD
Ministre de la Santé Publique
et des Affaires Sociales
N'Djamena
Republique du Tchad

COSTA RICA
Ministerio de Salud Public
San Jose, Costa Rica

CYPRUS
Director
Pharmaceutical Services
Ministry of Health
Republic of Cyprus
Nicosia, Cyprus

CZECHOSLOVAKIA
Office for Standards and Measures
Department for International Cooperation
Urad pro normalizaci a mereni
Oddeleni mezinarodne spoluprace
Vaclavske nam. 19
110 00 Praha 1, Czechoslovakia

- D -

DENMARK
National Board of Health
1, St. Kongensgade
DK-1264 Copenhagen K
Denmark

DOMINICAN REPUBLIC
Secretaria de Estado de Salud
Publica y Asistencia Social[4]
Ensanche La Fe
Santo Domingo, Republic Dominicana
Hispaniola Island, West Indies

- E -

ECUADOR
Ministerio de Salud Publica
Directora Nacionale de Control
Sanitario
Juan Larrea No. 444-Quito
Ecuador

EGYPT
Director General of
Importation Department
6 El-Shwarby St.
Cairo, Egypt

EL SALVADOR
Presidente del Consejo
Superior de Salud Publica
Avenida Espana No 736
San Salvador, El Salvador, C.A.

ETHIOPIA
Third Secretary[1]
Embassy of Ethiopia
2134 Kalorama Road, N.W.
Washington, D.C. 20008

- F -

FIJI
Permanent Representative
Fiji Mission to the United Nations
One United Nations Plaza
(26th Floor)
New York, New York 10017

FRANCE
Ministère de la Santé et de la
Secuité Sociale
1, place de Fontenoy
75700 Paris, Republique Française

FINLAND
Chief of the Office
Office of Technical Affairs
National Board of Health
Box 220
SF-00531 Helsinki, Finland

- G -

GABONESE REPUBLIC
Embassy of Republic of Gabon[1]
2034 20th Street, N.W.
Washington, D.C. 20009

GERMANY (WEST)
Ministerialrat[5]
Referat Va 5
Bundesministerium für Arbeit und Sozialordnung
Rochusstrasse 1
5300 Bonn-Duisdorf
Federal Republic of Germany

GERMANY (WEST)
Ministerialrat
Dr. Karl F. M. Feiden
Augustastr. 33
D-5300 Bonn-Bad Godesbert
Federal Republic of Germany

GHANA
Director of Medical Services
c/o P.O. Box M.44
Accra, Ghana

GREAT BRITAIN
(see United Kindgom of Great Britain)

GREECE
First Secretary
Ministry of Social Services
International Relations Division
17 Aristotelous Str.
Athens, Greece

GRENADA
Permanent Secretary
Ministry of Health
St. George's Grenada
Windward Island, West Indies

GUATEMALA
Embassy of Guatemala
2220 R Street, N.W.
Washington, D.C. 20008

GUINEA
Embassy of Guinea[1]
2112 Leroy Place, N.W.
Washington, D.C. 20008

GUINEA-BISSAU
Minister of Health and Social Welfare
Bissau
Guinea, Bissau
Africa

GUYANA
Government Analyst
Commissioner of Food and Drugs
19/21 Lyng and Evans Streets
Georgetown, Guyana

- H -

HAITI
Ministère Santé Publique[4]
Port-Au-Prince, Haiti
West Indies

HONDURAS
Director, General de Salud
Tegucigalpa, D.C., Honduras

HONG KONG
Hong Kong Government
Medical and Health Department
Sunning Plaza, 4th-13th Floors
10 Hysan Avenue, Causeway Bay
Hong Kong

HUNGARY
Eng., MSEE, B. Sc. Econ.[1]
Commercial Secretary
Hungarian Embassy
150 E. 58th Street
New York, NY 10022

- I -

ICELAND
Embassy of Iceland[1]
2022 Connecticut Avenue, N.W.
Washington, D.C. 20008

INDIA
Embassy of India[1]
2107 Massachusetts Avenue, N.W.
Washington, D.C. 20008

INDONESIA
Information Officer[1]
Indonesian Embassy
2020 Massachusetts Avenue, N.W.
Washington, D.C. 20036

IRAN
Embassy of Iran[1]
3005 Massachusetts Avenue, N.W.
Washington, D.C. 20036

IRAQ
Commercial Officer[1]
Iraq Embassy
1801 P Street, N.W.
Washington, D.C. 20036

IRELAND
The Food and Drugs Division
Department of Health
Room 503, Hawkins House
Dublin 2
Republic of Ireland

ISREAL
Embassy of Israel[1]
1453 International Drive, N.W.
Washington, D.C. 20008

ITALY
General Director
Pharmaceutical Division
Via della Civilta Romana, 7
I-00144 Rome, Italy

IVORY COAST (WEST AFRICA)
Ministere de la Sante Publique de la
Population et des Affaires Sociales
01 B.P. V4
Abidjan 01, Ivory Coast
Africa

- J -

JAMAICA
Acting Chief Medical Officer
Ministry of Health and
Social Security
10 Caledonia Avenue
Kingston 5, Jamaica
West Indies

JORDAN
Ministry of Health
Department of Pharmacy and Supplies
P.O. Box 86
Amman, Jordan

JAPAN
Director
Evaluation and Registration Division
Pharmaecutical Affairs Bureau
Ministry of Health and Welfare
1-2-2 Kasumigaseki
Chiyoda-ku, Tokyo
Japan

- K -

KENYA
Permane Secretary
Ministry of Health
P.O. Box 30016
Nairobi, Kenya

KOREA (SOUTH)
Director General
Bureau of Medical Administration
Ministry of Health and Social Affairs
77 Sejongro, Chongroku
Seoul
Republic of Korea

KUWAIT
Secretary to the Administrative Officer[1]
Embassy of Kuwait
2940 Tilden Street, N.W.
Washington, D.C. 20008

- L -

LAOS
Counsellor In Charge of Affairs[1]
Laos Embassy
2222 S Street, N.W.
Washington, D.C. 20008

LEBANON
Economic Attache[1]
First Secretary
Embassy of Lebanon
2560 28th Street, N.W.
Washington, D.C. 20008

LESOTHO
Permanent Secretary
Ministry of Health
P.O. Box 514
Maseru 100, Lesotho

LIBERIA
Research and Information Officer[1]
Embassy of Liberia
Liberian Information Center
1050 17th Street, N.W.
Suite 330
Washington, D.C. 20036

LITHUANIA
Lithuanian Legation[1]
2622 16th Street, N.W.
Washington, D.C. 20009

LUXEMBOURG
Ministère de la Santé
Division de la Pharmacie et des Medicaments
10, rue C.M. Spoo
Luxembourg

- M -

MADAGASCAR
Directeur des Pharmacies et
Laboratories
Ministere de la Sante Publique
Ambohidahy-Antananarivo
Madagascar Island
Malagasy Republic
Madagascar

MALAWI
The Secretary for Health
Ministry of Health Headquarters
P.O. Box 30377
Capital City
Lilongwe 3, Malawi
Africa

MALAYSIA
Assistant Director of Medical
Services (Equipment)
Hospital Division
Ministry of Health
Jalan Pahang
Kuala Lumpur 02-14, Malaysia

MALI
Embassy of Mali[1]
2130 R Street, N.W.
Washington, D.C. 20008

MALTA
Chief Pharmacist of the
Department of Health
Government Medical Stores
G'Mangia, Malta
Sovereign State of Malta

MAURITANIA
Directeur
Direction de la Santé et des
Affairs Sociale
Nouskchott, Mauritania
Island Republic of Mauritania

MAURITIUS
The Permanent Secretary
Ministry of Health
Government House
Port-Louis
Mauritius Island
Mauritius

MEXICO
Dr. Mario Lieberman[6]
Director General de Insumos para
la Salud
Secretaria de Salud
Mazaryk 490, 10o piso
11560 Mexico, D.F.
Mexico

MEXICO
Direccion de Control de Importaciones
y Exportaciones[7]
Direccion General de Control de
Alimentos Bebidas y Medicamentos
Secretaria de Salubridad y Asistencia
Liverpool #80, Esquina con Haure
Mexico 6, D.F.

MEXICO
Commercial Counselor[8]
1101 15th Street, N.W.
Suite 505
Washington, D.C. 20005

MOROCCO
Ministre de la Sante
Rabat, Morocco

MOZAMBIQUE
Chief of Pharmaecutical Department
Ministry of Health
People's Republic of Mozambique

- N -

NEPAL
Royal Nepalese Embassy[1]
2131 Leroy Place, N.W.
Washington, D.C. 20008

NIGER
Ambassador[1]
Embassy of Niger
2204 R Street, N.W.
Washington, D.C. 20008

THE NETHERLANDS
Head of the Central Division for Medicines, Medical Appliance and Infectious Diseases
Dr. H. A. M. Suykerbuyk
Ministry for Welfare, Health and Cultural Affairs
P.O. Box 5406
2280 H. K. Rijswijk
The Netherlands

NEW GUINEA
Economic Officer[1]
Papua New Guinea Embassy
Suite 631
1800 K Street, N.W.
Washington, D.C. 20006

NEW ZEALAND
The Manager
Medicines and Benefits
Department of Health
P.O. Box 5013
Wellington, New Zealand

NICARAGUA
Ing. Marcos Wheelock
Vice Ministro de Salud
Apartado Postal No. 107
Ministerio de Salud
Managua, Nicaragua

NIGERIA
Assistant Director (Drug Controls)
Food and Drug Administration
Federal Ministry of Health
Federal Secretariat
Ikoyi Lagos
Federal Republic of Nigeria

NORWAY
Director[9]
Directorate of Health
Akersgaten 42, Oslo
P.B. 8128 DEP.
N-0032 Oslo 1, Norway

NORWAY
Control Department[10]
National Institute of Public Health
Geitmyrsveien 75
0463 Oslo 4, Norway

- O -

OMAN
Second Secretary[1]
Embassy of Oman
2342 Massachusetts Avenue, N.W.
Washington, D.C. 20008

- P -

PAKISTAN
Deputy Director General/Drugs Controller
Government of Pakistan Health Division
Islamabad, Pakistan

POLAND
Ministry of Health
ul. Miodowa 15
00-923 Warszawa, Poland

PORTUGAL
Ministerio da Industria e Tecnologia[11]
Servicos de Normalizacao
Rua José Estevao, 88-A-29
Lisbon, Portugal

PANAMA
Director de Salud de Adultos
Salud del Canal
Ministerio de Salud
Direccion del Programa de
Salud de Adultos
Apartado 2048
Panama 1, Republic of Panama

PARAGUAY
Director, Instituto Nacional de
Tecnologia y Normalizacion
Av. Gral. Artigas y Gral. Roa
Asuncion, Paraguay

PERU
Ministerio de Salud
Director Superior
Avenida Salaverry, 8 ava Cuadra
Jesus Maria
Lima, Peru

PORTUGAL
Ministerio de Industria e Energia[4]
Direccao-Geral da Qualidade
Rua Jose Estevao, 83A
1199 Lisboa Codex, Portugal

PORTUGAL
Ministerio dos Assuntos Sociais[12]
Servicos de Instalacao e Equipamentos
da Secretaria-Geral
Rua de Arraios, 97
Lisbon, Portugal

PHILIPPINES
Administrator
Republic of the Philippines
Ministry of Health
Food and Drug Administration
San Lazaro Hospital Compound
Rizal Avenue
Metro Manila, Philippines

- Q -

QATAR
Director General of the
Ministry of Health
Ministry of Public Health
P.O. Box 42
Doha, Qatar,
Arabian Gulf

- R -

ROMANIA
Education Officer[1]
Embassy of Romania
1607 23rd Street, N.W.
Washington, D.C. 20008

RWANDA
Ministere de la Sante Sante Publique
B.P. 84, Kigali, Rwanda
Africa

- S -

SAUDI ARABIA
Ministry of Health
Riyadh, Saudi Arabia

SAINT LUCIA
The Minister
Castries, St. Lucia
West Indies

SWAZILAND
Director of Medical Services
Ministry of Health
P.O. Box 5
Mbabane, Swaziland

(the following eight addresses are for various health ministry offices in SWEDEN)

SENEGAL
Press Counsellor[1]
Embassy of Senegal
2112 Wyoming Avenue, N.W.
Washington, D.C. 20008

SIERRA LEONE
Education Attache[1]
Embassy of Sierra Leone
1701 19th Street, N.W.
Washington, D.C. 20009

SINGAPORE
Head (Inspectorate)
Director of Medical Services
Ministry of Health
55 Cuppage Road
Cuppage Center #09-00
Republic of Singapore 0922

SOMALIA
Somalia Embassy[1]
2600 Virginia Avenue, N.W.
Suite 508
Washington, D.C. 20037

SOMALI
Embassy of Somali[1]
600 New Hampshire Avenue, N.W.
Suite 710
Washington, D.C. 20037

SOUTH AFRICA
Director-General
Department of Health and Welfare
Pensions Building
Private Bag X63
0001 Pretoria
Republic of South Africa

SPAIN
Commercial Attache[1]
Embassy of Spain
2558 Massachusetts Avenue, N.W.
Washington, D.C. 20008

SRI LANKA
Information Officer[1]
Embassy of Sri Lanka
2148 Wyoming Avenue, N.W.
Washington, D.C. 20008

SWEDEN
Head, Hospital Division[13]
Socialstyrelsen
The National Board of Health and Welfare
S-106 30 Stockholm, Sweden

SWEDEN
Socialstyrelsen[14]
Lakemedelsavdelningen
Box 607
S-751 25 Uppsala, Sweden

SWEDEN
The National Swedish Industrial Board[15]
Energy Division
Electrical Safety Section
I Nilsson, GL
S-117 86 Stockholm, Sweden

SWEDEN
Statens Stralskyddsinstitut[16]
Karolinska sjukhuset
Box 60204
S-104 01 Stockholm, Sweden

SWEDEN
Produktkontrollnamndem[17]
Statens naturvardsverk
Box 1302
S-171 25 Solna, Sweden

SWEDEN
Box 3295[18]
S-103 66 Stockholm, Sweden

SWEDEN
SPRI[19]
Box 27310
S-102 54 Stockholm, Sweden

SWEDEN
Apoteksbolaget AB[20]
Centrallaboratoriet
Box 3045
S-171 03 Solna, Sweden

SUDAN
Government Analyst
Chemical Laboratories
Ministry of Health
P.O. Box 287
Khartoum, Sudan

SURINAM
Ministerie van Volksgezondheid
Gravenstraat 64
Paramaribo, Suriname

SWITZERLAND
Federal Office for Foreign[21]
Economic Affairs
Embassy of Switzerland
2900 Cathedral Avenue, N.W.
Washington, D.C. 2008

SYRIA
Minister of Health
Ministry of Health
Parliament Street
Damascus, Syria

- T -

TAIWAN
Department of Health
Executive Yuan
Bureau of Drugs
P.O. Box 91-103
Taipei, Taiwan
Republic of China

TANZANIA
The Principal Secretary
Ministry of Health
P.O. Box 9083
Dar Es Salaam, Tanzania

THAILAND
The Food and Drug Administration
Ministry of Public Health
Attn: The Secretary-General
Samsen Road
Bangkok 10200, Thailand

TOGO
Ministre de la Santé Publique
Lome, Togo
Africa

TRINIDAD and TOBAGO
Embassy of the Republic of Trinidad
and Tobago[1]
1708 Massachusetts Avenue, N.W.
Washington, D.C. 20036

TUNISIA
Ministère de la Santé Publique
Pharmacie Centrale de Tunisie
51, Avenue Charles Nicolle
Tunis, Tunisia

TURKEY
Saglik ve Sosyal Yardim Bekanligi
Donatim Genel Mudurlugu
Ankara, Turkey

- U -

UGANDA
The Permanent Secretary
Ministry of Health
P.O. Box 8
Entebbe, Uganda

UNION OF SOVIET SOCIALIST
REPUBLICS
Medical Attache[1]
Soviet Embassy
1125 16th Street, N.W.
Washington, D.C. 20036

UNITED KINGDOM OF GREAT BRITAIN
Director
Scientific and Technical Services
Department of Health and Social Security
14 Russell Square
London WC1B 5EP, England

UPPER VOLTA
Embassy of Upper Volta[1]
5500 16th Street, N.W.
Washington, D.C. 20011

UNITED ARAB EMIRATES
Undersecretary
Ministry of Health
P.O. Box 848
Abu Dhabi
United Arab Emirates

URUGUAY
Embassy of Uruguay[1]
1918 F Street, N.W.
Washington, D.C. 20006

- V -

VENEZUELA
Adjunto a la Direccion de Salud Publica
Ministerio de Sanidad y Assistencia Social
Edificio Sur, Centro Simon Bolivar
Caracas, Venezuela

- Y -

YEMEN
Minister[1]
Embassy of Yemen Arab Republic
600 New Hampshire Avenue, N.W.
Suite 860
Washington, D.C. 20037

YUGOSLAVIA
Jugoslovenski Zavod Za Standardizaciju
11000 Beograd, Cara Urosa 54
Socialist Republic of Yugoslovia

- Z -

ZAIRE
Embassy of Zaire[1]
1800 New Hampshire Avenue, N.W.
Washington, D.C. 20009

ZAMBIA
Executive Secretary
Food and Drugs Control Board
Ministry of Health
Republic of Zambia
Cairo Road
P.O. Box 30205
Lusaka, Zambia

---

[1]Present embassy contact pending the foreign government's designation of an official liaison to work with FDA on the enforcement of medical device laws and regulations.

[2]Belgium's foreign liaison responsible for handling all matters regarding medical devices except importation authorization.

[3]Belgium's authorizing official responsible for approving the importation of medical devices into Belgium.

[4]Present unofficial foreign liaison pending the foreign government's designation of an official liaison to work with FDA on the enforcement of medical device laws and regulations.

[5]Applications for the letter required by FDA should be addressed to Dr. Krebs and should be either written in German or accompanied by a German translation. Applications can only be considered if they are accompanied by detailed descriptions, which will permit the German authorities to evaluate the conformity of that device with German laws and regulations. Such descriptions should also be in German.

[6]The office in Mexico that authorizes the importation of medical devices and serves as the Center for Devices and Radiological Health foreign liaison.
[7]No medical devices may be sold in Mexico unless approved by and registered with the Office of Chemistry and Pharmaceutical Industry of the Secretaria de Patrimonia y formento Industrial located at this address. Information concerning these requirements may be obtained from the Secretaria de Salubridad y Aisstencia.
[8]The Department of Commerce of Mexico that is in charge of all import and export licensing.
[9]Norway's office responsible for serving as liaison on behalf of the Committee for Medical Devices for all devices except sterile medical devices for single use.
[10]Norway's office responsible for serving as liaison on behalf of the Committee for Medical Devices for sterile medical devices for single use.
[11]Portugal's office responsible for handling medical device regulatory matters.
[12]Portugal's office responsible for handling the manufacture and distribution of medical devices.
[13]Sweden's authorizing official responsible for approving the importation of medical devices into Sweden.
[14]Sweden's foreign liaison responsible for handling matters regarding industrially sterilized single-use medical devices.
[15]Sweden's office responsible for regulating electrical equipment.
[16]Sweden's office responsible for regulating equipment concerning radiation.
[17]Sweden's office responsible for regulating chemicals.
[18]Sweden's nongovernment office responsible for developing and enforcing standards.
[19]Sweden's office responsible for planning and handling rationalization programs.
[20]Sweden's office responsible for regulating preventatives and diabetic tests.
[21]Switzerland's office responsible for serving as the medical device foreign liaison and providing the letter of acceptance to U.S. device manufacturers requesting import authorization.

part twelve

Responding as Organizations to International Issues

59

Industry Role in International Standards

ROBERT C. FLINK *Medtronic, Inc., Minneapolis, Minnesota*

Increasingly, medical devices are sold in a global market. And access to the market is increasingly regulated. An important opportunity to impact these regulations is provided by active participation in international standards.

I. THE SIGNIFICANCE OF INTERNATIONAL STANDARDS

Industry's interest in international standards stems from the regulatory use of standards. Countries around the world currently use international standards in various degrees to regulate medical devices. Of particular interest is the current program in the European Economic Community (EEC) to "complete the internal market." This includes eliminating technical barriers to trade by harmonizing requirements. Reference to European or international standards will be the basis of harmonization.

United States medical device manufacturers and industry groups no longer see the large domestic market as an adequate goal. For substantial growth, they are finding ways of entering developed foreign markets, competing with established companies, and introducing products into Third World countries where markets are just being opened up. As these Third World countries increase purchase of these products, they are becoming more interested in standards among other regulatory controls.

II. INDUSTRY OBJECTIVES IN INTERNATIONAL STANDARDS

In this environment, what does an industry group or a company member of an industry group have as objectives for the standardization process? Five basic objectives can be identified.

Predictability. The process of standardization must be such that the final standard emerges in a clear way. Substantial requirements are not added or changed just before final approval.

Practicality. Practical products require careful tradeoffs. A standard requirement must be achievable and not maximize one or two features that can constrain and distort other important design priorities.

Performance. Standards should contain *performance* requirements rather than locking in specific designs. This well-established concept merits continual reemphasis because writing design standards is often the most convenient approach to a problem, albeit the most restrictive.

Punctuality. The standard process should have defined milestones and be managed so that participants can plan and organize their contributions in a businesslike fashion.

Precision. Problems of interpretation always emerge. New designs can easily present features and performance not anticipated by the framers of the standard.

Let's look at these five points in more detail.

A. Predictability

The conceiving of products to meet the marketplace 2-3 years in the future has basic risk. The market can change: new ideas can enter the market and change the focus and the desires of the customers. To avoid aggravating this situation, the standard must proceed in a deliberate manner so that the standard requirements do not change substantially in the process.

Standards become more predictable when the requirements in the standard are rooted clearly in practical clinical needs. In contrast, requirements based on hypothesis and speculation, devoid of practical roots, are easily replaced by other speculations, causing substantial changes in the process.

These concerns point to the importance of a proper process, which starts with clinical problem definition and analysis that continuously narrows down specific standards requirements that solve or mitigate the problems initially identified.

B. Practicality

As mentioned earlier, the standard must solve real clinical or patient problems. This requires the continuous input of clinicians practicing in the field. If only technologists and scientists are involved in planning and writing the requirements, the requirements easily become an exercise out of touch with reality. Strong industry representation by itself will not provide assurance that this problem will be avoided, as industry specialists may not have current clinical experience.

The test methods chosen should be as simple as possible. Simplicity tends to bring with it the benefits of repeatability and reproducibility. Such test methods should be written for use by a person who understands the basic science and technology associated with the product and with the test.

To keep standards requirements practical, insist on a companion rationale document. In the author's experience, a number of instances have occurred in international medical device standards where the simple process of writing down the rationale showed a requirement to be more practically and simply approached in a different way, or totally unneeded. In this regard, requirements that bear on safety have a more straightforward rationale than requirements that bear on medical effectiveness, an area of much more ambiguity. In most cases, the regulators' interest in standards is focused on the need for a safe device.

C. Performance

The preference for performance versus design standards is well established in standard work. If performance is carefully described, then a variety of design approaches can be used to accomplish this end and design is not unduly constrained. Development of a performance standard can, however, be much more difficult than a design standard. It should be noted that performance requirements, because they are written with the knowledge of existing technology and design approaches, can be design-restrictive in unforeseen ways. For example, attempts of the past to describe electrocardiograph (ECG) performance requirements based on the frequency response of analog amplifiers are not applicable to modern digital ECG equipment.

D. Punctuality

For the participating manufacturer, standards work has some of the aspects of product planning and development. Data must be developed and reviewed, technical problems must be solved, and test criteria must be developed. Objectives and milestones are needed so that available resources are allocated in a cost-effective manner. The well-managed standards project also has a publication date that is dependable so that products can be offered that conform to the standard.

The timing of standards is of particular concern in Europe. The European Economic Community (EEC) has set as its goal elimination of technical barriers to trade by 1992. A large number of safety standards must be available by that time to support harmonization; that is, compliacne with one standard will be all that is necessary to meet the regulatory needs of the EEC countries (and, possibly, in non-EEC European countries as well).

E. Precision

It is implicit that the standards be precise. If there is too much room for interpretation and no guidance is given, the manufacturers' and the regulators' different views will needlessly delay approval of products. In this regard, the standards writers should look at the document from the standpoint of the standards user.

The "user-friendly" standard is written to minimize the burden of interpretation when used by a qualified person. Such a qualified person is knowledgeable about the product's use, technology, and test technique, similar to the experts who have participated in the standardization process.

Standards written to accommodate a person who does not understand the technical and scientific fundamentals of what he is dealing with may have been useful in the past. For example, basic electrical equipment presents rather straight forward concerns of shock hazards and fire safety. However, with the rapid pace of technical and scientific change in modern products and test techniques, such "cookbooks" will take longer to write, be more inflexible, and require more frequent revision.

In the committee process, precision is one aspect of quality that can be lost for the sake of compromise and comfort, much as the horse designed by a committee resembles a camel.

Each standards committee needs to consciously and continually consider the "standard" of quality of their "product," the standard itself. Quality has been defined as conformance to requirements. Requirements, in this case, come from the immediate standards users: purchasers, regulators, and manufacturers. As the standard has a profound impact on products available in the marketplace, the ultimate beneficiary or victim of good or bad quality in a standard will be the product user.

III. ORGANIZING FOR INTERNATIONAL STANDARDS

For effective international standards work, it is essential for industry to organize to gain the leverage that comes from representing a group of people and the efficiency of collective work. Depending on the industry organization's capability and staff, the following list is offered in order of increasing resource outlays and impact on principles, timing, and content of the standard.

1. Monitor what is happening in standardization and inform members.
2. Collect the comments of members and submit them through appropriate national groups to the international work.
3. Develop and advocate proposals (defining and taking a proactive stance toward addressing safety problems). To be effective this must be done with consistent, long-term representation to the committee and working groups involved.
4. Host meetings of the international committees (members of the association can readily attend the meetings to get a feeling for the background of the documents they will be living with in the future).
5. Volunteer staff or representatives of member companies to administer international committees and standards drafting groups.
6. Make a commitment to the broad support of international standardization. Through the American National Standards Institute (ANSI), the trade associations can provide support for the International Standards Organization (ISO) and the International Electrotechnical Commission (IEC). Organizations in other countries in which your company does business also merit consideration.

IV. THE OPPORTUNITY OF INTERNATIONAL STANDARDS

For the most part, in industry, we live within regulatory rules not of our choosing. It is not necessary to accept this as our lot in international standards. An active segment of industry can have a role in writing a large part of the international technical rules to which their products will be subjected. Through international standards and standards work, regulators worldwide will be educated to your ideas. Your increased understanding of, and credibility with, the regulatory authorities will be an asset as issues and interpretations arise.

This opportunity carries a price. Influencing the future involves borrowing resources from the present and using the lessons from the past. To effectively influence international standards, a long-term commitment of talent and support resources is needed.

60

Networking Internationally

GEORGE T. WILLINGMYRE *American National Standards Institute, Washington, D.C.*

I. INTRODUCTION

Trade associations exist to serve the needs of their members. As medical device issues have arisen over the years in various countries, trade associations have sprung up to address or solve the problems on behalf of their members. Historically, most of the issues and problems have been product-specific and local. This environment has led to many national associations throughout the world that are organized around product sectors.

In Western Europe, these national bodies have further organized themselves to address European regional issues. Thus, the section in this report dealing with Europe summarizes the activities of four associations that are confederations of national associations in four major product sectors. In Japan, three associations serve the interests of the Japanese medical device and diagnostic products industry. Another committee under the American Chamber of Commerce addresses U.S. company concerns in Japan. In Australia, a single association represents the entire spectrum of medical device and diagnostic manufacturers, much as is the case in North America for the Canadian and U.S. trade associations.

These national and regional organizations exchange views and information periodically; however, at the time of the writing of this chapter there was no organization or structure with a mandate to deal with strategic worldwide issues. These national and regional organizations may organize themselves in the future to address such issues as regulatory initiatives in developing countries, the work of the World Health Organization, and multilateral trade negotiations. Or it may be that new organizational structures may be formed to address these issues. The remainder of this chapter provides summary information on the major regional and national associations.

II. EUROPE

The four pan-European associations representing medical product manufacturers are essentially confederations of the national associations in four product sections. The European Confederation of Medical Suppliers Associations (EUCOMED) is an umbrella group representing principally the interests of manufacturers of sterile devices, although its charter provides for a broader scope. The Coordination Committee of the Electromedical and Radiological Industries (COCIR) in the beginning of its activity represented the X-ray field and now serves the interests of the whole electromedical devices sector. The INternational Association of Medical Prostheses Manufacturers (IAPM) began as an organization to represent the interests of the pacemaker industry but has now expanded to include the interests of heart valve manufacturers as well. The European Diagnostic Manufacturers Association (EDMA) is a confederation of the national associations for *in vitro* diagnostic reagent and instrument manufacturers. COCIR, EUCOMED, and EDMA hold regular liaison meetings to coordinate their programs.

A. European Confederation of Medical Suppliers Associations (EUCOMED)

EUCOMED is an organization of national and international trade associations representing European producers and suppliers of nonpharmaceutical health care products. Some examples of the products manufactured by members of EUCOMED are anesthetic disposables and accessories, catheters, dressings, bandages and plasters, general surgical instruments, incontinence, ostomy and colostomy products, infusion/transfusion sets, nonwoven protective clothing and drapes, sterilizing equipment, surgical implants, artificial limbs, pacemakers, and syringes and needles. EUCOMED aims to promote the diverse interests of its members and to represent and safeguard their interests internationally.

The organization was founded in 1981. The specific objectives of EUCOMED are:

1. To assure permanent contact among its members in exchange of information.
2. To promote and encourage among its members ethical principles and practices voluntarily agreed upon.
3. To cooperate with governments and similar authorities whether national, European, or international through its members as appropriate.
4. To work towards harmonization of standards and regulatory requirements within the area of nonpharmaceutical health care products.
5. To study and deal with all matters of common interest.

It has a permanent secretariat located in London, and correspondence may be addressed to Alan Berton (Secretary, EUCOMED, 551 Finchley Road, Hampstead, London NW37BJ, United Kingdom; Tel: 011 441 431-2187 or 435-2087; Fax: 794-5271). Currently 13 European countries are represented in the confederation and 22 member associations are formal members. Combined, these 22 member associations represent over 800 manufacturers and distributors of medical devices. The organization is directed by a board consisting of two representatives from each member association. The board meets two times yearly. The EUCOMED bureau conducts day-to-day business and is composed of a chairman, a vice president of technical and a vice president of commercial operations, one treasurer, and one secretary.

Working groups focus on specific issues:

1. Legal and regulatory
2. Sterilization
3. Good manufacturing practices
4. Packaging
5. Bar code
6. Labeling
7. Public relations
8. Ethylene oxide

The Sterilization Working Group has prepared draft sterilization guidelines and responded to local issues in the United Kingdom and the Netherlands. EUCOMED is also developing liaisons with the Health Industry Business Communications Council in the United States. The Packing Committee deals with papers used in medical packaging and pack-seal testing. The Ethylene Oxide (EtO) panel has addressed risk benefit ratios, workplace and environmental safety considerations, and studies guidelines on EtO limits, toxicity, and safety. The Public Relations Group issues press releases and publications and sponsors EUCOMED conferences. The Legal and Regulatory Committee has responded to national regulatory developments in Denmark, Germany, and Belgium and has focused principally in 1988 on the European Commission (EC) initiative to harmonize the regulatory schemes for medical device regulations by 1992. EUCOMED has, in 1988, submitted its recommendations for an EC directive on medical devices. The GMP group has prepared advice on European GMPs and is working on subcontractor guidelines and a manufacturers' GMP checklist.

EUCOMED current members include the following organizations:

AIII
Mr. P. Vidal
AIII
59 Route de Paris
F-69260 Charbonnières-les-Bains
France
Tel: (78) 871853
Tlx: 370015 AIII

ASSOBIOMEDICA
Mr. Carlo Mambretti
ASSOBIOMEDICA
Via Accademia 33
20131 Milano
Italy
Tel: (2) 6362 249
Tlx: 332488 FECHIM I
Fax: (2) 6362310

AUSTROMED
(Arbeitsgemeinschaft der Hersteller Medizinischer Bedarfsartikel Osterreichs)
Mr. O. Heissenberger
Semperit AG, 1/TKF 1
A-2632 Wimpassing
Austria
Tel: (2630) 8355
Tlx: 1661 SEMPWP
Fax: (2630) 8355479

BVM
(Bundesvereinigung Verbandstoffe und Medizinische Hiffsmittel)

	Dr. L. Vedder BVM Viktoriastrasse 45 6200 Wiesbaden West Germany	Tel: (6121) 307088/9 Tlx: 4186443 BVM Fax: (6121) 374779
DHAEMAE	Mr. A. L. Berton 551 Finchley Road London NW3 7BJ United Kingdom	Tel: (a) 431 2187 Tlx: 923753 MONREF Fax: (1) 794 5271

DUFO
(Danish Utensils Manufacturers Organization)

	Mr. K. F. Christensen DUFO Meadox Surgimed Havremarken 3 3650 Olstykke Denmark	Tel: (2) 17 78 00 Tlx: 42545
EDANA	Mr. G. Massenaux EDANA 51 Avenue de Cerisiers 1040 Brussels Belgium	Tel: (2) 734 9310 Tlx: 26634 NWASSO Fax: (2) 733 3518
EMETA	Mr. Jean F. Luyckx HOSPITHERA sprl Rue Emile Feron 70 1060 Brussels Belgium	Tel: (2) 538 9050 Tlx: 23604 LUDECO Fax: (2) 537 6992
ESDREMA	Dr. M. Barth Lohmann GbmH & Co KG P O Box 120110 D 5450 Neuwied 12 West Germany	Tel: (2631) 786394 Tlx: 867883 LOMA Fax: (2631) 786467
FENIN	Mr. E. Cocero de Corvera General Secretary Zurbano 92-6°, Izda. 28003 Madrid Spain	Tel: (1)-441-7333/7444 Fax: (1) 441-7555

FICI
(Federation of Irish Chemical Industries)

	Mr. N. Buckley FICI 13 Fitzwilliam Square Dublin 2 Eire	Tel: (01) 765116 Tlx: 33212 FICI Fax: (01) 613821

IAPM
(International Association of Medical Prostheses Manufacturers)

	Mr. Gordon Higson IAMP 38 Rue de Lisbonne	Tel: (1) 4563 0310 Tlx: 641424/650742 Fax: (1) 4563 8627

75008 Paris
France

MEDISPA
(Medical Sterile Products Association)

Mr. C. G. Grenshaw
c/o Portex Ltd.
Hythe CT21 6JL
Kent

Tel: (303) 66863
Tlx: 96165
Fax: (0303) 66761

NEMEPRA
[Nederlandse Vereniging van Fabrikanten, Importeurs en Exporteurs van en Handelaren in Medische Disposables en andere (nietpharmaceutische) Hulpmiddelen]

Dr. M. F. A. van Diessen
NEMEPRA
Postbus 90154
5000 LG Tilburg
The Netherlands

Tel: (13) 654342
Tlx: 52384 VASPA
Fax: (13) 636715

NEVERBA
(Nederlandse Vereniging van Verbandstoffen fabrikanten)

Mr. H. Makaske
NEVERBA
c/o Paul Hartmann BV
Postbus 26
6500 AA Nijmegen
The Netherlands

Tel: (80) 774044
Tlx: 48083
Fax: (80) 778284

NLM

Mr. C. Heffermehl
NLM
Stortingsgt 30
Oslo 1
Norway

Tel: (2) 423247/ 414194

SAI-LAB
(Suomen Tukkukauppiaiden Liiton Jasenhysistys)

Mr. D. Metclfe
SDMA
Robinson & Sons Ltd.
Montonfields Road
Monton, Eccles
Manchester M30 8AT
United Kingdom

Tel: (61) 789 5341
Fax: (61) 787 7276

SGOP
(Syndicat General des Ouates et Pansements)

Mr. P. Rouard
SGOP
12 Rue d'Anjou
75008 Paris
France

Tel: (1) 4265 4180
Tlx: 640969 TEXTIL
Fax: (1) 4266 1381

SLF
(The Swedish Association of Suppliers of Hospital Equipment)

Mr. A. Hultman
SLF
Sveavagen 17, 7tr
S-111 57 Stockholm
Sweden
Tel: (8) 240700
Fax: (8) 218496

SNITEM
(Syndicat National de l'Industrie des Technologies Medicates)

Mr. P. Breitburd
SNITEM
10 Avenue Hoche
75382 Paris
Cedex 08
France
Tel: (1) 4563 0200
Tlx: 280900 FEDEMEC
Fax: (1) 4563 5986

UNAMEC

Ms. S. Vandermeersch
UNAMEC
Leuvensestraat 29
1800 Vilvoorde
Belgium
Tel: (2) 251 0509
Fax: (2) 252 4398

B. Coordination Committee of the Electromedical and Radiological Industries (COCIR)

COCIR was established in 1959 with the principal aim of studying and coordinating problems common to manufacturers of radiological and electromedical equipment. COCIR's principal objectives are:

1. To promote and represent the various interests of its members.
2. To develop, through national standards organizations, appropriate standards for the safety and proteciton of patients and users.
3. To encourage free worldwide trade.
4. To implement a European quality system assessment scheme.
5. To cooperate with governments and EC authorities in the achievement of the above objectives.

COCIR establishes working groups charged with drafting reports and statements on specific issues. Current working group activities cover:

1. Quality system assessment scheme
2. Servicing of medical electric equipment
3. Cooperation with European Commission authorities in the development of directives for medical equipment
4. Magnetic resonance imaging
5. Guide on good manufacturing practice (GMP)
6. Point-of-departure proposals for international standardization

Currently seven national trade associations comprise the COCIR membership. A chairman and secretary serve periodic terms and rotate among the members. The COCIR council is the managing body for current activities and the establishment of manufacturer recommendations. All COCIR recommendations are subject to ratification by the national manufacturers associations. The council comprises delegates from the national member associations. The chairman and secretary general serve for 2-year terms. Close links are established by COCIR with European and international radiological congresses. COCIR closely follows the work of the International Electrotechnical Commission TC62 as well as CENELEC covering electrical equipment in medical practice.

COCIR has also established good relations with sister European associations maintaining regular liaisons in order to coordinate general and horizontal issues relating to future European legislation on health field.

COCIR has produced several recommendations and guidelines voer the years. They include:

1. COCIR-R 11, June 1988—Aspects on the Servicing of Medical Electrical equipment (second edition)
2. COCIR-R 12, June 1987—Standardized Reporting of Test Results
3. COCIR-R 12/7, November 1987—Standardized Reporting of Test Results: Particular Report for Generators of Diagnostic X-rays
4. COCIR-R 13, June 1987—Guide to Application of Quality Systems to the Manufacture of Energized Medical Devices
5. COCIR-R 14, June 1988—Member Companies
6. COCIR-R 15, June 1987—Short Summary of COCIR Aims and Activities
7. COCIR-R 16, June 1987—The Market for Medical Electrical Equipment in Japan

The major current focus of COCIR is the European Commission initiative to harmonize the regulatory schemes for medical devices. On this matter, COCIR has organized a multidisciplinary experts panel to advise the Commission in Brussels and is preparing recommendations on a draft directive. The current Secretary General of COCIR is Mr. Fiore Pennone (c/o ANIE Gruppo 17, Via Alessandro Algardi, 2, 20148 MILANO Italy; Tel: 39-2-326.42.41; Telex: 321616 ANIE I; Telefax: 39-2-326.42.12). The Chairman of COCIR is Mr. Antonio Leone of Kontron Instruments SPA. Current members of COCIR are the following national trade organizations:

Belgium

FABRIMETAL
Federation des Enterprises de l'Industrie des Fabrications metalliques, mechaniques, electriques, electroniques et de la transformation des matieres plastiques
Mr. Arnold Rambout
21, rue des Drapiers—B-1050 Bruxelles
Tel: 32-2-51.02.311
Telex: 046-21078
Telefax: 32-2-51.02.301

FABRIMETAL Group 15/2 "Medical hospital and laboratory equipment" has 23 member companies with a total turnover of about ECU 175 million in 1986. The total number of employees is about 2,000.

France

SNITEM
Syndicate National de l'Industrie des Technologies Medicales
Mr. George Norkevicius
10, avenue Hoche
F-75381—Paris Cedex 08

Tel: 33-1-45.63.02.00
Telex: 280900 F FEDEMEC
Telefax: 33-1-45.63.59.86

SNITEM is the result of merger between SEMRAD and FACOMED. The new unity entered in force since January 1987. Member companies are now 130 with 11,000 employees, and a total turnover (in 1986) of ECU 880 million, export representing 35% of turnover.

Germany

ZVEI
Division Medical Engineering
Mrs. Ellen U. Meyer-Schulke
Postfach 701261
D-6000 Frankfurt (Main)-70
Tel.: 49-69-63.02.206/207
Telex: 411035 zvei d
Telefax: 49-69-63.02.317

ZVEI-Division Medical Engineering has aorund 100 members, providing high-technology medical devices to the health-care field in the areas of diagnostic imaging equipment and radiation therpay equipment, equipment used in the field of physiotherapy, and hydrotherapy devices for cardiac pacing and cardiovascular euqipment. The total turnover by its members in Germany is estimated to be approximately ECU 2150 million in 1986.

Italy

ANIE
Mr. Fiore Pennone
Gruppo XVII "Apparecchi Elettromedicali"
Via A. Algardi, 2
1-20148 Milano MI
Tel.: 39-2-32.64.227/241
Telex: 32.16.16 anie i
Telefax: 39-2-32.64.212

ANIE-Division Electromedical Equipment has 19 members, providing high-technology medical devices to the health-care field in the areas of diagnostic imaging equipment and radiation therpay equipment, equipment used in the field of physiotherapy, and hydrotherapy devices for cardiac pacing and cardiovascular equipment. The total turnover by its members in Italy is estimated to be approximately ECU 300 million in 1986. The total number of employees is about 2400.

The Netherlands

FME/FARON
Groep Fabrieken van Rontgen—en andere elektromedische apparatuur in Nederland
Mr. Peter H. Elskamp
Bredewater 20 - Postbus 190
NL-2700 AD Zoetermeer
Tel: 31-79-53.11.00
Telex: 32157 FME/NL
Telefax: 31-79-53.13.65

FARON (Dutch manufacturers association on medical electrical equipment) is a product division of the RME (association of the mechanical and electrical engineering industry), an independent, private association of companies.

FARON has 12 members, providing high-technology medical devices in the health-care field in the areas of diagnostic imaging equipment and radiation therapy equipment, equipment used in the field of physiotherapy and hydrotherapy, devices for cardiac pacing and cardiovascular equipment, and anaesthesia equipment. The total turnover by its members in the Netherlands is estimated to be approximately ECU 920 million in 1986. Employees total 4155.

Sweden

SLF
Svenska SjukvBrdsleverantorers Forening
Mr. Anders Hultman
Sveavagen 17, 6 tr
S-111 57 Stockholm
Tel: 46-8-24.07.00
Telefax: 46-8-21.84.96

SLF-The Swedish Association of Suppliers of Hospital Equipment is an organization consisting of companies—manufacturers, wholesalers, and agents—supplying the Swedish health and care service with medicine-technical installations, apparatus and instruments, disposable products, technical aids for the disabled, equipment for rehabilitation, etc. The total turnover by its members in Sweden is estimated to be approximately ECU 650 million in 1986. The SLF member companies number 57, with about 3000 employees.

United Kingdom

AXrEM
Association of X-Ray Equipment Manufacturers
Mr. John W. Christopher
Leicester House 8, Leicester Street
UK London WC2H 7BN
Tel: 44-1-43.70.678
Telex: 263536 elect london
Telefax: 44-1-43.74.901

AxrEM has 34 members, providing high-technology medical devices to the health-care field in the areas of diagnostic imaging equipment and radiation therpay equipment, equipment used in the field of physiotherapy and hydrotherapy, devices for cardiac pacing, and cardiovascular equipment. The total turnover by its members in the United Kingdom is estimated to be approximately ECU 400 million in 1986. Employees total 7000.

C. International Association of Medical Prosthesis Manufacturers (IAPM)

The IAPM is an organization of manufacturers. Established in 1978 by cardiac pacemaker companies, its scope broadened in 1986, when its activities were extended to medical prostheses, beginning with cardiac valves. The association pursues the following goals:

1. To establish a continuous dialogue with the medical community.
2. To collaborate with all interested parties in establishing international norms and standards.
3. To cultivate regular contacts with governmental authorities.

Today, its broader charter includes:

1. Formulation and coordination of EEC directives.
2. Direct representation in major bodies such as CEN/CENELEC (European Standards and Electrotechnical Standards Committee).
3. Continuous interface with the European medical and regulatory communities.

The overall objective is to promote the development of medical prostheses in line with ethical rules.

The IAPM has succeeded in establishing itself in the eyes of the Commission of the European Communities as a sufficiently important and responsible body to propose a draft directive for active implantable electromedical devices (A.I.E.M.D.), the first of its kind in Europe. Also, the association's input and support was instrumental in shaping the European norm "Safety of Implantable Cardiac Pacemakers." At present, the IAPM has the following members: BIOTEC, BIOTRONIK, C.P.I., INTERMEDICS, MEDTRONIC, SHILEY, SIEMENS-PACESETTER, TELETRONICS AND CORDIS PACING SYSTEMS, and VITATRON.

The Secretary General for IAPM is Mr. Gordon Higson, who succeeded Mr. Joseph Cordonnier as of August 1, 1988. He may be reached at 38, Rue de Lisbonne—P-75008, Paris; Tel: (1) 45 63 03 10; Telex: 641424/650742.

IAPM's working groups are:

1. Technical/Legal Working Group for Active Implantable Devices
2. Technical/Legal Working Group for Non-Active Prostheses

Ad hoc working gorups were created on the following subjects: IS 1, EMI, GMP, European Pacemaker Registration Card, European Directive, PMA, Clinical Evaluation, Resue, and CEN/CENELEC.

Apart from regular publications such as "Stimulation," newsletters, and regulatory information, IAPM has published a book on Europe entitled "Towards European Integration—88-92—The Completion of a Single Market: Consequences and Perspectives for Medical Community and Industry."

D. European Diagnostic Manufacturers Association (EDMA)

The European Diagnostic Manufacturers Association (EDMA) is the association of the national associations in Europe representing manufacturers and distributors of *in vitro* diagnostic products. The association's primary objectives as set out in its Statutes are:

1. To promote and encourage among its members ethical principles and practices voluntarily agreed upon.
2. To study and deal with all matters of common interest, for example, in the fields of health legislation, science, technology, and research.

3. To contribute expertise to and to cooperate with national, European, and international organizations, governmental, having aims and objects similar to those of the association or whose activities affect the interests of the members of the associations.

While the association has standing committees on Regulatory Affairs and on Economic Affairs, much of the work of the association in the scientific and technical field is carried out by specialized working parties. EDMA's "Guide to Good Manufacturing Practice for In-Vitro Diagnostic Products" was prepared by the association's GMP Working Party and adopted by the association in the hope that its observance by those who produce *in vitro* diagnostic products and its recognition by those who purchase or use them will contribute significantly to the maintenance of high standards of quality and to the free movement of such goods within Europe.

Current members of EDMA are the following:

Austria
Verband del Osterreichischen
Diagnostica-und Diagnostica-Gerateehersteller Fach
1 A-1101
Vienna
Austria

Belgium
Fabricants et Importateurs de Diagnostics
c/o AGIM a.s.b.l.
Square Marie Louise 49
B-1040 Brussels
Belgium

Denmark
Dansk Diagnostika Forening Borsen
DK-1217
Copenhagen k
Denmark

Finland
Sairaala-jaLaboratoriotarvikkei-den jalaitteiden Valmistrajat
SAVA ry
c/o General Industry Group (YTR)
Kaupiankatu 7 A
00160 Helsinki 16
Finland

France
Syndicat des Fabricants de Reactifs de Laboratoires
6, rue de la Tremoille
75008 Paris
France

Germany
Verband der Diagnostic-und Diagnosticagerate-Hersteller
Karlstrasse 21
6000 Frankfurt am Main 1
West Germany

Ireland
Federation of Irish Chemical Industries
13, Fitzwilliam Square

Dublin 2
Ireland

Italy

Assobiomedica—Federchimica
Via Accademia, 33
20121 Milan
Italy

Netherlands

Diagnostica Associatie Nederland
P.O. Box 80523
2508 GM
The Hague
Netherlands

Norway

Elektronik Importer Foreningen
Haakon VII's, gt 2
Oslo 1
Norway

Spain

Federacion Nacional de Empresas de Instrumentacion Cientifica, Medica, Tecnica y Dental
Zurbano, 92-6°
28003 Madrid
Spain

Sweden

Swedish Diagnostic Trade Association
S-751 82
Uppsala
Sweden

Switzerland

Swiss Society of Chemical Industries
Nordstrasse 15
Postfach
CH-8035 Zurich
Switzerland

United Kingdom

The Diagnostic Register
GAMBICA
Leicester House
8 Leicester Street
London WC2H 7BN
England

The Secretary of EDMA is Mr. Gordon C. Tuck (c/o Miles Ltd., Stoke Court, Stoke Poges, Slough SL2 4LY, England; Tel: (0) 2814-5151; Telex: 848337 Miles Lab).

III. FAR EAST

Four groups in Japan and one in Australia represent manufacturers' interests in the Far East. The American Chamber of Commerce in Japan (ACCJ) provides a forum for U.S. manufacturers of medical supplies and equipment to address common issues. There are three major association groupings representing Japanese manufacturers of medical devices and diagnostic products in Japan. The Japan Association for the Advancement of Medical Equipments (JAAME) is a government/private sector organization focused on advancement of medical technology. The second is the Japan Federation of Medical Devices Associations (JFMDA) which is a confederation of 14 associations with individual product sector interests. Another is a product sector organization, the Japan Association of Clinical Reagents Industries (JACR), dealing with *in vitro* diagnostic issues. The Australian Medical Devices and Diagnostics Association (AMDADA) represents the combined interests of all medical device manufacturing and importing companies on the Australian continent.

A. The American Chamber of Commerce in Japan (ACCJ)

The American Chamber of Commerce in Japan (ACCJ) sponsors a Medical Supplies and Equipment Subcommittee. This subcommittee provides an opportunity for U.S. manufacturers of medical devices and diagnostic products to consider common interest and issues, particularly with regard to the regulatory and import requirements of the Japanese government. The ACCJ subcommittee was very active during the 1985-1986 Market-Oriented Sector-Selective (MOSS) negotiations between the U.S. and Japanese governments addressing barriers to the sale of U.S. products in Japan. The ACCJ Committee also had a close relationship with the Health Industry Manufacturers Association in the United States during these talks. The current focus includes continuing work with the MInistry of Health and Welfare regarding implementation of the agreements resulting from the MOSS talks. The ACCJ Committee and HIMA coordinated industry advice to the U.S. government during follow-up discussions with the government of Japan in April 1988. The current Chairman of the ACCJ Medical Supplies and Equipment Subcommittee is T.E. Pierce of Technicon Asia-Pacific (Hazama Building, 2-5-8 Kita-Aoyama, Minato-ku, Tokyo 107, Japan; Tel: 03 405-7311). The ACCJ Committee works in close cooperation with the U.S. Embassy in Tokyo. Kyle E. Murphy, Assistant Commercial Attache at the U.S. Embassy, has provided U.S. government staff support to the ACCJ efforts and government negotiations with the Ministry of Health and Welfare. The phone at the U.S. Embassy in Tokyo is 03 244-5000.

B. The Japan Association for the Advancement of Medical Equipment (JAAME)

The JAAME was formed in 1985 to promote closer cooperation between the industry and the health and welfare governmental administration. Its purpose is to promote comprehensive solutions to problems in research and development, production and distribution, import and export, distribution and use of medical equipment, and training of clinical engineers. The JAAME provides a structure for participation of clinical experts, medical equipment enterprises, and other medical personnel to plan and execute projects concerning medical equipment. The projects are intended to promote the sound development of the medical equipment industry and the promotion of the health of the Japanese people. The JAAME has a Managing Director, Mr. Toshio Utsunomiya (JAAME, Ochanomizu Kimura Building, 6F, 2-19-3, Soto-Kanda, Chiyoda-ku, Tokyo 101; Tel: 03-255-9361). JAAME engages in the following activities:

1. Research and development of, and experiments concerning, medical equipment, and the provision of subsidies for the foregoing.
2. Investigation, study, and collection of information concerning research and development, production, export and import, distribution, layout, and use of medical equipment.
3. Guidance and other necessary technical support to medical organs and medical equipment enterprises related to research and development, etc.
4. Training of medical equipment operators and engineers.
5. Conducting qualification examinations for government-licensed clinical engineers.
6. Issuing publications and sponsoring lecture meetings.
7. Coordination and cooperation with overseas Japanese agencies and organizations relating to medical equipment.
8. Any other activities required to achieve the objects in the preceding Article.

The following projects were to be carried out in fiscal year 1987.

1. Entrusted Projects

The following are carried out as projects entrusted by the Ministry of Health and Welfare:

Streamlining Data Banks. Systems will be streamlined for the research and development of medical care equipment and for the gathering and offering of relevant information.

Surveys and Studies on the Application of Medical and Equipment for Medical Care. The reliability and safety of medical care equipment will be appraised, their medical effectiveness assessed, and technology assessments (TA) will be made with special reference to cost-effectiveness.

Fact-Finding Surveys on the Distribution and Use of Medical Care Equipment. Fact-finding surveys will be conducted on the distribution and use of therapeutical equipment.

Measures for Assurance of Safety in Anesthesia. Comprehensive safety measures for anesthetic facilities and equipment will be studied.

2. General Projects

a. Judging from trends in medical care outlays, JAAME presumes that hospitals and other medical installations will shift to a medical care system of outreach programs in the future. Studies will be performed on medical examination equipment that will be required for these programs.
b. As a project to control the use of medical care equipment, the maintenance and use of medical care equipment by medical care institutions will be surveyed and analyzed, and a system formulated for the maintenance, checking and daily control of medical care equipment.
c. As an education and training project, technical lecture meetings will be conducted for workers associated with the medical care euqipment at medical care institutions, for business corporations, and for other organizations.

d. As a survey and assessment project, medical care euqipment will be assessed in institutional terms, such as the frequency of maintenance and checks, and also in medical care terms, including the actual use of medical care equipment.
e. As a standardization project, international standards and criteria will be surveyed and studied primarily in an attempt to standardize the safety of medical care equipment.
f. As an overseas cooperation project, JAAME will train workers from developing countries in dealing with medical care equipment. This project is part of the State's project for international technical cooperation and is conducted in cooperation with the Japan International Cooperation Agency, which is the entity entrusted to perform the training.

In fiscal year 1988, in addition to the continuing projects outlined above, licensing of clinical engineers took place. A newly issued government act in 1987 requires registration of clinical engineers who can legally participate in blood dialysis, forced breathing, high-tension therapy, and other life-supporting techniques. Four thousand candidates are expected to apply in 1988.

C. Japan Federation of Medical Devices Association (JFMDA)

The JFMDA is a confederation of 14 separate product-specific associations representing various sectors of the medical device market. JFMDA was established in 1984, and taken together its members represent virtually the entire Japanese medical manufacturing establishment.

JFMDA's principal objectives are:

1. To promote the vairous common interests of its members.
2. To cooperate with government for the safety and effectiveness of users and patients.
3. To study legal and regulatory matters and work toward international harmonization of standards or regulatory requirements.
4. To study servicing of medical equipments and systems.
5. To coolaborate with related other parties including foreign countries.

Current working group activities cover:

1. GMP (quality system for manufacturers of medical devices).
2. Review of the maintenance service and inspection system.
3. Review of the current technical classification of medical devices.
4. Replies to inquiries from the industries to the Japanese government.

The Secretariat of JFMDA is now in the Office of the Japan Industries Association of Radiation Apparatus (1-6-2 Yuchima, Bunkyo-ku, Tokyo 113; Tel: 03-818-2310; Fax: 03-818 8920). Mr. Yoshisuke Iwai is Chairman of JFMDA.

The member associations of JFMDA are as follows:

1. Japan Industries Association of Radiation Apparatus (JIRA)
   1-6-2 Marunouchi, Chiyoda-ku, Tokyo 100
   Tel: (03) 816-3450

2. Electronic Industries Association of Japan (EIAJ)
   3-2-2 Marunouchi, Chiyoda-ku, Tokyo 100
   Tel: (030) 211-2765
3. Japan Association of Medical Equipment Industries (JAMEI)
   Eikouhaibu-Hongo Building, 3-32-6, Hongo, Tokyo 113
   Tel: (03) 816-5575
4. Japan Federation of Medical Trading & Manufacturing Association
   3-39-15, Hongo, Bunkyo-ku, Tokyo 113
   Tel: (03) 811-6761
5. Japan Industrial Society Artificial Organs (JISAO)
   Happy-Hongo Building, 402, 3-4-8, Hongo, Bunkyo-ku, Tokyo 113
   Tel: (03) 815-2602
6. Japan Association of Disposable Medical Device Industries (JADMI)
   Benisuzume Building, 1-4-6, Ginza, Chuo-ku, Tokyo 104
   Tel: (03) 567-6246
7. Japan Association of Physical Therapy Industries (JAPTI)
   Yahagi Building, 7-2-6, Hong, Tokyo 113
   Tel: (03) 811-8200
8. Japan Analytical Instruments Manufacturers Association (HAIMA)
   Taimei Building, 3-22, Ogawa-machi, Kanda, Chiyoda-ku, Tokyo 101
   Tel: (03) 292-0642
9. Japan Medical Optical Equipment Industrial Association (JMOIA)
   c/o Tokyo Kogaku Kikai K.K., 74-12, Hasunuma-cho, Itabashi-ku, Tokyo 174
   Tel: (03) 967-2693
10. Contact Lens Manufacturers Association of Japan (CLMAJ)
    c/o Menicon, 3-21-19, Aoi, Naka-ku, Nagoya 460
    Tel: (052) 935-1515
11. Japan Dental Machine Manufacturers Association (JDMMA)
    2-16-14, Kojima, Taito-ku, Tokyo 111
    Tel: (03) 851-6123
12. Japan Dental Materials Manufacturers Association (JDMA)
    Kyodo Building, 517, 2-18-17, Higashi-Ueno, Taito-ku, Tokyo 110
    Tel: (03) 831-3974
13. Japan Surgical Dressings Industrial Association (JSDIA)
    3-36-12, Takata, Toshima-ku, Tokyo 171
    Tel: (03) 971-1903
14. The Japan Health-living Apparatus Industrial Association (HAPI)
    Urban Ebis Building, 1-6-11, Ebisu-Minami, Shibuya-ku, Tokyo 150
    Tel: (03) 793-3121

D. Japan Association of Clinical Reagent Industries (JACR)

The JACR was founded in November 1983 and currently represents 150 businesses involved in the *in vitro* diagnostic reagents industry. The Chairman is Yasuaki Kobayashi, who may be reached at Yu Building, 5-7 Horidome-cho, 1-chome, Nihonbashi, Chuo-ku, Tokyo 103, Japan; Tel: 03-669-9101. The major objective of JACR is to develop and strengthen *in vitro* diagnostic business interests on a global basis, by encouraging high-quality products and contributing toward advanced health care programs. There are working committees on:

1. General administration and planning
2. Regulatory
3. Technical I, chemistry
4. Technical II, biology
5. Raw material specification
6. GMP
7. Distribution
8. Medical insurance
9. Technical investigation
10. Training and education

With regard to regulatory issues JACR was quite active in resolving the following issues in close cooperation with ACCJ and Japan Ministry of Health and Welfare:

1. Labeling for *in vitro* diagnostic products
2. Acceptance by the Japan Ministry of Health and Welfare of foreign clinical data in support of product approvals
3. The time clock for regulatory approvals of *in vitro* diagnostics
4. The adequate reimbursement by the health-care coverage and insurance system for payment of *in vitro* diagnostic products

As one of its long-term goals, JACR is lobbying for separation of *in vitro* diagnostic products from their present classificaiton as a drug. GMP requirements for *in vitro* diagnostic products were to be in effect as of December 1, 1988.

The JACR membership consists of 83% domestic Japanese suppliers, but JACR also maintains close ties to the American Chamber of Commerce in Japan (ACCJ) Medical Supplies and Equipment Subcommittee representing U.S. interests.

E. Australian Medical Devices and Diagnostics Association Inc. (AMDADA)

The Australian Medical Devices and Diagnostics Association Inc. (AMDADA) was formed in 1980 to represent the technical and regulatory interests of importers, manufacturers, and distributors of medical devices and diagnostic reagents. The association represents approximately 100 companies.

AMDADA has a permanent Secretariat, and the Chief Executive Officer may be contacted at the following address: AMDADA, Level 2, 77 Berry Street, North Sydney 2060, Australia; Tel: 61-02-9221157; Fax: 02-959-4860. A staff contact is Ms. Janice E. Hirshorn, Executive Secretary.

Policy is determined by an Executive Committee of 12 elected members.

AMDADA has effective liaisons with the Australian Department of Community Services and Health and represents members' interests with regard to registration of medical devices, GMP inspection of manufacturers, problem reporting, and other regulatory matters. AMDADA also provides an industry liaison to the Standards Association of Australia (SAA).

The association endeavors to ensure uniformity of approach between federal and state governments on relevant regulatory matters, providing input through expert working parties. It aims to minimize delays caused by regulation and urges adoption of overseas standards where applicable, in lieu of writing Australian standards.

The current work of AMDADA focuses on:

1. Therapeutic goods legislation
2. Customs and excise legislation
3. Guidelines for premarket evaluation of designated devices and other registration matters
4. Recall procedures and problem reporting
5. Proposed government regulation of diagnostic products, including AIDS testing
6. Assessment of Australian, ISO, and IEC standards
7. General industry issues

IV. NORTH AMERICA

Two associations on the North American continent representing manufacturers of medical devices and diagnostic products are the Health Industry Manufacturers Association (HIMA) in the United States and Medical Devices Canada (MEDEC) in Canada. Both associations provide forums for all manufacturers of medical devices and diagnostic products.

A. The Health Industry Manufacturers Association (HIMA)

As indicated in the association's bylaws, the objectives and purposes of the Health Industry Manufacturers Association (HIMA) are:

1. To represent the manufacturers of medical devices and diagnostic products . . . respecting activities of federal, state and local government agencies and legislative bodies as may pertain to or affect the medical device and diagnostic products industry and to take such action on behalf of the industry as may be necessary, proper, and in the public interest.
2. To develop programs and activities relating to medical and scientific matters.
3. To gather and disseminate information as to those activities of government agencies, legislative bodies, as well as industry or scientifically related organizations which may affect the industry and to maintain effective liaison with the foregoing.
4. To gather and disseminate information respecting international developments in regulatory, scientific, or standards-making fields which may affect the industry.
5. To undertake such other activities as may be proper to enhance or promote the welfare of the industry, including the conducting of scientific and educational seminars and programs, gathering and disseminating statistics and the industry information, and maintaining liaison with health-care delivery groups and service organizations.

HIMA in 1988 represented 330 corporate entities. HIMA's members account for more than 90% of the medical devices, diagnostics, and health-care information systems sold in the United States. The members include both large and small companies. The association is governed by a 30-member Board of Directors. The elected officers of the Association are the chairman, chairman-elect, secretary, and treasurer. The Association currently employs a staff of 50 administrative and professional personnel. The President is Frank E. Samuel, Jr.

Most of the work of the association is conducted by member company volunteers within three sections, as follows.

1. *Science and Technology (SAT) Section*

The Science and Technology section addresses problems and issues having a high technical or scientific component. The current focus includes use of various sterilization techniques, *in vitro* diagnostic product regulations, electromedical safety standards, responding to regulatory issues having a high technical component such as good manufacturing practices, and medical device software regulations, user education, AIDS, and various product-specific sector task forces. The seciton also addresses international trade issues such as foreign trade barriers and U.S. export controls on high technology products. In 1988, the section produced a strategic plan for 1989-1992, covering the expected impact of new technologies and environmental influences on HIMA programs.

2. *Legal and Regulatory Section*

The Legal and Regulatory section addresses legal issues and problems arising from the regulations issued by the Food and Drug Administration, Health Care Financing Administration, Occupational Safety and Health Administration, Environmental Proteciton Agency, and other federal and state governmental agencies. The section is currently heavily involved in reviewing draft legislation revising the Medical Device Amendments of 1976.

3. *Government and Public Affairs Section*

The Government and Public Affairs seciton deals with National Health Care Policy questions, matters of the structure of health-care payment and reimbursement, technology assessment, product liability and tort reform, and issues affecitng the Medicare budget and payment for medical technology.

All three sections are managed by Steering Committees composed of representatives of members. Various committees and task forces under the sections carry on the major business of the association. The association sponsors numerous educational activities, including annual meetings for members and special meetings on regulatory, technical, scientific, and economic issues. The association also publishes educational material on topical problems. The association has an effective relationship with the U.S. Food and Drug Administration, Health Care Financing Administration, Congress, and the Administration in the United States. The Health Industry Manufacturers Association may be reached at 1030 15th Street, N.W., Washington, D.C. 20005; Tel: (202)452-8240; Telefax: (202)289-1978.

B. Medical Devices Canada (MEDEC)

Medical Devices Canada (MEDEC) is the association representing the medical device sector's interest in Canada. Its major purposes are:

1. To assist the industry in delivering safe, effective, best technology medical devices to all Canadians and

2. To seek a business environment that encourages the growth of the industry and maximizes value added to products and services in Canada. Further, the association's specific goals are:

    a. To plan an active role in the development of policies and regulations that will promote the development, manufacture, and distribution of safe, economical and effective medical devices.
    b. To achieve full representation and the active participation of every firm in every sector of the Canadian industry in a structure designed to meet the needs of the special interest groups and sectors in the membership.
    c. To be recognized as a responsible partner in the Canadian health-care delivery.
    d. To increase the industry's contribution to the Canadian economy by fostering the growth, development, and Canadian value-added capabilities of member firms.

The association is governed by a Board of Directors. The officers and directors include chairman, chairman-elect, first vice-chairman, second vice-chairman, secretary, treasurer, and an immediate past chairman. The permanent staff includes several full-time employees: Phillip T. Nance, President, and Margaret Guerrier, Director of Regulatory Affairs and Standards. The Secretariat is located at 10 Four Seasons Place, Suite 405, Etobicoke (Toronto), Ontario, M9B 6HS Canada; Tel: (416)620-1915. Most of the business is conducted within six product-oriented sections: (1) assistive devices, (2) diagnostics, (3) hospital equipment, (4) implants, (5) medical imaging/therapy, and (6) surgical/medical.

MEDEC has an effective and close working relationship with the Health Proteciton Branch (HPB) of the Ministry of Health in Canada and is increasing its attention to provincial economic development and regulatory matters. MEDEC and HIMA have a cooperative relationship on regulatory issues arising in Canada and often coordinate submissions and comments to the Canadian HPB on regulatory matters. MEDEC and HIMA also shared perspectives during the 1987 negotiations between the U.S. and Canadian governments on a free-trade agreement.

V. THE FUTURE

The structures that have developed over the years have addressed the local and specific concerns of manufacturers in particular countries reasonably well. Language, customs, local markets, and cultures have all combined to make country-specific and product-sector structures the most logical organizations to address these historical problems. Such local organizations, however, may not be the best means to address strategic international and worldwide issues. For example, the European Commission initiatives for harmonization of the regulatory systems in Europe are currently structured along the four separate product sectors represented by the four trade associations described in Section II. At some points, an integration of the four sectors into a unified scheme of regulations seems to be in order. In the United States, for example, the single unifying impact of the Medical Device Amendments of 1976 led to HIMA's creation and current umbrella structure representing the breadth and spectrum of medical device manufacturers and health-care information systems.

There are other worldwide strategic developments that may not be most effectively addressed on a country-by-country or even regional response. These include, for example, proposals by the World Health Organization (WHO) to consider research and development activities forusing on the needs of developing countries. Opportunities for harmonizing regulatory schemes on a global scale may also fall between the cracks without an organization considering these issues on a strategic global scale. Opportunities for harmonization present themselves in such discussions as the multilateral trade negotiations (MTN) whereby governments of developed countries are discussing means and procedures to promote free trade. Meetings of medical device regulatory authorities such as occurred in Washington, D.C., in May 1986 also present issues and implications of a strategic global nature.

Currently the major trade associations described in this chapter have modest liaisons and commitments for exchange of information. In addition, the European Trade Associations meet occasionally to coordinate their efforts and programs, and HIMA hosts an annual meeting of all of these groups each year at its major membership event.

This summary review chapter must be considered only as a snapshot in time of the trade association organizations that have come into being through the decades of the 1970s and 1980s in response to the historical country and product-specific issues. While the present structures may suffice for the time being, the reader should be aware of the potential and opportunities for new structures, which may be collections or confederations on a global basis or structures that may be directly supported by manufacturers on a regional or global basis.

61

Economics of U.S. Trade in Medical Technology and Export Promotion Activities of the U.S. Department of Commerce

MICHAEL C. FUCHS *Office of Microelectronics and Instrumentation, U.S. Department of Commerce, Washington, D.C.*

Health-care expenditures are among a nation's largest social costs and have increased rapidly over the past 20 years. In 1960, the 24 countries belonging to the Organization for Economic Cooperation and Development spent an average of 4.2% of gross domestic product (GDP) on health care. By 1984, these expenditures had reached almost 8% of GDP. Such health-care expenditures have contributed to the growth of the industries that provide medical products and services, including health-care providers, pharmaceutical manufacturers, and medical equipment suppliers. The focus of this chapter is to analyze the role that international trade plays with respect to medical equipment and to provide insight into the export promotion activities of the U.S. Department of Commerce.

I. WORLD PRODUCTION AND CONSUMPTION OF MEDICAL EQUIPMENT

A. The World Market

For the purpose of this analysis, medical equipment consists of X-ray and electromedical equipment, surgical and medical instruments, surgical appliances and supplies, and dental instruments and supplies. Historically, this group of industries has enjoyed substantial growth. During the 10-year period ending in 1982, U.S. production of medical equipment grew at an average annual rate of 17.3% measured in current dollars. From 1983 to 1988, as public- and private-sector cost containment measures in the United States sought to slow escalating health-care costs, production nonetheless grew at an average of 8.5%.

The leading manufacturer and end user countries of medical equipment are the United States, Japan, and West Germany. United States companies in particular have driven world medical equipment production. The U.S.

share comprises about 60% of the $36.1 billion world total (excluding the Soviet Union and certain other nonmarket economies). Japanese and West German firms accounted for an additional 25% of world output in 1988 (see Fig. 1). These countries have a long-standing tradition in medical technology, having begun experimentation with X-rays around the turn of the century. As a result, a highly sophisticated medical electronics industry has evolved. Current products include computed tomography scanners that noninvasively image internal body organs and shock-wave lithotripters that crush kidney stones through the use of ultrasound.

Development of state-of-the-art medical technology has made advanced health-care systems possible. Most of these are in use in the same leading countries. The United States, Japan, and West Germany together consume over 75% of world production of medical equipment. Each of these "big three" manufactures more than it consumes and is therefore a net exporter.

The majority of medical equipment trade occurs among developed countries. For the United States, The main destinations for exports are Japan (14.4%), Canada (13.9%), and West Germany (9.9%) (see Table 1). In fact, developed countries purchase the bulk of U.S. medical equipment exports (79%). This situation is also true for West Germany and Japan. They supply over one-half of U.S. imports of medical devices; developed countries as a group account for over 90% of the total.

United States trade in medical equipment in 1987 focused predominantly on X-ray and electromedical equipment. This product area reflects the in-

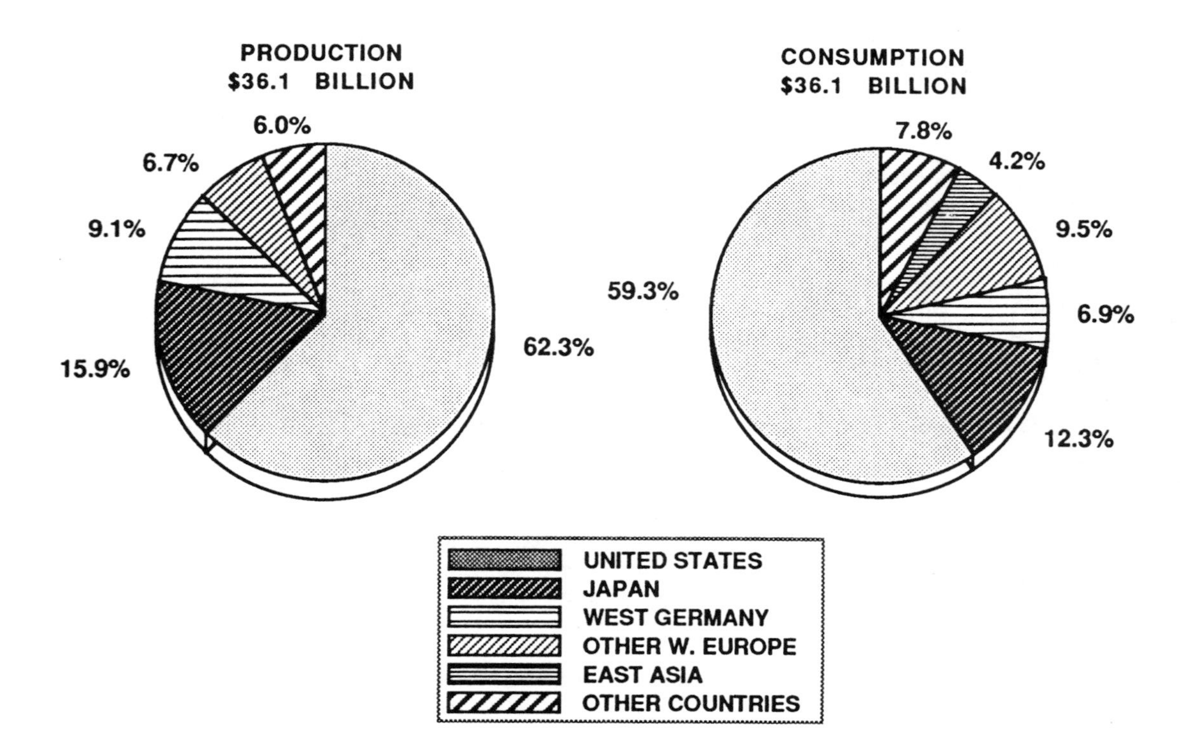

FIGURE 1 World production and consumption of medical equipment, 1988. (From *U.S. Industrial Outlook, 1988*, U.S. Department of Commerce, ITA.)

TABLE 1 U.S. Trade in Medical Equipment, 1987 ($ Million)

	Exports	Imports	Balance
Western Hemisphere	767	218	549
Canada	445	79	366
South, Central America, Caribbean	322	139	183
Western Europe	1453	1441	12
France	197	117	80
West Germany	318	683	(365)
Netherlands	196	123	73
United Kingdom	222	173	49
Eastern Europe	5	1	4
East Asia	675	743	(68)
Japan	459	611	(152)
Africa, Near East, South Asia, Oceania	297	104	193
Total	3198	2507	691

Source: U.S. Department of Commerce, ITA, derived from Bureau of Census data.

creased technological sophistication required to perform new and ever more advanced procedures. Medical instruments—mainly nonelectric diagnostic and therapeutic devices—accounted for another 27% of U.S. medical equipment trade. The surgical appliances and supplies industry, which manufactures a diversity of products ranging from surgical sutures and dressings to certain types of durable medical equipment, accounted for 15%. Dental equipment made up the remaining 5% of the total.

B. Internationalization of the Medical Equipment Industry

As the world market for medical equipment has grown, so too have the international linkages. Foreign medical students study in the United States, U.S. doctors attend medical conventions in West Germany, and manufacturers exhibit their products in Tokyo. With the prevalence of such ties, it is no surprise that manufacturers of medical equipment have also become more internationally oriented. For small to medium-sized companies this means more trade. For the large multinationals, this translates into trade as well as assembly operations and entire manufacturing facilities abroad. Reasons for expansion into the international arena include the following:

1. *Achieving economies of scale.* Economies of scale are critical to minimizing production production costs. Even though the United States is the world's largest market, U.S. firms have to expand outside U.S. borders simply to remain competitive with foreign firms, who must export to survive. Firms such as Baxter, Pfizer, and GE, to name just a few, have understood this concept and have been actively involved in foreign mar-

kets for a number of years. These firms have forged international linkages abroad, including distribution, assembly, and full-fledged manufacturing.

2. *Stronger foreign demand.* As standards of living rise outside the United States, the demand for more specialized medical procedures grows. Consequently, the demand for medical equipment is rising faster in certain foreign countries than in the United States. Because it is difficult to supply their own health-care needs domestically, countries must turn to imports.

   Demand factors vary from case to case. Public- and private-sector expenditure levels on health care, population demographics, predominance of various types of diseases, medical insurance reimbursement for use of medical equipment, and acceptance of medical products in the health-care system generally influence demand. Other determinants may also prove significant in specific instances. For example, in some Western European countries the ratio of magnetic resonance imaging (MRI) equipment to population is lower than in the United States, due to the high unit cost of this equipment. As these countries seek to upgrade their health care, demand for these devices should grow at rapid rates in the late 1980s and early 1990s.

3. *Foreign clinical research.* While the United States is recognized as having the best clinical research environment, U.S. firms often seek to complement their domestic research by conducting some of it abroad. This serves as an effective means to develop products that have applicability in both the U.S. and foreign markets.

   Japan, for example, is known for its low-cost medical electronics products. This is partially due to the Japanese health insurance system that reimburses end users of sophisticated medical technologies such as computed tomograph scanners and magnetic resonance imagers at rates approximately one-third of those in the United States. This has forced Japanese manufacturers to develop low-cost imaging systems that not only meet the needs of the Japanese health-care system, but are especially applicable in third countries. Many U.S. manufacturers have seized on this idea to develop products in Japan or have formed joint ventures with Japanese firms to develop or refine medical technologies.

C. International Trade Trends

While internationalization of the medical device industry is driven by the large multinationals, small to medium-sized firms have also been active in the international arena. While these firms may not have the resources to set up operations abroad, they have opted for arrangements (such as distribution agreements) with foreign companies for expanding their market share.

Trade plays an increasingly important role to the medical equipment industry, particularly for high-tech medical electronics. In 1977, X-ray and electromedical equipment exports accounted for 21.6% of product shipments. Ten years later, these exports comprised almost 30% of shipments. Table 2 shows historical growth trends for U.S. medical equipment imports, exports, and trade balances.

As Table 2 indicates, both exports and imports have increased steadily. However, the growth rates have been cyclical, especially on the export side. While exports grew at double-digit rates through 1981, their growth rates

TABLE 2 U.S. Exports, Imports, and Trade Balances for Medical Equipment, 1975-1987 ($ Million)

Year	Exports	Imports	Balance
1975	710	300	410
1976	827	349	478
1977	1009	416	593
1978	1209	499	710
1979	1410	579	831
1980	1663	621	1042
1981	2070	729	1341
1982	2148	867	1281
1983	2213	1095	1118
1984	2296	1382	914
1985	2417	1690	727
1986	2760	2149	611
1987	3198	2507	691

Source: U.S. Department of Commerce, ITA, derived from Bureau of Census data. Note: Exports are free along side; imports exclude insurance and freight.

dropped dramatically from 1982 to 1985, only then to rebound again. Fluctuations in international exchange rates have played a strong role in U.S. medical equipment export levels.

In years when the U.S. dollar depreciated (such as between 1975 and 1980), U.S. exports of medical equipment grew at double digit rates. The weak dollar effectively lowered prices of U.S. medical products to foreign customers while making imports more expensive. From 1981 to 1985, the opposite took place: the U.S. dollar appreciated sharply against the currencies of our major trading partners, resulting in sluggish export growth to those countries.

Another major shift in exchange rates occurred in 1986 with the U.S. dollar depreciating sharply against currencies of our major trading partners. This shift may stimulate exports again, as in the previous "weak" dollar period. There is evidence that this effect is beginning to take place. The balance sheets for medical equipment manufacturers in 1986 and 1987, especially those with strong foreign positions, attributed strengthening financial performance in part to increased international sales.

The reasons for this are quite clear. The price differential for medical equipment has varied drastically in a short period of time. In February 1985, a U.S.-made MRI with a price tag of $2 million sold for DM 6.9 million in West Germany. In December 1987, as the U.S. dollar dropped against the West German mark, the cost for the same machine was only DM 3.3 million, a decline of roughly 53%. The opposite has occurred to foreign medical electronics manufacturers, which earmark a substantial portion of their export sales to the United States and other dollar denominated countries. During this time period, they reproted declines in profitability, reduction in revenue, and layoffs.

The effects of the value of the dollar paint a picture of strong demand for U.S. medical products abroad, competing most effectively when they are price competitive. However, exchange rates do not adequately explain the continued surge in U.S. imports. Unlike the cyclical growth trends of exports, U.S. medical equipment imports have grown consistently, at an average annual rate of 19% from 1975 to 1987. United States imports accounted for nearly 12% of medical equipment consumption in 1987 and about 27% of X-ray and electromedical equipment consumption. The trade surplus that the United States enjoyed in 1981 of $1.3 billion in this sector was less than one million dollars in 1986.

United States imports have increased for a variety of reasons:

1. *Largest medical equipment market.* The United States is the world's largest market for medical equipment, with health-care expenditures accounting for 10.7% of GDP. In addition, it is an open market for these products. While medical device regulation is very stringent in the United States, the regulatory framework is open with interested parties able to provide input into the process. This has made the United States the market of choice for foreign producers.
2. *Faster acceptance of medical technology.* Trends for expensive medical products such as computed tomographs and magnetic resonance imaging equipment indicate that these technologies are adopted at faster rates in the United States. Many foreign countries take a more cautious approach in investing in new types of equipment and replacing existing medical technologies.
3. *Increased foreign competition.* Foreign competition has also increased sharply domestically and abroad. In many cases foreign-made goods have raised their technological sophistication to the extent that foreign quality equals and sometimes surpasses that of American products. According to the United Nations, the trade surplus for electromedical equipment increased $452 million for Japan, $297 million for West Germany, and $69 million for the Netherlands from 1982 to 1986. By contrast, for the United States the trade surplus in this sector shrank by $555 million in this period.
4. *Effect of exchange rates.* In order to avoid the full impact of fluctuating exchange rates, foreign firms are setting up manufacturing facilities in the United States to supply this market. Often components and parts are sourced from their home countries.
5. *Cost containment efforts.* Efforts by both the public and private sector to limit health-care expenditures in the United States may have contributed to accelerated imports. The prospective payment system in effect since 1983 has created a more competitive environment for medical devices by spurring the growth of competitive bidding arrangements and capitated reimbursement arrangements. This is focusing competition increasingly on price. As indicated earlier, imports were very price competitive during the 1981-1985 period.

The result of these rapidly increasing imports has been a deterioration of the U.S. trade surplus in medical equipment. This surplus peaked in

TABLE 3 Annual Growth Rates of U.S. Exports and Imports of Medical Equipment Compared to Change in the Value of the U.S. Dollar, 1975-1987

Year	Exports (%)	Imports (%)	Value of U.S. dollar[a]
1975	—	—	98.5
1976	16.5	16.3	105.7
1977	22.0	19.2	103.4
1978	19.9	20.0	92.4
1979	16.6	16.0	88.1
1980	17.9	7.3	87.4
1981	24.5	17.4	103.4
1982	3.8	18.9	116.6
1983	3.0	26.3	125.3
1984	3.8	26.2	138.2
1985	5.3	22.3	143.0
1986	14.2	27.2	112.2
1987	15.9	16.7	96.9

[a]Index of weighted average exchange rate values of U.S. dollar against currencies of other G10 countries: Belgium, Canada, France, Italy, Japan, the Netherlands, Sweden, Switzerland, United Kingdom, and West Germany.
Source: U.S. Department of Commerce, ITA, and Federal Reserve Board.

1981, registering $1.3 billion. By 1986, it had fallen to $611 million and rebounded to $691 million the following year (see Table 3).

For future U.S. trade in medical equipment the questions arise: Will U.S. exports for medical equipment remain dependent on the fluctuations of exchange rates? Or, will manufacturers be able to supply equipment at competitive prices during periods of a strong dollar? Will imports driven by the various factors (United States being the largest market, faster acceptance of medical technologies, increased competition, exchange rate fluctuations, and cost containment efforts) continue at historical rates? Or, can U.S. firms meet the competition head on in the domestic market?

While price is only one component of successful sales overseas, quality and after sale service are other important considerations. Producing quality medical devices not only helps manufacturers with domestic problems such as product liability, but also enhances the image of the U.S. industry as being a supplier of quality products. With renewed interest in exporting medical equipment, firms are also attempting to improve quality. They are designing better products to avoid safety problems, beginning to use new FDA "preproduction" quality assurance guidelines, and using automation to identify more accurately any defects in the production line.

In summary, to maintain competitiveness and increase exports of medical devices over the long term, price, technological improvements, quality, and after-sale service are all essential.

II. EXPORT PROMOTION ACTIVITIES OF THE U.S. DEPARTMENT OF COMMERCE

With regard to medical equipment, the U.S. government is involved in two types of activities: removal of obstacles to U.S. business abroad, and the promotion of U.S. exports.

A. Removal of Obstacles to U.S. Business Abroad

International negotiations involving foreign barriers to trade are led by the U.S. Trade Representative (USTR) or the U.S. Department of Commerce. The Tokyo round of multilateral trade negotiations, conducted under the auspices of the General Agreement on Tariffs and Trade (GATT), led to development of the Standards and Government Procurement Codes in 1979. These codes are currently being refined at the follow-up summit meeting in the 1987 Uruguay round of trade negotiations. The Standards Code is an agreement among signatory countries, including most major export markets for U.S. medical equipment, not to use standards as nontariff barriers. International standards and technical regulations, which have proliferated in the 1980s, are the primary trade barrier for U.S. medical device manufacturers.

Government activities in standards that relate to medical equipment are also addressed through bilateral negotiations. For example, the U.S. government's market-oriented sector-selective discussions with Japan led to the resoluation of many medical device trade barriers in 1986. Among these barriers were the lack of acceptance of foreign clinical test data, lengthy processing for the approval of new medical devices, restrictions on changes in import agents, and complex procedures for approving minor device changes that did not affect health and safety. In 1988, negotiations were reopened with Japan to further modify these issues as well as to address new problems experienced by the U.S. industry.

B. The Promotion of U.S. Exports

Most federal government export promotion activities are centered in the Department of Commerce. These include export counseling, search for foreign agents/distributors, and organizing for overseas trade missions, trade shows, and trade fairs. A full listing of the International Trade Administration's (ITA) export services that are geared toward our export oriented industries, including medical equipment, is provided in Section III.

Within the ITA, export promotion services are carried out by Trade Development (TD), the U.S. & Foreign Commercial Service (U.S. & FCS) and International Economic Policy (IEP).

1. U.S. & Foreign Commercial Service

The U.S. & Foreign Commercial Service is the first stop for new-to-export medical device companies. Its 48 district offices and 19 branch offices in cities throughout the United States and Puerto Rico provide companies immediate access to information and services available from the ITA. The district office can also direct U.S. companies toward other government agencies and private-sector export services, if necessary.

Each district office is staffed by trade specialists that can assist a company gain a basic understanding of profitable opportunities in exporting and help them evaluate their market potential overseas. The district office can give information to medical equipment companies on:

Trade and investment opportunities abroad
Foreign markets for medical equipment
Services to locate and evaluate overseas buyers and representatives
Financing aid for exporters
International trade exhibitions
Export documentation requirements
Medical equipment trade statistics
United States export licensing requirements
Esport seminars and conferences

The international activities of the U.S. & FCS are found in the 126 offices operating in 66 countries, which are staffed by about 175 commercial officers and 500 foreign nationals. This staff can provide a wide range of services to help U.S. companies sell overseas, which include background information on foreign companies, agency-finding services, market research, business counseling, assistance in making appointments with key buyers and government officials, and representation on behalf of companies experiencing trade barriers.

2. *International Economic Policy*

International Economic Policy houses the country desk officers, who specialize in a particular country's overall economy, trade policies, and overall political situation. They support both small and large U.S. companies' international activities by providing country-specific information.

Desk officers keep up-to-date on the economic and commercial conditions in their assigned countries. They collect information on country regulations, tariffs, business practices, economic and political developments, trade data and trends, market size, and growth. This provides U.S. companies with good resources on a foreign market's potential for U.S. products, services, and investment.

Country desk officers seek to remove obstacles to U.S. commercial activities in their assigned countries. They are actively involved in trade and inventment agreements, drawing on their country expertise. They also work with the ITÀ district offices in counsleing business and undertaking outreach programs to publicize trade opportunities in their assigned countries.

3. *Trade Development*

The Trade Development unit promotes the trade interests of industry and offers information on markets and trade practices wroldwide. The organization is grouped into seven sectors: Aerospace, Automotive Affairs and Consumer Goods, Basic Industires, Capital Goods and International Construction, Science and Electronics (where medical equipment falls), Services, and Textiles and Apparel. A cross-sectoral unit, Trade Information and Analysis, develops generic data that can be useful in export promotion.

International trade specialists promote exports of their industries through marketing seminars, foreign buyers groups, executive trade missions, business counseling, and information or market opportunities. They work together with their industries, trade associations, and state development agencies to plan international marketing programs and develop policy positions aimed at removing trade barriers and opening up markets. Industry specialists in TD offer the greatest concentration of industry expertise. They are the one spot in the U.S. government that offers a full combination of industry knowledge, trade policy, and an understanding of international business.

For medical equipment, the Science and Electronics unit (S & E) brings together export promotion and industry analysis staff to take a coordinated approach toward fostering competitiveness of the medical device industry and stronger participation in international markets. Science and Electronics annually prepares a chapter on medical and dental instruments and supplies in the *U.S. Industrial Outlook*. This document reviews developments of this industry during the past year and gives predictions for trade, production, and employment for the coming year. A competitive assessment detailing trands in the U.S. diagnostic imaging industry's ability to compete domestically and abroad was also published in 1988.

Export promotion activities for medical equipment sponsored by S & E strike a healthy balance between large markets in industrialized countries and small but rapidly growing markets in the developing world. In 1987, these activities included operating a business information office at "Medica," the world's largest medical equipment trade show held in Dusseldorf, West Germany, as well as showcasing U.S. medical equipment at the U.S. pavilion "Medic Asia" in Singapore. In 1988, an executive-level trade mission went to Beijing, Tianjin, and Shanghai in the People's Republic of China. This was the first time that ITA had sponsored a major medical equipment event to that country.

4. *Other Department of Commerce Export Services*

Other Department of Commerce export services relevant to medical device companies are provided in the areas of export licensing, foreign technical specifications, and export trading companies.

Export Licensing Assistance. Given by the Bureau of Export Administration, this provides high-level direciton for national export control policy and administration, strong enforcement of export laws, and improved service to U.S. business. United States export control laws prevent the unauthorized transfer of high technology that would harm America's national security by contributing to military capabilities of our adversaries. Business services of the Bureau of Export Administration include:

Electronic export license applications
Status updates on license applications
Export licensing educaiton seminars
Guidelines on preparing license documents

Foreign Technical Specifications for U.S. Products. Medical device companies must deal with a whole array of foreign national requirements, standards, testing, and certification requirements when seeking to penetrate foreign markets. The National Center for Standards and Certification Information in the National Bureau of Standards is the government's central repository for standards-related information. The NBS is also the U.S. focal point for notification of foreign standards under auspices of the General Agreement of Tariffs and Trade (GATT).

Export Trading Companies. The Office of Export Trading Company Affairs promotes the formation of export trading companies (ETC) and export management companies (EMC) by sponsoring conferences, workshops, and presentation. It provides a program for registering suppliers and ETCs and EMCs to help registrants identify and contact potential business partners. The office also administers the Export Trade Certificate of Review program under Title III of the Export Trading Company Act, which extends antitrust proteciton for joint exporting ventures. Both ETCs and EMCs are important vehicles for smaller firms to move quickly into export markets. The joint approach involved in ETCs helps participants to reduce up-front costs and share international expertise.

A primary mission of the U.S. Department of Commerce is to help American firms to export. Its programs are structured to address the most common needs of U.S. industry as it pursues the international market place. The department also has the experience and expertise to respond to the most unique and specialized situations. Exports represent the greatest untapped opportunity for American business today.

III. EXPORT SERVICES OF COMMERCE'S ITA

Export counseling. Trade specialists are available at ITA district and branch offices for individualized export counseling and may schedule appointments with department officials in Washington, D.C.

Agent/distributor service. A customized search for interested and qualified foreign representatives will identify up to six foreign prospects who have examined the U.S. firm's literature and expressed interest in representation.

Commercial News USA. A monthly magazine that promotes the products or services of U.S. firms to more than 110,000 overseas agents, distributors, government officials, and end users.

Comparison shopping. A custom-tailored search that provides firms with key marketing and foreign representation information about their specific products. Commerce Department staff conduct interviews abroad to determine nine key marketing facts about the product, such as sales potential in the market, comparable products, distribution channels, going price, competitive factors, and qualified purchasers.

Foreign buyers programs. Exporters can meet qualified foreign purchasers for their products and service at trade shows in the United States. The Commerce Department promotes trade shows worldwide to attract foreign buyer delegations, manages an international business center, counsels participating firms, and brings together buyer and seller.

Trade opportunities program. Provides companies with current sales leads from overseas firms seeking to buy or represent their product or service. These "leads" are available electronically from the Commerce Department and are redistributed by the private sector in printed or electronic form.

World traders data report. Custom-tailored reports that evaluate potential trading partners. Includes background information, standing in the local business community, credit worthiness, and overall reliability and suitability.

Overseas catalog and video catalog shows. Companies may gain market exposure for their product or service without the cost of traveling overseas by participating in a catalog or video catalog show sponsored by the Department of Commerce. Provided with the firm's product literature or promotional video, the department's US & FCS will send an industry expert to display the material to select foreign audiences in several countries.

Overseas trade missions. Officials of U.S. firms may participate in a trade mission that will give them an opportunity to confer with influential foreign business and government representatives. Commerce Department staff will identify and arrange a full schedule of appointments in each country.

Overseas trade fairs. United States exporters may participate in overseas trade fairs, which will enable them to meet customers face-to-face and also to assess the competition. The Commerce Department creates a U.S. presence at international trade fairs, making it easier for U.S. firms to exhibit and gain international recognition. The department selects international trade fairs for special endorsement, called certification. This cooperation with the private show organizers enables U.S. exhibitors to receive special services designed to enhance their market promotion efforts.

Matchmaker events. Matchmaker trade delegations offer introductions to new markets through short, inexpensive overseas visits with a limited objective: to match the U.S. firm with a representative or prospective joint venture/licensee partner who shares a common product or service interest. Firms learn key aspects of doing business in the new country and meet in one-on-one interviews the people who will help them be successful there.

part thirteen

Responding as Companies to International Issues

62

Using Partnerships and Other Joint Ventures

ROBERT G. PINCO and ANN K. POLLOCK *Baker & Hostetler, Washington, D.C.*

I. INTRODUCTION

A. Scope

The market for medical devices is an international market, particularly in light of fast-paced technological developments.[1] One simple way to build an international market is to license the medical device to a local company for distribution in that country or in a narrow region. However, the purpose of this chapter is to explain the use of a vehicle alternative to licensing—partnerships and other joint ventures—to expand the market for and perhaps, in part, continue research and development into a medical device. The chapter is practically oriented, providing initial guidance to companies contemplating use of such a structure to enter new markets and to continue research and development. Substantial issues, such as tariffs, import-export regulations, and antitrust considerations, are not reviewed. Nonetheless, for medical device inventors or manufacturers, this chapter provides a basic overview.

B. Components of a Transaction

The underlying reason why medical device firms often enter into mutual enterprises is to accomplish something they could not otherwise do through a licensing arrangement. For example, a hypothetical Swedish device manufacturer, which we will call Devical Company, may have an entire line of sophisticated diagnostic devices currently marketed in Sweden. As a successful company, Devical has the capital, the technology, the products, and the management to expand its markets; however, the company does not have the

---

[1]For the purposes of this chapter, the term "medical device" includes everything encompassed by the term "device" as it is defined in Section 201(h) of the Federal Food, Drug, and Cosmetic Act, 21 U.S.C. Section 321(h).

present capability to obtain foreign registrations and approvals or to market its products in foreign countries. Devical may also need additional facilities to manufacture its diagnostic devices at home and in foreign locations in sufficient quantities to expand its markets. A joint venture partner with a distribution network, a sales force, technical expertise in obtaining approvals, and compatible manufacturing facilities in the foreign country would be an ideal joint venture partner for Devical Company.

As a rule of thumb, there are six major common ingredients that the parties to a medical device joint venture should bring together:

1. The product and accompanying technology, as well as any patents, know-how, and other intellectual property rights
2. Technical expertise in obtaining product registrations and approvals needed under regulatory laws
3. Sufficient capital
4. Adequate management
5. Manufacturing facilities
6. Sales, marketing, and distribution capabilities

How these elements will be brought to the transaction and who will bring them dictates in large part how the joint venture is structured. In this chapter, the two basic business forms commonly used in the United States for joint ventures—partnerships and corporations—are first explained generally. Second, the six ingredients listed above are discussed in terms of how they may be considered in the joint venture. Third, because not all joint ventures are fruitful, this chapter will conclude with a discussion of the options for termination of the joint venture.

II. BUSINESS FORM OPTIONS

Under U.S. law, general partnerships, limited partnerships, and corporations are different business forms, which are treated differently under the law. The term "joint venture" is a broad, general term; in this context, a joint venture could be a partnership or a corporation. General partnerships, limited partnerships, and joint ventures are similar in many respects. Usually, all are equity ventures formed under, or not contrary to, the relevant state law in the United States by contracts where the parties to the contract may share control and divide the profits. Wherever a joint venture is formed, a new legal entity is created. There are two primary differences between the basic business forms to consider in deciding whether or not the joint venture to develop the medical device should take the form of a partnership or a corporation.

A. Differences in Liability

The first consideration, and perhaps the most important particularly for an equity venture marketing medical devices, is the consideration of exposure to potential product liability.[2] Under the laws of the individual U.S. states,

[2]Of course, as in any undertaking, there are other different potential liabilities involved aside from product claims, such as claims by creditors, or, in the case of joint ventures, claims by shareholders or partners.

the general partners of a partnership are jointly and severally liable[3] for any losses or other liabilities incurred by the partnership. For example, let us assume that Devical and a hypothetical U.S. corporation, Right, Inc., form a general partnership to develop and market Devical's diagnostic devices. If a diagnostic device manufactured and marketed by the partnership allegedly causes harm to a person, the partners, Devical and Right, will be jointly and severally liable for the loss to the injured third party, regardless of any contractual provisions in the partnership agreement allocating the risk of loss among the partners. For example, Right might agree to indemnify and hold harmless its partner, Devical, provided insurance could be obtained for all product liability claims. On the other hand, if Devical and Right form a corporation and the corporation manufactures and markets a diagnostic device, the corporation, not its shareholders, Devical or Right, will be the entity exposed to any liability for the loss.[4] In contrast, in a limited partnership, only the general partner has direct exposure to liability for product liability claims.[5]

Attorneys representing persons allegedly injured by medical devices will often bring actions against all parties in the "chain of distribution," from the retailer that sold the medical device to the parties owning the joint venture. Doctors and hospitals are also commonly parties to the litigation. In sum, the corporate form of the venture can be an important and successful first line of defense to limit exposure to product liability claims. The least advantageous form from the standpoint of minimizing liability is the general partnership, while the limited partnership is the hybrid giving the parties the option to limit the liability of certain partners.

B. Differences in Taxation

Second, tax consequences are often a major consideration in determining the form of the joint venture. Under U.S. law, for example, a general or limited partnership itself is not an entity subject to federal tax, and partnership income is taxed to the partners whether or not it is distributed to the partners in cash or its equivalent [1]. In a general or limited partnership, profits and losses are allocated for tax purposes directly to the partners. Corporations, on the other hand, have the problem of double taxation when profits are distributed as dividends to shareholders.[6] The corporation, as a separate entity, must pay taxes on its income. If the income is distributed to the shareholders as a dividend, the shareholders pay tax on the dividends as ordinary income. The corporation does not receive a deduction for the payment. A partnership is often the preferable business form selected by

---

[3]Liability is said to be joint and several when the plaintiff may sue one or more of the partners separately, or all of them together at his option.

[4]This statement assumes that the corporation is adequately financed, observes corporate formalities, and meets other criteria. If the corporation is a captive company or a shell, a court may "pierce the corporate veil" and find the individual corporate directors, officers, or employees liable.

[5]The limited partner, by law, trades off loss of control in management of the partnership for the insulation against exposure to liability (Uniform Limited Partnership Act, Section 17).

[6]If no dividends are declared, and profits are "expensed out" in management, employment, and other contracts, double taxation is avoided.

two corporations entering into a joint venture, at least from the standpoint of taxation. Moreover, in international transactions involving companies from different countries, issues of taxation can be even more complex in that income may be subject to taxation by more than one country, and issues regarding foreign tax credits need to be reviewed carefully [2]. To the extent that a U.S. partnership is involved in such international transactions, its partners are automatically caught up in tax complexity.

Related to the issue of taxation is the consideration of profit allocation. In a corporation, profits generally must flow in direct proportion to shareholder interests. For example, if Devical owned 51% of the stock of the joint venture corporation and Right owned the remaining 49%, Devical would always receive 51% and Right 49% of the dividends declared (and, presumably 51% and 49% of the proceeds from a sale of all the stock of the joint venture). In a partnership, however, there is greater flexibility because profits can flow disproportionately to invested capital, since they may be allocated in a manner generally determined in the partnership agreement. (United States tax laws should be carefully checked, however, to make sure that any such allocations do not trigger unintended tax consequences.)

C. Conclusion

The foregoing discussion has focused on the principal U.S. joint venture business forms; however, the considerations addressed must also often be carefully weighed in deciding which foreign business form the parties might use. In Japan, the corporate vehicle used has typically been a Kabushiki Kaisha, a joint stock corporation that is similar to the U.S. corporation [3]. In Germany, the Gesellschaft mit beschrankter Haftung (GmbH) has limited liability and would be used most often in German joint ventures in preference to the German public stock corporation. In the Netherlands, there is the Naamloze Vennootschap (N.V.) or the publicly held company and the Besloten Vennootschap (B.V.) or the privately held company. Most local laws permit various types of partnerships and several types of limited liability companies. Insofar as the major legal systems of Western countries are concerned, many similarities among the corporate forms exist [4]. This observation is less true in non-Western cultures. Therefore, while these general considerations to the selection of a business form are important, counsel familiar with the local laws of the foreign country should be consulted.

III. CONSIDERATIONS IN THE FORMATION OF AN ENTERPRISE

The joint venture agreement is the primary document in the formation of any joint venture. This contract sets forth the operating goals and procedures for the joint venture and identifies the various terms, conditions, and other documents needed for the transaciton. The task will be to get the product or technology transferred into the venture, create the ownership and control mechanisms, and allocate the economic benefits and burdens. Thus, whatever the form of the venture, the following general considerations should be taken into account.

A. The Product, Technology, and Patents

A medical device in final form or under development is the basis of the enterprise. The agreement will take into account the stage of development of the device. If the device has neither been tested nor approved in the country in which it is to be marketed, a research and development agreement will be necessary. If the device is in the very early stages of development, the parties may wish to consider making the formation or continuation of the joint venture contingent on the attainment of various research and development milestones. This is a standard protective provision for the developer or owner of the product or technology.

A research and development agreement may include provisions on budget, personnel, quality control, indemnification, and the scope of the research envisioned. The parties may decide on research plans to be developed or on specific protocols in the agreement itself. The parties should also consider who will conduct investigational animal and human studies on the device, to the extent that such studies are required for market approval. The safety and ethical standards under which the studies are to be conducted should be specified and agreed on by the parties. If the device is to be marketed in the United States and requires premarket approval (PMA) or if it must comply with performance standards, the studies must be conducted in accordance with the FDA regulations set forth in 21 CFR Part 812. If independent firms or investigators are used to conduct such studies, the parties must decide who will have the responsibility for supervising and monitoring the studies. Often, independent researchers, such as universities, engaged to conduct clinical research will want the sponsor to provide partial or complete indemnity for their activities, including minimum levels of insurance.

Two other important issues are (1) whether the joint venture will own the medical device as well as any developments or improvements and (2) whether the joint venture itself will become a party to any research and development agreement.

Usually as a condition to third-party financing of the joint venture, ownership rights to the device and any developments or improvements to the device are transferred to the joint venture, and the joint venture becomes a party to any research and development agreements. However, specific financial, liability, or tax considerations may dictate a different arrangement. For example, in our hypothetical situation, Devical may not wish to transfer ownership of its device to the joint venture if it is anticipated that the device has enormous profit potential. Devical may conduct the research and development itself and then enter into a licensing agreement that provides Right, Inc., with royalty payments. On the other hand, Devical may wish to transfer ownerrship of the device to the joint venture if it is anticipated that there are substantial potential product liabilities associated with its use. There are a myriad of potential arrangements between the parties that may be developed.

In any event, proprietary technology and secrets not patented should be protected through confidentiality agreements between the parties and between the parties and the new venture. These protections must be in place prior to negotiations. The confidentiality of any submissions to obtain government approvals should also be assured.

B. Product Registrations and Approvals

The parties must also decide who will be responsible for obtaining product approvals in the countries in which they plan to market the medical devices. Depending on the device, obtaining approval may be uncomplicated and swift, or, in some countries, approval may be complex and require a significant investment of time and money. Other government approvals that may be required include assurances relating to taxation and import and export restrictions. The joint venture agreement should specifically assign the responsibility of obtaining government approvals and may provide for termination in the event approvals are not obtained.

In the United States, if a premarket approval (PMA) or a 510(k) (premarket notification) has already been obtained by one of the parties to the joint venture transaction, the approval may need to be transferred to the new entity. In order to accomplish transfer in the United States, two informal letters should be sent together to FDA's Center for Devices and Radiological Health if a PMA is transferred [5]. The first letter, from the entity transferring the approval, need merely inform the agency of the transfer and the effective date. The second letter from the new entity should, in addition to announcing the transfer, inform the agency that it will manufacture the device in compliance with the conditions of approval and applicable FDA regulations. The new entity must also certify that it has a complete copy of the PMA. Supplements to the approval must be filed by the new entity in the event of a change in the labeling or if there is a new manufacturing facility. For a 510(k), notification of a change in ownership under the establishment registration regulations will suffice [6].

C. Capital

Each party to a joint venture will contribute capital to start the joint venture. The capital may take the form of cash, property (including intangible property), and, in some cases, services. One standard method to determine the amount of capital necessary is for the parties to prepare a business plan where pro forma financial projections for the first several years of the business are projected in detail.

Various cost and revenue assumptions are common, although cost analysis is usually the more predictable aspect. These costs will include expenses incurred in clinical and human testing of the device and obtaining government approvals for the device, the costs of manufacturing, distribution, and marketing, and any other expenses the parties reasonably anticipate. In the area of medical devices, equity ventures are most commonly employed where the projects involved are capital intensive and involve more than one medical device. Therefore, if the projected costs are small or if only one medical device is envisioned, the parties may wish to consider means other than an equity venture, such as a licensing agreement, to achieve their goals.

Once the cost projections are settled, often only one party to the venture puts up the majority of the front-end money; in return, that party can bargain for the lion's share of the profits of the equity venture. The parties investing capital must take into consideration the applicable tax consequences of their contribution. For example, under the U.S. Internal Revenue Code, a credit in an amount equal to 20% of the excess of the qualified research expenses for the year over the average base period research expenses for that year may be subtracted from the party's income tax return.[7] Qualified research expenditures include most research and experimental costs except expenses for foreign research, research in the social sciences, arts, or humanities, or subsidized research [7]. Moreover, the tax laws of the various countries involved in international transactions will dictate many provisions with regard to capital contributions. For example, a new venture should be formed in a country with a network of tax treaties to avoid double taxation.

D. Management

The parties must decide who will control the joint venture partnership or corporation, as well as who will operate the venture. With a partnership, the partners generally select a management committee and a managing general partner. The managing general partner usually has the responsibility for the day-to-day operation of the partnership in consultation with the management committee. The management committee may be provided with veto rights over key or specific material issues, depending on the degree of their equity participation in the partnership.

Multiple variations for dispute settlement mechanisms exist in the partnership format. In contrast, the management structure of a corporation is set forth by statute in state corporate laws. A Board of Directors and, at least, officers to exercise the duties of the offices of President, Secretary, and Treasurer must be elected by law. These officers, pursuant to directions by the Board of Directors and in accordance with the law and the bylaws, manage the corporation. Often, particularly in a closely held corporation, a shareholder's agreement dictates the affiliation and, frequently, the identity of the directors constituting the Board.

Nonmanagement personnel contributed from a joint venture participant often remain on their original employer's payroll, while the joint venture reimburses the participant for the costs of its personnel assigned to the joint venture. The joint venture may also subcontract for additional personnel.

E. Manufacturing the Device

The parties must decide where the medical device should be manufactured and by whom. The joint venture may manufacture the device itself or contract that function to a specialized manufacturer, perhaps even to one of the venturers. Obviously, the availability of medical facilities, technicians, and raw materials may automatically dictate the choice. However, the parties should also consider whether the country of manufacture has any export restrictions on technology. The amount of any tariffs that will be assessed and any shipping charges if the device is exported must be taken into consideration. In addition, joint ventures with manufacturing facilities located in Puerto Rico are permitted to exclude income from Puerto Rican sources from taxation [8], which is an attraction the parties should consider in making a decision to manufacture in or near North America.

---

[7] Internal Revenue Code Section 41. The Omnibus Budget Reconciliation Act of 1989, Pub. L. No. 101-239, extends the 20% tax credit through December 31, 1990.

Regardless of where the device is manufactured, the parties should consider having close ties between the manufacturing function and the adverse-reaction medical device reporting (MDR) function. This is helpful to insure that device malfunctions, attributable to either design or manufacturing defects, are resolved at the source. Quality control must also be assured.

The party with the physical manufacturing and warehouse facilities generally leases the facilities to the joint venture. If neither party has facilities to contribute to the joint venture, at least a local partner may have expertise in acquiring facilities and ensuring that they comply with building codes, environmental and safety regulations, and other required local clearances.

F. Sales, Marketing, and Distribution

In many countries, the joint venture will have to deal with many layers of wholesalers, suppliers, and retailers through exclusive and nonexclusive contracts. Because such dealings can often become complex, a joint venture partner with an existing distribution system through which the devices may be marketed is highly desirable. If neither partner has a distribution system, the joint venture may have to subcontract this function to a third party. However, for many sophisticated medical devices that require careful maintenance and service, subcontracting is often impractical because it creates an intermediary that may be poorly informed about the device and the technology unless properly instructed and supervised.

IV. WITHDRAWAL OR WINDING UP OF THE JOINT VENTURE

Regardless of the optimism of the parties at the commencement of a joint venture, adequate provision should be made for the termination of the relationship.

A. Withdrawal from a Corporation

1. Liquidation

Of course, it is possible to provide in the contract under which the joint venture is formed that it will be dissolved and liquidated given any irreconcilable conflict between the parties regarding material business strategy or upon the happening of any number of specific events, such as the bankruptcy of one party. However, more often than not, the joint venture will have more value as an ongoing entity and therefore will not be liquidated.

2. Buy-Sell Option

One very common way of providing for withdrawal from a joint venture is to create an option for either party to buy the other party's interest in the venture. The buy-sell option may take many different forms. First, the contract may provide for a formula by which either party's interests may be valued upon a triggering event. For example, the buy-out price could be determined in accordance with a formula based on book value or earnings or some variations of either or both. Or the contract may provide that an arbitrator be allowed to set the price to be paid to the withdrawing party.

One method to resolve disputes is known as the "shotgun." Under the shotgun approach, if a party to the joint venture wishes to withdraw, this party must offer to purchase the other party's interest at a specified price. This offer permits the other party the response option of either accepting the offer and selling, or purchasing the shares of the initiating party at the same price and under the terms and conditions as the party making the offer. The party forcing the sale is placed at risk in the sense that he may be forced to buy the entire entity. Consequently, this formula is widely believed to result in a fair price [9]. This formula is not often used in major corporate joint ventures because the parties desire preliminary certainty as to value.

The buy-sell option may also take the form of a "put" or a "call." With a put, the party owning the put has the right to force the other party to purchase his interest in the venture on a certain date or upon the happening of certain events at a price determined in accordance with a formula or at a price set by an arbitrator. Conversely, the party owning a call has the right to purchase the interest of the other party in the venture after a fixed period or upon the happening of one or more events at a price determined in accordance with the formula or at a price set by an arbitrator. If the contract tains a put or call or other provision that forces one venturer to buy out the other venturer, for the relief of the would-be purchaser, a provision compelling the venture's sale to a third party or forcing its liquidation may be included. If an eventual sale to a third party is envisioned as a possibility, the parties may at the outset consider whether they wish to restrict that right and, at least, provide each other with a right of first refusal. Tax considerations will also often play a large part in determining which entity will assume the buy-out obligation [10].

3. Miscellaneous Provisions Relating to Withdrawal

Provisions for periodic renegotiation of the contract under which the joint venture is formed may avoid termination of the joint venture [11]. Termination of the venture may also be avoided in the event the joint venture "goes public," enabling the withdrawing party to sell his shares in the public market.

B. Winding Up of a Partnership

A partnership form is more dependent on the parties themselves than a corporation. For example, under U.S. law if a party to the partnership goes bankrupt, the partnership may automatically be dissolved [12]. The partnership agreement should make provision for the reformation of the partnership in the event of changes. Termination should be avoided when the partnership continues to be profitable. The parties to a partnership may assign their partnership interests to a third party. However, in the United States the partners are protected by law from being forced to accept a new partner without their consent [13].

V. CONCLUSION

The formation of a joint venture to develop a medical device is a complex, time-consuming, and capital-intensive project. If the right business form is selected, a partner with complementary resources and assets is chosen, and all the appropriate business considerations and a variety of other factors are taken into account, such a venture may be successful and very profitable. Similarly, the "down-side" risk of forming a joint venture can be limited if the parties make appropriate provisions for termination. This chapter should provide a general practical guide for achieving these goals.

REFERENCES

1. Internal Revenue Code Section 701, Reg. Sections 1.701-1, 1.702-1.
2. Richard E. Cherin and James J. Combs, Foreign Joint Ventures: Basic Issues, Drafting, and Negotiation, *Business Lawyer 38*:1033 (May 1983) (discussion of foreign tax credit issues).
3. Rosser H. Brockman, Subsidiary or Joint Venture: Choosing an Entry Vehicle, *East Asian Executive Reports* 7:9 (1985).
4. Robert W. Hillman, Providing Effective Legal Representation in International Business Transactions, *International Lawyer 19*:3, 5 (Winter 1985).
5. Premarket Approval Manual, FDA, p. II-18 (October 1986).
6. *Fed. Reg. 42*:42520, 42523 (Aug. 23, 1977).
7. Internal Revenue Code Section 174.
8. Internal Revenue Code Section 936.
9. This approach is discussed in more detail by Alexander Konigsberg, Practical Legal Aspects of International Franchising, *International Business Lawyer 297* (July-August, 1985).
10. Harold A. Segall and Michael S. Sirkin, Providing for Withdrawal from a Joint Venture, *The Practical Lawyer 28*:75, 79 (January 15, 1982).
11. M. K. Gavin, Protecting the Entrepreneur: Special Drafting Concerns for International Joint Venture Contracts, *Corporate Practice Commentator 25*:573, 592 (Winter 1984).
12. Uniform Partnership Act Sections 29,31.
13. Uniform Partnership Act Section 18(g).

part fourteen

Current and Emerging Issues in the International Environment

63

An Overview of the European Community's 1992 Program

JEAN RUSSOTTO *Oppenheimer Wolff & Donnelly, Brussels, Belgium*

I have been asked by the Health Industry Manufacturers Association (HIMA) to undertake a preliminary review of the general impact on the health-care technology industry of the European Community's (EC) program to complete the internal market by the end of 1992. I have considered it appropriate, first, to highlight the background, principal features, and major objectives of the EC's internal market initiative. To this base are added an evaluation of the broad effects that the internal market is expected to have on U.S. industry at large; selected—but not comprehensive—observations relating to the internal market's potential impact for the health-care industry; and, finally, I have briefly described the major changes that may be anticipated in the EC's legislative framework and business environment following completion of the internal market. Obviously, therefore, this memorandum does not represent an exhaustive study of the EC internal market; nor is it a detailed analysis of the impact that specific EC legislative measures may have on the health-care industry and its individual sectors.

I. 1992: DEFINING THE EUROPEAN COMMUNITY'S INTERNAL MARKET

A. The Significance of 1992

The year 1992 is expected to be a milestone in the calender of the European Community. By the end of that year, the EC—12 sovereign states working together in close cooperation—will have made significant progress toward being a truly unified economic region, in which the vast potentials of a 320-million-strong home market can be realized by producers and consumers alike.

In the internal market of post-1992, the "four freedoms" originally set forth in the Treaty of Rome [1]—freedom of movement for goods, persons, capital, and services—are to be realized. It must be emphasized that December 31, 1992, is not a legally binding deadline. Thus, the Community's economic and business environment will not experience a sudden economic or political metamorphosis as of that date. But it is clear that change is already underway and will continue long after 1992.

B. The Origins of 1992

In order to appreciate the significance of the Community's drive toward achieving its internal market, a brief historical review of the background for this initiative is in order. In the early years of the Community, attention focused on the creation of a common external customs tariff. This was achieved in 1977, when all industrial product tariffs disappeared between the (then) nine Member States.[1] However, many nontariff barriers to trade remained, obstructing the free movement principles contained in the Treaty.

During the 1970s, the political momentum to achieve an integrated market slowed, due largely to national issues taking precedence over "European" issues, against a background of growing domestic economic difficulties and increased competition for world markets. By the mid-1980s, it had become evident that a fresh initiative was necessary if a barrier-free Europe were ever to become a reality.

During the "European Council" meeting (of EC heads of state and government) of March 1985, it was decided to give top priority to the achievement of a "single large market by 1992" with the purpose of creating a more favorable environment for stimulating enterprise, competition, and trade. The Council called upon the Commission (the EC's executive body) to draw up a detailed program of action and a specific timetable to achieve this goal. This initiative signaled a new-found commitment on the part of all EC member states to the achievement of stronger European integration.

C. The Commission's White Paper on Completing the Internal Market

The EC Commission responded with a White Paper on completing the internal market, published in June 1985 [2]. It contains proposals for approximately 300 legislative measures to be adopted between 1985 and 1992 along with a detailed timetable indicating the adoption dates for each item. These measures, intended to eliminate existing barriers to free movement in the European Community, are grouped under three headings: physical, technical, and fiscal.

The largest number of proposals listed in the White Paper relate to the removal of technical barriers to trade, and are grouped under seven headings:

1. Free movement of goods
2. Public procurement
3. Free movement for labor and the professions
4. Common market for services

[1] In 1973, Denmark, Ireland, and the United Kingdom joined the European Community, followed by Greece (1981) and Portugal and Spain (1986).

5. Capital movements
6. Creation of suitable conditions for industrial cooperation
7. Application of Community law

Physical and fiscal barriers act no less as a hindrance to efficient commercial operations in the EC and impose an unnecessary burden on industry. The Commission has set as a goal not the mere simplification of existing customs procedures, for example, but the total abolition of physical frontier controls. The removal of fiscal barriers lies partially in the harmonization of indirect taxation, that is, value-added tax (VAT) and excise duties.

II. IN PURSUIT OF THE INTERNAL MARKET: THE FORCES BEHIND ITS COMPLETION

A. The EC's Institutional Framework

Considering the size of the European Community, the diverse traditions and backgrounds of its 12 member states, and the broad spectrum of activities administered or supervised at Community level, the EC's governing institutions are relatively small, employing approximately 23,000 individuals. The four major institutions are the Commission (executive branch), the Council of Ministers (legislative branch), the European Parliament (supervisory/advisory body, directly elected), and the European Court of Justice. Appendix 1 gives a description of the responsibilities of these institutions.

Successful completion of the internal market requires close and sustained cooperation between these institutions, but most specifically between the Commission, the Council, and the European Parliament. Such cooperation has not always been present, however, often due to political differences between the Member States. Failure to reach agreement in the Council of Ministers during the late 1970s and early 1980s resulted in lengthy delays in the adoption of legislation and seriously threatened the progress of European integration. To remedy this situation, the Council of Ministers adopted, in December 1985, the Single European Act (the "Act"), bringing a number of significant amendments to the Treaty of Rome. Following ratification by all of the Member States, the Act came into force on July 1, 1987.

B. The Single European Act and the Internal Market

A principal objective of the Act is to facilitate the achievement of the internal market by, among other things, streamlining the adoption of legislation by the Council of Ministers. Whereas decisions to harmonize national legislation throughout the EC had increasingly been taken on the basis of a unanimous vote in the Council, the Act has substantially reduced the number of instances in which unanimity is required. Instead, the majority of measures relating to harmonization (of technical standards and company law, for example) may now be adopted by qualified majority voting. This is essentially a weighted voting system in which the member states are allocated votes according to their size.[2] A qualified majority requires a minimum of 54 out of

---

[2] Germany, France, Italy, and the United Kingdom have 10 votes each; Spain has 8; Belgium, Greece, The Netherlands, and Portugal have 5 each; Denmark and Ireland have 3 each; and Luxembourg has 2 votes.

the possible 76 votes. Thus it is no longer possible for one member state, claiming "vital national interest," to block passage of measures essential to the completion of the internal market.[3]

The Act has also increased the role of the European Parliament in the Community's legislative process. A cooperation procedure is now in place, which involves the review of selected proposed legislation by the Parliament and the Council of Ministers in two steps (as opposed to the previous single-step process). Fields of legislation now subject to the double review procedure include social policy, research and technological development, and the harmonization of legislation, that is, the majority of measures related to the internal market. The Community's legislative process pursuant to the "co-operation procedure" is diagrammed in Fig. 1.

C. The Impetus Behind the Internal Market

The progress made to date on achieving the internal market would not have been possible without strong support from all of the Community member states. Although there have been some well-publicized exceptions, the political will to cooperate is remarkable, and clearly in contrast to the political climate of the 1970s. The commitment of EC governments to move forward toward 1992 was reinforced at the European Council meeting held in Hannover in June 1988, where, among other things, agreement was reached to renew the appointment of Jacques Delors as President of the Commission.

European industrial and business sectors have been an equally important source of pressure to open the EC's internal boundaries and permit freedom of movement. Manufacturers of goods and services have long recognized the benefits to be gained from a "domestic market" of 320 million people. The advantages of the internal market were set forth in the Cecchini Report, prepared at the request of the EC Commission and published in March 1988 [3].

The report (which comprises 15 individual sector-oriented reports prepared by private consulting companies) not only quantifies the heavy cost paid by producers and consumers because of the many barriers that fragment the Community's economy into 12 separate markets; it also calculates the value of the opportunities that completion of the internal market will open up. Achievement of the internal market as projected in the Commission's White Paper would result in a total potential economic gain of approximately 200 billion ECU to the EC as a whole (figure expressed in 1988 prices).[4] In addition, the report concludes that a true internal market would result in a reduction of consumer prices by an average of 6% and would improve the balance of public finances by an average equivalent to 2.2% of gross domestic product.

The report is careful to point out, however, that these benefits will accrue slowly. Initially, the removal of barriers will produce narrow, technical, and

---

[3]Although well over half of the measures relating to the internal market may now be decided by this qualified majority voting process, unanimity is still required for the adoption of fiscal measures, those relating to the free movement of persons, and those relating to the rights and interests of employed persons.

[4]ECU (European Currency Unit): US $1.00 = 1.121 ECU (October 12, 1988).

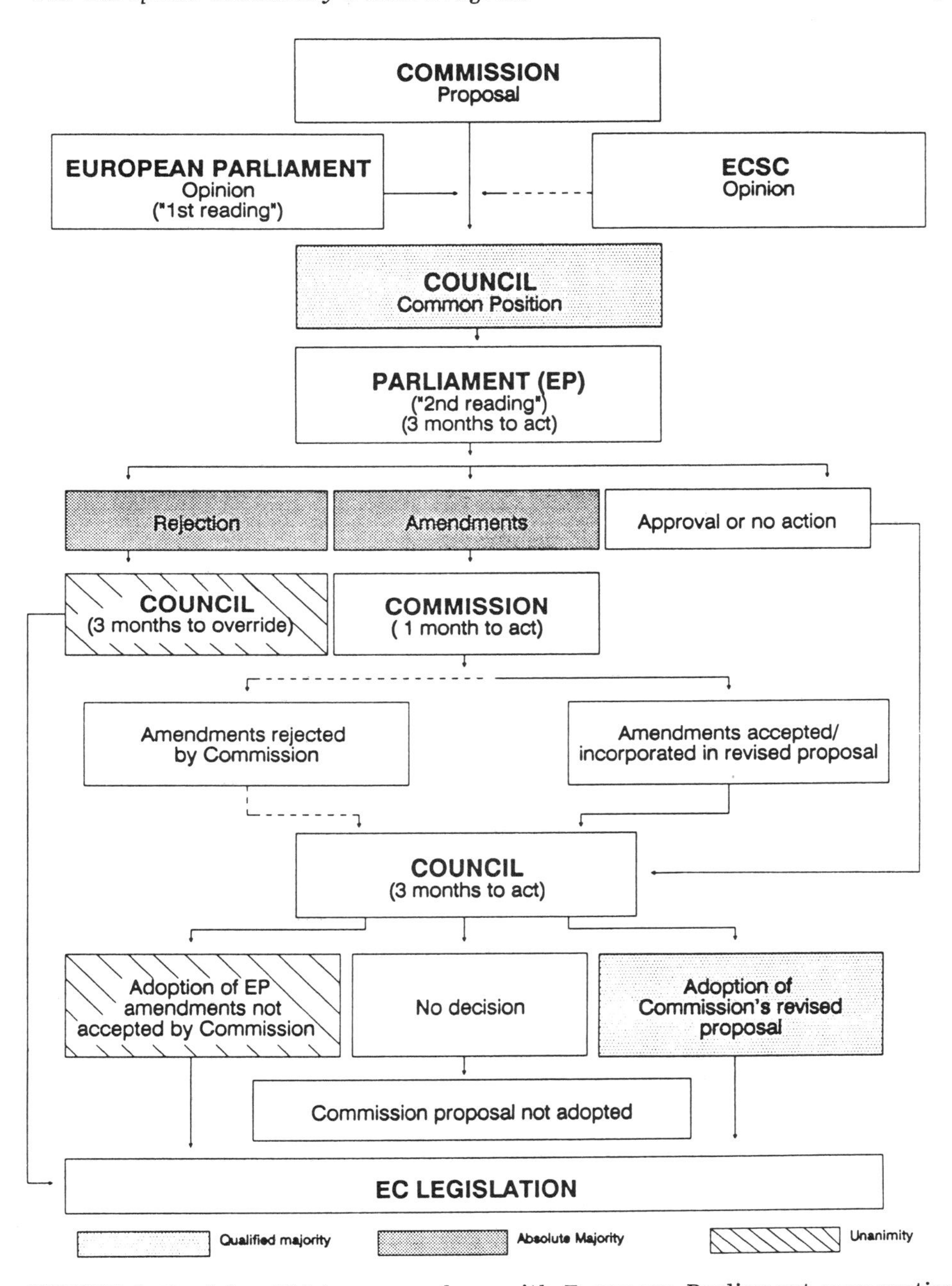

FIGURE 1 Decision-Making procedure with European Parliament cooperation.

short-term gains, but as the impact of market integration grows, the benefits will become much bigger and more widely felt. The report also stresses that all trade barriers must be removed if the beneficial effects of the internal market are to be fully realized.

III. PROGRESS TO DATE ON COMPLETING THE INTERNAL MARKET

A. Annual Commission Progress Reports

Since the adoption of the White Paper, the Commission has published annual reports summarizing the status of all legislative proposals. The third such report was adopted by the Commission in March 1988 and transmitted to the Council of Ministers and the European Parliament [4].

This progress report, which examines the status of all legislation up to March 15, 1988, did not contain very encouraging statistics, and the Commission expressed its disappointment at the number of proposals that were still pending in the Council of Ministers. For example, as of March 15, 1988, only 69 proposals contained in the White Paper had been finally adopted by the Council; an additional six measures have been partially adopted. The report attributes this slow progress to two major factors: first, the delayed entry into force of the Single European Act (6 months later than originally scheduled) and second, the Council of Ministers' reluctance to delegate wider implementing powers to the Commission as originally envisaged.

On a more positive note, the Commission stressed the successful adoption of several important measures that will have a wide and immediate impact on the internal market. These include directives relating to the insurance sector; the liberalization of capital movements; and standardization and harmonization measures relating to the environment, pharmaceuticals, and high-technology medicines. Also note that very recent figures on progress toward completing the internal market show that by the end of September 1988, 98 proposals have been either fully or partially adopted [5].

Of the three Community institutions involved in the decision-making process, the Commission has been able to act with the greatest speed. Lord Cockfield, Commissioner responsible for the Internal Market, announced the Commission's intention to present virtually all internal market proposals before the end of 1988. This action should have the effect of increasing both the pressure and momentum toward completing the internal market. Equally important, it will leave a space of 4 full years for the adoption of the proposed legislation.

The Commission's third progress report concluded by listing a number of important areas in which the Commission expects "significant progress" before the end of 1988. These include: (1) the second stage of the liberalization of capital movements; (2) non-life insurance; (3) direct tax measures designed to improve the fiscal environment for cross-border operations; (4) approximation of indirect taxation; and (5) the second banking directive.

As stipulated in the Act (Treaty of Rome Article 8B), before the end of 1988 the Commission will submit to the Council of Ministers a major report on progress toward the 1992 deadline. The report is to be discussed at the meeting of the European Council of Heads of State and Government in Greece during December 1988. In addition to listing the legislative proposals that have been adopted and those awaiting action, this report further confirms the accelerated pace of the Community's integration process and emphasizes future Community priorities in the context of the internal market, such as the social dimension, monetary policy, and fiscal harmonization.

B. Obstacles Still to Be Overcome

The Commission's White Paper lists three major types of barriers to movement within the European Community: physical, technical, and fiscal. The majority of proposals in the White Paper relate to the removal of technical barriers. Although legislation has been adopted to reduce barriers in each category, some real obstacles remain. For example, the elimination of internal border controls raises the sensitive issues of security against terrorism and the movement of illicit drugs. A number of the present veterinary and phytosanitary controls that the White Paper hopes to eliminate are perceived as providing protection to the individual member states' farming sector, which is already suffering from overproduction and inefficiencies. Removal of such controls is encountering considerable opposition.

Efforts to align member states' fiscal policies have likewise encountered substantial opposition, in particular from the United Kingdom and France.[5] Where the Commission's proposals involve the harmonization of *non*fiscal legislation, the Community's decision-making process should, in principle, be accelerated as a result of the practice of qualified majority voting in the Council of Ministers. But delays are still likely, as many harmonization proposals concern sensitive sectors of national economy and industry.

The Cecchini Report stresses that the benefits of the internal market will emerge slowly, and are likely to be accompanied by initial job losses of at least 500,000. With this prospect in view, the Commission is turning its attention to the internal market's "social dimension," and steps are being taken to propose measures that would, among other things, afford greater social protection for workers, increase retraining opportunities, and involve labor unions and employers' organizations more fully in the EC legislative process. Failure to take into account the effects of the internal market on the Community's social environment could result in labor unrest and hence jeopardize the success of the entire initiative.

Social unrest could also result if the Community is unable to correct the present (and growing) imbalance between its richer and poorer regions. In this context, some of the more prosperous northern member states have expressed concern that the free movement of labor envisaged in the White Paper will trigger a flow of workers (skilled and unskilled) from the less-industrialized southern regions, thus putting even greater strain on employment and social policies.

The drive to complete the internal market has raised once again the issues of nationalism and national sovereignty and tests the willingness of

---

[5]France and the United Kingdom reacted strongly to Commission proposals for closer alignment of national value-added taxes. France's objections relate, in the first instance, to the proposed implementation period for such alignment. The United Kingdom has put forward an alternative plan, which does not include the alignment of VAT at EC level, but rather as a function of market forces.

member states to cede decision-making authority to the EC institutions. Earlier this year, Commission President Delors predicted that within a decade, 80% of what are now national decisions taken by member state governments will be taken at Community level. His statement, that "80 percent of economic, perhaps even social and tax, policy will be of Community origin" prompted a strong reaction from U.K. Prime Minister Margaret Thatcher, who firmly opposes any such transfer of power. In a major policy speech outlining the United Kingdom's views on European integration and the European Community, Mrs. Thatcher responded, "working more closely together does *not* require power to be centralized in Brussels or decisions to be taken by an appointed bureaucracy . . . . We have not successfully rolled back the frontiers of the state in Britain, only to see them reimposed at a European level, with a European super-state exercising a new dominance from Brussels [6].

Despite these strong words, Mrs. Thatcher's speech should not be construed as signifying a reversal of the U.K. government's support for the internal market and European integration. Careful perusal of Mrs. Thatcher's remarks at Bruges shows that she took great pains to stress the British commitment to Europe. The criticisms contained in her speech were directed primarily at what she perceives to be an overzealous tendency toward concentrating authority and decision-making in Brussels, rather than at the concept of European integration. In this context, it should be recalled that in 1985 Mrs. Thatcher had been opposed to the holding of an Intergovernmental Conference that paved the way for the Single European Act, and she was likewise against the Single European Act itself. In both instances she eventually supported these initiatives. Finally, as regards substantive progress toward completing the internal market, it should be noted that 56 decisions were taken during the last 6 months of 1986 when the United Kingdom held the EC Council presidency.[6]

The present obstacles to completion of the internal market are both tangible and abstract. In January 1989, a new EC Commission, headed again by President Delors, begins its 4-year term of office. (Appendix 2 lists the members of the current EC Commission; Appendix 3 shows the composition of the Commission to be installed in 1989.) Lord Cockfield, the EC Commissioner responsible for the internal market—and considered by many as the author and driving force behind this initiative—is not part of the new Commission. Although assured of many competent senior European civil servants, the new Commission can expect a major challenge in seeing the internal market to fruition. Regardless of the commitment of the EC Commission to this goal, the most important ingredient necessary for the realization of the internal market remains the political will of the member states to consult and cooperate. With minor exceptions, EC leaders have clearly demonstrated, through statements and nation-wide campaigns to support the internal market, their desire to cooperate toward achieving this goal.

C. A Prognosis

Statistics such as those contained in the Commission's Third Progress Report on Completing the Internal Market and the September figures referred to

[6]This figure represents the largest number of internal market related decisions taken during a single EC Council presidency, with the exception of the German Council presidency (January-June 1988) which takes credit for 58.

above indicate that progress has been slower than anticipated. Nonetheless, there is little doubt that the EC will reach its goal, if not in 1992, then soon thereafter; 1992 does not represent a firm, legally binding deadline, and the Commission President has cautioned against transforming the date into a "sacred cow." It has, however, assumed great symbolic importance, not only in Europe but also in the rest of the world, as evidenced by the increased efforts of a multitude of corporations and businesses to prepare their organizations and strategies for post-1992 Europe.

IV. THE LIKELY IMPACT OF 1992 ON U.S. INDUSTRY

A. The Reactions of U.S. Industry

When first launched in 1985, the EC's internal market initiative attracted little attention in U.S. business circles. Today's situation is radically different. The increased media coverage reflects the sensitivity of U.S. firms to the potential advantages and disadvantages of Europe's internal market. Further evidence may be found in the number of companies—European as well as American and Japanese—that are creating in-house task forces or establishing a liaison office in Brussels to monitor "1992" developments.

There is clearly a concern—shared by U.S. and Japanese business alike—that post-1992 Europe will be "Fortress Europe," in which internal barriers have been removed and replaced by a protective wall around the entire Community.[7] United States reaction seems to center on one of two main themes: first, the increased opportunities that the internal market represents, and second, concern that the internal market will transform Europe into a protectionist bloc. Both reactions are understandable, given that the EC is the United States' largest trading partner.[8]

Achievement of the internal market should increase consumers' demands for goods and services, cut manufacturing costs, and enhance efficiency in production and distribution. On the negative side, the internal market is expected to result in sharpened competition, as companies reach across national borders to new markets. It must be borne in mind that European companies—with some notable exceptions—traditionally operate along national lines. At present, Germany is the largest home market: 61 million people. Following 1992, the Community's "home" market will be over 320 million. To serve this market will require companies of a size and capacity hitherto scarce in Europe.

European industry is responding to this challenge in a number of ways, most visibly perhaps in the number of mergers and takeovers that have occurred over the past 2 years. For example, there were 97 cross-border acquisitions in Europe reported by French companies in 1987; for Germany, the figure was 86. The first 4 months of 1988 witnessed more than twice the number of takeover bids by U.K. companies in the rest of Europe than dur-

---

[7]In a recent letter to the EC Commission, U.S. Trade Representative Clayton Yeutter expressed the U.S. position: "We welcome your efforts at completing the internal market—but not if a single internal market leads to increased external barriers or discrimination against U.S. firms' interest in the Community." [7]

[8]United States exports to Europe, combined with shipments of U.S. subsidiaries established in Europe, account for over $500 billion.

ing the previous year.[9] A major aim of this corporate alliance building is to achieve economies of scale, establish a corporate presence in new markets, and diversify into new (but generally related) industy sectors. In this process, a number of small, less efficient companies are expected to disappear, either as a result of absorption by larger competitors or simply due to their inability to compete.

To anticipate the post-1992 business environment, companies are also decentralizing their operations to enhance flexibility; undertaking management and staff retraining to focus on a "European" approach; increasing market research activities; and establishing manufacturing and/or sales subsidiaries in new European markets. Finally, it is expected that achievement of the internal market will be accompanied by an initial rise in unemployment, as companies seek to streamline operations by introducing more high-technology methods requiring less personnel.

These efforts on the part of European industry, if successful, will result in fewer but larger, more efficient European companies capable of competing with their U.S. and Japanese counterparts on a scale that has not yet been seen. There is, therefore, justification for the close attention which these developments are receiving in the United States.

United States concern also stems from uncertainty surrounding the issue of *reciprocity*, which has been raised most frequently in discussions concerning areas of economic activity not covered by the GATT, such as the provision of services. Several proposed EC measures, notably in the financial services sector, would permit the establishment of third-country companies (e.g., credit institutions) in the Community only if comparable EC companies were granted reciprocal treatment in the third country. A sub-issue raised in this connection concerns the possibility of Community demands for retroactive reciprocity.

Concern has also been voiced about possible transitional rules that may apply to businesses operating in the Community. Such rules have traditionally been used to provide a lead time during which member states may adjust their domestic legislation to EC-wide measures. However, the duration of this lead time may vary, giving rise to uncertainty.

Uncertainty has also been voiced as to what measures the EC may take to protect particularly vulnerable industrial sectors, including telecommunications and aeronautics, which are especially sensitive to outside competition.[10]

B. The EC's Response

The EC Commission invariably insists that the goal of 1992 is competitiveness, not protectionism. The Commission sees new business opportunities for EC and U.S. firms alike. But, it points out, whether the opportunities are fully

---

[9]These figures are contained in a study carried out early this year by the management consultants Booze Allen & Hamilton.

[10]U.S. Secretary of Commerce C. William Verity has recently expressed assurances that exports and investment in Europe should be "significantly improved" by the EC's 1992 program. At the same time, he cautions that companies must inform themselves of internal market developments and remain alert, both to the opportunities as well as to the potential difficulties which these developments may create.

exploited will depend on the "dynamism" of individual firms. With a fully integrated economy, the EC intends to play a more prominent role, commensurate with its economic weight. However, the Commission offers assurances that the EC's leading position in world trade would not result in actions that would conflict with its international obligations under the GATT.

On October 19, 1988, the Commission made its first formal policy statement on 1992 and its effects on trade with third countries. The statement, part of a yet-unpublished Commission document, is basically a reaffirmation that the intention of the internal market is not to create a "Fortress Europe," but rather a "Partnership Europe." Further, the Commission emphasizes its intention to honor the obligations laid down by GATT on free and open trade, and that the benefits of the removal of internal barriers in the EC should accrue to third countries as well as EC Member States.

On the issue of reciprocity, the Commission report indicates that in sectors not covered by international trade agreements (e.g., financial services and certain sectors of public procurement), the EC would "reserve the right" to demand reciprocity in granting access to its market, although it did not indicate what specific form such reciprocity would take.[11] Banking was singled out in the statement as a sector in which retroactive reciprocity agreements would not be sought. Subsidiaries of foreign banks already established in the member states would be treated as EC banks. The Commission did not, however, exclude the possibility of demanding retroactive reciprocity in service sectors other than banking.

The Community still intends to maintain import restraints on products from a small number of key sectors. Those most frequently mentioned include automobiles, textiles, and selected imports from the Eastern Bloc and developing countries. At the moment, imports are regulated by individual EC Member States; these national restrictions would be replaced by a Community-wide regime.

In evaluating the Community's stance vis-à-vis third countries, one should not lose sight of the fact that the principal goal of the internal market is to elevate the competitiveness of European industry and enable it to perform on an equal footing with its world trade partners.

C. An Evaluation

It would be premature to predict what specific effects the internal market will have on American industry. As this is written, 1992 is still a full four years ahead and, as the previous discussion has illustrated, there are several key areas in which the EC must still reach agreement (e.g., monetary and fiscal policy, border controls) before legislation can be adopted.

The elimination of barriers to intra-Community trade will create a large sales market. Especially for companies that are already selling across EC national borders—and doing so at a major cost—the removal of these barriers should bring about substantial savings. Products will, for example, no longer need to be differentiated to meet different national standards, and transportation should be both quicker and cheaper. Further, companies that have concentrated sales in particular national markets will likely begin extend-

---

[11]"Reciprocity" is defined by the Commission as a "guarantee of similar—or at least nondiscriminatory—opportunities" for EC enterprises to operate in foreign markets on the same basis as local companies.

ing their activities throughout Europe, and become increasingly competitive in the process. Distribution and pricing policies may have to change considerably. Already companies in many sectors are beginning to adapt their corporate structures to take advantage of the coming integrated market, as can be seen by some of the recently reported takeover attempts, mergers, and joint ventures.

In short, as pointed out above, increased competition is certain to be a prominent feature of the post-1992 European Community. European companies are projecting to experience important gains in productivity, economies of scale, and much faster new-product development due to the restructuring taking place. For example, 73% of French companies interviewed by the French Employers Federation have indicated that they intend to restructure their companies in light of the removal of trade barriers envisioned in the 1992 plans. The challenge of European companies will therefore be real.

As regards the Community's external trade, completing the internal market will mean, among other things, the phasing out of national restrictions on imports; future trade measures would instead be negotiated and executed at the Community level. Following this principle, if the EC could justify the need for Community protection from imports in a given sector, the resultant trade measure would be EC-wide legislation rather than national legislation adopted by an individual Member State.[12] Distinct from the benefits for individual companies of marketing or producing in the EC following completion of the internal market, the benefits of 1992 for certain sectors are projected to be very significant.[13]

V. IMPLICATIONS FOR THE HEALTH-CARE INDUSTRY

A. The Present Community Environment

The Community's pharmaceutical and health-care technology industry is characterized by the dominance of large, multinationally oriented firms, which operate on a world-wide basis. Sixty such firms operate in the EC; of these, 33 are Community-based, with German and British companies being the leaders, followed by French and Italian companies. Despite this international character, the marketing of health-care products, equipment, and services tends to closely follow national lines, due largely to national differences in attitudes to drugs and in traditions of medical practice [8].

The industry is highly regulated at the national level. Within the European Community, there are 12 national authorities or government agencies charged with approving the safety, efficacy, and high quality of the prod-

[12]Particularly sensitive sectors that could be the object of such measures include automobiles and textiles. The implications especially for non-EC producers of these "sensitive products" will be that they may find it advisable to create or expand production facilities within the Community, in order to be properly positioned to benefit from the "free movement" rules of the integrated market.

[13]See the Cecchini Report, which includes analyses of how certain manufacturing industries (e.g., pharmaceuticals, telecommunications, motor vehicles) will be affected. The report describes the main economic factors involved (e.g., economies of scale, restructuring, incidence of competition on corporate behavior, etc.).

ducts and equipment released on to their respective national markets. Registration of health-care products also takes place at the national level.

In the Community's view, a measure of uniformity within the EC has been attained through EC legislation adopted since 1965, especially concerning pharmaceutical products. Particularly noteworthy was the adoption in 1986 of four EC Directives aimed at harmonizing national legislation concerning the testing of veterinary and proprietary medicinal products and the marketing of high-technology medicinal products, for example, primarily those derived from biotechnology. These measures were the first to be adopted in the pharmaceutical sector within the context of the White Paper on completing the internal market. An important feature of the Directive concerning the marketing of high-technology medicinal products is that it introduces a marketing application procedure whereby an application is referred for an opinion to the competent Community-level body prior to the taking of any national decision. Nonetheless, discrepancies in important areas remain.

Also as a result of Community legislation, some convergence has been achieved concerning the requirements imposed by national regulatory authorities relating to the approval and registration of health-care products. There remain nonetheless significant differences, which stem, for example, from varying methods of evaluation.[14] It has been estimated that the delays in product registration represent 0.5-0.6% of industry costs within the EC. Conversely, unified and more rapid registration procedures in the EC would be expected to yield savings in direct costs of operation to U.S. firms in the range of 65.6-105.6 million ECU [8]. Although relatively small when compared to costs related to the approval of a single drug, these savings estimated by the Commission are noteworthy.

Community efforts to streamline the approval and registration procedure, notably through the EC's Committee for Proprietary Medicinal Products, have had little success. The search for a more unified registration system has focused on the mutual recognition of registration between member states or, alternatively, an EC-wide registration agency, patterned after the U.S. Food and Drug Administration. On this issue, the Commission intends to submit a proposal by the end of 1989 to revise the present registration system. It is not yet known, however, whether the proposal will be based on a system of mutual recognition, a centralized European agency, or perhaps a combination of the two. There are considerable difficulties with either approach, and an early solution to the present situation is not foreseen.

Given the financial involvement of EC governments in the provision of health care, most governments play an important role in the setting of prices for health-care products and equipment. Also, in some member states the prices that are set favor national products. Two major difficulties that are encountered by health-care manufacturers in the EC relate to, first, the approval and registration of their products and second, the pricing.

For the manufacturers of health-care equipment, an additional problem lies in the public purchasing policies followed by individual EC member state governments. Although existing EC legislation covering public supply and public works contracts directs government departments and local authorities

---

[14] As regards pharmaceuticals, EC legislation provides for a 120-day limit on the processing of applications for national approval. In practice, however, the Community average is 18-24 months.

to open up their tender procedures to firms throughout the EC, the practices persist of buying "national," employing discriminatory specifications, and the failure to advertise contracts properly, inter alia.

While there is a cluster of EC legislation relating to the manufacturing and/or marketing of pharmaceuticals in the Community, legislation covering medical equipment is more limited, addressing, for example, the electrical safety of equipment or the tax regime applicable to equipment transported for temporary use from one member state to another. In 1984 the Council adopted a Directive to harmonize member states' laws concerning electro-medical equipment.[15] The deadline for transposing this Directive into national legislation was September 26, 1986, yet there are still member states who have failed to do so. Pursuant to EEC Treaty Article 169, the Commission may, in such instances, initiate legal proceedings against the member state for failure to fulfill its Treaty obligations.[16] Such proceedings have begun concerning failure to apply this 1984 Directive.

In 1987 the Commission also initiated legal proceedings against Spain for its use of a system of national registration for medical devices that the Commission considers incompatible with the Community law relating to the free movement of goods [9].

B. Anticipated Changes in the Post-1992 European Community

It is undisputed that the completion of the internal market will have a substantial impact on the way in which companies do business in the European Community. There is a growing understanding on the part of corporate management that 1992 will eventually force a reappraisal of corporate strategy. Company pricing policies, in particular, will require adjustments. At present, due to market segmentation, the prices charged for identical products can vary widely between countries, leading in some instances to the practice of parallel imports. The post-1992 unified market will likely see an increase in the alignment of prices Community-wide, as hitherto compartmentalized national markets disappear. Not only do companies realize that they will be affected by the new business environment, but consumers are also becoming increasingly aware that the internal market is a very material reality. The year 1992 will also see the emergence of larger corporate customers, with their implicitly greater bargaining power. These are but two examples of the changes that are foreseen following achievement of the internal market.

1. *The General Impact*

At the general level, daily business operations will be affected by, for example, the harmonization of Community fiscal legislation, the transborder

---

[15]Council Directive of 17 September 1984 on the approximation of the laws of the Member States relating to electro-medical equipment used in human or veterinary medicine (see *EC Official Journal* No. L 300 of November 19, 1984).
[16]Article 169 regulates the procedure for bringing before the European Court of Justice and alleged failure of a member state to comply with its obligations under the Treaty. This Article is applicable only when the Commission is the complainant; neither other EC institutions nor private persons can bring proceedings of this kind. However, they may be able to instigate action by the Commission.

shipment of goods, and the way companies maintain their internal accounts. In the field of company law, proposed legislation on cross-border mergers would synchronize the examination of mergers by national authorities and ensure that shareholders, employees, and other third parties would be adequately protected [10]. To be noted in this context is that the Commission is also currently working on a competition regulation relating to mergers with a total value of over 1000 million ECU. Qualifying mergers will have to be notified to the Commission before being carried out. The Commission would like to adopt the block exemption regulation by the end of 1988, but sometime in 1989 seems a more realistic date.

Legislation has also been proposed regulating intercompany dealings, such as takeover bids. A draft measure now being discussed by the Commission details how much information a bidding company must disclose during takeovers and specifies a code of conduct. Other draft legislation under consideration concerns the voluntary liquidation of companies, and measures to harmonize national requirements concerning the publication of accounts by branches of foreign limited liability companies.

New Community instruments for industrial cooperation will also facilitate cross-border cooperation. European Community legislation that enters into force on July 1, 1989, provides a legal company framework known as the European Economic Interest Grouping (EEIG), which will allow businesses to cooperate more freely across borders. This represents the first example of a corporate body under Community—as opposed to national—law. Once formed and registered, an EEIG will have the capacity in its own name to acquire rights, incur obligations, make contracts, sue, and be sued. The EEIG is not intended to generate profits itself, but is intended as a vehicle for the business of its members.

The EEIG sets a precedent for the European Company Statute, a 1970 Commission proposal that has been recently revived. This proposal provides for the creation of "European companies" that would be wholly subject to a specific legal system that is directly applicable in all member states, thus freeing this form of company from any legal tie to a particular country. However, the size and complexity of this proposal—plus its provisions relating to taxation and worker participation—suggest that progress toward adoption of a definitive measure will be slow.

As regards financial services, the White Paper acknowledges the important role that this sector plays in the Community's economy. The Commission, recognizing that the competitiveness of Europe's manufacturing industry depends on cheap and competitive financing, has put forward proposals that aim to create a more closely knit financial market and provide for the free mobility of capital. In June of this year, the adoption of a Council Directive on the free movement of capital will greatly facilitate, for example, the repatriation of dividends and transfers of funds from one member state to another [11]. Other measures in this sector would include the proposed Second Banking Directive, which provides for a "single license" enabling a bank established in one member state to set up a branch in another member state without first seeking authorization from that member state [12]. It is in this connection that the issue of "reciprocity" has been raised. The Commission has also proposed legislation on the publication of annual accounting documents of foreign branches of banks.

In the field of public procurement, EC legislation to ensure greater transparency in the award of public supply contracts worth more than 200,000 ECU was adopted in March 1988 and becomes effective on January 1, 1989 [13].

This directive amends a 1977 Council Directive on the award of public supply contracts [14]. Modifications contained in the new directive will improve the information available for potential suppliers and provide for more stringent rules regarding the contents of invitations to tender. Further, in order to prevent public authorities from demanding standards that favor national suppliers, the directive will require wider use of existing European standards. Annex I to the new directive is a list of legal persons governed by public law and bodies corresponding thereto, whose public purchasing procedures are governed by the directive. At the most general level, this includes "associations governed by public law or bodies corresponding thereto formed by regional or local authorities." Named in particular are, for example, the U.K. National Health Service Authorities.

Substantial progress has also been made toward revising EC legislation concerning the awarding of public works contracts. In the revised legislation (which has been approved by the Council of Ministers and is now awaiting review by the European Parliament) an improved information procedure would enable EC firms to prepare their bids under better conditions and, in principle, benefit from equal access to public works contracts totalling approximately 150 billion ECU annually.

Manufacturers in all industry sectors will be affected by the 1985 EC legislation on product liability, which harmonizes national product liability measures and affords increased protection to consumers. Similarly, proposed environmental legislation, such as the proposed directive on civil liability for damage caused by waste, will have a broad impact on Community manufacturers. These are but a few examples of the broad effects that 1992 will have on companies doing business in the European Community.

2. *Sector-Specific Effects*

In addition to its general impact on industry, the Commission's White Paper contains a number of proposals that will affect individual sectors, such as health care. As regards pharmaceuticals, these would include legislation relating to the transparency of the pricing of pharmaceutical products, and the extension of existing legislation to medicinal products not yet covered, such as human immunological products, products derived from human blood, radiopharmaceuticals, and generic products.

For medical products, several legislative proposals are in various stages of preparation within the EC Commission. The general White Paper heading, "Medical equipment: electro-medical implantables," does not indicate the full range of this activity. It includes currently separate proposals on:

a. Active implantables (e.g., pacemakers and heart valves)
b. Electromedical equipment (e.g., monitoring and imaging devices)
c. *In vitro* diagnostic products
d. Sterile hospital products and other products not covered by the other three proposals

We understand that several European trade associations are actively working on these proposals with EC Commission staff.

The proposals are a result of the Commission's recognition that present EC legislation governing medical devices, such as the 1984 Directive, which lays down common rules relating to safety and inspection procedures in respect of a range of equipment used in human or veterinary medicine, cover only a small portion of the medical devices now available in the Community.[17]

Each of the above proposals falls under the Commission's "new approach" to technical harmonization, adopted in 1985 [15]. Pursuant to this new approach, Community legislation lays down essential safety requirements, while the European standards institutes (e.g., CEN and CENELEC[18]) are entrusted with drawing up the technical specifications, taking into account the current level of technology.

VI. CONCLUSIONS

Progress toward the completion of the internal market is now accepted as an irreversible process. It must also be acknowledged that the target date of "1992" is optimistic, and indeed may arrive before all measures envisaged in the Commission's White Paper have been adopted. However, the Community's overall commitment to achieving the internal market is not in question, and there is little doubt that the momentum will be sustained. Moreover, the Community does not lack the resources to achieve its goal, as long as the commitment remains firm.

The benefits that industry stands to gain from the "1992" initiative would appear clear; it is now the task of industry to critically but rapidly review its strategy and adapt, where necessary, its corporate plans in order to remain competitive in the enlarged, integrated EC market.

Major multinationals—among which are many U.S. companies—now operating in Europe may find that their size and experience give them an advantage in adjusting to the new post-1992 business climate. However, size and experience alone will not be sufficient to ensure continued success post 1992. Many companies are already taking a critical look at the "new" European market and their own level of preparedness. Such corporate self-evaluation raises a number of questions worth considering, such as:

1. Do the EC's 1992 objectives figure in the company's present business plan? If yes, have they been expressed in specific or general terms?
2. If, on the other hand, the company's business plan does not specifically refer to a 1992 target date, should it be modified accordingly?
3. Assuming the existence of a large market, should the company's marketing approach be reconsidered (in the light of, for example, consumer preferences and purchasing patterns)?
4. In a large single market, where presumably production might be increased, will product prices be affected, and if so, how?
5. Will the free circulation of workers impact positively or negatively on the labor costs?
6. How can/should the distribution network be best adapted to service the new market?
7. Should additional production/distribution centers be considered to service the enlarged market?

---

[17]See footnote 15.

[18]CEN, European Committee for Standardization; CENELEC, European Committee for Electrotechnical Standardization.

8. Because of the enlarged market, should the company's existing corporate structure remain unchanged? For example, are 12 marketing subsidiaries necessary? Should a company rather operate through distributors or agents? Or are cooperative ventures a more desirable vehicle?
9. In view of what might be a new corporate organization, will the existing management structure be altered? Are additional management-level personnel required to supervise the extended corporate activities? Or should the company reconsider reducing the level of its management? What adjustments are necessary for the present management in a new business environment (new languages, management approaches)?

For companies that do business—but do not have an established presence—in the Community, the prospect of a barrier-free EC internal market raises additional questions. In order to protect the continuation of their present activities, such companies should be exploring how to strengthen their position. Should a company subsidiary or branch be created in the EC? Should direct acquisition of an existing EC company be considered? If so, in which country? This question alone raises considerations of present national legislation regarding corporate establishment, labor laws, taxation, and social security provisions, to name a few.

Whether a company is already established in the EC or simply reviewing its options for entry, a corporate reappraisal in the light of a post-1992 internal market will test a company's flexibility and cohesion. As the aim of a corporate strategy is not to divine but to devise a precise and pragmatic plan of action, a reappraisal of the impact of the post-1992 market appears to be not only in order but indispensable.

REFERENCES

1. Treaty establishing the European Economic Community. Signed in Rome on March 25, 1957, by the original six Member States: Belgium, France, Federal Republic of Germany, Italy, Luxembourg, and The Netherlands.
2. *Completing the Internal Market.* White Paper from the Commission to the European Council. COM (85) 310 final, June 14, 1985.
3. *Research on the "Cost of Non-Europe."* Report prepared under the direction of Paolo Cecchini, former Director-General of the EC Commission, 1988.
4. Third Report from the Commission to the Council and the European Parliament on the implementation of the Commission's White Paper on completing the internal market. COM (88) 134 final, March 21, 1988.
5. Countdown 1992, *Business Journal*, No. 84, p. 28, October 1988.
6. Speech delivered at the opening ceremony of the 39th academic year of the College of Europe, Bruges, Belgium on September 20, 1988.
7. Laying the foundation for a great wall of Europe, *Business Week*, August 1, 1988, p. 16.
8. The "Cost of Non-Europe" in the Pharmaceutical Industry, Executive Summary, Economists Advisory Group Ltd., January 1988.
9. Written question no. 107/87, National regulations covering the import of medical devices. *EC Official Journal* No. C 23 of January 28, 1988.
10. Proposal for a Tenth Council Directive based on Article 54(3)(g) of the Treaty concerning cross-border mergers of public limited companies. *EC Official Journal* No. C 23 of January 25, 1985.

11. Council Directive for the implementation of Article 67 of the Treaty. *EC Official Journal* No. L 178 of July 8, 1988.
12. Second proposed directive on the coordination of banking legislation. *EC Official Journal* No. C 84 of March 31, 1988.
13. Council Directive of 22 March 1988 amending Directive 77/62/EEC relating to the coordination of procedures on the award of public supply contracts and repealing certain provisions of Directive 80/767/EEC. *EC Official Journal* No. L 127 of May 20, 1988.
14. Council Directive of 21 December 1976 coordinating procedures for the award of public supply contracts. *EC Official Journal* No. L 13 of January 15, 1977.
15. *Technical harmonization and standards: a new approach.* Communication from the Commission to the Council and to the European Parliament. COM (85) final, January 31, 1985.

APPENDIX 1 THE MAJOR INSTITUTIONS OF THE EUROPEAN COMMUNITY

The Commission

Composition	Seventeen members: two from each of the five largest member states, one from each of the remaining seven. Appointed for renewable 4-year terms by common agreement of the member governments.
Responsibilities	The executive arm of the EC: responsible for initiating Community policies, drafting legislation and assuring the proper application and enforcement of Community legislation. Individual Commissioners assigned specific areas of responsibility (e.g., external relations, internal market and industrial affairs, science, research and development). The day-to-day functions are carried out by 22 Directorates-General and a number of Specialized Services.

The Council of Ministers

Composition	Ministers or Government-level representatives of the 12 EC Member States; an intergovernmental rather than a supranational body. Membership varies with the subjects to be discussed at individual Council of Ministers' meetings (e.g., agriculture, economic affairs, labor).
Responsibilities	The EC's decision-making body. Responsible for adopting or rejecting legislative proposals submitted by the Commission. Also responsible for taking broad policy decisions, which may then be transformed into specific legislative proposals by the Commission. Council presidency held for a 6-month period by each EC member state on a rotating basis, according to alphabetical order. Assisted by a Committee of Permanent Representatives (COREPER) comprised of permanent representatives or ambassadors of the member states.

The European Parliament

Composition	A 518-member Assembly, directly elected by citizens of the EC Member States. Meets in plenary session once each month. Members of the European Parliament (MEPs) form cross-national political groups.
Responsibilities	To supervise the activities of the Commission and the Council of Ministers; to render its opinion on all proposals for Community legislation; to adopt the Community annual budget.

The European Court of Justice

Composition	Thirteen judges appointed for 6-year renewable terms by agreement among member state governments.
Responsibilities	To interpret and apply Community law. Decisions of the Court may not be appealed.

APPENDIX 2 RESPONSIBILITIES OF THE MEMBERS OF THE EC COMMISSION (1988)

Office	Member	Responsibilities
President	Jacques Delors (F)	Secretariat General Legal service Monetary affairs Spokesman's service Joint interpreting and conference service Security office
Vice-President	Lorenzo Natali (I)	Cooperation and development
Vice-President	Karl-Heinz Narjes (FRG)	Industrial affairs Information technology Research and science Joint Research Centre
Vice-President	Frans Andriessen (NL)	Agriculture Forestry
Vice-President	Lord Francis Arthur Cockfield (UK)	Internal market Customs union service Taxation Financial institutions
Vice-President	Henning Christophersen (DK)	Budget Financial control Personnel and administration
Vice-President	Manuel Marin (E)	Social affairs and employment Education and training

Office	Member	Responsibilities
Member of the Commission	Claude Cheysson (F)	Mediterranean policy North-South relations
Member of the Commission	Grigoris Varfis (GR)	Coordination of structural instruments Consumer protection
Member of the Commission	Willy De Clercq (B)	External relations and trade policy
Member of the Commission	Nicolas Mosar (L)	Energy Euratom supply agency Office for official publications
Member of the Commission	Stanley Clinton Davis (UK)	Environment Nuclear safety Transport
Member of the Commission	Carlo Ripa Di Meana (I)	Institutional questions Problems concerning a Citizens' Europe Information and communication policy Cultural affairs Tourism
Member of the Commission	Peter Sutherland (IRL)	Relations with the European parliament Competition
Member of the Commission	António Cardoso E Cunha (P)	Fisheries
Member of the Commission	Abel Matutes (E)	Credit, investments, and financial instruments Policy on small and medium-sized enterprises
Member of the Commission	Peter Schmidhuber (FGR)	Economic affairs Regional policy Statistical office

APPENDIX 3 COMPOSITION OF THE EC COMMISSION IN OFFICE IN JANUARY 1989

Country	Commissioner
Belgium	Karel Van Miert
Denmark	Henning Christophersen
Federal Republic of Germany	Martin Bangemann Peter Schmidhuber
France	Jacques Delors[a] Christiane Scrivener
Greece	Vasso Papandreou
Ireland	Ray MacSharry
Italy	Filippo Pandolfi Carlo Ripa di Meana
Luxembourg	Jean Dondelinger
Netherlands	Frans Andriessen
Portugal	Antonio Cardoso e Cunha
Spain	Manuel Marin Abel Matutes
United Kingdom	Sir Leon Brittan Bruce Millan

[a]President.

64

ANSI—Its Role in the Rapidly Changing Marketplace

RUSSELL BODOFF and MARILYN HERNANDEZ *American National Standards Institute, Washington, D.C.*

I. INTRODUCTION

The rapidly expanding and increasingly interlinked world economy has placed the United States at a new threshold of global competition. As international economic rivalries create a fiercer competitive environment, standards technology is gaining greater recognition for its strategic importance. This increased awareness, generated in part by Europe's internal market initiative, has helped foster a significant rise in program growth and interest in medical device standards activities.

At the same time, challenges threaten the U.S. voluntary standards system—a seventy-year-old system that continues to perform as the most effective and efficient means to develop voluntary consensus standards in the free world. More so than at any other time in its history, it is important that this system continue to forge ahead and respond to the challenges placed upon it with confidence and commitment.

The U.S. voluntary system of standards, testing, and certification mirrors our nation's culture and commitment to free enterprise by allowing the market to determine the optimum allocation of resources devoted to work in both the national and international arenas. For over seven decades this system has been successfully administered by the private sector under the auspices of the American National Standards Institute (ANSI) and through the efforts of ANSI's diversified federation which includes corporations, government agencies, professional and technical societies, consumer groups, and educational institutions. It is important to note that the private sector is the "stake holder" with the dominant investment of resources in the system.

II. ANSI'S ROLE

ANSI, founded in 1918, is a nonprofit, privately funded membership organization whose functions include:

Serving as the national coordinator for voluntary standards activities through which standards developers cooperate in establishing, improving, and approving standards based on a consensus of all interested parties

Furthering the voluntary standards movement as a means of advancing the national economy; benefiting the public health, safety and welfare; and facilitating domestic and international trade and commerce

Establishing, promulgating and administering procedures and criteria for the recognition and approval of standards as American National Standards so as to encourage existing organizations to prepare and submit such standards for approval

Representing the interests of the United States in international nontreaty standardization and certification organizations

The private sector voluntary standards system is unique. No other nation produces as many quality standards as quickly and at such relatively small cost as does the United States. Although the Institute does not develop standards, it has facilitated the development of tens of thousands of national and international consensus standards by qualified organizations—a feature that is one of the great strengths of the system. Of the 270 standards developing organizations in the United States that have active ongoing programs, 93% are ANSI members.

The Institute assures that voluntary standards submitted to it for approval as American National Standards are developed through due process and that such standards reflect a true consensus of all materially interested parties. It encourages the development of new standards where needed, provides the required organizational structure, and also offers a forum for voluntary resolution of disputes. American National Standards provide dimensions, ratings, terminology and symbols, test methods and performance and safety requirements for products ranging from bicycle helmets to electric motors to computers, robots and medical equipment. Standards exist in all industries, including telecommunications, safety and health, information processing, petroleum, banking, household appliances, and medical devices—to name a few.

III. INTERNATIONAL LEADERSHIP

The United States, through the ANSI federation of member organizations, has succeeded in assuming a proper leadership role in international standards efforts. The international efforts of the United States are unparalleled in its production of standards. The Institute has long served as the U.S. member body to the two nontreaty international standards organizations, the International Organization for Standardization (ISO), and the International Electrotechnical Commission (IEC) via the U.S. National Committee. In addition, ANSI is the United States representative to the Pacific Area Standards Congress (PASC), which is dedicated to improving the participation of Pacific Rim standards organizations in international standards groups (see Figure 1).

The leadership role of the United States in the international arena is unmatched by any country. On behalf of the United States, the ANSI federation

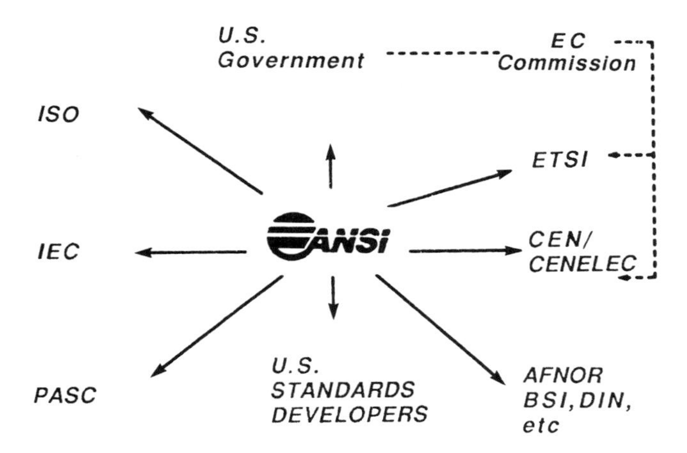

FIGURE 1 ANSI's global relationships. ANSI represents U.S. government and private sector interests to ISO, IEC. It is recognized as the interface with CEN/CENELEC as well as other regional and foreign national standards organizations. The U.S. government plays a role in direct relations with the EC Commission on matters such as tariffs and trade.

is active in most ISO and IEC committees and holds secretariats directly or by delegation to others, in sectors where U.S. interests are most affected by international standards and have great economic importance. While ANSI may delegate the administration of a secretariat to another member organization, the Institute, as the member body of ISO and IEC (via USNC), is nevertheless responsible for ensuring the proper functioning of U.S. secretariats.

Representatives to ISO and IEC committees are drawn from the Institute's various member body organizations and participate in Technical Advisory Groups (TAGs). These participants represent a broad cross section of 10% government agencies, 15% technical and trade organizations, and 75% companies working to ensure U.S. interests are well served in standards developed by over 2,200 international committees and subcommittees. On other committees where U.S. interests are not as great, the ANSI federation holds observer status. The Institute coordinates the activities of the several thousand participants who develop U.S. positions for ISO and IEC standards. seeking to:

Harmonize international standards with domestic standards
Ensure that the TAGs represent the interests of all materially affected parties
Verify that ISO and IEC representatives possess the required technical expertise

With the European community's initiative to create a single market by 1992 steadily advancing, the Institute's relationship with the European standardizing bodies ensures U.S. interests are appropriately considered. ANSI maintains access and liaison with the European private sector standards organizations—the European Committee for Standardization (CEN) and the European Committee for Electrotechnical Standardization (CENELEC). In most cases the European Commission relies on CEN and CENELEC to coordinate standards throughout the Community. Both these organizations, to ensure efficiency, will defer to ISO and IEC standards, except in sectors where these international bodies have failed to promulgate standards. Since many of the ISO and IEC standards mirror American National Standards, the United States, via the ANSI federation, is in a key position to influence the development of standards that will be used by the European commission.

Responding to its constituency's needs, the Institute opened a Brussels office that would provide increased U.S. presence in Europe, through 1992. With its newest location, ANSI is at the forefront of EEC developments in standards and certification and is able to provide the most up-to-date information to its members. To further enhance member knowledge of international and European standards, ANSI maintains copies of, and sells standards initiated by, the major international and regional standards developers, such as ISO, IEC, CEN, CENELEC, and individual countries who are major trading partners with the United States.

ANSI has been very active in developing closer working relationships officials from all over the world. The Institute regularly meets with representatives of ISO, IEC, CEN/CENELEC and many of the major foreign national standards bodies such as the British Standards Institution (BSI), the French Standards Association (AFNOR), and the German Institute of Standards (DIN). These meetings have led to new understandings and agreements to open more channels of communication and of mutual cooperation. At the same time, ANSI is also working to increase communications with U.S. government as well as with the Institute's member councils and U.S. standards developers to improve understanding as Europe moves to create a single market by 1992.

ANSI publishes a biweekly newsletter, Standards Action, which includes a listing of CEN/CENELEC draft standards available for review and comment by the United States. Comments are also solicited on national and international draft standards and information is provided on the initiation of standards projects. Reports on newly approved and published standards are included as well. Special publications, such as ANSI Global Standardization Report provide information on the latest standards issues taking place in Europe. Also, in addition to providing information on policy actions of ANSI, ISO, IEC, and national and international standards issues affecting the marketplace, the Institute's monthly newsletter, the ANSI Reporter, also includes updates on European developments.

Concurrent with the growth in the global economy and the need for U.S. industry to compete more effectively in the international arena, ANSI has experienced an unprecedented period of membership growth from within the medical device community, with many major corporations lending their support to the voluntary standards system. ANSI now holds secretariats for the following medical device and medical-related activities within the ISO and IEC:

ISO Technical Committee Secretariats
- ISO/TC 42—Photography

ISO Subcommittee Secretariats
- ISO/TC 106/SC 2—Dentistry/Prosthodontic materials
- ISO/TC 121/SC3—Anaesthetic and respiratory equipment/Lung ventilators and related equipment
- ISO/TC 150/SC 2—Implants for surgery/Cardiovascular implants

IEC Subcommittee Secretariats
- IEC/TC 62/SC 62—Electrical equipment in medical practice/Electromedical equipment

The Institute also maintains active participation through ANSI TAGs administered by organizations such as the Association for the Advancement of Medical Instrumentation (AAMI), American Society for Testing and Materials (ASTM), National Electrical Manufacturers Association (NEMA), and the American Dental Association (ADA).

U.S. Participation in ISO Technical Committees
- ISO/TC 42—Photography
- ISO/TC 76—Transfusion, infusion, and injection equipment for medical use
- ISO/TC 106—Dentistry
- ISO/TC 121—Anesthetic and respiratory equipment
- ISO/TC 150—Implants for surgery
- ISO/TC 150/SC 2—Cardiovascular implants
- ISO/TC 157—Mechanical contraceptives
- ISO/TC 168—Prosthetics and orthotics
- ISO/TC 172—Optics and optical instruments
- ISO/TC 173/SC 1—Wheelchairs
- ISO/TC 173/SC 3—Aids for ostomy and incontinence
- ISO/TC 194—Biological evaluation of medical and dental materials and devices

U.S. Participation in IEC Technical Committee
- IEC/CE/TC 62—Electrical equipment used in medical practice
- IEC/SC 62A—Common aspects of electrical equipment used in medical practice
- IEC/SC 62B—X-ray equipment operating up to 400 kV and accessories
- IEC/SC 62C—High-energy radiation equipment for nuclear medicine
- IEC/SC 62D—Electromedical equipment

The large number of organizations involved in the development of medical device standards and their participation in the many related international activities can create an environment for redundancies and duplications of effort between organizations. To prevent these problems and to provide leadership and coordination to the myriad of activities, the ANSI Medical Device Standards Board (MDSB) was established. The Board has representation of materially interested parties from all of the key trade associations, professional societies, government agencies, and even major companies who cooperate to make this coordination possible. The excellent relationship that exists between the private sector organizations and government agencies on the MDSB sets an example of the overall private sector/public partnership that is needed in standards activities to best represent a consensus of U.S. activities at all times. To advance this goal of a unified government/industry voice, ANSI opened a

Washington office in 1989 to promote enhanced interaction between the Institute and its Washington-based membership as well as with the administration and congress.

In 1984, President Reagan awarded ANSI with a Private Sector Initiative Commendation for its administration and coordination of the voluntary standards effort in the United States. In 1986, a presidential commission advised the government to adopt private sector standards whereby it would substitute more commonly available commercial products for expensive, custom-made items. Today, an increasing number of nongovernment standards are being used by government agencies for regulatory and procurement purposes. Moreover, the voluntary standards system with tens of thousands of participants is also a major success story in President Bush's advocacy of volunteerism.

IV. CHALLENGES FOR TOMORROW

The American voluntary standards system has withstood the test of time. Yet as strong and effective as the ANSI federation is, it is occasionally challenged. Every now and then a federal agency seeks to either regulate or replace ANSI without the support of the federal government, and without the support of the system's primary users and beneficiaries, the private sector. These attempts do not focus on a voluntary, consensus system but instead arise with the belief that governmental control is necessary because the private sector does not have the ability to regulate itself.

Added to the threat of the regulators is an overall trend in many U.S. industries to cut back on standards activities as a means of trimming budgets. CEOs in Japan and Germany possess a greater awareness of the importance of standards technology than do American CEOs. Japanese and German CEOs, while recognizing that they may not reap the benefits of their investments for many years, understand that financing standards pays a long term dividend. Unfortunately, in the United States, too many CEOs tie their expenditures to one or two year returns. This makes it easy to cut back on standards efforts—especially financial support of international participation when budget trimming is in order. But these executives are sacrificing a part of America's future international competitiveness for an easy solution today. While many industries are at fault, certainly credit should be given to the senior executives in the medical devices, information technology, and other industries who are significantly increasing their commitment to, and financial support for, standards activities.

As it embarks on the decade of the nineties, the Institute's vision is of a world in which expanding trade, along with healthy competition and rapid technology change, creates unparalleled opportunity and global prosperity. It is a world where goods, services, and capital move freely satisfying global economic and human needs. There is universal recognition of the essential role of standardization technology in advancing global commerce. And, as a world leader in standardization technology, the United States is a key player in satisfying the needs of the global marketplace. Moreover, the United States' process for standards development mirrors our national culture and commitment to free enterprise. It is a vision where the U.S. system prospers as a private sector administered voluntary standards system funded by the participants and devoid of market distortions. Central to the system and U.S. leadership internationally, is the ANSI federation that enjoys the strength of unity resulting from a partnership of all materially interested parties (e.g., standards developers, industry, consumers, academia, and government). This federation serves to foster U.S. standardization interests globally. And, the best interests of the Institute's "stakeholders" are pursued with courage and conviction.

Index